W9-AGM-413

REMOTE SENSING OF SEA ICE AND ICEBERGS

WILEY SERIES IN REMOTE SENSING

Jin Au Kong, Editor

Tsang, Kong, Shin • THEORY OF MICROWAVE REMOTE SENSING

Hord • REMOTE SENSING: METHODS AND APPLICATIONS

Elachi • INTRODUCTION TO THE PHYSICS AND TECHNIQUES OF REMOTE SENSING

Szekielda • SATELLITE MONITORING OF THE EARTH

Maffet • TOPICS FOR A STATISTICAL DESCRIPTION OF RADAR CROSS SECTIONS

Asrar • THEORY AND APPLICATIONS OF OPTICAL REMOTE SENSING

Steinberg and Subbaram • MICROWAVE IMAGING TECHNIQUES

Curlander and McDonough • SYNTHETIC APERTURE RADAR: SYSTEMS AND SIGNAL PROCESSING

Haykin and Steinhardt • ADAPTIVE RADAR DETECTION AND ESTIMATION

Haykin, Lewis, Raney, Rossiter • REMOTE SENSING OF SEA ICE AND ICEBERGS

REMOTE SENSING OF SEA ICE AND ICEBERGS

Simon Haykin
McMaster University

Edward O. Lewis
Department of Fisheries and Oceans, Canada

R. Keith Raney
Department of Natural Resources, Canada

James R. Rossiter
Canpolar Inc.

A WILEY-INTERSCIENCE PUBLICATION
JOHN WILEY & SONS, INC.
New York • Chichester • Brisbane • Toronto • Singapore

This text is printed on acid-free paper.

Library of Congress Cataloging in Publication Data:

Remote sensing of sea ice and icebergs / [edited by] Simon Haykin . . .
[et al.].
p. cm.—(Wiley series in remote sensing) (Adaptive and
learning systems for signal processing, communications, and control)
"A Wiley-Interscience publication."
Includes bibliographical references and index.
ISBN 0-471-55494-4 (acid-free)
1. Sea ice—Remote sensing. 2. Icebergs—Remote sensing.
I. Haykin, Simon S., 1931– . II. Series. III. Series: Adaptive
and learning systems for signal processing, communications, and control.
GB2401.72.R42R46 1994
551.3′42′028—dc20 94-6252

Printed in the United States of America

10 9 8 7 6 5 4 3 2 1

CONTRIBUTORS

BRIAN W. CURRIE, Communications Research Laboratory, McMaster University, Hamilton, Ontario, Canada

JOHN C. FALKINGHAM, Ice Branch, Atmospheric Environment Service, Environment Canada, Ottawa, Ontario, Canada

DAVID M. FARMER, Institute of Ocean Sciences, Department of Fisheries and Oceans, Sidney, British Columbia, Canada

CAREN GARRITY, Microwave Group—Ottawa River, Dunrobin, Ontario, Canada

SIMON HAYKIN, Communications Research Laboratory, McMaster University, Hamilton, Ontario, Canada

J. SCOTT HOLLADAY, Aerodat Limited, Mississauga, Ontario, Canada

VYTAS KEZYS, Communications Research Laboratory, McMaster University, Hamilton, Ontario, Canada

EDWARD O. LEWIS, Bayfield Institute, Department of Fisheries and Oceans, Burlington, Ontario, Canada

CHARLES E. LIVINGSTONE, Canada Centre for Remote Sensing, Department of Natural Resources, Ottawa, Ontario, Canada

RAYMOND T. LOWRY, Intera Information Technologies (Canada) Limited, Calgary, Alberta, Canada

DENNIS M. NAZARENKO, RADARSAT International Inc., Richmond, British Columbia, Canada

R. KEITH RANEY, Canada Centre for Remote Sensing, Department of Natural Resources, Ottawa, Ontario, Canada

JAMES R. ROSSITER, Canpolar Inc., Toronto, Ontario, Canada

IRENE G. RUBINSTEIN, Institute for Space and Terrestrial Science, York University, North York, Ontario, Canada

SATISH SRIVASTAVA, RADARSAT Program Office, Canadian Space Agency, Ottawa, Ontario, Canada

SEBASTIAN TAM, MPB Technologies Inc., Dorval, Quebec, Canada

JOHN WALSH, Faculty of Engineering, Memorial University, St. John's, Newfoundland, Canada

YUNBO XIE, School of Earth Sciences, University of Victoria, Victoria, British Columbia, Canada

FOREWORD

If Sir John Franklin had remotely sensed ice information in 1845, he never would have left England to sail the Northwest Passage, because all Canadian Arctic waters were still in the grip of the Little Ice Age. Climate moderation allowed CCGS *Labrador* in 1956 and USCGS *Manhattan* in 1969 to make the Passage.

Because Canada has the longest coastline of any country and most of that coastline is ice-bound and in darkness for much of the year, remote sensing of the environment is essential to Canada. In Canada, conventional data collection and analysis methods are hindered by the environmental extremes, geographical vastness, and logistical expenses involved in polar research. Remotely sensed data will play a critical scientific and operational role in the Canadian Arctic.

Remote sensing may be defined as the body of techniques that allow non-intrusive observation of an object from a distance; for example, a camera is a passive sensor. Remote sensing in this case is the science of acquiring, processing, and interpreting images obtained from satellites and aircraft that record the interaction between electromagnetic (EM) radiation and material; it is a basic assumption that all energy can be reduced to electromagnetic radiation.

Virtually all satellite and radar measurements of the atmosphere or of a surface are strongly dependent upon the characteristics of the sensor; they are not pure functions of the state of the atmosphere or surface that they are measuring. We must embrace with excitement possible new tools. In some applications there has not been much progress in basic sensors and image analysis methods, but sea ice applications have progressed and evolved from the simple days of 1947/48 experiments in which brighter returns on a radar screen merely indicated greater resistance to navigation through the ice. An early problem of aircraft altimeters identifying the bottom of an ice cap, instead of the ice surface which the pilot thought he was seeing, led to several crashes. There is also the longstanding problem of designing a system which can be mounted on a ship to reveal thickness and physical properties of ice ahead of the vessel in real time, as opposed to reliance upon airborne systems. Present radars cannot reliably distinguish a hundred-ton growler (iceberg fragment) from countless waves which rise to a similar height above the sea surface.

Radar was first used in the ice environment in 1943 but it was only in the early 1980s that radar equipment was modified or designed to operate specifically in that environment. Ice Island T3 was remotely sensed by photography in 1947 and only in 1979 was it sensed remotely by synthetic aperture radar. I recall the proceedings of a seminar on thickness of floating ice by remote sensing held at Defense Research Establishment Ottawa (DREO) in October 1970 "with the aim of bringing together those working on the difficult problem of remote measurement of ice thickness"; it attempted to be a summary of the previous 20 years of work. The book in your hand is a comprehensive summary of the work of the following 20 years: many of the "difficult problems" posed in 1970 have been solved by Canadian researchers.

Remote Sensing of Sea Ice and Icebergs is laid out very logically; from early experiments to present, from surface to satellite, from simple to complex, from passive to active, from cheap to expensive. The topics are well explained by recognized authorities, building on basic data and theory developed 30 to 40 years ago and earlier. There are excellent reference lists for those readers who may wish to pursue greater detail. Past accomplishments are brought into focus with the reality of today. There are good discussions about tactical ice information; how to observe and identify ice from various platforms, how and why various radar images are seen in varied sea and ice conditions, how a ship may sail in effectively open water conditions in spite of almost complete ice cover, and how to detect glacial and old ice within first-year pack ice. Specific attention has been devoted to topics that are often difficult to appreciate by novice users of radar data.

Several new systems that are just becoming operational are covered. For example, *RADARSAT*, the Canadian radar satellite to be launched in 1995, is given special attention in the book. This satellite is designed to produce data within four hours of scanning by satellite and will bring the operational use of remote sensing to a new plateau. Most of the Canadian Arctic will be imaged every day. Canada now has the ability, sought since the early 1970s, to monitor its northern and offshore regions.

The requirements for new knowledge of the surface environment have led to rapid expansion of research, and Canadians are leaders because there is a high level of cooperation between government, university, and industry researchers. Supercomputers have permitted rapid simulations of processes and increasingly realistic models, but the costs of such research are high. With a fundamental understanding and capability in this field, we will all be well placed in the future to make effective use of the latest remote sensing technology for resource and environmental monitoring. As for operational use, it is essential to be able to ingest, analyze and deliver reliable and timely ice information to users. However, it is still a major problem to discern icebergs in open water driven by high sea states; there is much work to be done.

"What's new in Remote Sensing?"—here it is! Here is a practical textbook with the required theory and the applications. Here are summaries of significant advancements in research and development of remote sensing equipment in Canada. All authors are experienced in the recognition and identification of the various signals and the physical mechanisms responsible for those signals.

The organization that wants facts and information instead of guesses and opinions finds the well organized library the most efficient research tool and the greatest economy of time and effort. A library contributes to the three processes basic to civilization: the discovery of knowledge, the conservation of knowledge and the transmission of knowledge. This book is a worthy addition to any library.

GEORGE D. HOBSON,
February 21, 1994

PREFACE

Ice forms an integral part of the Canadian ocean environment for most of each year. Its presence, be it in the form of sea ice (year-round in the Arctic and at more southerly latitudes in the winter months) or icebergs off Canada's east coast, represents a major potential hazard to navigation and other marine activities. Ice is becoming recognized increasingly as an important component in scientific studies of ocean–atmosphere interaction and global climate. It is, therefore, fitting that the remote sensing of ice has received significant scientific and engineering attention in Canada over the past two decades.

This effort was remarkably successful: techniques from acoustic to passive microwave radiometry, to several different forms of active radar (such as, over-the-horizon, marine, side-looking airborne, and synthetic aperture), to electromagnetic induction, as well as data processing, display, and dissemination, moved from conceptual ideas to (for the most part) full operational status. Surface, airborne, and spaceborne platforms are all being used.

In 1990, the urge to document these technical achievements occurred to the four of us. Our vision was to coedit a book, written by a select group of Canadian scientists from coast to coast, devoted to the effective use of remote sensing techniques for the detection, classification, and measurement of sea ice and icebergs. We hope that the book will appeal to readers who are involved in ice-related research and operations, as well as to those who are keen to gain an appreciation for the wider use of remote sensing techniques.

The task of assembling the large group of authors and coordinating their individual contributions into a cohesive, readable, and informative work has been challenging. The fact that many of the authors have now moved on to new projects meant that for many this was indeed a labor of love, done on their own time. Now that the project is complete, we will miss the conference calls which we held regularly and with increasing frequency over the past two years, during which the many details associated with the book were discussed and resolved.

The production of this book would not have been possible without the dedicated effort and enthusiasm throughout from Lyn Arsenault, our Technical Editor. Lyn brought not only her superlative editing skills to the book but also her deep knowledge and appreciation for the topic, based on years of practical experience. Lyn's meticulous adherence to a set of guidelines is responsible for the cohesive manner in which the different chapters have been integrated. Her influence, in one form or another, can be found on every page. All the coeditors and authors owe Lyn a debt of gratitude.

Hamilton, Ontario, Canada	SIMON HAYKIN
Burlington, Ontario, Canada	ED LEWIS
Ottawa, Ontario, Canada	KEITH RANEY
Toronto, Ontario, Canada	JAMIE ROSSITER

August 1994

ACKNOWLEDGMENTS

A few words of acknowledgement are in order, which I would like to record in my capacity as the senior co-editor, Simon Haykin, on behalf of my co-editors and co-authors of this book.

First, we would like to thank George Hobson for writing the Foreword to the book: George Hobson has degrees in Math and Physics, and Geophysics and an honorary D.Sc. He has conducted geophysical studies throughout the Canadian Arctic, and from 1972 until 1988 he was director of the Canadian Polar Continental Shelf Project, a position providing exposure to much of the Arctic research carried out in Canada. He is a fellow of the Arctic Institute of North America, and the Royal Canadian Geographical Society, and has served as a director of the Science Institute of the North and the Geographical Society.

We are all most grateful to Ed Lewis for his tireless effort throughout the past two years, working behind the scenes to put the manuscript for the entire book together. We are also indebted to our editor, George Telecki, for his patience, understanding and support, and to the staff of John Wiley & Sons, Inc. for their help in the production of the book.

CONTENTS

CHAPTER 8
SURFACE-BASED RADAR: NONCOHERENT 341

Edward O. Lewis, Brian W. Currie, and Simon Haykin

ACRONYMS, TERMS

NOTATION[1]

α	Depth of EM penetration into a medium
α_m	Antenna main beam efficiency
β_h	Azimuth antenna beamwidth (two-way power, equivalent rectangle norm)
β_r	Angular resolution, radians
γ	Scattering per unit projected illuminated area; $\sigma^0/\sin\theta_i$
γ	Ratio of surface tension to water density
Γ	Scattering coefficient, complex, dimensionless
δ	Loss tangent: $\tan\delta = \epsilon_r''/\epsilon_r'$
δd_h	Horizontal separation between interferometer antennas
$\delta G/G$	Rms gain fluctuation and drift
δT	Radiometer sensitivity
$\delta(\ )$	Dirac (delta) distribution, an impulse of unity area
η	Antenna radiation efficiency
θ	Independent variable, angle, in the vertical (elevation) plane
θ_B	Brewster (incidence) angle
θ_i	Incidence angle, with respect to the vector normal to the earth's geoid
θ_t	Local incidence angle, with respect to the vector normal to the local slope of the reflecting facet
ϵ	(Absolute) electrical permittivity of a material, farad m^{-1}
ϵ_r''	(Relative) dielectric loss, dimensionless
ϵ_r'	(Relative) dielectric constant, dimensionless
ϵ_0	Dielectric constant of free space, farad m^{-1}
ϵ_r	Relative permittivity $\epsilon_r = \epsilon/\epsilon_0 = \epsilon_r' - j\epsilon_r''$, dimensionless

[1]Notation for Chapter 3 and Chapter 5 may be found at the end of chapters.

μ_0	Magnetic permeability of free space
ρ	(Ice) density, kg/m^3
σ	Radar cross section (RCS), discrete object, m^2
σ	Electrical conductivity
σ_{DC}	Conductivity, direct current
σ^0	Reflectivity density, nominal ground plane, dimensionless
σ_{st}	Ice strength
σ_x	Standard deviation of the random variable x
$\sigma_{R(G,2N)}$	Width parameter of the Rayleigh (Gaussian, chi-squared) distribution
τ	Length of transmitted pulse, coded or uncoded, seconds
τ	Integration time
τ	Optical opacity
τ_0	Length of compressed pulse, seconds
τ_r	Radiometer integration time, seconds
Φ_w	Potential function, gravity waves
$\Phi_0(\ )$	Compressed azimuth impulse response
ϕ	Independent variable, angle, in the azimuth (cross range) plane
φ	Phase, radians
χ	Ellipticity of a polarized wave
ψ	Orientation of (elliptically) polarized wave
Ω	Solid angle, steradians
a_τ	Atmospheric opacity
$a_{H(V)}$	Amplitude of H-polarized (or V-polarized) signal
a_T	Amplitude of the transmitted (T) pulse
A	Attenuation rate, EM wave into conducting material, dB m^{-1}
A_R	Effective area, receiving antenna, m^2
B	Bandwidth, Hz
B_f	Black body spectral brightness, Wm^{-2}sr^{-1}Hz^{-1}
B_n	Receiver noise bandwidth (Hertz), equivalent rectangle power
B_p	Pulse bandwidth, Hz
C	Total ice concentration
C	Fluid phase velocity, ω/k_w
C_A	Azimuth compression ratio
C_R	Range compression ratio
d	Depth of water
$d_{v(h)}$	Antenna linear dimension in vertical (horizontal) plane
$ds_{t,i}^{\omega_0}$	Signal increment at the radar, at carrier frequency, from i^{th} transmission

E[ ]	Expectation (average) operator
$E_{(i,r)}$	Electric component (incident or reflected), scalar or vector form, of an EM field
EM	Electromagnetic wave
e	Emissivity, T_B/T
F_n	Receiver noise figure, dB
f	Frequency (Hz = cycles per second)
f_p	Pulse repetition frequency, Hz
G	Free space antenna gain, power, one way. May have angular variation. (Upper case subscripts denote transmit, receive, or polarization)
G_p	Azimuth processing gain, coherent radars
G_0	One-way peak power antenna gain
GR	Gradient ratio
g	Gravitational constant
$g(\theta, \phi)$	One-way antenna power pattern, normalized to unity peak power
g_T	One-way voltage gain of the transmitting antenna
g_1	Two-way voltage amplitude coefficient, range compressed received signal
h	Sensor altitude
h_{fb}	Freeboard, ice surface above mean sea level
$\hbar$	Planck's constant
I	Inphase component, coherently demodulated signal
I_f	Radiation energy flux ($Wm^{-2}Hz^{-1}$)
$I(M, r)$	Detected image: azimuth line number M, range position r
$i^C(M, r)$	Complex image: azimuth line number M, range position r
k	Wave number in free space; $k = 2\pi/\lambda$
k_B	Surface Bragg wavenumber
k_w	Wave number, fluid surface gravity waves
K	Boltzman's constant
L	System losses, dB
m	Mean (average) of a distribution
N	Number of statistically independent samples
N_T	Number of transmitted pulses summed in signal processing
P_f	Spectral power
P_n	(Receiver) noise power
$\overline{P_T}$	Average transmitted power, w
PR	Polarization ratio
$p(t)$	Range pulse (slant range), normalized to unity peak value

$p_0(t)$	Range impulse response (slant range), normalized to unity peak value
$p_0(r)$	Range impulse response, spatial domain (slant range), normalized, voltage
Q	Quadrature component, coherently demodulated signal
r	Independent variable, range (slant)
R	Range (usually a nominal parameter, not an independent variable)
r_D	Depolarization ratio $\sigma^0_{HV}/\sigma^0_{HH}$
r_P	Copolarization ratio $\sigma^0_{HH}/\sigma^0_{VV}$
r_a	Azimuth resolution
r_{ah}	Spatial resolution, meters, for altimetry or radiometry from altitude h
r_{gr}	Ground range resolution
r_{sr}	Slant range resolution
S	Ice salinity
$S(\omega)$	Wave-height spectrum, water waves
T	Receiver noise temperature, K
T_A	Antenna radiometric temperature, K
T_B	Brightness temperature, K
T_{Bsp}	Brightness temperature of a surface target, "p" polarization
T_{Bf}	Brightness temperature per unit bandwidth, K/H
T_C	Temperature, °C
T_D	Debye relaxation time
T_p	Interpulse period, $1/f_p$
T_R	Receiver noise temperature, K
T_S	Scene brightness temperature, K
T_{sys}	System noise temperature, K
TBP	Time-bandwidth product (of a distribution, such as a pulse)
t	Independent variable of time, range ("fast" time)
U_w	Wind speed, ms^{-1}
V_{br}	Brine volume, percent
η_{RF}	Group velocity (RF or DC), ms^{-1}

REMOTE SENSING
OF SEA ICE AND ICEBERGS

1

INTRODUCTION

EDWARD O. LEWIS
Bayfield Institute
Department of Fisheries and Oceans
Burlington, Ontario, Canada

JAMES R. ROSSITER
Canpolar Inc.
Toronto, Ontario, Canada

1.1 REMOTE SENSING OF ICE IN CANADA

1.1.1 Background

Over the past quarter century, study of the earth's polar regions has grown dramatically, driven by a variety of interests. An increased awareness of the importance of the polar regions to global climate has led to studies of ice cover and heat exchange between the ocean and the atmosphere. The discovery of valuable natural resources in the Arctic has encouraged significant exploration activity and planning for production. The need to move people, resources, and material safely into and out of polar regions has required development of new ice detection systems. Fundamental scientific interest in sea ice, icebergs, and the oceanography of cold oceans has fostered development of techniques to observe these phenomena effectively.

Not only are the polar regions vast, making up some 25% of the earth's area, but they are remote to most of the world's population, hostile for much of the year, dark for significant portions of each year, and often shrouded in cloud. Under these conditions, remote sensing techniques offer the only practical means of observation for many parameters. Until recently, however, very little was known about the response from ice and oceans for most types of remote sensors. Moreover, few sensors had been designed specifically to observe sea ice or to detect icebergs.

Remote Sensing of Sea Ice and Icebergs, Edited by Simon Haykin, Edward O. Lewis, R. Keith Raney, and James R. Rossiter.
ISBN 0-471-55494-4

Canada has taken a major part in the development and use of more effective sensors, driven by the large Arctic oceanic area of Canada and the substantial hydrocarbon exploration activity there during the 1970s and 1980s. Because new sensing systems were required, the remote sensing of ice became the focal point for leading-edge research into sensing technologies, including data processing, interpretation, and display techniques.

The research and development activities carried out were not centered in a particular region, agency, or sector, but were dispersed throughout the country in industry, government laboratories, and universities. However, the people involved were well aware of each other's work and were able to critique work in progress, encourage exciting developments, and build upon each other's successes. Now that these technologies are being implemented operationally, it is timely to document the major technical achievements of this group of scientists, engineers, and researchers.

This book is written for several quite different types of reader:

- those who are interested in sea ice and icebergs and how to detect or observe them
- those who wish to learn about state-of-the-art remote sensing techniques
- those who are involved with observation or monitoring of ice-covered oceans.

It has been written by Canadian contributors to both ice research and to remote sensing systems' development, many of whom are well known for their contributions to both fields. It was the goal of the editors that the book be useful to readers whose interests lie within this wide span of disciplines.

1.1.2 Canada—An Arctic Nation

Canada is considered to be a polar nation, with vast tracts of Arctic land and ocean areas. Armstrong et al. [1] made the following observation:

> "Geographically, Canada is certainly a northern country. Economically, culturally and politically, however, Canada reflects the attitudes of her population, the great majority of which lives within 200 miles (320 km) of the east-west border with the United States, . . ."

Nonetheless, Canada has more northern territory than any other country but Russia (Fig. 1.1) and has the greatest area, more than 2.5 million square kilometers (or 25%) north of the tree line (a common definition for Arctic regions). After Russia, Canada has by far the longest coastline bordering the Arctic Ocean. More than 170,000 km of shoreline are north of 60°N and a further 6800 km of maritime shoreline are affected by ice for some part of the year, which results in more than 90% of Canada's maritime shoreline being affected by ice.

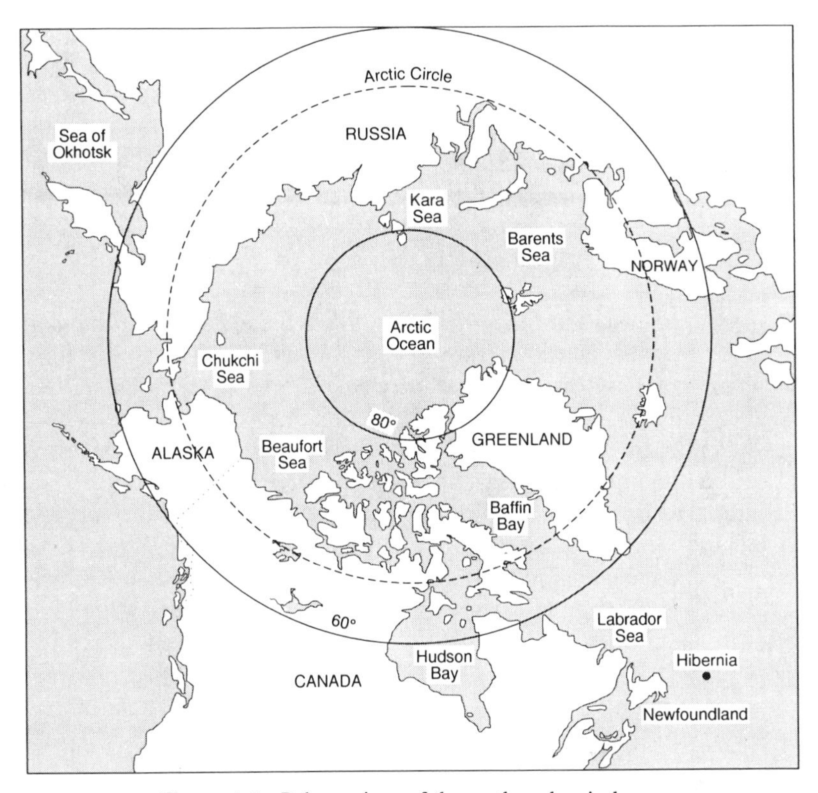

Figure 1.1 Polar regions of the northern hemisphere.

The Arctic and subarctic regions of the northern hemisphere cover more than 41 million square kilometers—about 8% of the surface of the planet—including 15% of the land area and 5% of the ocean area. At its maximum seasonal extent, sea ice covers about 15 million square kilometers in the northern hemisphere and 54 million square kilometers in the southern hemisphere. Although always conceded as important, recent research findings have established the Arctic and Antarctic regions as increasingly important to the world's climate and atmospheric circulation, and have given important warnings about ozone depletion, climate change, and global warming.

Although Canadian Arctic regions are sparsely populated and relatively underdeveloped, they contribute more than $3 billion (all $ are Canadian dollars) to Canada's gross domestic product annually (Indian Affairs and Northern Development [2]). Almost half of this value comes from mineral production, most of which is transported from the Arctic by sea. A total of more than 300,000 metric tons of ore, primarily lead and zinc concentrate from the Polaris and Nanisivik mines, are shipped out on 15 to 18 voyages each year. This ore represents more than 5% of the world's production of lead and zinc. Hydrocarbon exploration has contributed also to operations in ice-covered waters.

Arctic oil and gas exploration peaked in 1985 with annual expenditures of over \$1 billion. Current (1993) estimates of Mackenzie Delta/Beaufort Sea and Arctic Island reserves are 2 billion barrels of oil and 25 trillion cubic feet of gas. Since 1985, there has been an average of two shiploads (150,000 barrels per shipload) of crude oil transported from Bent Horn on Cameron Island in the high Arctic to local markets in the Arctic and to eastern Canadian refineries. These shipments, together with Arctic community resupply, account for more than 95 voyages each year into Arctic areas which are at least seasonally ice covered.

The navigational season for properly ice-strengthened shipping into and out of the Arctic extends from late May to early November. During much of this period, many ports and passages are heavily ice covered, making ship access potentially dangerous and costly. With daily operating costs in excess of \$30,000 for ice-strengthened carriers, effective navigational and ice information is essential for both the safety of the ship and for economical operation.

In addition to economic activities in the Arctic, the Hibernia oil field off Newfoundland will be operating in an area of seasonally ice-covered waters. The oil field and the shipping route to the mainland are in an area with iceberg counts as high as 2000 pear year.

Ice information is required to support offshore development, marine transportation, fisheries operations, weather forecasting, and research activities. This ice information requirement takes many forms, ranging from long-range strategic needs for route planning and on-ice operations to close tactical avoidance of ice hazards. Ice information needs also range from regional coverage to site-specific details. Equally important to the sensing requirement is the distribution of ice information in a timely manner.

1.1.3 Arctic Research Drivers

Although research and exploration activities in the Canadian Arctic date from British Admiralty days, the discovery of oil at Norman Wells in 1921 began the modern development and exploitation of the Arctic. The discovery of gold at Yellowknife in the 1930s and the construction of the Alaskan highway in 1942 continued this trend.

In the postwar period, Canada has invested heavily in the Arctic, first in construction of the DEW Line Air Warning System and then in the establishment of permanent communities, primarily to draw the population together for educational purposes.

However, it was not until the 1970s and 1980s that major research and exploitation activities began, prompted by hydrocarbon finds first off Alaska in 1968 and subsequently in the Beaufort Sea and the Arctic Islands in the 1970s. Rich mineral deposits on Little Cornwallis and Baffin Island have led to the opening of mines and the transportation of concentrated ore out of the Arctic. Western Arctic transportation was enhanced with the construction of the Dempster Highway from the Yukon to Inuvik which crossed the Arctic Circle in 1977.

The reemergence of Arctic sovereignty as an issue has led to further research and occupation activities in the north. The HMCS *Labrador*, the first Canadian icebreaker capable of sustained operations in the Arctic, undertook extensive surveys in the 1950s and was the first icebreaker to navigate the Northwest Passage. The icebreaking voyages of the supertanker MV *Manhattan* through the Northwest Passage in 1969 and of the U.S. icebreaker *Polar Sea* in 1985 created renewed interest in supporting Canada's Arctic sovereignty.

Moreover, hydrocarbon exploration and potential development off the Canadian east coast, in seasonally ice-covered waters, has led to further significant ice and cold water research and development.

1.1.4 Canadian Cold Regions' Research

Canada has a long history of research and development related to the Arctic, through such diverse activities as the National Research Council's Building Research program, the Atmospheric Environment Service's northern weather stations and instrumentation, the Polar Continental Shelf Project of Energy, Mines and Resources, and the research projects of many universities and of several other Canadian government departments.

On land, Canada has made major contributions to permafrost research, Arctic road and building construction techniques, and the use of utilidors for distributing services in northern communities. Canada has led in the development and use of specialized Arctic vehicles, ranging from the Skidoo to larger track vehicles, such as the Nodwell and Arktos.

Canadian industry also has pioneered the development of offshore drilling techniques for Arctic areas, including artificial islands, caisson-retained island structures, specialized floating platforms, the building of ice roads, and the construction of thickened ice drilling platforms. Canada has led in the development and application of novel icebreaking techniques: for the bulk carrier MV *Arctic* of Canarctic Shipping Ltd. and for icebreakers, including the MV *Kigoriak* which was built for use in the Beaufort Sea by Dome Petroleum Ltd.

It is not surprising that development and application of remote sensing tools in Canada has included extensive investigations into their use in Arctic areas and, in particular, their effectiveness in observing sea ice and icebergs in pursuit of both operational and scientific objectives.

One of the major success stories in Canadian remote sensing is the development of airborne ice-mapping sensors including side-looking airborne radar (SLAR) and synthetic aperture radar (SAR) systems. Research and experimentation undertaken by the Canada Centre for Remote Sensing since the 1970s has yielded operational, state-of-the-art SAR systems which are currently being flown in support of marine transportation in the Arctic and in mapping activities in developing countries around the world.

Canadian research into enhanced marine radars—rig-mounted radars and display systems in the Arctic, experiments on drillships off the east coast, and shore-based research in the Arctic—has led to a variety of technologies for surface-based applications. These advances include enhanced, dual-polarized shipborne radars, sophisticated radar processing and display systems, integration of airborne and surface-sensed data, and experimental Doppler radar detection of small targets in sea clutter.

Investigations into airborne ice thickness technologies have resulted in quasi-operational through-ice bathymetry systems and ice thickness systems. Research on passive satellite and airborne sensors and on high-frequency, over-the-horizon radars is bringing them near to operational status. Acoustic and seismic systems will start to play a role in Arctic remote sensing in the near future.

The goal of this book is to document these remote sensing technologies from the Canadian perspective. Remote sensing of ice has been at the forefront of remote sensing research over the past two decades, so it is our belief that this will be valuable not only for those working in ice-covered environments, but also for those with a wider interest in remote sensing.

1.1.5 The Research and Development Process—Philosophical Discussion

This book describes in detail the development of a variety of sensors, technologies, and operational infrastructure for the surveillance of ice-infested waters. These activities cannot be described as anything but successful, both for their technical achievement and for their diversity. It is of interest to investigate how a nation, with a relatively small population (26 million) and a relatively low expenditure on research and development as a percentage of gross domestic product, undertook these activities. This section examines the research and development process with a view to understand better how these developments took place.

The first point to be noted is that there was no master plan or program to undertake the development of offshore surveillance capability. There was no central agency, such as in the American NASA program, nor was there a central source of dedicated funding. The development activities took place at a wide range of institutions, including government research laboratories, university research groups, and industry participants, both resource companies and consulting firms.

To examine how this process occurred, it is important to recall the economic milieu of the 1970s. The world appeared on the brink of a hydrocarbon resource shortage and it was generally accepted that oil and gas would need to be located and developed in frontier regions. Canada not only appeared to have substantial resources off its east coast and in its Arctic, but was concerned about exercising sovereignty over its huge offshore domain. The American MV *Manhattan* traverse of the Northwest Passage in 1969 reinforced this latter concern. It was understood that Canada needed to put in place far better systems for the management of its northern and offshore regions. The single, most identifiable aspect of northern operations was the ability to detect and monitor ice. Ice was recognized as the crucial factor that differentiated arctic and subarctic operations from other regions of the world.

Similarly, it was recognized that relatively little was known about ice formation, location, decay, forces, movement, and so on. Few tools were available to study this natural phenomenon scientifically and repetitively. In addition, ice exists under an enormous range of conditions, from the calm, cold Arctic Archipelago to the highly active Beaufort Gyre, to the extremely turbulent and storm-tossed eastern seaboard. Ice conditions vary from thin (and relatively benign) first-year ice to thick (and hazardous) old ice, to substantial ice ridges, and to potentially disastrous icebergs.

In this economic and political environment, it was clear that Canada had to invest both in the tools to study these frontier regions and in deploying the scientific and technical resources to understand these areas better. Therefore, although there was no master plan for development of these sorts of technologies, it was agreed generally that they had to be developed and that the results of doing so would be worthwhile. There was a tacit understanding amongst funding agencies and Arctic operators that much more needed to be done. This situation resulted in two key factors:

1. Funding was available from a variety of sources to undertake new and interesting research.
2. The intellectual challenge of a rich and diverse problem could attract some of Canada's finest scientists and engineers.

Therefore, although there was no overall scheme for development of the technologies that we have been describing, there was certainly an environment conducive to their development. Researchers from a variety of organizations examined the problem with their own particular expertise assured that their funding supporters would look favorably upon such activities and that their efforts could prove worthwhile.

The result of these initiatives was the emergence of an "intellectual soup" of activity from coast to coast. Researchers at universities and government laboratories and within industry all developed projects for enhancing offshore surveillance technologies. Funding, likewise, came from a wide variety of sources. The Canadian federal government had put in place a very generous scheme for encouraging offshore oil and gas exploration called the Petroleum Incentive Program. Under this program, \$6.3 billion was invested in various aspects of offshore exploration from 1981 to 1987. In addition, the government financed a variety of intramural activities through the Panel on Energy Research and Development. Funds were deployed through a variety of federal agencies and departments both to governmental laboratories and to contractors. The Natural Science and Engineering Research Council further encouraged and supported research activities along these lines.

Quite quickly a variety of projects emerged, centered in different locations. For example, Intera Technologies Ltd. started development of a commercial airborne SAR, backed by Dome Petroleum Ltd. through a long-term surveillance contract. The intellectual basis for this development had been undertaken previously by the Sursat program of Energy, Mines and Resources. Ice thickness and over-the-horizon radar programs were initiated at the Centre for Cold Ocean Resources Engineering (C-CORE) in Newfoundland. The Atmospheric Environment Service undertook investigation of passive microwave sensing, and started to plan for a fully operational and highly computerized operational Ice Centre. The foundations for RADARSAT, a Canadian spaceborne SAR, were laid around the same time.

Although there was an uncoordinated aspect to these activities, it was recognized that a group of well-informed scientists and engineers were aware of all these activities. This group met regularly throughout the 1970s and early 1980s, both through ad hoc working groups (such as the Canadian Advisory Committee on Remote Sensing) and at regular scientific meetings at which results of ice studies were presented. There was also the opportunity to review and critique each other's progress, both at informal meetings and as peer review adjudicators. Therefore, although there was a gentle rivalry between various groups, there was also a much deeper recognition of the complementary nature of various activities because of the highly diverse nature of the problem and because of the quite different rates of development of different technologies.

It was, therefore, clear that although the activities were not orchestrated by a single source, there was no substantial duplication of effort and Canada had a good chance, through this diversity, of achieving much greater overall success. The emergence of these activities became self-sustaining as a cadre of students and young professionals could see career opportunities in these fields. In Chapter 13 we will examine where this critical mass of intellectual capability is now headed.

1.1.6 Global Applications

Global applications for ice-related technology and capability developed in Canada are twofold: first, for other Arctic or Antarctic regions, and second, for applications not associated with ice. Canadian ice sensing technology has global application for polar (Arctic and Antarctic) transportation, offshore oil and gas exploration, coastal surveillance, and for scientific studies, all in ice-infested waters.

There is significant application for ocean and ice surveillance with over-the-horizon radar in many ice-covered areas of the world. Japan, China, Scandinavia, Russia, and the USA all have offshore areas where a long-range offshore surveillance system would have application for navigation, fisheries, and development monitoring in ice-covered waters.

These same countries also can make use of Canadian-developed surface-based radar, airborne SLAR and SAR, and satelliteborne sensor technologies for ice applications. Visits to China have confirmed their need for surface-based radar, and for airborne and spaceborne SAR for operational assistance to offshore oil production in ice-covered waters in the Bohai Sea. Discussions in Japan have shown the opportunity for use of Canadian-developed ice-thickness sensing technology and surface-based radar technology in the Sea of Okhotsk off northern Japan. The Russian and Baltic states have a large maritime fleet that operates year-round in ice-covered regions. There is considerable offshore oil and gas exploration in the Barents Sea and potential future offshore hydrocarbon development in Russia which would benefit from Canadian technology.

In addition, there are many other areas of the world where both long-range and site-specific ocean surveillance systems would prove invaluable. Immediate applications include anti-smuggling surveillance and drug enforcement, territorial limit surveillance, fisheries enforcement, and so on. Intera's SAR system is currently being used for onshore mapping applications in third world countries, particularly in areas where frequent cloud cover or poor visibility makes conventional visual mapping techniques impractical.

1.2 SENSING TECHNIQUES

Remote sensing may be defined as the body of techniques that allow nonintrusive observation of an object from a distance; ranging from as close as a few meters (as in acoustic sensing) to as far as several thousand kilometers (as in satellite sensing). In this book, the object of interest is usually floating ice, although many of the concepts presented are applicable to other scenes. Detection is performed using an appropriate sensor, as shown in Fig. 1.2. In most general terms, the sensor is made up of a *transmitter* (or *source*) and a *receiver* (or *detector*), mounted in or on a suitable platform, which can be a shore

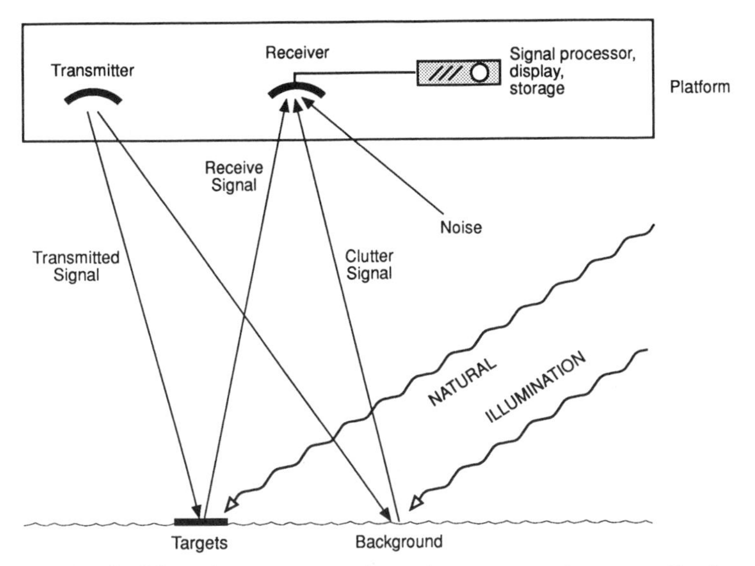

Figure 1.2 Sketch of the major components of an active remote sensing system. Passive systems do not transmit a signal but receive signals that are emitted naturally from the target.

station, a ship, a helicopter, an airplane, or a satellite. The source emits a signal which travels through some path to the target. There it interacts with the target and some portion of it is returned to the receiver, where it is detected. After suitable processing to separate the returned signal from various kinds of noise and to enhance particular features of interest, it is displayed for interpretation by an operator.

All sensible signals must derive from a source. *Active sensors* provide their own signal, whereas *passive sensors* rely on natural means of illumination, such as the sun. For example, the eye is a passive sensor, whereas radar is an active device. There are important examples of passive microwave sensors for sea-ice observation.

Most of the sensors described in this book use electromagnetic (EM) radiation. The EM spectrum is shown in Fig. 1.3 and runs from very low-frequency

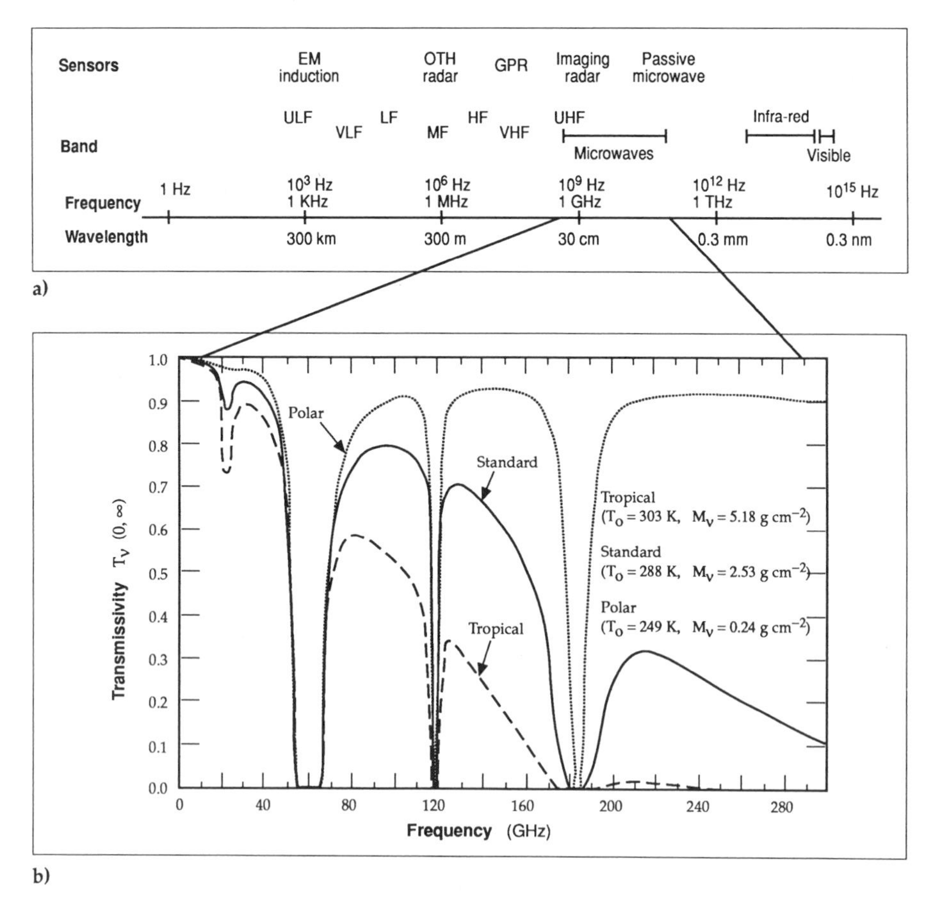

Figure 1.3 (a) EM spectrum showing correspondence between frequency and wavelength. The approximate location in the spectrum of the EM systems discussed in this book is shown. Acoustic systems typically operate in the 1 Hz to 1 kHz frequency range, with wavelengths (in water) of 1.5 km to 1.5 m, respectively. (b) Transmissivity of the atmosphere at different latitudes in the microwave portion of the EM spectrum.

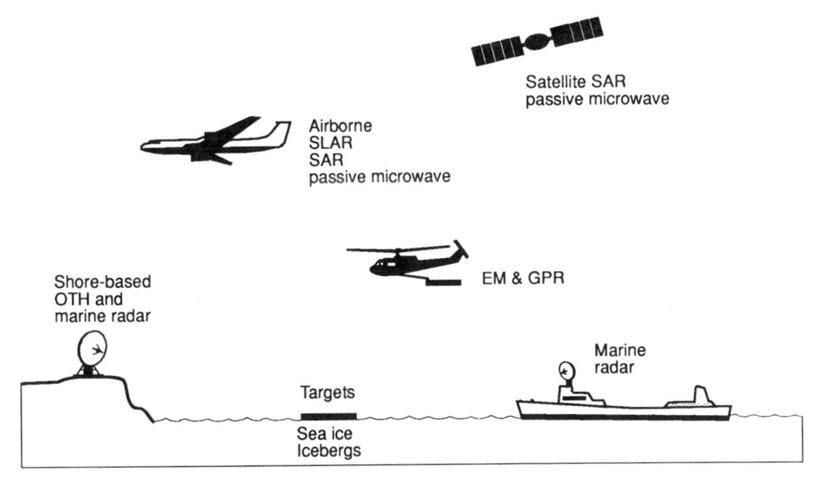

Figure 1.4 Sketches of different EM sensors discussed in this book.

magnetic fields through the various radio frequency bands, up through the millimeter bands, to the infrared, visible, and ultraviolet bands. A sketch of various EM sensors is given in Fig. 1.4. As most sensor systems look through at least some portion of the earth's atmosphere, they are designed to operate at frequencies that travel in a relatively undisturbed manner through the medium. For this reason, most of the sensors described in this book operate at frequencies below the visible band; although the visible spectrum is a useful band for many applications, it is severely hampered by cloud, fog, and darkness.

Note that a particular EM frequency can be described either by its frequency in Hertz (Hz), the standard term for cycles per second, or by its wavelength. For EM radiation the frequency and wavelength are determined by:

$$f\lambda = c \tag{1-1}$$

where

f = frequency (Hz)
λ = wavelength (m)
c = speed of light (m/s)

In vacuum (and approximately in air), the speed of light is 3×10^8 m/s. Therefore,

$$\lambda = \frac{3 \times 10^8}{f} \text{ m}$$

The other major sensor type described in this book is acoustic, a class that includes seismic sensors. Many of the same principles apply to acoustic sensors but the radiating energy (whether active or passive) is mechanical instead of electromagnetic. In water, the speed of acoustic signals is somewhat variable but is about 1500 m/s. With this change, (1-1) still applies and the analog with EM radiation is almost perfect, except that compressional sound waves propagate without polarization. More subtle relationships apply within the ice itself, depending upon the type of elastic wave under consideration, as discussed in Chapter 3.

1.2.1 Resolution and Related Issues

A number of important concepts are introduced concerning the resolving power of remote sensing systems. For most sensing devices, resolution is limited ultimately by the wavelength of the sensor, so systems tend to operate at the highest possible frequency. However, there are other limitations, such as the opacity of the atmosphere (as shown in Fig. 1.2), the wave interaction with the surface, or hardware considerations, which determine the frequency band which is employed.

The *beamwidth* of an antenna system is determined by its *aperture* relative to the wavelength of the radiation used in the aperture. The larger the aperture, the narrower the beam that can be sent out or received from an antenna. For a given antenna size, d, a longer wavelength λ (lower frequency) will have a wider beamwidth, β, which is the Rayleigh diffraction limit according to $\beta = \lambda/d$. Hence, for many *real-aperture* systems, lateral or *azimuthal* resolution is determined by the aperture of the antennas employed which, in turn, is often constrained by the platform available.

For active systems, a finite *pulse length* can be transmitted, which allows measurement of two-way travel time of the pulse to determine distance to the target. The shorter the pulse length, the finer the time-delay measurement which, when scaled for pulse propagation velocity, determines range resolution. Pulse length is controlled by a number of factors, including frequency (again!) and a number of electronic considerations. Passive systems do not have the capability of determining distance to target, except for the case of dispersive seismic waves trapped in the ice.

Many of the sensing techniques developed for ice must operate at lower frequencies than traditional optical remote sensing systems, so significant effort has gone into improving resolution through signal processing. The advent of low-cost computing capability has led to a number of techniques becoming practical and effective. For example, certain active systems use complex coded pulses which, when decoded by signal processing, may have a compressed length that is much shorter than the length of the original pulse.

Synthetic aperture radar uses the property of a moving radar to observe more distant targets for a longer time than nearer targets. This additional information can be used to produce an image in which the azimuthal resolution is kept constant, regardless of the aperture size, the wavelength used, and the distance to the target. Advanced signal processing techniques have made this technique practical, even from space. For the sensors described in this book, signal processing is often critical to the success of the techniques.

1.3 PHILOSOPHY AND ORGANIZATION OF THE BOOK

1.3.1 Approach

This book has been structured so as to connect early experiments with present capability, joining state-of-the-art developments with proven technologies, and blending theoretical elegance with operational practicality. Development grows from a theoretical and/or historical foundation to discussion of current developments. Current practice is described, and evolution into commercial usage with future directions indicated.

There are other ways to look at the pattern of chapters. One is from the ground up. From the ice properties chapter, one moves to surface (acoustic) sensing, through near-contact methods (helicopters and shore-based systems), to aircraft systems, and finally to a satellite radar system—a discussion which includes a sophisticated ice information ground segment, thus completing the logical bridge.

Another view is from simplicity to complexity. For sensing techniques, one progresses from passive systems to active, from long wavelength systems to short (with the exception of Chapter 5 on passive microwaves), from rudimentary laboratory apparatus to a half-billion-dollar satellite system.

Finally, one may look at the book as an evolution from first principles to practice. Early chapters reemphasize the theoretical treatments more than the later ones. Chapter 10 on operational airborne radars and Chapter 12 on RADARSAT and ice information are largely descriptive, although practical results may be found in all chapters. It seems that remote sensing of sea ice is an eclectic discipline, and this book reflects that characteristic.

To keep the book within practical bounds, several topics of interest have had to be excluded. The ice of primary interest is floating ice, both sea ice and icebergs. Landfast glacial ice, in particular the ice sheets covering Greenland and Antarctica, is not included, nor is freshwater ice such as may be found on the Great Lakes (although the sensing techniques may be applied). The remote sensing techniques covered include neither optical nor infrared methods because of the practical restrictions too often imposed on these sensors by cloud and darkness. The reader may note the absence of important instruments such as the space radars in the Kosmos 1500 series operated by the CIS. Our approach is largely technique-oriented rather than device-specific. The topic of inclusions and exclusions is developed further in the closing chapter of this book.

1.3.2 Overview

Chapter 2 introduces the subject matter of the book: sea ice and glacial ice, together with their natural neighboring elements, snow and water. Ice has a remarkable variety of forms, depending on the conditions of its formation, its change in structure and physical properties with time and temperature, and its regional location. The chapter provides the fundamental background required for understanding abilities and limitations of the various sensing technologies. It introduces the "problem," and the remaining chapters are intended to address specific aspects of a "solution." Of course, each of the technique chapters contributes additional material about the ice regime.

As is well known by anyone who has spent time near a large frozen body of water, or by a submariner who may have depended on the phenomenon for the concealment of the vessel, ice can be very noisy. Chapter 3 introduces the field of acoustic and seismic study of ice. Both passive and active methods are employed. Passive methods rely on quantitatively sensing the ambient noise ensemble generated by the ice itself. Active methods use sonar and other acoustic generators to create sonic probes of the ice environment. Mathematical models are being developed to relate acoustic data to ice conditions. However, contrary to many of the EM methods discussed, acoustic and seismic sensing of ice is still a very young science. Much work has been done, primarily for defense purposes, in characterizing ambient sound caused by ice. Apart from the work described in Chapter 3, however, little has been done on the inverse problem of using the signal for remote sensing purposes.

One of the most important ice parameters, at least from a practical point of view, is thickness. Determining the ice thickness, however, poses one of the most challenging ice-measurement problems. Chapter 4 discusses the importance of ice thickness and describes recent progress in its measurement. The most successful technique uses long-wavelength electromagnetic (EM) radiation, with specialized equipment mounted on a helicopter. This EM-induction technique, developed in Canada, offers a means of operational remote ice-thickness measurement, with application to Coast Guards and commercial shipping as well as being an active research field.

Perhaps the most powerful method of wide-scale remote sensing of sea ice uses passive radiometry. Just as ice is "noisy" in the acoustic domain, ice, like all other matter on earth, emits a low-level electromagnetic noise in the microwave spectrum. The art of exploiting this property is the field of microwave radiometry. Chapter 5 provides an introduction to the basic principles of passive microwave sensing, whereby the "brightness" of the natural microwave emissivity is measured. Ice may be detected and, within limits, classified using radiometers, as demonstrated both from aircraft and from spacecraft. One constraint on radiometry is that, in general, the spatial resolution that can be achieved is rather coarse, on the order of tens of kilometers for a spaceborne system. This is an advantage, however, for studies of ice distribution and its change on a global scale.

Chapter 6 is aimed at the description of the salient features of active radar systems in a unified model. This description is particularly important because most radars used in ice sensing depend on two-dimensional signal processing in the complex signal domain. Specific cases of the model include the principal active radar systems considered in subsequent chapters of the book. The basic radar equation in each case is introduced.

Sea ice often assumes an enhanced economic importance when it is near inhabited coastal regions, particularly in navigable waters. Therefore, a shore-based technique for offshore ice observation is attractive, as recognized by researchers in Newfoundland before 1980. Chapter 7 describes the technique of employing high-frequency EM radiation to observe ocean surface features beyond the optical horizon. These ground-wave radars have matured, both in theory and in practice, to the point that they offer a practical solution to the problem of over-the-horizon ice monitoring. The chapter includes a description of the prototype operational system installed at Cape Race, Newfoundland, in 1990.

The classic "radar" that most easily comes to mind is perhaps a standard navigation radar on board a ship with its telltale antenna mounted on the mast, rotating several times each minute, scanning for other ships, buoys, shoreline features, or ice hazards. Such radars are not always reliable devices for detecting ice. Chapter 8 examines these radars and describes a variety of developments in Canada that have improved their performance in an ice environment. Specialized techniques range from optimization of basic parameters, such as frequency and polarization, to advanced data analysis methodology (such as neural network pattern analysis). Controlled experiments verify that significant improvement is possible. Implications of these results on marine radar design are discussed in Chapter 8.

There is an important distinction between noncoherent and coherent radars. In a coherent radar, each transmitted pulse is controlled so that the phase of the received signal sequence may be measured. This capacity allows a significant additional measurement that may be exploited to further enhance the detectability of ice by surface-based radars and leads to better performance than from noncoherent systems. The important concept of Doppler is introduced through the use of a coherent radar and Chapter 9 discusses experimental results using Doppler detection, enhanced through an analytic model of the sea's background reflectivity variations. These radars may be improved through clever use of the polarization signature of the received signal to improve ice target detection, for which advanced signal processing techniques are useful.

Ice observation by airborne radars has a long history. Chapter 10 looks at the development of the operational capability of these systems. Whereas ice had been observed on the rudimentary scanning aircraft radars in use at the end of World War II, serious ice observation using radars followed the development of side-looking imaging radars in the 1960s. The field has progressed, with dedicated ice surveillance missions flown routinely by high-performance systems, using real-time image generation and data analysis. The chapter describes important commercial and governmental operations. The subject of reliable iceberg detection from airborne radars is reviewed; it remains an elusive goal.

High-resolution radar imaging at aircraft (or satellite) altitudes requires a special class of coherent radar. Synthetic aperture radar (SAR) is sufficiently important as an ice-observation remote sensing technique that it merits a full chapter. Chapter 11 reviews the basic principles of SAR operation, which relies on the integration of a *virtual* aperture in the processor rather than attempting to build, and fly, a real aperture of sufficient size to yield the same resolution. The Canada Centre for Remote Sensing (CCRS), Ottawa, has pioneered in the development of SAR technology and its application to ice observation. A wealth of experimental data for sea ice has been collected by the CCRS radars, a summary of which is included in this chapter.

Based on positive experience in radar technology and applications development, Canada has embarked on a major endeavor: the RADARSAT Project. RADARSAT, to be launched in the mid-1990s, will carry an SAR as its only imaging instrument. Unlike other radar satellites that will operate in this decade, RADARSAT will have a variety of modes well-suited to ice observation and it will be able to observe the entire ice sheet of Antarctica. As one of the forces behind Canadian national support for RADARSAT was the need for wide-scale and timely ice surveillance, the ice information system of the Ice Centre, Environment Canada, takes on special significance. Through this system, radar data from the satellite will be processed into ice maps, combined with other ice environment data, and rapidly disseminated as a government service. Chapter 12 describes RADARSAT and the ice information service.

In Chapter 13, we raise unexplored topics of multisensor data fusion, radar vision, and interferometry. Areas for future research and development are identified and spin-off benefits of current developments also are described briefly.

1.4 GENERAL BIBLIOGRAPHY

The publications listed below provide a general overview of the material covered in each chapter and are listed in chapter order. At the end of each chapter, a list of references is given for documentation cited in the chapter.

Chapter 2: Properties of Snow and Ice

Yu. P. Doronin and D. E. Kheisin (1975) *Sea Ice*, Gidrometeoizdat Publishers, Leningrad. (Also published for the Office of Polar Programs and the National Science Foundation, Washington, DC by Amerind Publishers Co. Pvt. Ltd., New Delhi, 1977).

C. Mätzler (1987) "Applications of the interaction of microwaves with the natural snow cover," *Remote Sensing Rev.*, **2**, 259–387.

W. F. Weeks and S. F. Ackley (1982) *The Growth, Structure and Properties of Sea Ice*, USA Cold Regions Research and Engineering Laboratory, Monog. 82-1.

Chapter 3: Acoustic and Seismic Sensing Techniques

M. Mellor (1986) *Mechanical Behaviour of Sea Ice*, N. Untersteiner (Ed.), NATO ASI Ser., Plenum Press, New York, 165–281.

E. R. Pounder (1965) *Physics of Ice*, Pergamon Press, London.

F. Press and W. M. Ewing (1951) "Propagation of elastic waves in a floating ice sheet," *Trans. Am. Geophys. Union* **32** (5), 673–678.

Chapter 4: Ice Thickness Measurement

A. Kovacs, N. Valleau, and J. S. Halladay (1987) "Airborne electromagnetic sounding of sea-ice thickness and sub-ice bathymetry," *Cold Regions Sci. Tech.* **14**, 289–311.

J. R. Rossiter and D. P. Bazeley (1980) *Proceedings of the International Workshop on the Remote Estimation of Sea Ice Thickness*, St. John's, Nfld., 1979, C-CORE Tech. Pub. 80-5, St. John's Nfld.

J. R. Rossiter and L. A. Lalumiere (1988) *Evaluation of Sea Ice Thickness Sensors*, Transport Canada Rept. TP-169E, Transportation Development Centre, Montreal.

Chapter 5: Passive Microwave Systems

F. T. Ulaby, R. K. Moore, and A. K. Fung (1981) *Microwave Remote Sensing, Active and Passive*, Vol. I, Addison-Wesley Publishing Co., Reading, MA.

F. T. Ulaby, R. K. Moore, and A. K. Fung (1982) *Microwave Remote Sensing, Active and Passive*, Vol. II, Addison-Wesley Publishing Co., Reading, MA.

F. T. Ulaby, R. K. Moore, and A. K. Fung (1986) *Microwave Remote Sensing, Active and Passive*, Vol. III, Artech House, Dedham, MA.

Chapter 6: Active Microwave Systems

M. Born and E. Wolf (1959) *Principles of Optics*, Pergamon Press, Macmillan, New York.

W. M. Brown (1963) *Analysis of Linear Time-Invariant Systems*, McGraw-Hill, New York.

W. B. Davenport and W. L. Root (1958) *An Introduction to the Theory of Random Signals and Noise*, McGraw-Hill, New York.

M. I. Skolnik (1962) *Introduction to Radar Systems*, McGraw-Hill, New York.

F. T. Ulaby, R. K. Moore, and A. K. Fung (1982) *Microwave Remote Sensing, Active and Passive*, Vol. II, Addison-Wesley Publishing Co., Reading, MA.

Chapter 7: Over-the-Horizon Radar

J. Walsh, B. J. Dawe, and S. L. Srivastava (1986) "Remote sensing of icebergs by ground wave Doppler radar," *IEEE J. Ocean. Eng.* **OE-11** (2), 276–284.

J. Walsh and R. Donnelly (1987) "A new technique for studying propagation and scatter for mixed paths with discontinuities," *Proc. Royal Soc., London,* **A412** 125–167.

Chapter 8: Surface-Based Radar: Noncoherent

S. Haykin, B. W. Currie, E. O. Lewis, and K. Nickerson (1985) "Surface based radar imaging of sea ice," *Special Issue on Radar. Proc. IEEE,* **73** (2), 233–251.

E. O. Lewis, B. W. Currie, and S. Haykin (1987) *Detection and Classification of Ice*, John Wiley, New York.

Chapter 9: Surface-Based Radar: Coherent

B. W. Currie, S. Haykin, and C. Krasnor (1990) "Time-varying spectra for dual-polarized radar returns from targets in an ocean environment," *IEEE Int. Radar Conf.*, Arlington, VA, 365–369.

S. Haykin (1991) "Adaptive multidimensional signal processing for ocean surveillance," *IEEE Pacific RIM Conf.*, Victoria, B.C., 1–4.

S. Haykin, C. Krasnor, T. J. Nohara, B. W. Currie, and D. Hamburger (1991) "A coherent dual-polarized radar for studying the ocean environment," *IEEE Trans. Geosci. Remote Sensing,* **29** (1), 189–191.

C. Krasnor, S. Haykin, B. W. Currie, and T. Nohara (1989) "A coherent dual-polarized radar for ice surveillance studies," *Int. Conf. Radar*, Paris, 438–443.

Chapter 10: Operational Airborne Radars

M. Skolnik (1970) *Radar Handbook*, McGraw-Hill, New York.

G. W. Stimson (1983) *Introduction to Airborne Radar*, Hughes Aircraft Co., El Segundo, CA.

F. T. Ulaby, R. K. Moore, and A. K. Fung (1981) *Microwave Remote Sensing, Active and Passive*, Vol. I, Addison-Wesley Publishing Co., Reading, MA.

F. T. Ulaby, R. K. Moore, and A. K. Fung (1982) *Microwave Remote Sensing, Active and Passive*, Vol. II, Addison-Wesley Publishing Co., Reading, MA.

F. T. Ulaby, R. K. Moore, and A. K. Fung (1986) *Microwave Remote Sensing, Active and Passive*, Vol. III, Artech House, Dedham, MA.

Chapter 11: Synthetic Aperture Radar

S. R. Brooks (1979) "Synthetic Aperture Radar: An introduction to space borne systems." *Marconi Rev.*, **213**.

J. P. Fitch (1988) *Synthetic Aperture Radar*, Springer Verlag, New York.

S. A. Hovanessian (1980) *Introduction to Synthetic Array and Imaging Radars*, Artech House, Dedham, MA.

G. W. Stimson (1983) *Introduction to Airborne Radar*, Hughes Aircraft Co., El Segundo, CA.

K. Tomiyasu (1978) "Tutorial Review of Synthetic Aperture Radar," *Proc. IEEE,* **66** (5), 563–583.

Chapter 12: RADARSAT and Operational Ice Information

W. M. Brown and L. J. Porcello (1969) "An introduction to synthetic aperture radar," *IEEE Spectrum,* **6**, 52–62.

J. C. Falkingham (1991) "Operational remote sensing of sea ice," *Arctic,* **44** (1).

R. K. Raney, A. P. Luscombe, E. J. Langham, and S. Ahmed (1991) "RADARSAT," *Proc. IEEE,* **79** (6), 839–850.

Chapter 13: Discussion

S. Haykin (1990) ‘‘Radar Vision,’’ *Int. Radar Conf.*, McLean, VA.

REFERENCES

[1] T. E. Armstrong, G. W. Rogers, and G. Rowley (1978) *The Circumpolar North: A Political and Economic Geography of the Arctic and Sub-Arctic, Methuen, London.*

[2] Department of Indian Affairs and Northern Development, Ottawa, Canada (1990) *Northern Indicators*.

2

PROPERTIES OF SNOW AND ICE

EDWARD O. LEWIS
Bayfield Institute
Department of Fisheries and Oceans
Burlington, Ontario, Canada

CHARLES E. LIVINGSTONE
Canada Centre for Remote Sensing
Energy, Mines and Resources
Ottawa, Ontario, Canada

CAREN GARRITY
Microwave Group—Ottawa River
Dunrobin, Ontario, Canada

JAMES R. ROSSITER
Canpolar Inc.
Toronto, Ontario, Canada

The physical properties of a remotely sensed target determine the techniques that will be effective in detecting that target and the strategies that must be used to be successful. Often it is these very properties which are being studied, and therefore their range needs to be known. In this chapter, the physical properties of ice and snow are given for two reasons:

- to summarize succinctly the physical properties for the reader who may not be familiar with them;
- to discuss those properties which have the largest influence on remote sensing.

Remote Sensing of Sea Ice and Icebergs, Edited by Simon Haykin, Edward O. Lewis, R. Keith Raney, and James R. Rossiter.
ISBN 0-471-55494-4

Ice and snow are ever-changing natural materials, with wide changes in their physical properties daily, through the season, and with geographical region. In this chapter the major properties of sea ice, glacier ice (and therefore icebergs), and snow are discussed. The electromagnetic properties, which control most remote sensing techniques, and radar signatures by season are described in some detail.

2.1 BULK PROPERTIES OF SEA ICE

Sea ice is an anisotropic material with wide spatial and temporal variability. The properties of the ice are a function of its history, temperature, salinity, density, and age, amongst other factors.

The ice cover can be divided into four major categories: young, first-year, second-year, and old ice. Young and first-year ice form within the current ice year (late October to approximately May or June in the northern hemisphere). The young ice category includes all sea-ice growth stages leading to a relatively stable form, known as first-year ice in the nomenclature of the World Meteorological Organization [116]. Any of the first-year ice that survives the summer melt season becomes second-year ice at the start of the next ice growth season, and increases in thickness throughout that season unless it is destroyed by a melt event. Old ice has survived more than one melt season. Within each category, ice properties vary, depending on diurnal, seasonal, and ambient conditions.

2.1.1 Ice Growth

The formation of sea ice is a complex process that begins at $-1.8°C$ in typical sea water. When the salinity of sea water exceeds 24‰, its maximum density is reached at temperatures below the freezing point. The increasing density of the chilled surface layers results in an unstable vertical density distribution which produces vertical convection in the surface layer of the ocean. No ice forms until this "mixed layer" is cooled to the freezing point. In the Arctic, this mixed layer can reach a thickness of approximately 50 m.

Ice formation begins at the sea surface with the formation of a suspension of small ice crystals known as frazil. These crystals are pure water ice and take the form of small needles or small spheres and circular platelets as described by Kumai and Itagaki [62]. The rejection of salts from the forming ice crystals enhances the vertical instability of the water column and results in the formation of frazil throughout the upper few meters of the ocean.

When the ice forms in a mechanically quiet environment (calm seas and little or no wind), the buoyant frazil crystals rise to form an unconsolidated mat of randomly oriented ice crystals on the sea surface known as grease ice (Fig. 2.1). This grease ice can be up to 10-cm thick, is only 20% to 40% ice crystals by volume, and acts to stabilize the sea surface by suppressing the

Figure 2.1 Grease ice.

formation of capillary waves in the presence of wind. Continued freezing results in a smooth, thin (up to 10-cm thick), elastic, weakly consolidated ice known as dark nilas. Consolidation progresses by water crystallizing in the interstitial brine within the ice sheet, with a resulting increase in salinity of the remaining liquid. Some of the brine is rejected (forced out of the ice mass) to the sea beneath the growing ice sheet and to the surface of the ice where it forms a thin, high-salinity film. The remainder of the brine is trapped within the ice volume in small, initially spherical, brine pockets. Continued ice growth occurs on the bottom of the ice sheet as congelation ice.

When the ice forms in mechanically active environments produced by strong winds and/or waves, the initial formation results in different ice forms than those observed in quiet environments. Grease ice, instead of forming a large sheet, is blown into streamers and wind rows. Wave action causes the ice to coalesce into lumps or slob ice and then to form small rounded floes called pancakes 30 cm to 3 m in diameter (Fig. 2.2). Wave-deposited frazil and slush created by interacting pancakes freeze onto the edges of these floes to create raised rims, sometimes several centimeters in height. During calm, wave-free periods, the frazil and slush surrounding the pancakes will freeze, bonding them together into an ice sheet (Fig. 2.3). Continued freezing under quiet conditions will result in consolidated ice, with growth by congelation at the bottom of the sheet. Recurrent wave action without long periods of consolidation will refracture the ice sheet to create composite pancake floes in a mobile ice pack. Wave and wind-driven pack compression can raft the pancakes to several layers in thickness, as is often observed in marginal ice zones.

The surface layer of new ice consists of randomly oriented crystals inde-

Figure 2.2 Formation of pancake ice.

pendent of its formation environment. However, as the congelation ice continues to grow below this layer, a preferred orientation of crystal growth develops. When the ocean surface becomes covered with ice, the growth rate is substantially reduced by the insulation of the sea water from the colder air. The growth rate of the ice is controlled thereafter by the temperature gradient in the ice cover.

Figure 2.3 Consolidation of a complete ice cover.

The growth of ice crystals in a preferred orientation is a complex subject which is well developed in Weeks and Ackley [114] and will not be repeated here. In summary, however, crystals which are aligned with their *c*-axis horizontal have the maximum rate of growth in the vertical direction. These faster growing crystals in effect choke out other slower growing crystals, resulting in an ice structure in which crystals with the *c*-axis aligned horizontally predominate (Perey and Pounder [94]).

This transition from random orientation to the preferred *c*-axis horizontal orientation takes place within a vertical distance of between 5 and 10 cm, although the transition may take place at up to 1 m from the surface in the case of a thick frazil layer. Not surprisingly, this layer is known as the transition zone. The ice below the transition zone is columnar in composition with *c*-axis orientation which is very close to horizontal. The thickness of this seasonal ice is typically between 1 and 2 m.

As the ice sheet grows, individual crystals project downward into the sea water. As more water freezes, much of the salt is expelled, increasing the salinity of the water at the ice-water interface. However, some of the brine is trapped within the growing ice, both within the crystal structure itself, but more predominantly in the interstitial zones between the crystals. The salt thus trapped is present both as liquid and solid inclusions. As the ice continues to grow, the ice crystals bridge together, trapping the liquid brine in elongated cells (brine pockets), as shown in Fig. 2.4. A brine pocket is a bubble of brine surrounded by ice that has either limited or no communication with other brine masses (Niedrauer and Martin [82]). The amount of brine trapped is a function of the salinity of the sea water and the freezing rate. The faster the freezing rate, the more brine that is entrapped. In frazil ice, the brine pockets are largely spherical (Campbell et al., [9]), while in columnar ice the brine pockets are elongated in the vertical direction.

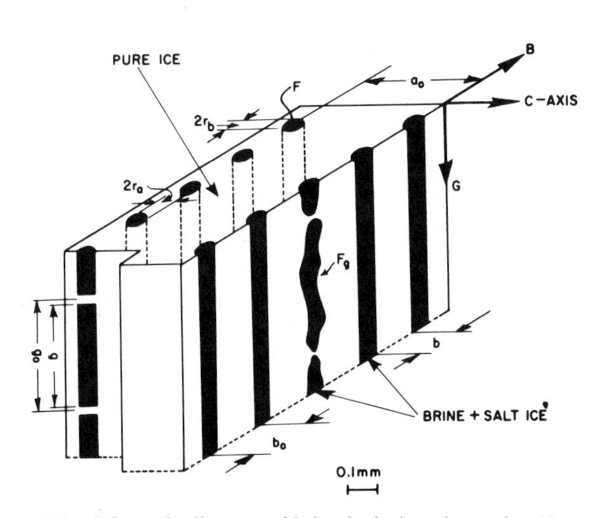

Figure 2.4 Schematic diagram of brine inclusions in sea ice (Assur [5]).

Addison [1] notes that the diameter of the brine pockets is about 0.05 mm, while the length varies from 2 to 3 cm. The spacing between brine inclusions, a_0, is usually 0.5 to 0.6 mm, although there is a gradual increase in the spacing with depth. Rather than being directly related to depth, it appears that a_0 is related to the ice growth rate, which is itself generally related to depth. Results by Nakawo and Sinha [80] show that the a_0 is inversely proportional to growth rate. Thus slow growth rate produces large a_0 values.

The process of growth is a process of bulk desalination. As the ice grows and ages, its freeboard (the height of the ice surface above the mean sea level) increases and there is a continuing reduction in the entrapped salts by several mechanisms known as brine drainage. Starting from the dark nilas mentioned earlier, increasing ice thickness is accompanied by drainage of some of the trapped brine within the ice sheet and its replacement by air. The air-filled voids change the visual appearance of the ice from dark gray (almost black) for dark nilas to medium gray for light nilas to light gray for gray ice to white for all thicker ice forms. Accompanying the changes in visual appearance are changes in the electromagnetic and mechanical properties of the ice. Brine drainage mechanisms include brine pocket migration, brine expulsion, and gravity drainage. By far the most significant mechanism is gravity drainage, which is the only one that will be discussed here; Weeks and Ackley [114] give a more complete description of known mechanisms.

Under the force of gravity, the liquid brine drains through channels in the ice known as brine drainage channels. These channels are like the branches of a tree, with the smallest channels closer to the surface of the ice, coalescing and widening with increasing depth in the ice. Near the bottom of the ice, the channels have a horizontal spacing of about 15 to 20 cm and a diameter of about 1 cm (Schwartz and Weeks [98]). In addition to these larger channels, Lake and Lewis [63] also reported smaller tubes near the ice water interface, with diameters of 0.3 to 0.5 mm and a density of 42 tubes/cm^2.

As the temperature of the ice decreases, the liquid inclusions in the brine pockets begin to freeze and the salts begin to precipitate, thus reducing the liquid brine volume. At −8.2°C, sodium sulphate decahydrate starts to precipitate (Anderson [3]), while at −22.9°C sodium chloride dehydrate begins to precipitate, causing a large change in brine content. Other salts precipitate at still lower temperatures, including potassium chloride at −36°C and calcium chloride hexahydrate at −54°C.

The ice surface, far from retaining the relatively smooth structure of original formation, undergoes significant and continual change. Under forcing from wind, waves, and ocean currents, adjacent ice sheets can impact under intense pressure. While the bulk salinity is large (during the dark nilas, light nilas, and gray ice growth stages), the ice is sufficiently elastic to first buckle under compression and then fracture so that portions of adjacent sheets ride over each other to form ice rafts. The most common rafting pattern appears as a row of interlaced fingers of over-riding ice called finger rafting. Rafting can occur repeatedly and several layers of horizontal, rafted ice can accumulate.

Compression rafting of flat-lying ice sheets seldom occurs in ice thicker than 30 cm.

In marginal ice zones, wave action can raft pancake floes to thicknesses of several layers. In these regions, the rafting process can leave upended blocks on top of the ice layer to produce meter-scale roughness on the ice surface. Rafted ice floes with individual floe thickness up to 90 cm have been observed in the Labrador Sea marginal ice zone. If the rafted ice is not further disturbed mechanically and the conditions remain cold, the ice will consolidate into a solid sheet of nonuniform thickness. Under these conditions, sea water pockets trapped between the subsurface layers will freeze to generate transitional, and at times locally, columnar crystal structures between the layers of frazil ice.

When the bulk salinity of the ice decreases sufficiently (gray–white and first-year growth stages), the mechanical properties of the ice change sufficiently so that compression failure of the ice sheet will occur along local lines of weakness and the ice will fracture in a brittle manner to form piles of rubble blocks above the ice surface (pressure ridge sails) and below it (pressure ridge keels). As in the case of rafting, the majority of the failed ice will be found below the water surface. Pressure ridge sail heights vary from less than 1 m for gray–white ice (30–50-cm thick) to more than 10-m thick in first-year ice (up to 2m) and in old ice.

In addition to the mechanical deformation processes discussed above, the surface of the growing ice undergoes a series of transformations determined by the environment and the growth process. One distinguishing property of very young, high-salinity ice is the thin film of high-salinity surface brine resulting from the consolidation process. Under cold conditions (air temperatures below −5°C), the high vapor pressure of this film favors the growth of vertical hoar frost crystals, known as frost flowers (Drinkwater [25]). These structures, although fragile, can grow to a few centimeters in height. The surface brine film wicks up the hoar crystals to produce a high-salinity structure that is a very efficient scatterer of vertically polarized, high-frequency, microwave radiation. As the ice thickens beyond the gray ice stage, brine drainage depletes the surface brine film and frost flower growth is no longer supported.

Snow cover forms the thermal interface between the atmosphere heat sink (or source) and the ice heat source (or sink) and plays a very significant role in the evolution of sea ice. Snowfall during the initial ice-formation stage increases the ice crystal content in the sea water suspension of frazil and promotes the formation of grease ice.

Snow accumulation on a consolidated ice sheet is a mechanical load and acts to reduce the freeboard of the ice. In regions which have heavy snowfalls, such as the Antarctic Ocean and the East Greenland marginal ice region, snow loads are often sufficient to flood the ice surface layers with sea water (Tucker et al. [102], Lang and Eicken [65]).

Snow crystals on the surface of young ice create a porous layer that transports surface brine upward into the snow pack, thus reducing the surface salinity of the ice and creating a layer of saline snow.

The thermal conductivity of compacted snow is approximately one-tenth that of sea ice (Nakawo [81]) and uncompacted, fresh snow is an even better thermal insulator. Even a very thin snow layer (2 to 4 cm) can have a large impact on the heat transfer through the ice cover and thus will affect the ice growth rate, especially for thin ice. The combination of the thermal insulation provided by the snow and the brine that has migrated into the snow base creates conditions favorable to snow-base recrystallization and to the growth of depth hoar crystals at the snow–ice interface. The coarsely recrystallized snow or "snow–ice" contributes to the ice surface roughness at scales of importance to microwave sensors. The thermal insulation provided by the snow maintains the snow–ice interface at temperatures different from the ambient air temperature and thus influences the evolution of both the ice and the snow. Under cold conditions, the warmer ice temperature accelerates brine drainage, drainage channel formation, and ice surface transformations beyond those expected from the air temperature. When atmospheric temperatures and/or surface solar heating provide sufficient energy input at the snow surface to cause local melting or a significant increase in the partial pressure of water vapor, the colder, underlying ice drives recrystallization processes within the snow volume. Extensive snow base transformations including the formation of "superimposed ice" (Onstott and Gogineni [85]) and depth hoar formation are observed.

At the end of the growth (winter) season, increasing air temperatures and increased solar input to the snow and exposed ice surfaces drive a melting and restructuring process in which the snow is a very important factor. The decomposition and melting process progresses through several stages (Livingstone et al. [70]) as the "early melt," "melt onset," and "advanced melt" seasons. The early melt season, characterized by an increase in water vapor and free water within the snow cover, causes extensive crystal metamorphosis within the snow volume and at the ice surface. Crystal growth is driven by the cold underlying ice. Melt onset is characterized by a significant accumulation of free water (free water content approaches or exceeds 3%) within a snow pack that is rapidly becoming isothermal, and a free water accumulation at the snow–ice interface (slush layer growth). While the snow pack becomes saturated with water, the snow surface becomes increasingly rough. During advanced melt, the snow pack ablates rapidly, the ice surface floods and drains, and ice decomposition occurs. During the melting–decay process, the snow cover contributes a granular surface layer to the ice and supplies a volume of fresh, or relatively fresh, water to the ice surface. Drainage occurs through holes and cracks and through the formation of open drainage channels in the ice sheet (Holt and Digby [52]). The absorption of solar radiation by the snow–ice system is governed by its surface albedo.

Since the albedo is approximately 0.85 for snow, 0.6 for dry ice, and 0.2 for water-covered ice, the snow cover is thought to control the rate of melt in a local sense. Where the snow is thinnest, the melting snow will become fully saturated, thus decreasing the albedo and accelerating the rate of melt. When low patches of level ice are exposed, these melt faster than surrounding snow-covered areas to form melt ponds in which the rate of melt accelerates to up to 5 cm/day (Carsey [11]) because of the even lower albedo of open water. Higher relief areas and regions of thicker snow cover (drifts for example) lose melt water by drainage to surrounding lower areas, retain a higher albedo, and melt more slowly. On level first-year ice, the ice surface is quickly covered by saturated snow and melt water (Holt and Digby [52] report 75% to 85% coverage with melt water). Young ice forms which have shallow snow cover quickly disintegrate. Much first-year ice develops thaw holes which mechanically weaken the ice and speed its breakup.

Some first-year ice survives the summer. On this ice, the differential melting results in hummocks and weathered ridges topped by recrystallized snow base remnants (Onstott and Gogineni [85]) and melt ponds and drainage channels. The melt and drain process removes salts from the ice both by surface flushing (Weeks and Ackley [114]) and by expanded drainage channels within the ice sheet. The layer of ice formed from the frazil ice is completely recrystallized and in low-lying areas the columnar ice is exposed. This modified first-year ice becomes second-year ice at the beginning of freeze-up. In second-year ice and also in old ice, the top layer (approximately the first meter) of the ice has been transformed to hard, clear, bubbly, low-salinity ice over most of the floe. In these ice types, ridge and hummock tops are large-grained, porous ice with little residual brine (Tucker et al. [103]). Frozen melt ponds are nearly freshwater ice containing small air bubbles, and thaw holes regrow first-year ice with a very thin frazil layer. Second-year and old ice growth commences at the bottom of the ice sheet as columnar ice and total ice thicknesses in the range of 2 to 5 m are not uncommon by the end of subsequent ice growth seasons; thicknesses in excess of 10 m have been observed (Walker and Wadhams [110]).

The role of snow in the life history of second-year ice and old ice parallels that of the first-year ice. During the initial stages of advanced melt, water collects in melt pond depressions to depths up to 20 cm (Gogineni et al. [37]) while the porous ice layers on ridge and hummock tops and snow drift sites are enhanced by the remaining snow base. In general, surface melting will result in the older, topmost layers of old floes, ablating so that the oldest ice could be 10 to 12 years old.

In addition to contributing to reduced salinity in the ice column, the melt season also causes substantial smoothing and rounding of surface features. Pressure ridges, rubble fields, and rafted zones are all rendered less angular. Figure 2.5 shows a typical old floe with significant relief but with well-rounded features. This same process causes pressure ridges to melt and then refreeze in a more consolidated form, resulting in the ridges becoming much stronger and more difficult to break.

The melting and thawing process produces pools of low-salinity water on the floe surface and the remaining water subsequently freezes, producing smooth areas of nearly freshwater ice. The structure of the resulting old floe is shown diagrammatically in Fig. 2.6.

Figure 2.5 Old-ice floe with significant vertical relief.

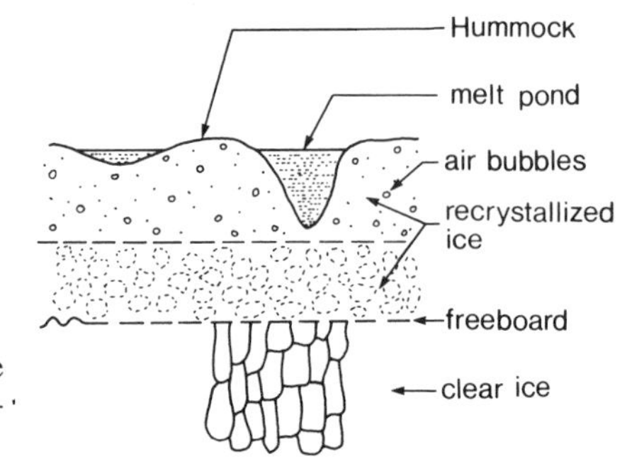

Figure 2.6 Schematic diagram of old-ice floe structure (Vant et al. [107]). Reprinted by permission of the American Institute of Physics.

2.1.2 Salinity

The salinity of sea ice is related directly to its rate of growth, the ice and atmospheric temperature, and age. As discussed earlier, as the ice ages, brine drainage takes place and the ice becomes less saline. Although the salinity of sea water in the open ocean is on the order of 35 parts per thousand (35‰), the initial salinity of the ice can be considerably higher.

During formation, the salinity of the thin layer of new ice can be greater than 50‰ (Martin [74]), although 25‰ seems like a more usual value. There is an initial rapid decrease in salinity with time, but for cold Arctic ice the salinity reaches a quasi-stable value which then decreases very slowly during the rest of the growing season. Nakawo and Sinha [80] reported a decrease in salinity from 25‰ on 2 December to 8.5‰ on 9 December in first-year ice (Fig. 2.7). This rapid initial decline as well as the slower decrease in salinity

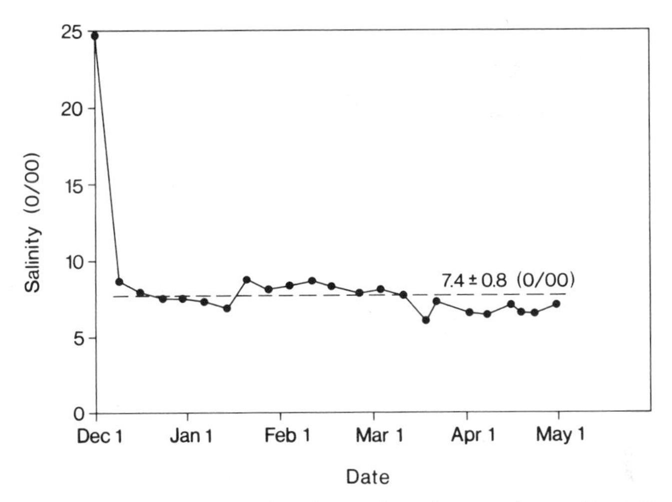

Figure 2.7 Variation of salinity with time for northern first-year ice at 40-cm depth. Broken line indicates average salinity excluding initial high values (Nakawo and Sinha [80]). Reproduced by courtesy of the International Glaciological Society from *Journal of Glaciology* (**27** (96), 315–329).

over the remaining period of the ice growth are shown. Nakawo and Sinha also noted that desalination increased rapidly once the warmer weather of May and June arrived.

In the warmer conditions off Labrador, Weeks and Lee [112] observed the same rapid decline in salinity shortly after ice formation (Fig. 2.8), but also

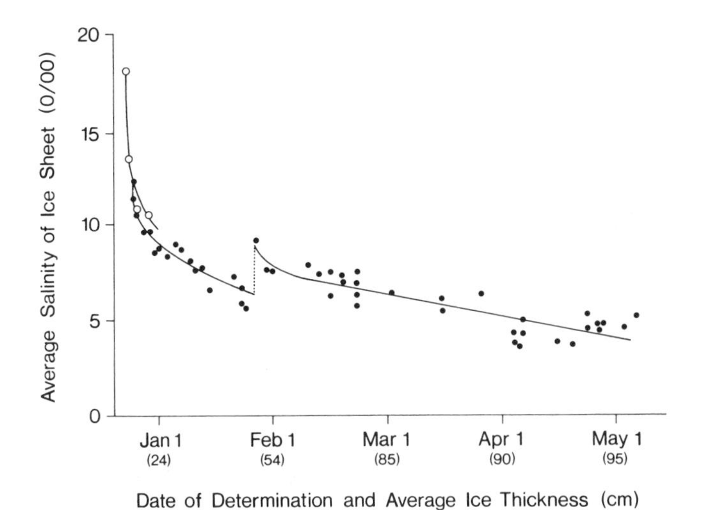

Figure 2.8 Average salinity of ice sheet versus time for Labrador coast ice (Weeks and Lee [112]). Reproduced by permission of the Arctic Institute of North America from *Arctic* (**11** (3), 135–156).

noted a continuing decline in salinity. The warmer ice temperatures at this site led to continuing desalination.

Typical salinity values for Arctic first-year ice, after initial rapid desalination, appear to be about 10 to 20‰ near the ice surface, 4 to 6‰ in the bulk of the ice sheet, and rise to 10 to 30‰ at the ice–water interface. Figure 2.9 shows a typical salinity versus depth plot for first-year ice.

When the ice goes through the melt season as previously discussed, significant brine drainage takes place. Figure 2.10 shows the reduction in salinity and the change in the shape of the salinity profile during the rapid desalination period. Not only does the net salinity decrease, but there is a substantial reduction in salinity near the ice surface and near the ice–water interface.

Figure 2.11 shows an average profile of salinity for old ice. The salinity is near 0 at the ice surface, increasing to 2 to 3‰ in the bulk. These measurements agree favorably with more recent results (Campbell et al. [9]; Cox and Weeks [18]). Figure 2.12 shows contours of salinity for an old floe where the salinity of the ice above the mean sea level is generally low. However, the salinity in the ice depressions can remain quite high as shown on the left and right sides

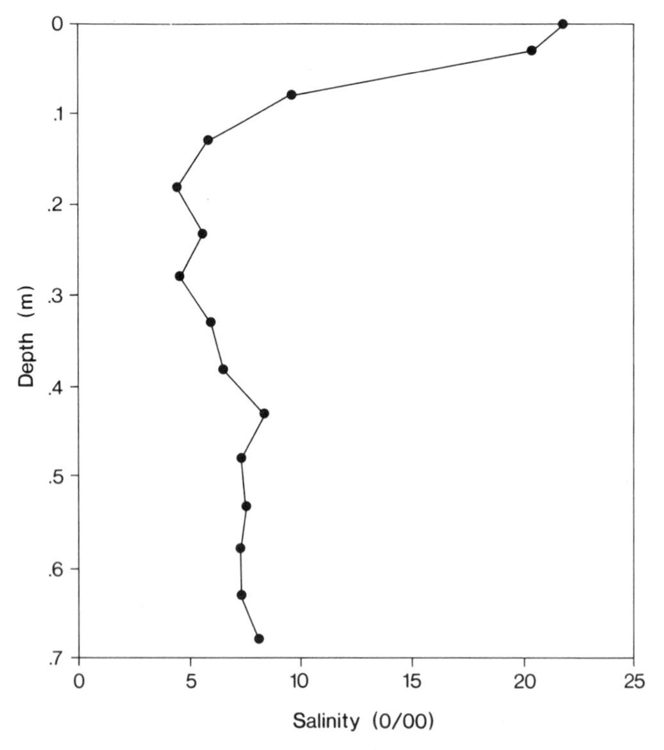

Figure 2.9 Typical salinity versus depth profile for first-year arctic sea ice (Martin [74]). Reproduced by courtesy of the International Glaciological Society from *Journal of Glaciology* (**22** (88), 473–502).

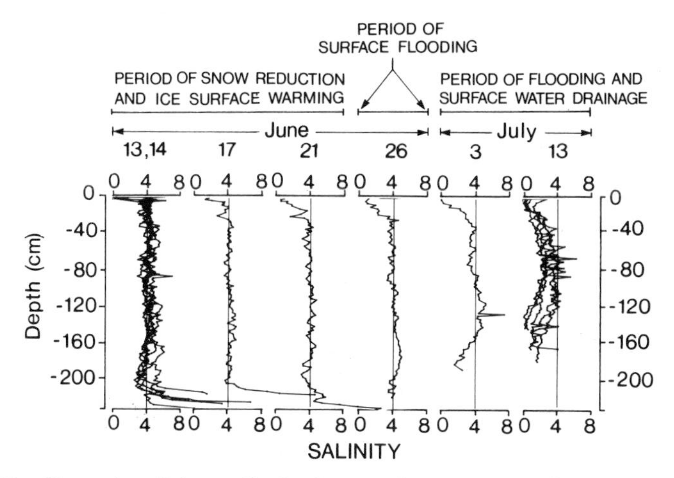

Figure 2.10 Change in salinity profile for first-year ice during rapid desalination period (Holt et al. [52]). Reproduced by permission of the American Geophysical Union.

of Fig. 2.12, whereas the salinity on the hummock in the center is considerably lower.

2.1.3 Temperature

Typically, the temperature (at equilibrium) in ice increases linearly from the atmospheric temperature at the ice–air interface to about $-1.8°C$ at the ice–

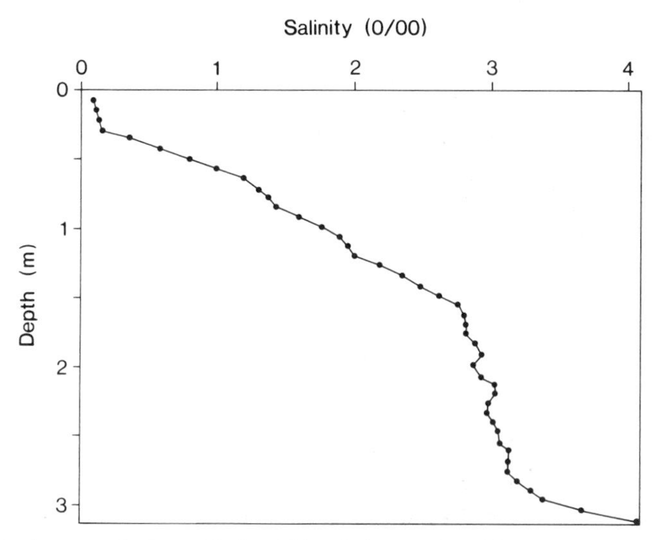

Figure 2.11 Average salinity profile for old ice (Schwarzacher [99]). Reproduced by permission of the American Geophysical Union.

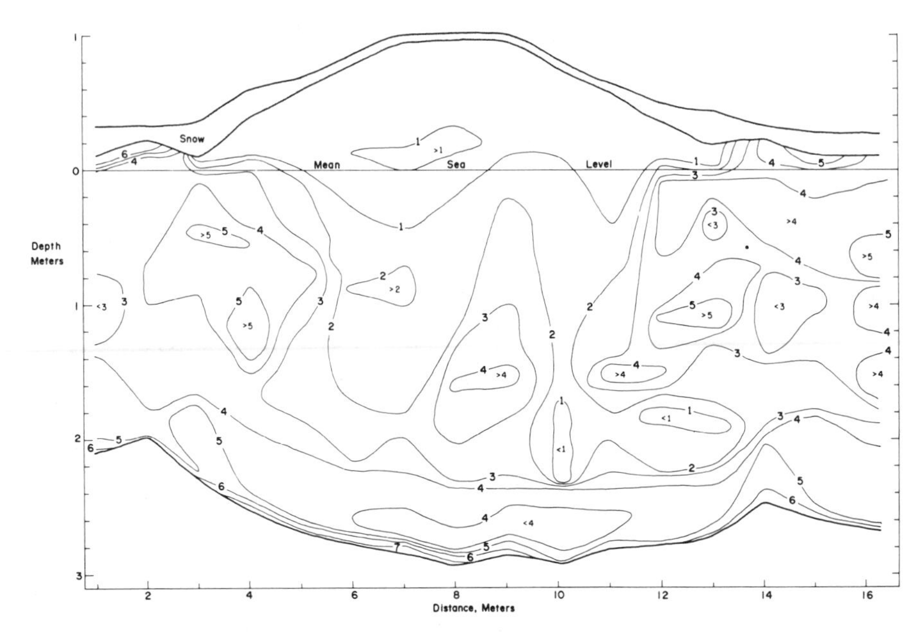

Figure 2.12 Salinity contours in old-ice floe. Contour lines are at 1‰ intervals (Cox and Weeks [18]). Reproduced by courtesy of the International Glaciological Society from *Journal of Glaciology* (**13** (67), 109–120).

water interface. Plot A of Fig. 2.13 shows the temperature structure for first-year ice while Plot B shows a similar temperature structure for old ice.

2.1.4 Density

The density of sea ice decreases with age as the brine pockets are replaced by ice (Weeks and Lee [112]). Figure 2.14 shows the density of ice as a function of temperature and salinity. Weeks and Lee [112] reported densities of newly formed ice at Hopedale, Labrador, as high as 0.945 g/cm^3. Schwartz and Weeks [98] and Evans [31] reported densities of first-year ice in the range of 0.910 to 0.920 g/cm^3, while values for old ice are quoted as 0.910 to 0.915 g/cm^3. Evans [31] notes that for old ice, the higher the freeboard, the lower the destiny, as given by:

$$\rho = -194 f_{fb} + 974$$

where ρ is the density in kg/m^3 (note, however, that density is generally expressed as g/cm^3) and f_{fb} is the freeboard in meters. The freeboard is the height of the ice surface above the mean sea level. These density values compare to a density of sea water at 1.026 g/cm^3 at 5°C.

For the top 10 to 20 cm of very porous second-year ice, Sinha (verbal communications) has found densities as low as 0.65 g/cm^3. Cox et al. [19]

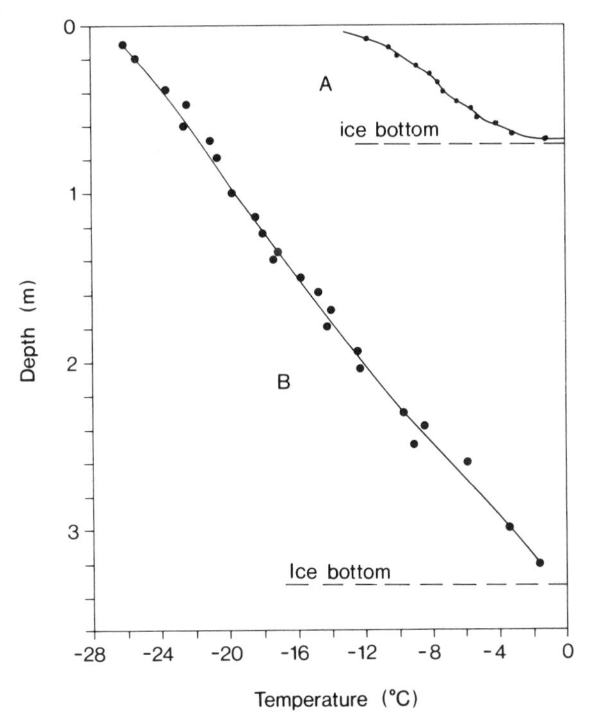

Figure 2.13 Temperature profile for first-year (A) and old (B) ice. (Plot A from Martin [74]; Plot B from Cox and Weeks [18].) Reproduced by courtesy of the International Glaciological Society from *Journal of Glaciology* (**22** (88), 473–502, and **13** (67), 109–120).

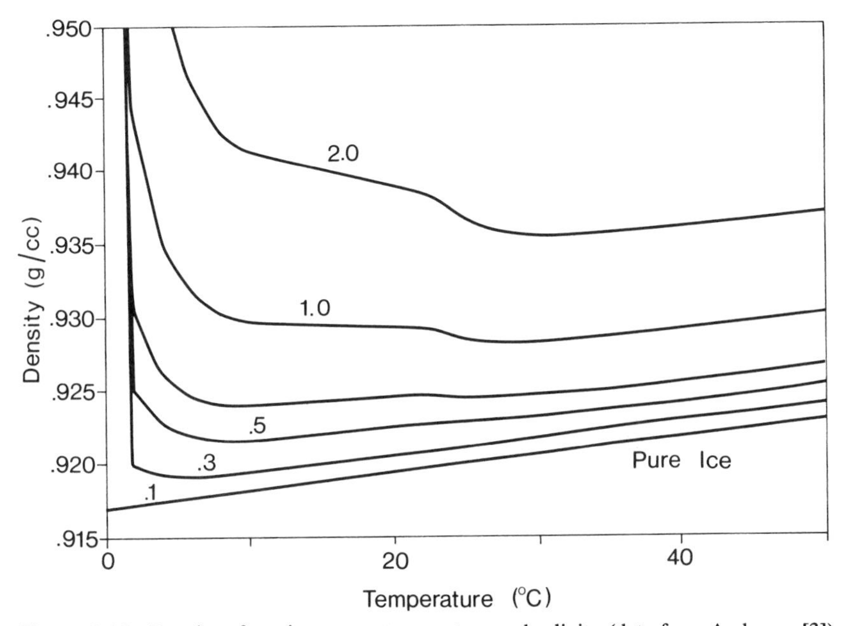

Figure 2.14 Density of sea ice versus temperature and salinity (data from Anderson [3]).

reported average densities of 0.875 g/cm^3 in pressure ridge sails and 0.899 g/cm^3 in the keels. These low densities, not atypical for the top surface of old ice, are the result of a great number of relatively large air bubbles and voids. Sinha reported that near the surface the bubble diameters are in the range of 4 to 6 mm, while at the 10-cm depth they are about 3 mm. There is a sharp decrease in size to 0.4 mm at 15 cm, with most voids in the lower columnar ice being in the range of tenths of millimeters. Vant et al. [107] assume spherical bubbles of 5-mm diameter for old ice for modeling purposes.

There are very few reports of bubbles in first-year ice, although Nakawo [81] reported values of 0.5 to 1.0 mm as typical. Vant et al. [107] assumed values of bubble sizes in first-year ice as 0.5 mm.

2.1.5 Brine Volume

The salinity of the ice is a measure of its salt content and, in combination with the ice temperature, determines the brine volume within the ice. The brine volume not only affects the electrical properties of the ice, but also significantly affects the structural properties and, in particular, ice strength (see Subsection 2.1.6). Anderson [3] reports that "under natural conditions, young sea ice is generally only one-third to one-half as strong as fresh ice but increases in strength as it freshens during subsequent winters."

Frankenstein and Garner [32] have formulated an empirical relation for brine volume in parts per thousand (‰) for the temperature range of −0.5° to −22.9°C:

$$V_{br} = S\left(-\frac{49.185}{T_c} + 0.532\right)$$

where S is the salinity of the ice in ‰, and T_c is the temperature in centigrade, for $-0.5 \leq T_c \leq -22.9$.

2.1.6 Mechanical Properties of Ice

Mellor [78], based on work over the last 30 years, gives a very comprehensive description of the mechanical properties of sea ice. Generally speaking, sea ice is a very complicated material in terms of its mechanical properties; its elastic parameters are functions of temperature, salinity, brine volume, crystal size, crystal orientation, time, etc. Brine volume (see Subsection 2.1.5) is defined by Pounder [95] as the fraction of the volume of a piece of sea ice occupied by fluid, brine, or air. Pounder suggested that sea ice, to a first order of approximation, be considered as a viscoelastic material and its mechanical behavior be described by a rheological model known as a Maxwell unit. Although this is a simple model, it can be used to illustrate the mechanical

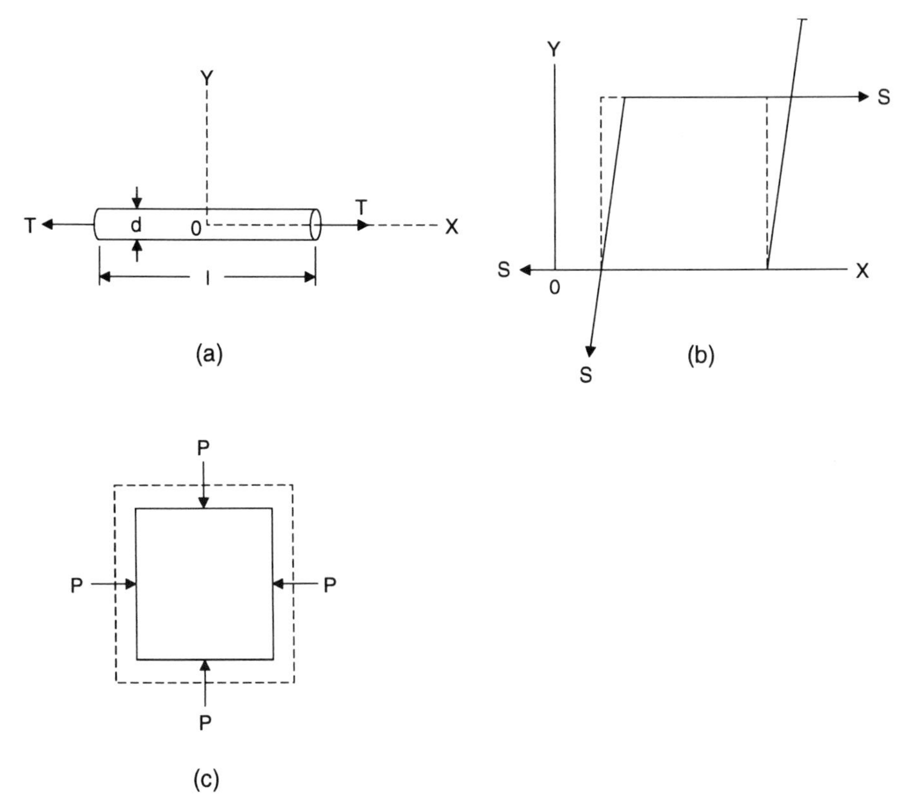

Figure 2.15 Sketch illustrating three types of stresses (after Pounder [95]). (a) Pure tensile stress on an ice cylinder. (b) Pure shear stress on an ice block; the dotted rectangle indicates an unstrained block of ice. (c) Compressional pressure on an ice block.

behavior of sea ice. Therefore, it is worthwhile presenting the basic ideas of the Maxwell unit.

Neglecting the anisotropic nature of sea ice, for any kind of deformation one can decompose the stresses applied to the ice into pure tensile stress, illustrated in Fig. 2.15(a) by an ice cylinder, pure shear stress [Fig. 2.15(b)] by an ice block, and pure compression or dilatation as shown in Fig. 2.15(c) by a piece of ice immersed in a fluid. The relationships between strain, e_{ij}, and stress, s_{ij}, for the three cases are given as follows:

$$s_{xx} = E_y e_{xx}$$

$$s_{xy} = \mu e_{xy}$$

$$p = k \frac{\Delta v}{v}$$

and

$$\sigma_P = -\frac{e_{yy}}{e_{xx}}$$

where

E_y = Young's modulus of ice
μ = the shear modulus or rigidity of ice
k = the bulk modulus of ice and
$\Delta v/v$ = the fractional decrease in volume
p = hydrostatic pressure
σ_P = Poisson's ratio

Let e_e, e_v, e_t be the elastic, viscous, and total strains, respectively. Assume a sinusoidally varying shear stress $s = s_0 \sin \omega t$ is applied to a block of viscoelastic ice. For the elastic part of the strain, by Hooke's law,

$$\mu e_e = s \tag{2-1}$$

For the viscous part of strain, using the Newtonian flow model,

$$\eta \frac{de_v}{dt} = s \tag{2-2}$$

where η is the coefficient of viscosity of the ice. Since $e_t = e_e + e_v$,

$$\frac{de_t}{dt} = \frac{de_e}{dt} + \frac{de_v}{dt} \tag{2-3}$$

Substituting (2-1) and (2-2) into (2-3), one has:

$$\frac{de_t}{dt} = \frac{1}{\mu}\frac{ds}{dt} + \frac{1}{\eta}s \tag{2-4}$$

This is the mathematical expression for the Maxwell unit shown in Fig. 2.16.

In Fig. 2.16, a spring and dash pot have been connected in series to simulate the viscoelastic property of sea ice indicated by (2-4); that is, the total strain

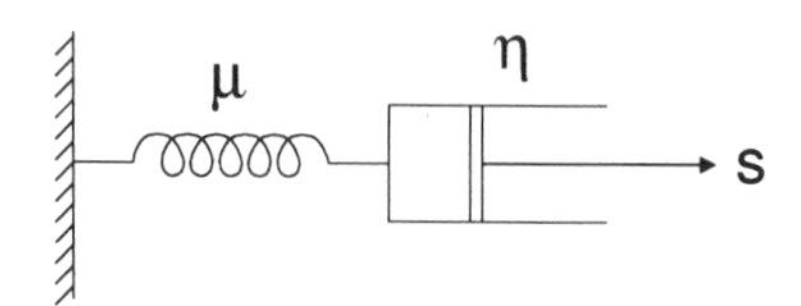

Figure 2.16 A rheological model for sea ice: the Maxwell unit.

in the ice is the sum of strains due to elastic and viscous deformations, respectively. A more thorough discussion of ice properties based on the Maxwell unit can be found in Pounder [95].

Driven by the harmonic stress $s = s_0 \sin \omega t$, (2-4) leads to a response

$$e_t = \frac{s_0}{\mu} \sin \omega t + \frac{s_0}{\omega \eta} (1 - \cos \omega t) \tag{2-5}$$

From (2-5), it is seen that if $\omega >> \mu/\eta$, the first term is dominant and the strain is almost entirely elastic. Taking $\mu = 2.5 \times 10^9$ Pa (a reasonable value for pure ice), and $\eta = 1.0 \times 10^9$ Pa $\cdot$ s/cm^2 (a reasonable figure for temperature ≈ -5°C and stress $\approx 9.8 \times 10^4$ Pa), one has $\omega >> 2.5$ rad/s. That is to say, any periodic force with a period much less than 1 s will result only in elastic deformations of ice. The viscosity of ice affects its elastic behavior if the ice is subject to a slowly varying force.

It should be pointed out that the Maxwell unit is the simplest model for ice. Apart from elastic and viscous behavior, ice has plastic and creeping properties. With creeping, ice can actually memorize stresses. In other words, it can delay its deformative response to a stress being applied to it for a long time. All these characteristics can be modeled in a way similar to the derivation of the Maxwell unit. More comprehensive models are summarized in Mellor [78].

In general, the strain–stress curve of ice has three different ranges. They are:

1. the elastic region where Hooke's law holds;
2. the plastic region where the relationship between strain and stress is nonlinear and after the stress is removed ice cannot revert to its original shape;
3. the yield point, or the ultimate stress point, at which ice breaks.

However, the distribution of these ranges varies greatly with the rate of stress application, or strain rate. If the stress is applied rapidly to ice, and for a short period of time only, the elastic range of the ice will be quite large and the plastic range very small, that is, stress only slightly above the elastic limit will lead to fracture. These characteristics were illustrated by Karlsson [57] in his numerical model. Figure 2.17 shows the results.

Sea ice grows out of sea water and forms a layered structure, from which sea ice inherits a so-called profile property; i.e., the mechanical properties of the ice cover vary with depth. Generally, the Young's modulus in the surface layer of an ice sheet is greater than that in the lower layer. Cox and Weeks [20] reviewed a wide variety of field studies conducted in different polar regions. They summarized data sets collected by a number of people and derived vertical distributions of temperature and salinity for ice sheets of different thickness. Using a least squares fitting method, Cox and Weeks obtained empirical

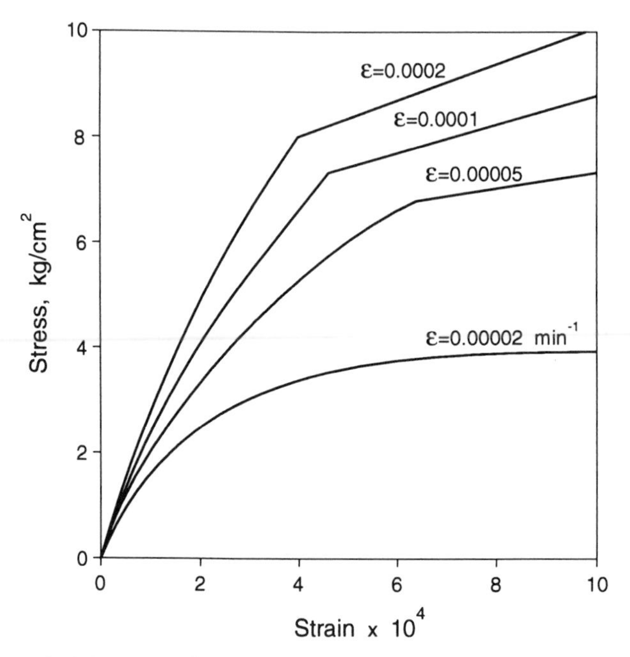

Figure 2.17 Uniaxial stress-strain curves for viscoelastic-plastic model at various strain rates indicated by $\dot{\epsilon}_0$.

formulae for Young's modulus and Poisson's ratio in terms of brine volume profiles in ice sheets. Figure 2.18 shows profiles of Young's modulus for ice sheets of 10 different thicknesses.

Physical Process of Ice Failure

In studying the strength of an elastic plate, Griffith [45] first postulated a criterion for growth of a crack of length $2C$. The basic idea is that the change of potential energy in the plate is equal to the change of surface energy in the crack as it grows in length. Mathematically this idea can be expressed as:

$$\frac{\partial}{\partial C}\left(\frac{\pi s^2 C^2}{E}\right) = \frac{\partial}{\partial C}(4C\gamma)$$

where γ is the specific surface energy of the material; s is the applied stress (tensile stress perpendicular to the long axis of the crack) at which crack growth occurs, and E is the surface energy in the crack. Therefore, the minimum applied stress necessary for crack growth is:

$$s = \left(\frac{2E\gamma}{\pi C}\right)^{1/2} \tag{2-6}$$

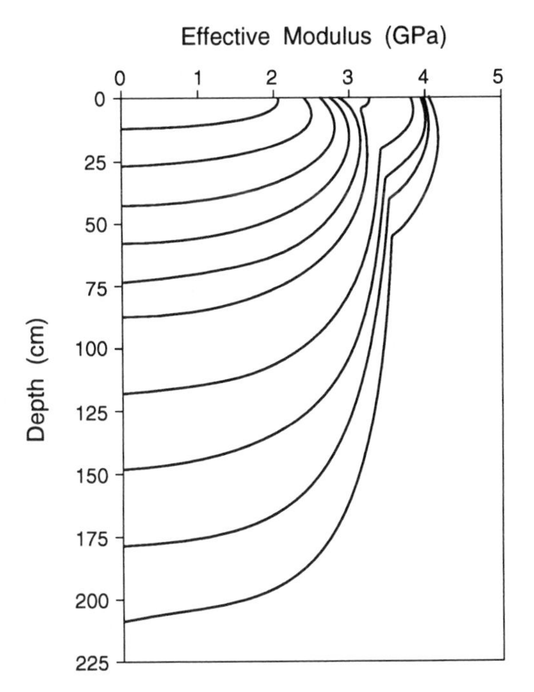

Figure 2.18 Young's modulus profiles for 10 different ice thicknesses (after Cox and Weeks [20]). Each profile intersects the vertical axis at the depth corresponding to the bottom of the ice and thus shows the model profile for ice of that thickness.

However, the predicted values from (2-6) have been found to be too low for the actual strength of most materials, including ice samples. This discrepancy is explained by the fact that once a crack occurs, a large amount of stress will concentrate at the crack's tip, which has the effect of blunting the crack. Hence, further growth of the crack requires an extra amount of work to overcome this barrier or "plastic region."

Orowan [90] modified Griffith's model by including a specific energy for plastic working γ_p in (2-7). Since $\gamma_p >> \gamma$, (2-6) becomes:

$$s = \left(\frac{2E\gamma_p}{\pi C}\right)^{1/2} \tag{2-7}$$

This result is important because it emphasizes the fact that as the crack lengthens (C increases), the stress required for further extension of the crack decreases.

Although the above model gives a nice picture of failure processes in an elastic plate, it is hard to use this theory to get a quantitative interpretation of sea ice cracking processes. Sea ice has a more complicated structure than an elastic plate and, as mentioned above, its elastic parameters vary with temperature, salinity, and other factors. Therefore, the discussion of ice properties in terms of these parameters is of limited value. It is almost impossible to get an analytical form relating plastic parameters with these factors.

Based on a laboratory study, Dykins [27] concluded that:

1. Neither the crystal grain size nor the spacing of platelets (smallest elements of ice crystals) have much influence on the tensile strength.
2. Orientation of the crystal grain structure in relation to the stress field has a significant influence. That is, the vertical tensile strength is 2 to 3 times greater than the horizontal one.
3. Temperature has a linear effect on ice strength while salinity has a nonlinear effect. The influence of temperature and salinity could be converted into a single function—brine volume.

Based on field data obtained from sites scattered throughout the Arctic, Cox and Weeks [20] derived empirical formulae for the tensile (*TS*), shear (*SS*), and flexural strength (*FS*) of sea ice in terms of its brine volume V_{br}. These are:

For horizontal tensile stress tests: $$TS = 0.816 - 0.0689\sqrt{V_{br}}$$

For vertical tensile stress tests: $$TS = 1.54 - 0.0872\sqrt{V_{br}}$$

For shear stress tests: $$SS = 1.68 - 0.118\sqrt{V_{br}}$$

For flexure stress tests: $$FS = 0.959 - 0.0608\sqrt{V_{br}}$$

where strength is in megapascals (MPa) and V_{br} is in parts per thousand (ppt).

It is a matter of common observation that sea ice strength increases as its temperature drops. Assur [5] also gave a theoretical model for sea ice structure and related it to the strength of sea ice. The basic idea is very simple. He assumed that the reduction in the strength of sea ice was proportional to the reduction in ice volume due to the existence of rows of brine cylinders in the ice. The strength for ice with this structure is

$$s = s_0\left(1 - 2\sqrt{\frac{\zeta}{\pi\frac{b_0}{a_0}}}\sqrt{V_{br}}\right)$$

where

s and s_0 = the ultimate tensile strengths with and without brine, respectively

ζ = g/g_0 with g being the average length and g_0 the average separation of brine pockets along the growth axis, respectively

V_{br} = the brine volume
a_0 = the average platelet thickness
b_0 = the average brine cylinder separation

As revealed by the above equation, the strength of sea ice decreases as the brine content increases.

2.2 GLACIER ICE

Glacial ice, unlike sea ice, has its origin on land. Glaciers spawn billions of tons of icebergs each year from tongues which reach the sea. The icebergs drift over an area of 70,000,000 km^2, or more than 20% of the Earth's ocean area, and pose a serious threat to navigation and offshore activity in many areas of the world. The "unsinkable" ocean liner MV *Titanic* collided with a west Greenland-spawned iceberg in the north Atlantic in April 1912, with the loss of 1513 lives. More recently, the Danish ice-breaking passenger cargo ship MV *Hans Hedtoft* sank off Greenland in 1959 after collision with an iceberg with a loss of 95 lives.

2.2.1 Morphology and Density

Glaciers form where precipitation in the form of snow exceeds its loss due to evaporation and/or melting. The formation of glacial ice starts with snowfall on the high altitude areas of existing glaciers. The new-fallen snow has a low density (0.06 to 0.08 g/cm^3) because of the air trapped between the hexagonal snow crystals.

The length of time for the snow to be converted to ice largely depends on temperature. In temperate glaciers (glaciers where the ice is at the pressure-melting point throughout), the delicate points of the snow crystals begin to melt quickly, the snowflakes are converted to spherical particles, the snow settles, and the density increases. The density can increase to 0.2 g/cm^3 in a few days. This more dense snow is granular in form. However, in subpolar and polar glaciers, this melting and compression process may take years. The compaction process is accompanied by, and is a result of, the ice crystals growing larger, joining together, and thus eliminating the air spaces between the crystals. As the snow continues to become more dense through compression, recrystallization, and freezing and thawing, it reaches an intermediate step on route to becoming glacier ice. This intermediate step, known as firn or neve, occurs when the snow reaches a density of between 0.4 and 0.55 g/cm^3.

As the density increases further, mostly by recrystallization, and the interconnecting air passages between ice grains are sealed off, the firn is transformed into glacier ice. The density at which this occurs is around 0.8 to 0.85 g/cm^3. For Greenland glaciers this process may take from 150 to 200 years from the original snowfall. As the firn is transformed into ice, the air that was trapped in the firn is either forced out through cracks or fissures in the ice or is trapped in bubbles. The further increase in density of the ice is due to the compression of these air bubbles. The pressure in air bubbles within Greenland glacial ice has been measured as high as 10 to 15 atmospheres (Scholander and Nutt [97]). The glacial ice is impermeable to air or water.

The visual "whiteness" of glacial ice is caused by the tiny air bubbles distributed throughout the ice. These bubbles give glacial ice a 4 to 9% volume of air. Scholander and Nutt [97] reported that for Greenland icebergs, in general, the bubbles are elongated with diameters of 0.02 to 0.18 mm, and lengths of up to 4 mm. They also reported a lesser number of spherical bubbles with diameters ranging from microscopic up to 2 mm. Gow [39], in a detailed analysis of bubble content in Antarctic glaciers, reports a range of spherical bubble sizes of 0.49 to 0.33 mm, decreasing with depth. The density of the bubbles, however, was relatively independent of depth. Gammon et al. [33] report round or ellipsoidal bubbles, with elongation on the order of a few tenths of millimeters. It would seem that typical values for bubble sizes in glacial ice would be on the order of 0.1 to 0.5 mm. The occasional piece of bluish glacial ice indicates ice which is free of bubbles.

The final density of glacial ice can exceed about 0.9 g/cm^3. As sea water has a density of approximately 1.0 g/cm^3, about six-sevenths of the mass of floating glacier ice will be below the sea surface, while the freeboard of individual icebergs will depend on their shape.

2.2.2 Icebergs and Their Fragments

Under the pressure of its own weight, the glacier flows from its high altitude source towards sea level. Glacier flow in Greenland has been measured at up to 20 m/day. In many instances the glacier continues to flow out over the sea, often reaching distances of 40 to 60 km offshore. Under the continuing action of the tides and ocean level changes due to winds and swell, large pieces of ice break off glacier tongues or ice shelves.

Depending on their size, these ice features are classified as either ice islands or icebergs. Ice islands break away from ice shelves and are large pieces of ice which in the Arctic (Untersteiner [104]) usually protrude about 5 m above sea level, are 30 to 50 m thick, and from a few thousand square meters to 500 km^2 or more in area. Ice islands are relatively rare in the Arctic but are common in the Antarctic. In the Antarctic (Husseiny [53]), ice islands and massive icebergs measuring 200–250 m in thickness with freeboards of 35–55 m and lengths of many kilometers (but averaging on the order of 1 km) break off from the floating ice shelves. Icebergs break away from glacier tongues, ice shelves, and ice islands, and are pieces of ice higher than 5 m above water level, but cover less area than an ice island.

Each year, west Greenland and, to a lesser extent, eastern Canadian Arctic glaciers produce about 10,000 icebergs. The icebergs are trapped by sea ice in

TABLE 2.1 Classification of Icebergs by Size

Type	Height Above Waterline (m)	Waterline Length (m)	Physical Area Above Waterline (sq m)	Relative Size	Mass (tonnes)
Iceberg	>5	>30	300	Merchant ship	180,000
Bergy bit	1–5	10–30	100–300	Small house	up to 5400
Growler	<1	<10	<100	Grand piano	up to 120

Icebergs are further classified by size as follows:

Small:	Height less than 16 m, length less than 65 m
Medium:	Height 16 to 48 m, length 65 to 130 m
Large:	Height 48 to 70 m, length 130 to 225 m
Very Large:	Height greater than 70 m, length greater than 225 m

the Arctic fjords until they are released during breakup. The icebergs then begin a long and circuitous route which can take from 2 to 5 years, until their final disintegration, typically east of Newfoundland where the cold Labrador Current meets the Gulf Stream.

As the icebergs drift and begin to deteriorate, they fracture and break, or calve off smaller pieces. Pieces which are from 1 to 5 m above sea level are referred to as bergy bits, while smaller pieces, up to 1 m above sea level, are called growlers. Table 2.1 lists the representative height above water level, waterline length, and physical area for icebergs, bergy bits, and growlers.

Arctic icebergs are swept along by the counterclockwise Baffin Bay circulation into northern Baffin Bay, and then flow southward off Baffin Island, driven by the relatively narrow but rapidly flowing Baffin current. Icebergs tracked by satellite (Marko et al. [73]) have shown long-term drift rates of up to 12 km/day. However, the icebergs routinely make intrusions into Lancaster Sound, on occasion venture into Navy Board Inlet, regularly become trapped in landfast ice, or run aground. Although Marko et al. tracked four icebergs which traveled from Lancaster Sound to Davis Strait in only 8 to 15 months, they suggested that the most likely duration for an iceberg to drift from Lancaster Sound to Northern Labrador (66°N) would be in excess of 3 years.

The incidence of icebergs decreases farther south: Fig. 2.19 shows that there is an almost linear decrease in icebergs from northern to more southerly latitudes. The number of icebergs that cross 48°N annually and enter the Grand Banks area is shown in Fig. 2.20 (data from the International Ice Patrol). The number has been as high as over 2000 in 1984 and as low as none in 1966. The relatively large number of icebergs reported in recent years, however, is partially the result of better surveillance and detection techniques.

The deterioration of icebergs is slow in Baffin Bay and along the Labrador

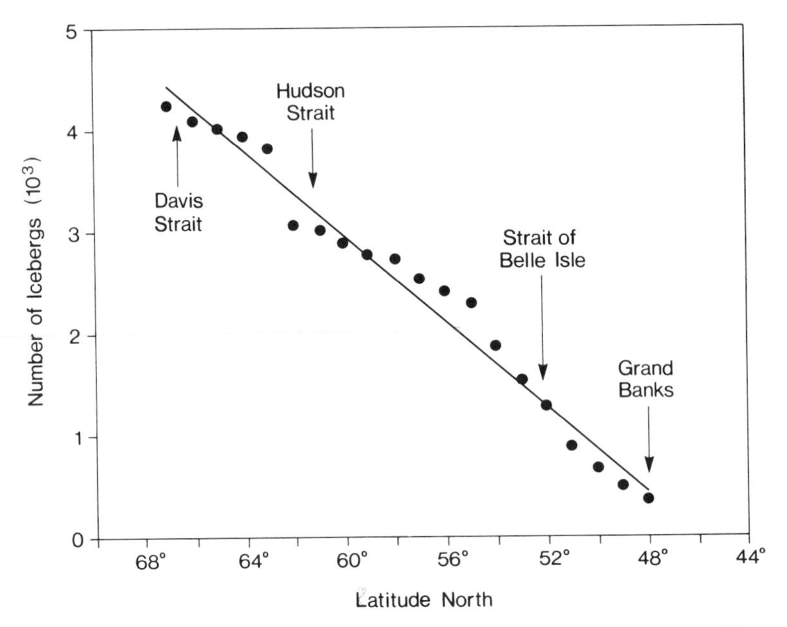

Figure 2.19 Annual average number of icebergs (dots) crossing selected latitudes (Ebbesmeyer et al. [29]).

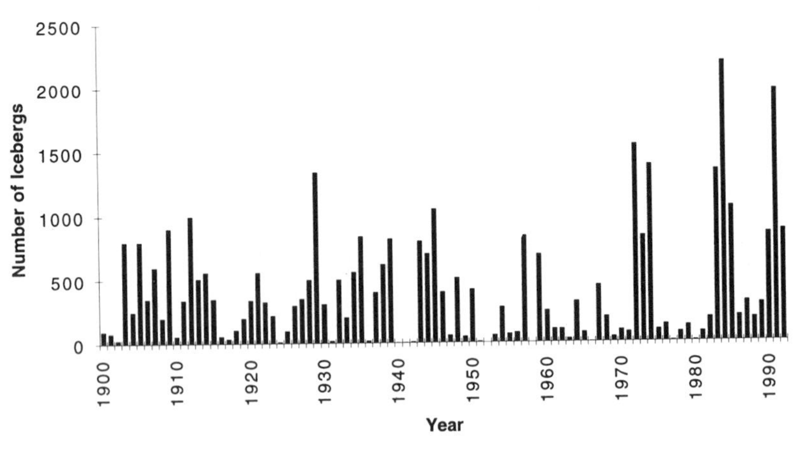

Figure 2.20 Number of icebergs crossing 48°N (data from International Ice Patrol).

coast because of the near-freezing sea water temperatures. However, as the icebergs move farther south into warmer water, they deteriorate more rapidly. When the icebergs reach the vicinity of 48°N (northern Newfoundland), they have lost more than 85% of their mass (Davidson and Denner [22]). Few icebergs pass south of 44°N and icebergs are rarely reported south of 40°N.

2.3 SNOW

In order to understand the microwave characteristics of sea ice, the overlying snow layer must also be understood. Like sea ice, snow is complex, and this complexity influences the microwave characteristics of snow-covered sea ice. Snow is also highly and rapidly changeable. Significant changes in the physical properties of snow on sea ice occur during the onset of melt in the Arctic and Antarctic. These changes have been observed by Garrity [34] for the Arctic (Greenland and Barents Sea) and Antarctic (Weddell Sea) based on 318 snow pit measurements.

A classification scheme has been developed for snow on land by the research group at the University of Bern (Schanda et al. [96]) and is also applicable to physical properties of snow observed on sea ice in the Arctic and Antarctic. Spring snow is one class from this scheme with two sub-groups: wet spring and dry spring snow. Wet spring snow is defined: "The snowpack surface consists of thick (at least several centimeters deep) firm layers of wet, quasi-spherical ice crystals (1–3 mm in diameter) formed during the day at temperatures above the freezing point and usually associated with either the passage of warm fronts or sunny, clear-sky conditions." The definition for dry, or refrozen spring snow is: "The surface of the snowpack consists of a layer of refrozen, firm snow that forms during clear, cold nights and is several centimeters thick." Subsection 2.3.4 goes a step further by presenting descriptive snow models for early, mid-, and late spring in the Arctic and for the Weddell Sea based on different ice regimes. Such descriptive snow models are needed in order to understand the interaction of microwaves with the surface of sea ice.

The onset of melt is defined here as the spring period during which the snow cover experiences changes in free-water content throughout its depth due to the redistribution of the water. The average free-water content of the snow cover is between 0% and 3% for this period. This average wetness does not include the slush at the snow–ice interface.

2.3.1 Morphology

Snow grain size and shape govern the snow density and amount of free water within a snow cover. Often, snow grain size is used to describe the texture and roughness of the snow. Snow grains are composed of an aggregation of molecules with a definite internal and external solid form enclosed by symmetrically arranged plane faces. Since the molecular structure of the snow is not discussed here, the term "grain" will be used to describe the size and general external shape of snow units.

From the time a snowflake forms until its destruction, it undergoes metamorphism. Even a dry, cold snow cover will change due to absorption of ionic impurities or to elastic strain (Colbeck [14, 15]). Grains of wet snow are found in tightly packed grain clusters for a snow wetness of less than 7% since the free water coheres individual snow grains together through surface tension. For a higher liquid-water content, snow grains are larger due to the growth of larger grains at the expense of smaller ones, and surface tension cannot hold the larger grains together as tightly (Langham [66]). Snow with a low liquid-water content typically has snow grains clustered in groups of two to six. Tetrahedral clusters are common. Snow regimes can be defined by the types of interfaces present. If there are three interfaces: ice–gas, ice–water, and water–gas, then the snow is in the pendular regime. If there are two interfaces: ice–water and water–gas, then the snow is in the funicular regime (Colbeck [12]) which corresponds to a relatively high snow wetness (Denoth [23]). If there is only a water–ice interface, snow is in the saturated regime.

Depth hoar has a variety of shapes, such as cups, plates, and needles. It is generally observed near the snow–ice interface where a temperature gradient often exists (Colbeck et al. [16]). Larger snow grains are common on the snow surface due to surface cooling from radiation heat loss through a clear atmosphere during cold nights, resulting in a fast growth rate of the grains. These grains are typically round in shape for dry snow and rounded and connected forming irregular shapes for wet snow. When there is slow grain growth, even for dry snow, the equilibrium shape of snow is not necessarily well defined (Colbeck [14, 15]).

Free-water distribution in snow adds to its complexity as a material. The amount of water held by a snow cover is a function of grain size, shape, and temperature. Based on field measurements, Lemmela [67] reported a percent wetness by volume of 2% to 3% at which free water starts to redistribute itself in the snow cover. The force of gravity causes free water to move lower in the snow or onto the surface of the layer below. However, based on laboratory measurements, Ebaugh and DeWalle [28] set this value at 1% to 2%. Denoth [23] observed a value of 7%, and Gerdel [36] observed a value as high as 10% for fresh snow.

2.3.2 Density

The density of snow depends primarily on its compaction. The density of wet snow is similar to that of dry snow since the density of water (1.00 g/cm^3) is similar to that of ice (0.92 g/cm^3).

Observations by Garrity [34] gave the average spring snow density in the Greenland and Barents Sea as 0.36 g/cm^3 (71 samples from 1987, 1989, and 1990) over both first-year and old ice. The average density during the summer was 0.30 g/cm^3 (27 samples from 1990 and 1991). The snow grains were loose, having a texture similar to sugar, with the grain size varying from millimeters to centimeters, giving a large range in density. Also, the snow–ice interface was not well defined, with snow grading to ice.

In the Weddell Sea, Antarctic, during spring 1989 the average snow density over first-year ice was 0.28 g/cm^3 (90 samples) compared to 0.37 g/cm^3 (22 samples) for that over second-year ice. The higher density for second-year ice was probably due to the thicker snow (44 cm vs. 15 cm) over second-year ice causing increased compaction lower in the snow. It appears that the range of snow density over first-year ice and old ice in the Antarctic may be greater than in the Arctic.

2.3.3 Wetness

Free water in snow is any form of liquid water which can be found around snow grains and as a meniscus between the grains. The amount of free water in a snow cover will influence its dielectric properties and thus its microwave characteristics.

The resometer developed by Mätzler at the University of Bern, Switzerland (Denoth et al. [24]) provides a tool for obtaining quantitative free-water measurements. The resometer is an instrument which measures the resonance characteristics of a material using a microwave cavity. Nondestructive measurements can be made of the snow permittivity and hence of its liquid-water content. Using a resometer on the snow surface, the surface sensor can limit the measurement to a layer of about 5 mm in thickness. The measurement is conducted by placing the sensor onto the snow cover, just touching, but not penetrating it. The weight of the sensor exerts enough pressure to make good contact with the snow surface. The small penetration depth of the sensor permits measurements of the vertical distribution of free water through a snow cover by repeating the measurements at different layers. In order to reflect incoming radiation, the sensor is white and put in the shade to avoid any melting of the snow. A single measurement consists of positioning the sensor and recording the resonance frequency in the range of 300 to 1300 MHz. There is an approximate linear relationship between the resonance frequency and the permittivity of the snow layer (Mätzler [75]). Once the permittivity and density of the snow are obtained, the free-water content in the snow can be calculated as a percent by volume using an empirical relationship developed by Ambach and Denoth [2].

The wetness of the snow surface is often highly variable due to topography on the surface. Therefore, a number of snow wetness measurements should be made of the surface. A measurement should also be made for each distinct layer in the snow cover. If the snow cover is homogeneous, a measurement should be made at the surface, middle, and near the snow-ice interface. The average of all snow wetness measurements from the surface and throughout the vertical snow profile will give a representative percent wetness of the snow cover.

The error of a snow wetness measurement using the resometer is mainly due to error in the density measurement. A good estimate in the error for density measurements is ± 0.05 g/cm^3, based on 80 density measurements which were compared using two different snow samplers and weighing devices (Garrity and Burns [35]). This error is at the upper bound, since the snow samples were weighed on site using a balance, and wind often made the balance unstable. It is advantageous to use an electronic balance protected from the wind. Using this upper bound gives an error in snow wetness of $\pm 0.4\%$ for a snow permittivity of 2.

Using a resometer Garrity [34] measured the snow wetness in the Arctic and Antarctic. The average snow wetness during spring in the Arctic was similar for first-year and old ice at 0.8% (84 samples). However, in the Antarctic, the difference was large: 1.1% (89 samples) over first-year ice and 0.5% 922 samples) over second-year ice. During the summer in the Arctic, the average snow wetness of 1.2% (26 samples) was not much larger than during the spring.

2.3.4 Evolution of Snow Cover—Arctic

In the Arctic, there is a progressive change in the snow cover properties as the spring period advances from early, to mid-, to late spring, as shown in Figs. 2.21(1) through 2.21(3), respectively.

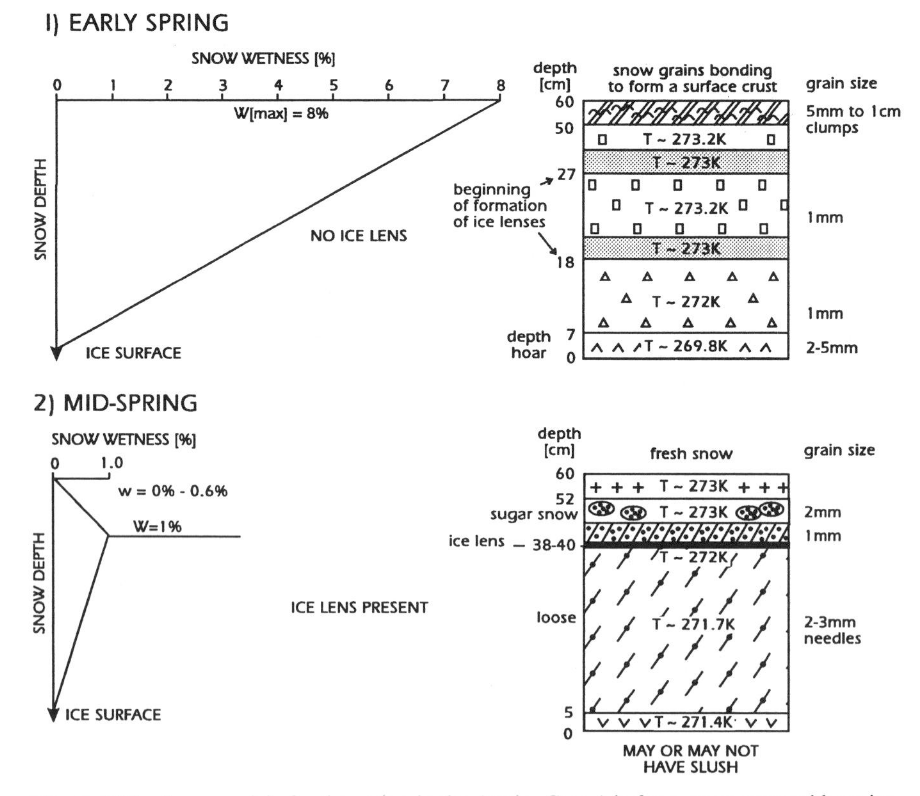

Figure 2.21 Snow models for the spring in the Arctic. Case 1 is for a snow cover with no ice lenses which applies to the early spring; Case 2 is for the mid-spring when ice lenses are present in the snow cover; Case 3 is for the late spring when there is slush at the snow–ice interface. The snow classification symbols were obtained from the Hydro-Tech Taylor field book (Seattle, Washington).

TYPICAL SNOW MODELS
SPRING - APRIL to JUNE
HIGH ARCTIC

3) LATE SPRING

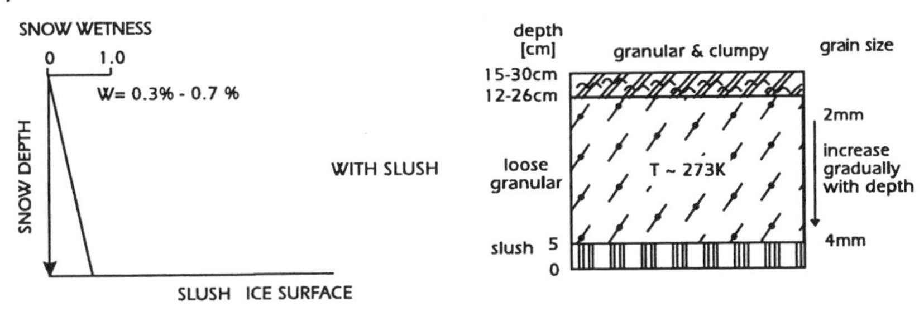

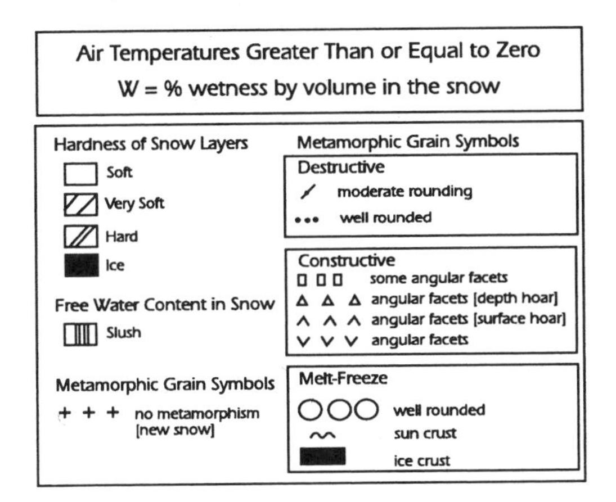

Figure 2.21 *(Continued)*

These snow evolution models are based on 127 distinct vertical snow profiles (snow pits) consisting of 350 individual measurements obtained during 1987 and 1989 in the Greenland Sea plus 1989 and 1990 in the Barents Sea (Garrity [34]).

In general, during early spring, the upper portion of the snow cover is warmer than the lower by about 4°C, whereas for the mid-spring the temperature difference is reduced to 2°C. There is no difference by the late spring when the snow cover is isothermal and near the melting point.

The descriptive early spring model (Fig. 2.21(1)) shows a decrease in snow wetness with snow depth. This is characteristic of a snow cover on sea ice during the onset of melt. After the winter months, this is the start of free water in the upper part of the snow cover which can be as high as 8% due to the passage of a warm weather front. However, 1% is more representative for the snow surface wetness. Some typical properties of a snow cover are also shown in Fig. 2.21(1), such as decreasing snow temperature with depth. This type of snow cover generally has no ice lenses; there are indications of structural changes, however, which may lead to the formation of ice lenses. A surface crust exists and can be described as being composed of a granular layer which has many snow crystals bonded together forming crystal aggregates (Colbeck [13]; Colbeck et al. [16]) and can be described as clumps. The average snow grain diameter is 5 mm, the size of the clumps about 1 cm. The larger grains in the surface layer are a result of the increased amount of incoming radiation during the spring period which enhances melt at the grain boundaries and impurities. Refreezing of water on the grains in the surface layer during the night increases their size. The remainder of the snow cover has a grain size of 1 mm except for the formation of 2–5-mm depth hoar crystals at the snow–ice interface due to an upward vapor flux (Colbeck et al. [16]).

Often thin layers of very hard snow or ice with a thickness of only a few millimeters to centimeters are in a snow cover. Refreezing of water in a snow cover can produce flat horizontal ice lenses in less than 24 hours. This was observed in Resolute Passage, Canada, during June 1990 (Garrity [34]) when a layer of solid ice 4-cm thick formed in a snow cover overnight after a day when the air temperature increased from below freezing to 2°C. The snow was about 30-cm thick and the ice lens formed 10 cm below the surface.

By mid-spring [Fig. 2.21(2)], very often there is at least one well-defined ice lens present within the top 10 to 20 cm of the snow cover. An ice lens will form when downward motion of water in the snow is prevented from percolating further by a less permeable snow layer, such as a previously wind-blown surface covered by a new snowfall, followed by freezing snow temperatures. A decrease in the snow wetness at the surface during early spring, which can be as high as 8%, is due to redistribution of free water in the snow cover followed by either refreezing to form an ice lens, or possibly accumulation as a slush layer at the snow–ice interface. If a slush layer was present, the snow depth generally was reduced compared to the early spring. The ice lens is an obstacle causing a decrease in the amount of free water below its location in the snow cover. Snow wetness just above the ice lens is often significant, for example 1%. The soft snow layer above the ice lens could be described as "sugar" snow which is advanced in its metamorphism compared to that of snow grains in early spring. The snow grains are rounded, about 2 mm in size, and very loose, forming a medium that has the texture of large raw sugar grains. This provides an excellent medium for free water to move through. Below the ice lens, there is commonly a layer containing elongated grains, which are moderately rounded and 2 mm to 3 mm in size. This layer facilitates the free-water percolation through the snow to form slush at the snow–ice interface.

The late spring (or a period of prolonged onset of melt in the snow cover) is characterized by a snow cover that increases in snow wetness with depth. The snow wetness is not necessarily large, ranging from 0.3% near the surface to 1% above a slush layer. The snow depth has decreased from the previous stage and the snow cover has become more homogeneous. A surface crust exists composed of a layer only a few millimeters in thickness, with an average grain size of 2 mm. The increasing intensity of solar radiation, approaching a maximum in the afternoon, can cause the amount of free water to increase at the snow surface. The free water around a snow grain can cause another grain to connect to it due to the increased surface tension of the water-covered grain. Grain size increases gradually with depth, reaching a maximum size of approximately 4 mm just above the slush layer. The bulk of the snow can be described as loose and granular with a temperature near 0°C. The water is able to move through the bulk of the snow with ease due to the looseness of the snow and its large grains producing channels for water movement. This often results in a slush layer at the snow–ice interface and also has been observed extensively by Grenfell and Lohanick [44] in the Beaufort Sea in 1982 during the late spring. The slush layer is usually 1 to 5-cm thick consisting of a mixture of snow and melted snow, and sometimes rotting ice where slush on first-year ice is saline. Slush that is formed from flooding with sea water after submergence of the ice by a heavy snow load will always be saline, reaching a salinity close to that of sea water.

The various cases given above are for a snow cover on both first-year and old ice; however, it was observed that a snow cover on first-year ice can be further advanced in the metamorphic process for a given period. For example, in 1987 the snow cover on old ice was similar to that shown in Fig. 2.21(1), whereas the snow cover on first-year ice had advanced to that of Fig. 2.21(2) for the same period. There are two major causes for the different snow metamorphic stages over first-year compared to old ice. The snow cover on first-year ice can be thinner than on old ice (Tucker et al. [101]; Perovich et al. [93]; Garrity [34]). It is likely that the density of snow, hence its thermal conductivity, is less during the early growth period (Nakawo and Sinha [80]), emphasizing the thinner snow cover on first-year compared to old ice. Since snow is a good insulator, it takes longer for the air temperature and solar radiation to have an influence on the thicker snow cover on old ice. During the early spring, the snow–ice interface temperature is also less for old than first-year ice (Meeks et al. [77]) which is partly due to the better thermal insulation provided by the snow.

Although no descriptive snow model has been included for the summer, a few words will describe the snow cover observed during the summer, from July to August 1990 in the Greenland Sea and August through September 1991 in the Arctic Ocean (Garrity [34]). It is surprising, but understandable, that the snow wetness is not high (1.2%) and similar to the spring. The snow cover is thin, averaging 3.1 cm, with a loose snow layer over a hard, almost ice, layer. The snow–ice interface is not distinguishable, thus an accurate snow depth is difficult to measure. Since the snow grains are distinct units, about 0.5–2 cm, water was able to drain freely from the snow cover into drainage channels on the ice surface. This results in a snow cover which has largely disappeared, leaving a grainy texture and a rough ice surface.

2.3.5 Evolution of Snow Cover—Antarctic

During the Weddell Sea experiment in September and October 1989, 191 distinct snow profiles (384 individual measurements) were obtained from first-year and old ice (Garrity [34]). The old ice present in the western area of the Weddell Sea was mainly second-year ice, which was also observed near the ice edge. The average air temperatures decreased from −3°C near the ice edge to less than −5°C very close to the Antarctic continent. Table 2.2 summarizes the different ice regimes near the ice edge, which have characteristic snow covers. Thicker snow covers were found near the ice edge, consistent with measurements made in 1986 (Wadhams et al. [109]). The snow depths generally ranged from 50 cm to 1 m on the second-year ice and to a maximum of 50 cm (level ice) on first-year floes which were sometimes rafted and/or deformed near the ice edge. The large snow depths caused loading of the ice, and the negative freeboard allowed for slush to infiltrate the snow–ice interface. These substantial snow depths were characteristic of second-year and first-year ice for the western and eastern Weddell Sea areas, respectively. First-year ice near the ice edge also exhibited slush at the snow–ice interface. The substantial 50-cm snow cover over the first-year ice, along with the deformation process, caused cracks in the ice, providing a path for water to flood the snow–ice interface. The observed slush thickness for both the eastern and western Weddell Sea was generally 6 cm. The slush extended over the entire ice floe for both the second- and first-year ice. Undeformed first-year ice often did not have slush at the snow–ice interface.

The presence of slush was visible from the air as a gray area on floes in July to September 1986 (Wadhams et al. [109]). During September to October 1989, the slush layer was hidden by a snow cover (Garrity [34]). There was in fact no visible melt on the ice or snow surface during the spring of 1989 in the Weddell Sea, which is opposite to spring periods in the Arctic. Most of the melt in the Weddell Sea is from the bottom of the ice (Andreas and Ackley [4]).

Snow profiles are also considerably different from those observed in the

TABLE 2.2 Predominant Ice Regimes Near the Ice Edge in the Weddell Sea

Ice Type	Location	Snow Depth (cm)	Comments
Second-year	Western Weddell Sea 64°19′S 46°44′W and 65°25′S 40°17′W	50–100	Slush present/negative freeboard
		50	Dry snow–ice interface
First-year	East/West Weddell Sea 63°47′S 02°23′W	50	Slush present/deformed ice
		50	Dry snow–ice interface

Arctic. For example, it is common to have a surface crust in the Arctic caused either by refreezing of the snow surface after a warm period or by strong winds. However, if there is a surface crust in the Antarctic (where there is a lack of distinct surface melting), it is only millimeters in thickness and is formed primarily by the wind. The wind crust has small snow grains ranging from 0.5 to 1 mm in size. This is a medium capable of retaining water with a larger surface area for a given volume. In general, the vertical structure of the snow cover is more homogeneous in the Antarctic than in the Arctic.

For air temperatures ranging from −12°C to −5°C, a descriptive model for a snow cover without the presence of slush is shown in Fig. 2.22(1a). The snow wetness ranges from 0.1% to 0.4% and is constant throughout the snow

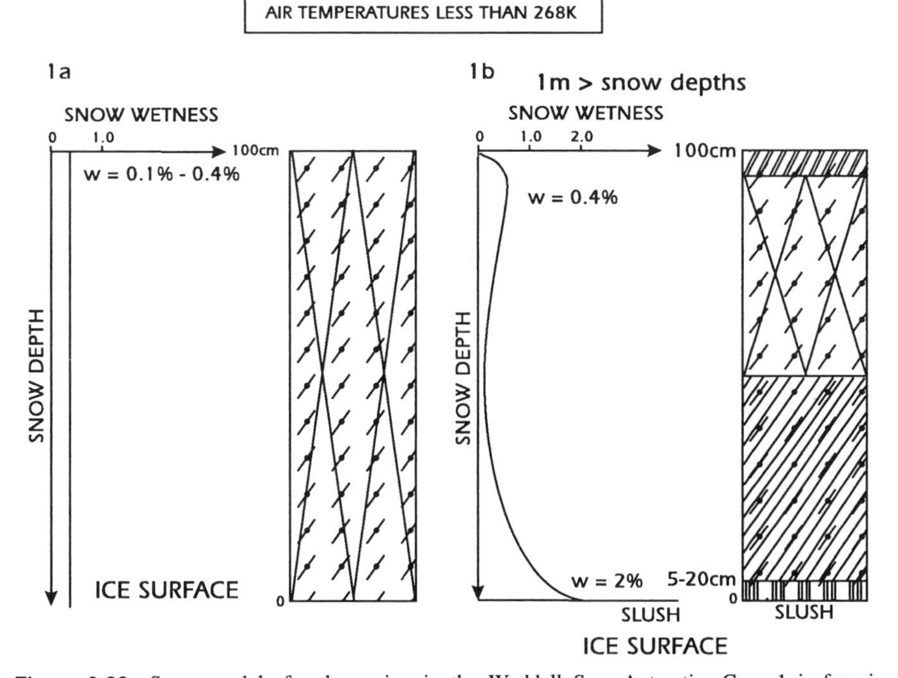

Figure 2.22 Snow models for the spring in the Weddell Sea, Antarctic. Case 1 is for air temperatures less than 268K: Case 1(a) is without slush at the snow–ice interface and Case 1(b) with slush. Case 2 is for an air temperature between 268 K and 273 K. The snow wetness distribution changes significantly for Case 3, when the air temperature is greater than 273 K. Cases 1(a)–(b) and 2(a) are representative snow covers found at the western Weddell Sea ice edge, whereas Cases 2(b) and 3 are found at the eastern ice edge. Case 3 is characteristic of the interior of the Weddell Sea. There is no descriptive snow model close to the Antarctic continent presented in this figure.

cover. The snow wetness distribution in the presence of slush is shown in Fig. 2.22(1b). The absorption of water from the slush layer at the snow–ice interface is analogous to a sponge soaking up water. This soaking up, or wicking, of water caused the snow wetness to increase to 2% just above the slush layer. These descriptive models are for snow on second-year ice in the western Weddell Sea.

For air temperatures between −5°C and 0°C, there is often significant free water occurring in the snow cover. The resulting snow physical properties are shown in Figs. 2.22(2a) and 2.22(2b) (slush present). Figure 2.22(2a) shows snow characteristics observed in the western ice edge areas where there is a gradual increase in snow wetness with depth, reaching a maximum of 2% near the snow–ice interface. Wicking is shown again [Fig. 2.22(2b)] above the slush layer which is generally present on rafted first-year floes in the eastern ice edge area. The snow surface layer is wetter than shown in Fig. 2.22(1b) due to the warmer air temperatures.

There were only a few snow measurements made for air temperatures above freezing in the eastern ice edge and interior areas of the Weddell Sea. This case is shown in Fig. 2.22(3). The surface was wetter than the immediate

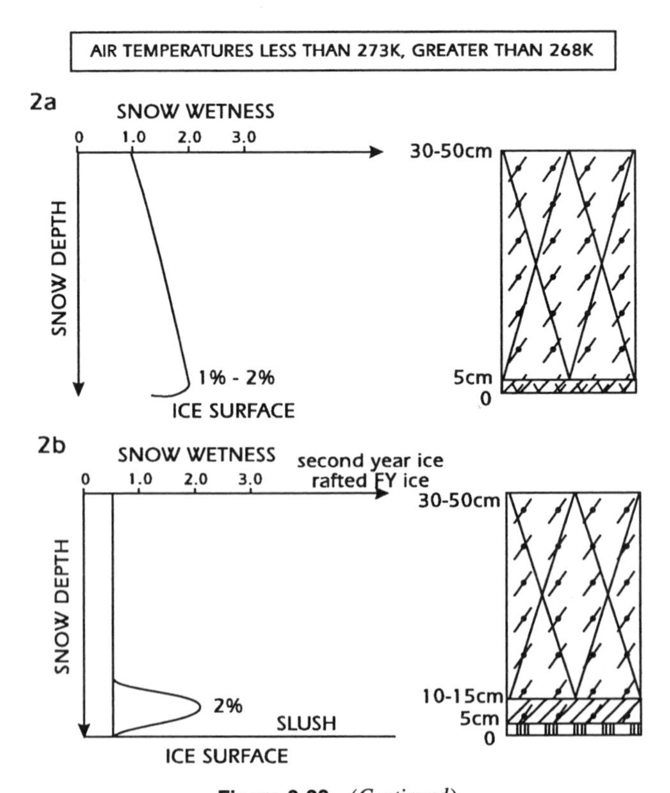

Figure 2.22 *(Continued)*

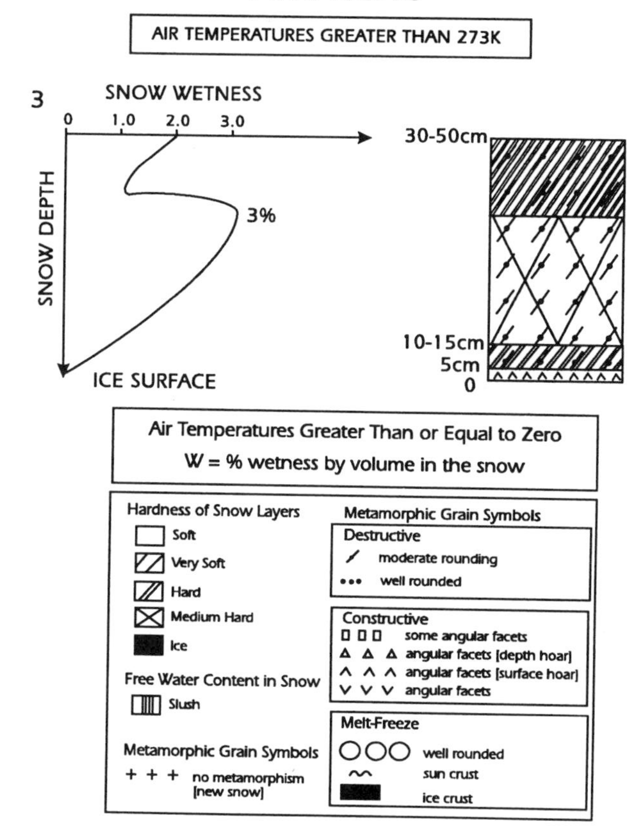

Figure 2.22 *(Continued)*

underlying layer and snow wetness increased near the middle of the snow cover. The wetness decreased gradually to 0% at the snow–ice interface. There was also more structure in the snow cover as compared to that shown in Fig. 2.22(2a)–(2b), mainly due to the higher air temperatures.

2.4 ELECTROMAGNETIC PROPERTIES OF ICE AND SNOW

2.4.1 Electrical Properties

It is essential to understand the electrical properties of a material that is to be studied by electromagnetic waves since these properties, as well as the structure of the material, greatly affect the return signals. At frequencies lower than about 1 MHz, the dominant electrical property is the bulk conductivity of the ice, σ_{DC}, while at higher frequencies, the complex dielectric permittivity, ϵ_r, is used.

The electrical properties of ice display a smooth, continuous variation over the frequency range from 1 kHz to 10 GHz and above. However, they are normally characterized in somewhat different ways for lower frequencies (quasistatic case) and for higher frequencies (propagating wave case).

In the quasistatic approximation and considering the problem in the time domain, the dc conductivity, σ_{DC}, of the sea ice is the principal factor controlling both propagation velocity, v_{DC}, and attenuation of the EM signal in the ice, although usually the attenuation is almost negligible at frequencies below about 500 kHz:

$$v_{DC} = \frac{c}{\sqrt{\pi \sigma_{DC} \mu_0 t}}$$

where

c = speed of light in vacuum (3×10^8 m/s)
μ_0 = magnetic permeability of free space
t = time

Note that v_{DC} is effectively the group velocity of the current system as it travels down into the ice: at the ice–seawater interface, the velocity will drop dramatically (by as much as 50 times for old ice) due to the much higher conductivity of sea water. As sea ice is typically only a few meters thick and is quite resistive, it is more practical to sense the transit time of the signal as a frequency-domain phase shift than as a time-domain delay.

The electrical property of ice that is most important at higher frequencies is the relative complex permittivity (defined with respect to the permittivity of free space, ϵ_0):

$$\epsilon_r = \epsilon_r' - j\epsilon_r''$$

where ϵ_r' is the dielectric constant and ϵ_r'' is the dielectric loss. The speed of a propagating wave through the material, v_{RF}, is controlled by ϵ_r':

$$v_{RF} = \frac{c}{\sqrt{\epsilon_r'}}$$

The loss tangent:

$$\tan \delta = \frac{\epsilon_r''}{\epsilon_r'}$$

is directly related to the attenuation of the radar signal in the medium, α, given by (Vant et al. [107]):

$$\alpha = 868.6 \left(\frac{2\pi}{\lambda_0}\right) \left\{\frac{\epsilon_r'}{2} (\sqrt{1 + \tan^2 \delta} - 1)\right\}^{1/2}$$

where λ_0 is the free space wavelength.

According to Morey et al. [79], there are two mechanisms that cause dissipation of the radar energy as heat: namely, relaxation and conduction. Conduction, which causes losses at low frequencies, is caused by electron and ion movement in an electric field. At higher frequencies, typically above about 1 GHz, relaxation losses due to the drag of oscillating bipolar water molecules constitute the major losses. For ice, relaxation losses are virtually entirely dependent on free-water content.

The value of ϵ_r' for freshwater ice has been measured extensively: 3.17 at 10 GHz (Lamb [64]); 3.15 at 9.375 GHz (Cumming [21]); and 3.17 at 10 GHz (Von Hippel [108]). The value of ϵ_r' for saline first-year sea ice is higher, and is highly dependent on the temperature and salinity of the ice. First-year sea ice is a complex mixture of pure ice and brine. In a laboratory experiment conducted by Stogryn and Desargant [100], measurements are reported over the frequency range 7.5 to 40 GHz of the complex dielectric constant of fresh water and of sea water brine in equilibrium with sea ice. The brine dielectric constants were measured over the temperature range 0°C to −25°C, as was the ionic conductivity, C_{ic}. It was found that the dielectric constant, ϵ_r', of brine in thermal equilibrium with sea ice could be expressed in terms of the Debye relationship:

$$\epsilon_r' = \epsilon_{inf} + \frac{(\epsilon_s - \epsilon_{inf})}{1 - j2\pi T_D} + \frac{jC_{ic}}{2\pi\epsilon_0 f} \tag{2-8}$$

where ϵ_s and ϵ_{inf} are the limiting static and high frequency values of the real part of ϵ_r', f is the electromagnetic frequency, T_D is the Debye relaxation time, $j = (-1)^{1/2}$, ϵ_0 is the permittivity of free space (8.85419) $\times 10^{-12}$ farad/m), and C_{ic} is the ionic conductivity (mho/m).

The Debye constants (ϵ_s, ϵ_{inf}, and T_D) in (2-8) were empirically determined as:

$$\begin{aligned}
\epsilon_s &= (939.66 - 19.086T_c)/(10.737 - T_c) \\
\epsilon_{inf} &= (82.79 + 8.19T_c^2)/(15.68 + T_c^2) \\
2\pi T_D &= 0.10990 + 0.13603 \times 10^{-2}T_c + 0.20894 \\
&\quad \times 10^{-3}T_c^2 + 0.28167 \times 10^{-5}T_c^3
\end{aligned}$$

where T_D is the relaxation time in nanoseconds and T_c is the temperature in °C.

The ionic conductivity (mho/m) of seawater brine is:

$$C_{ic} = -T_c \exp(0.5193 + 0.8755 \times 10^{-1} T_c) \text{ if } T_c >= -22.9°\text{C}$$

$$C_{ic} = -T_c \exp(1.0334 + 0.1100 T_c) \qquad \text{if } T_c < -22.9°\text{C}$$

Although there is only a small quantity of brine present in the ice, its large dielectric constant of approximately 80 has a significant influence on the resulting dielectric properties of the ice–brine mixture. Figure 2.23 shows the apparent dielectric constant and brine volume for first-year ice as a function of depth. Kovacs and Morey [61] reported a typical value of 4.6 for ϵ_r' for first-year ice. Figure 2.24 shows the relationship between ϵ_r' and temperature. The value of ϵ_r' for old ice is lower than that for first-year ice and is much less temperature sensitive. Kovacs and Morey [60, 61] reported a value of 3.7 for old ice.

The loss tangent for pure ice has been measured by several investigators:

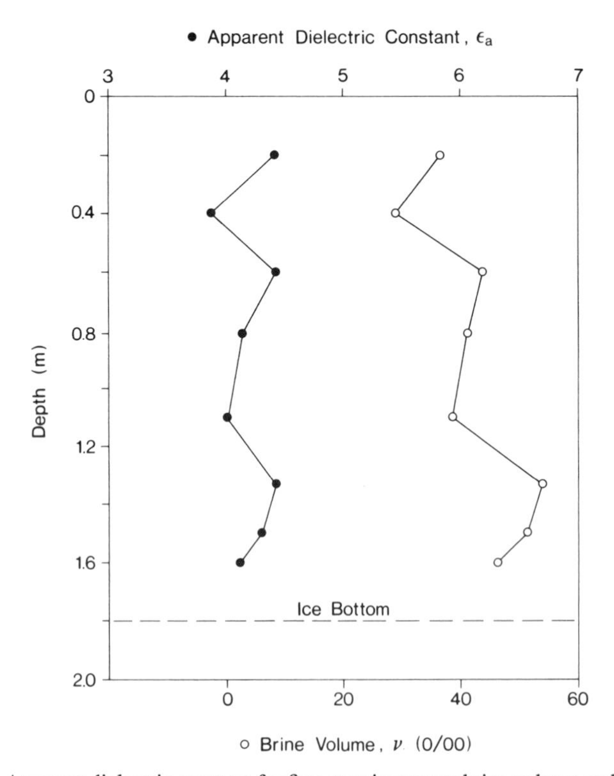

Figure 2.23 Apparent dielectric constant for first-year ice versus brine volume and depth (Morey et al. [79]).

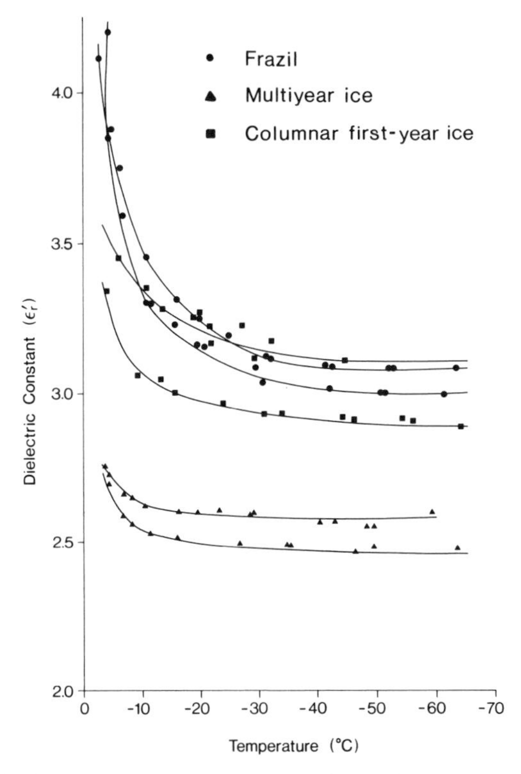

Figure 2.24 Dielectric constant versus temperature for sea ice (Vant et al. [106]). Reprinted by permission of the American Institute of Physics.

20×10^{-4} (Vant et al. [106]), 12×10^{-4} (Lamb [64]), 27×10^{-4} (Cumming [21]), and 7×10^{-4} (Von Hippel [108]), all at slightly different temperatures.

Unlike the dielectric constant, which is relatively independent of frequency above about 1 MHz, the dielectric loss for sea ice is strongly dependent on frequency. This condition results from the fact that water inclusions exist in sea ice and because the dielectric properties of water are strongly dependent on frequency in the range of 10^9 to 10^{11} Hz (Hoekstra and Spanogle [51]). Figure 2.25 shows this clearly. In the range of 3 to 10 GHz (*S*- to *X*-band radar), the dielectric loss is at a minimum.

As in the case of ϵ_r' the dielectric loss for sea ice is also strongly temperature dependent. Figure 2.26 shows the dielectric loss versus temperature for several salinities. The strong increase in dielectric loss as the temperature increases is a result of the precipitated salts going back into solution. In the normal range of ice temperatures encountered ($-30°$ to $0°$C), the dielectric loss increases very substantially with increasing temperature. The dielectric loss is much lower for old ice because of its lower salinity, and is much less temperature sensitive. Figure 2.27 shows the variation of ϵ_r'' with temperature for old ice.

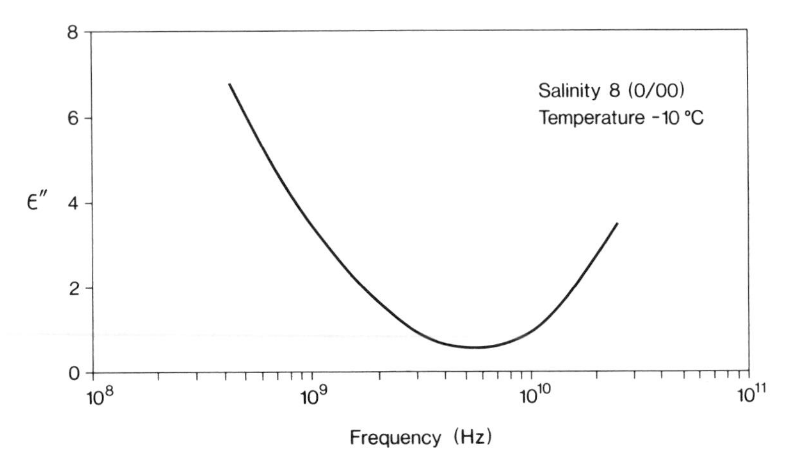

Figure 2.25 Dielectric loss versus frequency for first-year ice (Hoekstra and Spanogle [51]). Reprinted by permission of the Cold Regions Research and Engineering Laboratory.

As in the case of the strength of ice, the dielectric properties can be shown to be related to the brine volume. Hoekstra and Spanogle [51] showed that at 9.8 GHz, ϵ_r' can be determined by:

$$\epsilon_r' = \frac{(\epsilon_r')_{\text{ice}}}{1 - 3\ V_{br}}$$

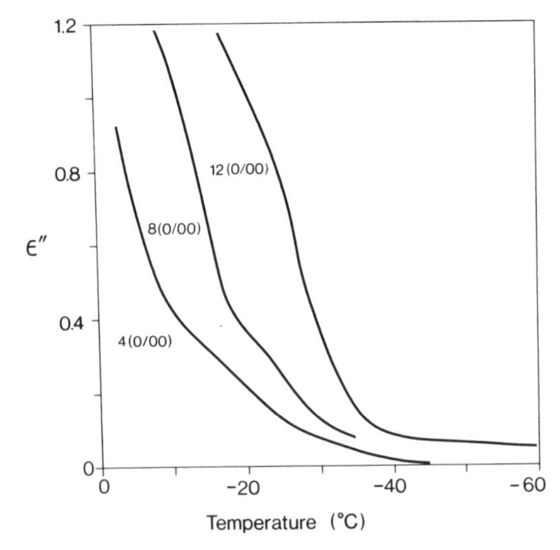

Figure 2.26 Dielectric loss versus temperature for several salinities of sea ice (Hoekstra and Spanogle [51]). Reprinted by permission of the Cold Regions Research and Engineering Laboratory.

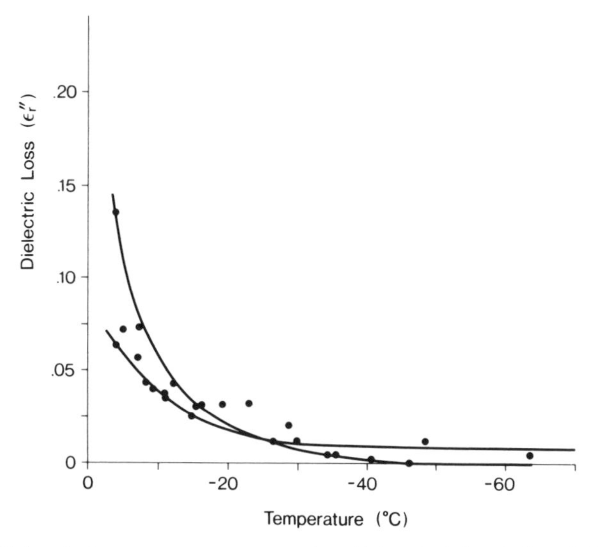

Figure 2.27 Dielectric loss versus temperature for old ice. Salinity 0.61 to 0.70 ppt (Vant et al. [106]). Reprinted by permission of the American Institute of Physics.

where $(\epsilon_r')_{\text{ice}}$ is the dielectric constant for pure ice, and V_{br} is the percent brine volume. Vant et al. [107], also at 10 GHz, developed the relationship:

$$\epsilon_r' = a_0 + \frac{a_1}{1 - 3\,V_{br}}$$

where the best fit to the data for columnar first-year ice, frazil first-year ice, and old ice showed that a_0 does not equal 0 and a_1 does not equal 3.14. These results were interpreted to mean that all the brine was not in spherical inclusions, but that perhaps the brine was held in elliptical inclusions angled to the electrical field. Vant et al. [107] also showed that ϵ_r'' is directly related to V_{br}. Through the use of a theoretical model and correlation with real data, they showed that ϵ_r' and ϵ_r'' varied with the angle of inclination of the assumed elliptical brine pockets. Figure 2.28 plots the theoretical model and real data, and shows the best agreement when the brine pockets are oriented at 35° to 45° to the vertical.

Vant et al. [107] also made empirical fits to their results of the form:

$$\begin{Bmatrix} \epsilon_r' \\ A \end{Bmatrix} = a_0 + a_1 V$$

where

ϵ_r' = measured dielectric constant of the sample

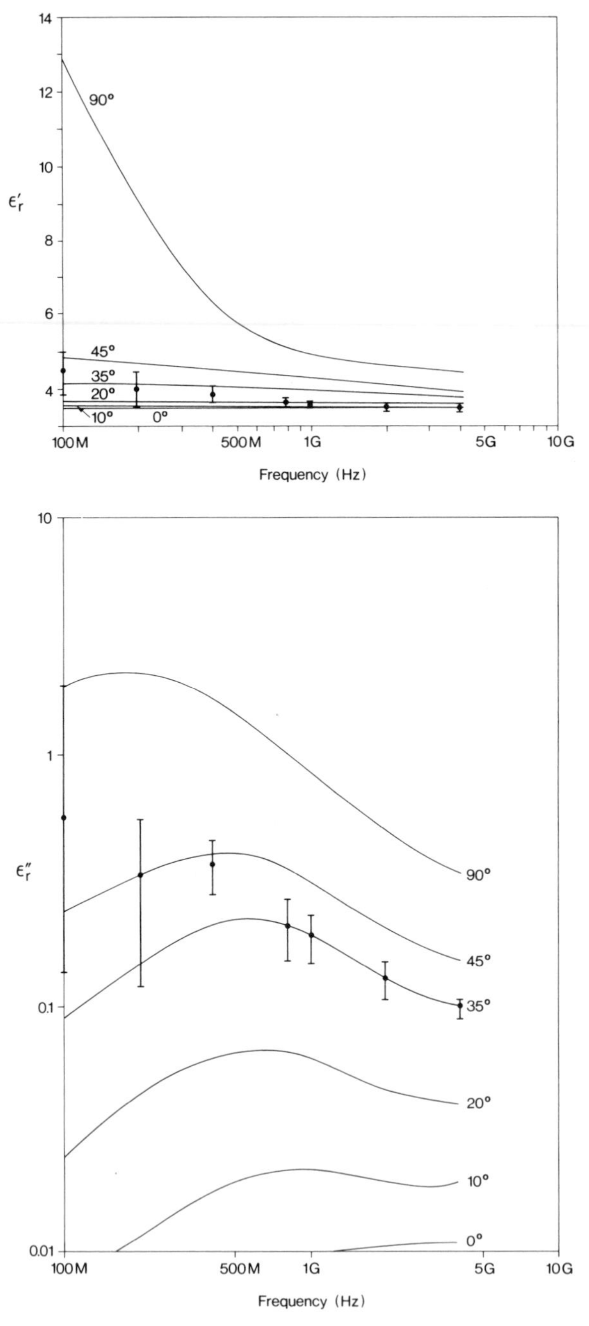

Figure 2.28 Comparison of theoretical model (solid line) with experimental data for ϵ_r' and ϵ_r'' (Vant et al. [107]). Reprinted by permission of the American Institute of Physics.

A = the attenuation rate of the sample (dB/m)
V = brine volume of the sample
a_0, a_1 = frequency-dependent fit parameters

For 43 samples of first-year ice, at 100 MHz they found $a_0 = 3.22$, $a_1 = 20.6$ for dielectric constant; and $a_0 = 0.97$, $a_1 = 51.1$ for attenuation rate. At 40 GHz (47 samples), the parameters were $a_0 = 3.05$, $a_1 = 7.2$ for dielectric constant; and $a_0 = 13.37$, $a_1 = 1369.8$ for attenuation rate.

There has been research interest in the horizontal anisotropy of sea-ice crystals. Although it has been known for some time that the *c*-axis of ice crystals is oriented in the horizontal plane, it has been found that widespread areas of sea ice also have a preferred orientation in the horizontal plane (Weeks and Gow, [111]. The preferred orientation has been related to ocean currents beneath the ice. This phenomenon also affects radio sounding of sea ice (Campbell and Orange, [8] and echoes with the electric field perpendicular to the *c*-axis orientation are very much weaker than those parallel to it (Kovacs and Morey [60, 61]; Morey et al. [79]). Therefore, with a linearly polarized radar antenna, it is possible to obtain no echoes in one orientation and useful sounding measurements with a 90° antenna rotation.

More rigorously, the penetration depth of an electromagnetic wave into sea ice is governed by two processes, absorption of electromagnetic energy as described above, and scattering of the wave by inhomogeneities within the ice. The effective wave penetration is defined in terms of an extinction coefficient, K_e:

$$K_e = K_s + K_a$$

where K_s describes the energy lost to the reflected wave due to scattering in the ice and K_a is the absorption coefficient described above.

The effective wave penetration depth is $1/K_e$ which is smaller than or equal to the classical skin depth. The classical skin depth and absorption coefficients are most commonly reported in the literature. The scattering contribution can be estimated by means of various scattering models (Kim et al. [58]; Hallikainen and Winebrenner [46]) but is not easily observed experimentally. It is expected to be most important in low-loss, inhomogeneous materials such as old ice. From the viewpoint of electromagnetic wave interactions with sea ice it is most useful to define the penetration depth in free space wavelength units, "λs."

Table 2.3 shows the skin depth as a function of loss tangent at *X*-band (9.4 GHz, 3.2-cm wavelength).

It can be seen that the higher losses in first-year ice significantly reduce penetration into the ice, while the lower losses in old ice allow significant penetration. This is noted by Weeks and Ackley [114] who define young and first-year ice as a "high-loss" and old ice as a "low-loss" dielectric material.

TABLE 2.3 Skin Depth at X-Band as a Function of Loss Tangent

Loss Tangent (tan δ)	Ice Type	Skin Depth (cm)	Skin Depth (λs)
0.001	Glacier	541.0	169.0
0.005	Glacier	108.0	33.8
0.010	Glacier/Old	54.1	16.9
0.020	Glacier/Old	27.1	8.5
0.04	Glacier/Old/FY	13.5	4.2
0.06	Glacier/Old/FY	9.0	2.8
0.08	First-year	6.8	2.1
0.10	First-year	5.4	1.7
0.15	First-year	3.6	1.1
0.20	First-year	2.7	0.8
0.30	First-year	1.8	0.6
0.40	First-year	1.4	0.4
0.50	First-year	1.1	0.3

Electrical Properties of Icebergs

The relative dielectric constant for glacial ice is approximately 3 at 10 GHz (Kirby and Lowry [59]). Measurements by Pearce and Walker [92] at 30 MHz for an ice density of 0.92 g/cm^3 yielded a dielectric constant of 3.31 $\pm$ 0.04, while Jezek et al. [56] reported values between 3.09 and 2.89. Since glacial ice is not saline, it is a low-loss dielectric. Page and Ramseier [91] reported that the loss tangent had values less than 2×10^{-3} for all frequencies greater than 1 GHz.

Electrical Properties of Snow

The physical structure of snow on arctic sea ice can be modeled as a volume distribution of ice crystals, air, and water (Evans [31]). The physical parameters which determine the electromagnetic properties of snow at microwave frequencies are temperature, density, ice crystal size distribution, liquid water fraction, and the size distribution of liquid water inclusions. The size distribution of ice crystals in a sea-ice snow cover is dynamic, even at low temperatures, due to recrystallization processes driven by heat transfer through the ice. Salinity of the snow base and underlying ice is a factor in this process. From the viewpoint of low-frequency microwave radar, the snow cover on sea ice can be treated as a bulk medium in which reflections from layers are determined by changes in the dielectric constant of the snow and absorption losses are determined by the imaginary part of the dielectric constant. When the dielectric constant increases slowly with wavelength-normalized depth, the effect of the snow is to decrease the dielectric discontinuity at the ice surface and to reduce the surface component of the radar reflectivity of the ice. At radar frequencies for which a significant number of ice crystals or water inclusions are large enough at a wavelength scale, volume scattering and scattering contributions to the extinction coefficient of penetrating radiation must be considered.

The bulk electromagnetic properties of snow have been studied extensively since the 1940s and are summarized in a detailed review article by Mätzler [76]. The following discussion of the bulk properties of snow are based on Mätzler's paper.

When the ionic impurities in the snow cover are sufficiently small and no liquid water is present, the Debye relaxation frequency of ice crystals is a few kilohertz. Thus, the imaginary part of the dielectric constant of snow crystals varies slowly with frequency. Empirical results quoted by Mätzler are summarized as:

$$\epsilon''_{ice} = A/f + Bf^C$$

where

$$0.00035 < A < 0.0013$$

$$0.000036 < B < 0.00012$$

$$1.2 > C > 1.0$$

(the first limit is for pure ice and the second is for low-concentration ionic impurities). The coefficients A, B, C are temperature dependent.

The real part of the snow crystal dielectric constant is independent of radar frequency and is weakly dependent on temperature. Mätzler quotes Wegmüller's result;

$$\epsilon_\infty = 3.1884 + 0.91 \times 10^{-3}\,(T - 273)$$

Snow which has <0.01‰ salinity contains no significant liquid water at temperatures below −1°C and is thus "dry." The upper snow layers on old sea ice fall in this category during the winter. In the bottom layers of a sea ice snow pack, the snow salinities can vary over a wide range depending on the ice and snow history. This snow can contain liquid brine down to temperatures at which the brine freezes to a eutectic solid. The Debye relaxation frequency of water is approximately 8.8 GHz at 0°C with the result that contributions to the snow dielectric constant from water will be frequency dependent over the commonly used range of radar frequencies (1 GHz to 30 GHz). Mätzler reports that empirical results for saline snow support the relationship:

$$\epsilon'' \approx \epsilon''_{pi} + D\,S$$

where

ϵ''_{pi} = the imaginary part of the dielectric constant for pure ice
D = 0.05 at T = −5°C
f = 5 GHz
S = snow salinity in ‰

The relationship between salinity and the imaginary part of the dielectric constant is claimed to be valid over the range of 0‰ to approximately 25‰.

The model of snow as a mixture of ice crystals, water inclusions, and air suggests that the dielectric constant of snow can be modeled by a mixing formula which combines the dielectric constants of the constituents according to their population. Suitable population parameters are the snow density and the snow wetness (volume fraction of liquid water).

For dry snow, Mätzler quotes Tiuri's results for the real part of the dielectric constant as $\epsilon_d' = 1 + 1.7\rho + 0.7\rho^2$ where ρ is the snow density in g/cm^3 and the data supporting the empirical relationship span the range of $0 < \rho < 0.5$. Comparisons of these results with theoretical calculations show good agreement and show that the results are independent of snow type. The imaginary part of the snow dielectric constant was shown to obey the relationship: $\epsilon_s'' = \epsilon_{\text{ice}}''(0.52\rho + 0.62\rho^2)$ at 2 GHz.

As snow becomes moist, water begins to accumulate at ice grain boundaries. Between liquid water (volume) fractions of 0% and 11 to 15%, water can be treated as a volume dispersion of droplets (water inclusions in an air–ice matrix) for the calculation of dielectric constant. At the 11% to 15% saturation limit, the snow structure changes and calculations must be based on air inclusions in an ice–water matrix. Typical liquid water volumes in unsaturated, wet snow range between 0% and 10%. For this water concentration range and microwave frequencies below 30 GHz, Mätzler proposes the relationship:

$$\epsilon_{ws} = \epsilon' + j\epsilon'' = \epsilon_d + \frac{23W}{1 - \dfrac{jf}{f_0}}$$

where

ϵ_d = the dielectric constant of dry snow
W = the volume fraction of liquid water
f = the microwave frequency
$j = \sqrt{-1}$
f_0 = 10.0 GHz

This model is considered by Mätzler to be an interim approximation and further research is required.

Model calculations of microwave emissivity, scattering cross section, and extinction coefficients for snow cover over sea ice require knowledge of the size distribution functions associated with ice crystals and water inclusions in addition to the bulk properties outlined above. Although the specific case of snow cover on sea ice has been studied, most of the current knowledge of the behavior of scattering effects in snow packs has been obtained at terrestrial test sites. Mätzler reviews much of this work but his models and results are beyond the scope of this chapter. Some of the principles underlying the scattering of electromagnetic waves are outlined in the following sections of this chapter. Passive microwave effects in snow packs have been studied by Garrity [34] and may be applied to some of the scattering effects in snow packs.

Summary

As noted above, the electrical properties of ice and snow are quite complex, depending on frequency, temperature, ice type or age, and mode of formation for ice and on frequency, temperature, density, and moisture content for snow. A summary is given in Table 2.4.

2.4.2 Reflection Coefficient

As stated by Page and Ramseier [91], "radar reflections occur when there is an abrupt change in the value of ϵ_r, i.e., where significant changes in ϵ_r take place within the distance of a wavelength of the radio wave." Budinger et al. [7] related the reflection coefficient Γ to the dielectric constant as:

$$\Gamma = \frac{1 - \sqrt{\epsilon_r'}}{1 + \sqrt{\epsilon_r'}}$$

We can, therefore, expect reflections from air–ice interfaces, ice–water interfaces (if the losses in the ice are low enough to permit penetration), ice–air interfaces (for low losses) from the back surface of the ice, and from inhomogeneities in the ice.

Active radar returns from sea ice are determined by the electromagnetic fields scattered from the snow surface, the ice surface, and from the volume of the snow–ice medium back into the radar antennas. The physics of the scattering process is governed by the interaction of electromagnetic waves with matter and is rigorously (but not practically) described by the solutions to Maxwell's equations over the entire, actual dielectric structure of the surfaces and throughout the volumes of the ice and snow being observed. Since the actual structures are a priori unknown, models based on the statistics of these structures and their complex dielectric constants have been developed to provide tools for the inference of the physical structures present from electromagnetic measurements and from physical measurements of a few examples. The models may be empirically based, theoretically based, or semi-empirical. All current electromagnetic scattering models for sea ice and snow make use of the statistics of scatterer (ice, air, or brine inclusions) size distributions, axial ratios, and correlation lengths. No existing model accounts for the observed spatial inhomogeneity of ice and snow structures in an effective manner. The literature on sea ice and snow scattering models is extensive and will not be reviewed here; however, a comparative review of the main families of scattering models can be found in Winebrenner et al. [115].

TABLE 2.4 Summary of Electrical Properties of Ice and Snow

Ice Type	Typical Thickness (m)	σ_{DC} (S/m)	ϵ_r'	A(dB/m)	ϵ_r'	A (dB/m)
Comments	Approx.	Slight temperature dependence	At 100 MHz (somewhat temperature and frequency dependent)	At 100 MHz (strongly temperature and frequency dependent)	At 10 GHz	
Young	<0.5	>0.05	6–?	>100	5–?	$\gg 100$
First-year	0.5–2.5	0.05–0.01	4–6	5–10	3.5–4.5	>100
Old	1–10	$<.001$	3.5–4.5	0.1–10	3.5–4.5	1–50
Icebergs	>10	$<.001$	3.2	0.1	3.2	0.1
Snow	≤ 1	10^{-7}–10^{-9}	2–3.5	0.01–0.1	2.5–3	1–10
Sea water	N/A	4–5	80	10^2	~ 50	10^4

It is within the scope of this chapter to sketch a few of the key concepts underlying electromagnetic scattering. Any monostatic scattering process requires an electromagnetic wave interaction with a dielectric discontinuity in a medium or with a surface relief element on a surface. Two factors combine to determine the strength of the wavelet reflected from a single scattering element, the dielectric constant contrast at the surface of the element, and the ratio, r, between the "scale dimension," s, of the element and the radar wavelength, λ. Formally:

$$r = 2\pi s/\lambda \text{ or } r = ks$$

where $k = 2\pi/\lambda$ is the wave number of the electromagnetic wave.

If $r >> 1$ the scattering event is determined by the large-scale geometry of the scatterer, geometric optics solution apply, and the scattering efficiency varies with the scatterer area (r^2). If r is approximately equal to 1 resonant (Mie), scattering occurs and the back-scattered field strength depends on the shape of the scattering element in a complex manner. If $r < 0.5$, the scatterer shape is unimportant, Rayleigh scattering applies, and the scattering "efficiency" varies as r^4. Describing the scattering efficiency as a function, $f(r)$, whose behavior is outlined above, the "strength" of an individual scatterer can be defined as $\Gamma f(r)$. The radar return from an ensemble of scatterers, whose "population" density is $N(r)$, is thus determined by the spatial distribution of $N(r)\Gamma f(r)$. For example, the plate- and needle-like hoar frost crystals in frost flowers can be a few centimeters long but only a few millimeters in thickness, are nearly vertically oriented, and are brine coated. Frost flowers covering a sheet of smooth, young ice yield high radar returns to vertically polarized radiation at C-band and higher frequencies, lower returns to horizontally polarized radiation in the same frequency range, and very low returns to L-band radiation at any polarization. Following similar reasoning, the ensembles of millimeter-scale ice crystals within dry snow form a significant radar target at 37 GHz and higher radar frequencies but contribute very little scattered energy below 13 GHz.

One of the major problems in the formulation of realistic scattering models is the definition of a reasonable statistical description of $N(r)\Gamma f(r)$. $N(r)$ is very difficult to measure, propagation losses through the medium must be accounted for, and physically realistic descriptions of $f(r)$ must account for multiple scattering. Surface measurements (dielectric constant, salinity, density, and crystal structure) of sea ice and its snow cover suggest that the spatial variability of $N(r)$ is large.

For surface scattering events, wavelength-normalized correlation lengths can be used to statistically describe the ice surface. The surface is said to be smooth (slightly rough) when its surface correlation length $r_{surf} > 6$ and when its rms slope $m < 0.25$. In this case, the surface scattering can be modeled by a scalar approximation to the Kirchhoff surface integral. When $r_{surf} > 6$ and the surface-height standard deviation, ρ, yields $r_h > 1.6/\cos\theta$ (for incidence angle θ), the surface is said to be rough and a stationary phase approximation to the Kirchhoff integral must be used. The conditions $r_{surf} > 6$ and $s^2/(\rho\lambda) > 2.76$ are the basic assumptions for the Kirchhoff model. When the Kirchhoff model is not valid ($r_{surf} < 6$ and/or the scatterer radius of curvature is smaller than a wavelength), more complex descriptions of the scattering process must be used. As in the more general case described above, the spatial variability of ice surface parameters is large and the roughness statistics of the surface may not be stationary.

In all cases, the scatterer size distribution (number density of scale dimensions) and spatial correlation lengths are needed to determine the expected values of the scattered fields. Some effects of varying correlation functions are discussed by Eom [30]. Local measurements of surface roughness parameters have been made successfully for a small number of test sites and some success has been achieved in modeling mean surface scattering effects (Kim et al. [58]; Livingstone and Drinkwater [72]).

At each layer interface, the Fresnel reflection and transmission coefficients (reflections and transmissions at a plain dielectric interface) are used to model the bulk effects of the layer structure. For an interface between a medium with a mean, complex permittivity ϵ_1 and another medium with mean, complex permittivity ϵ_2, and for a wave propagating from medium 1 to medium 2, the Fresnel reflection coefficients are:

$$\Gamma_H = \frac{\sqrt{\epsilon_1}\cos(\theta_i) - \sqrt{\epsilon_2}\cos(\theta_t)}{\sqrt{\epsilon_1}\cos(\theta_i) + \sqrt{\epsilon_2}\cos(\theta_t)} = \frac{\sin(\theta_i - \theta_t)}{\sin(\theta_i + \theta_t)}$$

$$\Gamma_V = \frac{\sqrt{\epsilon_2}\cos(\theta_i) - \sqrt{\epsilon_1}\cos(\theta_t)}{\sqrt{\epsilon_2}\cos(\theta_i) + \sqrt{\epsilon_1}\cos(\theta_t)} = \frac{\tan(\theta_i - \theta_t)}{\tan(\theta_i + \theta_t)}$$

where

Γ_H = the Fresnel reflection coefficient for horizontally polarized radiation
Γ_V = the coefficient for vertically polarized radiation
θ_i = the incident angle of the wave at the boundary
θ_t = the corresponding propagation angle of the transmitted wave in medium 2

$$\theta_t = \sin^{-1}\left(\sqrt{\frac{\epsilon_1}{\epsilon_2}}\sin(\theta_i)\right)$$

Since the medium permittivities are complex, the reflection coefficients are complex, so the phase of the wave is modified upon reflection. The transmission coefficients, $T_{H,V}$ are defined in terms of the reflectivities, $\Gamma_{H,V}\Gamma^*_{H,V}$ by:

$$T_{H,V}T^*_{H,V} = 1 - \Gamma_{H,V}\Gamma^*_{H,V}$$

If medium 2 has a much larger dielectric constant than medium 1, as is true for the interface between air and wet ice for example, the reflection coefficients approach 1, and very little energy is coupled into medium 2. Surface scattering dominates. If the dielectric constants of media 1 and 2 are similar in magnitude, as for a fine-grained, dry snow surface under cold conditions, the reflection coefficients are small and surface scattering at the interface is unimportant, and most of the incident energy couples into medium 2. If the interface is not plane but varies slowly in slope (with respect to the wavelengths), as with old ice ridges and hummocks, these formulations of the reflection coefficients still can be used if the incidence angle is defined with respect to the interface normal. If the interface geometry is highly convoluted spatially at wavelength scales, the reflection coefficient formulations used here may not be valid and more complex relationships may be required.

When some of the other concepts in this section are collected together and are examined from the viewpoint of radar scattering, a number of other useful, general observations can be made. When electromagnetic wave energy is coupled through a medium interface, if the penetration depth in the medium exceeds 1 wavelength, and if the scale dimensions of the scattering elements within this medium are significant with respect to the radar wavelength (there is a large population of scatterers which have $r > 0.05$), the radar returns from the medium will have a significant volume scattering component (Kim et al. [58] and many others). This is the case for cold, old ice at radar frequencies above 1 GHz. If the penetration depth in the medium is of the order of the radar wavelength or less, little or no radar returns will be observed from the volume of the medium and only surface scattering at the upper interface will dominate the radar returns. This is observed for first-year ice at radar frequencies above 1 GHz. If the penetration depth in the medium exceeds the thickness of the medium, the next lower layer interface will receive sufficient energy to contribute to the radar returns. This case is common for dry snow over sea ice under winter conditions at radar frequencies below 37 GHz and for icebergs at L-band (Gray and Arsenault [43]). Strong cross-polarized radar returns (one radar polarization was transmitted and the orthogonal polarization received) are seen for volume-scattering targets in which multiple scattering occurs as with snow and cold old ice, and for geometric reflections at dihedral interfaces as for rafts, fractures, and ridge blocks. Damp snow packs with volume moisture contents near 3% have small penetration depths at most radar frequencies, may have low surface reflection coefficients, and can appear as microwave absorbers.

2.5 RADAR SIGNATURES OF SEA ICE

In this section, the linkages between surface observations of the detailed structure and properties of the ice cover and some of the associated electromagnetic scattering properties of the ice that are important for large area observations are explored by an examination of experimental results obtained from surface and near-surface measurements made by nonimaging sensors. The parameter measured by most of the scattering sensors discussed is the normalized radar cross section of the ice, $\sigma^0(\theta)$, which depends on the electromagnetic properties of the ice and is a strong function of the incidence angle, θ.

2.5.1 Parameter Dependence

Incidence Angle Dependence

Two classes of experiment have been used to characterize the angular dependence of sea ice scattering: surface and near-surface measurements using short-range high-resolution scatterometers, and airborne survey measurements using scatterometers and synthetic aperture radars.

The surface and near-surface scattering measurements (Onstott et al. [83, 84, 86]; Onstott and Gogineni [85]; Gogineni et al. [38]; Collins et al. [17]) have been necessarily local in nature and have been conducted in conjunction with detailed surface measurements of the physical properties of the ice. Typical $\sigma^0(\theta)$ results are shown in Fig. 2.29. Because of the close association of

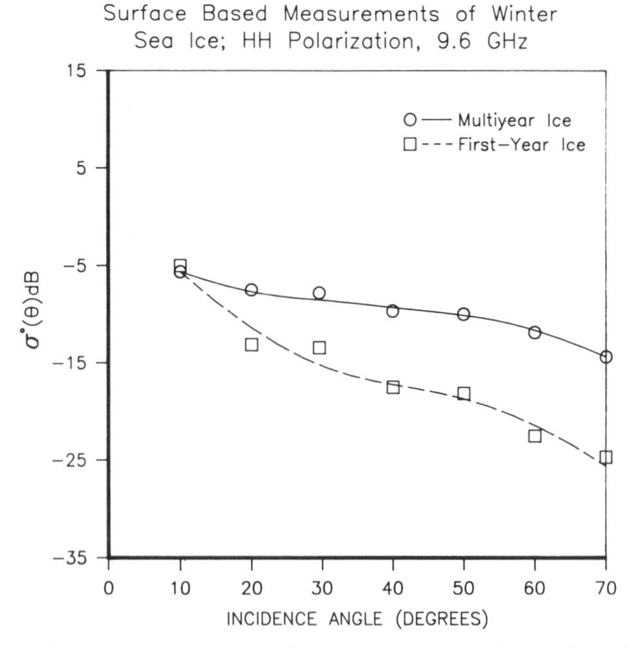

Figure 2.29 Surface-based radar cross-section measurements provide precise values for small spatial areas. As a result, only a limited subset of the surface conditions which define the scattering cross-section variance of the ice type ensemble are contained in these measurements. This limitation is usually offset by companion data which describe the material properties of the area measured, but generalizations of surface measurements to ensemble properties must be made with caution.

the scattering measurements with physical measurements of the structure of the ice and snow cover, these studies have provided essential insights into the sea ice scattering problem and have been responsible for many of the recent advances in electromagnetic scattering models for sea ice. Although surface-based scattering experiments provide very precise knowledge of the scattering properties of thoroughly documented ice and snow structures, they are limited to a small number of ice samples. In the previously cited work, profiling scattering measurements made from low-flying helicopters have been used to extrapolate results from surface measurement sets and to interpolate between surface measurement sites so that the properties of the ice in the study area are more completely known.

The airborne survey measurements conducted with fan beam scatterometers and SARs (Gray [41]; Gray et al. [42]; Hawkins et al. [47–49]; Livingstone et al. [68–71]; Livingstone and Drinkwater [72]; Drinkwater et al. [26]) all suffer from a more remote coupling between the electromagnetic scattering and the surface structure measurements; however, they do permit an examination of the ensemble statistics of the scattering properties of sea ice classes. Coincident measurements using near surface and survey instruments (Livingstone et al. [71]) provide one cross-reference example between the two experiment classes. Sea ice research with a fan-beam scatterometer shows the seasonal evolution of $\sigma^0(\theta)$ for sea ice type ensembles found in the Beaufort Sea. It is shown that at 13.3 GHz, under cold conditions, there are characteristic $\sigma^0(\theta)$ relationships for a wide range of sea ice types. When the ice surface is bare and wet under summer conditions, $\sigma^0(\theta)$ reduces to a tightly clustered set of curves with no clear ice-type dependence.

Looking at a surface which has an arbitrary roughness distribution, the incidence angle dependence of the scattering cross section can be subdivided into three regimes:

1. *The ''normal incidence'' regime.* The normal, or vertical, incidence angle regime spans the incidence range from 0° to approximately 10° for ''smooth'' ice surface and to approximately 20° for rough ice surfaces. In this regime, σ^0 is dominated by returns from those surface elements whose local (surface normal) incidence angles lie closest to 0°. For very smooth or very regular surfaces, the phase structure of the signal is preserved on reflection and the surface reflections are specular.
2. *The ''diffuse scattering'' regime.* Over the incidence angle range from approximately 20° to approximately 80°, σ^0 is dominated by diffuse scattering in which all surface elements contribute elements to the backscattered field according to the surface roughness statistics. Here, the radar cross section varies relatively slowly, but with a decreasing trend with increasing incidence angle.
3. *The ''grazing angle'' or ''shadowing'' regime.* For incidence angles larger than approximately 70° for rough ice or larger than approximately 80° for relatively smooth ice, some parts of the surface are shadowed by

adjacent ice and thus do not contribute to the scattered energy. Most surfaces approach specular forward scattering with θ_i approaching 90°, so that the backscatter component becomes very small. In this region, σ^0 decreases more rapidly with incidence angle than in the diffuse scattering region and large-scale surface deformation features (ridges, blocks, rubble piles, etc.) become increasingly significant. At incidence angles greater than 85°, the deformation features may totally dominate the radar returns from the ice.

Under cold conditions, Livingstone et al. [69] and Kim et al. [58] show that the diffuse scattering regime of $\sigma^0(\theta)$ can be modeled by an ice type and frequency-dependent relationship of the form:

$$\sigma^0(\theta) \approx A\theta + B \text{ dB}$$

where A and B are determined by a χ^2 fit to $\sigma^0(\theta)$ (θ is in degrees) data over the range $10° < \theta < 50°$. Livingstone et al. [69] explore the use of this A–B plane for ice-type classification and the clustering of ice types found in the Beaufort Sea during a winter 1979 experiment as shown in Fig. 2.30. Although the incidence angle dependence of the scattering cross section shows some promise as an ice classification feature for winter ice conditions, this dependence is not seasonally robust and the measurements needed to provide the classification data are not routinely available from commonly used sensors. The dependence of radar frequency on A and B is not well known; however, some data derived from published results is shown in Table 2.5.

Consider the scattering from an exposed or thinly snow-covered ice surface under cold conditions, for ice whose brine volume is sufficiently high that the radar penetration depth is less than a wavelength. The scattering in this case is localized at the physical surface of the ice; snow volume effects and volume scattering within the ice can be ignored.

When the wavelength is normalized, horizontal correlation length of the surface, $r_{surf} > 1$, and the rms slope of the surface, m, is less than 0.25, the surface is electromagnetically smooth and behaves as a slightly rough mirror. In this case, the radar returns (and thus σ^0) at normal incidence ($\theta = 0°$) will be large, and as θ increases, σ^0 will decrease rapidly since most of the radar energy scatters into the forward wave direction. Taking an example from Kim et al. [58], for an ice surface with $m = 0.2$, the model yields $\sigma^0(0°) = 1$ dB and $\sigma^0(40°) = -35$ dB.

In contrast, when the rms slope exceeds $m = 0.4$, the surface is "rough" and the variation of scattered energy with the incidence angle is much reduced. Again referring to Kim et al. [58], at $m = 0.4$, $\sigma^0(0°) = -5$ dB and $\sigma^0(50°) = -12$ dB. A factor of two change in "surface roughness" can produce a factor of 800 change in incidence angle sensitivity.

The form of the surface roughness correlation function also influences the incidence angle sensitivity of surface scattering from sea ice. Gaussian (or

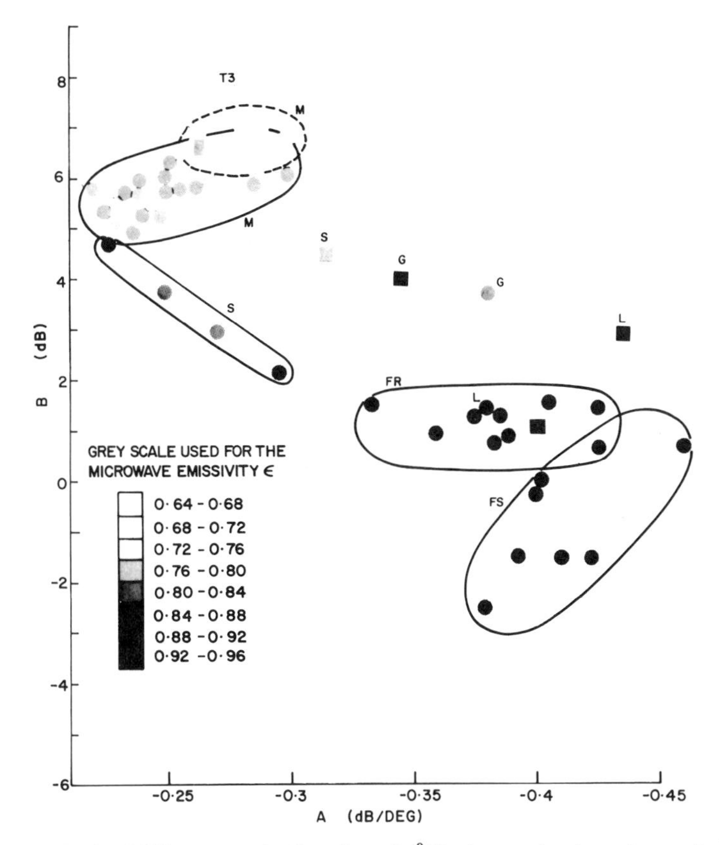

Figure 2.30 In the "diffuse scattering" region of $\sigma^0(\theta)$, the angular dependence of σ^0 (dB) is approximately linear and can be fit to a simple linear equation. The coordinate axes in the diagram above are the line intercept, B, and the line slope, A. Each point represents the *HH*-polarized data (at 13.3 GHz) from a single contiguous sample of an ice type (which is identified by a letter code). The gray-scale of the points is the *H*-polarized emissivity of the ice measured at 19.4 GHz. Sea-ice types are seen to form clusters in A–B space and thus feature spaces of this type have potential for ice type classification of winter sea ice. The ice type codes are: T3—an ice island (fresh water ice); M—old ice; S—second-year ice; FR—first-year rough ice; FS—first-year smooth ice; L—gray-white ice; G—gray ice.

TABLE 2.5 Incidence Angle Dependence of $\sigma^0_{HH}(\theta)$ for Winter Conditions at Different Frequencies ($\sigma^0_{HH}(\theta) = A\theta + B$, in dB)

Frequency	Ice Type	B	A	Reference
13.3 GHz	First-year	−0.6	−0.37	[72]
5.3 GHz	First-year	−11.0	−0.35	Note (a)
13.3 GHz	Old	−5.5	−0.25	[72]
9.6 GHz	Old	−5.0	−0.11	[90]
5.2 GHz	Old	−8.0	−0.16	[59]

Note (a): Livingstone et al. (1991, unpublished)

similarly shaped) correlation functions produce greater incidence angle sensitivities than do exponential correlation functions (or other cusped functions such as power laws). Comparisons between scattering surface model calculations for single scattering conditions (multiple scattering or scattering of the scattered wave can be ignored) and in situ measurements are discussed by Kim et al. [58]. Scattering cross-section measurements of wet, rough, Beaufort Sea ice at 13.3 GHz are well matched to rough-surface scattering model calculations based on surface measurements which yielded an exponential surface correlation function with 7.7 cm for correlation length and a height standard deviation of 0.72 cm. The bulk dielectric constants required in this case were in the range 3 to 5 and are compatible with measurements made on similar ice by other experimenters.

Volume scattering effects from scattering centers in the ice and snow cover are only weakly dependent on incidence angle over the 10° to 70° incidence angle range. The degree to which volume scattering is a significant contributor to σ^0 is dependent on the extinction coefficient of the material, which determines the penetration depth of the incident radiation, and thus on both salinity and temperature. Seasonal effects are large. Under cold conditions, the snow cover is a weak volume scatterer and volume scattering is found primarily in old, low-salinity ice (second-year ice and older) whose surface has not been flooded with sea water (Carsey [11]). Combinations of surface and volume scattering from typical snow cover and sea ice parameters are modeled by Kim et al. [58] and a detailed summary of modeling results can be found in Carsey [11].

Frequency Dependence

Some intriguing results come from multi-frequency scatterometer measurements performed by Onstott et al. [83–84], [86] and Onstott [88–89] in the late 1970s and early 1980s over the frequency range 1 to 18 GHz, and from their use in the formulation of semi-empirical scattering models by Kim et al. [58]. When the reported scattering cross sections are reduced from logarithmic form (dB, as published) to linear form, the following relationships can be extracted from the published data:

1. First-year ice cross section at *HH* polarization and 42° incidence angle varies with frequency as $\sigma^0(42°) = 2.23 \times 10^{-3} f^{1.02}$.
2. Gray ice cross section at *HH* polarization and 40° incidence angle varies with frequency as $\sigma^0(40°) = 10^{-3} f^{1.25}$.
3. Old ice cross section at *HH* polarization and 40° incidence angle varies with frequency as $\sigma^0(40°) = 1.26 \times 10^{-3} f^2$.
4. Empirically scaled, pure volume scattering model for sea ice at *HH* polarization and 30° incidence angle varies with frequency as $\sigma^0(30°) = 3.98 \times 10^{-5} f^{3.4}$.

In all cases above, the radar frequency is expressed in GHz. The data used represent a small spatial sample of each ice type at various sites and under a range of conditions (all cold) and thus the degree to which the above results are representative is unknown. Assuming that the $\sigma^0(\theta, f)$ relations given above have some general validity, their incidence angle dependence remains to be determined, although results given in the previous section suggest that incidence angle effects will be well behaved as a function of ice type. Similarly, a dependence on polarization is expected, but is not known at this time.

It is instructive to note that the old ice scattering represented by point (3) above is known to be produced by a combination of surface scattering represented by (1) and (2) and volume scattering represented by (4). The volume-scattering contribution to the scattering cross section of old sea ice is seen to decrease rapidly with radar frequency for frequencies below 4 GHz. As free water accumulates within the snow cover and within the surface layers of the ice (starting at the early melt season) volume scattering ceases to be a contributor to the radar returns and all ice types become surface scattering media. During freeze-up conditions in the fall of the year, old ice signatures will approach their winter levels as the freezing horizon in the ice volume descends to sufficient depth.

Although these comments do not represent fully established results at this time, they present sufficient evidence to identify a potentially fruitful area for further research. Firm results could have a major impact on the future development of electromagnetic scattering models.

Polarization Sensitivity

Radar polarization describes the orientation of the electric field vector of the radar wave in space. By convention, the wave is said to be horizontally polarized if the electric field vector lies in a horizontal plane at the radar and is said to be vertically polarized if the electric field vector lies in the vertical plane which contains the propagation vector. At normal incidence, all fields are horizontally polarized. For the remainder of this chapter, let us consider measurements made by radars that transmit either horizontal, *H*, or vertical, *V*, linear polarizations and receive either or both polarizations. When one radar polarization is transmitted and two are received, the received signals will have like-polarized, *HH* or *VV*, and cross-polarized, *HV* or *VH*, components. By convention, the first letter designates the transmitted polarization.

When radar systems are used to measure the scattering properties of sea ice surfaces, the scattering cross sections σ^0_{HH} and σ^0_{VV} are not necessarily equal. Two classes of phenomena must be considered:

1. The Fresnel reflection coefficients in Subsection 2.4.2 are different functions of incidence angle. Γ_V has the property that it is reduced as the Brewster angle of the medium ($\theta_B = \tan^{-1}(\epsilon^{1/2})$ for ϵ equal to the real part of the dielectric constant) is approached. The surface component of *VV* polarized radar returns are correspondingly reduced when energy is scattered from a smooth surface.
2. Spatially coherent orientations of elongated scatterers can preferentially interact with radar polarizations for which they exhibit efficient scattering. Elements of this type are found in hoar frost (frost flower) arrays on the surface of smooth, young ice. Low frequency radars whose signals penetrate the columnar ice layers can be influenced by the brine channel orientation.

Cross-polarized returns from sea ice surfaces are generated by multiple scattering, either within the ice or snow volume or at large surface features (e.g., edges of rafted ice and ridges).

For rough ice that is dominated by surface scattering processes it is important to note that the Fresnel reflection coefficient subscripts are referenced to the local surface normal at each point on the surface. As the surface normals for rough surfaces are uncorrelated with the radar range vector, the reflection coefficient averaged over a radar resolution cell has no preferred polarization. It is expected that $\sigma_{HH} = \sigma_{VV}$ for this ice. This idea is supported by polarization ratio observations of very rough first-year ice in the Labrador Sea marginal ice zone, Fig. 11.31. In the $\sigma^0_{VV}/\sigma^0_{HH}$ image in this figure, rough first-year ice is displayed as the dark tones on the right-hand side of the scene.

The surface of cold old ice found in the Arctic is covered by hummocks and frozen melt ponds. The surface salinity of the ice is low, the penetration depth is large, and the bulk dielectric constant is not a tensor (to first order). The ice at, and underlying, the hummock surfaces is both coarse grained and porous and is thus an effective surface scattering medium and an effective volume scattering medium at many radar frequencies. Melt pond surfaces are smooth, the trapped air bubbles are small and these features are, thus, inefficient radar scattering media. At radar resolution cell scales, the hummocks dominate the radar returns from this ice. The arguments of the previous paragraph can be applied to individual scattering centers on and within the ice with the result that σ^0_{HH} is expected to be equal to σ^0_{VV} for old ice. This has been confirmed experimentally by surface scatterometer measurements at 10 GHz during the CEAREX program (Onstott [87–89]).

When the ice is saline and the surface is smooth, the behavior of the Fresnel reflection coefficients controls the copolarized scattering from the ice surface. The Brewster angle minimum of the V-polarized reflection coefficient corresponds to a peak in the transmission coefficient of the surface. The V-polarized incident fields are thus more tightly coupled to the internal structure of the ice than are H-polarized incident fields (these are efficiently scattered in the forward direction). It is thus expected that the V-polarized back-scattered fields will be larger than the H-polarized back-scattered fields. The polarization sensitivity of smooth saline ice was investigated for artificially grown sea ice in the CRRELEX experiment. Results quoted in Onstott [89] show that $\sigma^0_{HH}/\sigma^0_{VV}$ at 10 GHz for this ice ranges from 0.6 at 20° incidence angle to about 0.2 at 60°.

The depolarization ratio of sea ice is not constant, but varies with ice type and deformation structure. The cross-polarized radar cross section $\sigma^0_{HV} = \sigma^0_{VH}$ thus contains information not found in σ^0_{HH} or σ^0_{VV} at least some of the time. From studies of electromagnetic scattering models (Winebrenner et al. [115]), the volume scattering process generally results in a cross-polarized radar return component. Studies of the polarimetric response of simple targets (for example, van Zyl et al. [105]) have shown that dihedral reflectors can be efficient depolarizing targets. There is evidence that volume scattering and even bounce multiple reflections both play roles in determining the depolarization ratio for sea ice. Ensemble statistics for sea ice scattering at 13.3 GHz (Livingstone et al. [70]) show that the depolarization ratio $r_D = \sigma^0_{HV}/\sigma^0_{HH}$ varies from approximately 0.25 for volume scattering media (cold old and second-year ice) to approximately 0.13 for rough surface scattering ice (rough first-year ice) to approximately 0.04 for very smooth young ice. Unweathered ice fracture features such as ridges or rafted edges (large-scale multiple-surface scattering) can have depolarization ratios exceeding 0.4 (Livingstone et al. [69]) as can be deduced from the foregoing and from Fig. 2.31. Weathered ridges on old ice (Fig. 2.32) have depolarization ratios similar to the background ice.

2.5.2 Seasonal Variations

In the earlier parts of this chapter, the physical processes that take place during sea-ice growth, aging, and decay are described. The electrical properties of sea ice are tightly coupled to these physical processes and thus it is reasonable to propose that sea-ice seasons can be defined in terms of the evolution of the measurable properties of the ice. The ice season hypothesis was supported by available ice surface measurements in the Beaufort Sea, Crozier Channel, and Davis Strait and by the ensemble (cluster) statistics of ice-type scattering cross section at 13.3 GHz and brightness temperature at 19 GHz. A feature space representation of these signatures at 45° incidence angle is shown in Figs. 2.33 to 2.37. In these figures, the one standard deviation ellipsoid of each cluster is used to represent the cluster boundary.

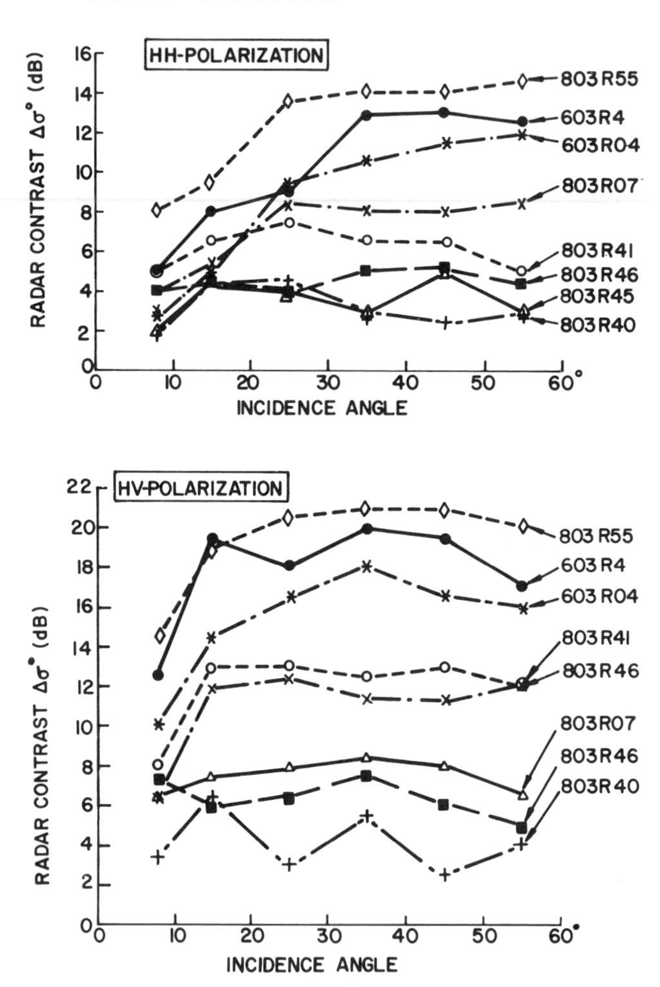

Figure 2.31 Fan beam scatterometer measurements (at 13.3 GHz) of sea-ice ridge returns for winter, first-year, sea ice in the Beaufort sea were used to estimate the contrast between first-year ice ridges and undeformed first-year ice, shown above. The ridge widths were estimated from simultaneous photographic imagery and ridge heights were estimated from ridge shadows. No significant correlations between ridge dimensions and radar cross section were observed. As is expected from the detailed geometry of ice ridge structures, the radar cross section of the ridges is extremely variable and changes more slowly with incidence angle than the cross section of undeformed first-year ice.

RADAR CONTRAST BETWEEN MULTI-YEAR RIDGES AND MULTI-YEAR ICE

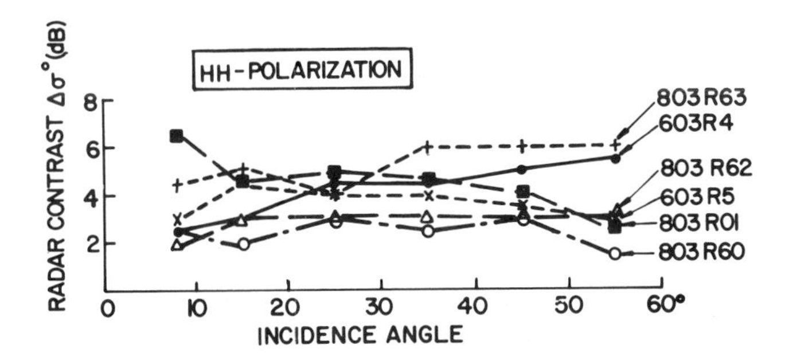

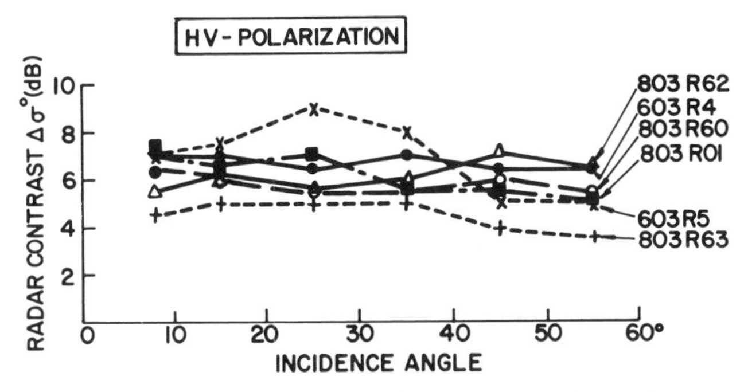

Figure 2.32 As part of the study mentioned in the previous figure, the contrast between old ice ridges and the background ice was also measured. Ridges on old ice floes have been extensively weathered and have characteristics similar to hummocks on the ice surface. The graphs above show that the old ridge contrasts at 13.3 GHz are much more tightly clustered than the corresponding first-year ice ridge contrasts and are considerably smaller.

Rigorously, the generality of these results is constrained to the regions and conditions measured. Results from other experiments, where the physical characteristics of the ice and the environmental conditions were similar to those used to define the feature space, yield compatible results (Carsey [11]). Results from experiments where the processes which determine the ice-formation history differ markedly from those found in the Arctic regions studied to define the feature space show that signatures for other ice types may be required. Examples are: young and first-year ice that have grown in mechanically disturbed conditions such as are found in the Labrador Sea (Livingstone and Drinkwater [72]), ice islands (Jefferies and Sackinger [55]), and ice that has been subjected to sufficiently heavy snow loads to cause surface submergence during growth or aging such as is commonly found in the Antarctic (Gow et al. [40]) and occasionally found in the Fram Strait (Tucker et al. [102]). The

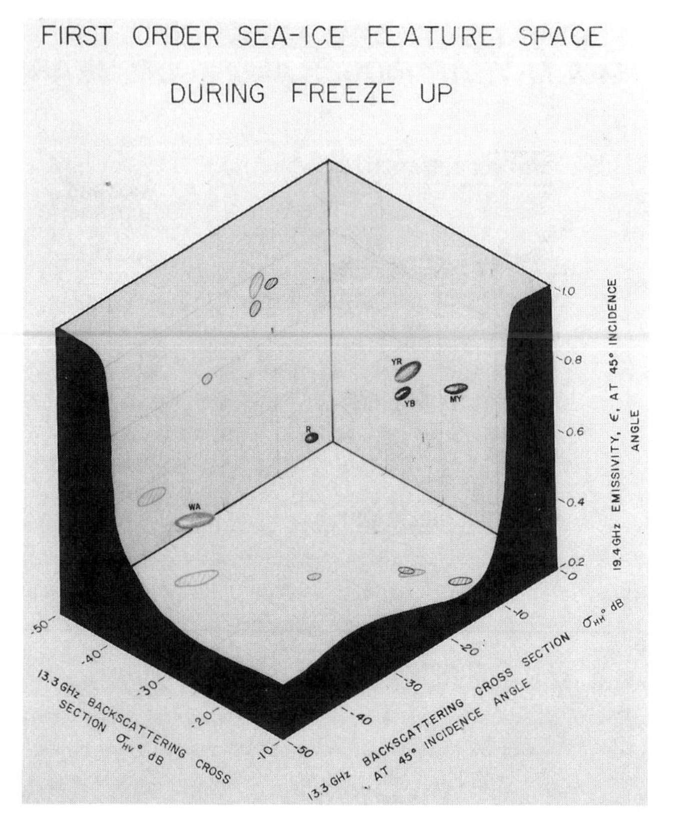

Figure 2.33 The position and first standard deviation of open water, WA; grease ice, R; young, rough ice, YR; young, broken (rubbled) ice, YB; and old ice, MY; in a feature space defined in terms of the *HH*- and *HV*-polarized scattering cross section at 13.3 GHz and the *H*-polarized, 19.4 GHz emissivity are shown all at 45° incidence angle. The data used to generate the clusters were acquired in the Beaufort Sea in the later portion of freeze-up (this is a temporal snapshot of a continuous process). At the time and frequencies presented, the signatures are approaching their winter positions in this space.

feature space shown in this section is not unique in its ability to provide models for sea-ice classification (Livingstone et al. [70]) and was originally devised to illustrate the value of the quantitative use of multi-sensor data sets for this function.

The sea-ice season concept that provides the framework for the feature space is physically based and may prove valuable in the long-term as a tool for grouping measurement sets and ice classification algorithms. As presented here, the defined ice seasons are probably not a complete set. Analysis of passive microwave satellite imagery of summer sea ice (Carsey [10]) shows that the variability of ice signatures is large during the melt season. Since the details of the melt water source and the ice drainage channels differ between the period

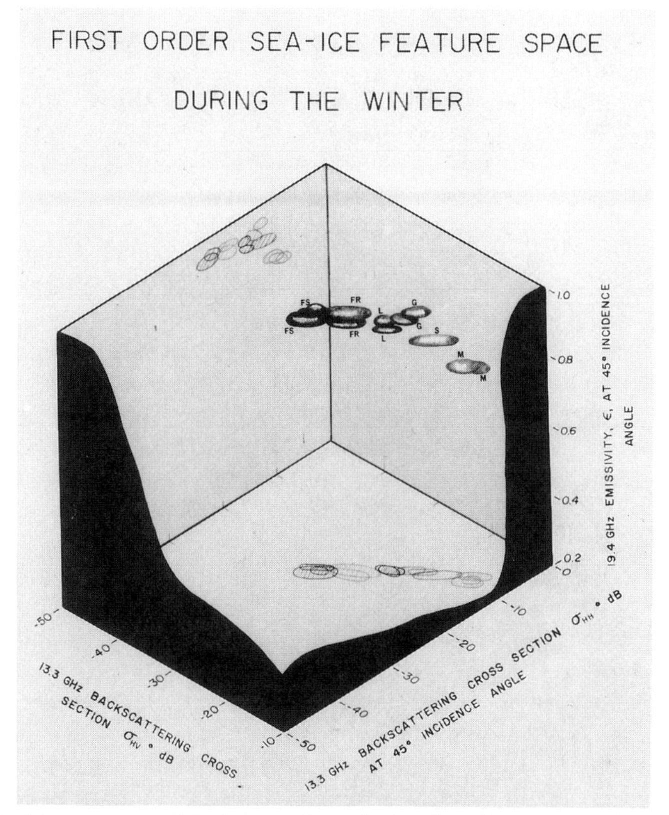

Figure 2.34 The signatures of the ice type shown in the winter feature space: first-year smooth, FS; first-year rough, FR; gray-white, L; gray ice, G; second-year, S; and old, M. These were compiled from large numbers of samples collected in two different years and demonstrate the expected stability.

preceding and during the peak melt (ice flooding) and the well-drained period in the later part of the summer, it may be reasonable to redefine the "melt season" as two distinct ice seasons. This has not yet been done, except for investigations of snow cover.

The seasonal cycle can be considered to begin with freeze-up. This season is physically characterized by the beginning of ice formation in open water, by the formation of a freezing horizon in existing ice, and by the start of snow cover accumulation. At the start of freeze-up, any of the previous season's first-year ice that has survived the summer melt becomes second-year ice. During freeze-up, the mean temperature is consistently below freezing, young ice growth commences, and the freezing horizon in old ice descends from the ice surface into the volume of the ice with a consequent loss of the liquid fraction in the upper layers of the ice. The transition of old ice scattering signatures from their variable summer states to their stable winter states follows the descent of the freezing horizon as a function of frequency (winter signatures

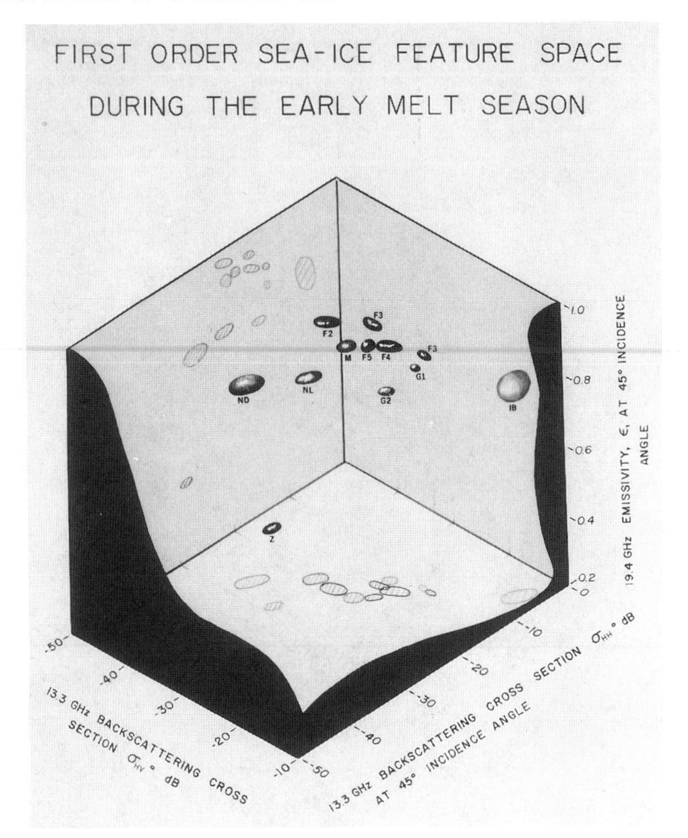

Figure 2.35 The ice classes represented are: frazil, Z; dark nilas, ND; light nilas, NL; gray, G; first-year, F; old, M; and icebergs, IB. The numbers attached to some classes designate identifiable (from auxiliary data) sub-classes of the ice type.

appear first at high-radar frequencies, *Ku*-band and higher, and last at low-radar frequencies, *L*-band and lower). Freeze-up can be considered to end with the growth of first-year ice. During the freeze-up season, ice signatures (Fig. 2.33) migrate from their end-of-summer configuration to their winter form.

The onset of the winter ice season is marked by the appearance of recognizable first-year ice. Available data indicate that microwave sea-ice signatures are quite stable throughout this season in the Arctic. During the arctic winter the temperatures on the pack ice vary from cold to very cold and few transitions above freezing are seen. Young ice grows in leads that open in the pack due to wind forcing and their growth rates (and thus, signatures) are relatively stable. This ice forms and ages in a predictable manner in this season and the microwave signatures of the ice evolve in a consistent manner. Arctic snow cover tends to be thin and dry over the winter season. Snow is not a major contributor to sea-ice signatures at radar frequencies below 18 GHz. Figure 2.34 shows a feature-space representation of winter sea-ice signatures.

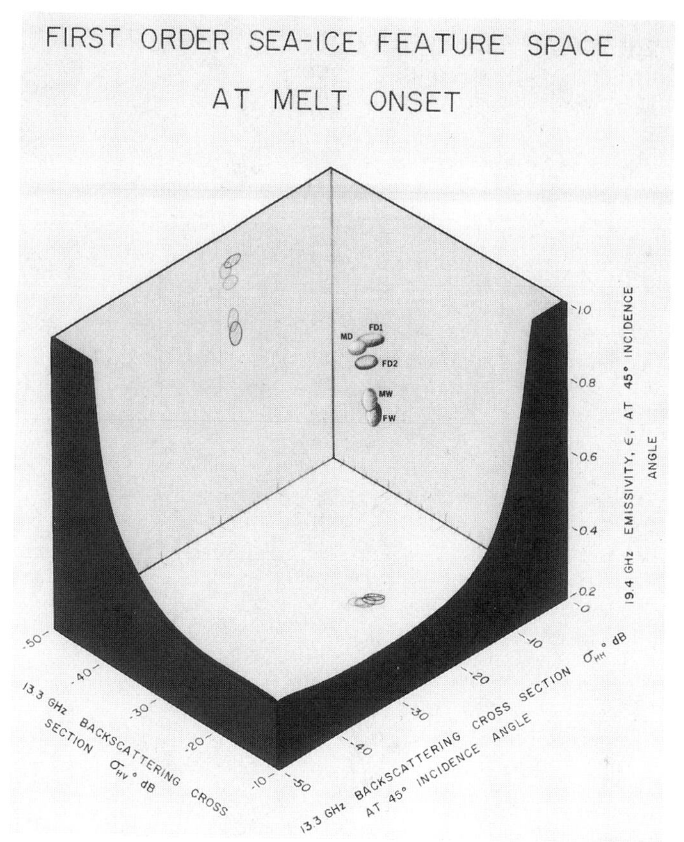

Figure 2.36 Feature space scattering signatures for the melt onset season cluster into a small segment of the feature space and the emissivity signatures provide an estimate of free water on the ice and in the snow pack. The ice classes identified from aerial photography and surface measurements are: first-year wet, FW; first-year dry, FD; old wet, MW; and old dry, MD. The "wet" and "dry" descriptors indicate the presence of free water on exposed surfaces.

At winter's end, when the diurnal temperature cycle peaks exceed the freezing point, moisture begins to accumulate in the snow pack and a period of extensive snow-pack recrystallization begins. In the early melt season, diurnal heating and occasional warming events increase the partial pressure of water vapor within the surface layers of the ice and snow cover. The bulk of the ice is cold and recrystallization occurs. In this early melt season, the snow cover contributions to the sea-ice scattering signatures become stronger as scatterers within the snow pack increase in size and as accumulating moisture within the upper snow layers increases the loss tangent of the snow. Feature-space clusters which had remained stable throughout the winter begin to migrate within the space as shown in Fig. 2.35.

As moisture continues to accumulate in the snow cover, water begins to collect at the ice surface. In the early stages of this process, the ice is a heat

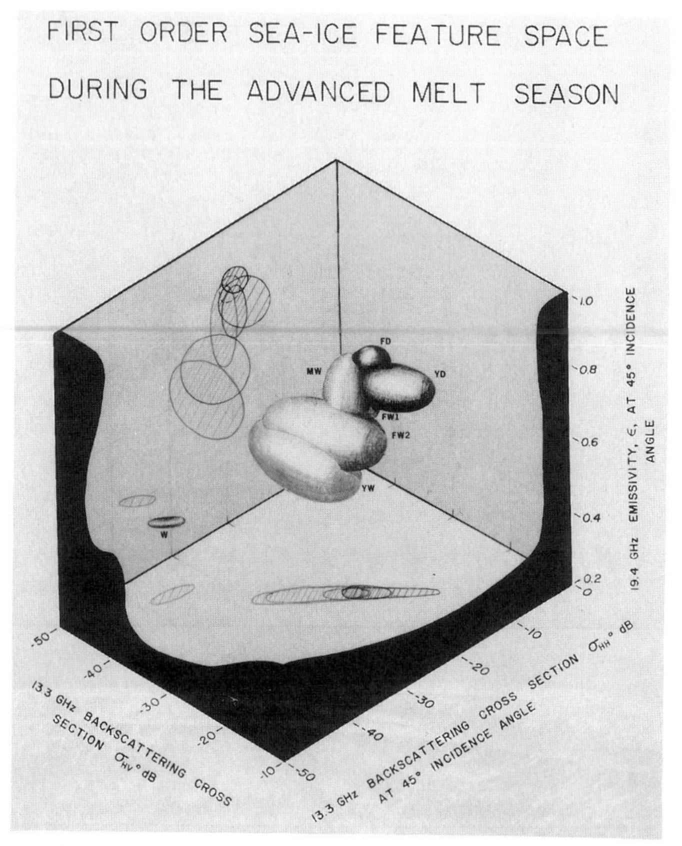

Figure 2.37 The advanced melt season feature space representation shown contains sample blocks from both rapid melt and freezing conditions in the Beaufort Sea and the Arctic islands. The ice classes shown are: young wet ice, YW; young dry ice, YD; first-year wet ice, FW; first-year dry ice, FD; old wet ice, MW. The "wet" and "dry" descriptors refer to the ice surface appearance in aerial photographs and to conditions reported by surface parties.

sink, well below its freezing point, and extensive recrystallization of the snow base creates a layer of rough "superimposed ice" (Onstott and Gogineni [85]). In places where the snow was thin, it erodes completely, exposing rough, wet ice surfaces. Remaining snow often has free-water concentrations exceeding 5% and becomes optically thick. This condition has been called "melt onset" and is shown as a features space diagram in Fig. 2.36. Here, all ice types in the scene have become localized in the radar cross-section plane and only emissivity differences are observed. Scattering is dominated by the snow cover and by bare, wet ice. Measurements of the dependence of scattering cross section on incidence angle show surface scattering characteristics only.

The summer melt season is characterized by continuous or intermittent ablation of the ice surface as the local temperature cycles through the melting point of the ice. Where a continuous ice sheet has been present, the melt season is often initiated by a series of flood-drain cycles as the rapidly melting snow surface floods the ice with melt water, which then drains through fractures and eventually through opened brine drainage channels within the ice volume. Subsequent melting and ice ablation is interspersed with freezing periods, depending on the Arctic region being examined. Ice that is not completely melted by the end of the melt season is largely desalinated, has a coarse-grained, bubble-filled upper surface on its high points (hummocks), and has melt ponds filled with nearly fresh water in its drainage basins. Measurements from the early part of the melt season in the Beaufort Sea were used to create the feature space representations shown in Fig. 2.37. In this diagram it is seen that the clusters for the all-ice types in the data set have very large standard deviations and are extensively overlapped in the scattering plane. The radar is observing ice projections of varying roughness and geometry on a smooth, wet background.

Onstott et al. [86] examine accumulated results from sea-ice measurements made in the Fram Strait and trace the complex evolution of scattering and emissivity signatures through the melt season. They suggest that the two dimensional texture patterns associated with ice type may be of use in classification. Barber and LeDrew [6] suggest that texture-based classifications may be more difficult than was originally envisaged.

2.6 SUMMARY

First-year sea ice is relatively saline (5 to 20%) and thus is lossy and difficult to penetrate with microwave energy. This ice is usually columnar in structure, with a possible thin layer of frazil ice on the surface. The surface topography is generally sharp and angular with moderate relief.

Old ice is a low-loss dielectric substance which permits significant penetration by microwave energy. The top layer (about 1 m) of the ice is porous with large (2 mm) air voids. The surface is undulating with smooth and rounded features, often with significant vertical relief.

Glacial ice is also a low-loss dielectric material which permits significant microwave penetration. It usually contains a large number of small (0.2 to 0.5 mm) air bubbles. Surface features of the glacial ice on the ocean (icebergs, bergy bits, and growlers) vary from angular and square to smooth and well rounded.

Quantitative descriptive snow models have been developed for the Arctic (Greenland and Barents Seas) as well as the Antarctic (Weddell Sea) during the onset of the melt period. Understanding the metamorphic processes in a snow cover described in these descriptive snow models allows changes in snow conditions to be related to changes in observed, remotely sensed characteristics.

The problem of spatial inhomogeneities in ice and snow regimes is discussed, and linkages between ice and snow characteristics and electromagnetic scattering properties for large area observations are explored.

Much of the early work in designing effective sensors for the remote detection and surveillance of ice entailed gaining a better understanding of its physical properties. However, the physical properties of ice and snow vary enormously, by ice type, age, season, time of day, and region, and this variability posed difficulties in accurate characterization. We now have a good understanding of the properties of most kinds of ice, even though there is still work to be done for some types of ice, such as conditions near freeze-up and breakup. In subsequent chapters of this book, the properties of ice and snow and particularly the electromagnetic properties will be the underlying factors in almost all the discussions.

ACKNOWLEDGMENTS

The authors wish to acknowledge the work on the properties of ice and snow undertaken by a large cadre of international researchers as referenced throughout the chapter. The authors also wish to acknowledge the program and financial support of the Department of Fisheries and Oceans, the Canada Centre for Remote Sensing, the Atmospheric Environment Service, and the Office of Energy Research and Development (OERD).

REFERENCES

[1] J. R. Addison (1969) "Electrical properties of saline ice," *J. Appl. Phys.*, **40** (8), 3105–3114.

[2] W. Ambach and A. Denoth (1980) "The dielectric behaviour of snow: A study versus liquid water content," NASA Workshop on the microwave remote sensing of snowpack properties, Fort Collins, CO, May 20–22, 1980, NASA CP-2153.

[3] D. L. Anderson (1960) "The physical constants of sea ice," *California Institute of Technology, Research*, **13** (8), 310–318.

[4] E. L. Andreas and S. F. Ackley (1982) "On the differences in ablation seasons of Arctic and Antarctic sea ice," *J. Atmos. Sci.*, **39** (3).

[5] A. Assur (1958) "Composition of sea ice and its tensile strength," in *Arctic Sea Ice*, U.S. National Academy of Sciences–National Research Council, Pub. 598, 106–138.

[6] D. G. Barber and E. F. LeDrew (1991) "SAR sea ice discrimination using texture statistics: A multivariate approach," *Photogrammetric Eng. Remote Sens.*, **57** (4), 385–395.

[7] T. F. Budinger, R. P. Dinsmore, P. A. Morill, and F. M. Soule (1960) "Iceberg detection by radar," *Int. Ice Patrol Bull.*, **45**, United States Coast Guard, 49–97.

[8] W. J. Campbell and A. S. Orange (1974), "The electrical anisotropy of sea ice in the horizontal plane," *J. Geophys. Res.*, **79**, 5059–5063.

[9] W. J. Campbell, J. Wayenberg, J. Ramseyer, and R. O. Ramseier (1978) "Microwave remote sensing of sea ice in the Aidjex main experiment," *Boundary-Layer Meteorology*, **13**, 309–337.

[10] F. D. Carsey (1985) ''Summer arctic sea ice character from satellite microwave data,'' *J. Geophys. Res.*, **90** (C3), 5015–5034.

[11] F. D. Carsey (1992) Ed., *Microwave Remote Sensing of Sea Ice*, Geophysical Monogr. 68, American Geophysical Union, Washington, DC.

[12] S. C. Colbeck (1975) ''Grain and bond growth in wet snow,'' *Proc. Grindelwald Symp.*, IAHS Publ. 114.

[13] S. C. Colbeck (1979) ''Grain cluster in wet snow,'' *J. Colloid Interface Sci.*, **72** (3), 371–384.

[14] S. C. Colbeck (1982) ''The geometry and permittivity of snow at high frequencies,'' *J. Appl. Phys.*, **53** (6), 4495–4500.

[15] S. C. Colbeck (1982) ''An overview of seasonal snow metamorphism,'' *Rev. Geophys. Space Phys.*, **20**, 45–61.

[16] S. Colbeck (Chair), E. Akitaya, R. Armstrong, H. Gubler, J. Lafeuille, K. Lied, D. McClung, and E. Morris (1988) ''The international classification for seasonal snow on the ground,'' Prepared by the Working Group on Snow Classification.

[17] M. J. Collins, R. O. Ramseier, and S. P. Gogineni (1990) ''Active/passive signatures of springtime Barents Sea ice,'' *Proc. IGARSS'90*, 1517–1520.

[18] G. F. N. Cox and W. F. Weeks (1974) ''Salinity variations in sea ice,'' *J. Glaciol.*, **13** (67), 109–120.

[19] G. F. N. Cox, J. A. Richter-Menge, W. F. Weeks, and M. Mellor (1984) ''Mechanical properties of multiyear sea ice; Phase 1 test results,'' U.S. Army Corps of Engineers, Cold Regions Research and Engineering Laboratory, Hanover, NH, Rep. 84–9.

[20] G. F. N. Cox and W. F. Weeks (1988) ''Profile properties of undeformed first-year sea ice,'' CRREL Rep. 88–13.

[21] W. A. Cumming (1952) ''The dielectric properties of ice and snow at 3.2 centimeters,'' *J. Appl. Phys.*, **23** (7), 768–773.

[22] L. W. Davidson and W. W. Denner (1982) ''Sea ice and iceberg conditions on the Grand Banks affecting hydrocarbon production and transportation,'' *Proc. IEEE Oceans '82*, 1236–1241.

[23] A. Denoth (1980) ''The pendular-funicular transition in snow,'' *J. Glaciol.*, **25** (91), 93–97.

[24] A. Denoth, A. Fogar, P. Weiland, C. Mätzler, H. Aebischer, M. Turi, and A. Sihvola (1984) ''A comparative study of instruments for measuring the liquid water content of snow,'' *J. Appl. Phys.*, **56**, 2154–2160.

[25] M. R. Drinkwater (1988) ''Important changes in microwave scattering properties of young, snow covered sea ice as indicated from dielectric modelling,'' *Proc. IGARSS'88*, 793–797.

[26] M. R. Drinkwater, R. Kwok, D. P. Winebrenner, and E. Rignot (1991) ''Multifrequency polarimetric synthetic aperture observations of sea ice,'' *J. Geophys. Res.*, **96** (11), 20679–20698.

[27] J. E. Dykins (1969) ''Tensile and flexural properties of saline ice,'' N. Riehl, B. Bullemer, and H. Engelhardt (Eds.), *Proc. Int. Symp. Phys. of Ice*, Plenum, New York, 251–270.

[28] W. P. Ebaugh and D. R. DeWalle (1977) ''Retention and transmission of liquid water in fresh snow,'' *Proc. 2nd Conf. Hydrometeorology*, American Meteorological Society, Toronto, 255–260.

[29] C. C. Ebbesmeyer, A. Okubo, and H. J. M. Helset (1980) "Description of iceberg probability between Baffin Bay and the Grand Banks using a stochastic model," *Deep-Sea Res.*, **27A**, 975–986.

[30] H. J. Eom (1982) "Theoretical scatter and emission models for microwave remote sensing," Ph.D. thesis, University of Kansas.

[31] S. Evans (1965) "Dielectric properties of ice and snow—A review," *J. Glaciol.* **5** (42), 773–792.

[32] G. Frankenstein and R. Garner (1967) "Short note: Equations for determining the brine volume of sea ice from −0.5 to −22.9°C," *J. Glaciol.*, **6** (48), 943–944.

[33] P. H. Gammon, R. E. Gagnon, W. Bobby, and W. E. Russell (1983) "Physical and mechanical properties of icebergs," *15th Ann. Offshore Tech. Conf. 4459*, Houston, TX, 143–149.

[34] C. Garrity (1991) "Passive microwave remote sensing of snow covered floating ice during spring conditions in the Arctic and Antarctic," Ph.D. dissertation, York University, CRESS Dept., North York, Canada.

[35] C. Garrity and B. A. Burns (1988) "Electrical and physical properties of snow in support of BEPERS-88," Tech. Rept. MWG 88-11.

[36] R. W. Gerdel (1948) "Physical changes in snow-cover leading to runoff, especially floods," *Int. Assoc. Hydrological Sci.*, *Proc. General Assembly of Oslo*, Commission of Snow and Ice, **2**, 42–51.

[37] S. Gogineni, R. G. Onstott, R. K. Moore, and J. Chancellor (1984) "Intermediate results of radar backscatter measurements from summer sea ice," Remote Sensing Laboratory, Univ. Kansas, Center for Research, Inc., Lawrence, KS, Tech. Rept. 3311-2.

[38] S. Gogineni, R. K. Moore, Q. Wang, A. Gow, and R. G. Onstott (1990) "Radar backscatter measurements over saline ice," *Int. J. Remote Sensing*, **11** (4), 603–615.

[39] A. J. Gow (1968) "Bubbles and bubble pressures in Antarctic glacier ice," U.S. Army Corps of Engineers, Cold Regions Research and Engineering Laboratory, Hanover, NH, Res. Rept. 249.

[40] A. J. Gow, S. F. Ackley, W. F. Weeks, and J. W. Govoni (1982) "Physical and structural characteristics of Antarctic sea ice," *Ann. Glaciol.*, **3**, 113–117.

[41] A. L. Gray (1981) "Microwave Remote Sensing of Sea Ice," in *Oceanography from Space*, Marine Science Ser., Plenum, New York, **13**, 785–800.

[42] A. L. Gray, R. K. Hawkins, C. E. Livingstone, L. D. Arsenault, and W. M. Johnston (1982) "Simultaneous scatterometer and radiometer measurements of sea-ice microwave signatures," *IEEE J. Ocean. Eng.*, **OE-7** (1), 20–33.

[43] A. L. Gray and L. D. Arsenault (1991) "Time-delayed reflections in L-band synthetic aperture radar imagery of icebergs," *IEEE Trans. Geosci. Remote Sensing*, **29** (2), 284–291.

[44] T. C. Grenfell and A. W. Lohanick (1985) "Temporal variations of the microwave signatures of sea ice during the late spring and early summer near Mould Bay, NTW," *J. Geophys. Res.*, **90** (C3), 5063–5074.

[45] A. A. Griffith (1920) "The phenomena of rupture and flow in solids," *Philosophical Trans.*, *Royal Society of London*, **A221**, 163–198.

[46] M. Hallikainen and Dale Winebrenner (1992) "Chapter 3: The physical basis for sea ice remote sensing," in *Microwave Remote Sensing of Sea Ice*, F. D. Carsey (Ed.), Geophys. Monog. 68, American Geophysical Union, Washington, DC.

[47] R. K. Hawkins, A. L. Gray, C. E. Livingstone, and L. D. Arsenault (1981) "Seasonal effects on the microwave signatures of Beaufort Sea ice," *Proc. 15th Int. Symp. Remote Sensing of the Environment*, 239–257.

[48] R. K. Hawkins, A. L. Gray, C. E. Livingstone, and L. D. Arsenault (1982) "Microwave remote sensing of sea ice," *J. Int. Soc. Photogrametry and Remote Sensing*, 879–886.

[49] R. K. Hawkins, A. L. Gray, C. E. Livingstone, and L. D. Arsenault (1982) "Microwave remote sensing of sea ice," *Proc. Int. Symp. ISPRS Comm. 7*, 789–799.

[50] P. Hoekstra and P. Cappillino (1971) "Dielectric properties of sea and sodium chloride ice at UHF and microwave frequencies," *J. Geophys. Res.*, **76** (20), 4922–4931.

[51] P. Hoekstra and D. Spanogle (1972) "Radar cross-section measurements of snow and ice," U.S. Army Corps of Engineers, Cold Regions Research and Engineering Laboratory, Hanover, NH, Tech. Rept. 235.

[52] B. Holt and S. A. Digby (1985) "Processes and imagery of first-year fast ice during the melt season," *J. Geophys. Res.*, **90** (C3), 5015–5034.

[53] A. A. Husseiny (1978) *Iceberg Utilization*, Permagon Press, New York.

[54] B. Holt, S. Digby, and A. Usan (1985) "Processes and imagery of first-year fast sea ice during the melt season," *J. Geophys. Res.*, **90** (C3), 5045–5062.

[55] M. O. Jefferies and W. M. Sackinger (1990) "Ice island detection and characterization, with airborne synthetic aperture radar," *J. Geophys. Res.*, **95** (4), 5371–5377.

[56] K. C. Jezek, J. W. Clough, C. R. Bentley, and S. Shabtaie (1978) "Dielectric permittivity of glacier ice measured in situ by radar wide-angle reflection," *J. Glaciol.*, **21** (85), 315–328.

[57] T. Karlsson (1971) "A viscoelastic-plastic material model for drifting sea ice," *Proc. Int. Conf. on Sea Ice*, Reykjavik, Iceland.

[58] Y. S. Kim, R. K. Moore, and R. G. Onstott (1984) "Theoretical and experimental study of radar backscatter from sea ice," Remote Sensing Laboratory, Univ. Kansas Rept. RSL TR 331-37.

[59] M. E. Kirby and R. T. Lowry (1979) "Iceberg detectability problems using SAR and SLAR systems," *5th Ann. Pecora Symp.: Satellite Hydrology*, Sioux Falls, SD.

[60] A. Kovacs and R. M. Morey (1978) "Radar anisotropy of sea ice due to preferred azimuthal orientation of the horizontal *C* axes of the ice crystal," *J. Geophys. Res.*, **83** (C12), 6037–6046.

[61] A. Kovacs and R. M. Morey (1979) "Anisotropic properties of sea ice in the 50- to 150-MHz range," *J. Geophys. Res.*, **84** (C9), 5749–5759.

[62] M. Kumai and K. Itagaki (1953) "Cinematographic study of ice crystal formation in water," *J. Faculty of Science*, Hokkaido Univ., Hokkaido, Japan, Ser. II, **4**, 235–246.

[63] R. A. Lake and E. L. Lewis (1970) "Salt rejection by sea ice during growth," *J. Geophys. Res.*, **75** (3), 583–597.

[64] J. Lamb (1949) "The dielectric properties of ice at 1.25 cm wavelength," *Proc. Physical Soc. London*, Sect. B, 272.

[65] M. A. Lang and H. Eicken (1991) "The sea ice thickness distribution in the northwestern Weddell Sea," *J. Geophys. Res.*, **96**, 4821–4837.

[66] E. J. Langham (1974) "Phase equilibria of veins in polycrystalline ice," *Can. J. Earth Sci.*, **11**, 1280–1287.

[67] R. Lemmela (1973) "Measurements of evaporation-condensation and melting from a snow cover," *Proc.: The role of Snow and Ice in Hydrology*, UNESCO-WMO-IASH, 670–679.

[68] C. E. Livingstone, R. K. Hawkins, A. L. Gray, K. Okamoto, T. L. Wilkinson, S. Young, L. D. Arsenault, and D. Pearson (1981) "Classification of Beaufort Sea ice using active and passive microwave sensors," in *Oceanography from Space*, Marine Science Ser., Plenum, New York, **13**, 813–821.

[69] C. E. Livingstone, R. K. Hawkins, A. L. Gray, L. D. Arsenault, K. Okamoto, T. L. Wilkinson, and D. Pearson, (1983) "The CCRS/SURSAT Active-passive experiment 1978–1980: The microwave signatures of sea ice," Canada Centre for Remote Sensing, Data Acquisition Div. Rept.

[70] C. E. Livingstone, K. P. Singh, and A. L. Gray (1987) "Seasonal and regional variations of active/passive microwave signatures of sea ice," *IEEE Trans. Geosci. Remote Sensing*, **GE-25** (2), 159–173.

[71] C. E. Livingstone, R. G. Onstott, L. D. Arsenault, A. L. Gray, and K. P. Singh (1987) "Microwave sea ice signatures near the onset of melt," *IEEE Trans. Geosci. Remote Sensing*, **GE-25** (2).

[72] C. E. Livingstone and M. R. Drinkwater (1991) "Springtime C-band backscatter signatures of Labrador Sea marginal ice: Measurements versus modelling predictions," *IEEE Trans. Geosci. Remote Sensing*, **29** (1), 29–41.

[73] J. R. Marko, J. R. Birch, and M. A. Wilson (1982) "A study of long-term satellite-tracked iceberg drifts in Baffin Bay and Davis Strait," *Arctic*, **35** (1), 234–240.

[74] S. Martin (1979) "A field study of brine drainage and oil entrainment in first-year sea ice," *J. Glaciol.*, **22** (88), 473–502.

[75] C. Mätzler (1985) *Resometer Manual*, Univ. Bern, Switzerland.

[76] C. Mätzler (1987) "Applications of the interaction of microwaves with natural snow cover," *Remote Sensing Rev.*, **2**, 259–387.

[77] D. C. Meeks, G. A. Poe, and R. O. Ramseier (1974) "A study of microwave emission properties of sea ice—AIDJEX 1972," Aerojet Electrosystems Co., Univ. Washington, Final Rept. 1786FR-1.

[78] M. Mellor (1986) "Mechanical behaviour of sea ice," N. Untersteiner (Ed.), NATO ASI Ser., Plenum, New York, 165–281.

[79] R. M. Morey, A. Kovacs, and G. F. N. Cox (1984) "Electromagnetic properties of sea ice," *Cold Regions Sci. Tech.*, **9**, 53–75.

[80] M. Nakawo and N. K. Sinha (1981) "Growth rate and salinity profile of first-year sea ice in the high Arctic," *J. Glaciol.*, **27** (96), 315–329.

[81] M. Nakawo (1983) ''Measurements on air porosity of sea ice,'' *Ann. Glaciol.*, **4**, 204–208.

[82] T. M. Niedrauer and S. Martin (1979) ''An experimental study of brine drainage and convection in young sea ice,'' *J. Geophys. Res.*, **84** (C3), 1176–1186.

[83] R. G. Onstott, R. K. Moore, and W. F. Weeks (1979) ''Surface-based scatterometer results of Arctic sea ice,'' *IEEE Trans. Geosci. Electron.*, **GE-17**, 78–85.

[84] R. G. Onstott, R. K. Moore, S. Gogineni, and C. Delker (1982) ''Four years of low-altitude sea ice broadband backscatter measurements,'' *IEEE J. Ocean. Eng.*, **7**, 44–50.

[85] R. G. Onstott and S. P. Gogineni (1985) ''Active microwave measurements of sea ice under summer conditions,'' *J. Geophys. Res.*, **90** (C3), 5035–5044.

[86] R. G. Onstott, T. C. Grenfell, C. Mätzler, C. A. Luther, and E. A. Svendsen (1987) ''Evolution of microwave sea ice signatures during early summer and midsummer in the marginal ice zone,'' *J. Geophys. Res.*, **92** (C7), 6825–6835.

[87] R. G. Onstott (1990) ''Near surface measurements of arctic sea ice during the fall freeze-up,'' *Proc. IGARSS'90*, 1529.

[88] R. G. Onstott (1990) ''Polarimetric radar measurements of Arctic sea ice during the coordinated eastern Arctic experiment,'' *Proc. IGARSS'90*, 1531–1532.

[89] R. Onstott (1992) ''Chapter 5: SAR and scatterometer signatures of sea ice,'' in *Microwave Remote Sensing of Sea Ice*, F. D. Carsey (Ed.), Geophys. Monog. 68, American Geophysical Union, Washington, DC.

[90] E. Orowan (1950) *Fatigue and Fracture of Metals*, W. M. Murray (Ed.), Wiley, New York, 139–157.

[91] D. F. Page and R. O. Ramseier (1975) ''Application of radar techniques to ice and snow studies,'' *J. Glaciol.*, **15** (73), 171–191.

[92] D. C. Pearce and J. W. Walker (1967) ''An empirical determination of the relative dielectric constant of the Greenland ice cap,'' *J. Geophys. Res.*, **72** (22), 5743–5747.

[93] D. K. Perovich, A. J. Gow, and W. B. Tucker III (1988) ''Physical properties of snow and ice in the winter marginal ice zone of Fram Strait,'' *IGARSS'88*, ref. ESA SP-284, 1119–1123 (IEEE 88CH2497-6), Edinburgh, Scotland.

[94] F. G. J. Perey and E. R. Pounder (1958) ''Crystal formation in ice sheets,'' *Can. J. Phys.*, **36**, 494–502.

[95] E. Pounder (1965) *Physics of Ice*, Pergamon Press, London.

[96] E. C. Schanda, C. Mätzler, and K. Kunzi (1983) ''Microwave remote sensing of snow cover,'' *Int. J. Remote Sensing*, **4**, 149–158.

[97] P. F. Scholander and D. C. Nutt (1960) ''Bubble pressure in Greenland icebergs,'' *J. Glaciol.*, **3** (28), 671–678.

[98] J. Schwartz and W. F. Weeks (1977) ''Engineering properties of sea ice,'' *J. Glaciol.*, **19** (81), 499–530.

[99] W. Schwarzacher (1959) ''Pack-ice studies in the Arctic Ocean,'' *J. Geophys. Res.*, **64** (12), 2357–2367.

[100] A. Stogryn and G. Desargant (1985) ''The dielectric properties of brine in sea ice at microwave frequencies,'' Aerojet Electrosystems Rept. 7788 (prepared for USN, ONR).

[101] W. B. Tucker, III, A. J. Gow, and W. F. Weeks (1987) "Physical properties of summer sea ice in the Fram Strait," *J. Geophys. Res.*, **92** (C7), 6787–6803.

[102] W. B. Tucker, III, T. C. Grenfell, R. G. Onstott, D. K. Perovich, A. J. Gow, R. A. Schuchman, and L. L. Sutherland (1991) "Microwave and physical properties of sea ice in the winter marginal ice zone," *J. Geophys. Res.*, **96** (C3), 4575–4587.

[103] W. B. Tucker, D. K. Perovich, A. J. Gow, W. F. Weeks, and M. R. Drinkwater (1992) "Chapter 2: Physical properties of sea ice relevant to remote sensing," in *Microwave Remote Sensing of Sea Ice*, F. D. Carsey (Ed.), Geophys. Monog. 68, American Geophysical Union, Washington, DC.

[104] N. Untersteiner (1986) *The Geophysics of Sea Ice*, NATO ASI Ser. B: Physics, **146**, Plenum Press, New York.

[105] J. J. van Zyl, H. A. Zebker, and C. Elachi (1987) "Imaging radar polarization signatures: Theory and observation," *Radio Sci.*, **22** (4), 529–543.

[106] M. R. Vant, R. B. Gray, R. O. Ramseier, and V. Makios (1974) "Dielectric properties of fresh and sea ice at 10 and 35 GHz," *J. Appl. Phys.*, **45** (11), 4712–4717.

[107] M. R. Vant, R. O. Ramseier, and V. Makios (1978) "The complex-dielectric constant of sea ice at frequencies in the range 0.1–40 GHz," *J. Appl. Phys.*, **49** (3), 1264–1280.

[108] A. Von Hippel (1954) *Tables of Dielectric Materials*, Massachusetts Institute of Technology Press, Boston, MA.

[109] P. Wadhams, M. A. Lange, and S. F. Ackley (1987) "The ice thickness distribution across the Atlantic sector of the Antarctic Ocean in midwinter," *J. Geophys. Res.*, **92** (C13), 14,535–14,552.

[110] E. R. Walker and P. Wadhams (1979) "Thick sea-ice floes," *Arctic*, **32** (2), 140–147.

[111] W. F. Weeks and A. J. Gow (1978) "Preferred crystal orientation in the fast ice along the margins of the Artic Ocean," *J. Geophys. Res.*, **83**, 5105–5121.

[112] W. F. Weeks and O. S. Lee (1958) "Observations on the physical properties of sea ice at Hopedale, Labrador," *Arctic*, **11**, 135–155.

[113] W. F. Weeks and O. S. Lee (1962) "The salinity distribution in young sea ice," *Arctic*, **15**, 93–108.

[114] W. F. Weeks and S. F. Ackley (1982) "The growth, structure and properties of sea ice," U.S. Army Corps of Engineers, Cold Regions Research and Engineering Laboratory, Hanover, NH, Monog. 82-1.

[115] D. P. Winebrenner, J. Bredow, A. F. Fung, M. R. Drinkwater, S. Nghiem, A. J. Gow, D. K. Perovich, T. C. Grenfell, H. C. Han, J. A. Kong, J. K. Lee, S. Mudaliar, R. O. Onstott, L. Tsang, and R. D. West (1992) "Chapter 8: Microwave sea ice signature modeling," in *Microwave Remote Sensing of Sea Ice*, F. D. Carsey (Ed.), Geophys. Monog. 68, American Geophysical Union, Washington, DC.

[116] WMO (1985) WMO Sea-Ice Nomenclature, Suppl. #4, WMO #259.Tp.145, World Meterological Organization.

3

ACOUSTIC AND SEISMIC SENSING TECHNIQUES

DAVID M. FARMER
Institute of Ocean Sciences
Fisheries and Oceans Canada
Sidney, British Columbia, Canada

YUNBO XIE
Institute of Ocean Sciences
Fisheries and Oceans Canada
Sidney, British Columbia, Canada

School of Earth and Ocean Sciences
University of Victoria
Victoria, British Columbia, Canada

3.1 INTRODUCTION

Other chapters in this book describe the active and passive remote sensing of sea ice from above the surface using various forms of electromagnetic radiation. Although far-reaching and comprehensive in its coverage, remote sensing from above the ice cover inevitably yields a one-sided view. To study the ice from within and below, acoustic and seismic methods are being investigated. Acoustic and seismic sensing of floating ice is a young science and its full potential is hard to discern. Nevertheless, the rapid progress now being made in this field and the range of applications being investigated ensure that it will have an important role to play in ocean remote sensing at high latitudes. In this chapter, some of the recent developments in the field are considered, with emphasis on interpretation of the naturally occurring sound field.

In some respects, the use of *active* sonar devices to probe the ice field is similar to that used to map the sea floor. In general the instrumentation is not

Remote Sensing of Sea Ice and Icebergs, Edited by Simon Haykin, Edward O. Lewis, R. Keith Raney, and James R. Rossiter.
ISBN 0-471-55494-4

particularly sophisticated, and quite simple devices can yield useful data. What distinguishes such measurements is the observational platform that must be used. Except in the case of bottom-mounted or moored measurements, and very specialized local mapping carried out from the surface, sonar studies of the underside of the ice must be conducted from an autonomous underwater vehicle or a submarine. It is perhaps not surprising that relatively little has been published on this topic although what has appeared is of great interest. Some work has also been done on the problem of ice scattering effects on long-range acoustic propagation, but primarily from the acoustic propagation point of view. Only recently has the inverse problem been tackled; that is, the determination of ice properties and distribution from its influence on acoustic and seismic wave propagation.

The upward refractive sound-speed profile ensures that long-range acoustic propagation in the Arctic is sensitive to the properties of the bounding ice. This fact is currently being investigated with a view to using propagation over several thousand kilometers to study potential effects of global warming. For example, in the case of an acoustic path between Barrow Canyon and Svalbard, a 10-cm reduction in mean ice thickness is estimated to result in a 0.1 wave period decrease in propagation time at 30 Hz, and a 3-dB increase in received intensity (Mikalvesky, personal communication, 1992; see also Jin and Wadhams [9]). The increase in intensity occurs because of a change in ice roughness which is known to be correlated with ice thickness. The discussion will mention active acoustic methods only briefly, although it is clear from what has already been achieved that both the backscatter and propagation approaches have interesting potential.

In contrast to active sonar measurements, which have analogues in electromagnetic scattering, the interpretation of naturally occurring acoustic and seismic signals leads to a fundamentally different type of information. Ice radiates sound and seismic waves because it is a dynamic material, responding to changing stress fields imposed by the atmosphere and ocean. The interaction of adjacent ice floes and the cracking of ice sheets, even microscopic fracturing, can produce measurable acoustic and seismic radiation that reveals details of the response mechanisms and of ice properties. The distinction is important because no other remote sensing technique appears to offer such an intimate and dynamic view of ice processes.

In this chapter, three distinct theoretical approaches to the problem of acoustic and seismic radiation from an ice crack are examined, and then some experimental results and field observations that illustrate the theory are described. The first of these approaches may be described as the classical waveguide model and accounts for waves trapped in the ice. This energy may be detected with geophones on the ice surface, or with a hydrophone appropriately mounted within the ice. Guided modes in the ice do not travel very far, usually less than 10 km; some of the energy is scattered into the water by inhomogeneities in the ice and some, for example that from P waves, is acoustically coupled to the water.

The second approach is described as a *near-field* theory which describes the sound radiated into the water from the immediate vicinity of the crack. The theoretical model is able to account for some important properties of the radiated field when the cracking process is small enough that it does not alter the bulk properties of the ice; it is therefore especially appropriate to the description of thermal-stress cracking.

The third approach is appropriate for the description of cracks deep enough to affect bulk properties. This development, which is also defined as a *near-field* theory, is analogous to that used in earthquake mechanics. It is necessarily quite simplified and does not include reflections from the upper surface of the ice explicitly, it being just a solution to the Helmholtz equation for two adjacent layers. Nevertheless, it appears that the results contain useful insight on the sound-generation mechanism. Each of these three approaches is then illustrated with field measurements.

As will become evident, the study of naturally occurring or ambient sound and seismic signals is at present at the stage of single-event analysis. A long-range goal is to learn how to use the sound and seismic signals to infer the distribution of stresses over large scales. Such observations would be valuable for the interpretation of the larger scale stress field which plays an important role in determining ice morphology. The prospects for such studies form a concluding comment to this review.

3.2 ACTIVE ACOUSTIC METHODS

Before discussing the use of naturally occurring ocean sound in the Arctic, the use of active acoustic techniques is described briefly. Ice reflects acoustic radiation. Backscatter sonars are, therefore, effective tools for remotely sensing ice distribution and properties. A simple and effective technique is to use an inverted narrow beam echo-sounder and to record range to the lower surface of the ice. The inverted echo-sounder has been used successfully both from a submarine (Wadhams [30]) and, more recently, from a bottom-mounted device (H. Melling, Institute of Ocean Sciences, personal communication, 1992). In the latter case, horizontal measurement depends upon the drift of the ice cover. Such measurements require simultaneous pressure recordings to differentiate between effective sea-surface level and ice draft. If the ice cover is intermittent, an independent check can be made.

For inverted echo-sounders mounted on the sea floor, an interpretation of the scale of observed topography requires knowledge of ice motion. Whereas there are various ways of obtaining this information, bottom-mounted acoustic Doppler or correlation sonar appears to provide a viable option.

Inverted echo-sounder observations have already revealed several interesting properties of the ice cover, including evidence for thinning of the ice north of Greenland. Detailed morphological study is aided by use of side-scan sonar imaging (Wadhams [30]). These images have revealed patterns of narrow cracks on the smooth bottom surface of first-year ice. The cracks tend to be invisible from above due to the uniform snow cover and are most likely caused by mechanical stress exerted from nearby pressure ridges and surrounding rough ice, and by thermal stress. Old ice, in contrast, is very irregular. It is thought that melting at the lower surface is enhanced beneath surface melt pools, thus amplifying topographic irregularities. It is perhaps worth emphasizing the usefulness of active acoustic observations such as those described above because they provide views and measurements of the underneath of the ice cover which have hitherto been unavailable.

3.3 ORIGIN OF NATURAL ACOUSTIC AND SEISMIC RADIATION FROM ICE

Ice radiates acoustic and seismic waves whenever there is a sudden relief of stress through a fracture or collapse, or from the rubbing of ice edges against each other. Such waves may be generated by external processes also: wind effects, such as air flow over an irregular surface, wind driven snow, and wave action against exposed ice edges. Other sounds may also be present in the Arctic, including the voices of marine mammals and the noise of icebreaker and drilling operations. However, the signal of particular relevance to the mechanical properties and behavior of the ice cover is that from ice cracking, and this will consequently be the principle area of focus.

It is the dynamic origin of most energy radiated by ice that distinguishes it from the signals used in other remote sensing techniques. The naturally occurring acoustic and seismic radiation provides clues to processes occurring within the ice sheet. Moreover, such radiation depends not only on the stress responsible for its origin, but also on the material properties of the surrounding ice. Its interpretation, in turn, depends upon use of an appropriate model of the failure mechanism and the ice sheet's acoustic and seismic response.

Ice cracking occurs as a result of mechanical failure under stress. The stress may arise from a local boundary effect, such as that at the surface, or it may be driven on a large scale, as for example, by the action of wind or current. Thermal stress cracking is known to occur during periods of cooling when tension at the upper boundary is relieved by small-scale fracturing. The resulting acoustic and seismic response is distinguished by the continuing mechanical integrity of the ice sheet although it is quite possible that thermal cracks can on occasion penetrate more deeply. Deep cracks, on the other hand, alter the local ice response and must be treated quite differently. Moreover, a thermally induced microcrack might usefully be treated as an essentially motionless source, whereas larger cracks radiate sound from the ends of the crack which can travel at a speed comparable with that of a shear wave, thus introducing in effect a Doppler shift in the detected spectrum.

Whether the source is a superficial or deep fracture, some fraction of the radiated energy propagates as elastic waves in the ice sheet, while part is coupled directly into the water near the source. Energy trapped in the ice waveguide can leak into the water. Waveguide propagation is a classic problem and perhaps for this reason it is the part of ice acoustics and ice seismics that has received most attention. However, evidence is accumulating that direct acoustic coupling into the water close to the source is by far the dominant radiation mechanism. Energy is certainly trapped in the ice as seismic waves and is readily detectable with geophones. As a source of information on ice mechanics and properties, seismic waves may be even more useful than acoustic waves in the water, but they are attenuated rapidly and so are quite local.

The mechanics of ice fracturing has some analogies with earthquake mechanics and it may be useful to treat the problem as one of ice seismics. Thus, ice failure can occur as a tensile, shear, or compression fault. Tensile failure is typical of the response to thermal stress during cooling and to the effects of a divergent wind or current stress. Horizontal shear faults can be expected as a result of transmitted longitudinal stress from differential drag or current shear; for example, along a coast where there is landfast ice. Vertical shear failure occurs as a result of differential vertical loading; the tidal crack along the edge of landfast ice is one example. Compressive failure is responsible for pressure ridging, although active ridging is associated with many types of seismic sources, in particular bending failure, rafting of broken ice, and related processes. Pressure ridging is thought to be an important source of low-frequency sound in the Arctic.

Interaction of adjacent ice floes in moving pack ice can generate an acoustic and seismic response without actually involving ice failure at all. Although this type of noise has received little attention in the open literature, it appears that the radiated signal contains useful information for remote sensing of ice properties (Xie and Farmer [33]; Rottier [23]). Undoubtedly there are other acoustic and seismic sources generated by ice which await investigation. For example, it is known that seismic signals are a common feature of glaciers. Glacier activity at the head of fjords, especially iceberg calving, is an active acoustic source. It should be emphasized that acoustic and seismic remote sensing of ice is at a formative stage; some signals have been detected and are beginning to be interpreted, but the full range of signals available has yet to be determined.

3.4 MATHEMATICAL DESCRIPTION OF SOUND RADIATION FROM CRACKING ICE

Formulation of an adequate mathematical description of sound generation and propagation is the first essential step in the development of a remote sensing capability using passive acoustics and seismics. Fairly comprehensive numerical techniques, in particular the SAFARI model (Schmidt and Jensen [25]), have been developed and can be applied to ice acoustics and seismics (Stein [28]; Kim [12]; Miller [18]). The analytical formulation is described here because of the insight it provides into the underlying physics; moreover, it appears that particular solutions of the relevant equations have special application to the observations, as will be discussed in Subsection 3.4.6.

In this section the constitutive and wave equations are given in Subsection 3.4.1 and then the classical waveguide model of Press and Ewing [20] is summarized in Subsection 3.4.2, relevant to propagation within the ice sheet. This model is referred to as a *far-field* model, as it includes no explicit description of the sound-generation mechanism. The primary interest is in ice cracking; therefore some of the known properties of fracturing are summarized in Subsection 3.4.3. The emphasis is on fracturing over scales larger than the complex crystalline microstructure of the ice. A *near-field* theory relevant to thermal stress failure is given in Subsection 3.4.4, followed by finite-volume theory in Subsection 3.4.5. Both of these theories focus on the signal radiated into the water for a specific type of fracturing event.

3.4.1 Small-Amplitude Wave Equations in an Isotropic, Elastic Medium

The stress–strain relationship for small deformations in an isotropic, elastic medium can be described in terms of Lamé's constants, λ and μ (Ewing et al. [4]), in matrix form:

$$\begin{bmatrix} p_{xx} \\ p_{yy} \\ p_{zz} \\ p_{xy} \\ p_{yz} \\ p_{xz} \end{bmatrix} = \begin{bmatrix} \lambda + 2\mu & \lambda & \lambda & 0 & 0 & 0 \\ \lambda & \lambda + 2\mu & \lambda & 0 & 0 & 0 \\ \lambda & \lambda & \lambda + 2\mu & 0 & 0 & 0 \\ 0 & 0 & 0 & 2\mu & 0 & 0 \\ 0 & 0 & 0 & 0 & 2\mu & 0 \\ 0 & 0 & 0 & 0 & 0 & 2\mu \end{bmatrix} \begin{bmatrix} e_{xx} \\ e_{yy} \\ e_{zz} \\ e_{xy} \\ e_{yz} \\ e_{xz} \end{bmatrix} \tag{3-1}$$

where p_{ij} and e_{ij} are, respectively, stress and strain tensors. The constants λ and μ, sometimes referred to as *bulk modulus* and *shear modulus*, respectively, are related to Young's modulus, E, and to Poisson's ratio, σ, through

$$\lambda = \frac{\sigma E}{(1 + \sigma)(1 - 2\sigma)} \tag{3-2}$$

and

$$\mu = \frac{E}{2(1 + \sigma)} \tag{3-3}$$

For small amplitude deformation, the equation of motion is

$$\rho \frac{\partial^2 \vec{s}}{\partial t^2} = (\lambda + 2\mu)\nabla(\nabla \cdot \vec{s}) - \mu\nabla \times (\nabla \times \vec{s}) \tag{3-4}$$

where ρ is the density of the medium and $\vec{s}$ is total displacement caused by the disturbance.

In an isotropic, elastic solid, wave motion consists of both equivoluminal and irrotational deformations. Accordingly, $\vec{s}$ can be decomposed into two components in terms of a scalar potential, Φ, and a vector potential, $\vec{\Psi}$:

$$\vec{s} = \nabla\Phi + \nabla \times \vec{\Psi} \tag{3-5}$$

and

$$\nabla \cdot \vec{\Psi} = 0 \tag{3-6}$$

Substitution of (3-5) into (3-4) leads to two linear independent wave equations governing equivoluminal and irrotational motions, respectively. The propagation of equivoluminal motion is governed by

$$\frac{\partial^2 \Phi}{\partial t^2} = \alpha^2 \nabla^2 \Phi \tag{3-7}$$

and the propagation of irrotational motion is governed by

$$\frac{\partial^2 \vec{\Psi}}{\partial t^2} = \beta^2 \nabla^2 \vec{\Psi} \tag{3-8}$$

where α = compressional wave speed in solid; β = shear wave speed in the solid; t = time.

These speeds are related to λ and μ through the following expressions:

$$\alpha = \sqrt{\frac{\lambda + 2\mu}{\rho}} \tag{3-9}$$

$$\beta = \sqrt{\frac{\mu}{\rho}} \tag{3-10}$$

The Ewing et al. [4] theoretical model of wave propagation in an ice sheet is based on (3-7) and (3-8).

3.4.2 Classic Wave Propagation Model for an Ice Sheet

Ewing et al. [4] derived general solutions for an ice sheet bounded by a vacuum above and by a fluid below. For this purpose the vacuum representation is a very good approximation to the atmosphere. It is assumed that the shear stress vanishes at the two boundaries. At the upper boundary, the normal stress also vanishes but at the lower boundary, both normal stress and normal displacement are continuous. The general solution in this case is

$$\Phi_1 = [A \sinh(\xi z) + B \cosh(\xi z)]e^{ikx} \tag{3-11}$$

$$\Psi_{y1} = [C \sinh(\eta z) + D \cosh(\eta z)]e^{ikx} \tag{3-12}$$

The corresponding acoustic wave in the water is

$$\Phi_2 = Ee^{-\zeta z}e^{ikx} \tag{3-13}$$

where the harmonic factor $e^{-i\omega t}$ has been omitted and k is the horizontal wavenumber which, for a given frequency, is invariant both in ice and in water. The variables ξ, η, and ζ are defined as

$$\xi = \sqrt{k^2 - \left(\frac{\omega}{\alpha_1}\right)^2} \tag{3-14}$$

$$\eta = \sqrt{k^2 - \left(\frac{\omega}{\beta_1}\right)^2} \tag{3-15}$$

$$\zeta = \sqrt{k^2 - \left(\frac{\omega}{\alpha_2}\right)^2} \tag{3-16}$$

where α_1 = compressional wave speed in ice; β_1 = shear wave speed in ice; α_2 = sound speed in water; ω = circular frequency of waves.

The next step is to find eigenvalues for the horizontal wavenumber, k, or equivalently, phase speed ($c_p = \omega/k$) so that the boundary conditions can be satisfied by the solutions. Applications of the boundary conditions leads to a characteristic equation in terms of k for the ice sheet waveguide from which eigenvalues can be found giving various elastic waves in the ice. Unfortunately, this characteristic equation cannot be solved analytically for c_p eigenvalues except in two special cases. However, asymptotic roots of c_p can be expressed in a closed form for either longwaves (where the horizontal wavelength of the waves, λ_h, is much greater than the ice thickness, $\lambda_h >> H$) or shortwaves ($\lambda_h << H$).

These roots correspond to five types of elastic waves. For shortwaves, there are three kinds of interfacial waves at the upper and lower surfaces of the ice sheet:

1. Rayleigh wave at the air–ice interface with $c_p = c_r = 0.9194\beta_1$, where c_r is the Rayleigh wave speed.
2. Attenuated Rayleigh wave at the ice–water interface with $c_p = c_r(1 + i\frac{1}{4})$, where $i = \sqrt{-1}$.
3. Stonely wave at the ice-water interface with $c_p = 0.87\alpha_2$.

For waves that are long relative to the ice thickness, there are two types of waves in the ice:

1. Compressional wave or P wave with a dispersion relation

$$c_p = 2\beta_1(1 - \beta_1^2/\alpha_1^2)^{1/2}(1 + ib) \tag{3-17}$$

where $b \sim O(H/\lambda_h)^3$ and α_1 is the compressional wave speed in ice.

2. Flexural or F wave with a dispersion relation given by

$$c_p^2/\beta_1^2 = (8/3)(\rho_1/\rho_2)(kH)^3(1 - \beta_1^2/\alpha_1^2)(1 + 2kH\rho_1/\rho_2) \tag{3-18}$$

One of the only two analytical solutions to the characteristic equation is $c_p = \alpha_1$, which corresponds to the so-called Crary wave. The Crary wave is a well-trapped vertically polarized wave, or SV wave, which reflects totally from the ice sheet boundaries. The resulting normal mode frequencies are given by

$$f_n = (n + 1)/(4H\sqrt{\beta_1^{-2} - \alpha_1^{-2}}) \tag{3-19}$$

with $n = 1, 2, \ldots$. For Crary waves, it can be shown that the total reflection of SV wave energy at both surfaces of the ice is caused by the incidence of SV waves at an angle $= \cos^{-1}(\beta_1/\alpha_1)$ with the horizontal. At this angle, the interference of incident waves with reflected waves results in a zero displacement at the boundaries, thus preventing energy from radiating into the water for this particular mode.

In accordance with shear deformation in the horizontal direction, there will be horizontally polarized shear (SH) waves. SH waves are also called Love waves, and both surfaces of the ice plate remain pressure free (with respect to this type of wave). Therefore, the SH wave propagation problem is equivalent to finding a solution for wave motions in a plate bounded by a vacuum. The corresponding phase speed is

$$c_p = \beta_1/\sqrt{1 - (n\pi/kH)^2} \tag{3-20}$$

The normal mode frequencies are given by

$$f_n = n\beta_1/2(2H) \tag{3-21}$$

with $n = 0, 1, 2, \ldots$. The presence of water beneath the ice therefore does not affect the characteristics of the ice plate as a perfect waveguide for trapping SH waves.

The SH and Crary waves are important in the seismic study of elastic parameters of sea ice and its thickness. Although they are well trapped in an ideal ice plate, they may contribute to underwater ambient sound by leaking energy to the water once they interact with discontinuities at the ice–water interface (Milne [17]; Langley [13]). As the frequencies associated with SH and Crary waves are quite unique (discrete bands), together with the direct acoustic waves they can provide additional information on source positions and mechanisms.

As for the intermediate wavelengths, there will be an unbounded number of normal modes corresponding to the roots of the characteristic equation. In this case, it is impossible to obtain an analytical form of the dispersion relation for the characteristic equation. Stein [26] has carried out a detailed numerical investigation of this equation and gained some interesting dispersion curves for P and F waves for $\sigma = 0.33$ (a typical value for sea ice). Using a monopole model, Stein [27] was able to obtain empirical formulae of wave attenuations for flexural and compressional waves in an ice sheet of 3-m thickness. For the flexural wave, the attenuation due to absorption is

$$0.003f^{0.5}\ \mathrm{dB \cdot m^{-1}}$$

and for the compressional wave, the attenuation from absorption is

$$0.0002f\ \mathrm{dB \cdot m^{-1}}$$

Stein [27] estimates that for a 10-Hz wave signal, the flexural wave suffers a 10-dB transmission loss at 1-km range, and the compressional wave suffers the same loss after propagating in the ice for 10 km.

3.4.3 Properties of Ice Relevant to Cracking

In studying the strength of an elastic plate, Griffith [7] first postulated a criterion for the growth of a crack of length $2L$. The basic idea is that "the change of potential energy in the plate is equal to the change of surface energy in the crack as it grows in length." Mathematically this idea can be expressed as

$$\frac{\partial}{\partial L}\left(\frac{\pi s^2 L^2}{E}\right) = \frac{\partial}{\partial L}(4c\gamma) \tag{3-22}$$

where γ is the specific surface energy of the material and s is the applied stress (tensile stress perpendicular to the long axis of the crack) at which crack growth occurs. Therefore, the minimum applied stress necessary for crack growth is

$$s = \left(\frac{2E\gamma}{\pi L}\right)^{1/2} \tag{3-23}$$

However, the predicted values from (3-23) were found to be too low for the actual strength of most material, including ice samples. This discrepancy is explained by the fact that once a crack occurs, a large amount of stress will concentrate at the crack's tip, which has the effect of blunting the crack. Hence, further growth of the crack requires an extra amount of work to overcome this barrier or *plastic region*.

Orowan [19] modified Griffith's model by including a specific energy for plastic working, γ_p, in (3-23). As $\gamma_p >> \gamma$, (3-23) becomes

$$s \approx \left(\frac{2E\gamma_p}{\pi L} \right)^{1/2} \tag{3-24}$$

This result is important because it emphasizes the fact that less energy is required to lengthen a crack than is required to start one. It is seen that as the crack lengthens (L increases), the stress required for further extension of the crack decreases.

Whereas the above model gives a good description of failure processes in an elastic plate, it is difficult to use this theory for a quantitative interpretation of sea-ice cracking. Sea ice has a more complicated structure than an elastic plate and, as mentioned above, its elastic parameters vary with temperature, salinity, and other factors. Therefore, the discussion of ice properties in terms of these parameters is of limited value. It is almost impossible to obtain an analytical form relating elastic parameters with these factors. Based on field data obtained from sites scattered throughout the Arctic, Cox and Weeks [2] derived empirical formulae for the tensile (TS), shear (SS) and flexural strength (FS) of sea ice in terms of its brine volume (BV) as follows:

1. For horizontal tensile stress tests:

$$TS = 0.816 - 0.0689\sqrt{BV} \tag{3-25}$$

2. For vertical tensile stress tests:

$$TS = 1.54 - 0.0872\sqrt{BV} \tag{3-26}$$

3. For shear stress tests:

$$SS = 1.68 - 0.118\sqrt{BV} \tag{3-27}$$

4. For flexure stress tests:

$$FS = 0.959 - 0.0608\sqrt{BV} \tag{3-28}$$

where the strength is given in MegaPascals (MPa) and BV is in parts per thousand.

3.4.4 Near-Field Theory for the Acoustic and Seismic Response to a Surface Thermal-Stress Crack

A significant source of ice noise arises from thermal-stress cracking (Milne [16]; Farmer and Waddell [5]; Waddell and Farmer [29]). Figure 3.1, after

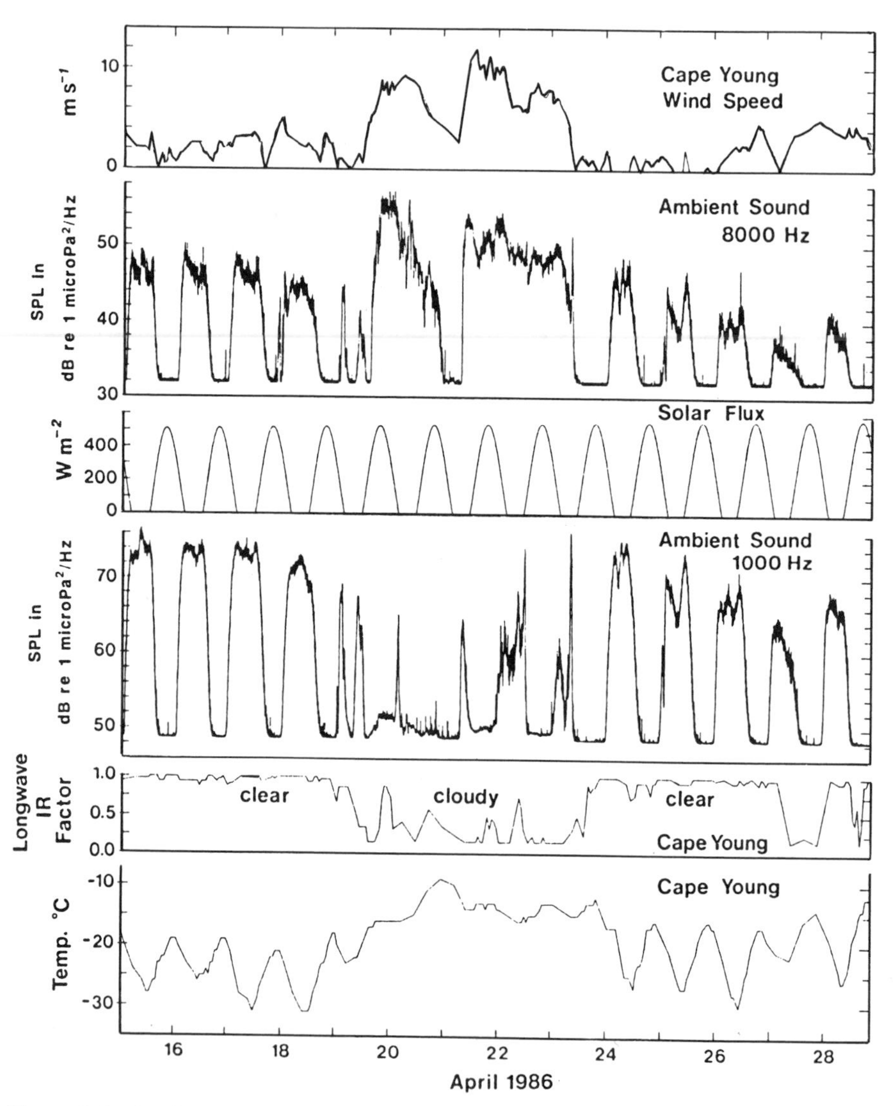

Figure 3.1 Ambient sound measured at Cape Young (8 kHz and 1 kHz), together with wind speed, solar flux, inferred infrared back-radiation estimated from cloud cover, and air temperature. Cloud cover inhibits radiant heat loss, and thus thermal stress cracking. After Farmer and Waddell [5]. Reprinted by permission of Kluwer Academic Publishers.

Farmer and Waddell [5], shows an example of the time series of radiated sound over the course of a clear night when there was strong radiation cooling. The signal rises rapidly to a maximum which it maintains until dawn, after which it drops precipitously. It is remarkable that during cloudy periods, when the back-radiation drops, the acoustic noise drops simultaneously, thus allowing the presence or absence of clouds to be detected from an acoustic sensor on the sea floor!

A thermal stress crack arises from the relief of a surface thermal stress. It is thought that most such cracks tend to be superficial, leaving the bulk of the ice undamaged. The theory for acoustic radiation from such a crack has been developed and discussed in detail by Xie and Farmer [32] and Xie [31]. In the following, a brief derivation of the theoretical model is provided.

The model consists of an ideal infinite plate of uniform thickness, H (see Fig. 3.2). The plate is bounded by two infinite media, air above and water below. Suppose an incident plane wave of unit amplitude, P_{oi}, insonifies the upper surface of the plate. As a result, a reflected wave (P_{or}) will be excited with the rest of the sound energy penetrating the plate, and a transmitted wave, P_t, will radiate into the water. Taking the x, z frame as shown in Fig. 3.2, the three waves can be expressed as follows:

$$P_{oi} = e^{ik_0(x\sin\theta_0 + z\cos\theta_0)} \tag{3-29}$$

$$P_{or} = Ae^{ik_0(x\sin\theta_0 - z\cos\theta_0)} \tag{3-30}$$

$$P_t = Be^{ik_2(x\sin\theta_2 + (z-H)\cos\theta_2)} \tag{3-31}$$

Therefore, the sound pressure in the upper space is

$$P_0 = P_{0i} + P_{0r} \tag{3-32}$$

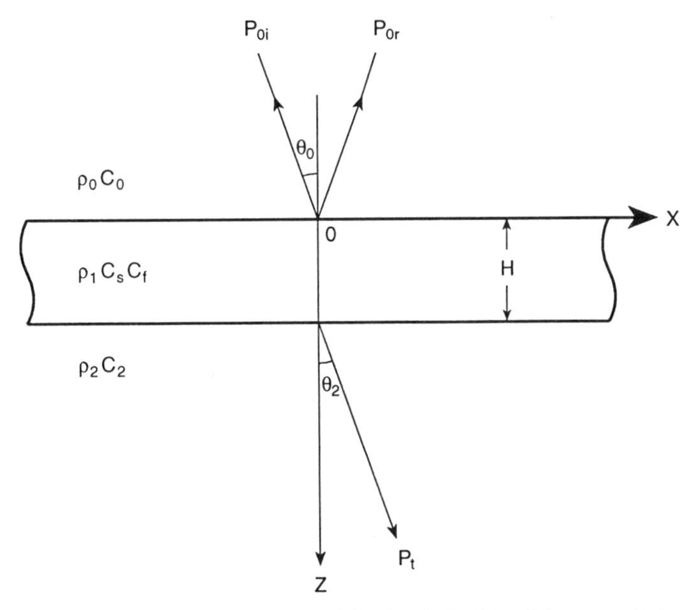

Figure 3.2 An (x, z) coordinate system used for the derivation of the transmission coefficient of an elastic plate. After Xie and Farmer [33]. Reprinted by permission of the American Institute of Physics.

and in the lower space,

$$P_2 = P_t \tag{3-33}$$

Here, A and B are coefficients of reflection and transmission, respectively, and k_0 and k_2 are the wavenumbers in air and in water, respectively. The harmonic time dependence $e^{-i\omega t}$ is omitted.

The excited motion in the plate is the superposition of two classes of eigenmodes causing a vibration of the plate in the z direction (see Brekhovskikh [1]). They are the *symmetric* and *antisymmetric* modes, which are illustrated in Fig. 3.3. The goal is to determine B, the transmission coefficient of the plate. This parameter is analogous to the system function of a linear system. B is expressed in terms of the acoustic impedance of each of the two modes.

By matching the normal displacement along upper and lower surfaces of the plate with that from the surrounding pressure fields in the air and water, one obtains

$$B = \frac{2Z_0^{(2)}(Z_s - Z_a)}{Z_s Z_a + Z_0^{(2)}(Z_s + Z_a)} \tag{3-34}$$

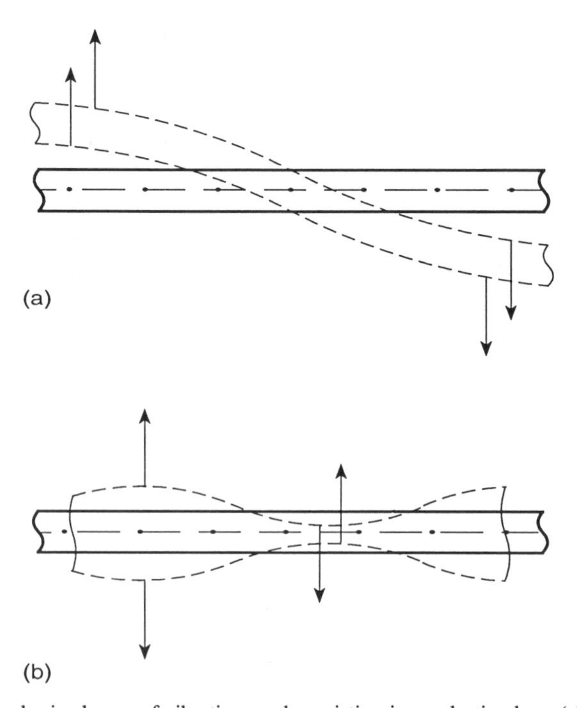

Figure 3.3 Two basic classes of vibration modes existing in an elastic plate. (a) Antisymmetric mode; (b) Symmetric mode. Arrows indicate displacement vectors at the two surfaces of the plate. After Xie and Farmer [33]. Reprinted by permission of the American Institute of Physics.

where

$$Z_s = \frac{2iE_1[1 - (C_s \sin \theta_2/C_2)^2]}{\omega H[1 - \sigma_1^2 - (C_s \sin \theta_2/C_2)^2]} \tag{3-35}$$

is the acoustic impedance of the thin plate due to symmetric mode motion;

$$Z_a = -i\omega M \left(1 - \frac{C_f^4}{C_2^4} \sin^4 \theta_2\right) \tag{3-36}$$

is the acoustic impedance of the thin plate due to antisymmetric mode motion;

$$C_s = \sqrt{E_1/\rho_1} \tag{3-37}$$

is the longitudinal wave speed in the plate where ρ_1 is the density of the plate.

$$C_f = \left(\frac{\omega^2 E_1 H^3}{12M}\right)^{1/4} \tag{3-38}$$

is the flexural wave speed in the plate.

E_1 = elastic modulus for the plate

H = plate thickness

M = mass per unit area

C_2 = sound speed in water

It is seen from (3-34) that B has a maximum when either $Z_s = 0$ or $Z_a = 0$. For $Z_s = 0$, from (3-35) (using the thin-plate approximation),

$$\theta_b = \sin^{-1}\left(\frac{C_2}{C_s}\right) = \sin^{-1}\left(\frac{C_2}{\sqrt{E_1/\rho_1}}\right)$$

or

$$E_1 = \rho_1 \left(\frac{C_2}{\sin \theta_b}\right)^2 \tag{3-39}$$

where θ_b is a frequency-independent radiation angle at which the plate becomes a broadband system and signals of all frequency components will pass through the plate at the same transmission rate. θ_b is called the broadband radiation angle.

When $Z_a = 0$, from (3-36),

$$\frac{C_f}{C_2} \sin \theta_2 = 1 \tag{3-40}$$

Substitution of (3-38) into (3-40) leads to

$$\omega_c = \frac{C_2^2}{\sin^2 \theta_2} \left(\frac{12\rho_1}{E_1 H^2} \right)^{1/2} \tag{3-41}$$

This result implies that if the phase speed of free flexural waves in the plate matches the speed of sound in the water, then the sound pressure caused by a cracking process in the ice will be coupled most effectively into the water. This relationship defines a characteristic frequency, ω_c, which is a function of the radiation angle. Consequently, an angular filtering phenomenon appears: the ice cover selects a ω_c corresponding to the frequency of maximum sound transmission.

Thus, the frequency content of a signal detected at a hydrophone depends upon the radiation angle at which the sound is projected into the water. If two signals arrive at the hydrophone from two different radiation angles, α_1 and α_2, the ratio between the two dominant frequencies, f_1 and f_2, is

$$R_T = \frac{f_2}{f_1} = \left(\frac{\sin \alpha_1}{\sin \alpha_2} \right)^2 \tag{3-42}$$

Together with the presence of broadband radiation at angle θ_b, this theoretical result provides a basis for inferring the flexural rigidity of the ice sheet.

The response of the plate to force concentrated at a single point is given in Xie and Farmer [32] where it is shown that maximum sound radiation occurs when either $Z_a = 0$ or $Z_s = 0$ is satisfied.

3.4.5 Acoustic Radiation from a Growing Finite-Volume Crack

In Subsection 3.4.4, acoustic radiation was considered from a small, superficial crack that does not penetrate deep into the ice. Radiation from larger cracks that result in a significant spreading of the crack faces results in an altogether different signal. In this case, the ice sheet can no longer be considered to be unaffected by the crack, and it is necessary to use the theoretical approach developed to describe tensile and shear failures in earthquake mechanics. The sound is radiated primarily from the ends of the crack, and is therefore subject to the effects of rapid motion of the source.

Figure 3.4 shows a cylindrical coordinate system (R, Z) used for derivation of the theoretical model. Assume the velocity potential, Φ, is harmonic in time,

$$\phi_{1,2}(R, Z, t) = \Phi_{1,2}(R, Z)e^{-i\omega t}$$

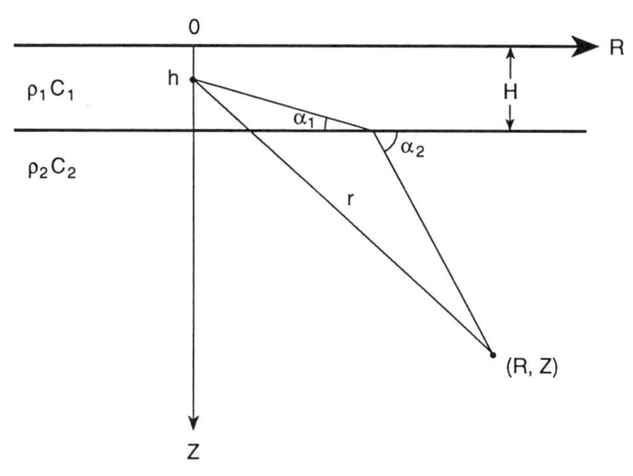

Figure 3.4 Cylindrical coordinate system (R, Z), with a slant range r used for a point source model. After Farmer and Xie [6]. Reprinted by permission of the American Institute of Physics.

and $\Phi_{1,2}(R, Z)$ is a function of radius R in the horizontal plane and depth Z (see Fig. 3.4), with the subscript 1,2 referring to the ice and water, respectively. For a time-harmonic point source of velocity potential per unit area located at $(R = 0, Z = h)$, $\Phi_{1,2}(R, Z)$ satisfies the following Helmholtz equations:

$$\frac{1}{R}\frac{\partial}{\partial R}\left(R\frac{\partial \Phi_1}{\partial R}\right) + \frac{\partial^2 \Phi_1}{\partial Z^2} + k_1^2\Phi_1 = -4\pi\delta(R, Z - h) \tag{3-43}$$

$$\frac{1}{R}\frac{\partial}{\partial R}\left(R\frac{\partial \Phi_2}{\partial R}\right) + \frac{\partial^2 \Phi_2}{\partial Z^2} + k_2^2\Phi_2 = 0 \tag{3-44}$$

where $k_{1,2}$ is the wavenumber in the ice and water, respectively. The right-hand term in (3-43) represents a point source of unit strength. The corresponding boundary condition at the ice surface, neglecting the small coupling to the atmosphere, is then

$$\Phi_1 = 0 \qquad \text{at } Z = 0 \tag{3-45}$$

which means that no waves are transmitted into the air. At the ice–water interface, both the vertical displacement and pressure should be continuous:

$$\left.\begin{aligned} \frac{\partial \Phi_1}{\partial Z} &= \frac{\partial \Phi_2}{\partial Z} \\ \rho_1\Phi_1 &= \rho_2\Phi_2 \end{aligned}\right\} \qquad \text{at } Z = H \tag{3-46}$$

At the source, Pekeris's source condition (Ewing et al. [4]) is incorporated:

$$\left.\frac{\partial \Phi_1}{\partial Z}\right|_{h^-} - \left.\frac{\partial \Phi_1}{\partial Z}\right|_{h^+} = 2 \int_0^\infty J_0(\eta R)\eta d\eta \tag{3-47}$$

where η is the horizontal wavenumber and $J_0(\eta R)$ is the zero order Bessel function.

Solutions of (3-43) to (3-47) are derived in Appendix A of Xie [31]. Only the direct acoustic propagation through the water is of concern, so only the solution for Φ_2 is needed:

$$\Phi_2 = \int_0^\infty S(\eta)e^{i\beta_2(Z-H)} J_0(\eta R)\eta d\eta \tag{3-48}$$

and

$$S(\eta) = \frac{2 \sin \beta_1 h}{m\beta_1 \cos \beta_1 H - i\beta_2 \sin \beta_1 H} \tag{3-49}$$

where $\beta_1 = (k_1^2 - \eta_2)^{1/2}$ and $\beta_2 = (k_2^2 - \eta_2)^{1/2}$ are vertical wavenumbers in the ice and water, respectively, and $m = \rho_2/\rho_1$. $S(\eta)$ given by the integrand of Equation (4-46) of Ewing et al. [4] is the same as (3-49) except for a sign difference in the denominator.

Using the method of stationary phase, (3-48) and (3-49) may be evaluated as

$$\Phi_2(r, t) \approx \frac{k_2|\sin \alpha_2|S(\eta_0)}{r} e^{ik_2 r - i\omega t} \tag{3-50}$$

where $\eta_0 = k_2 \cos \alpha_2$ and α_2 is the angle of refraction (see Fig. 3.4). It is worth noting that (3-50) represents only the principal value of the integral given by (3-48). This principal value describes the most important contribution of the monopole source to the subsurface sound field.

A harmonic-time-dependent point source of form $e^{-i\omega t}$ has been assumed. As the source will be of finite volume and frequency bandwidth for a real crack, the rate of volume change, Q, is

$$Q = \int_s U_s(t)\, ds = 2AU_s(t) \tag{3-51}$$

where $U_s(t)$ is the amplitude of the displacement velocity at the source and A is the area of the fault plane per unit length. The source model described by (3-51) incorporates the assumption that the sound is generated by a tensile crack acting uniformly over the area A centered at depth d. This result is now applied to a more realistic source model.

Motivated by the structure of these observations and their similarity to those observed in certain earthquake records, Haskell [8] is followed to take a sinusoidally roughened ramp function for the displacement of the fault planes, except that the roughness elements are allowed to be some fraction (ϵ) of the primary amplitude. The physical concept here is of a "slip-stick" shear fault, or of a tensile fault where the yield point advances in small jumps, determined by the thickness of the ice sheet. The normalized fault displacement is modeled as

$$G(t) = \begin{cases} 0 & t < 0 \\ \dfrac{1}{T} \quad t - \epsilon \cdot \dfrac{\sin (2n\pi t/T)}{2n\pi/T} & 0 \leq t \leq T \\ 1 & t > T \end{cases} \tag{3-52}$$

where T is the rise time of the displacement function and n is an integer equal to the displacement roughness elements in the fault formation. If $\epsilon = 0$, (3-52) reduces to a linear ramp function. Figure 3.5 shows the displacement function for a roughened fault ($n = 4$, $\epsilon = 0.5$), together with the corresponding displacement velocity:

$$\frac{dG(t)}{dt} = \begin{cases} 0 & t < 0 \\ \dfrac{1}{T}[1 - \epsilon \cdot \cos (2n\pi t/T)] & 0 \leq t \leq T \\ 0 & t > T \end{cases} \tag{3-53}$$

The slip or tear on the fault is presumed to be controlled by the narrowest dimension of the rupture surface, which is the ice thickness, H. In earthquake mechanics, it is generally assumed that, at any given point, the crack continues to widen until the rupture front is $H/2$ past that point (Savage [24]). Following this assumption, the tear at a point begins as the rupture front passes that point and continues until a time roughly equal to $T = H/2v_r$, where v_r is the rupture velocity.

For total fault displacement, D_0, the speed of the fault planes at the source is then $U_s = D_0 G'(t)$, and its Fourier Transform is

$$\hat{U}_s(\omega) = D_0 e^{i\omega T/2} \frac{\sin (\omega T/2)}{\omega T/2} \frac{1 - (1 - \epsilon)(\omega T/2n\pi)^2}{1 - (\omega T/2n\pi)^2} \tag{3-54}$$

The Green's function for a source of finite bandwidth at a point distant r from the source is then found to be (see also Farmer and Xie [6])

$$\Phi_2\left(t - \frac{r}{c}\right) = \frac{2Ak_2|\sin \alpha_2|S(\eta_0)}{2\pi r} \int_{-\infty}^{+\infty} \hat{U}_s(\omega) e^{ik_2 r - i\omega t}\, d\omega \tag{3-55}$$

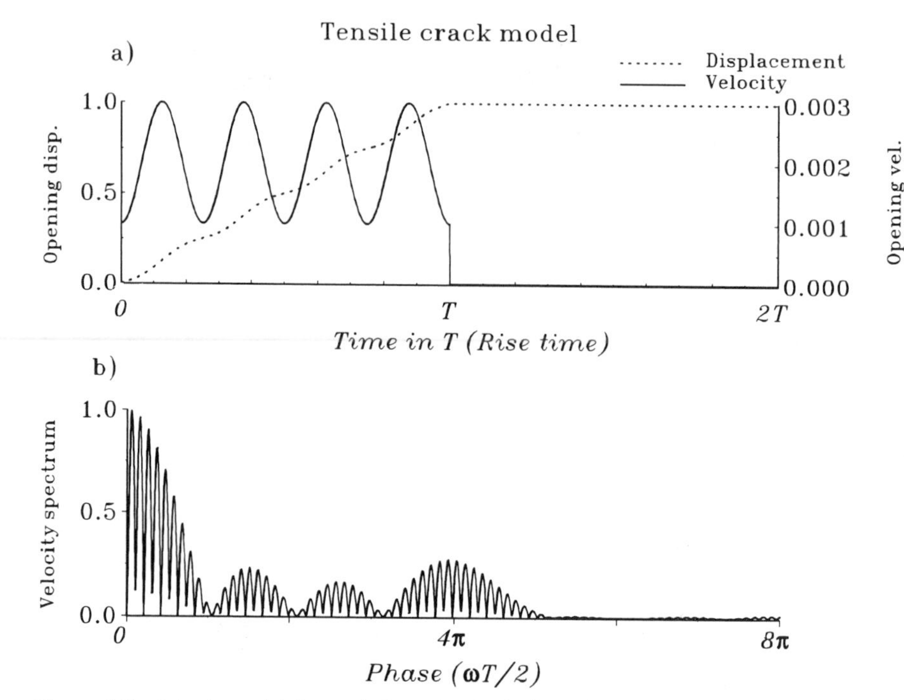

Figure 3.5 Seismic model for crack formation in the ice. Above, model displacement function (dotted line) and corresponding speed for $\epsilon = 0.5$, $n = 4$. Below, spectrum of model displacement speed for $f_h/f_b = 3$. f_h and f_b are related to depth and length of a crack through (2-29) and (2-30), respectively, in Farmer and Xie [6]. Reprinted by permission of the American Institute of Physics.

which is the delayed signal in the water from direct acoustic transmission from a crack of unit area A. Thus, Φ_2 is the Green's function for the moving source model, and c is the mean speed in water. The directly transmitted signal for a moving source is then found by integration of (3-55) over the length of the crack (Kasahara [11]).

Application of Huygens' principle for moving point sources, corresponding to the ends of the moving crack, then allows calculation of the full spectrum $|\hat{P}_2(\omega)|$:

$$|\hat{P}_2(\omega)| = \frac{m(\rho_1 V_0)k_2|\sin \alpha_2||S(\eta_0)|}{r} \cdot \left[\frac{\sin(\omega T/2)}{\omega T/2}\,\frac{1-(1-\epsilon)(\omega T/2n\pi)^2}{1-(\omega T/2n\pi)^2}\right]\left[\frac{\sin(\omega\tau/2)}{\tau/2}\right] \tag{3-56}$$

The quantity $V_0 = 2HLD_0$ can be interpreted as the volume of ice expelled from the crack during the process of sound generation; $\rho_1 V_0$ is the expelled mass, and $m = \rho_2/\rho_1$ is the density ratio.

Thus, the sound pressure spectrum is proportional to the product of two terms [in square parentheses in (3-56)]. The first represents the envelope of the source and the second shows the effect of the source motion. The source spectrum has the argument $\omega T/2$, where $T = H/2v_r$, which is determined by the time taken for the crack to propagate through the ice depth. The fault displacement speed is modeled as the product of a rectangular function and a sine wave given by (3-53). The corresponding argument of the source motion term is $\omega\tau/2$, where $\tau = L/v_r$ for $\cos\theta = 0$ is determined by the time taken for the crack to propagate over a path of length L. An identical result occurs for propagation to $-L$. The source motion introduces nulls in the spectrum at $\omega\tau/2 = 0, \pi, 2\pi, \ldots$.

The combination of large- and small-scale components of the crack, together with crack propagation, determine the form of the signal detected in the water. Fracture of the ice from its surface to a certain depth, d (which for larger cracks will be the ice thickness), takes the rise time T, which depends on the ratio d/v_r and is described by a rectangular modulation of the pulse of width T, as described by (3-52) and shown in Fig. 3.5. A rectangular pulse concentrates most of its energy at low frequencies and provides the necessary condition for the existence of a base frequency signal. The dominant component of the baseband signal, however, is determined by the modulation of the source spectrum resulting from its motion. If the source remained motionless, the dominant baseband signal would be determined by T but its motion at velocity v_r over the crack length L introduces a sequence of lobes, the first of which will be dominant. One would not necessarily expect identifiable nulls in the spectrum to occur as indicated in (3-56) because in a real environment the source will never be perfectly coherent. Source motion will alter the spectrum and ensure that the details of the detected signal will vary with hydrophone position, at least in the near field.

For given values of n and T, the envelope of the spectrum is defined and expressed by the first term in square parentheses in (3-56). However, the second term in square parentheses in (3-56) depends upon τ, and thus on the crack orientation with respect to the hydrophone. This term defines the fine structure in the spectrum because, typically, $\tau >> T$. In the near field, the angle θ and, hence, τ, will be a stronger function of hydrophone location than in the far field.

If $\tau > T$, two characteristic frequencies can be identified with which to describe the crack. This can be easily demonstrated by calculating the spectrum of the acceleration G'', as shown by Haskell [8]. The high-frequency peak occurs at $\omega = 2n\pi/T$ and at a slightly higher frequency in the velocity spectrum than that given by (3-56).

To a good approximation, the high frequency is centered at

$$f_h = n/T = 2nv_r/d \ (n > 1) \tag{3-57}$$

The baseband frequency is defined as the location of the first peak of sin

$\omega\tau/2$ or

$$f_b = 1/2\tau = v_r/2L \tag{3-58}$$

The relative contribution of the high-frequency component will be governed by the physics of the cracking process. These two frequencies, which may be obtained from direct observations, provide a basis for inferring important parameters of the crack: the microstructure parameter, n, and its coherent length, L.

3.5 EXPERIMENTAL RESULTS

An application of the foregoing theory is provided by experiments and observations obtained in the Canadian Arctic. Although the detailed hydrophone and geophone configuration differed in each case, the essential feature of these experiments was the use of a small hydrophone and geophone array in three dimensions, so as to allow both precise positioning of the source and detection of the acoustic and seismic response at different angles and ranges. In the following subsections, a few examples of acoustic and seismic signals are given from both artificial sources and natural ice fracturing.

3.5.1 Acoustic and Seismic Response of Ice to a Point Impact

An exploration of the impulse response of an ice sheet was obtained by observing the acoustic signal radiated from a sharp hammer blow delivered to the ice surface. As the bulk ice properties are not affected by the impact, a hammer blow should result in an acoustic response in the water approximating the Green's function (3-55) and will also generate waves trapped in the ice, as described by the classic theory outlined in Subsection 3.4.2. The following discussion follows Xie and Farmer [32].

The experiment was carried out on relatively uniform first-year ice, 1.75-m thick. An 8-lb hammer was used to provide a surface impact at a range of 100 m. The signal was detected on a vertical array of four hydrophones ranging from 0 to 70 m in depth, the uppermost of which was incorporated in the ice. Thus the sound received at each hydrophone corresponded to a radiation angle varying from 0° to 44.4°.

Figure 3.6 shows an example of the detected sound. The time series of sound pressure is given for each of the hydrophones. Saturation sometimes occurred for hydrophone A_0 located in the ice. At the 5-m hydrophone (A_5), one can see a weak precursor of about 140 Hz, identified as P in the figure. This precursor is the dilatational or P wave, excited by the blow as described in Subsection 3.4.2; it had a phase speed of about 2630 m/s in the ice and radiated into the water at an angle of about 33° (with the vertical). Following

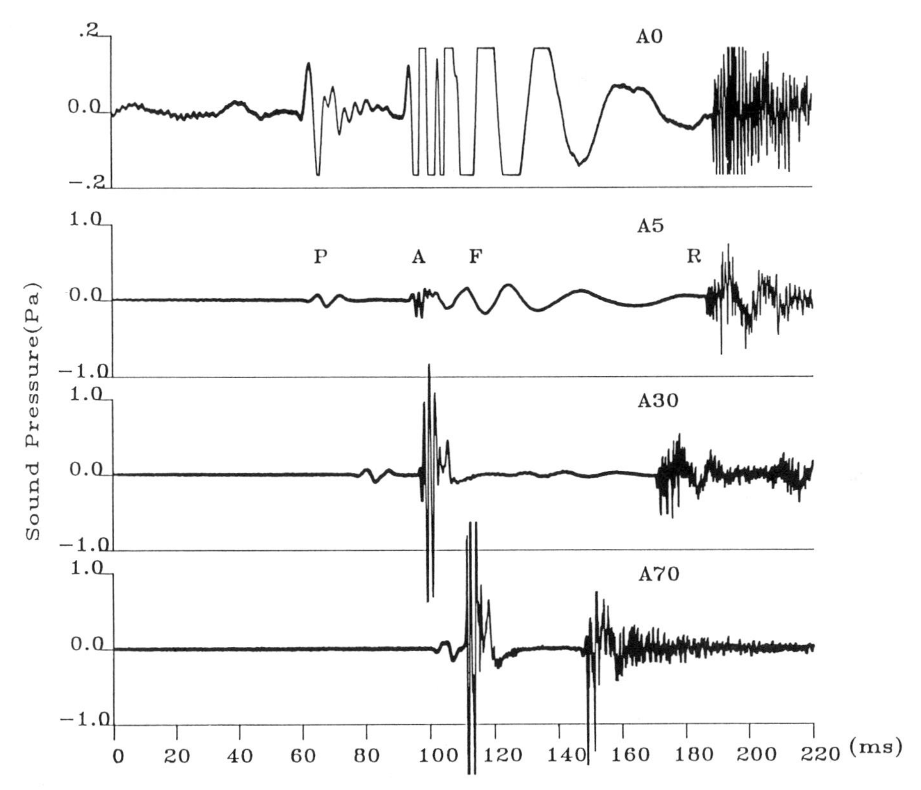

Figure 3.6 Sound-pressure time series recorded by four vertically spaced hydrophones for a hammer blow at 100-m range. Hydrophones at depths of 0 m (A_0), 5 m (A_5), 30 m (A_{30}), and 70 m (A_{70}). *P:* P wave; *A:* direct-path acoustic wave; *F:* flexural wave; and *R:* sea-floor reflected wave. After Xie and Farmer [33]. Reprinted by permission of the American Institute of Physics.

the P wave is the acoustic wave (identified as *A* in Fig. 3.6) with a frequency of 516 Hz. This wave arrived directly from the source through the water. Generation of this signal is described in Subsection 3.4.4. It was weak because of the low-radiation gain of the source function at this shallow angle with the horizontal. Radiation from thermal cracking events has been shown to approximate the dipole pattern closely (Zakarauskas and Thorleifson [34]; Greening, personal communication, 1991) and also is consistent with the theory of Subsection 3.4.2. Almost simultaneously, a low-frequency pulse arrived (identified as *F*) which was anomalously dispersive, with higher frequency (~60 Hz) components arriving first, followed by lower frequency components (down to 30 Hz). This feature is the SV wave, or for very long wavelengths as in this case, the flexural wave, F, which propagates as an inhomogeneous plane wave in water, because its maximum phase speed in ice is less than the sound speed in water [see (3-18)]. The evanescent property of the flexural wave is apparent from its rapid decay with depth: it was barely detectable at the 30-m hydro-

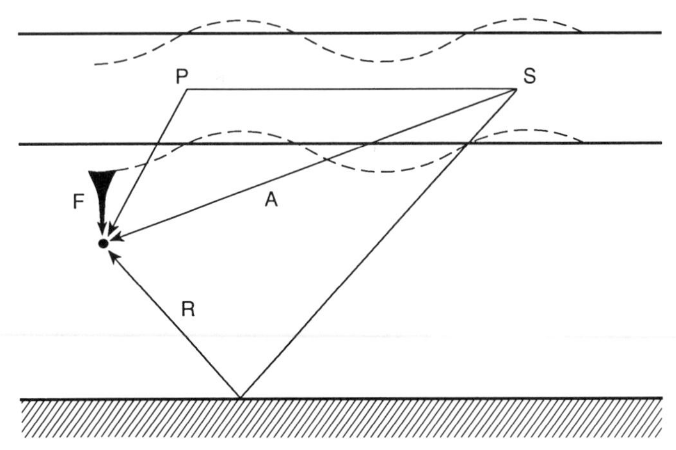

Figure 3.7 Paths showing the arrival of four waves at a hydrophone from a hammer blow source. After Xie and Farmer [33]. Reprinted by permission of the American Institute of Physics.

phone. Finally, a strong reflected acoustic wave (identified as R) arrived from the sea floor. Fig. 3.7 illustrates the paths of each of the four types of waves.

It is apparent that there is a shift in spectral peaks between direct and reflected signals, which is consistent with the radiation pattern defined by (3-41).

A spectrum of the pressure signal at 5 m is shown in Fig. 3.8: the spectrum has been prewhitened with a -14 dB/decade filter. It is emphasized that P and F waves are caused by radiation from the ice cover (i.e., waveguide propagation), or *far-field radiation*, whereas the A (acoustic) and R (reflected) waves radiate into the water from the local ice region at the source, or *near-field radiation*.

The experimental analysis of the ice response to a hammer blow provides an approximate solution to an impulse response of the air–ice–water system. It is interesting that even for a broadband source such as a hammer blow, the detected signals have relatively narrow bandwidths; it appears that the ice cover serves as a multiple band-pass filter as predicted by the theory of Subsection 3.4.4.

It is clear that the frequencies of spectral peaks associated with the P and F waves are determined by eigenvalues of a waveguide system described in Subsection 3.4.2. Although Press and Ewing's [20] theory predicts an infinite number of possible eigenmodes existing in the waveguide, this example shows that only two modes were excited. Other modes associated with both symmetric and antisymmetric forms of solutions may exist, but the data imply that in a real environment, the P and F waves are dominant.

To show how the acoustic response of an ice sheet can be related to its mechanical properties, the above example is used to evaluate elastic parameters of the ice sheet based on the difference in time of arrival of various elastic waves.

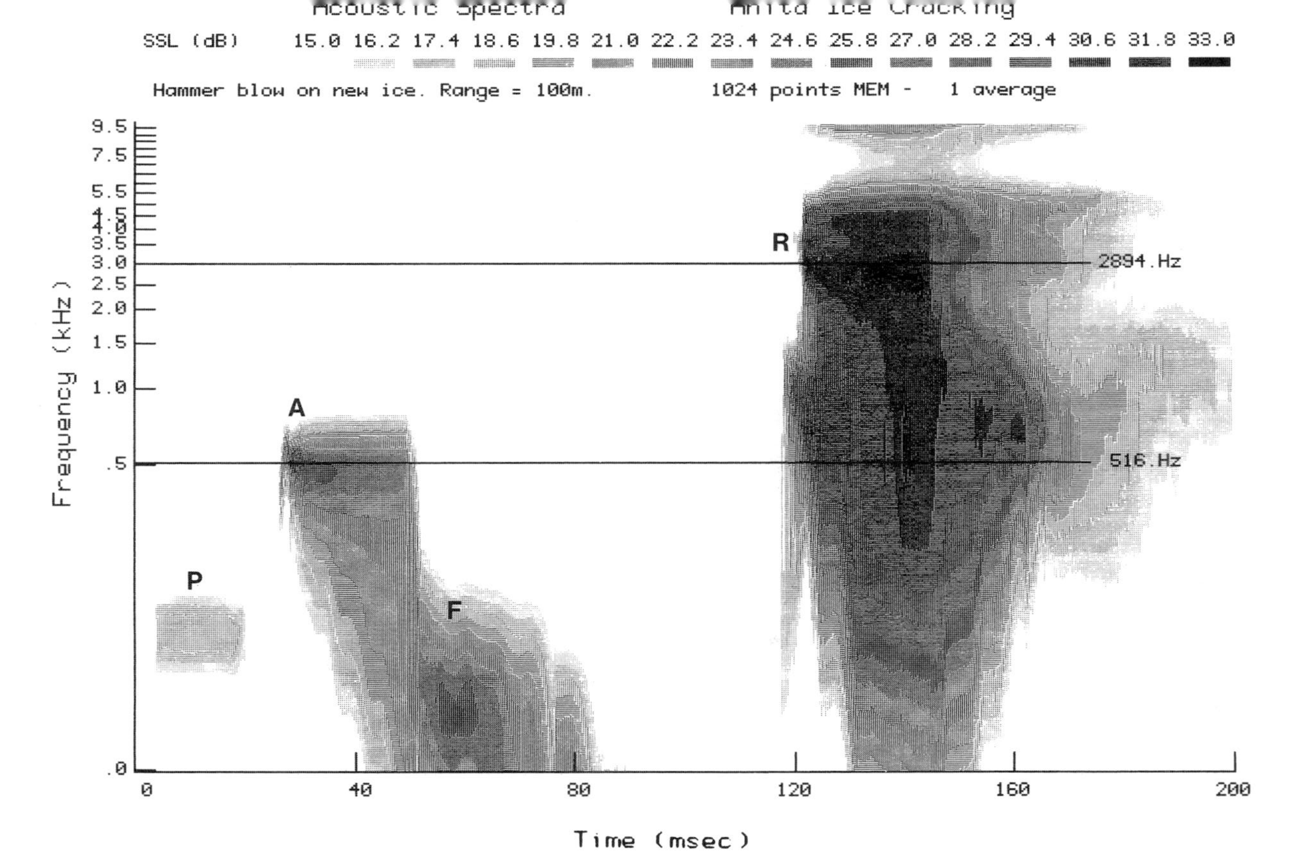

Figure 3.8 Spectral intensity as a function of time for a hammer blow, with *P:* P wave; *A:* direct-path acoustic wave; *F:* flexural wave; and *R:* sea-floor reflected wave. The data are based on record A_5 in Fig. 3.6. After Xie and Farmer [33].

The hammer blow excited four types of waves, as shown in Figs. 3.6 and 3.7. In this case, conductivity–temperature–depth (CTD) casts provided a precise profile of sound speed in the water column, allowing evaluation of elastic parameters for the first-year ice based on delay relationships among F, P, and acoustic waves.

3.5.2 Phase Speed of a P Wave

The signals received at A_0 can be used to calculate the phase speed, c_P of the P wave (see Fig. 3.9). As the group speed of F is very close to the sound wave speed in the water, it can be seen that the acoustic and F waves overlap to some extent. Therefore, the reflected acoustic wave R along a path r_R is chosen as a time reference for the travel-time estimates along the P-wave path r_P. Figure 3.9 illustrates the two paths.

The difference in travel time between the two waves is

$$\Delta t = \frac{r_R}{c_W} - \frac{r_P}{c_P} \tag{3-59}$$

where $c_W = 1438$ m/s, the mean speed of sound in the water column. With $\Delta t = 130$ ms, $r_P = 100$ m and $r_R = 241.66$ m, (3-59) yields

$$c_P = 2628 \text{ m} \cdot \text{s}^{-1} \tag{3-60}$$

Thus the radiation angle, θ_2, of the P wave from ice to water is

$$\theta_2 = \sin^{-1}(1436/2628) = 33.1°$$

where 1436 m/s is the sound speed near the ice-water interface.

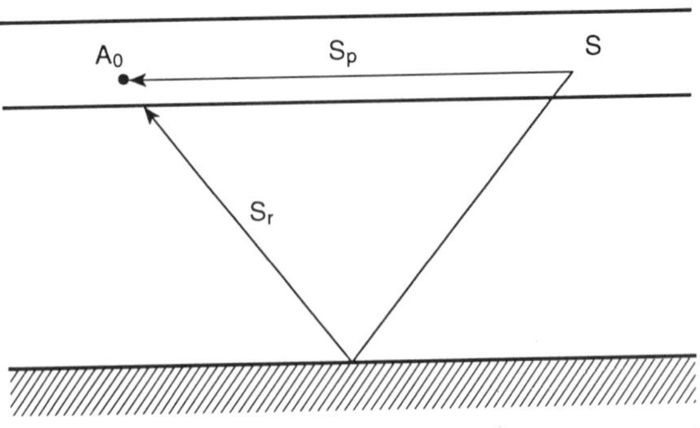

Figure 3.9 Paths illustrating the arrival of P and R waves at the A_0 hydrophone for a nearby hammer blow, where r_P and r_R denote paths for P and R waves, respectively.

3.5.3 Shear-Wave Speed

As mentioned above, the flexural wave is anomalously dispersive. Judging from the signals detected at A_5 (Fig. 3.6), it is estimated that the frequency of maximum group velocity for F was about 60 Hz, and the time delay between the P and F waves was approximately 32 ms at A_0. Therefore, the maximum group speed (U_{max}) for F is

$$U_{max} = 1428 \text{ m} \cdot \text{s}^{-1} \tag{3-61}$$

According to Ewing et al. [4], as in Fig. 6.7 where Poisson's ratio has been assumed to be 0.345, the shear-wave speed in the ice, β, is

$$\beta = \frac{U_{max}}{0.8} = 1785 \text{ m} \cdot \text{s}^{-1} \tag{3-62}$$

and the corresponding phase speed is

$$c_{Pmax} = 0.55\beta = 982 \text{ m} \cdot \text{s}^{-1} \tag{3-63}$$

The calculations of U_{max} and c_{Pmax} for the flexural wave are consistent with the observations (Fig. 3.6). The fact that U_{max} is close to the speed of sound in water implies that F and A waves arrive almost simultaneously at hydrophone A_5; the fact that c_{Pmax} is less than the speed of sound in water implies an evanescent property of the F wave.

3.5.4 Evaluation of Other Elastic Parameters

Knowing β, the compressional wave speed, α, of the first-year ice can be calculated. From Equation (1-12) in Ewing et al. [4], it follows that

$$\alpha = \beta \left[\frac{2(1 - \sigma)}{1 - 2\sigma} \right]^{1/2} \tag{3-64}$$

With $\beta = 1785$ m/s and $\sigma = 0.345$, (3-64) gives

$$\alpha = 3669.4 \text{ m} \cdot \text{s}^{-1} \tag{3-65}$$

The compressional, shear-wave speed and Young's modulus (E) can be expressed in terms of the Lamé constants, λ and μ (see (3-9) and (3-10) and Ewing et al. [4]). That is,

$$E = \frac{\mu(3\lambda + 2\mu)}{\lambda + \mu} \tag{3-66}$$

where the ice density is taken as 910 kg $\cdot$ m^{-3}. The inferred values of α and β are

$$\mu = 2.9 \times 10^9 \text{ Pa} \tag{3-67}$$

$$\lambda = 6.45 \times 10^9 \text{ Pa} \tag{3-68}$$

and

$$E = 7.8 \times 10^9 \text{ Pa} \tag{3-69}$$

There are some differences in the values of E obtained by different methods. It is probable that these were caused by profile properties of the inhomogeneous ice plate.

3.5.5 Thermal-Stress Cracks

In this subsection, examples are shown of naturally occurring sound caused by strong surface cooling at night, when the air temperature dropped to −23°C. The increased sound level associated with strong cooling was described first by Milne [16]. Figure 3.10 illustrates the sound of a thermal-stress crack re-

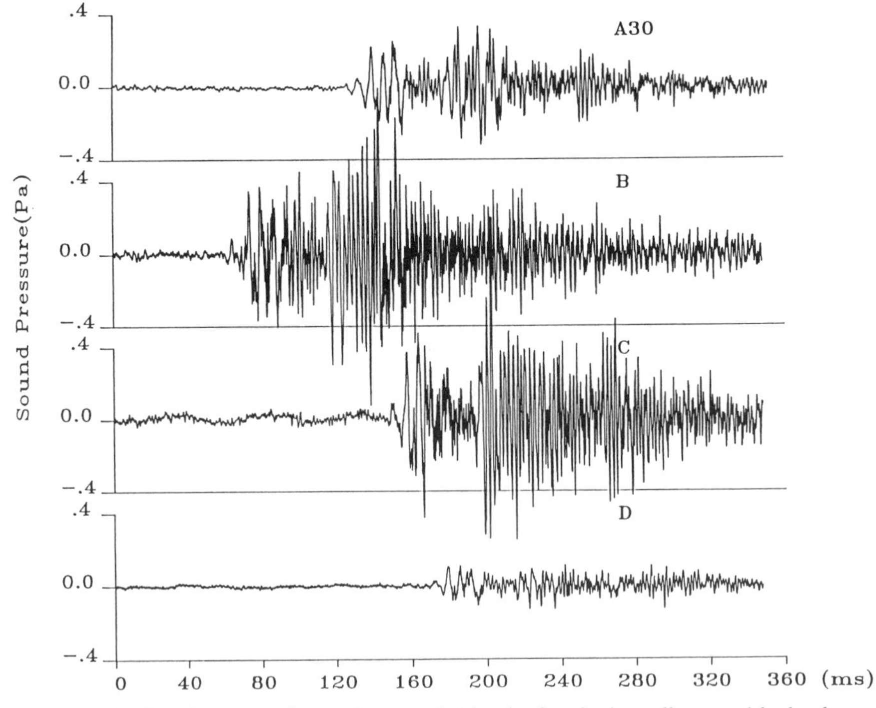

Figure 3.10 Sound-pressure time series recorded by the four horizontally spaced hydrophones for a cracking event occurring in an old ice field.

ceived on four horizontally spaced hydrophones. The event occurred on old ice where the rough surface topography ensured that there were snow-free surfaces well exposed to atmospheric cooling. As a result, many thermal cracks formed in these areas, which made the old ice field a very active source of acoustic noise.

The most interesting feature is that the bottom-reflected signal possessed higher frequency components than did the direct-path signal. The effect was also observed in the artificially generated signals shown in Fig. 3.6 but it is more apparent in Fig. 3.11 where the signal has been transformed into frequency space using the maximum entropy method (Press et al. [21]) and is presented as a "waterfall" plot. The direct path signal has a frequency range of 100–300 Hz, whereas the reflected signal is centered at 400–600 Hz. This result is fully consistent with the near-field radiation model described in Subsection 3.4.4, thus providing insight on the nature of this type of ice-failure mechanism and its contribution to ambient sound in the Arctic Ocean.

First-year ice in the vicinity was relatively quiet during the same cooling period because of the insulating properties of a uniform snow layer (a few centimeters) on the ice. Nevertheless, a few cracks did occur and Fig. 3.12

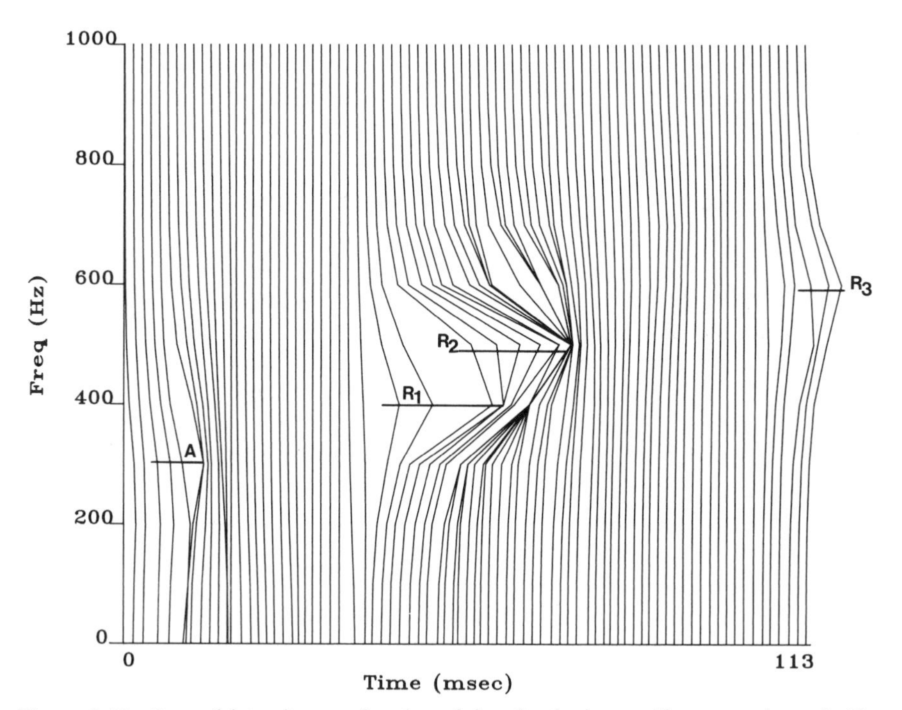

Figure 3.11 Spectral intensity as a function of time for the ice-cracking event shown in Fig. 3.10. Horizontal lines represent three predicted dominant frequencies. *A* is the direct path acoustic wave (300 Hz), R_1 is a first bottom-reflection (399 Hz), and R_2 is a first bottom-surface-bottom reflection (473 Hz). A third multiple reflection (R_3) is also apparent.

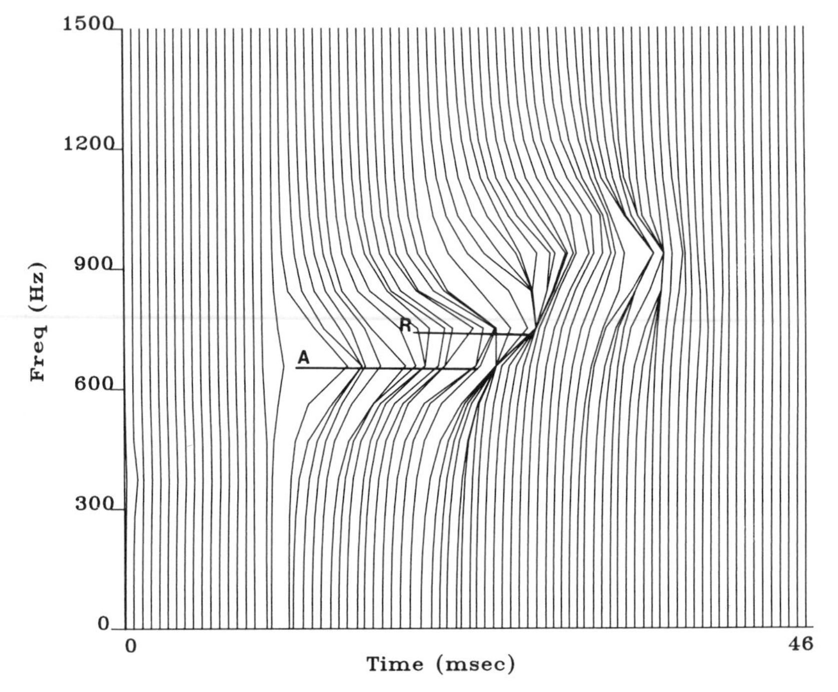

Figure 3.12 Spectral intensity as a function of time for a far-field thermal crack in new ice. The predicted frequencies are indicated by two lines for direct path acoustic wave *A* (622 Hz) and sea-floor reflected wave *R* (656 Hz).

shows a sequence of spectra of the acoustic signal caused by a crack at 1-km range in first-year ice.

The source is a short pulse centered on 622 Hz. The reflected path at this range partly overlaps the direct-path signal; the dominant frequency shift is from 622 to 668 Hz in the reflected portion. The theory predicts that this frequency shift is consistent with a local ice thickness at the source of 1.4 m, similar to but slightly less than that in the vicinity of the hydrophone array.

Crack radiation from old ice is more complicated because of highly variable ice topography. Xie and Farmer [33] described several examples of acoustic radiation from fractures in old ice. In general, the presence of an angular frequency dependence still seems to apply, but without the consistent properties observed for first-year ice. One curious example is shown in Fig. 3.13 where the signal changes frequency rapidly with time.

Radiation frequency as a function of radiation angle is shown in Fig. 3.14 for both first-year and old ice. It is clear from this that the angular frequency dispersion, which is consistent for first-year ice, is much more scattered for old ice. Nevertheless, it is apparent that for first-year ice the results are consistent, allowing flexural rigidity of the ice sheet to be inferred remotely from the sound of thermal-stress fracturing.

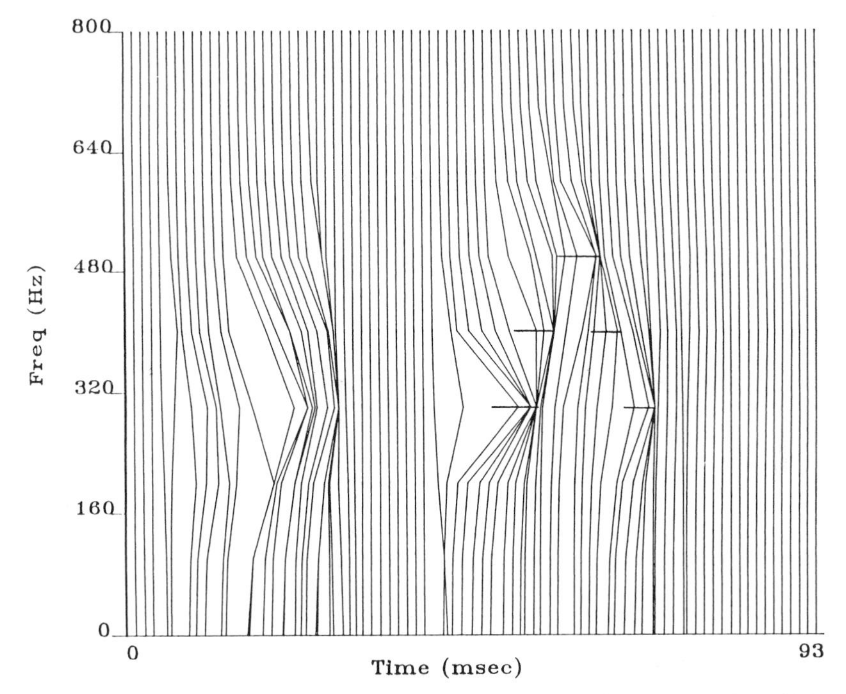

Figure 3.13 Spectral intensity as a function of time for a cracking event during which the primary frequency content of the sound changes rapidly. This unusual example originating from old ice consists of an up-chirp followed by a down-chirp.

3.5.6 A Finite-Volume Crack

Another example of the way in which acoustic radiation can be used to probe the structure of ice fracturing is provided by the sound of a finite-volume crack. Although the driving force for the failure process in this example is unclear, the event is chosen to provide a comparison between observed signals and the theory on acoustic radiations from a growing finite-volume crack. In this case, the crack was deep enough to alter the bulk properties of the ice materially, so that the theory of Subsection 3.4.2 is inappropriate. Therefore, the finite-volume theory of Subsection 3.4.5 is applied. A key result was the derivation of two spectral frequency components that provide insight on the crack properties.

A time-series plot of sound pressure from one of the hydrophones for a period of 220 ms is shown for this event in Fig. 3.15. Most acoustic energy in this event occurs in two frequency bands: a baseband pulse of frequency ~200 Hz with a much higher frequency superimposed over the first few cycles. This double-band feature is more clearly shown in the time-evolving power spectrum of Fig. 3.16.

Equations (3-57) and (3-58) allow derivation of the horizontal scale and fine-structure property. For the example shown, the fractional amplitude (ϵ) of the

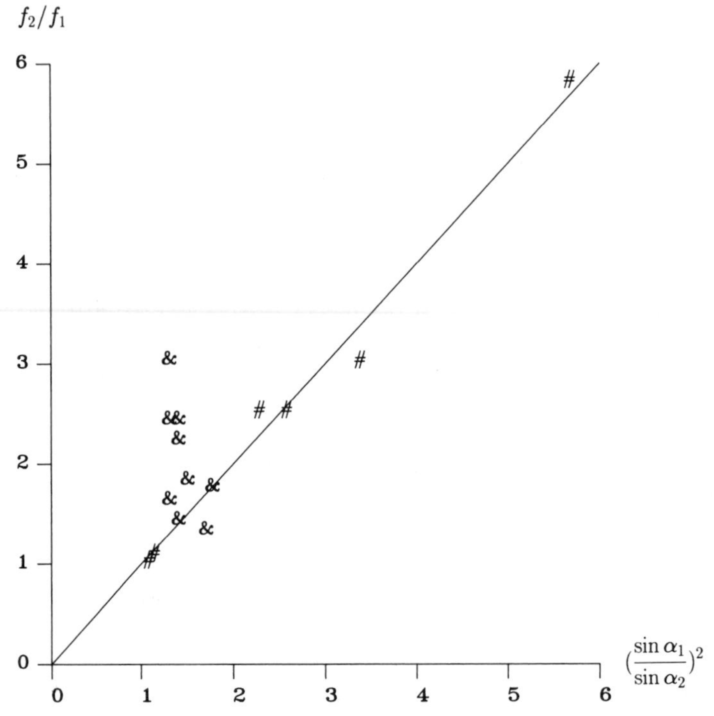

Figure 3.14 Ratio of primary frequencies between direct and reflected path signals of ice-cracking sounds in both first-year and old ice versus $(\sin \alpha_1/\sin \alpha_2)^2$ as defined in (3-42). Symbols # and & indicate first-year and old-ice events, respectively.

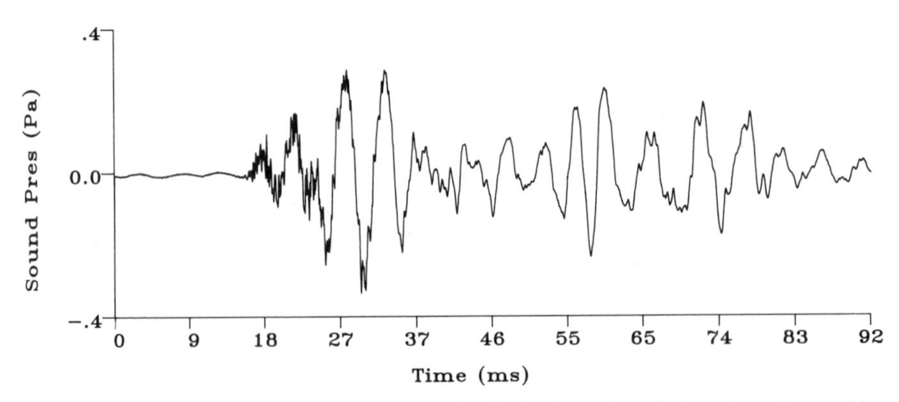

Figure 3.15 Sound-pressure time series of a 30-m hydrophone for a finite-volume ice-cracking event. After Xie [31].

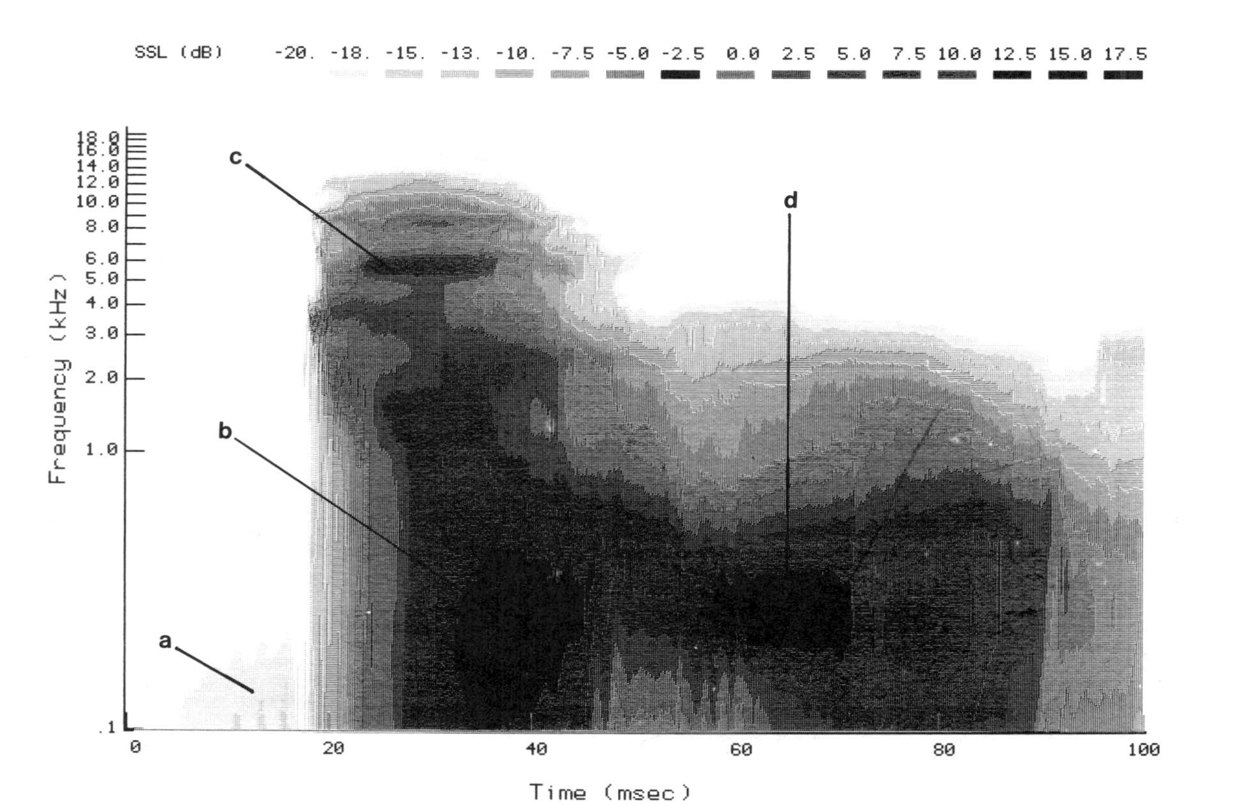

Figure 3.16 Spectral intensity (indicated by shades) of the signal shown in Fig. 3.15, showing (*a*) *P* wave, (*b*) direct-path, baseband signal, (*c*) direct path, high-frequency signal, and (*d*) reflected signals from the sea. After Farmer and Xie [6]. Reprinted by permission of the American Institute of Physics.

high frequency component is 0.12. The baseband frequency of 200 Hz implies a coherent length scale of 6.9 m. Farmer and Xie [6] discussed these and similar results in detail. There is some ambiguity in the interpretation of the high-frequency component because of the unknown value of n in (3-52) and (3-53). Moreover, the interpretation of the low-frequency component depends on the concept of coherence. The crack length calculated from the baseband frequency, therefore, represents the minimum length. The total length may be the sum of several coherent segments, which may be established from the duration of the direct-path cracking sound (found to be 20 ms on the basis of high-frequency radiation), and from independent estimates of crack propagation velocity. In this case, a total crack length of 27 m is estimated.

3.5.7 An Uplift Event

An example of the insight to be gained from simultaneous seismic and acoustic measurements is the analysis of a physical event that was tracked back to its origin in an experiment in Allen Bay, NWT, March 1992. Figure 3.17 shows the instrument deployment. Three 3-axis geophones were located 90 m apart on 1.5-m thick first-year ice. Five hydrophones were located at 30-m depth and 100-m spacing; a sixth was deployed at 5-m depth. A pressure ridge (Fig. 3.18) ran close by the array.

Events were mapped back to their origin in the field, using time-delay analysis of the geophone and hydrophone signals. The location of one such event is indicated in Fig. 3.17. The 5–8-cm snow cover was cleared from the surveyed location, revealing a well-defined segment of ice approximately $0.5 \times 3\ \text{m}^2$ that was raised 0.5 cm above the surrounding ice sheet (Fig. 3.19). Three-axis geophone records from G2 and the hydrophone record from H3, corresponding to this event, are shown in Fig. 3.20. The hydrophone [Fig. 3.20(d)] senses high-frequency radiation and is plotted on an expanded scale.

The three axes of the geophone record illustrate the highly anisotropic nature of the seismic signal trapped in the ice. The horizontal transverse component G2X [Fig. 3.20(a)] is dominated by the arrival of a strong horizontal shear wave. This can be used to evaluate the shear-wave speed based on the time of arrival of the corresponding acoustic wave recorded on H3 [Fig. 3.20(d)], which is found to be $1736\ \text{m} \cdot \text{s}^{-1}$. On the other hand, the longitudinal component G2Y [Fig. 3-20(b)] shows the arrival of the P wave which can be used to calculate $c_P = 3402\ \text{m} \cdot \text{s}^{-1}$. Finally, the vertical component G2Z [Fig. 3.20(c)] reveals the flexural wave, which at this range of 160 m is considerably dispersed [see (3-18)]. This dispersive property of the flexural wave is more clearly revealed in the time-evolving frequency display of Fig. 3.21. A strong vertical component can be expected from a vertical motion of the ice at the source, which is consistent with the physical observation (Fig. 3.19). The elastic constants can thus be found using the procedures discussed in Subsection 3.5.3. It is interesting to note that there is a weak coupling of the acoustic signal into the ice. The acoustic signal overlaps with the beginning of the

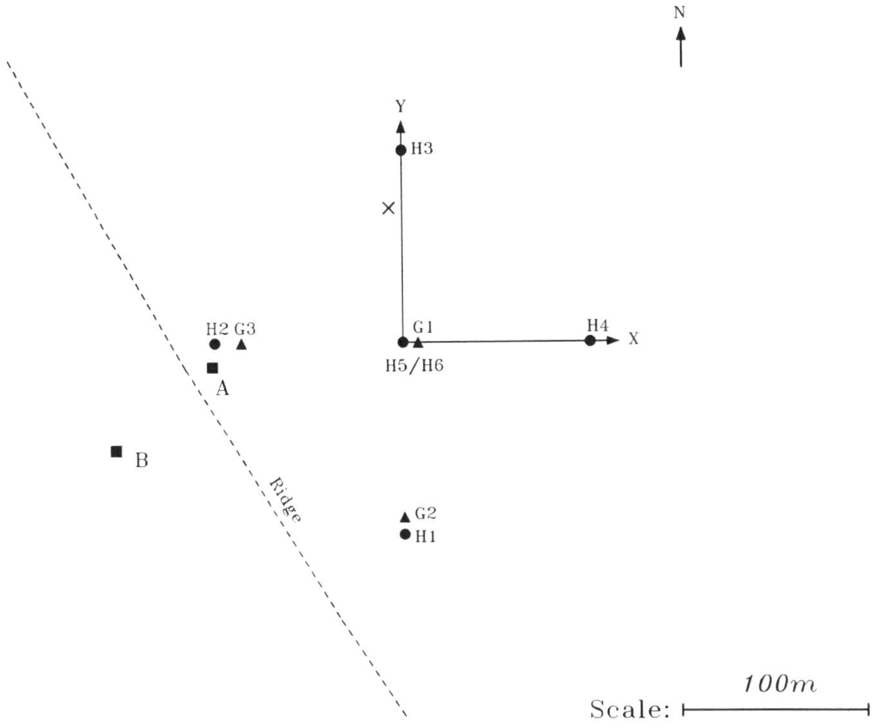

Figure 3.17 Configuration of acoustic and seismic array used an experiment in Allen Bay, NWT, in March 1992. Six hydrophones (H1–H6) and three 3-axis geophones (G1–G3) were deployed. A and B mark the locations of two lead-ball drops made on each side of the ridge; × marks the location of an uplift event, the result of which is shown in Fig. 3.19.

Figure 3.18 Photo showing a close-by ridge with an estimated mean height of 2 to 3 meters.

Figure 3.19 Photo showing an acoustically located uplifted "ice island" dimensions of which are: $3 \times 0.5 \times 0.005$ m^3.

flexural wave in G2Z and a trace of the bottom-reflected acoustic wave appears at 280 ms.

The acoustic signal is shown on an expanded scale in Fig. 3.20(d). Note that the initial sound signal is a pressure decrease. This is consistent with an upwards movement of the ice face as observed in Fig. 3.19, which results in

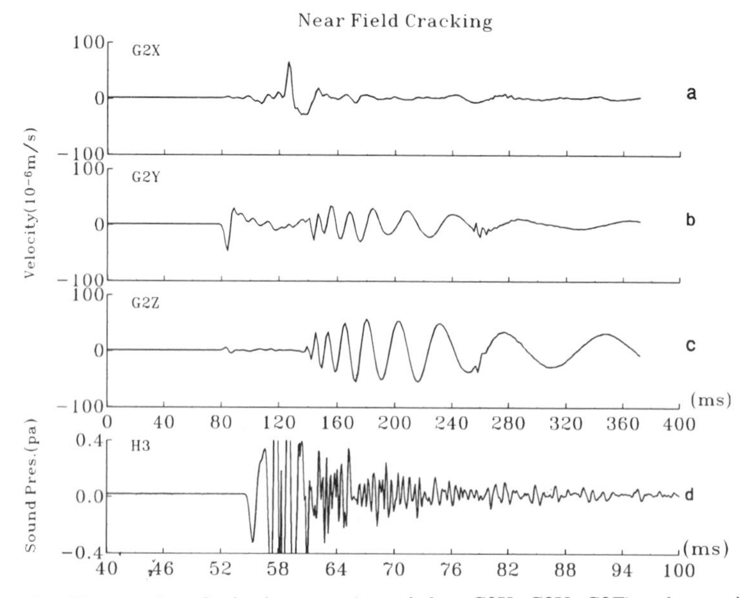

Figure 3.20 Times series of seismic waves (recorded on G2X, G2Y, G2Z) and acoustic wave (recorded on H3) generated by an uplift event.

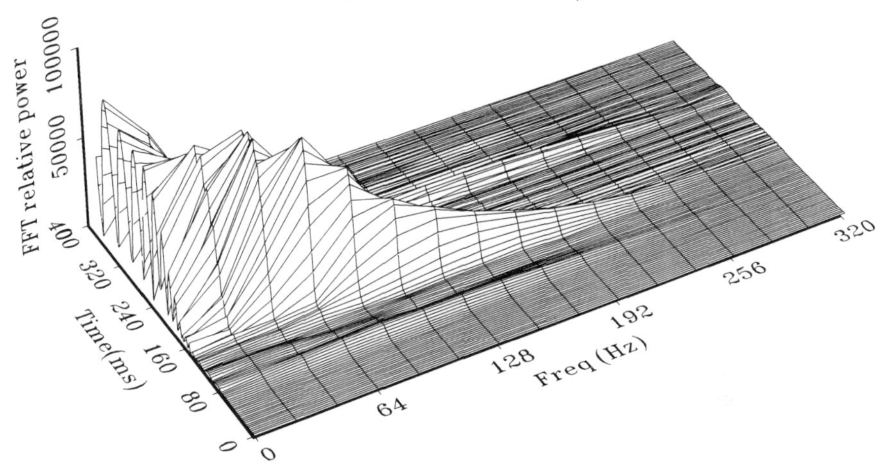

Figure 3.21 A 3-D display of the spectrum evolution of seismic waves generated by an uplift ice event. The raw signal is shown in Fig. 3.20 (G2Z channel). The dispersive character of the flexural wave is clearly evident.

a local reduction of acoustic pressure. This example illustrates the intimate relationship between the detected signals and their origin, which underlies the potential of this approach for probing the failure process.

3.5.8 Wave Propagation Through an Ice Ridge

One final example illustrates the potential for probing anisotropies in the ice properties. Since energy trapped in the ice is sensitive to its bulk properties, fractures, pressure ridges, and other features can be expected to leave their trace on the detected signal. An extreme example of this is provided by a comparison of seismic propagation from two closely spaced and essentially identical mechanical inputs in the experimental array shown in Fig. 3.17. A lead ball was dropped from a fixed height (1.32 m) onto the ice from which the snow had been cleared. Figure 3.17 shows the paths between two such sources, *A* and *B*. Figure 3.22 shows the vertical geophone component for each path, illustrating the arrival of both the flexural wave and the bottom-reflected acoustic wave. Although each path is approximately the same distance (120 m), the flexural wave is severely attenuated. Evidently the flexural-wave energy is either absorbed or scattered by the ridge.

Although this is an extreme example, it does serve to illustrate a principle that can be used to probe the existence of heterogeneity and ice faults that are less obvious. The three geophone components each provide information about the different mechanical properties of the ice along the path. A large number of such paths, using many source locations and a few geophones, allows mapping of the spatial distribution of ice properties by tomographic techniques.

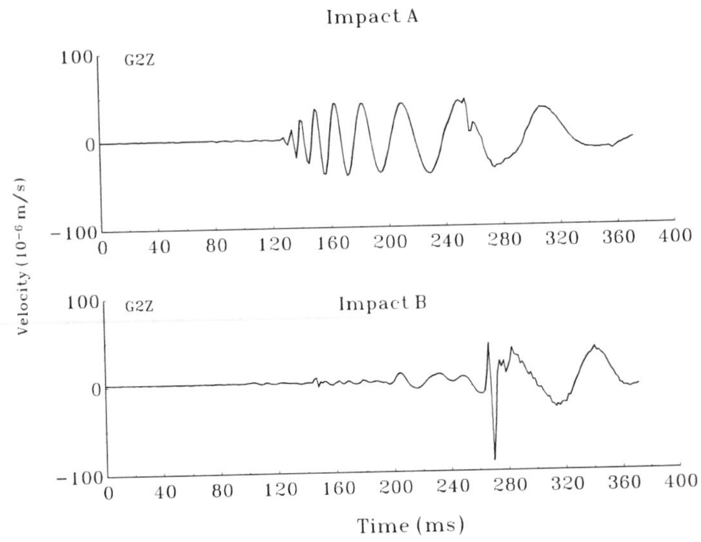

Figure 3.22 Time series of seismic waves (recorded on G3Z channel) generated by two lead-ball impacts made at site A and site B shown in Fig. 3.17. Impact B generates waves that must travel through the ridge resulting in noticeable attenuation, illustrating a principle that can be exploited in tomographic studies of ice structure.

3.6 FUTURE WORK

Passive acoustic and seismic sensing of sea ice is different from other remote sensing techniques described in this book, not just because it depends on underwater sound or elastic propagation in the ice sheet but because the signal arises from dynamic processes within the ice. For these reasons, the kind of information that can be gleaned relates to the mechanical conditions and properties of the ice, such as its flexural rigidity and the stress to which it is subjected and its structural inhomogeneities. As our understanding of this topic develops, such information will be especially relevant to the study of ice behavior, which is essential to the development and testing of predictive models of sea ice.

It must, however, be emphasized that passive acoustic and seismic sensing of sea ice is a young science. At this stage, scientists are learning how to identify the different acoustic and seismic signatures for different ice-cracking events. The interpretation depends on an adequate model of the sound source and thus motivates development of a suitable theoretical framework, some of which has been summarized. The models described are necessarily simplified. For example, the ice was treated as a vertically uniform and horizontally homogeneous elastic slab, whereas ice is known to be heterogeneous, marked by numerous cracks and fissures, and having a distinctive vertical profile (Cox and Weeks [2]). However, the simple models appear to provide a useful description and form a starting point for more complex models and data-inversion schemes.

Current work includes the simultaneous use of small-scale acoustic tomography, where acoustic transducers are embedded in the ice (S. Rajan, personal communication, 1992), together with acoustic and seismic observations of the type discussed to resolve spatially and temporally evolving properties, and analysis of propagation through ridges and other structures within the ice as discussed in Subsections 3.5.7 and 3.5.8.

The focus in this chapter has been on the physics of sounds and seismic waves generated by ice fracturing but this represents only part of the full range of ice noise sources in the Arctic, although it is one of the more thoroughly studied. For example, some distinctive signals have been detected near pressure ridges (Pritchard [22]). More general descriptions of ambient ice noise signals and their relation to environmental parameters may be found in Makris and Dyer [15], and Dyer [3] has proposed generation mechanisms of low-frequency sound.

When ice is not consolidated, for example, in the marginal ice zone, the interaction between adjacent ice floes can be a significant source of noise. These interactions have been the focus of recent numerical (Rottier [23]) and observational studies (Johannesen et al. [10]). The frequency of interaction is modulated by atmospheric or oceanographic effects such as waves and eddies. Detailed analyses of the sound generated by individual floe collisions have also been carried out. Xie and Farmer [32], for example, show that the rubbing of adjacent floes can cause narrowband squeaks associated with natural resonances in the ice. As in the case of ice cracking, analysis of the structure of the sound radiated by individual events serves as a probe of the physics of the event itself and of the local properties of the ice, whereas analysis of the global sound yields insight into the larger scale forcing field.

Attempts to integrate these signals into models of the general stress field are just beginning. Lewis and Denner [14] have calculated the stress induced by heat flux at the ice surface and experimental work is now underway to investigate the relationship between this stress and the sound radiated from a distribution of thermally induced fractures.

Opportunities may also exist for exploiting distinctive acoustic signatures of, for example, pressure ridging to help identify areas of ice cover posing a hazard to navigation. (This possibility has already been proposed for use in the Gulf of St. Lawrence.) Whatever the application, our ability to exploit these naturally occurring signals ultimately depends on our knowledge of the underlying physics.

Incorporation of these effects into numerical models of the sound generation over larger areas, and even over the whole Arctic basin, may eventually be required to explain the background sound field that is observed. Arrays of hydrophones can be used to focus on this sound in a highly directive way, implying the potential for mapping physical processes across large areas of the ice cover. These results will be of particular interest with the advent of high-resolution radar maps (from space or aircraft) using techniques described elsewhere in this book. The simultaneous application of more than one remote sensing technique to the study of particular phenomena is invariably enlightening. As our ability to interpret the ambient sound signal improves, acoustic and seismic sensing techniques are expected to play an increasing role in our understanding of the properties and behavior of sea ice.

ACKNOWLEDGMENTS

This work is supported by the Canadian Panel on Energy Research and Development, project 67136, and the U.S. Office of Naval Research under the administration of Dr. Thomas Curtin.

NOMENCLATURE

λ	Bulk modulus, Pascal
μ	Shear modulus, Pascal
σ	Poisson's ratio, dimensionless
E	Young's modulus, Pascal
E_1	Elastic modulus for a plate $[= E/(1 - \sigma^2)]$, Pascal
σ_1	Poisson's ratio for a plate $[= \sigma/(1 - \sigma)]$
$\vec{s}$	Displacement vector
Φ	Displacement scalar potential
$\vec{\Psi}$	Displacement vector potential
$P_{i,j}$	Stress tensors
$e_{i,j}$	Strain tensors
α	Compressional wave speed in ice, $\text{m} \cdot \text{s}^{-1}$
β	Shear wave speed in ice, $\text{m} \cdot \text{s}^{-1}$
c_p	Phase speed in horizontal direction
λ_h	Horizontal wavelength
k	Horizontal wavenumber, $k = 2\pi/\lambda_h$
ω	Circular frequency
H	Ice thickness, meters
f_n	Discrete eigenfrequencies, Hertz
L	Half length of a bilaterally growing crack
γ	Specific surface energy (meter · Pascal)
BV	Brine volume (parts per thousand)
TS	Tensile strength of sea ice, Pascal
SS	Shear strength of sea ice, Pascal
FS	Flexural strength of sea ice, Pascal
P_{oi}	Incident acoustic pressure
P_{or}	Reflected acoustic pressure
P_t	Transmitted acoustic pressure
P_2	Sound pressure in water
θ_0	Incident angle

θ_2	Transmitted angle
Z_s	Acoustic impedance of a thin plate due to symmetric motion
Z_a	Acoustic impedance of a thin plate due to anti-symmetric motion
C_s	Longitudinal wave speed in a thin plate, $m \cdot s^{-1}$
C_f	Flexural wave speed in a thin plate, $m \cdot s^{-1}$
C_2	Sound speed in water, $m \cdot s^{-1}$
U_{max}	Maximum group speed of flexural waves, $m \cdot s^{-1}$
θ_b	Broadband radiation angle
ω_c	Cutoff circular frequency
α_i	Radiation angle along path i
f_i	Dominant frequency of acoustic wave propagating along path i
R_t	Frequency ratio
Φ_i	Velocity potential of acoustic waves in medium i
h	Depth of a monopole source from the upper ice surface
η	Radial wavenumber
β_i	Vertical wavenumber in medium i
k_i	Wavenumber in medium i
Q	Volume change rate associated with an acoustic source
$U_s(t)$	Displacement velocity associated with a rupture
$\hat{U}_s(\omega)$	Spectrum of $U_s(t)$
D_0	Total displacement induced by a fault
$G(t)$	Normalized displacement due to a fault
T	Rise time of an opening induced by a fault
ν_r	Rupture velocity
ϵ	Fraction of the primary amplitude associated with a fault opening
n	Number of roughness elements associated with a fault formation
$\hat{P}_2(\omega)$	Spectrum of acoustic pressure in water
τ	Propagation time of a rupture over half of a coherent crack length
f_h	High frequency caused by roughened cracking
f_b	Baseband frequency related to the average speed of a crack opening
$G_{2x,y,z}$	Geophone sensors along longitudinal, transverse, and vertical directions

REFERENCES

[1] L. M. Brekhovskikh (1980) *Waves in Layered Media*, 2nd ed. (translated by R. T. Beyer), Academic Press, New York.

[2] G. F. N. Cox and W. F. Weeks (1988) ''Profile properties of undeformed first-year sea ice,'' CRREL Rep. 88–13.

[3] I. Dyer (1988) ''Speculations on the origin of low-frequency Arctic Ocean Noise,'' in *Sea Surface Sound*, B. Kerman (Ed.), NATO ASI Series, Kluwer Academic, Boston, 513–532.

[4] W. M. Ewing, W. S. Jardetzky, and F. Press (1957) *Elastic Waves in Layered Media*, McGraw-Hill, New York.

[5] D. M. Farmer and S. R. Waddell (1988) ''High-frequency ambient sound in the Arctic,'' in *Sea Surface Sound*, B. Kerman (Ed.), NATO ASI Series, Kluwer Academic, Boston, 555–563.

[6] D. M. Farmer and Y. Xie (1989) ''The sound generated by propagating cracks in sea ice,'' *J. Acoust. Soc. Am.*, **85** (4), 1489–1500.

[7] A. A. Griffith (1920) ''The phenomena of rupture and flow in solids,'' *Philosophical Trans., Royal Society of London*, **A221**, 163–198.

[8] N. A. Haskell (1964) ''Total energy and energy spectral density of elastic wave radiation from propagating faults,'' *Bull. Seis. Soc. Am.*, **54,** 1811–1841.

[9] G. L. Jin and P. Wadhams (1989) ''Travel time changes in a tomography array caused by a sea cover,'' *Progress in Oceanography*, **22**, 249–275.

[10] O. M. Johannesen, S. G. Payne, K. V. Starke, G. A. Gotthard, and I. Dyer (1988) ''Ice eddy ambient noise,'' in *Sea Surface Sound*, B. Kerman (Ed.), NATO ASI Series, Kluwer Academic, Boston, 599–605.

[11] K. Kasahara (1981) *Earthquake Mechanics*, Cambridge University Press, Cambridge, U.K.

[12] J. S. Kim (1989) ''Radiation from directional seismic sources in laterally stratified media with application to arctic ice cracking noise,'' Ph.D. thesis, Massachusetts Institute of Technology.

[13] A. J. Langley (1989) ''Acoustic emission from the Arctic ice sheet,'' *J. Acoust. Soc. Am.*, **85**, 692–701.

[14] J. K. Lewis and W. W. Denner (1988) ''Higher frequency ambient noise in the arctic ocean,'' *J. Acoust. Soc. Am.*, **84**, 1444–1455.

[15] N. C. Makris and I. Dyer (1991) ''Environmental correlates of arctic ice-edge noise,'' *J. Acoust. Soc. Am.*, **90** (6), 3288–3298.

[16] A. R. Milne (1972) ''Thermal tension cracking in sea ice: A source of underice noise,'' *J. Geophys. Research*, **77**, 281–327.

[17] A. R. Milne (1974) ''Wind noise under winter ice fields,'' *J. Geophys. Res.*, **79** (12), 803–809.

[18] B. E. Miller (1990) ''Observation and inversion of seismo-acoustic waves in a complex Arctic ice environment,'' M.Sc. thesis, Massachusetts Institute of Technology.

[19] E. Orowan (1950) *On Fatigue and Fracture of Metals*, W. M. Murray (Ed.), Wiley, New York, 139–157.

[20] F. Press and W. M. Ewing (1951) ''Propagation of elastic waves in a floating ice sheet,'' *Trans. Am. Geo. Union*, **32** (5), 673–678.

[21] W. H. Press, B. P. Flannery, S. A. Teukolsky, and W. T. Vetterling (1986) *Numerical recipes*, Cambridge University Press, Cambridge, U.K.

[22] R. S. Pritchard (1991) ''Arctic ocean background noise caused by ridging of sea ice,'' *J. Acoust. Soc. Am.*, **75**, 419–427.

[23] P. J. Rottier (1990) ''Wave/ice interactions in the marginal ice zone and generation of ocean noise,'' Ph.D. thesis, University of Cambridge, U.K.

[24] J. C. Savage (1972) ''Relation of corner frequency to fault dimensions,'' *J. Geophys. Res.*, **77** (20), 3788–3795.

[25] H. Schmidt and F. B. Jensen (1985) ''A full wave solution for propagation in multi-layered viscoelastic media with application to Gaussian beam reflection at fluid-solid interfaces,'' *J. Acoust. Soc. Am.*, **77**, 813–825.

[26] P. J. Stein (1986) ''Acoustic monopole in a floating ice plate,'' Ph.D. thesis, Massachusetts Institute of Technology.

[27] P. J. Stein (1988) ''Interpretation of a few ice event transients,'' *J. Acoust. Soc. Am.*, **83**, 617–622.

[28] P. J. Stein (1990) ''Predictions and measurements of the directivity of a monopole source in a floating ice plate,'' presented at NATO Workshop on Natural Physical Sources of Underwater Sound, Cambridge, U.K., 1990, ed. Bryan Kerman, Dordrecht: Kluwer Academic Publishers, 625–639.

[29] S. R. Waddell and D. M. Farmer (1988) ''Ice break-up: Observations of the acoustic signature,'' *J. Geophys. Res.*, **93**, 2333–2342.

[30] P. Wadhams (1988) ''The underside of Arctic sea ice imaged by sidescan sonar,'' *Nature*, *(London)*, **333**, 161–164.

[31] Y. Xie (1991) ''An acoustical study of the properties and behaviour of sea ice,'' Ph.D. thesis, University of British Columbia.

[32] Y. Xie and D. M. Farmer (1991) ''Acoustical radiation from thermally stressed sea ice,'' *J. Acoust. Soc. Am.*, **89** (5), 2215–2231.

[33] Y. Xie and D. M. Farmer (1992) ''The sound of ice break-up and floe interaction,'' *J. Acoust. Soc. Am.*, **91** (3), 1423–1428.

[34] P. Zakarauskas and J. M. Thorleifson (1991) ''Directivity of ice cracking events,'' *J. Acoust. Soc. Am.*, **89**, 722–734.

4

ICE-THICKNESS MEASUREMENT

JAMES R. ROSSITER
Canpolar Inc.
Toronto, Ontario, Canada

J. SCOTT HOLLADAY
Geonex Aerodat Inc.
Mississauga, Ontario, Canada

4.1 INTRODUCTION

As surface operations in polar regions and ice-covered oceans become more common, there are many needs for information on the thickness of sea ice over large areas, for both scientific and engineering requirements (Thorndike et al. [67]). From a scientific point of view, knowledge of sea-ice thickness is essential in order to understand the dynamics and the thermodynamics of the ice cover, the world heat budget and climate change, and the interactions between the atmosphere and the ocean. Engineers need to know ice thickness for planning transportation routes and for estimating maximum loads on structures in ice-covered waters. These needs for transport are both strategic and tactical, and apply to both over-ice vehicles, which must avoid dangerously thin ice, and to icebreakers, which try to find the thinnest and weakest ice.

The need to measure iceberg thickness relates to its draft and its potential to scour the seabed, upon which pipelines or cables may be deployed. If an iceberg is to be used as a source of fresh water, knowledge of the iceberg's thickness is required both to estimate its volume and to detect internal structural flaws before a decision is made to tow it to a distant location.

Remote Sensing of Sea Ice and Icebergs, Edited by Simon Haykin, Edward O. Lewis, R. Keith Raney, and James R. Rossiter.
ISBN 0-471-55494-4

4.1.1 Historical Overview

Although attempts have been made to estimate sea-ice thickness using passive visual and microwave techniques and other imaging sensors, these have not been highly successful because, in general, there is poor or ambiguous correlation between ice surface characteristics and ice thickness (Campbell et al. [9]; Page and Ramseier [51]; Rossiter and Bazeley [59]). Radio echosounding has been particularly successful over glaciers and ice sheets (Robin et al. [55]) and lake ice (Cooper et al [13]); however, typical radio echosounding equipment used for glacier sounding emits radar pulse lengths which are too long to be useful for sea ice thickness measurement. In the mid-1970s, impulse radar (now usually called ground-penetrating radar or GPR) sounding of sea ice was studied (Campbell and Orange [7]; Kovacs [28, 30]; Rossiter et al. [64]). These systems use a short pulse length at a relatively low frequency (typically near 100 MHz) in order to maximize resolution while achieving reasonable penetration in lossy materials (Annan and Davis [3]; Davis and Annan [14]). Although very successful for freshwater ice, the results for sea ice have been only partly successful due to the brine content of sea ice, which increases the conductivity of sea ice and thus severely limits the propagation distance of radio-frequency energy (Rossiter [65]).

A more practical technique for the routine measurement of sea-ice thickness from an airborne platform, based on electromagnetic (EM) induction, has been under development since the early 1980s. Helicopter EM-measurement systems are now reaching the end of validation testing (Rossiter and Lalumiere [58]; Kovacs and Holladay [33]; Rossiter et al. [56]; Prinsenberg et al. [53, 54]; Holladay et al. [21, 22]; Holladay [24]).

Radar sounding of icebergs has also been investigated (Kovacs [29]; Rossiter and Gustajtis [62]). In icebergs, which are made up of glacier ice with very low radar attenuation, penetration is not usually an issue. Successful radar sounding of icebergs appears to be related more to iceberg shape. Large, tabular icebergs are amenable to sounding using radar; whereas irregularly-shaped icebergs, more common off Canada's east coast, are difficult to sound accurately (Rossiter et al. [61]; Rossiter and Gustajtis [63]).

Other techniques have been used to obtain long profiles of sea ice thickness using upward-looking sonar from submarines and laser profilometry from fixed-wing aircraft (Wadhams and Comiso [68]), by examining the keel or sail, respectively, and assuming neutral buoyancy for the ice. Although these techniques are effective for inferring statistical information over a long track, they are not usually available for tactical, operational needs.

The electrical properties of sea ice that affect ice-thickness measurement are presented in Chapter 2. In this chapter, the development and experimental results of EM induction for ice-thickness measurement are discussed in some detail. Other ice-thickness measurement techniques, including radar and portable ice thickness sensors, which are useful in some circumstances, are described, and finally, future directions in this field are discussed.

4.2 EM-INDUCTION ICE-THICKNESS SENSOR

The use of EM-induction sounding for airborne ice-thickness measurement is an outgrowth of technology developed for mineral prospecting during the late 1960s. Since then, a variety of instruments has been developed to address problems ranging from airborne hydrography to mapping of toxic waste sites. These systems use frequencies in the 50- to 50,000-Hz range, although frequencies of up to 500,000 Hz may eventually prove useful. A set of transmitter and receiver coils are towed in a sensor platform, or "bird," at altitudes ranging from about 15 to 30 m, as shown in Fig. 4.1. The transmitted or "primary" field induces eddy currents in nearby conductive objects or regions, which in turn generate "secondary" EM fields. These secondary fields are detected at the receiver.

During sea-ice–thickness profiling, the electrical "conductor" generating most of the secondary field strength is the seawater beneath the ice sheet, as

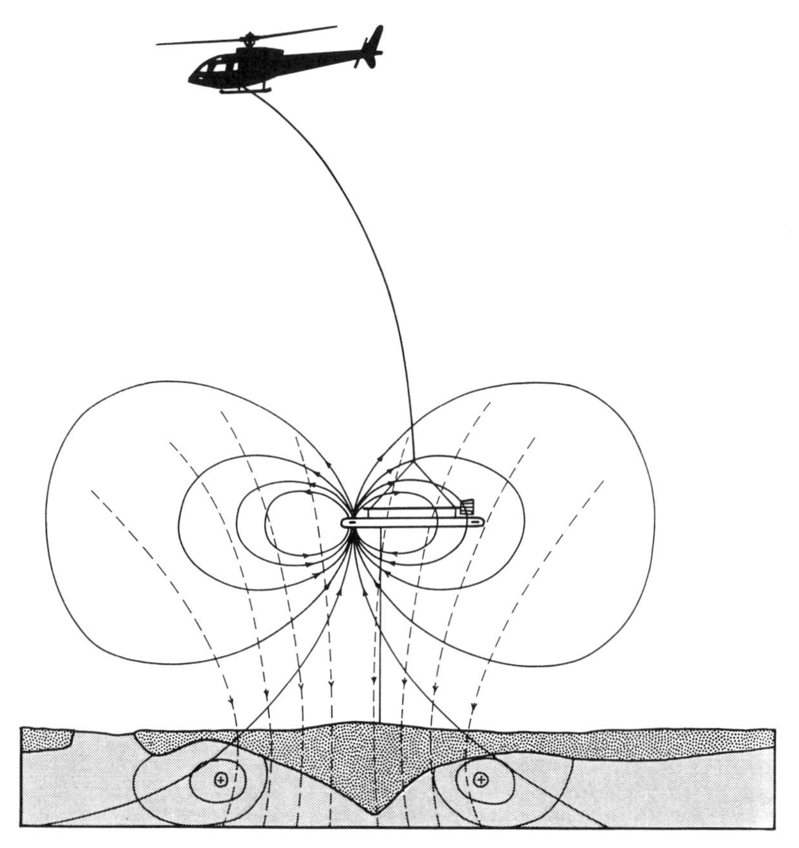

Figure 4.1 Sketch of Ice Probe and EM-induction ice-thickness measurement concept.

it is typically several orders of magnitude more conductive than the ice, as discussed in Chapter 2. By measuring the amplitude and phase of the secondary field (in parts per million of the primary field at the receiver coil location), the distance from the bird to the water can be estimated. A laser profilometer, also mounted in the bird, is used to measure the distance from the bird to the top of the snow and/or ice. The difference between the two measurements gives the thickness of the snow plus ice layer.

The approximate area of ice illuminated, or "footprint," of the EM system, although it depends somewhat on coil orientation (see below), has a diameter similar to the bird height; i.e., the profilometer's laser spot size is much smaller. Therefore, when lateral changes in ice thickness are rapid compared to the flying height, such as occur near ridges, the peak ice thickness tends to be underestimated.

4.2.1 Theoretical Background

Many of the topics in this book deal with the ice measurement or imaging at frequencies exceeding 100 MHz. At these frequencies, the EM fields propagate from the transmitter to the scatterer and back to the receiver according to the EM wave equation, which is a special case of the Helmholtz equations for the EM field

$$\begin{aligned} \nabla^2 \boldsymbol{E} + (\omega^2 \mu \epsilon - j\omega\mu\sigma)\boldsymbol{E} &= 0 \\ \nabla^2 \boldsymbol{H} + (\omega^2 \mu \epsilon - j\omega\mu\sigma)\boldsymbol{H} &= 0 \end{aligned} \tag{4-1}$$

where

$\boldsymbol{E}, \boldsymbol{H}$ = time-varying electric and magnetic fields
ω = angular frequency $2\pi f$
σ = electrical conductivity
μ = magnetic permeability (typically $\mu = \mu_0$, the permeability of free space)
ϵ = electrical permittivity
$j = \sqrt{-1}$

Under the wave approximation, the term $\omega^2\mu\epsilon$ is much greater than $\omega\mu\sigma$. In contrast, the EM-induction method exploits the properties of the EM-diffusion equation, which is also a special case of (4-1). In this case, $\omega\mu\sigma$ is dominant over $\omega^2\mu\epsilon$. The assumption that $\omega\mu\sigma >> \omega^2\mu\epsilon$ (typically valid at frequencies below 500 kHz) is known as the quasistatic approximation. Note that for conditions under which the quasistatic approximation holds, the conductivity σ also becomes essentially constant at σ_{DC}.

Mathematical models which predict the response of EM-induction sensors to nearby conductors have been constructed for a variety of model geometries.

A particularly useful model for airborne mapping purposes is the layered half-space model, which computes the response of a model earth, consisting of a series of plane layers of known thickness and conductivity, to an EM sensor at a given distance above the top layer. This model is well suited to the interpretation of EM-induction sounding data over floating sea ice. In such a model, the ice often can be approximated as one or more horizontal layers of electrically resistive material, overlying a good conductor which models seawater.

The EM-induction sensor used for sea-ice-thickness sounding is based on the helicopter-borne electromagnetic (HEM) method, which is described in detail in Palacky and West [52]. In its most basic form, the HEM system consists of three coil-type antennas: the transmitter, receiver, and bucking antennas. The transmitter, which is normally modeled as a magnetic dipole, generates a quasistatic primary magnetic field which induces eddy currents in nearby conductors. These eddy currents, which in this application are dominated by current flowing in seawater beneath the transmitter antenna, give rise to a secondary magnetic field that is detected by the receiver antenna, centered at a distance l from the center of the transmitter. The bucking antenna, which is often located at the midway point between the transmitter and receiver antennas, is used to cancel the intense primary field sensed at the receiver antenna so that the much weaker secondary fields may be measured precisely. The antenna array is normally housed in a rigid Kevlar tube, which ensures that the relative positions of the antennas remain constant: any relative motion generates noise in the EM measurement.

EM-induction systems typically measure the response of nearby conductors in the form of the secondary coupling ratio, which is the ratio of the secondary magnetic field strength observed at the receiver antenna (Rx) to the transmitted field strength measured at the same location. For a layered half-space in cylindrical coordinates centered on the transmitter antenna (Tx) location and in the wavenumber domain, the coupling ratio can be specified using the following integrals (Anderson [1]):

$$T_0(l, \boldsymbol{p}, f) = \int_0^\infty R_0(\boldsymbol{p}, f, \lambda) e^{-\lambda(h_1 + h_2)} J_0(\lambda l) \lambda^2 d\lambda$$

$$T_1(l, \boldsymbol{p}, f) = \int_0^\infty R_0(\boldsymbol{p}, f, \lambda) e^{-\lambda(h_1 + h_2)} J_1(\lambda l) \lambda^2 d\lambda$$

$$T_2(l, \boldsymbol{p}, f) = \int_0^\infty R_0(\boldsymbol{p}, f, \lambda) e^{-\lambda(h_1 + h_2)} J_1(\lambda l) \lambda d\lambda$$

where

$\boldsymbol{p}$ = vector of model parameters (layer conductivities, σ_i, and thicknesses, t_i)
f = frequency
l = transmitter-receiver antenna separation

λ = spatial wavenumber
J_0, J_1 = zero and first-order Bessel function, respectively
h_1, h_2 = height of transmitter and receiver antenna above surface
R_0 = $R_{0,n}$, a complex reflection coefficient, defined recursively (Koefoed [27]) for an n-layered model as

$$R_{i-1,n}(l, p, \lambda) = \frac{V_{i-1,i} + R_{i,n}e^{-2t_i\nu_i}}{1 + V_{i-1,i}R_{i,n}e^{-2t_i\nu_i}}$$

where

$$(R_{n,n} = 0)$$

$$\nu_i = \sqrt{\lambda^2 + k_i^2}$$

$$k_i = \sqrt{-j2\pi f\mu\sigma_i}$$

$$V_{i,k} = (\nu_i - \nu_k)/(\nu_i + \nu_k)$$

Coupling ratios for two typical antenna orientations, known as the horizontal coplanar and coaxial modes, are given by

$$\left(\frac{Z}{Z_0}\right)_{HCOP} = l^3T_0$$

$$\left(\frac{Z}{Z_0}\right)_{COAX} = \frac{l^2(T_2 - lT_0)}{2}$$

respectively. The Hankel transform integrals, T_0, T_1, and T_2, can be evaluated by a variety of means but to date the most suitable in terms of processing speed and flexibility has proven to be the Gauss-Laguerre method, a standard technique in numerical analysis (Conte and de Boor [12]).

The footprint size depends on the coil orientation. This issue has been examined in detail by Kovacs et al. [38]. They found, through theoretical and experimental analyses, the apparent footprint diameter to be 1.25 and 3.75 times the antenna height above the ice surface for coaxial and coplanar antenna orientations, respectively. Therefore, the coaxial coil arrangement is better in principle for assessment of two-dimensional structures; however, the signal-to-noise for this configuration is reduced by about a factor of four. As system drift and noise improvements are achieved, the coaxial configuration may become more useful.

4.2.2 EM Data Processing and Interpretation

Drift Removal and Calibration

There are three stages in the first phase of data reduction: transcription of the field data to a processing computer (data volumes range up to 12 Mbytes/hr at present); removal of residual baseline drift; and calibration of the data. The data-transcription operation is relatively straightforward and leaves the data in a format in which they can be manipulated easily in later stages of reduction.

Precise correction for instrument drift now requires that the sensor package be flown to high altitude (from 120 to 500 m, depending on the system characteristics) every 30 min or so during data acquisition to remove the sensor package from the effect of the highly conductive seawater. Correction of the variation observed between these high-altitude baseline measurements provides a reliable basis for removal of small amounts of baseline drift in the system, based on the assumption of linear drift between the high-altitude measurements. As EM-system drift has become smaller and more linear with equipment improvements, the permissible baseline-monitoring interval has become progressively longer, increasing survey efficiency. Furthermore, what drift remains has become more predictable, which permits estimation and approximate compensation for system drift in real time.

Once the data have been drift-corrected, they can be calibrated. Until 1990, all systems depended on calibration measurements made with an external tuned coil and a magnetically permeable, nonconductive rod to establish their sensitivity and phase characteristics. As these calibration techniques were found to be unreliable for quantitative measurements of this type, a new approach was developed. In this approach, the system is flown over deep, open water (or flat, lead ice over deep water) having a known conductivity-depth profile to well beyond the depth penetration of the EM system: an accurate calculation of the expected response is then compared with the observed response, and correction factors are calculated which are applied to bring the measured response into line with the calculated response of the known ice-water structure. This technique has become known as "surface validation calibration" and it has proved to be quite satisfactory under a variety of conditions. Recent instrumentation advances have introduced a stable, accurate internal calibration method, which means that surface validation calibration has become more of a verification tool than a universally applied calibration technique.

Inversion of Survey Data for Ice Thickness

Two methodologies are used for estimation of snow-plus-ice-thickness profiles from the airborne data. The original approach was to postprocess the data set after the flight, taking it through a series of data reduction and interpretation operations. This approach, while somewhat time-consuming, still yields the most accurate results. A more recent innovation has been the introduction of real-time processing, in which the data are reduced and fed to interpretation software in the helicopter. This approach provides thickness data within seconds of flying over an area of interest, and is well suited to operational use of the technology. The real-time interpretation results are also stored on disk and can be used to plot an ice-thickness map within 30 min of landing. The real-time approach uses the same processing steps as postprocessing, with the exception of drift removal which is replaced by the removal of a drift estimate predicted from the earlier behavior of the system.

The computation of the bird to seawater distance is performed from the measured EM response. The drift-corrected, calibrated data, which consist of the secondary field response both in-phase and out-of-phase with the transmitted or primary field, vary according to the altitude of the bird (and, to a small extent, upon its attitude) above the surface of the seawater underlying the ice. The response is not a simple function of altitude, but is affected in a nonlinear fashion by the conductivity of the seawater (at low frequencies) and of the ice (at high frequencies). Inverse models which directly calculate ice properties from measured data (i.e., without iteration) do not yet exist. However, forward computer models which can predict the EM response over a given layered structure, such as an ice–water model, have existed for many years (e.g., Anderson [1]).

A number of relatively complex methods, all based on forward numerical modeling of the EM response, have been developed to invert the observed responses approximately to yield the desired bird to seawater distance (and perhaps other parameters of interest, such as ice conductivity). So far, the most robust of these procedures uses the damped singular value method of nonlinear regression to seek iteratively the set of model parameters (layer thicknesses and conductivities) which minimizes (in a least-squares sense) the variance between the observed data and the response of a computed model. At each iteration, the misfit from the previous iteration is used, together with a predictor or Jacobian matrix, to select a set of changes in the model parameters that will reduce the model variance. The iteration ends when no significant improvement can be made in the variance or the model response fits the data to within an acceptable tolerance.

Consider a general, layered half-space response, $\boldsymbol{c}$, a known function of the model parameter vector, $\boldsymbol{p}$ (the parameters sought during inversion), and the known system parameter vector, $\boldsymbol{s}$ (which includes l, f, the average bird altitude, bird pitch and roll, etc.):

$$\boldsymbol{c} = \boldsymbol{c}(\boldsymbol{p}, \boldsymbol{s})$$

where, for example, $\boldsymbol{c}$ could be the horizontal coplanar response of the model. $\boldsymbol{c}$ is a nonlinear function, but can be decomposed into a Taylor series expansion in the model parameters $\boldsymbol{p}$ about a trial model $\boldsymbol{p}_0$:

$$\boldsymbol{c}(\boldsymbol{p} - \boldsymbol{p}_0) = \boldsymbol{c}(\boldsymbol{p}_0) + \frac{\partial \boldsymbol{c}(\boldsymbol{p}_0)}{\partial \boldsymbol{p}} \cdot (\boldsymbol{p} - \boldsymbol{p}_0) + \frac{\partial^2 \boldsymbol{c}(\boldsymbol{p}_0)}{\partial^2 \boldsymbol{p}} \cdot (\boldsymbol{p} - \boldsymbol{p}_0)^2 + \cdots$$

If the "true model" parameter vector, $\boldsymbol{p}_{\text{true}}$, which most closely represents the geoelectric properties of the ice-seawater system under observation, is "close" to $\boldsymbol{p}_0$, the Taylor series may be approximated by truncation to the first-order term:

$$\boldsymbol{c}(\boldsymbol{p} - \boldsymbol{p}_0) \simeq \boldsymbol{c}(\boldsymbol{p}_0) + \frac{\partial \boldsymbol{c}(\boldsymbol{p}_0)}{\partial \boldsymbol{p}} \cdot (\boldsymbol{p} - \boldsymbol{p}_0)$$

If the observed response of the EM system is $\boldsymbol{o}$, the residual response, $\Delta \boldsymbol{c}_0$, is the difference between the observed and calculated (for $\boldsymbol{p}_0$) responses,

$$\Delta \boldsymbol{c}_0 = \boldsymbol{o} - \boldsymbol{c}(\boldsymbol{p}_0)$$

The truncated Taylor series can, therefore, be represented as

$$\boldsymbol{A}\Delta \boldsymbol{p} = \Delta \boldsymbol{c}_0$$

where $\Delta \boldsymbol{p}$ represents a correction to the parameter vector, $\boldsymbol{p}$, and the matrix, $\boldsymbol{A}$, is given by

$$\boldsymbol{A} = \frac{\partial \boldsymbol{c}_i}{\partial \boldsymbol{p}_j} = \frac{\partial \boldsymbol{c}(\boldsymbol{p}_0)}{\partial \boldsymbol{p}}$$

The Jacobian matrix, $\boldsymbol{A}$, is both nonsquare and nonnegative definite, and thus cannot be guaranteed to be invertible, even in a theoretical sense. To construct an approximate matrix inverse while facilitating statistical analysis of the inversion process, the singular value decomposition (SVD) (Lanczos [41]) is used. This decomposition is expressed as

$$\boldsymbol{J} = \boldsymbol{U}\boldsymbol{\Lambda}\boldsymbol{V}^t$$

where

$\boldsymbol{U}$ = "data" eigenvector matrix
$\boldsymbol{V}$ = "parameter" eigenvector (not to be confused with $V_{i,j}$ discussed earlier)
$\boldsymbol{\Lambda}$ = diagonal matrix of singular values

The approximate inverse, $\boldsymbol{A}_d^{-1}$, is constructed by discarding rows and columns of $\boldsymbol{U}$ and $\boldsymbol{V}$ corresponding to zero singular values and damping as required to reduce the condition number of the resulting damped Jacobian $\boldsymbol{A}_d = \boldsymbol{U}\boldsymbol{\Lambda}\boldsymbol{V}$ to an acceptable level, then inverting the decomposed $\boldsymbol{A}_d$ matrix:

$$\boldsymbol{A}_d^{-1} \simeq \boldsymbol{V}\boldsymbol{\Lambda}^{-1}\boldsymbol{U}^t$$

An approximation to the parameter vector correction required to minimize the residual vector, $\Delta \boldsymbol{c}_0$, in a least-squares sense is given by $\Delta \boldsymbol{p}_0$:

$$\Delta \boldsymbol{p}_0 = \boldsymbol{A}_d^{-1}\Delta \boldsymbol{c}_0$$

which is used to correct the parameter vector, $\boldsymbol{p}_0$, to give the next approximation, $\boldsymbol{p}_1$. A new forward model, $\boldsymbol{c}_1$, is then computed, its residual, $\Delta \boldsymbol{c}_1$, formed, and so on. This iterative search procedure continues until the magnitude of the residual vector falls within an acceptable tolerance or fails to converge further. A few iterations of this algorithm usually provide estimates of the ice thickness that are accurate to within the range of accuracy of the laser altimeter for relatively level ice.

There are other approaches to the inversion problem for level ice: look-up table methods (Kovacs et al. [31]) permit rapid estimates of ice thickness and seawater conductivity, as does a method due to Bergeron [4], which uses complex image theory to approximate and invert EM-induction responses. Both of these approaches perform well over sea ice at moderate (10–30 kHz) EM frequencies, but tend to de-emphasize the relief of ice structures, such as ridge keels, even more than the SVD approach described above, and do not provide estimates of ice conductivity. It is also possible to use neural networks to estimate ice thickness and conductivity (Holladay et al. [22]), although this approach has not been applied to routine interpretation as yet.

Maximum thickness estimates for ice that has large lateral thickness variations over short horizontal distances (slopes $> 10°$) tend to be underestimated by this method: more suitable approaches have been developed to deal with these situations by Liu and Becker [45] and Liu et al. [46]. They describe two approaches to the problem; one, based on interpretation nomograms constructed with the help of forward modeling, and the other, an elaborate (and very slow) inverse modeling procedure based on the same forward model. A third technique, also based on forward modeling of ice structures such as ridges, uses neural networks to perform the two-dimensional inversion operation. This approach has been implemented successfully (Holladay et al. [22]) and shows promise as the optimal approach for fast routine interpretation.

4.2.3 Equipment Description

EM and Laser Subsystems

Prototype EM systems designed primarily for ice measurement have been described by Holladay et al. [21, 22], Kovacs and Holladay [32, 33], Rossiter and Lalumiere [58], and Kovacs et al. [31]. The sensor packages used in all field tests to date have included three principal elements: a multifrequency, helicopter-towed EM-sounding system; a laser profilometer; pitch and roll sensors mounted in the towed bird; and a downward-looking video camera mounted in the helicopter. The EM system included an EM signal processor unit and a data acquisition/logging unit mounted inside the helicopter. In survey work prior to 1990, all data were sampled at 0.1-s intervals before digital recording on magnetic tape: since that time, data rates have been increased to 20, and even 50, samples per second, and data have been recorded on removable-media disks. The video imaging system, mounted on the helicopter, monitored ice conditions below the helicopter, and the image was annotated with time and other information before being recorded on videotape using a videocassette recorder.

Typically, small single-engine helicopters were used for surveying over fast ice. Safety and range considerations mandated the use of a much larger, twin-engine helicopter for offshore work during the 1989 Labrador Ice Margin Experiment (LIMEX'89), whereas a smaller twin-engine helicopter was used for Canadian Coast Guard operational trials in February to April 1993.

Birds designed for ice sensing are typically about 4-m long and weigh about 125 kg, whereas for conventional HEM systems the birds are about 7-m long and weigh about 250 kg. Ice-measurement systems also operate at higher frequencies in order to sense ice conductivity more accurately: a Cold Regions Research and Engineering Laboratory (CRREL) fixed-frequency system has a top frequency of 50 kHz; a fixed-frequency sensor developed for Transport Canada operates up to 100 kHz; and the wideband CRREL system described in Kovacs and Holladay [33] can operate at up to 250 kHz.

System noise levels (expressed in terms of ice thickness error at 20-m bird altitude) are typically on the order of ± 0.1 m under normal flat ice survey conditions. Baseline drift in the EM measurements typically limits continuous data acquisition to 10- to 20-min segments, which in existing systems must be separated by a short (10 s) period of data acquisition at bird altitudes of 150 m or higher. Fortunately, drift rates have been reduced substantially in recent years so that longer periods of continuous profiling are now becoming possible. It is also possible to obtain estimates of surface roughness by high-pass filtering the laser profilometer results, although more work remains before continuous reconstruction of surface relief becomes practical.

Radar Snow-Thickness Subsystem

One of the parameters of interest both to mariners and scientists is the thickness of snow overlying the ice. Snow can offer substantial frictional resistance to a hull moving through ice. It can also provide a substantial thermal insulating layer over ice and can affect other remote sensing measurements (Chapter 2).

As described above, the EM-laser subsystem can obtain only the ice-plus-snow thickness, rather than ice and snow individually. To overcome this problem, a GPR subsystem was added (Holladay et al. [22]). The GPR echoes from both the top and the bottom of the snow can be used to estimate snow thickness by measuring the travel time in the snow layer.

The speed of the radar signal in dry snow varies from about 0.30 to 0.15 m/ns, depending primarily on the snow density (Evans [16]). Moisture in the snow decreases the signal speed somewhat, to values from 0.20 to 0.13 m/ns. As the density and moisture content may not be known exactly, the accuracy of this approach will be limited but still it offers a significant improvement over the EM-laser equipment alone.

The GPR system consists of a Geophysical Survey Systems, Inc. (GSSI) 500-MHz center-frequency antenna mounted in the bird and a separate control unit and recorder in the cabin. The pulse length used is approximately 3 ns, giving a minimum snow thickness that can be measured of about 0.2 m. The sampling rate is approximately 13 Hz. Data are recorded on tape or disk and processed off-line. A prototype automated algorithm has been written to detect the top and bottom snow echoes and to measure the travel time between them.

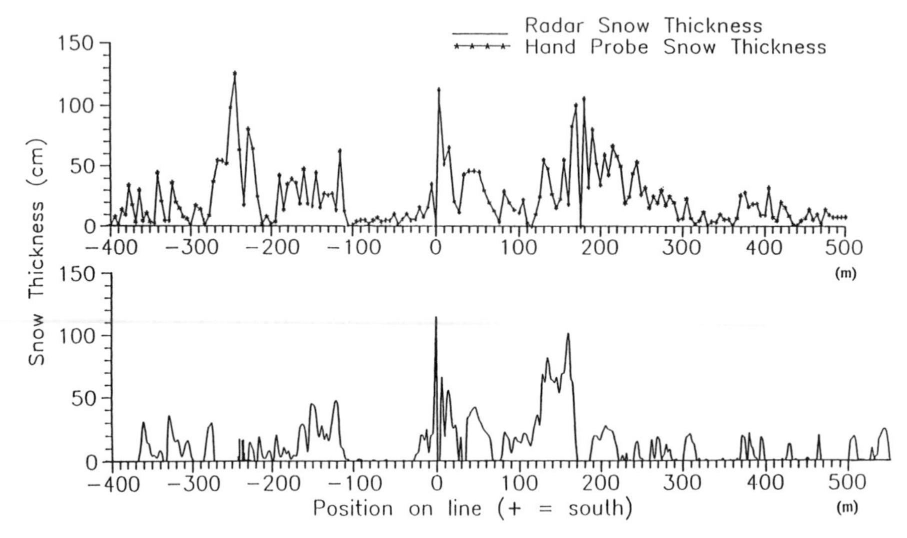

Figure 4.2 Snow-thickness measurements at Beaufort Sea site, April 1991. Upper panel—on-ice observations; lower panel—radar results (after Prinsenberg et al., [54]).

Tests have been carried out both in the Arctic and off Newfoundland (Prinsenberg et al. [53, 54]; Lalumiere [39]; Lalumiere et al. [40]). Arctic results are shown in Fig. 4.2, compared to surface measurements over the same line. For the east coast results (given in Fig. 4.3), measurements were made before and after a heavy rainfall. Acceptable agreement was achieved in both situations, although the radar signal speed used was decreased to allow for moisture in the snow. Other differences seem to be based on the wider area illuminated by the GPR dipole antenna, which effectively averages over a diameter of several meters. The presence of seawater spray does not seem to have affected attenuation through the snow seriously, but a subsurface slush layer does. In the 1993 survey, there was a 6-cm slush layer which reflected the radar signal and caused the observed snow depth to be less than the measured depth.

4.2.4 Experimental Results

A variety of experimental, airborne ice-measurement surveys have been conducted since 1985, using progressively more specialized equipment. The results of eight of these have been published to date: (a) four CRREL trials near Prudhoe Bay, Alaska, in 1985, 1987, 1990, and 1991 (Kovacs et al. [31]; Kovacs and Holladay [32, 33]); (b) trials off Newfoundland in 1989 as part of LIMEX'89 (Rossiter et al. [56]; Holladay et al. [21]); (c) a further series of trials near Tuktoyaktuk, Northwest Territories, in 1991 (Prinsenberg et al. [53]) and St. Anthony, Newfoundland, in 1992 (Holladay et al. [22, 23], Prinsenberg et al. [54]); and (d) a set of operational trials conducted by the Canadian Coast Guard near Charlottetown, Prince Edward Island, and from the icebreaker

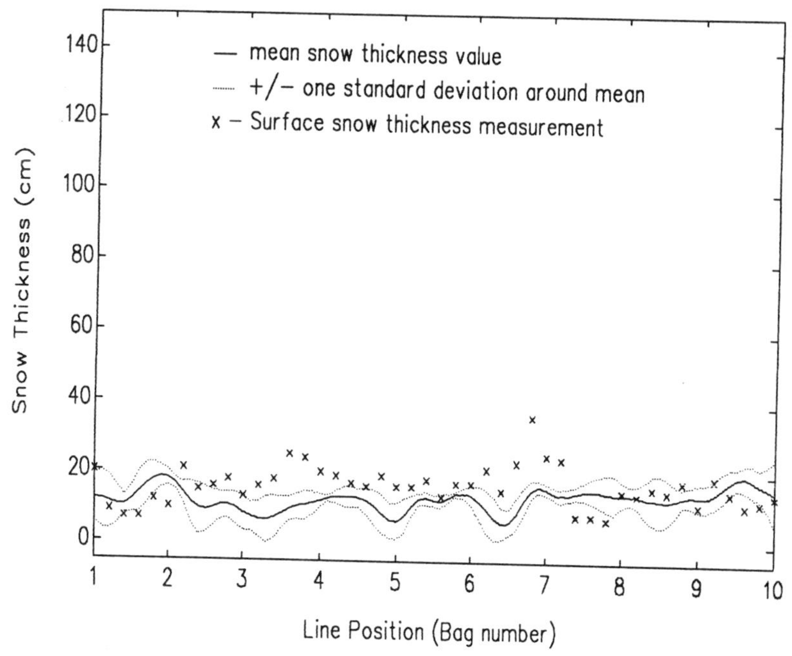

Figure 4.3 Snow-thickness measurements off Newfoundland, March 1992. Radar data are taken from five passes over the same line (after Lalumiere et al. [40]).

CCGS *Henry Larsen* in Northumberland Strait in February–April 1993 (Holladay [24]).

The 1985 CRREL work was a feasibility study which used a modified mineral-prospecting HEM system, whereas the 1987 CRREL trial was intended as a detailed assessment of the first dedicated, airborne EM ice-measurement sensor to be constructed. The 1990 and 1991 CRREL trials utilized a more sophisticated EM sensor, operating over a wider frequency range. The LIMEX'89 measurements also used a modified mineral-prospecting system and they were intended primarily to evaluate system performance over the Labrador Sea (marginal zone) ice. The Canadian trials of 1991–1993 tested a specialized sensor, known as Ice Probe, and were directed toward system validation, and eventual operational deployment, abroad an icebreaker.

During the 1987 Prudhoe Bay survey, data over marked grids on three old ice floes, a first-year pressure ridge, and a variety of extended survey lines were obtained with the EM system. Validation measurements consisted mainly of repeated passes over a set of clearly marked and densely drilled (5-m grid) surface sites on first-year and old ice floes near Prudhoe Bay. The snow and ice thicknesses were combined and averaged over swaths along the axis of each grid to generate an average snow-plus-ice thickness for comparison to the airborne results. An example from a multiyear floe is presented in Fig. 4.4: it includes the averaged snow-plus-ice thickness and four EM ice-thickness pro-

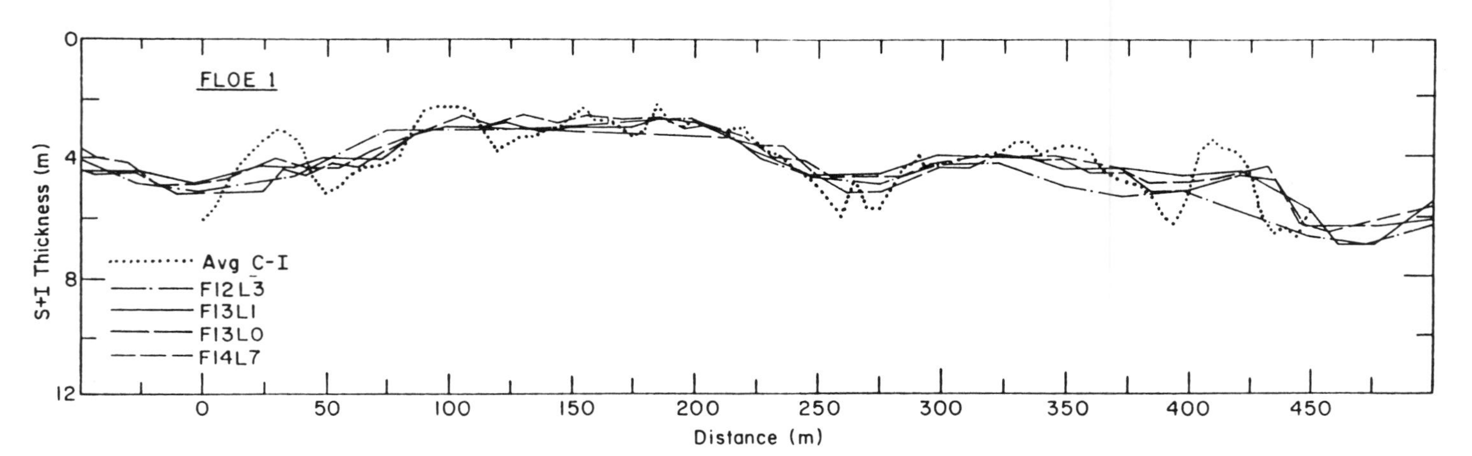

Figure 4.4 Ice-thickness measurements, Prudhoe Bay, 1987. Dotted line is average of snow plus ice-thickness measurements made by auger over a 30-m wide corridor of a multiyear floe. Other traces are various EM passes (after Kovacs and Morey [32].

files, as indicated in the legend. It is apparent that the EM-derived profiles follow the long-period thickness trends but not the short-period undulations. This result was expected due to the effective footprint of the EM system, which was on the order of 30–50 m. The average EM-derived floe thickness and the average borehole snow-plus-ice thickness differ by less than 10%.

The 1990 and 1991 Prudhoe Bay trials tested a new type of EM-induction sensor which used a simple set of transmit-bucking-receiving antennas but operated at multiple frequencies simultaneously. This was achieved by transmission of a square wave and Fourier analysis of the received waveform. Basic system operation was demonstrated in 1990, although postprocessing was required to obtain accurate ice-thickness profiles. The 1991 trials demonstrated accurate, real-time data inversion for ice thickness and yielded quantitative EM estimates of sea-ice conductivity. These conductivity measurements clearly, and repetitively, distinguished first-year sea ice from old ice for the first time.

The LIMEX'89 ice-thickness survey was made at two locations. The first effort was concentrated on fast ice in the Bay of Exploits, southwest of Twillingate, Newfoundland, and provided a check on system performance and further validation of the method, against surface measurements. A total of 16 augered holes were measured, along with conductivity profiles in the water to 2.25-m depth. Ice thickness varied from 38 to 59 cm, snow cover thickness varied from 0 to 21 cm; the water immediately under the ice had a conductivity of about 0.5 S/m, increasing sharply to 2.5 S/m at a depth of about 1.0 m.

Interpretation of the Bay of Exploits EM data was based on a two-layer model for the water, used to account for the brackish water layer under the ice. The agreement with auger measurements was well within ± 0.1 m (see Fig. 4.5). The EM results suggest a gradual thinning of the fast ice from 0.5 m to 0 m, as the line progressed offshore. Water depth was typically greater than 75 m, so that bias in the ice-thickness measurement due to finite water depths should be negligible.

The second EM sounding survey was along a 500-km flight from St. John's out to the M/V *Terra Nordica* (the LIMEX'89 project ship) and back. EM-sounding measurements were made over most of the flight line; i.e., about 300 km. Ice encountered varied from strips and patches of brash ice, to small cakes of first-year ice, to small floes of first-year ice around the ship. There were some rafted pieces of ice visible on the video flight record which were verified by others working on the ice. At the time of the flight, an ice motion sensor and an ARGOS beacon had been set up on the ice by other investigators and estimates of ice thickness were available at these two locations. These sites were clearly visible from the air.

A selected portion of data near the ship is shown in Fig. 4.6. The first-year pack ice was made up of many pans, typically 10 m in diameter and varying from 0.3 m to over 1.6 m in thickness. Ice salinity was typically 5 ppt and ice temperature around $-2°C$. Although there were few comparative points, data from two auger holes are shown (bars indicate range of auger measurements). The EM ice-thickness sounding results agree to within 0.05 m of augered ice

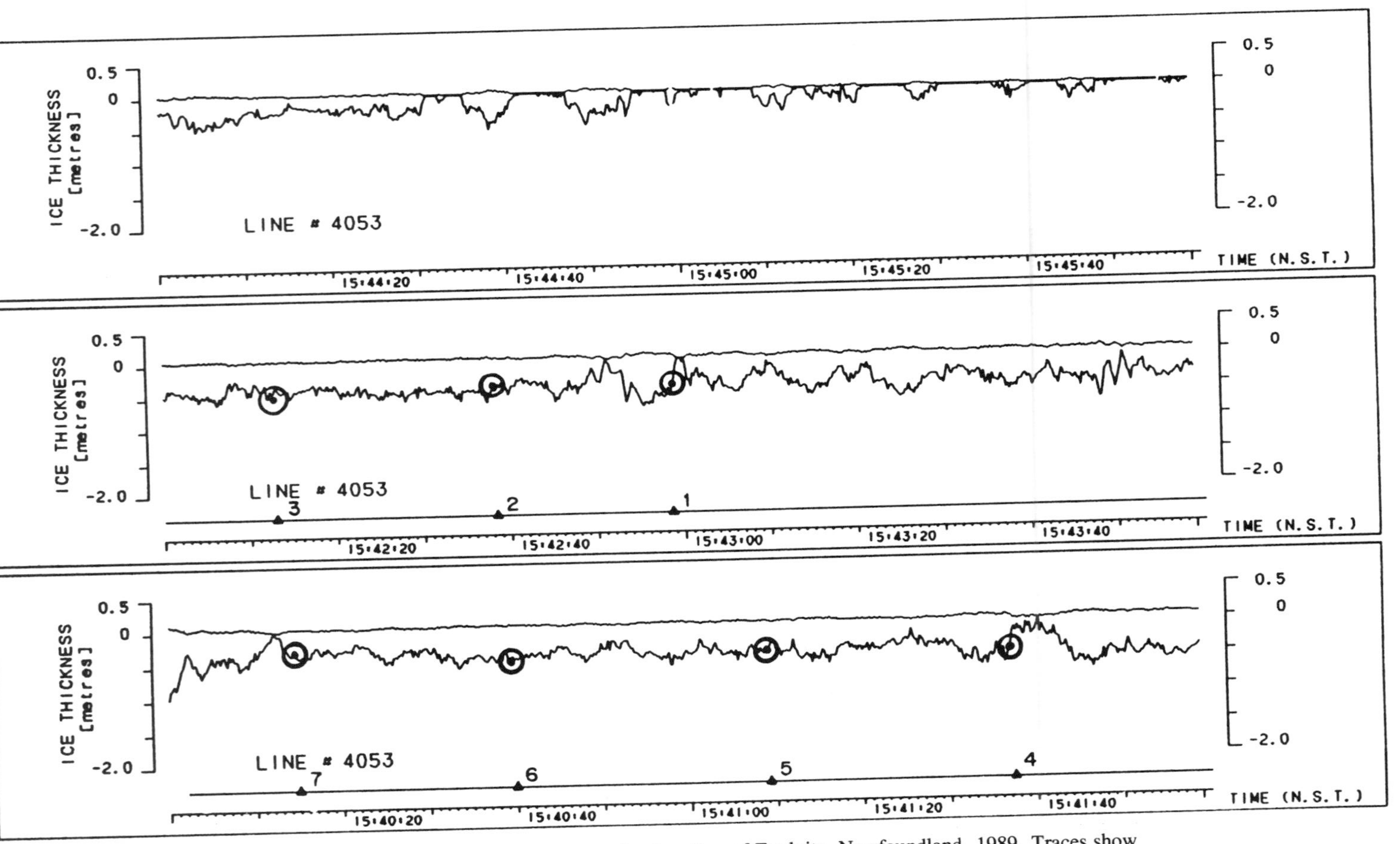

Figure 4.5 Ice-thickness measurements, fast ice, Bay of Exploits, Newfoundland, 1989. Traces show EM results; circles are auger measurements (after Holladay et al. [21]).

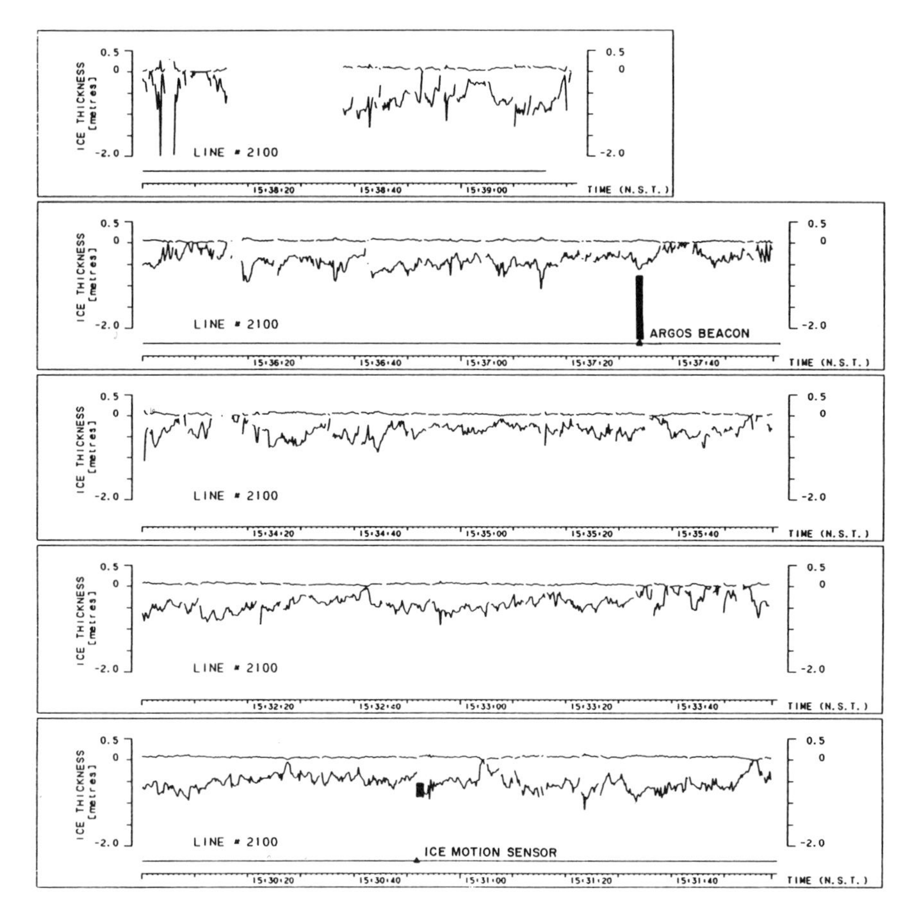

Figure 4.6 Ice-thickness measurements, first-year pack ice off Newfoundland, 1989. Traces show EM results; bars show range of auger estimates (after Holladay et al. [21]).

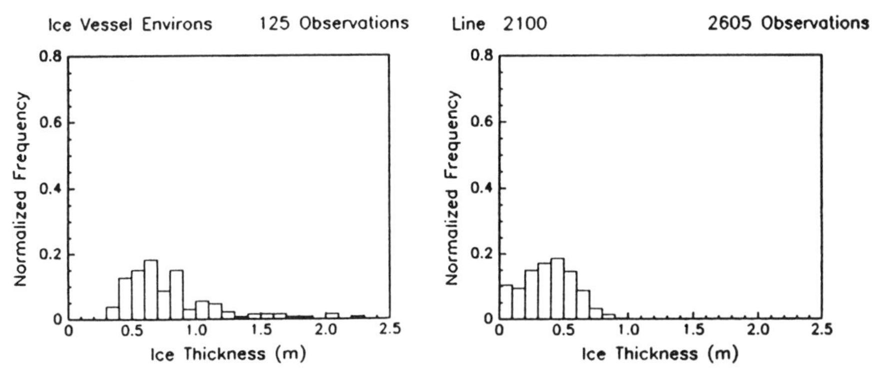

Figure 4.7 Histograms of ice thickness for first-year pack ice off Newfoundland, 1989, as measured by auger and using EM system for segment shown in Fig. 4.6 (after Holladay et al. [21]).

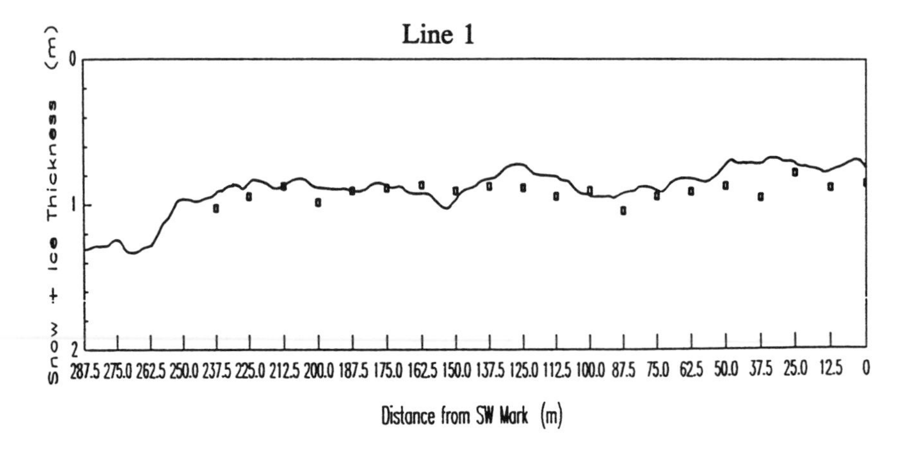

(a)

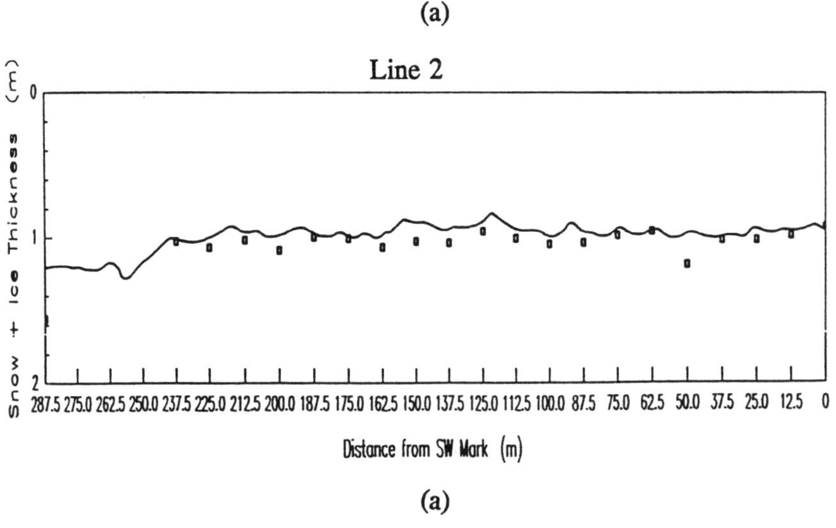

(a)

Figure 4.8 Snow plus ice thickness measurements over landfast ice, Newfoundland, 1992, measured with Ice Probe (continuous line) and auger (points). The thicker ice to the left is a zone of rafted ice (after Holladay et al. [22]). (a) 6 March results; (b) 13 March results, after 2 days of warm weather and rain. Note that the ice is approximately 0.13 m thinner than in (a).

thicknesses in the area of the ice motion sensor and to within 0.20 m at the ARGOS beacon (Fig. 4.6).

A histogram of ice thickness based on 125 "random" auger hole measurements made by investigators in the same area is shown in Fig. 4.7, together with the histogram computed from EM results along a line approximately 10 km long of the airborne survey (Fig. 4.6). The differences point out the effects of local variability in the ice at this site, since the surface measurements were not made along the EM line. However, the results also suggest two difficulties in comparing airborne and auger results: (a) surface measurements

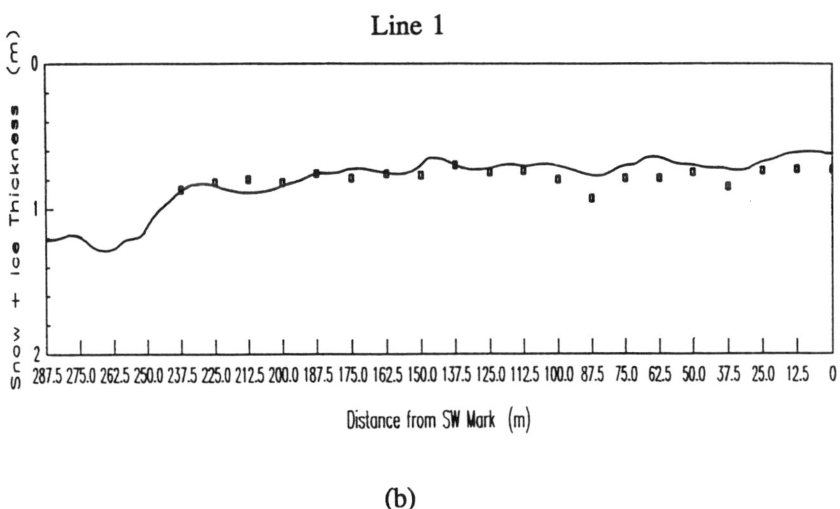

(b)

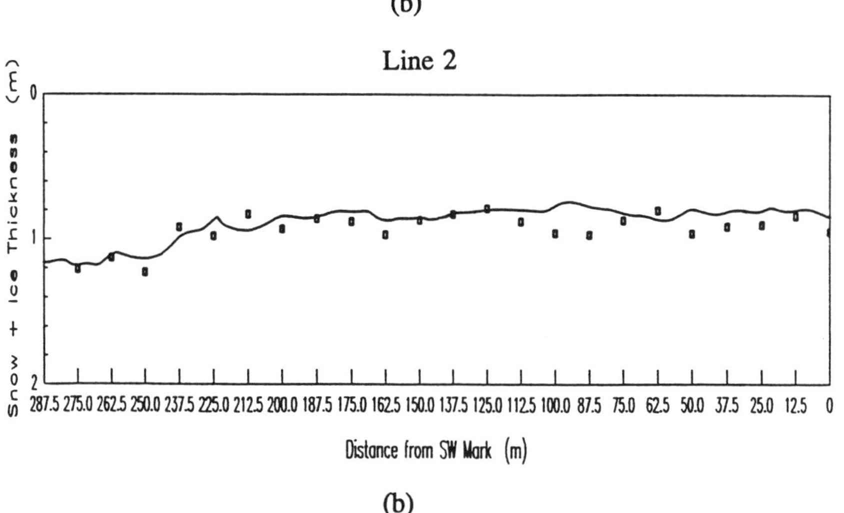

(b)

Figure 4.8 *(Continued)*

are taken only at specific points, whereas the EM system averages over an area of approximately 10^2–10^3 m^2; and (b) surface measurements are biased inevitably against thin ice (too thin to stand on) and/or thick ice (beyond reach of the auger stem available). These differences represent one of the important advantages of remote ice-thickness measurement for scientific applications, namely that more representative ice-thickness distributions can be obtained.

During the March 1992 trials of Ice Probe near St. Anthony, Newfoundland, located at the tip of the northwestern peninsula of Newfoundland, a series of validation flights was undertaken, as well as three data-collection missions about 200-km long, traversing pack ice, patches of thin ice, and fast ice near shore. Ice and snow thicknesses, ice salinity, and seawater conductivity were measured directly by auger, snow probe, and conductivity meter along a set of marked survey lines, which were then profiled repeatedly with the airborne system.

Figure 4.8(a) is a composite of EM ice-thickness estimates for passes over two lines on fast ice near St. Anthony, performed on 6 March, along with snow-plus-ice thicknesses observed on the surface at augered sites. At the time of this flight, cold, very windy winter conditions prevailed. The thicker ice at the left side of the figure is a rafted ice zone. On 11–12 March, the temperature rose sharply and rain fell for several hours, decreasing the average snow thickness by 0.13 m but having negligible effect on the ice thickness. Figure 4.8(b) displays EM snow-plus-ice thickness estimates and surface measurements obtained on 13 March along the same lines. Comparison of the surface measurements with the EM-derived snow-plus-ice thickness estimates indicates that the EM results are well within the target accuracy level of ± 0.2 m: the mean and standard deviations for differences between the EM and surface measurements are -0.06 m and 0.07 m, respectively.

Excellent correspondence is seen between passes executed at different speeds and altitudes and on different days over the validation lines, demonstrating the independence of flight conditions and stability of system calibration which are essential to an operational sensor. It is also evident from a comparison of Figs. 4.8(a) and (b) that the system has faithfully reproduced the 0.13-m average reduction in snow-plus-ice thickness caused by the rainfall and melting conditions of 11–12 March. The rafted section at the eastern end of the E–W lines is visible clearly on the profiles and its estimated thickness matches the 13 March surface measurements accurately.

The 1993 operational trials used a Canadian Coast Guard helicopter and pilots, and were intended to assess the utility of the system for routine reconnaissance at both the strategic and tactical levels. System accuracies were confirmed as better than ± 0.1 m during repeated passes over marked lines on fast ice in the approaches to Charlottetown harbor during a validation phase. Then the system was tested for operational effectiveness by performing a series of reconnaissance flights. Results from an icebreaker-based reconnaissance flight undertaken on 12 March 1993 are given in Fig. 4.9. Ice thicknesses were plotted in profile form along the flight-path trace, and the results are referenced to a latitude–longitude grid overlay. The results indicate unusually heavy ice for the Gulf of St. Lawrence, with ice thicknesses of up to 3 m in rafted areas and an average ice thickness of over 1 m being observed. These results were in agreement with visual estimates made by the Ice Specialist on board.

To summarize, experimental results obtained since the first survey work in 1985 have consistently indicated that the EM-induced sounding method yields useful ice-thickness estimates. The accuracies obtained have improved in recent years, to the point where results accurate to better than ± 0.1 m are routinely obtained over all types of level sea ice. It is equally significant that the most

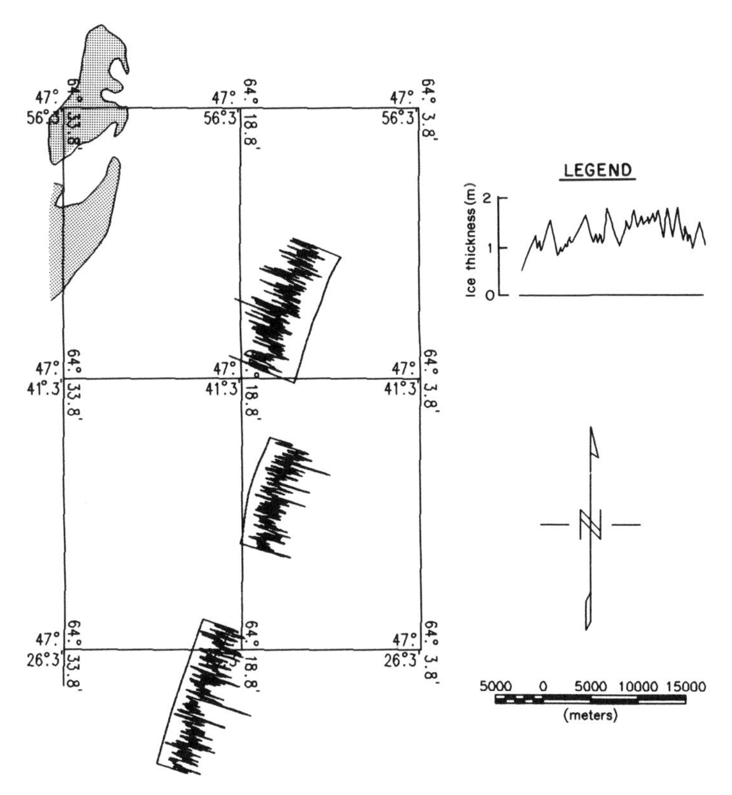

Figure 4.9 Ice-thickness profiles obtained using Ice Probe during an icebreaker-based reconnaissance flight in the Gulf of St. Lawrence, March 1993.

recent generation of systems are sufficiently reliable and easy to use that they can be operated effectively by non-specialist personnel.

4.2.5 Operational Implementation

Ice Probe consists of two principal components: a bird, which is suspended beneath a helicopter on a tow cable, and a helicopter package which includes an operator's console, a power-distribution unit, and a graphic recorder (see block diagram in Fig. 4.10). Two operating frequencies are used: a set of 2500 Hz coaxial coils and 94,000 Hz coplanar coils. The former provides better resolution of deformed ice and ridges; whereas, the latter provides higher resolution of thinner ice and estimates of ice conductivity. A video flight-path recording system, comprising camera, video annotation unit, monitor, and video cassette recorder, is also incorporated in this package.

The console itself has five subsystems: a main computer which interacts with the operator and controls the entire system; a secondary computer which

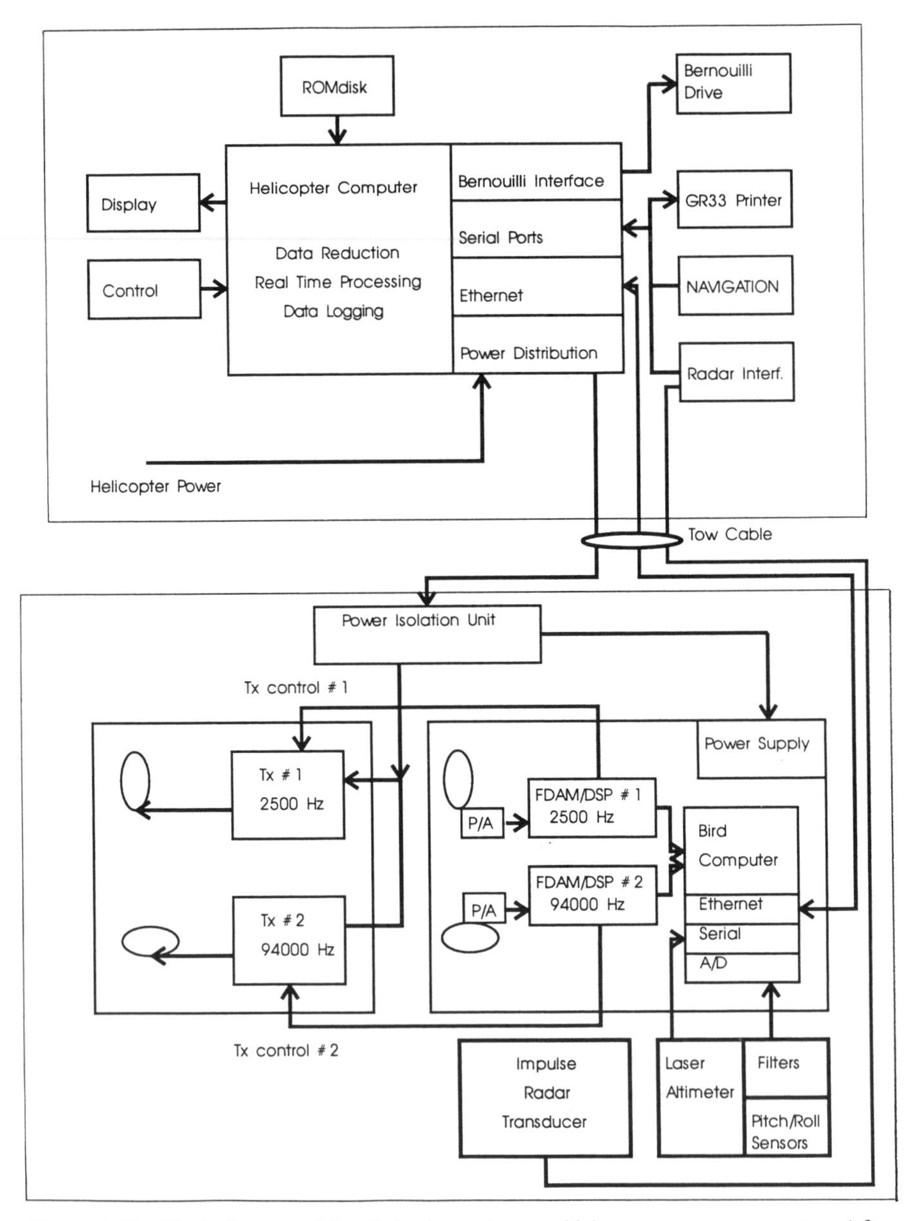

Figure 4.10 Block diagram of Ice Probe ice and snow thickness measurement system (after Holladay et al. [22]).

performs the inversion component of the real-time processing operation; an Ethernet interface for communication with the bird; an intelligent serial communications interface for control of, and communication with, the navigation unit, the laser altimeter display, and the graphic recorder; and a video overlay unit which annotates the video flight-path imagery with position, timing, and ice-thickness information. It should be noted that the video flight-path subsystem is principally required for validation and scientific purposes and would not necessarily form part of a production system. The weight of the system is about 80 kg for the helicopter electronics package and 125 kg for the bird. Photographs of the bird, the helicopter package, and the system in flight are shown in Figs. 4.11, 4.12, and 4.13, respectively.

A few special features make Ice Probe easier to work with while surveying and while operating over an icebreaker's flight deck: for instance, the bird-to-surface distance, as measured by the laser altimeter, is displayed for the pilot's guidance. In addition, the system has been equipped with a wheeled cradle into which the bird can be maneuvered during landings: once secured, it can be wheeled off the flight deck or landing pad to make room for the helicopter to land.

Another major component of Ice Probe is the system software. As the system incorporates many computers of different types, there are many software routines running simultaneously during normal operation. These include a Digital Signal Processor program used to perform the preliminary processing of the data in the bird; a more general program which controls all aspects of bird operation and communications, another program running on the secondary processor in the helicopter package which inverts the observed EM and laser altimeter data to yield the ice-thickness estimate; and the master program which

Figure 4.11 Photograph of Ice Probe sensor bird on an ice floe beside a Bell 206L helicopter.

Figure 4.12 Photograph of Ice Probe installation in a Bell 206L helicopter.

Figure 4.13 Photograph of Ice Probe bird in flight beneath a Bell 206L helicopter.

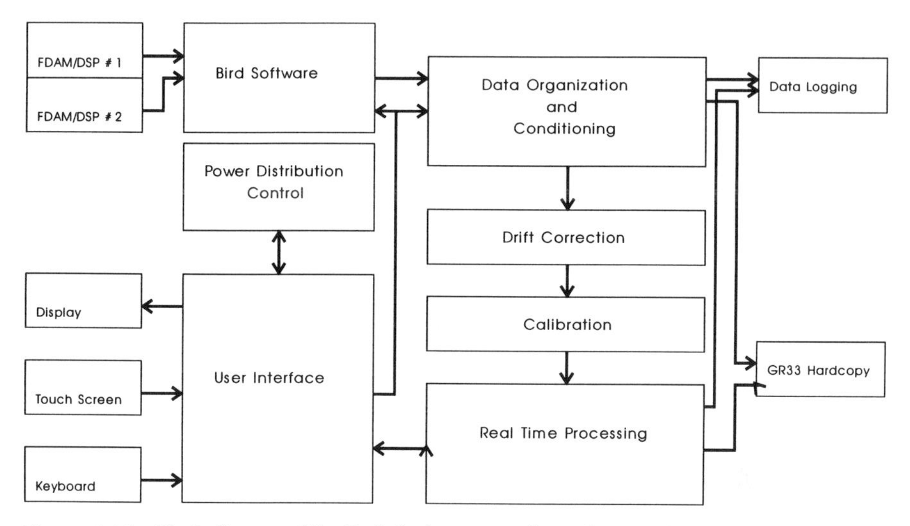

Figure 4.14 Block diagram of Ice Probe's data-processing software (after Holladay et al. [22]).

interacts with the operator, communicates with the bird and secondary processor, annotates the video, reads the incoming navigation data, and generates the hard copy record (see Fig. 4.14). The system generates ice-thickness estimates at 20 samples per second, although data acquisition rates of up to 40 samples per second or down to 10 samples per second have been used for various applications. At typical flight speeds of 60 to 80 knots (30 to 40 m/s), 20 samples per second corresponds to surface sampling distances of 1.5 to 2.0 m.

4.3 OTHER TECHNIQUES

4.3.1 Ice-Thickness Radars

Ground-penetrating radar (GPR), or impulse radar, has been used successfully for sounding many types of ice: freshwater ice of lakes (Kovacs [30]), rivers (Annan and Davis [22]), icebergs (Kovacs [29], Rossiter and Gustajtis [62], Rossiter et al. [61]), and ice shelves (Kovacs and Gow [36]); first-year ice (Campbell and Orange [7, 8]); old ice floes (Kovacs [28]); and a multiyear ridge (Kovacs [30]). Although many of the results were obtained on the ice surface, there have also been successful airborne measurements (Morey [48], Kovacs [28], Rossiter et al. [60, 64], Butt and Gamberg [6]). Successful radar sounding in the Soviet Union is reported by Bogorodskii and Tripol'nikov [5], Finkel'shteyn and Kutev [17], and by Chizhov et al. [10].

There have also been less successful reports. Penetration of ice at X-band frequencies (Iizuka et al. [26], Chudobiak et al. [11]) is limited by the very high absorption in sea ice at higher frequencies. Poor results have also been obtained over warm, porous pack ice of the Labrador Sea because of its high inherent attenuation (Goodman et al. [19], Rossiter et al. [64]). First-year pressure ridges pose problems because of brine pockets trapped between pieces of ice, which act as both attenuators and scatterers (Campbell and Orange [7]). Misleading results are obtained if there is a reflecting layer of brine between two pieces of rafted ice, so that the reflection comes not from the bottom of the ice but from the intermediate layer. It is not possible usually to sound sea ice with radar after it begins to melt in the spring because of the large reflection from surface meltwater which contains salts from the ice formation in the fall.

Most measurements in the Western world (to date) have used a commercially available GPR system (Morey [48]; Butt and Gamberg [6]) or a system built in England (Oswald [49, 50]). These instruments transmit broadband, near-monocycle pulses and most results have been obtained using a center frequency near 100 MHz. Data have also been collected at center frequencies of 80, 200, 400, and 625 MHz, with pulse lengths on the order of 2 to 20 ns. One of the practical advantages of these systems is that they display results as a continuous profile. As any individual echo can be of marginal quality, due to irregularities in ice surface and rapid changes in ice properties over short distances, it is essential to plot many sequential scans to obtain an interpretable record.

Airborne measurements suffer a number of degradations from surface sounding. First, the increased signal travel distance decreases the average signal power at the ice surface due to geometric spreading. Second, as there is a finite reflection from the ice surface, only a portion of the energy is transmitted into the ice. Third, there are scattering losses at a rough surface. Morey [48] has quantified the path loss considering these aspects, which limits the maximum flying height for any given system over a particular piece of ice. A fourth factor, and perhaps the most important, is the increased side-scatter echoes due to the large footprint illuminated.

Morey [48] reported flying 40 to 50 m above the ice surface but noticed degradation of the bottom echo above 40 m when flying over first-year ice near Tuktoyaktuk, NWT. Ice thickness was 1.5 to 2.0 m, but no salinities were reported. Kovacs [28] was able to sound first-year ice near Prudhoe Bay, Alaska, which was approximately 1.9-m thick and old ice up to 8-m thick at altitudes of up to 90 m. Rossiter et al. [64] reported airborne sounding off northeast Newfoundland from altitudes of 2 to 20 m for ice 40-cm thick, with salinities of approximately 2 ppt and an effective dielectric constant of 4.5 ± 0.5. They found two very different types of ice, indistinguishable from their surface appearance, intermixed with each other. Rossiter et al. [60] sounded the length of Lake Melville, Labrador, which was covered with 0.9- to 1.4-m thick brackish ice (salinities 0 to 2 ppt), flying at approximately 15-m elevation. They also reported sounding a piece of old ice more than 14-m thick and encountered refrozen melt pools which had a very characteristic signature.

Measurements have also been made from a Twin Otter fixed-wing aircraft, as shown in Fig. 4.15 (Rossiter et al. [60]). The flying height was usually about 30 m, although successful soundings of low-salinity, first-year ice were made from over 100-m elevation at air speeds of 50 to 70 m/s (100 to 140

Figure 4.15 Photograph of 200-MHz ground-penetrating radar antenna mounted under a de Havilland Twin Otter. This 1979 installation was the first reported use of a fixed-wing, sea-ice thickness sensor (after Rossiter et al. [60]).

knots). Measurements were made at a center frequency of 200 MHz. Analysis of 400 line-km of data over an area of old ice showed that the echoes were not consistent from floe to floe, or even within the same floe, suggesting a high degree of variability in the properties of the ice.

One of the novel features of the latter study was the collection of simultaneously imaged data, in this case aerial photography. Photographs were taken both at low level, from the aircraft containing the radar, and at 2440 m (8000 feet), from a second Twin Otter. As the radar and photography were taken simultaneously, the flight path of the radar aircraft could be plotted and the radar record correlated with ice features in the imagery (Fig. 4.16). This imagery allowed significant improvements to be made in the interpretation of the radar returns; in particular, identification of ice type from the nature of the radar echoes and an estimation of the thickness distribution of the old ice encountered. The thickest floe measured had a calculated thickness of 13.5 m.

In 1979 this system was used to profile ice island T3, which was situated at the time in the Beaufort Sea near 72°37′N, 131°34′W. A portion of the GPR record is shown in Fig. 4.17, going off the end of the island and over 2-m thick first-year ice. Although no simultaneous thickness measurements were made over T3, its thickness was about 25 to 30 m over most of the profile. This thickness agreed well with 1973 measurements of 30 m and an assumed ablation rate of 1 m/year. These results show clearly the effectiveness of GPR

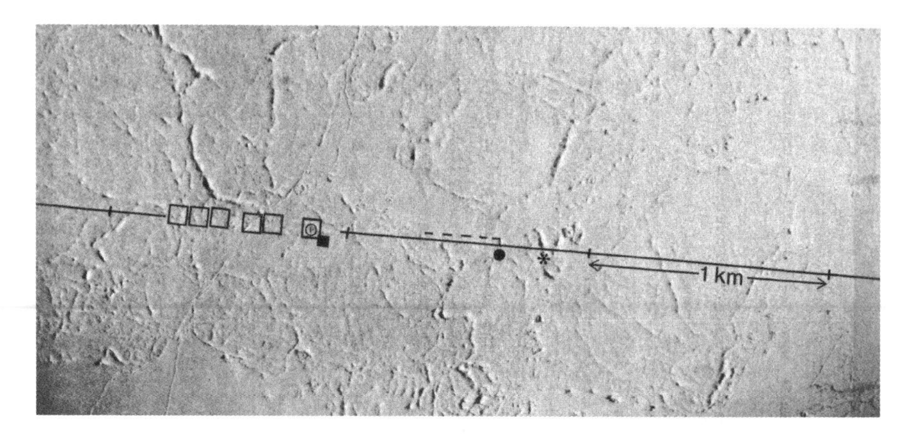

Figure 4.16 Aerial photograph of ice in the Beaufort Sea, March 1979, taken at an altitude of 2440 m. A second aircraft at 30-m altitude made simultaneous ice-thickness measurements using the system shown in Fig. 4-15, and this aircraft can be seen in the photograph. Its flight path, and symbols for various features, are also shown (after Rossiter et al., [60]).

sounding of freshwater ice, such as large icebergs, and this technique has promise for use in measuring the size of Antarctic icebergs (Rossiter and Gustajtis [63]).

Although it is usually not possible to estimate the thickness of first-year sea-ice ridges using GPR, Kovacs [30] was able to sound an old pressure ridge

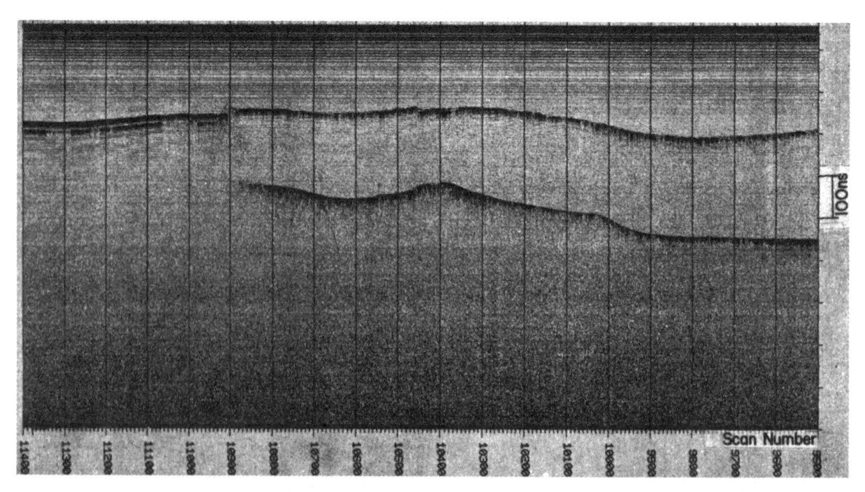

Figure 4.17 Airborne radar data (Fig. 4.15) over Ice Island T3 in the Beaufort Sea, March 1979. Data were collected at approximately 30-m elevation and 50-m/s ground speed. Horizontal distance is approximately 2.0 km. Vertical scale shown is based on a dielectric constant in the ice of 3.2. Note edge of ice island at scan 10,900 and sea ice to the left.

from its surface with very good agreement to augered measurements. The ridge had a maximum thickness of 14 m. Even when a return is not received from the bottom of a ridge, it is sometimes possible to estimate the ridge's width as the ice thickness starts to increase on each side of the ridge.

It is also possible to detect cracks and crevasses using GPR (Campbell and Orange [7]; Kovacs and Abele [37]). These features act as point targets which stand out clearly in a data record with a characteristic hyperbolic shape. It has not been possible to distinguish between wet or dry cracks, although cracks as small as 4- to 5-mm wide have been detected. Cracks appear to be less easily observed from the air.

Ice-thickness and pressure ridge statistics have been estimated using ice surface profiles collected with a laser profilometer, after removal of aircraft height variations (e.g., Lowry and Wadhams [47]). Similar statistics should be obtainable from radar sounding data because they also contain an ice surface profile. (Kovacs [28] did the inverse, calculating the ice freeboard from the estimated radar sounding thickness.) However, the effect of the larger radar footprint will have to be accounted for, and there are no reports of this step having been taken. Because of the scatter from ice rubble, one approach would be to include a high-frequency (i.e., narrow-beam) radar altimeter with an airborne sounding radar.

The major operational use of radar today for sea-ice–thickness measurement is from the ice surface. To ensure the safety of heavy equipment moving over the ice (often the most practical transportation route for remote areas), a light vehicle equipped with a GPR system leads the equipment train. Ice thickness is displayed in real time in the cab of this lead vehicle and detours are taken around dangerously thin areas. A GPR antenna with two crossed dipoles is used to avoid obtaining weak echoes from areas in which there is horizontal anisotropy in the ice (Weeks and Gow [69]). Each of the dipoles is fired in turn and the results are averaged. It is becoming mandatory to have this type of radar when moving heavy equipment over sea ice in the Canadian Arctic.

4.3.2 Portable Ice-Thickness Sensors

Although most of this book is devoted to remote sensing technologies, there is some interest in equipment that can be carried easily over the ice surface to reduce the need for augering. As described above, GPR is used routinely from vehicles traveling over the ice. In this section, we examine briefly two additional techniques, acoustic and EM.

Acoustic Sensors

In the mid-1980s work was done in Canada to develop a portable acoustic sensor (Hudson et al. [25]). One advantage is that the acoustic properties of ice appear to be less variable than the electrical properties of ice. The velocity of compressional waves in ice is not strongly dependent on temperature or frequency. The attenuation rate is more strongly frequency-dependent but below 100 kHz is not usually more than about 5 dB/m, (Langleben [42]).

To overcome earlier problems with single-shot impulse sources, the device developed by Hudson et al. used a continuous-wave system, swept over the frequency range of 150 to 12,000 Hz. The resonance peaks were located and used to calculate the ice thickness based on an assumed sound velocity. The system consisted of a linear 30-W acoustic source feeding a rugged, moving-coil loudspeaker source, three low-mass accelerometers, detectors, a receiver, and a portable computer-controller.

The device was operated by clearing away snow on the ice surface and freezing the sound source and the accelerometers to the ice with a small amount of fresh water. After a measurement cycle, the computer determined the frequencies of resonance using a Fast Fourier Transform and a correlation technique, and then calculated the ice thickness in real time. Although the portable computer available in 1984 made this step slow (9 min.), this would not be an issue today.

The results on several types of sea ice and freshwater ice were promising. The results at nine different sites, with augered ice thicknesses between 1 m and more than 2 m, gave acoustic readings accurate in all cases to $\leq \pm 0.3$ m. An assumption, or calibration, must be made concerning the sound velocity in the ice but this parameter probably does not change rapidly in a given area. No measurements were made near ridges and because the resonance approach assumes a layered medium, it would be expected that corrections would be needed near deformed ice.

Another approach to use of acoustic waves to infer ice thickness is reported by de Heering [15]. Bullets shot into the ice were used as the sound source. A microphone suspended approximately 0.30 m above the ice and approximately 10 m from the point of impact was used to detect the response. The received time-series response was autocorrelated to determine the resonant response of the ice sheet. The results, although preliminary, showed promise.

In summary, acoustic techniques have been shown to work under field conditions and may have some utility due to their inherent low cost, simplicity of operation, and portability.

Portable EM Sensor

Initial work using a surface-based EM induction system goes back to the 1970s (Sinha [66]) and showed the effectiveness of this approach in principle. Additional work led to a system based on the Geonics EM-31 ground conductivity meter (Hoekstra et al. [20]). This device is a 9-kg, portable EM unit operating at 9.8 kHz, with the transmitter and receiver at opposite ends of a rigid 3.66-m boom. Although attempts were made to market this approach, it did not achieve commercial success due to apparently erratic results.

More recently, Kovacs and Morey [34] reexamined the algorithms used to convert the EM-31 measured response into an ice thickness. They found several sources of error and, after correction, tested the device in the Arctic. The results were accurate to within $\pm 10\%$ over sea ice from 0.7- to 3.5-m thick. This device could be used in on-ice situations in which rapid measurements are required from the surface, in environments for which GPR could be ineffective due to relatively high attenuation at GPR-sounding frequencies. It should be noted that this device, unlike the acoustic sensor, will not work over freshwater ice because it depends on the conductivity of the underlying seawater for its response, although there has been a report of its use to estimate the extent of frazil ice in a river (Lawson et al. [43]).

4.4 FUTURE DIRECTIONS

Over the next few years it is expected that accurate sea-ice thickness measurement using airborne EM induction will become fully operational. Therefore, future research work in this field will concentrate on refinements and use of the technology, rather than development of the technology itself. In this section we suggest some of the areas that are likely to be of continuing research interest.

Although there has been substantial progress in development of real-time inversion procedures, this area is likely to evolve over many years. The continuing availability of inexpensive computers of rapidly increasing power will make implementation of more computationally complex varieties of real-time inversion possible. In addition, use of newer software approaches based on artificial intelligence are expected to be effective. Both expert systems (Lee et al. [44]) and neural networks (Rossiter et al. [57]) have shown promise.

In terms of equipment, the next stage appears to be development of fixed-wing implementation. This development will be technically difficult but promises rapid coverage of large areas using long-range aircraft. Although there have been a number of successful EM systems used from fixed-wing aircraft for geophysical exploration applications (Palacky and West [52]), there have been none for ice thickness that we are aware of, apart from some experiments carried out by the Finnish Geological Survey using an EM-processing system. The technical problems that will be encountered in implementing such a system relate to the relatively higher sensitivity required for this application as well as the relatively low-sensor altitude required to achieve a reasonably small footprint. It is likely that a towed-bird implementation of the technology will be most effective in this setting as well.

The most exciting aspect of the availability of a practical ice-thickness measurement tool will be its application to operational and scientific problems. It is expected that this technique will lead both to a better understanding of the ice regime and its interaction with its environment and to less expensive and hazardous ice navigation.

ACKNOWLEDGMENTS

Primary financial support for much of this work has come from the Transportation Development Centre and the Canadian Coast Guard, both of Transport

Canada. Mr. Maurice Audette of TDC has been successful in keeping the project moving through a number of years. More recently, Dr. Simon Prinsenberg, of the Department of Fisheries and Oceans, Bedford Institute of Oceanography, Dartmouth, has orchestrated several validation efforts. Additional support has come from the Ice Centre, Atmospheric Environment Centre, Environment Canada, Ottawa, from the Centre for Cold Ocean Resources Engineering, St. John's, from the Geological Survey of Canada, Energy, Mines and Resources, Ottawa, and from the Department of National Defence, Ottawa. Special reference should also be made to the Polar Continental Shelf Project of Energy, Mines and Resources, which provided logistical support for field work both at Tuktoyaktuk and Resolute over many years.

Mr. Austin Kovacs of the U.S. Army Cold Regions Research and Engineering Laboratory, Hanover, NH, has pursued similar goals for many years, and has provided substantial contractual support to Geonex Aerodat. He has been instrumental in advancing the state of the art, and we are pleased to acknowledge the many scientific discussions that we have enjoyed with him.

The support of our employers, Canpolar Inc. and Geonex Aerodat Inc., who have continued to believe that this activity has both technical and commercial merit, is also gratefully acknowledged.

REFERENCES

[1] W. L. Anderson (1979) ''Numerical integration of related Hankel transforms of order 1 and 2 by adaptive digital filtering,'' *Geophysics.*, **44**, 1287–1305.

[2] A. P. Annan and J. L. Davis (1977) ''Impulse radar applied to ice thickness measurements and freshwater bathymetry,'' Geol. Surv. Can. Paper, 77-1B, 63–65.

[3] A. P. Annan and J. L. Davis (1976) ''Impulse radar soundings in permafrost,'' *Radio Sci.*, **11**, 383–394.

[4] C. J. Bergeron, J. W. Ioup, and G. A. Michel II (1987) ''Theory of the modified image method for airborne electromagnetic data,'' in *United States Geologic Survey Bulletin 1925*, 65–74.

[5] V. V. Bogorodskii and V. P. Tripol'nikov (1974) ''Radar sounding of sea ice,'' *Sov. Phys. Tech. Phys.*, **19**, 414–415.

[6] K. A. Butt and J. B. Gamberg (1979) ''Technology of an airborne impulse radar,'' in *Proc. Int. Workshop on Remote Estimation of Sea Ice Thickness*, J. R. Rossiter and D. P. Bazeley (Eds.), C-CORE Publ. 80-5, 385–412.

[7] K. J. Campbell and A. S. Orange (1974) ''A continuous profile of sea ice and freshwater ice thickness by impulse radar,'' *Polar Rec.*, **17**, 31–41.

[8] K. J. Campbell and A. S. Orange (1974) ''The electrical anisotropy of sea ice in the horizontal plane,'' *J. Geophys. Res.*, **79**, 5059–5063.

[9] K. J. Campbell, R. O. Ramseier, W. F. Weeks, and P. Gloersen (1975) ''An integrated approach to the remote sensing of floating ice,'' in *Proc. Third Can. Symp. on Remote Sensing*, 39–72.

[10] A. N. Chizhov, V. G. Glushnev, and B. D. Slutsker (1977) ''A pulsed radar method of measuring ice-cover thickness,'' *Meteor Gidrol.*, **4**, 90–96.

[11] W. J. Chudobiak, R. B. Gray, R. O. Ramseier, V. Makios, M. Vant, J. L. Davis, and J. Katsube (1974) "Radar remote sensors for ice thickness and soil moisture measurements," in *Proc. Second Can. Symp. on Remote Sensing*, 417–424.

[12] S. D. Conte and C. de Boor (1972) *Elementary Numerical Analysis*, McGraw-Hill, Toronto.

[13] D. W. Cooper, R. A. Mueller, and R. J. Schertler (1976) "Remote profiling of lake ice using an s-band short-pulse radar aboard an all-terrain vehicle," *Radio Sci.*, **11**, 375–381.

[14] J. L. Davis and A. P. Annan (1989) "Ground-penetrating radar for high-resolution mapping of soil and rock stratigraphy," *Geophys. Prospecting*, **37**, 531-551.

[15] P. de Heering (1989) "An impact sound source useful for arctic remote sensing," *IEEE J. of Oceanic Engineering*, **14**, 166–172.

[16] S. Evans (1965) "Dielectric properties of ice and snow—a review," *J. Glaciology*, **5**, 773-792.

[17] M. I. Finkel'shteyn and V. A. Kutev (1974) "Probing of sea ice with a sequence of video pulses," *Rad. Eng. Elec. Phys.*, **17**, 1680–1684.

[18] G. Frankenstein and R. Garner (1967) "Equations for determining the brine volume of sea ice from −5° to −22°C," *J. Glaciology*, **6**, 943–944.

[19] R. H. Goodman, E. Outcalt, B. B. Narod, and G. K. C. Clarke (1977) "Radar techniques in the measurement of floating ice thickness," in *Proc. Fourth Can. Symp. on Remote Sensing*, 459–468.

[20] P. Hoekstra, A. Sartorelli and S. B. Shinde (1979) "Low frequency methods for measuring sea ice thickness," in *Proc. of Intl. Workshop on the Remote Estimation of Sea Ice Thickness*, J. R. Rossiter and D. P. Bazeley (Eds.), C-CORE Publ. 80-5, 313–330.

[21] J. S. Holladay, J. R. Rossiter, and A. Kovacs (1990) "Airborne measurement of sea ice thickness using electromagnetic induction sounding," in *Proc. Ninth Intl. Conf. Offshore Mech. and Arctic Eng.*, ASME, Vol. VI, 309–315.

[22] J. S. Holladay, I. R. St. John, V. Schoeggl, J. Lee, J. R. Rossiter, and L. Lalumiere (1992) "Airborne EM ice measurement sensor phase 1-2," Transportation Development Centre Rep., TP11282E.

[23] J. S. Holladay, J. Lee, I. St. John, J. R. Rossiter, L. Lalumiere, S. J. Prinsenberg (1992) "Real-time airborne electromagnetic measurement of sea ice," *62nd Annual SEG Meeting*, New Orleans, Expanded Abstracts with Biographies, 1992 Technical Program, Society of Exploration Geophysicists, Tulsa, 461–465.

[24] J. S. Holladay (1993) "Operational electromagnetic ice thickness sensor trial," *Can. Coast Guard Rept.* TP1160.

[25] R. Hudson, J. Pann, and T. Day (1985) "Development and testing of a portable hand carried ice thickness measuring device," Transportation Development Centre Rept. TP6816E.

[26] K. Iizuka, H. Ogura, J. L. Yen, V. Nguyen, and J. R. Weedmark (1976) "A hologram matrix radar," *Proc. IEEE.*, **64**, 1493–1504.

[27] O. Koefoed (1979) *Geosounding Principles*, Elsevier, New York.

[28] A. Kovacs (1977) "Sea ice thickness profiling and under-sea oil entrapment," in *Proc. 9th Ann. Offshore Techn. Conf.*, 547–554.

[29] A. Kovacs (1977) "Iceberg thickness profiling," in *Proc. Fourth Intl. Conf. on Port and Ocean Eng. under Arctic Conditions*, 766–774.

[30] A. Kovacs (1978) ''A radar profile of multiyear pressure ridge fragment,'' *Arctic*, **31**, 59–62.

[31] A. Kovacs, N. Valleau, and J. S. Holladay (1987) ''Airborne electromagnetic sounding of sea ice thickness and sub-ice bathymetry,'' CRREL Rept., 87-23.

[32] A. Kovacs and J. S. Holladay (1989) ''Airborne sea ice thickness sounding,'' in *Proc. 10th Intl. Conf. on Port and Ocean Eng. under Arctic Conditions*, 1042–1052.

[33] A. Kovacs and J. S. Holladay (1990) ''Sea ice measurement using a small airborne electromagnetic sounding system,'' *Geophysics*, **55**, 1327–1337.

[34] A. Kovacs and R. M. Morey (1991) ''Sounding sea ice thickness using a portable electromagnetic induction instrument,'' *Geophysics*, **56**, 1992–1998.

[35] A. Kovacs and R. M. Morey (1992) ''Estimating sea ice thickness from impulse radar sounding time of flight data,'' Geol. Surv. of Can. Paper, 90-4, 117–124.

[36] A. Kovacs and A. J. Gow (1975) ''Brine infiltration in the McMurdo Ice Shelf, McMurdo Sound, Antarctica,'' *J. Geophys. Res.*, **80**, 1957–1961.

[37] A. Kovacs and G. Abele (1974) ''Crevasse detection using an impulse radar system,'' *Antarctic J. of the U.S.*, **9**, 177–178.

[38] A. Kovacs, J. S. Holladay, and C. J. Bergeron Jr. (1993) ''Footprint size of a helicopter-borne electromagnetic induction sounding system versus antenna altitude,'' *CRREL Rept. 93-12*.

[39] L. A. Lalumiere (1992) ''Analysis of snow thickness data collected by impulse radar over the Beaufort Sea Shelf in 1991,'' Can. Contractor Rept. of Hydrog. and Ocean Sci. 43.

[40] L. A. Lalumiere, S. Prinsenberg, C. Coram, and J. R. Rossiter (1994) ''Analysis of airborne ground penetrating radar measurements of snow thickness, St. Anthony, Newfoundland, 1992,'' Can. Contractor Rept. of Hydrog. and Ocean Sci.

[41] C. Lanczos (1961) *Linear Differential Operators*, D. van Nostrand, New York.

[42] M. P. Langleben (1969) ''Attenuation of sound in sea ice, 10–500 kHz,'' *J. Glaciology*, **4**, 399–406.

[43] D. E. Lawson, E. F. Chacho Jr., B. E. Brockett, J. L. Wuebben, C. M. Collins, S. A. Arcone, and A. J. Delaney (1986) ''Morphology, hydraulics and sediment transport of an ice-covered river,'' CRREL Rept. 86–11.

[44] S. Lee, E. Milios, R. Greiner, and J. Rossiter (1992) ''An expert system for automated interpretation of ground penetrating radar data,'' Geol. Surv. of Can. Paper, 90-4, 125–131.

[45] G. Liu and A. Becker (1990) ''Two-dimensional mapping of sea-ice keels with airborne electromagnetics,'' *Geophysics*, **55**, 239–248.

[46] G. Liu, A. Kovacs, and A. Becker (1991) ''Inversion of airborne electromagnetic survey data for sea-ice keel shape,'' *Geophysics*, **56**, 1986–1991.

[47] R. T. Lowry and P. Wadhams (1979) ''On the statistical distribution of pressure ridges in sea ice,'' *J. Geophys. Res.*, **84**, 2487–2494.

[48] R. M. Morey (1975) ''Airborne sea ice thickness profiling using an impulse radar,'' U.S. Coast Guard Tech. Rept. CG-D-178-75.

[49] G. K. A. Oswald (1988) ''Geophysical radar design,'' *I.E.E. Proc.*, **135**(F), 371–379.

[50] G. K. A. Oswald (1992) ''Radar design for geophysical sounding,'' Geol. Surv. of Can. Paper, 90-4, 151–164.

[51] D. F. Page and R. O. Ramseier (1975) ''Application of radar techniques to ice and snow studies,'' *J. Glaciology*, **15**, 171–191.

[52] G. J. Palacky and G. F. West (1991) ''Airborne electromagnetic methods,'' in *Electromagnetic Methods in Applied Geophysics*, Vol. 2, Part B, Misac Nabighain (Ed.), Society of Exploration Geophysicists, Tulsa, 811–879.

[53] S. J. Prinsenberg, J. S. Holladay, J. R. Rossiter, and L. Lalumiere (1992) ''1991 Beaufort Sea EM/Radar ice and snow sounding project,'' Can. Tech. Rept. of Hydrog. and Ocean Sci. 139.

[54] S. J. Prinsenberg, J. S. Holladay, and L. A. Lalumiere (1993) ''Electromagnetic/Radar ice and snow sounding project over the Newfoundland Shelf in 1992,'' Can. Tech. Rept. of Hydrog. and Ocean Sci. 144.

[55] G. de Q. Robin, S. Evans, and J. T. Bailey (1969) ''Interpretation of radio echo sounding in polar ice sheets,'' *Phil. Trans. Royal Soc.*, **A265**, 437–505.

[56] J. R. Rossiter, J. S. Holladay, and L. A. Lalumiere (1992) ''Validation of airborne sea ice thickness measurement using electromagnetic induction during LIMEX '89,'' Can. Contractor Rept. Hydrog. and Ocean Sci. 41.

[57] J. R. Rossiter, J. S. Holladay, and L. A. Lalumiere (1991) ''Operational airborne sea ice thickness measurement system,'' Second WMO Operational Ice Remote Sensing Workshop, Ottawa, Ontario.

[58] J. R. Rossiter and L. A. Lalumiere (1988) ''Evaluation of sea ice thickness sensors,'' Transportation Development Centre Rept., TP9169E.

[59] J. R. Rossiter and D. P. Bazeley (Eds.) (1980) *Proceedings of the International Workshop on the Remote Estimation of Sea Ice Thickness*, C-CORE Publ. 80-5.

[60] J. R. Rossiter, K. A. Butt, J. B. Gamberg, and T. F. Ridings (1980) ''Airborne impulse radar sounding of sea ice,'' in *Proc. Sixth Can. Symp. on Remote Sensing*, 187–194.

[61] J. R. Rossiter, B. B. Narod, and G. K. C. Clarke (1979) ''Airborne radar sounding of Arctic icebergs,'' in *Proc. Fifth Intl. Conf. on Port and Ocean Eng. under Arctic Conditions*, 289–305.

[62] J. R. Rossiter and K. A. Gustajtis (1978) ''Iceberg sounding of impulse radar,'' *Nature*, **271**, 48–50.

[63] J. R. Rossiter and K. A. Gustajtis (1979) ''Determination of iceberg underwater shape with impulse radar,'' *Desalination*, **29**, 99–107.

[64] J. R. Rossiter, P. Langhorne, T. Ridings, and A. J. Allan (1977) ''Study of sea ice using impulse radar,'' in *Proc. Fourth Int. Conf. on Port and Ocean Eng. under Arctic Conditions*, 556–567.

[65] J. R. Rossiter (1980) ''Review of impulse radar sounding of sea ice,'' in *Proc. of Intl. Workshop on the Remote Estimation of Sea Ice Thickness*, J. R. Rossiter and D. P. Bazeley (Eds.) C-CORE Publ. 80-5, 77–107.

[66] A. Sinha (1976) ''A field study for sea ice thickness determination by electromagnetic means,'' Geol. Surv. of Can. Paper, 76-1C, 225–228.

[67] A. S. Thorndike, C. Parkinson, and D. A. Rothrock (Eds.) (1992) ''Report of the Sea Ice Thickness Workshop,'' Univ. Washington, Seattle.

[68] P. Wadhams and J. C. Comiso (1992) ''The ice thickness distribution inferred using remote sensing techniques,'' in *Geophysical Monograph 68: Microwave Remote Sensing of Sea Ice,* F. D. Carsey (Ed.), American Geophysical Union, 375–383.

[69] W. F. Weeks and A. J. Gow (1978) ''Preferred crystal orientation in the fast ice along the margins of the Arctic Ocean,'' *J. Geophys. Res.*, **83**, 5105–5121.

5

PASSIVE MICROWAVE SYSTEMS

IRENE GOLDHAR RUBINSTEIN
Institute for Space and Terrestrial Science
York University
North York, Ontario, Canada

DENNIS M. NAZARENKO
RADARSAT International Inc.
Richmond, British Columbia, Canada

SEBASTIAN TAM
MPB Technologies, Inc.
Dorval, Quebec, Canada

5.1 INTRODUCTION

All matter radiates electromagnetic waves as a result of the thermally induced, random motion of electrons and protons. These thermally induced waves contain components at all frequencies and can be described using Planck's theory for blackbody radiation. The observed radiation intensity depends, in general, on temperature, observation frequency, material composition, and the geometrical configuration of the target's surface. All remote sensing systems make use of information obtained from the processing and interpretation of electromagnetic radiation emitted and/or reflected from the earth's surface. Passive microwave radiometry is the measurement of incoherent electromagnetic radiation from 0.3 to 300 GHz (1 m to 1 mm in wavelength).

Polar regions, in addition to their vast extent, remote location, and inaccessibility, are shrouded in darkness and cloud for much of the year which creates a problem for sensors at optical wavelengths. Remote sensing at microwave

Remote Sensing of Sea Ice and Icebergs, Edited by Simon Haykin, Edward O. Lewis, R. Keith Raney, and James R. Rossiter.
ISBN 0-471-55494-4

frequencies has several attractions, primarily because it offers a means of sensing the character of the earth's surface through clouds, independent of daylight. Microwave signals are relatively free from atmospheric interference from clouds, particularly at the lower frequencies. Transmission characteristics of the atmosphere for the microwave portion of the spectrum are known, making it possible to tailor this technology to specific applications.

In the 1930s and 1940s, one of the earliest applications of radiometric techniques was the measurement electromagnetic energy of extraterrestrial origin. After two decades of outward-looking radioastronomical and atmospheric studies, scientists at the University of Texas (Straiton et al. [82]) pointed an antenna downward, from an airborne platform. Using this sensor, at 4.3-mm wavelength, they pioneered radiometric measurements of microwave-emitting properties of water, grass, and asphalt. In the last 25 years, microwave radiometry (or passive microwave remote sensing) has become an important tool for environmental observation. Satellite remote sensing using passive microwave systems has proven to be useful in providing long-term global monitoring.

The first spaceborne passive microwave observations (at 15.8- and 22.0-GHz frequencies), on 14 December 1962, were made by a two-channel microwave radiometer on the *Mariner 2* spacecraft. This sensor provided the first observations of the planet Venus (Barrett and Lilley [3]).

The study of the physical properties of the earth's surface and atmosphere using spaceborne microwave radiometry began in 1968 with the launch of *Cosmos 243* (Basharinov et al. [4]). Since 1968, nearly a dozen passive microwave sensors have been placed in space: see Table 5.1.

Microwave radiometry is particularly effective for monitoring seasonal and spatial changes in the sea-ice cover of the world's oceans because of the large contrast between microwave emissions from the ice-covered and ice-free sea surface. The physical variables which may be inferred from passive microwave sensors are the concentration of sea ice (that is, the percentage of the observed area covered with ice) and, with some seasonal and regional limitations, ice type (old, first-year, or young ice).

Single-frequency observations from 1972 to 1976 by the electrically scanning microwave radiometer (ESMR) on *Nimbus 5* provided, for the first time, a set of global observations that permitted sea-ice studies on a synoptic scale (Zwally and Gloersen [97]; Staelin et al. [75]; Zwally [95]). These 19.35-GHz measurements of microwave emissions were one of the primary sources of data used to produce U.S. Navy operational ice analyses (Wilheit et al. [93]); however, the single-frequency observations were of limited use in discriminating first-year from old ice.

Since June 1978, multifrequency, dual-polarized observations (Hollinger [41]) of the earth's microwave radiance have been used to monitor the extent and physical properties of sea ice and other geophysical parameters.

The scanning multichannel microwave radiometers (SMMR) on *Seasat-A* and *Nimbus 7* (both launched in 1978) measured microwave radiation with five different frequencies: 6.633, 10.69, 18.0, 21.0, and 37.0 GHz. Two radiometers were used for 37- and 18-GHz observations, one for each polarization. Four radiometers, alternating between polarizations at successive scans, were used for the other frequencies.

Over the past two decades, passive microwave sensors and algorithms for the interpretation of their microwave signals have increased in complexity and capabilities, leading to a wide range of applications for passive microwave data (Campbell et al. [8]; Anderson, [1]). Improved spectral resolution and the long life of the SMMR on *Nimbus 7* resulted in the development of improved algorithms for retrieval of ice-concentration and ice-type estimates (Cavalieri et al. [13]; Cavalieri and Zwally, [17]; Williams [94]; Walker [89]; Rubinstein [67]; Etkin [29].

Presently, spaceborne passive microwave data are available to the remote sensing research community from the special scanning microwave imager (SSM/I) on a Defence Meteorological Satellite Program (DMSP) satellite. The first SSM/I, built by Hughes Aircraft Company under the direction of the Air Force Space and the Naval Space Systems Division, was launched in 1987 and two additional sensors were placed in orbit in 1991. The program calls for the availability of at least one SSM/I sensor until the next generation of passive microwave sensors is launched in 1997. The SSM/I data are processed by the Naval Oceanography Command and the Air Force Weather Service for near-real time retrieval of environmental parameters, using sophisticated algorithms. Cloud cover, rain rates, water vapor over ocean, marine wind speed, extent, and age of sea-ice cover are generated for ocean areas. Land surface type, snow-cover parameters, and soil moisture information are obtained for terrestrial areas.

Several research teams have assessed the performance of sea-ice retrieval algorithms (Bjerklund et al. [5]) for passive microwave data. The Environment Canada Atmospheric Environmental Service (AES) team recommended the operational use of a Canadian algorithm (known as AES/York algorithm) instead of the prelaunch version of the Hughes algorithm (Hollinger et al. [40]) for SSM/I data. In addition, the AES team recommended the continuation of research activities dedicated to understanding seasonal and regional variabilities of the microwave properties of sea ice and the intervening atmosphere, aimed at reducing some of the Canadian algorithm's shortfalls.

In Canada, several important applications are being developed for passive microwave radiometry (Ramseier et al. [62]), ranging from ice information for strategic evaluation of ship traffic in ice to studies of regional and synoptic climate variability. Sea-ice charts produced with spaceborne passive microwave observations are used on a routine basis at several ice-forecasting centers and during sea-ice research campaigns. The coarse spatial resolution (on the order of 15 km or larger) of spaceborne microwave radiometers is a limiting factor for the application of these sensors.

Stimulated by the successful application of such satelliteborne sensors for sea-ice monitoring and the results of previous efforts in the United States with airborne sensors, a Canadian-built airborne imaging microwave radiometer

TABLE 5.1 Past, Present, and Future Passive Microwave Sensors, on Meterological Satellites

Spacecraft	Launch Year	Sensor	Frequency (GHz)	Swatch Width (km)	Resolution (km)
Mariner 2	1962	PMW	15.8, 22.2	Planetary	1,300
Cosmos 243	1968	PMW	3.5, 8.8, 22.2, 37.0	. . .	37
Cosmos 384	1970	PMW	3.5, 8.8, 22.2, 37.0	. . .	13
Nimbus-5	1972	ESMR	19.35 (H)	3,000	50
		NEMS	22.2, 31.4, 53.6, 54.9, 58.8	185	185
Skylab	1973	S-193	13.9	11–170	16
		S-194	1.4	280	115
Meteor	1974	PMW	37.0	. . .	. . .
Nimbus-6	1975	ESMR	37.0 (V&H)	1,270	20 × 43
		SCAMS	22.2, 31.6, 52.825, 53.8, 55.4	2,618	145 to 330
DMSP-5D	1978	SSM/T	50.5, 53.2, 54.35, 54.9, 58.4, 58.825, 59.4	1,600	175
TIROS-N	1978	MSU	50.3, 53.7, 55.0, 57.9	2,300	110
Seasat	1978	SMMR	6.63, 10.69, 18.0, 21.0, 37.0 (V&H)	600	149 × 87 to 16 × 27

Nimbus-7	1978	SMMR	6.63, 10.69, 18.0, 21.0, 37.0 (V&H)	800	148 × 151 to 27 × 32
MOS-1	1987	MSR	23.8 (H), 31.4 (V)	317	32, 23
MOS-1b	1990				
MOS-2	1992				
DMSP-5D1	1985	SSM/T2	91.5, 150, and 3 channels near 183.0	175–200	50
DMSP-5D2	1987	SSM/I	19.35, 37.0, 85.5 (V&H), 22.23 (V)	1,394	70 × 45
ERS-1	1991	ATSR-M	23.8, 36.5	500	22
NOAA 'NEXT' and EOS-A	1991 1998	AMSU-A	23.9, 31.4, 12 channels of 50.3 to 57.29, 89.0	2,300	50
GOES 'NEXT'	1991	PMW	92, 118, 150, 183, 230	500	35 to 15
TOPEX/ Poseidon		PMW	18.0, 21.0, 37.0	1,400	51, 40, 27
DMSP	1997	SSM/IS	19.35, 21.0, 37.0, 92.0, 150	. . .	50 to 25
EOS-A	1998	MIMR	6.8, 10.65, 18.7, 23.8, 36.5, 90 (all V&H)	1,400	60 to 5

(AIMR) was completed in 1989. It was designed with dual frequency (37 and 90 GHz) and vertical and horizontal polarization capabilities.

The objective of this chapter is to present an overview of the current status of sea-ice applications of passive microwave remote sensing. Section 5.2 outlines the theoretical basis for microwave radiometry and reviews the performance and characteristics of passive microwave sensors, both spaceborne and airborne. Section 5.3 reviews the radiometric properties of sea ice and Section 5.4 describes interpretation techniques used to infer sea-ice information from spaceborne, multispectral passive microwave observations. Current applications of the retrieved information and future plans for the use of passive microwave data are discussed in Sections 5.5 and 5.6, respectively.

5.2 PASSIVE MICROWAVE RADIOMETRY

The term radiometry means the measurement of intensities of naturally emitted electromagnetic radiation. The terminology and units used to describe microwave radiometric measurements were adapted from microwave engineering. A passive microwave radiometer consists of an antenna, a wideband receiver, and a recording device for measuring radiant electromagnetic energy. A block diagram of a typical radiometer can be seen in Fig. 5.1. The receiver, essentially a heterodyne receiver, converts the microwave signal power into a direct current voltage output. It detects minute amounts of incident radiation energy which can be much smaller than the thermal noise generated within the receiver itself. It is, therefore, important to understand how the radiometer works and to quantify the relative magnitudes of the different contributions to the radiometer input.

5.2.1 Fundamental Concepts

The concept of blackbody radiation is fundamental to understanding radiometry. A blackbody is an idealized, perfectly opaque material that absorbs all

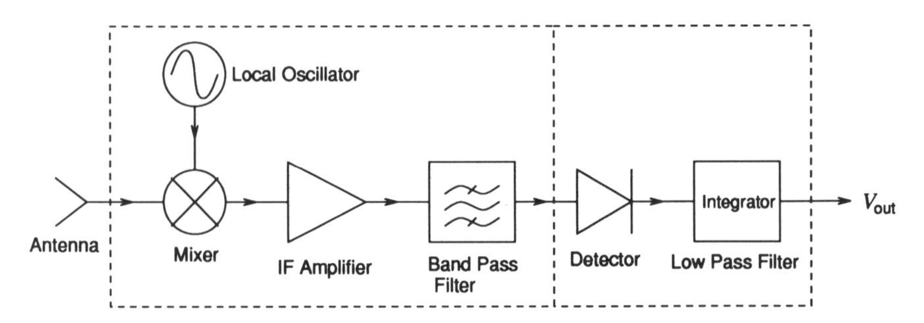

Figure 5.1 Block diagram of typical total power radiometer. Radiation incident upon the antenna is converted into an IF signal which is then detected and integrated over time to produce a voltage to be recorded.

incident radiation at all frequencies and, therefore, reflects none. In order that it may remain in thermodynamic equilibrium with its surroundings, a blackbody also must be a perfect emitter of radiation, otherwise the energy it absorbs would increase its temperature indefinitely. Max Planck, using a quantum-mechanic model, derived an expression to describe the behavior of the intensity of radiation emitted from a blackbody as a function of frequency (or of wavelength) and the temperature of the blackbody. According to this expression, a blackbody radiates uniformly in all directions with a spectral brightness, B_f, given by (Fowles [31])

$$B_f = \frac{2hf^3}{c^2}\left(\frac{1}{e^{hf/kT} - 1}\right)$$

where

B_f = blackbody spectral brightness in $\mathrm{Wm^{-2}sr^{-1}Hz^{-1}}$
h = Planck's constant
k = Boltzmann's constant
f = frequency in Hz
T = absolute temperature in K
c = velocity of light

Planck's radiation formula was originally derived from I_f of a blackbody, I_f being the radiation energy flux density ($\mathrm{Wm^{-2}Hz^{-1}}$) per unit solid angle normal to the emitting surface. However, it can be shown by simple geometry that I_f is numerically equivalent to B_f, although the two differ in definition (Charton [18]).

Plots of brightness as a function of frequency for two distinct temperatures, as in Fig. 5.2, reveal that as the temperature is increased, the overall level of spectral brightness increases, and that the frequency at which the brightness is maximum increases with temperature. At temperatures in the normal terrestrial environment range, the maximum brightness is in the infrared frequency range. With increasing temperature, the maximum brightness traverses the visible and tends toward the ultraviolet and X-ray frequencies. At temperatures in the vicinity of 300 K, the maximum brightness in the microwave region of the electromagnetic spectrum is about six orders of magnitude less than the maximum brightness in the infrared, corresponding to an amount of blackbody radiation far less in the microwave region than in the infrared. The technology required to detect this small amount of radiation presents a design challenge especially at millimeter wave frequencies beyond 60 GHz.

In the microwave region, where $(hf/kT) \ll 1$, Planck's Law can be approximated by what is known as the Rayleigh-Jeans approximation:

$$B_f = \frac{2hf^2kT}{c^2}$$

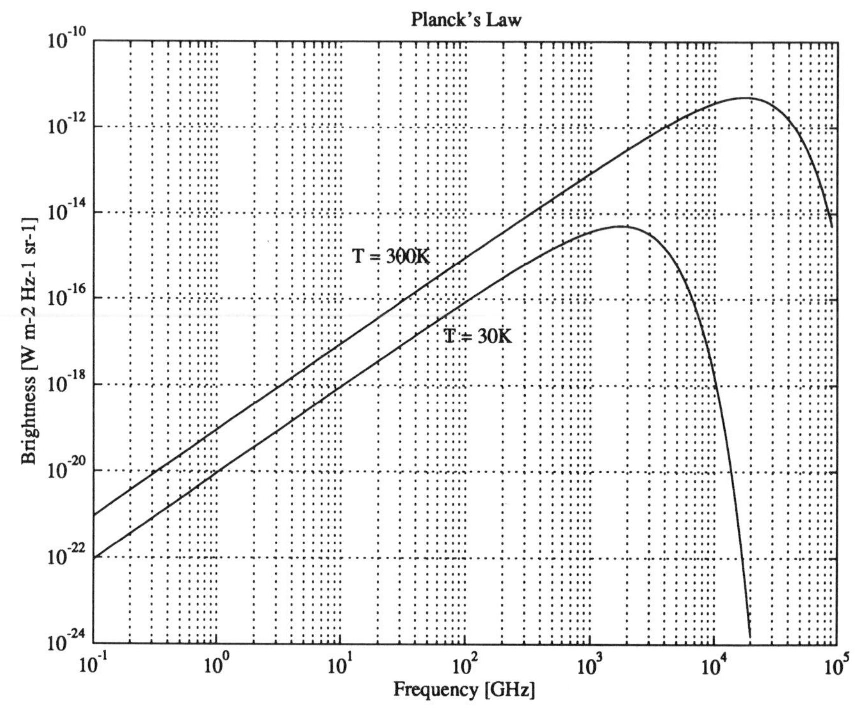

Figure 5.2 Blackbody brightness plotted as a function of frequency for two temperatures T.

No ideal blackbodies are known to exist in nature. Certain materials exhibit blackbody behavior for limited ranges of frequency and temperature. Soot, for example, behaves very much like a blackbody in the infrared and visible ranges whereas graphite-impregnated epoxy approaches the behavior of a blackbody in the microwave range. Most materials, however, even over a limited frequency range, emit more and absorb less radiation than a blackbody and are termed gray bodies. The brightness, B_{gray}, of a gray body relative to that of a blackbody, B_{bb}, at the same temperature, then, is defined as the emissivity, e, such that

$$e = \frac{B_{gray}}{B_{bb}} = \frac{T_B}{T}$$

Since B_{gray} is always less than or equal to B_{bb}, the emissivity is always less than or equal to unity, which reveals in turn that the brightness temperature of the gray body, T_B, is always less than or equal to its physical temperature, T.

The emissivity of a material, measured at frequency f, is dependent on its shape or surface roughness, its composition, and its permittivity and permeability. It is independent of the material's temperature unless the material composition is temperature-sensitive. For example, a wet soil surface has a much lower emissivity than the same soil when frozen. Models that relate emissivity to physical parameters are discussed in Ulaby et al. [86–88]. The emissivity of a surface generally

1. decreases with an increase in the reflectivity, or the real component of the dielectric constant;
2. increases with an increase in the absorptivity, or the imaginary component of the dielectric constant;
3. increases with increasing surface roughness; and
4. is greater in the case of vertical polarization than of horizontal polarization.

A parameter related to the emissivity of a material is its reflectivity. A blackbody has a reflectivity of zero, as it absorbs all incident radiation. The sum of the reflectivity and emissivity of an opaque object is unity. A poor emitter is thus a good reflector, and vice versa. Metallic surfaces, for example, are good reflectors for frequencies up to the ultraviolet.

Antenna Temperature

A microwave radiometer antenna receives electromagnetic energy radiated by the observed scene. The spectral power, dP_f, received by an antenna (Fig. 5.3)

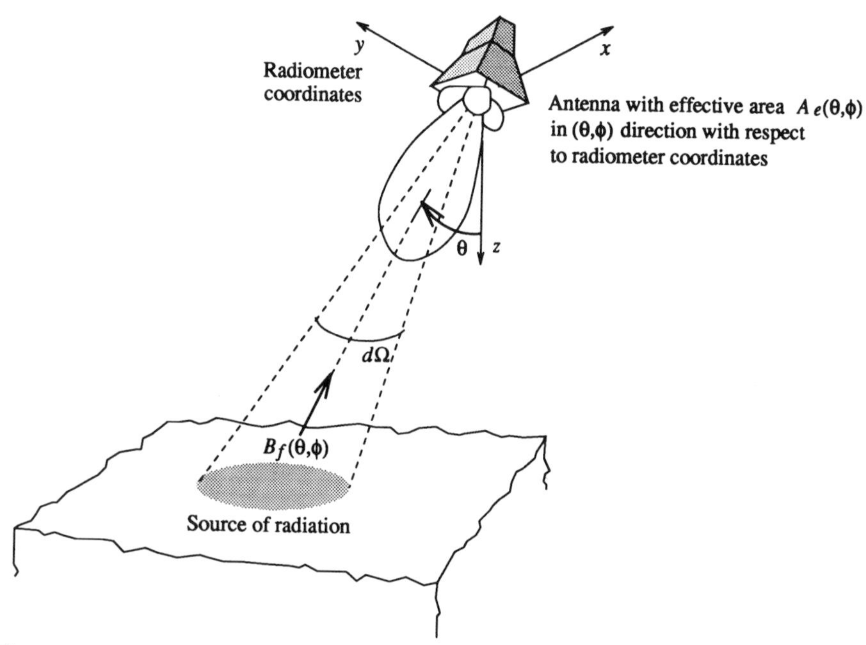

Figure 5.3 Geometry of airborne/spaceborne radiometer measuring thermal emission from terrain.

is given by

$$dP_f = A_e(\theta, \phi)B_f(\theta, \phi)\, d\Omega$$

where

$A_e(\theta, \phi)$ = effective antenna aperture in the (θ, ϕ) direction (Johnson [45])
$B_f(\theta, \phi)$ = spectral brightness in power flux density
$d\Omega$ = the unit solid angle subtended by the emitting surface

The spectral power received by an antenna is then

$$P_f = A_r \iint_{4\pi} B_f(\theta, \phi)F_n(\theta, \phi)\, d\Omega$$

where

A_r = effective antenna temperature
$F_n(\theta, \phi)$ = normalized radiation of the antenna
$\iint_{4\pi} F_n(\theta, \phi)\, d\Omega = \lambda^2/A_r$

For a blackbody, the corresponding spectral power, $P_{bb,f}$, received by a lossless antenna place inside a blackbody enclosure is

$$\begin{aligned} P_{bb,f} &= \tfrac{1}{2} A_r \iint_{4\pi} B_f(\theta, \phi)F_n(\theta, \phi)\, d\Omega \\ &= \frac{kTA_r}{\lambda^2} \iint_{4\pi} F_n(\theta, \phi)\, d\Omega \\ &= kT \end{aligned}$$

This spectral power is analogous to the noise power from a resistor at temperature T. The factor $\frac{1}{2}$ allows for the fact that a linearly polarized antenna receives half of the total power from unpolarized radiation. Thus, for microwave radiation for which the brightness is proportional to the brightness temperature, it is convenient to define a radiometric temperature T_i, as

$$P_{i,f} = kT_i$$

and

$$T_i = \frac{\iint_{4\pi} T_{Bi}(\theta, \phi)F_n(\theta, \phi)\, d\Omega}{\iint_{4\pi} F_n(\theta, \phi)\, d\Omega}$$

where $T_{Bi}(\theta, \phi)$ is the brightness temperature in connection with the energy incident upon the antenna from the *i*th source. The direct linear relationship between power and temperature is significant in microwave remote sensing because it allows a rather interchangeable use of the two terms.

Radiation incident upon a real antenna pointing at an object or surface may contain components originating from several different sources. The temperature detected by the antenna is a combination of the brightness temperature of the object and the radiation reflected from the object and the surrounding scene. If the object does not fill the antenna main beam, there are contributions also from the emission and the reflection of the background. The temperature detected by the antenna is called the apparent temperature of the object or surface, a value generally not equal its brightness temperature. Thus, the total spectral power received by a radiometric antenna is given by $P_{r,f} = kT_r$. T_r consists of the antenna radiometric temperature, T_A, and various other effects. The antenna temperature in the main lobe of the antenna is given by (Gagliano [32]):

$$T_A = \frac{1}{L_{atm}} (T_B + T_{SC}) + T_{UP}$$

where the radiometric temperatures T_B, T_{SC}, and T_{UP} are obtained from the corresponding brightness temperatures $T_{B(B)}(\theta, \phi)$, $T_{B(SC)}(\theta, \phi)$, and $T_{B(UP)}(\theta, \phi)$, associated with the terrain emission, the scattered radiation, and the atmospheric upward emission, respectively; they are integrated over only the main beam. L_{atm} allows for the loss due to atmospheric attenuation. Then, the radiometer's received temperature, T_r, present at the antenna's output is given by

$$T_r = \eta \alpha_m T_A + \eta(1 - \alpha_m) T_{SL} + (1 - \eta) T_0$$

where

η = radiation efficiency of the antenna
α_m = main beam efficiency of the antenna
T_{SL} = sidelobe antenna temperature
T_0 = physical temperature of antenna

The last term, $(1 - \eta)T_0$, represents the antenna's self-loss. For precision measurements, one needs to improve the antenna's efficiency and reduce the antenna loss and antenna sidelobes.

System Noise Temperature

All practical receiver systems have internal noise which limits the minimum detectable signal at the input. Imagine a receiver whose input part is connected to a resistor which is matched to the receiver input impedance. If the resistor is cooled to absolute zero while the receiver is maintained at the ambient temperature $T_0 = 290$ K, the output, P_{n0}, of the receiver will be due entirely to noise generated in the receiver. If the receiver is now cooled to absolute zero (which would make the receiver lossless) and the resistor temperature is raised until the receiver output is once again equal to P_{n0}, then the equivalent input noise temperature, T_{rec}, is the temperature of the resistor:

$$P_{no} = kT_{rec}BG$$

where

k = Boltzmann's constant = 1.38×10^{-23} J/K
B = system bandwidth in Hz
G = receiver power gain

When the receiver and the resistor are both maintained at ambient temperature T_0, the output becomes

$$P_{no} = kT_0BG + kT_{rec}BG$$

The noise figure, F, is defined as the ratio of (1) the input signal-to-noise ratio with the input connected to a matching resistor at $T_0 = 290$ K to (2) the signal-to-noise ratio at the receiver output (Charton [18]). F is a measure of the degradation in signal-to-noise ratio by the receiver as a result of its internal noise. As the output signal equals the input signal multiplied by the gain G, F can be written as

$$\begin{aligned} F &= \frac{\text{Output noise power}}{\text{Input noise power} \times G} \\ &= \frac{kT_0BG + kT_{rec}BG}{kT_0BG} \\ &= 1 + \frac{T_{rec}}{T_0} \end{aligned}$$

This gives

$$T_{rec} = (F - 1)T_0$$

When a receiver consisting of N stages is cascaded, the output is given by

$$\begin{aligned} G_1 \cdots G_{N-1}T_{rec} = T_{E1}G_1 \cdots G_{N-1} + T_{E2}G_2 \\ \cdots G_{N-1} + T_{E3}G_3 \cdots G_{N-1} + \cdots \end{aligned}$$

where T_{Ei}, G_i are, respectively, the equivalent input noise temperature and the gain of the ith stage, $i = 1, 2, \ldots, N$. The overall equivalent noise temperature T_{rec} is given by

$$T_{rec} = T_{E1} + \frac{T_{E2}}{G_1} + \frac{T_{E3}}{G_1 G_2} + \cdot \cdot \cdot + \frac{T_{EN}}{G_1 G_2 \cdot \cdot \cdot G_{N-1}}$$

This result shows that if the signal power can be amplified in the first few stages, e.g., G_1, $G_2 >> 1$, the noise contribution from later stages can be reduced significantly. However, if the earlier stages are lossy, e.g., G_1, $G_2 << 1$, then the noise contribution from these stages will be significant. Thus, it is extremely important that the loss from the antenna to the receiver input port should be reduced to a minimum, or a low-noise amplifier should be used immediately after the antenna, if possible.

In terms of the noise figure,

$$F_{rec} = F_1 + \frac{F_2 - 1}{G_1} + \frac{F_3 - 1}{G_1 G_2} + \cdot \cdot \cdot + \frac{F_N - 1}{G_1 G_2 \cdot \cdot \cdot G_{N-1}}$$

Here F_i is the noise figure of the ith stage, $i = 1, 2, \ldots, N$. The system noise temperature of a radiometer is given by $T_{sys} = T_r + T_{rec}$, with T_r being the antenna noise temperature. Using the results from statistical analysis of random processes, one can show that the temperature sensitivity due to thermal noise for the ideal radiometer is given by (Ulaby et al. [86])

$$\Delta T_N = \frac{T_{sys}}{\sqrt{B\tau}} \tag{5-1}$$

where B is the predetection bandwidth of the radiometer (i.e., the bandwidth over which the emissivity of the source of the antenna temperature is being measured) and τ is the integration time.

The ideal radiometer case assumes that the receiver's system gain remains constant. In practice, gain variations are unavoidable because of supply voltage changes and ambient temperature fluctuations. The temperature uncertainty due to gain variations alone is given by

$$\Delta T_G = T_{sys} \left(\frac{\Delta G}{G} \right) \tag{5-2}$$

where ΔG is the rms value of the gain variation.

The radiometer sensitivity, i.e., the combined uncertainty, ΔT, assuming that the respective contributions to the uncertainty due to noise and gain variations are statistically independent, is given as

$$\Delta T = T_{sys} \sqrt{\frac{1}{B\tau} + \left(\frac{\Delta G}{G} \right)^2} \tag{5-3}$$

The relative significance of the two sources of measurement uncertainty can be shown by the following example. Consider a total-power radiometer operating at a center frequency of 1.4 GHz. It is characterized by the following parameters: $T_{rec} = 600$ K, $B = 100$ MHz, $\tau = 0.01$ s, $\Delta G/G = 0.01$, and $T_r = 300$ K. Then $\Delta T_N = 0.9$ K, $\Delta T_G = 9$ K, and $\Delta T = 9.05$ K, showing that the radiometer sensitivity is dominated by gain variations. The desired sensitivity in remote sensing observation usually is on the order of 1 K or less. Using modern receiver components and frequent receiver calibration, it is possible to eliminate the effect of gain variations so that ΔT is essentially equal to ΔT_N. In radiometry, the objective is to measure T_A and its variation, ΔT_A, over different scenes of observation but the precision of such measurements is dictated by the sensitivity, ΔT. The accuracy of the measured T_A depends on the calibration procedure to be described later.

5.2.2 Radiometer Systems

Radiometers are highly sensitive receivers designed to measure thermally generated electromagnetic radiation from finite or extended scenes. A typical microwave radiometer consists of three subsystems: antenna, receiver, and processor. The antenna subsystem receives incoming radiation. The wideband receiver and electronics subsystem detects and amplifies the received radiation within a specific frequency band. The control and data processing subsystem maintains the antenna and receiver subsystems and processes and records the radiometric data. A good introduction to microwave and millimeter wave radiometer systems for remote sensing purposes may be found in Ulaby et al. [86], whereas a more applied discussion of some systems is given in Skou [74]. Presented here is a review of the operation of the microwave radiometer receiver, the performance characteristics of several different types of radiometers, and some calibration techniques.

The microwave radiometer is a calibrated microwave receiver designed to measure the amount of power received at the antenna or the antenna radiometric temperature, T_r. As the receiver has internal or system noise, the radiometer output contains a component which can be represented by an equivalent receiver noise temperature, T_{rec}, at its input.

The power output of the radiometer can be given by

$$P = kBG(T_r + T_{rec})$$

where B is the bandwidth and G is the receiver gain.

The radiometric resolution or radiometric sensitivity, ΔT, is the minimum detectable change in the radiometric antenna temperature of the observed scene. The spatial resolution of the radiometer is expressed usually as the angular resolution of the radiometer. It is the smallest angle at which two discrete sources can be distinguished.

Many different hardware configurations have been used for microwave and millimeter wave radiometers. Each design is the result of trade-offs between parameters that must be measured, most significantly the temperature (radiometric) and spatial resolutions of the radiometer, the physical size and weight limits of the platform that will carry the radiometer, and the financial resources. Other parameters that must be considered include the power consumed by the radiometer and the data processing, storage, and display requirements.

Total-Power Radiometer

The simplest radiometer configuration is that of the total-power radiometer for which a block diagram is shown in Fig. 5.1. The components include an antenna, a mixer and local oscillator, a band-pass filter, an intermediate frequency amplifier, a square-law detector, and a low-pass filter before the output. Square-law detectors are used because the output voltage from the detector is proportional to the input power to the detector.

If the antenna is connected to the receiver through a lossy waveguide section, then the equivalent noise temperature, T_{rec}, at the waveguide input is obtained by assuming the radiometer to be a two-stage cascaded system, so that

$$T_{rec} = (L - 1)T_w + T'_{rec}L$$

Here, T'_{rec} is the equivalent noise temperature of the receiver at the waveguide output; T_w is the waveguide physical temperature; and L is the loss factor of the waveguide. The term $(L - 1)T_w$ is the equivalent noise temperature of a lossy waveguide with a loss factor L. Since $L > 1$, it is evident that the waveguide loss increases the receiver noise temperature and has to be reduced to a minimum.

The system temperature of the radiometer is given by:

$$T_{sys} = T_r + T_{rec}$$

which includes the radiation, T_r, received by the antenna. The radiometric sensitivity due to thermal noise, ΔT_N, of the radiometer is reproduced here from (5-1):

$$\Delta T_N = \frac{T_{sys}}{\sqrt{B\tau}}$$

The output voltage from the total-power radiometer is given by

$$V_{out} = (T_r + T_{rec})G$$

where G is the total system gain taking into account amplifications, conversion losses, and path losses.

The radiometric sensitivity, ΔT, can be lowered (improved) by increasing the postdetection-integration time, τ, or by decreasing the postdetection bandwidth. The upper limit to the integration time in a scanning system is determined by the amount of time it takes the beam to move by one antenna beamwidth, otherwise the system could not properly detect differences in the brightness temperatures of adjacent footprints.

As seen in previous discussions, gain variations also affect the radiometric sensitivity of the total-power radiometer. If ΔG is the root mean square value of the system power gain variation, the minimum temperature resolution of the radiometer is given by (5-3),

$$\Delta T = T_{sys}\sqrt{\frac{1}{B\tau} + \left(\frac{\Delta G}{G}\right)^2}$$

In practice, the temperature resolution of total-power radiometers which are not calibrated frequently is dependent mostly on gain variations.

Dicke Radiometer

The Dicke radiometer was the first to overcome the gain stability problem to a great extent. Its construction is shown in Fig. 5.4. The Dicke differs from a total-power radiometer in having a switch (the Dicke switch) before the input to the receiver. The switch can send either the antenna signal to the receiver or the signal from a reference noise source whose noise temperature, T_R, is known and is stable. After the detector is a synchronous demodulator which can multiply the output from the detector by either 1 or -1. The switching rate, F_s, is chosen to be fast enough so that the amplifier gain is constant over

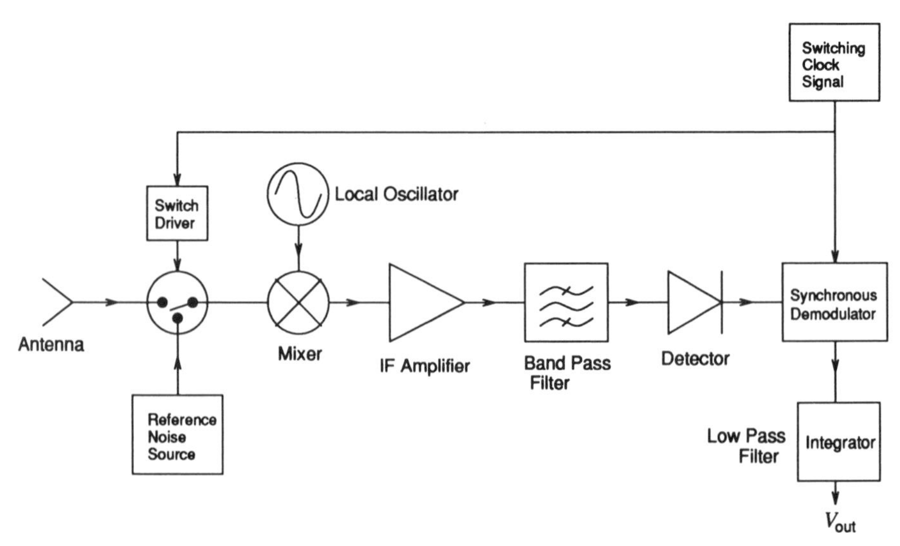

Figure 5.4 Block diagram of Dicke radiometer.

one switching cycle. The output signal, when the input is the antenna, is proportional to the sum of $T_A + T_{rec}$. When the input signal is the reference, the output is proportional to $T_R + T_{rec}$. By setting the multiplier to -1, while the input is the reference, the signal received at the integrator over one switching cycle is proportional to $T_A - T_R$. In this scheme, the dependence of the output on the receiver temperature is eliminated whereas the effect of the system gain variation is reduced because the output voltage of the Dicke radiator is

$$V_{out} = (T_A - T_R)G$$

with $(T_A - T_R) << (T_A - T_{rec})$.

As the Dicke radiometer is looking at the antenna for half of the time compared to the total-power radiometer, the temperature sensitivity of the Dicke radiometer is worse than the temperature sensitivity of the total-power radiometer by a factor of two (if gain fluctuations are not considered) such that

$$\Delta T_{\text{Dicke}} = 2\,\frac{T_{rec} + T_A}{\sqrt{B\tau}}$$

Noise-Injection Radiometer

The noise-injection radiometer is a Dicke radiometer with a directional coupler added before the switch, as shown in Fig. 5.5. A variable noise source (or a noise source with a variable attenuator) is attached to one port of the coupler. A feedback loop is also added to the system. The output of the Dicke radiometer is proportional to $T_A - T_R$. The reference is chosen always to be greater than the antenna temperature. If additional noise, of temperature T_i, is injected into the antenna through the directional coupler, so as to make $T_A + T_i = T_R$, the output of the Dicke radiometer will be zero. The feedback loop is set so as to increase or decrease T_i to keep the output voltage from the Dicke radiometer at zero. The output of the system to the data-collection computer is then the setting of the feedback loop gain (or the injected noise temperature, T_i).

The advantage of the noise-injection radiometer over the ordinary Dicke radiometer is that all dependence of the temperature sensitivity on the system gain has been removed. Several variations of the feedback loop exist. In one other commonly used design, the noise source is pulse modulated with a pin diode switch, at a frequency much higher than the switching rate of the Dicke switch.

Two-Reference–Source Radiometer

Other radiometer configurations exist which differ from the three previously discussed. One type is similar to the Dicke radiometer but it uses two reference sources, one hotter and the other colder than the antenna. The system operates by looking first at the hot reference, then at the antenna, and lastly at the cold reference. The antenna temperature is found by interpolations using the two known references. The temperature sensitivity of this type of radiometer also

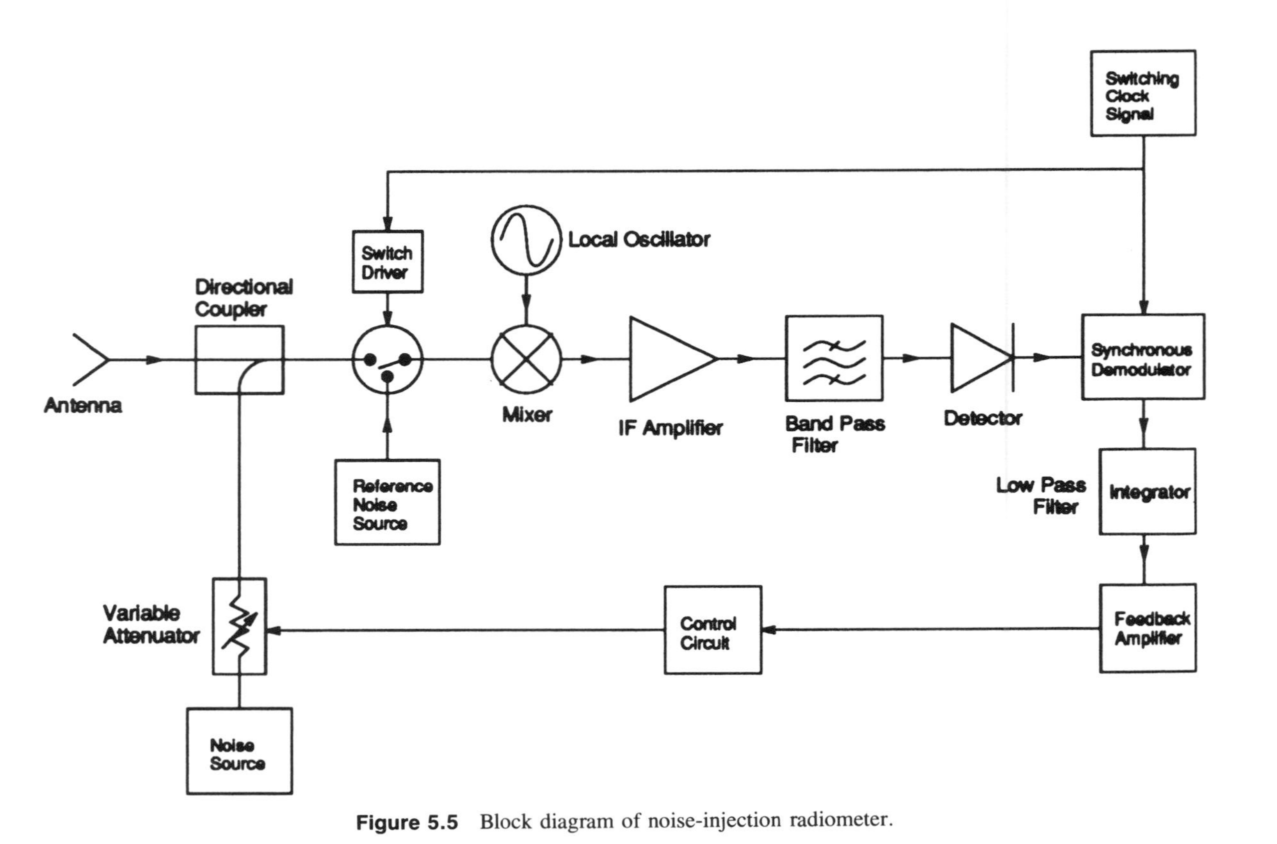

Figure 5.5 Block diagram of noise-injection radiometer.

is independent of system gain variation if the switching cycle between the references and antenna is done fast enough so that the amplifier gain is essentially constant.

The six-channel AIMR built for AES by MPB Technologies Inc., in Montreal, was of the two-reference–load type. Both the "hot" and "cold" loads were heated, temperature-stabilized waveguide terminations. A five-port, three-junction ferrite switch assembly, specifically designed for use in radiometers, was used to cycle the radiometer input between the load and the antenna, as shown in Fig. 5.6.

One disadvantage with radiometers which use internal waveguide-type reference sources, such as noise diodes or thermally stabilized waveguide terminations, is that the actual noise temperature seen by the radiometric receiver usually is not equal to the noise temperature of the noise source. This situation occurs because there is loss through the waveguides and junctions and through the switch. Thus, the internal noise sources cannot be used to give an absolute calibration to the radiometer. Periodically the radiometer must be calibrated externally as will be discussed under "Calibration."

Imaging Radiometers

Imaging radiometers are designed to scan in some manner so as to give a two-dimensional radiometric image of a scene. They are the type of radiometer used most commonly aboard satellites and aircraft. The axis of motion parallel to the flight direction gives one dimension of the image while either the radiometer or a reflector of some kind is moved to give the other dimension of the scan.

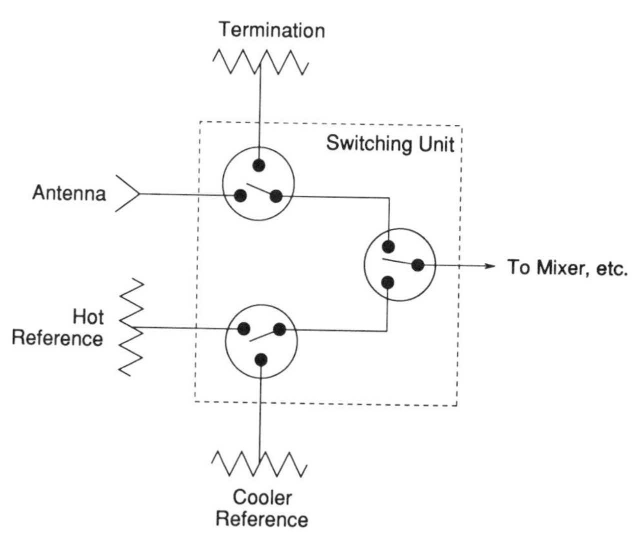

Figure 5.6 Reference noise source and antenna configuration of MPBT six-channel radiometer built for AES.

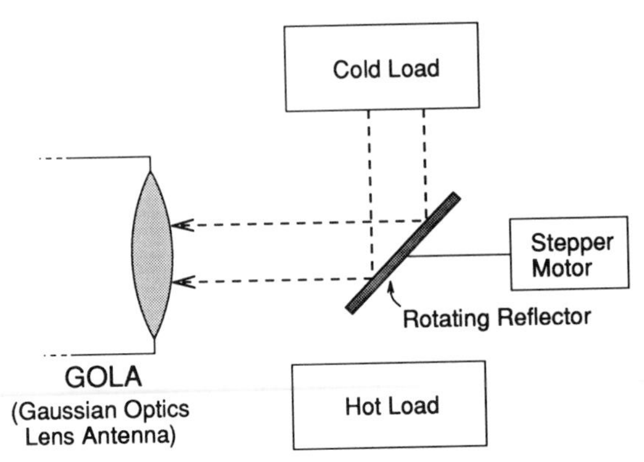

Figure 5.7 Scanning mirror and calibration load configuration of the airborne imaging microwave radiometer.

As a specific example, AIMR used a mirror mounted at a 45° angle to its axis of rotation, as shown in Fig. 5.7. This mirror allowed ±60° of the scene to be scanned in one dimension while the forward aircraft motion gave the other dimension of the scan. Mounting the AIMR, or some other similar scanning radiometers, on an elevation positioner would also permit imaging of an extended area.

One other important feature of the AIMR was the configuration of its calibration loads, as shown in Fig. 5.7. With each scan of the mirror, the radiometer looked first at a hot load, at the ±60° of the scene, and then at a cold load. The loads were designed carefully (and were thoroughly tested) to be blackbody emitters at the frequencies of concern. The temperatures of the loads also were monitored with care. With each scan of the mirror, the signal from the scene was compared with the known blackbody emissions of the loads, allowing the scene brightness temperature to be determined. This design allowed the radiometer to be recalibrated with each scan of the mirror. Gain variations and noise fluctuation effects on the systems radiometric resolution were minimized.

Calibration

The radiometric resolution, or the minimum temperature change that can be detected by a radiometer, is an essential quantity when specifying a radiometer's performance. Another important feature is how close the radiometer's output is to the actual brightness temperature of the scene, or the absolute accuracy of the radiometer. The performance of both mechanical and electronic components will tend to change over time because of thermal, chemical, electrical, and mechanical effects. Hence total-power and Dicke radiometers, and all radiometers that employ internal noise sources to reduce system gain fluctuations, must be recalibrated to allow their absolute accuracy to remain within some specified limit. Modern components and design techniques are allowing radiometers to operate for longer periods before recalibration.

A typical recalibration procedure involves the use of two or more blackbody sources of known temperature. Usually one of them is very cold, e.g., a liquid nitrogen load, whereas the other is at ambient temperature. Other techniques rely on pointing the radiometer at a natural source of known brightness temperature, such as a clear sky or space. The radiometer antenna is pointed at the cold load and the radiometer output corresponding to the cold-load brightness temperature is recorded. The radiometer output corresponding to the brightness temperature of the warmer load is recorded similarly. From these two points, the radiometer is adjusted to give an output corresponding to the true brightness temperature seen by the antenna. In modern radiometers, this adjustment is usually accomplished with software.

One of the advantages of a scanning radiometer design, such as that employed by the AIMR, with free-space loads used as reference loads instead of waveguide loads, is that the brightness temperature seen by the antenna is the true brightness temperature of the load. This design removes the necessity for conducting frequent calibrations.

5.2.3 Spaceborne Radiometers

In 1987 the U.S. military launched the first in a series of spacecraft that carry the special scanning microwave imager (SSM/I). DMSP spacecraft are placed in a sun-synchronous, near-polar orbit at an altitude of about 883 km with an inclination of 98.8° and an orbit period of 102.0 min. With a swath width of 1400 km and 14.1 full-orbit revolutions per day, the SSM/I provides nearly global coverage every day. It was the first passive microwave sensor to be used operationally in near-real time. This sensor measures both horizontally (H) and vertically (V) polarized radiances at 19.4, 37.0, and 85.5 GHz and V-polarized radiance at 22.2 GHz. The effective field of view of the SSM/I is frequency dependent, with the highest resolution at 85.5 GHz (15 km to 13 km) and the lowest at 19 GHz (69 km to 43 km). A detailed description of the sensor calibration, geolocation of the observed variables, antenna pattern correction, and translation of raw data to brightness temperatures is presented in Hollinger et al. [40].

The antenna system used for both the SMMR (on the earliest *Seasat-A* and *Nimbus 1* satellites) and SSM/I contains a parabolic reflector fed by corrugated, broadband, multiport horn antenna. Scanning is achieved by oscillating the reflector about an axis coincident with the axis of the feedhorn. Figure 5.8 illustrates the instantaneous field of view of the SSM/I during the scan. The separation between successive scans is 12.5 km along the in-track direction, almost equal to the resolution of the 85-GHz beams.

Calibration temperatures for the SSM/I are provided by cold-sky radiation and a hot reference absorber following each scan (i.e., about once every 2 s). The feedhorn is passed beneath a hot-load blackbody radiator at a nominal temperature of 250 K and a small mirror reflecting the cosmic background radiation of 3 K into the feedhorn's field of view. Five measurements of microwave emission from each calibration load are recorded. A voltage value at

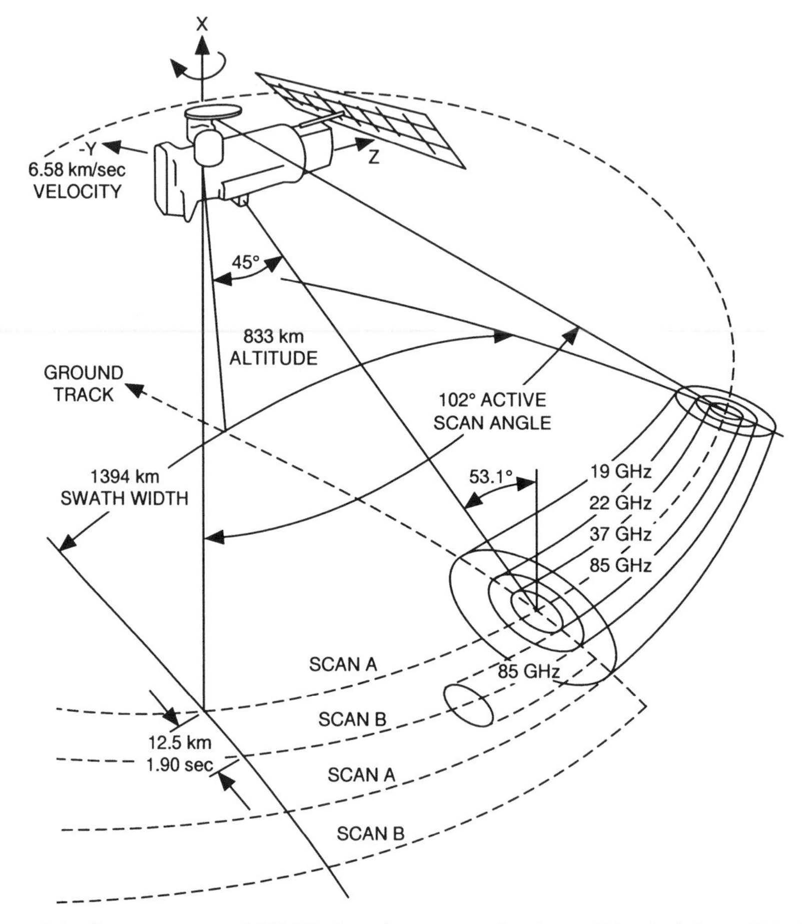

Figure 5.8 Scan geometry of SSM/I. Scan A: scene stations/scan 128, pixels/scan 576; Scan B: scene stations/scan 128, pixels/scan 256; scene stations/orbit 404,224 and pixels/orbit 1,373,728.

the output of the SSM/I receiver associated with each of these measurements is quantized and stored, using a digital unit of measure called "counts" (Hughes Aircraft [43]). The counts values C_{cold} and C_{hot} correspond to the SSM/I measurements of the cold and hot calibration loads, with known brightness temperatures, T_{cold} and T_{hot}, respectively. The calibration loads emit a constant amount of microwave emission during the time (0.1 s) of measurement. The subsequent counts value (C) is converted to a calibrated value of T_A:

$$T_A = aC + b$$

where

$$a = \frac{T_{hot} - T_{cold}}{C_{hot} - C_{cold}}$$

and

$$b = T_{cold} - aC_{cold}$$

The frequent radiometric calibration and the use of low-frequency noise amplifiers and detectors result in extremely small receiver-gain drift over the calibration period.

Verification of the long-term stability of the SSM/I was achieved by evaluating the repeatability of the absolute brightness temperatures for a number of selected target areas. Diverse surface types were selected, such as the Sargasso Sea, Congo Basin, Amazon Basin, Libyan Desert, and the Kalahari Desert.

The absolute calibration of the SSM/I was achieved using two techniques. In the first method, brightness temperatures were compared with those derived from measurements by SSM/I-simulator radiometers mounted on aircraft underlying the satellite. A total of 18 underflights were made. Data from flights over calm seas (representing a large homogeneous region) away from the shoreline were used to determine the standard error of the SSM/I absolute calibration. The second method involved comparison of measured brightness temperatures with model-predicted values. Model calculations were made for three different surface types: (1) clear, calm ocean areas which exhibit coldest brightness temperatures at 85 GHz, (2) Amazon rain forest, and (3) Arabian desert. These targets were selected because of homogeneous surface characteristics and the existence of tested models for calculations of the brightness temperatures. The expected uncertainties in modeled values are lowest for oceanic calculations and highest for desert surfaces calculations. The best agreement with SSM/I observations was for the calm ocean and the Amazon rain forest. The conclusion reached was that the standard error of the absolute calibration of the SSM/I, using both methods, is on the order ± 3 K (Hollinger [41]). Even for the coldest targets (e.g., 70 K for severe storm centers), the relative error is less than $\pm 5\%$.

During processing of the SSM/I data used by the research community at Fleet Numeric Oceanographic Centre (FNOC), ephemeris information from the U.S. Space Command is used as an input to the registration algorithm developed by Hughes Aircraft Company [43] so that SSM/I data can be geolocated. The SSM/I data are used for routine mapping of sea-ice concentration, ice-edge delineation, sea-surface wind speed, integrated water vapor, liquid water and rain rate, snow cover on land, and soil moisture.

Although DMSP is a military satellite system, the SSM/I data have been made available to the research community and private sector through the Cryospheric Data Management System in Boulder, Colorado (historical data), and the Civilian Navy/National Oceanic and Atmospheric Administration (NOAA) Oceanographic Data Distribution System.

5.2.4 Airborne Imaging Radiometer

Stimulated by the successful use of satelliteborne sensors for sea-ice monitoring and the results of previous efforts in the United States with airborne platforms, development of a Canadian-built airborne passive microwave sensor was undertaken by AES, Environment Canada. A 37- and 90-GHz airborne millimeter-wave radiometer, AIMR, was built by MPB Technologies Inc., in Dorval, Quebec, and delivered to the AES Ice Branch in February 1989. This system was designed as both a scientific tool and an operational system for ice surveillance. The AIMR can process data and display images in real time for operational requirements and the raw data can be stored in an on-board computer for further analysis.

AIMR consists of two major subsystems, the external payload assembly (EPA) and the cabin-mounted equipment (CME). Affixed underneath the aircraft fuselage, the EPA collects radiometric data from a swatch extending 60° on each side of the nadir (the path directly below the aircraft). The computer system in the EPA controls the scanner, collects data samples, reads the temperatures in the reference loads, and communicates with the CME via an optical link. At the CME, the data are processed and stored on digital tapes. After correction for aircraft motion and geometric distortion, image data can be displayed in real time on a monitor.

When AIMR is used for ice surveillance, the radiometric temperature of a swath of the surface below the aircraft is recorded with the use of a scanning mirror. This mirror sweeps the narrow antenna beam of the AIMR over the surface, collecting data points every 0.5° of mirror movement. The beamwidths for the 37- and 90-GHz channels are 2.4° and 1°, respectively. The scanning rate is adjusted constantly to maintain 50% overlap of adjacent sweeps, to a maximum rate of about 6 rps.

The External Payload Assembly

The EPA consists of: antenna system, scanning system, calibration loads, millimeter-wave radio-frequency front end, analogue electronics, preprocessor, and temperature-control system. All these systems are contained in a weatherproof housing which is shielded against electromagnetic interference. A block diagram of the EPA arrangement is shown in Fig. 5.9. The antenna system consists of a Gaussian optics, beam-forming lens which receives the millimeter-wave radiation and selects the proper frequency and polarization to be fed to each of the four channels (see Fig. 5.10).

The scanning system consists of a mirror driven by a stepper motor which is controlled by the preprocessor. During each complete rotation, the mirror scans up to a maximum rate of 6 rps through the hot and cold calibration loads and the scene consecutively. This arrangement allows the radiometer to be calibrated every revolution, thus relaxing the receiver-gain stability problem. In the AIMR, the hot load is maintained about 100 K above the temperature of the cold load. The latter is left at ambient temperature. The hot- and cold-load temperatures are monitored continuously with thermistor sensors embedded in the loads. The calibration loads are designed to simulate a blackbody, with emissivity close to unity. They are made by molding the high-temperature microwave absorber into special shapes and have less than −40-dB reflectivity when measured at 37 and 90 GHz.

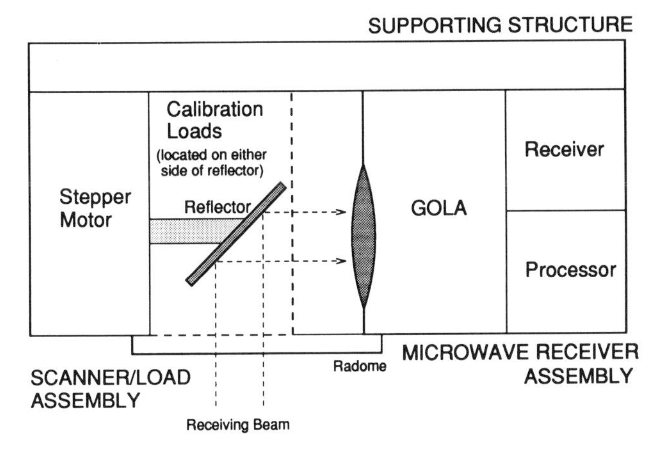

Figure 5.9 Layout of the external payload assembly including rotating 45° reflector for scanning.

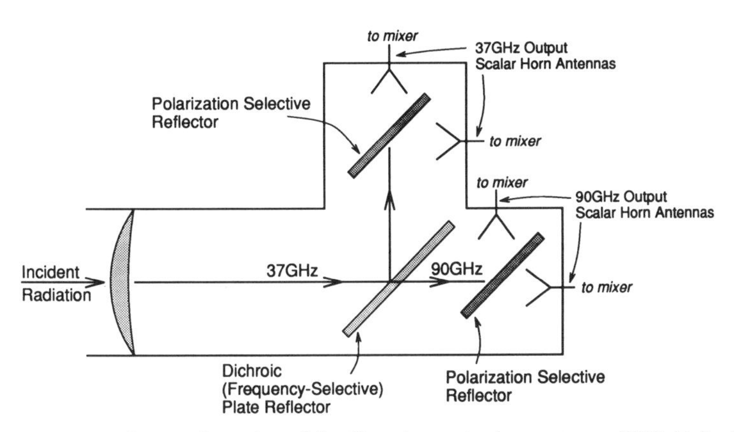

Figure 5.10 Design configuration of the Gaussian optics lens antenna (GOLA) System.

The EPA preprocessing system also monitors the load temperatures and other critical temperatures, controls channel gains, and multiplexes the four channels of data into a serial bit stream for transmission on the fiber optic link to the CME.

Cabin-Mounted Equipment

The CME consists of the following rack mounted units: power supply, processor, control unit, digital recorder unit, red-green-blue (RGB) color monitor, NTSC encoder, and videocassette recorder (VCR). There is also a small, rack-mounted power-control panel.

The processor is a 32-bit VME bus system using a 16.7-MHz 68020 processor and a 6888 floating-point coprocessor. Its operating system is booted from the control unit which has a rack-mounted video terminal. Control of the system is done through menu-type displays. The processor receives data from the EPA through the optical link, and navigation data from the aircraft management computer. In addition to recording such data on the digital recorder, the processor converts the angular information for each pixel to a ground position corrected for velocity, altitude, and aircraft roll and drift. It also converts the radiometer output to brightness temperatures.

At the operator's command, the processor generates and stores either "raw" images of apparent brightness temperature maps or "derived" images. The latter are estimates of the fraction of ice of different types as derived from algorithms incorporated in the software. These algorithms utilize the radiometric characteristics (i.e., the brightness temperatures at the two frequencies and the two polarizations) to estimate the percent concentration of the different ice types. The images can be displayed in real time on a monitor for quick review. Hard-copy, higher quality but static, color images are produced on a ground-based analyzer.

AIMR has logged about 600 hr of successful airborne operations. At the present time, plans call for implementing software and hardware upgrades and using the AIMR (mounted on the National Centre for Atmospheric Research aircraft) in future sea-ice research campaigns.

5.3 RADIOMETRIC SIGNATURES OF SEA ICE AND ICEBERGS

The sea-ice classification approved by the World Meteorological Organization (WMO) is based on the visible albedo and morphology of ice. The ice categories, in general, correspond to different stages of ice growth: (a) young ice, less than 30-cm thick; (b) first-year ice, between 30 cm and 2 m in thickness; and (c) old ice, with thickness exceeding 2 m which has survived at least one summer season.

Because sea ice is a mixture of ice, salt, brine pockets, and air bubbles, it is structurally complex. The dielectric constant of this mixture is strongly influenced by the number of brine inclusions. The shape and the concentration of these brine inclusions depends on the ice growth rate.

Microwave emissions from sea ice are a function of several factors, including ice thickness, dielectric constant and temperature profile of the ice layer, thickness of any snow layer that may be present on top of the ice, and degree of inhomogeneity and anisotropy of the ice volume (Comiso [19]; Comiso et al. [21]). The modeled and observed dependence of sea-ice emissivities for several frequencies indicate that for frequencies greater than 10 GHz, the ice layer behaves electromagnetically although it is infinite in depth, i.e., the emission originates from layers near the surface (see Chapter 2).

At the present time, only the brightness temperatures of open-ocean and first-year ice without snow cover are well understood. In addition, the spatial resolution of sensors on different platforms may limit discrimination of the number of ice-type classes. While surface sensors provide point measurements of radiometric properties of a single ice floe, satellite sensors collect radiation from a wider area which may contain several different ice types. Observations from aircraft platforms provide information on a finer scale than satellite sensors, allowing not only comparison between signatures of different ice types but adding to our understanding of the relationships between such features as ridges, leads, hummocks, and meltponds and observed brightness temperature.

In the following subsections, the radiometric characteristics of sea ice and icebergs are presented. A detailed discussion of temporal and regional radiometric properties of sea ice can be found in Carsey [10]. Analysis of sea-ice brightness temperatures measured at SMMR frequencies has shown (Rothrock et al. [64]) that only three channels can be used as independent sources of information. The addition of the 85-GHz frequency to the SSM/I provides more information but its sensitivity to atmospheric changes reduces the use of this channel in retrieval algorithms. As yet no procedure has been developed for correcting atmospheric effects at 85 GHz that could be used for unsupervised ocean surface classification (or processing of data with algorithms). Our discussion will be limited to newly formed, first-year ice and to old-ice radiometric properties because the existing algorithms can differentiate only a limited number of ice surface conditions. At this stage, considering the coarseness of the SSM/I resolution, the routines have not been developed for monitoring icebergs. Very large icebergs observed with the SSM/I in the Weddell Sea had brightness temperatures typical of mixtures of old and first-year ice (Carsey [10]).

5.3.1 Formation and Growth

Changes in the dielectric properties of water as it freezes and as new ice develops result in corresponding changes in microwave emissivity. The microwave emissivity is a function of ice growth rate, distribution of the brine trapped in the ice, and during the initial stages of growth (up to 0.3 m), on the weather conditions (Stogryn and Desargent [79] Cavalieri et al. [14], Stogryn, [80]). The mechanical and thermodynamical influences of the ocean and atmosphere on the roughness, chemical composition, and the mechanical properties of sea ice and, therefore, its dielectric properties, are very complex. Hence, empirical relationships are used to relate the electromagnetic properties (i.e., microwave emissivity) and the formation stages (Grenfell and Comiso [35]). Changes in the salinity of ice as it ages, accumulation and metamorphosis of the snow cover, and deformation of the ice sheet each contribute to the observed variance in radiometric brightness temperatures (the physical and dielectric properties of sea ice and snow are described in Chapter 2).

A smooth water surface is radiometrically colder than land or sea ice and its microwave emission is highly polarized. Laboratory and field experiments suggest that observed variations in radiometric brightness temperatures of freezing water and forming sea ice can be explained by considering the physical changes of a surface composed of ice and water. For example, it was observed

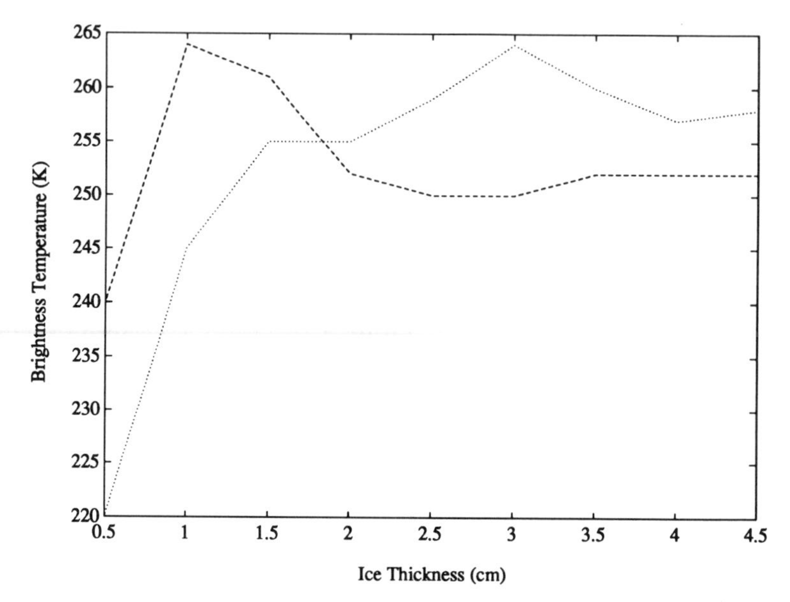

Figure 5.11 Vertical polarization brightness temperature versus ice thickness. Symbols used: --- 37 GHz, · · · 19 GHz. (After Grenfell et al. [36]).

(Eppler et al. [28]; Grenfell et al. [36]) that at 33.6 GHz, the brightness temperature of water is between 135 K and 155 K (depending on the wind-induced surface roughness). Newly formed first-year ice has brightness temperatures between 230 K and 260 K, depending on the physical temperature and thickness of the ice (Fig. 5.11). The potential effect of different factors in sea-ice structure on the observed brightness temperature is illustrated in Fig. 5.12 (Eppler et al. [28]).

Discrimination of newly formed ice from open water relies on the contrast in the emitting (or reflecting) properties. Unfortunately, similarities in the signatures of new ice and open water, across the electromagnetic spectrum, limit their discrimination by passive microwave airborne and spaceborne observations. At the present time, there is no reliable, single-sensor technique for determination of thin ice within the field of view. In addition to resolving some of the problems involved in using single-sensor radiometry for thin-ice detection, a multisensor approach may remove some of the ambiguities of ice-type discrimination.

Passive microwave sensors on satellites may not be capable of detecting small areas of new ice but they have greatly improved our knowledge of the seasonal extent and variability of sea-ice zones. With daily satellite observations of polar regions, the areas with newly formed ice are monitored through different stages of ice growth. For example, a large-scale ice formation in the Greenland Sea, known as "Oden", has been documented by both SMMR and SSM/I. During several research campaigns, it was determined that this tongue

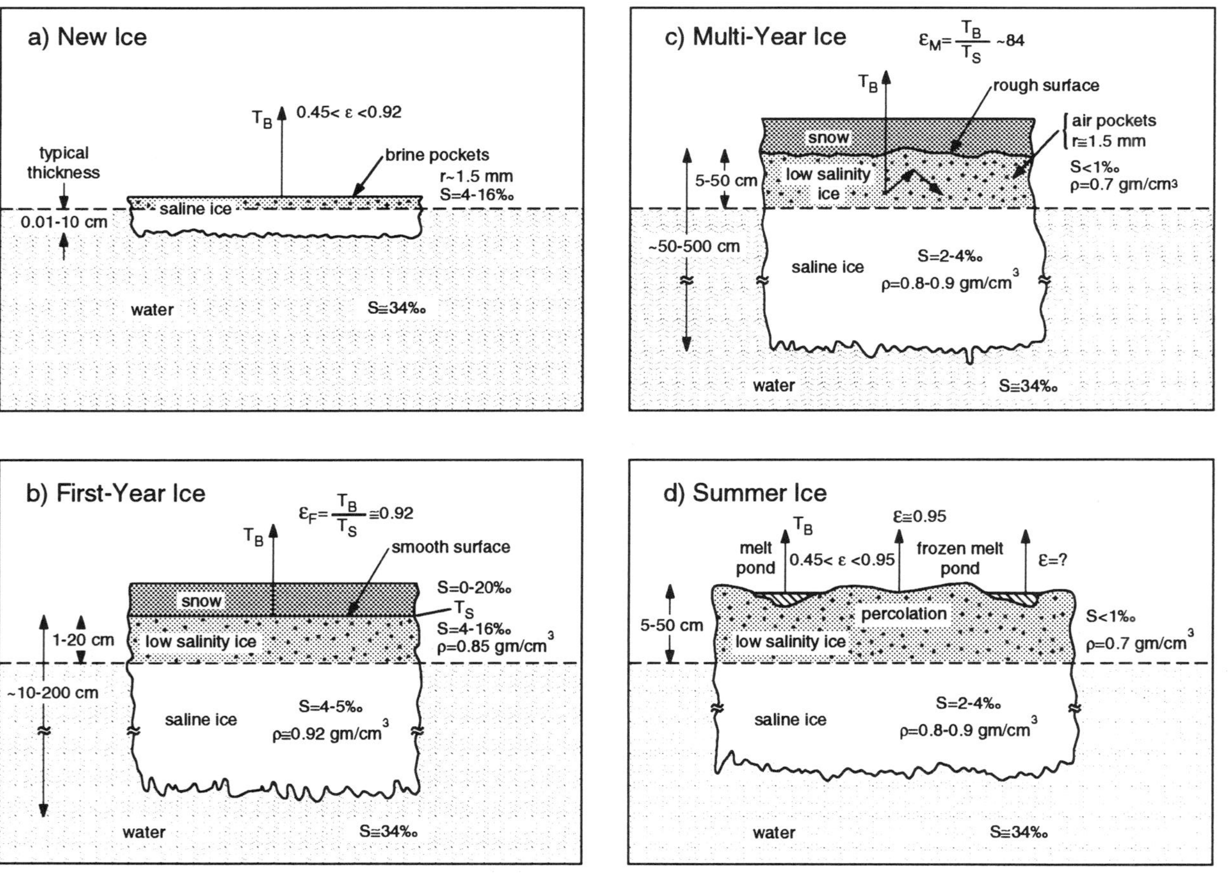

Figure 5.12 Factors that have potential effect on apparent brightness temperature for a new, first-year and old ice. (From Zwally et al. [96]).

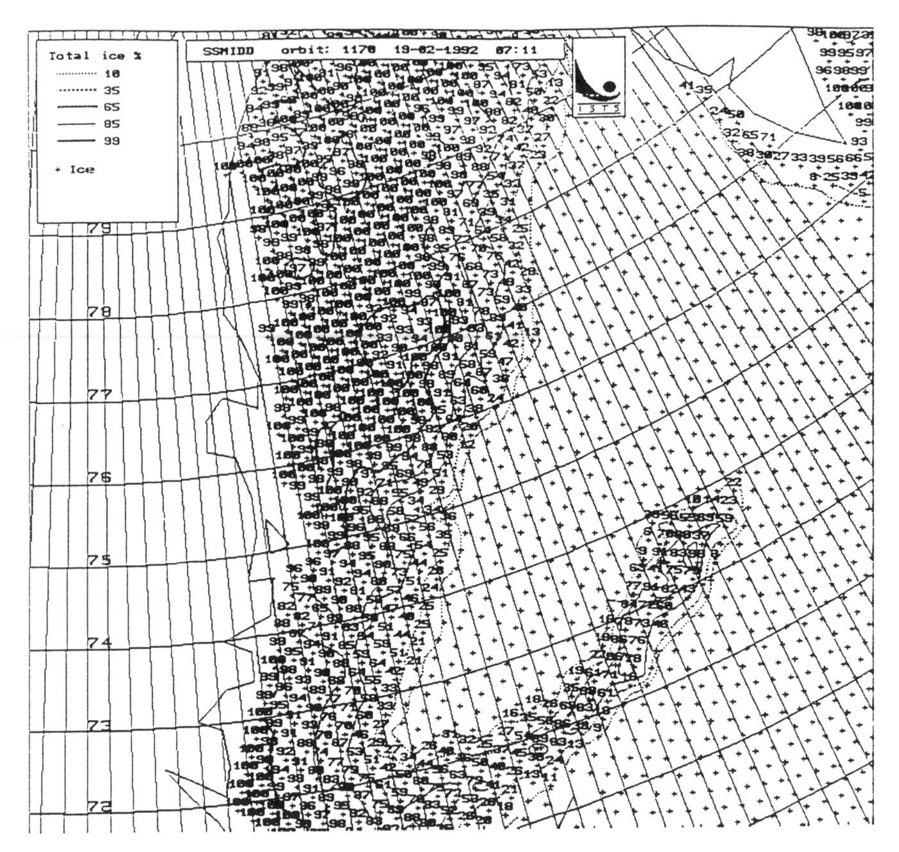

Figure 5.13 Sea ice chart of ice conditions in the Greenland Sea produced using SSM/I data. Numbers indicate percentage of the pixel area covered with ice. The ice field protruding to the right from the pack-ice is known as "Oden." Eastern Greenland Sea is on the left and Spitzbergen is at upper right.

of ice extending east from the main ice pack contained mainly thin ice types (Figs. 5.13, 5.14). The brightness temperatures of sea ice during the growth stage (up to 10 cm in thickness) increase rapidly (Grenfell and Comiso [35]). Surface-based and remote sensors showed that higher frequency radiometric measurements yield greater sensitivity to initial changes in the ice thickness, and the difference between measurements with vertical and horizontal radiation decreases with the increase in ice thickness.

The AES ice chart for 24 January 1992 (Crawford and Madej [25]) classifies ice in the southern area as 70% gray ice (less than 15-cm thick) and 20% gray-white ice (less than 30-cm thick). Figure 5.15 shows an ice chart of retrieved information. The northern study area contained ice of which 70% was less than 30-cm thick and 20% was between 30- to 70-cm thick. The differences in first-year ice signatures (more than 70-cm thick) and less compact, thinner ice can

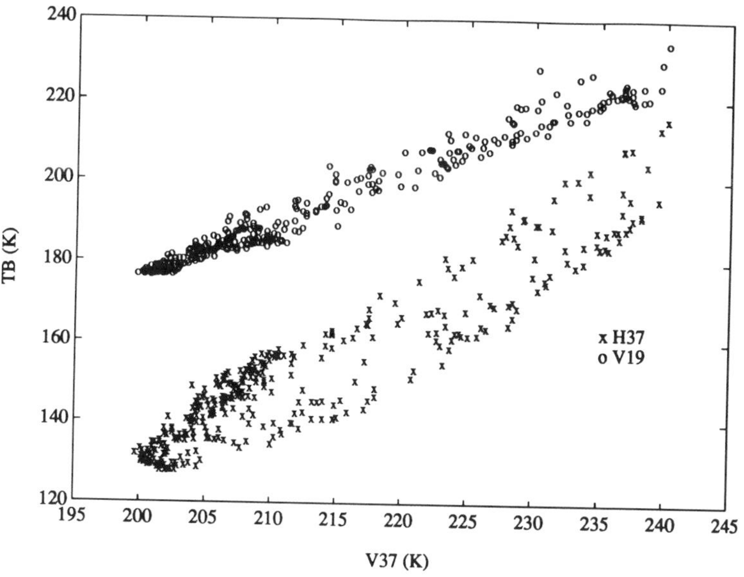

Figure 5.14 Scatterplot of differences in brightness temperatures over the Greenland Sea "Oden." Data obtained by DMSP SSM/I at 19 and 37 GHz for 71.5° to 74.0°N and 14°W to 10°E, 19 February 1992. Most of the ice in "Oden" was at the new to young ice stage.

be observed by comparing Figs. 5.16 and 5.17. Figures 5.16 and 5.17 illustrate changes in the brightness temperatures of ice with time as it thickens. Note the differences in the SSM/I brightness temperatures between the Labrador Sea and eastern Newfoundland waters (49° to 51° N, 46° to 55° W and 52° to 54° N, 46° to 55° W).

5.3.2 First-Year and Old Ice

As microwave emissions from first-year ice originate near the surface, the observed brightness temperatures for snow-free ice areas show very weak dependence on the passive microwave (Carsey [10]). The regional and temporal variability in first-year brightness temperatures can be attributed to scattering and emissions from the snow cover (Comiso [19]). This variability is illustrated in Figs. 5.18 and 5.19. The transect along 130°W from 68° to 86°N contains data for first-year ice south of 72°N, mixed pixels of first-year and old ice from 72° to 76°N, and predominantly old ice north of 76°N.

Typical signature changes of sea ice in seasonal ice zones are illustrated in Fig. 5.20. Fluctuations in brightness temperature that occur in ice greater than 50 cm in thickness can be attributed to the accumulation and metamorphosis of the snow cover on ice. Evaluation of attenuation by the snow requires information on snow-pack structure and the amount of free water (Ulaby et al.

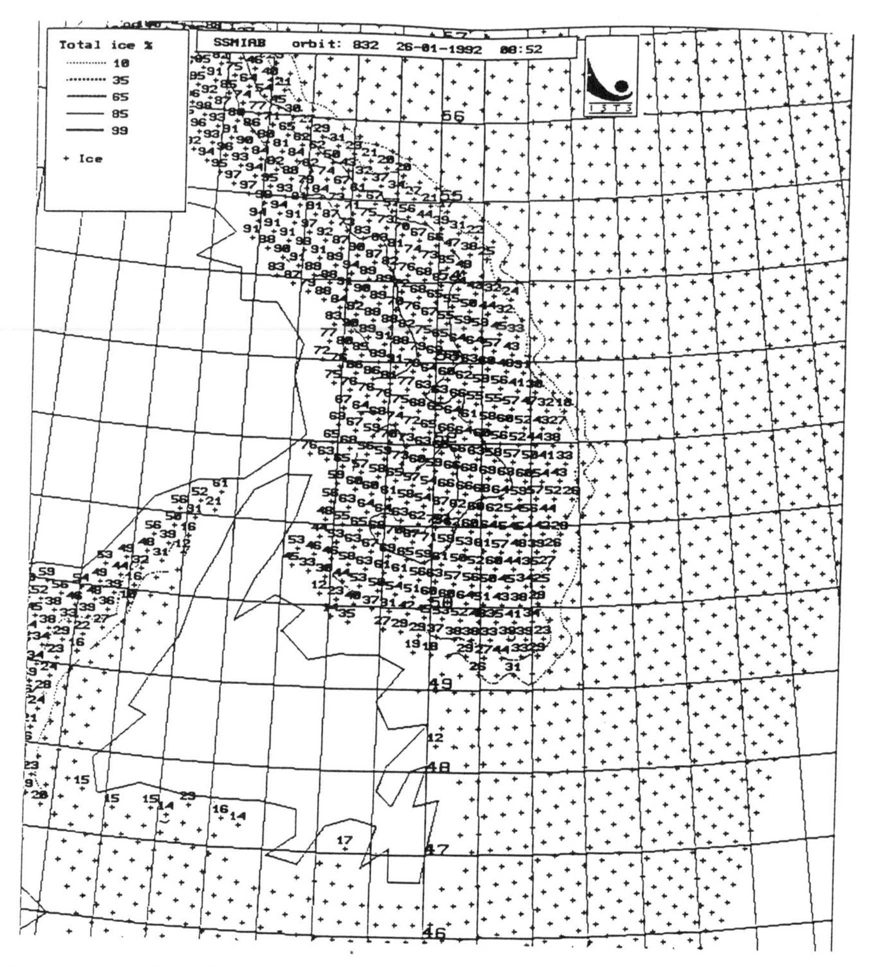

Figure 5.15 Chart of ice conditions in the Labrador Sea produced using SSM/I data for 26 January 1992. The AES/York algorithm was used to calculate ice concentrations.

[86]). The physical temperature of the ice surface and, therefore, the intensity of microwave radiation from snow-covered ice vary depending on the thickness of the snow cover. The effect of snow accumulation on observed brightness temperatures of first-year ice is illustrated in Figs. 5.21 and 5.22. The results of microwave measurements of sea-ice and snow cover can be found in numerous publications (Livingstone et al. [49]; Tucker et al. [85]; Carsey [10]).

The increase in snow-cover depth (i.e., increase in the thickness of an insulating layer and load on the ice surface) is one of the causes for brine release from the layers of ice below the surface. Hence, the lowest layers of snow can become transformed into slush which changes the emitting properties of the surface. In addition, the weight of snow can push the ice below the water

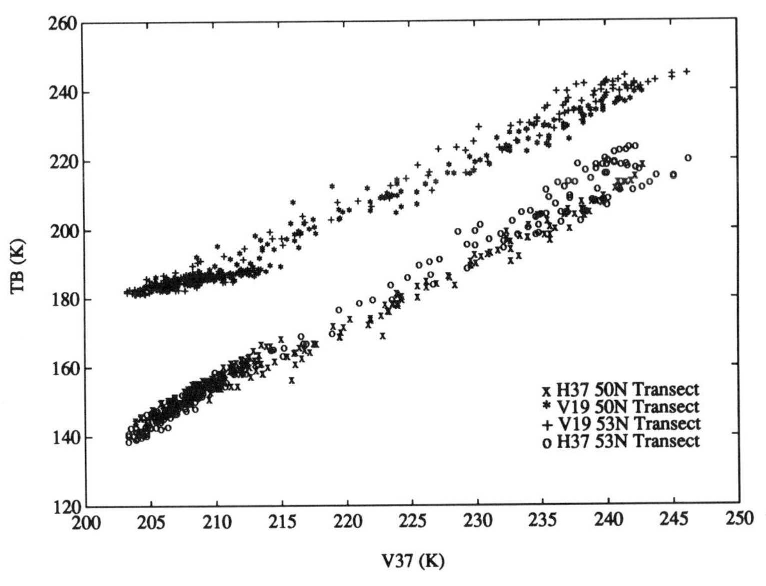

Figure 5.16 Scatterplot of the differences in brightness temperatures of young sea ice. Vertical 19 GHz and horizontal 37 GHz data are compared with the vertical 37 GHz observations. Data obtained by DMSP SSM/I off eastern Newfoundland (50° transect), and Labrador (53° transect), 26 January 1992. The sea ice was predominantly grey ice off Newfoundland and predominantly grey-white ice off Labrador.

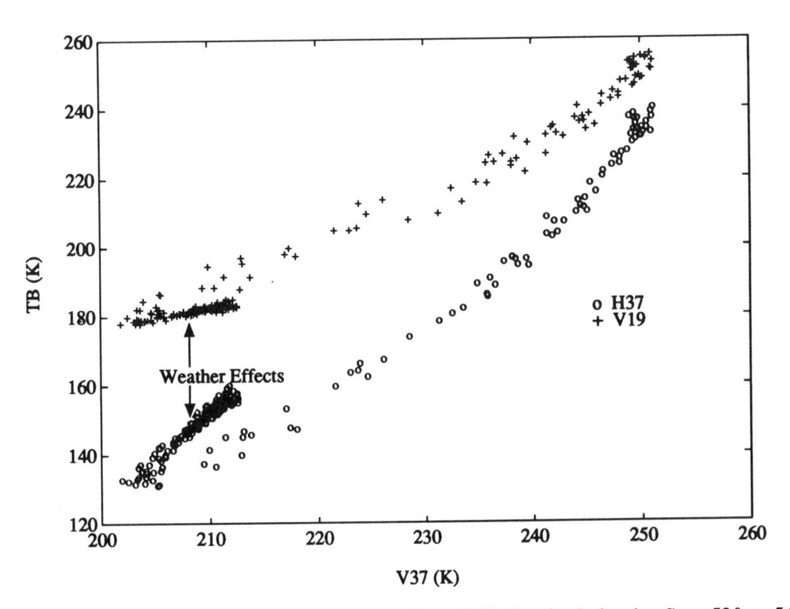

Figure 5.17 Scatterplot of V19 vs V37, and H37 vs V37, for the Labrador Sea, 52° to 54°N, 46° to 55° West, on 5 February 1992 SSM/I data.

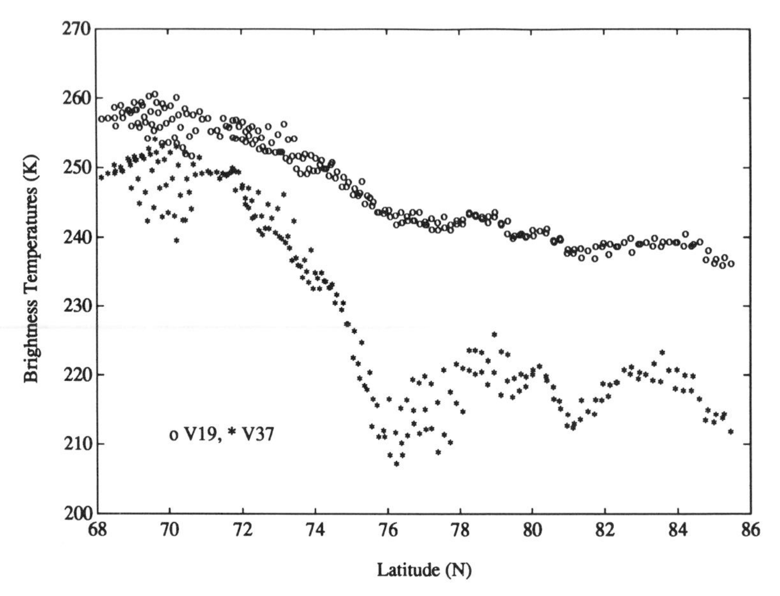

Figure 5.18 Winter brightness temperatures (Arctic Ocean). Symbols used: * for 37 GHz vertical polarization, ○ for 19 GHz vertical polarization.

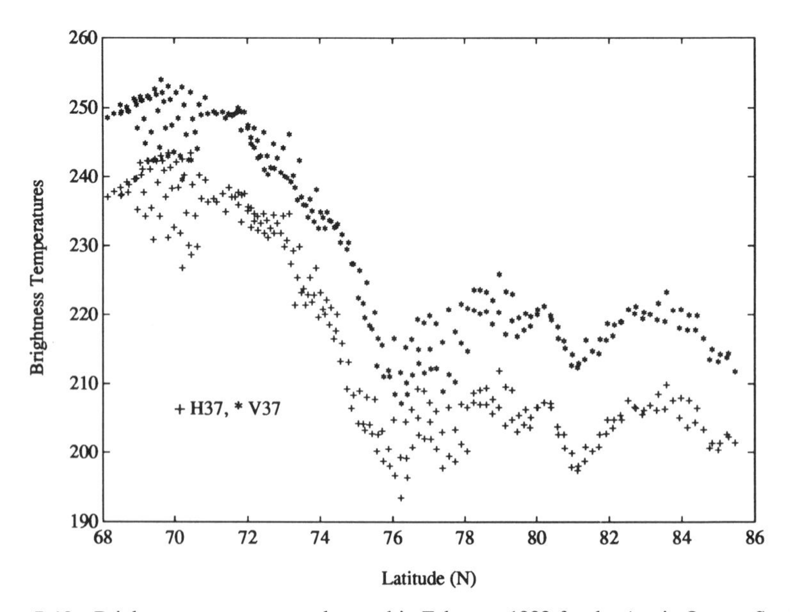

Figure 5.19 Brightness temperatures observed in February 1992 for the Arctic Ocean. Symbols used: * 37 GHz Vertical polarization, + 37 GHz Horizontal polarization.

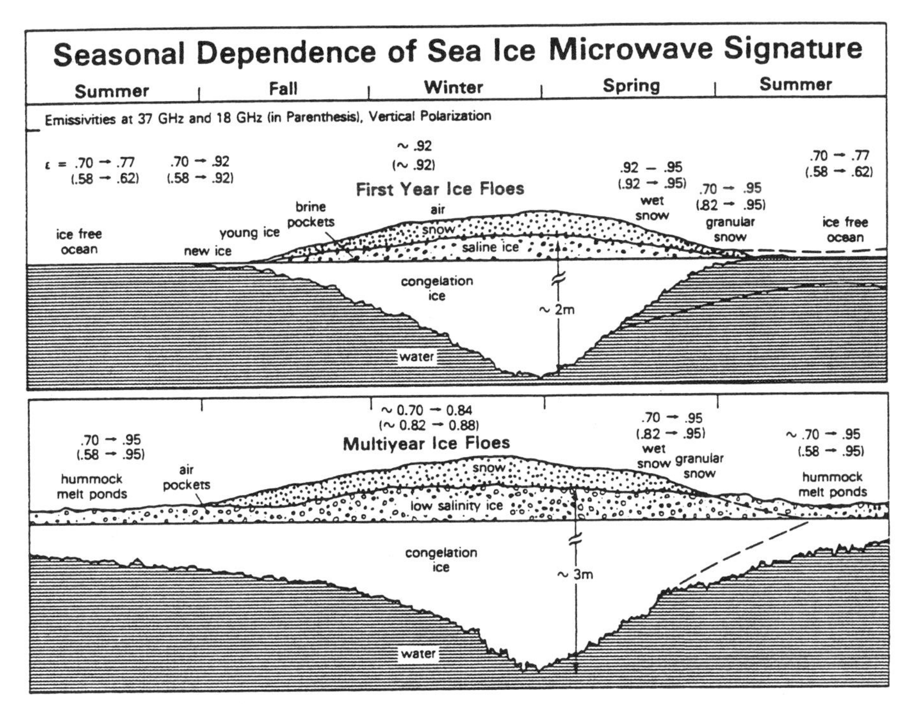

Figure 5.20 Seasonal Dependence of sea-ice emissivity (Arctic) at 18 and 37 GHz. (Courtesy of Comiso, [19]).

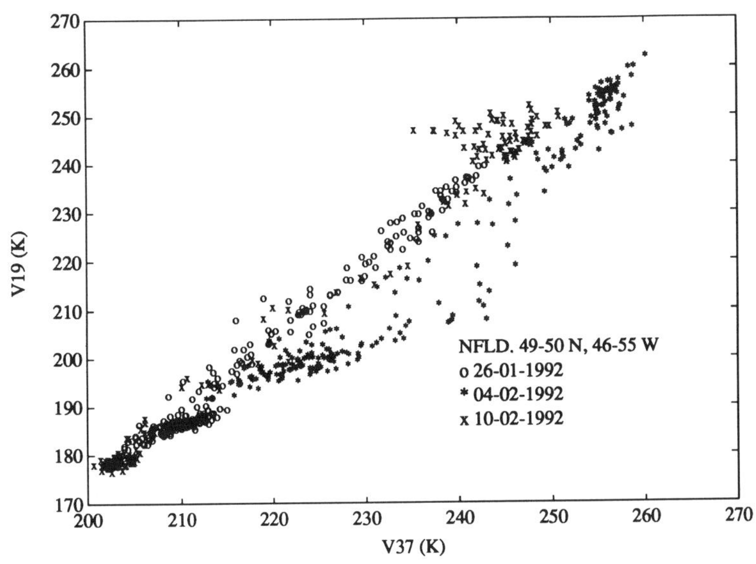

Figure 5.21 The effect of snow cover on first-year ice brightness temperatures. Note changes in brightness temperature at V37 GHz (10 February 1992 data) as compared with V19 GHz observations. Two snow storms (with up to 25 cm of snow) were reported for Newfoundland between 3 February and 10 February.

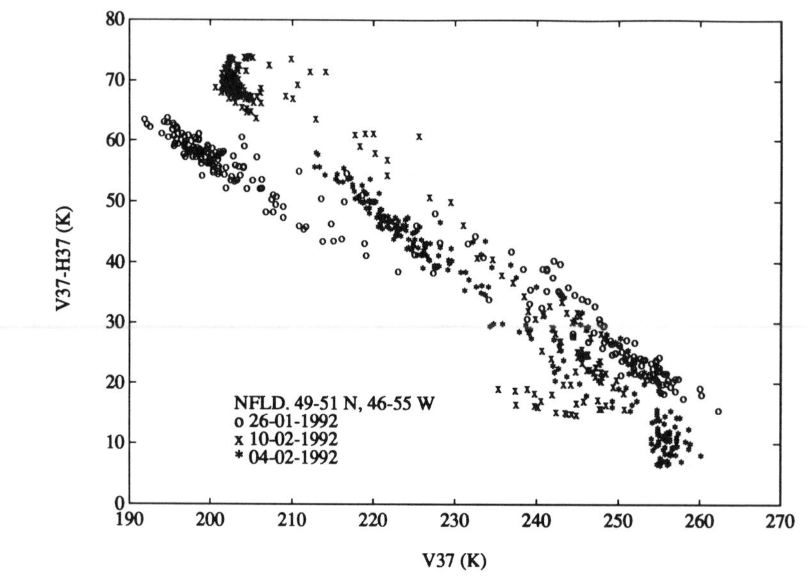

Figure 5.22 The effect of snow accumulation on a polarization difference at 37 GHz for first-year sea ice. The Canadian Atlantic Storm Project reported two snow storms within the period (Crawford and Madej, [25]).

surface, creating a water layer at the snow–ice interface (Comiso et al. [22]; Tucker et al. [85]). The onset of snow melt increases in the free water content which transforms the snow cover into a blackbody emitting surface (see Chapter 2 for an explanation of emitting-depth dependence on temperature) and, this, in turn leads to an increase in observed brightness temperatures. During melt, brightness temperatures of old ice become similar to those of first-year ice (Grenfell and Lohanick [37]; Carsey [10]) (Figs. 5.23, 5.24).

The melt process, refreezing, and consolidation of any remaining snow form a rough layer of superimposed ice; complete melt of the next winter's snow will expose this rough surface. The melting of the ice surface generates pools of water on the surface and the salinity of the ice decreases (Livingstone et al. [49]). Meanwhile, melt water draining from the ice creates a layer of fresh water under the ice. When the ice refreezes with dropping temperatures, it is usually rough and hummocky. Such change of the surface conditions and salinity of the sea ice during melt season processes reduces the radiometric contrasts between the different ice types but the radiometric brightness returns to values typical of cold conditions in late summer or early fall (Cavalieri et al. [11]).

Radiometric signatures of ice that has survived more than one melt season are more complex than signatures of first-year ice and show greater regional variability. One of the factors contributing to this variance is the formation of

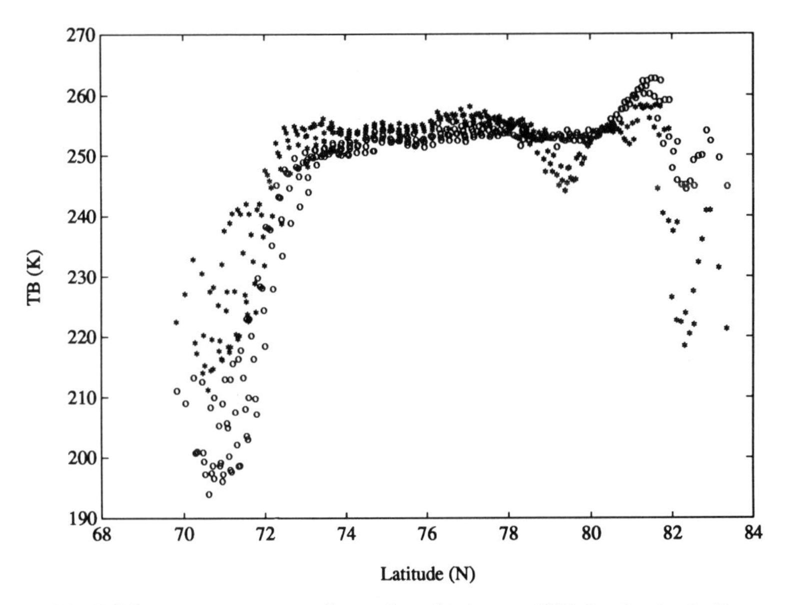

Figure 5.23 Brightness temperatures observed on 14 August 1992 for the Arctic Ocean. Symbols used: * 37 GHz Vertical polarization, ○ 19 GHz Vertical polarization.

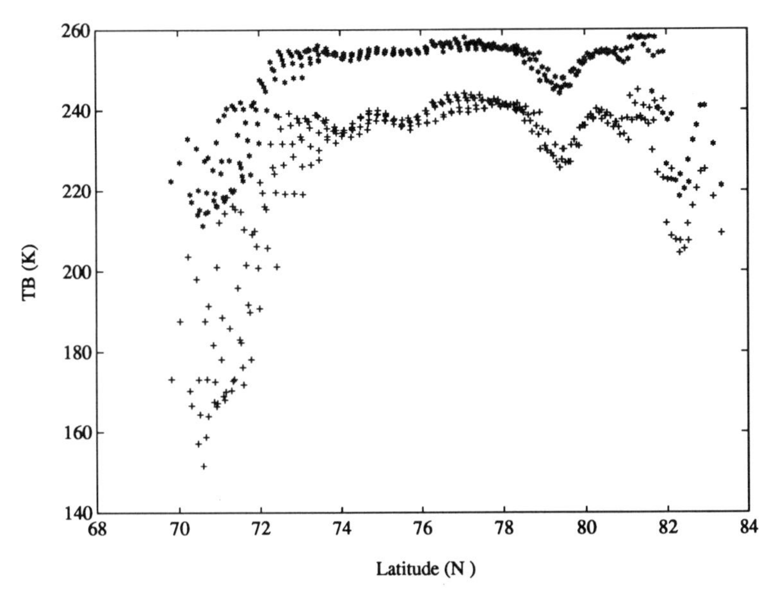

Figure 5.24 Vertically and horizontally polarized 37 GHz brightness temperatures on 14 August 1992 for the Arctic Ocean. Symbols used: ○ for V37, + for H37.

a low-density, porous layer as sea ice melts. The increase in volume scattering (Grenfell [34]) causes reduction in emissivity. The contrast in radiative properties of first-year and old ice is a useful factor in retrieval algorithms. The spatial variability of remotely sensed passive microwave brightness temperatures reflects regional differences in the history and formation of pack ice. For more detailed information on the radiometric properties of different sea-ice types, the reader is referred to Carsey [10].

5.3.3 Icebergs

Icebergs, being fragments of glacial (fresh water) ice, are radiometrically colder than first-year sea ice. When locked within a field of first-year ice, they can be easily detected with an airborne passive microwave (Seling and Nance [72]). When moving in ice-free waters, they may generate a trail of ice that forms when the less saline melt water they shed freezes. This results in an effective area as much as three times larger than the horizontal cross section of the iceberg which contains new and brash ice and, thus, is radiometrically different than sea water.

For detection from space, the icebergs must be very large to fill adjacent pixels of satellite-borne passive microwave sensors. Several icebergs measuring about 20 × 50 km have been observed in the Antarctic Ocean with the SSM/I (Stone [81]) (Fig. 5.25).

5.4 SEA-ICE PARAMETER RETRIEVAL TECHNIQUES

5.4.1 General Description

There are two approaches to the interpretation of remote sensing data. One approach is image oriented because it capitalizes on the pictorial aspects of the data. It uses analysis techniques developed for generation of images. The second approach is numerically oriented because it emphasizes the inherently quantitative aspects of the data, treating the data abstractly as a collection of measurements. In this approach, an image is not thought of as data, but rather as a convenient mechanism for viewing the data. The retrieval of geophysical parameters from multispectral passive microwave data is based on the second approach.

The set of numerical procedures to retrieve physical parameters from the collection of measurements is called an algorithm. One of the most common mathematical techniques used in the development of algorithms is to treat measurements as components in a multidimensional space, i.e., N measurements by a multispectral sensor (e.g., seven channels) can be represented as N points in seven-dimensional space. When several such measurement vectors are plotted, all vectors representing the same data type, but not necessarily in the same geographic location, will plot as a localized cluster, or cloud, of points.

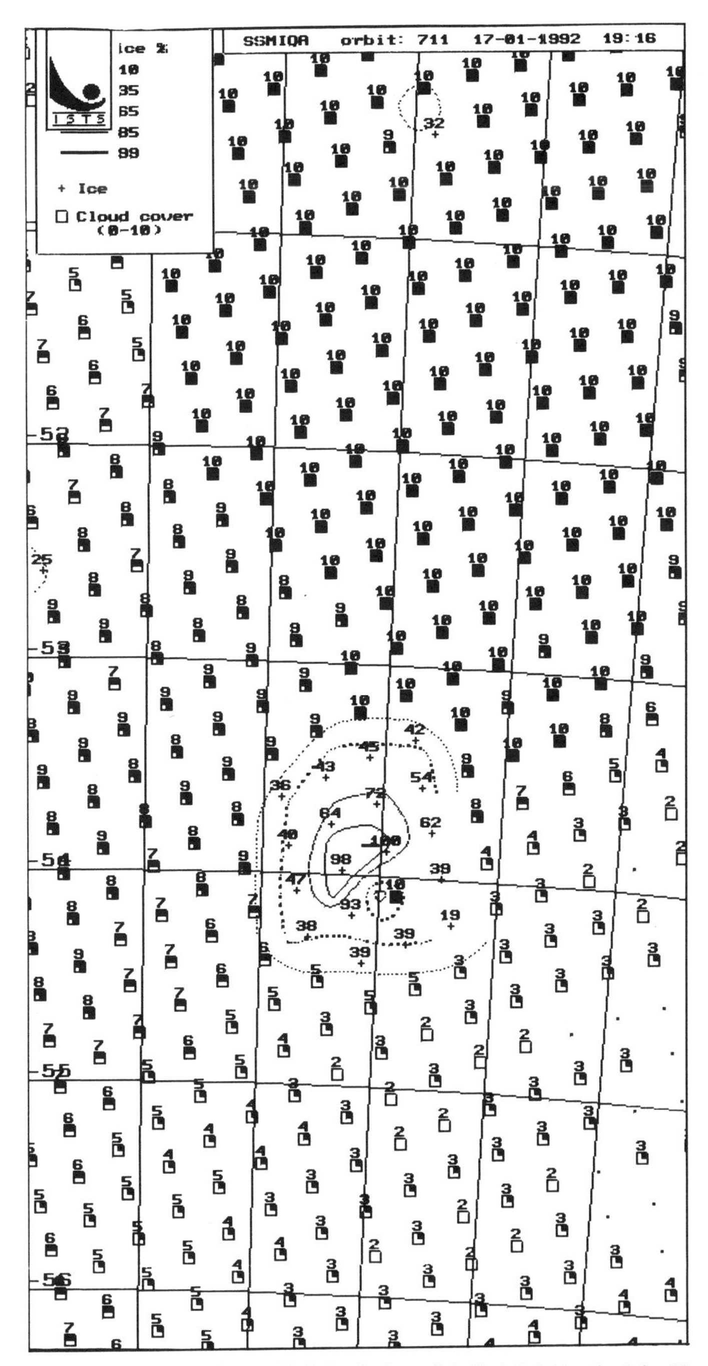

Figure 5.25 SSM/I detection of a Weddell Sea iceberg (labelled A-24 by the Joint Navy/NOAA ice centre) location on 17 January 1992. AES/York algorithm was used for interpreting SSM/I data. Numbers next to + symbols show calculated percentage area of the 37 GHz footprint covered with ice.

The first task of a classifying algorithm is to include criteria (or discriminant functions) for dividing the measurement space into decision regions, separating clusters corresponding to a specific discriminable class. Although it is sometimes possible to determine discriminant functions on the basis of either theoretical consideration or knowledge about the physical problem, most often these functions are derived using information obtained from a set of measurement vectors for targets with known identity. In many remote sensing applications, the spectral responses of the classes of interest overlap and the discriminating process then must include additional information, e.g., statistical distribution. In such cases, the use of statistical patterns allows classifications which are "most probably" correct.

Microwave radiation from the earth is a complex function of the temperature, physical composition, and properties of the earth's surface, altered by absorption, emission, and scattering from the atmosphere. Therefore, the quantitative determination of environmental parameters of interest is obtained from a limited set of microwave radiation measurements that are relatively noisy. In general, these measurements are not sufficient for unique determination of the environmental parameters without some prior empirical knowledge and mathematical models of the relationship between these parameters and measured radiometric temperatures. The empirical information is used to impose limits within which these parameters can vary. The accuracy of the retrieved information is, therefore, affected by the noise in the empirical data and the uncertainties in the assumptions used in model equations.

The main application of spaceborne passive microwave observations is global monitoring of sea ice. The contrast in the emitting properties of the ice-free as compared to the ice-covered ocean is used in all current algorithms. These interpretive techniques for algorithms were developed using signature brightness temperatures from areas identified as containing predominantly one ice type (Cavalieri et al. [14]). The selection of areas of signature brightness temperatures (called tie-point brightness temperatures in some publications) was made by a sea-ice algorithm working group (Cavalieri et al. [13]) and these values were used in deriving sea-ice algorithms for SMMR measurements. Table 5.2 is a list of these brightness temperatures and emissivities, derived from spaceborne observations and ice-surface physical temperature. The ice-surface temperature was obtained from a thermal infrared sensor on Nimbus 7.

After the launch of the SSM/I radiometer, the brightness temperatures from the same geographical areas were used to tune and improve the retrieval algorithms for sea ice and other geophysical parameters.

Algorithm Development

The starting equations for all algorithms are obtained from an approximation to the radiative transfer equations. The assumptions are that at frequencies used in algorithms, the atmospheric effects (excluding precipitating conditions) can

TABLE 5.2 Brightness Temperatures (T_B) and Emissivities (e) for Sea Ice and Ocean Observed with the SMMR

Frequency (GHz)	6.5	10.7	17.6	21.0	37.0
Wavelength (cm)	4.6	2.8	1.7	1.4	0.81
First-Year Ice:					
T_B H	221.43	225.12	225.12	234.29	228.47
T_B V	238.57	239.25	238.51	237.26	240.06
e H	0.865	0.879	0.880	0.915	0.892
e V	0.932	0.934	0.931	0.927	0.932
Old Ice:					
T_B H	211.78	208.33	193.19	195.73	166.57
T_B V	231.17	224.57	206.96	199.57	176.93
e H	0.851	0.837	0.776	0.787	0.669
e V	0.929	0.902	0.832	0.802	0.711
Seawater:					
T_B H	79.26	88.71	97.38	127.11	130.34
T_B V	140.67	149.11	156.31	170.15	190.52
e H	0.282	0.316	0.347	0.452	0.464
e V	0.501	0.531	0.556	0.606	0.678

be evaluated using seasonal or regional values for the amounts of water vapor and liquid droplets in the atmosphere. The brightness temperatures measured by any of the sensor's channels contain contributions from the earth, atmosphere, and space (Fig. 5.26). Assuming a nonprecipitating atmosphere, the observed brightness temperature can be expressed as

$$T_B = T_{Bs} \exp(-\tau) + T_1 + (1 - e)T_2 \exp(-\tau) + (1 - e)T_c \exp(-2\tau) \tag{5-4}$$

where

T_{Bs} = surface radiation
τ = atmospheric opacity
T_1 = atmospheric upwelling radiation
e = surface emissivity
T_2 = downwelling atmospheric radiation
T_c = cosmic space contribution

The atmospheric components, under the assumptions stated above, can be expressed in terms of an atmospheric mean temperature T_a. In polar regions, for the frequencies used in the algorithms, the atmospheric opacity is small but

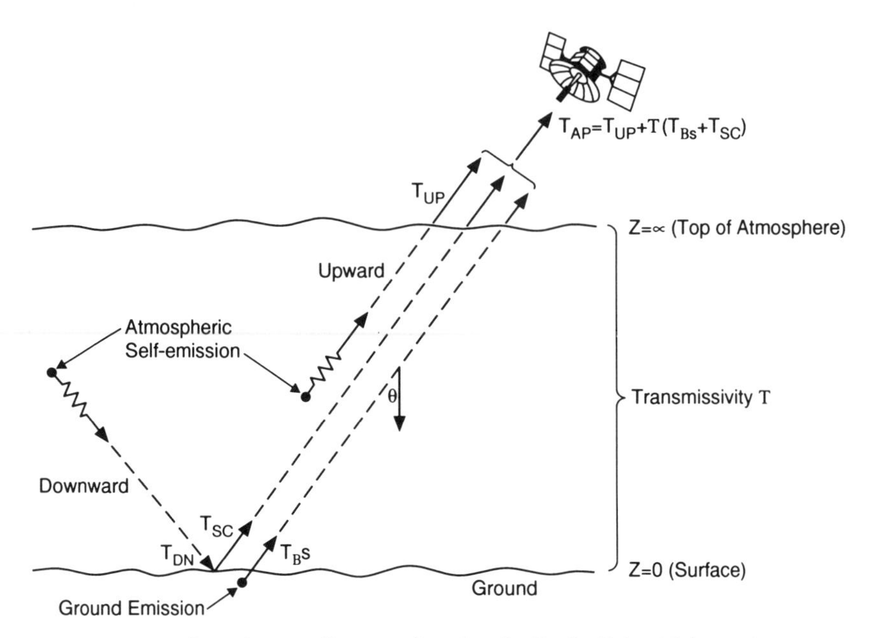

Figure 5.26 Space borne radiometer observing the Earth. (After Ulaby et al. [88].)

not negligible. Due to the noncoherent nature of the natural microwave emission from the surface, the surface radiation can be written as a sum of contributions from different ice types and open water. Equation (5-4) can be rewritten for the following scenarios:

1. Single-channel, dual-polarized observations;
2. Dual frequency, same polarizations for both channels.

The surface radiation term, if C represents the fraction of the footprint covered with ice, can be written as

$$T_{Bsp} = CT_{BIp} + (1 - C)T_{BWp} \tag{5-5}$$

where

subscript p = vertical or horizontal polarization
T_{BIp} = ice signature brightness temperature
T_{BWp} = open-ocean signature brightness temperature

The surface emissivity also can be written as a sum of the emissivities from the ice fraction and open ocean:

$$e_p = Ce_{Ip} + (1 - C)e_{Wp} \tag{5-6}$$

where e_{Ip} is the emissivity of ice, and e_{Wp} is the emissivity of open ocean. Substitution of (5-5) and (5-6) into (5-4) yields:

$$\begin{aligned} T_{BV} &= \exp(-\tau)[CT_{IV} + (1 - C)T_{BWV} + (Ce_{WV} - Ce_{IV})T_2] \\ &\quad + T_1 + (Ce_{WV} - Ce_{IV})T_{sp} \exp(-2\tau) \\ T_{BH} &= \exp(-\tau)[CT_{IH} + (1 - C)T_{BWH} + (Ce_{WH} - Ce_{IH})T_2] \\ &\quad + T_1 + (Ce_{WH} - Ce_{IH})T_{sp} \exp(-2\tau) \end{aligned} \tag{5-7}$$

The existing algorithms (Pedersen [59]; Carsey [10]) use either the difference temperature for two polarizations at a single frequency or the difference in brightness temperatures for two frequencies at the same polarization (19 and 37 GHz is the most commonly used combination). The polarization difference divided by the sum of brightness temperatures, for example $PR = (V37 - H37)/(V37 + H37)$ (polarization ratio) and $GR = (V37 - V19)/(V37 + V19)$ (gradient ratio), is used in an algorithm known as the NASA Team algorithm (Cavalieri et al. [13]).

Figure 5.27 illustrates the changes in the gradient and polarization ratios calculated along a transect (51°N) of the Labrador Sea in February 1992. The ice-edge location, according to the AES ice charts, was at 51°N, 50°W and cloud-free sky covered the transition area of open water to ice cover. Taking the difference between the equations in (5-7) and neglecting the contribution

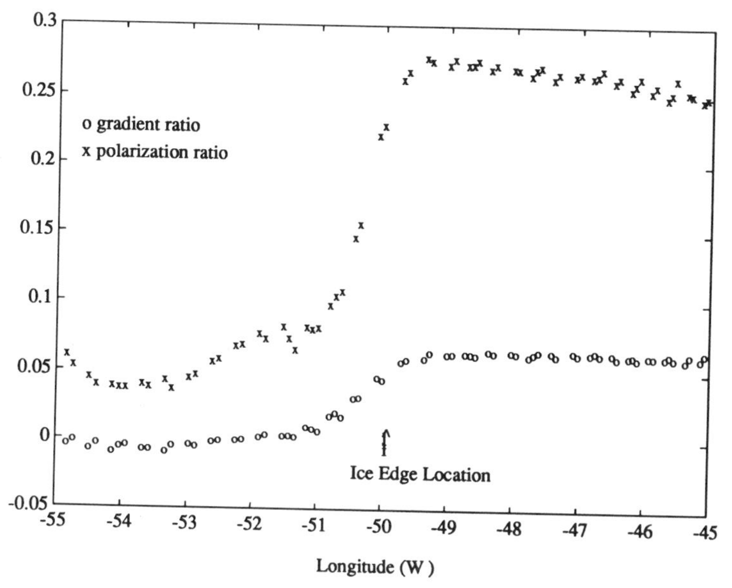

Figure 5.27 Gradient and polarization ratios calculated for 16 February 1992 SSM/I brightness temperatures observed along 51°N, Labrador sea. Ice edge was located at 51°N, 50.0°W.

from space yields the following expression for C:

$$C = [\exp(\tau)\Delta TB - \Delta T_{BW}]/[\Delta T_{BI} - \Delta T_{BW} + (\Delta e_W - \Delta e_I)T_2] \quad (5\text{-}8)$$

where

$$\Delta TB = T_{BV} - T_{BH}$$

$$\Delta T_{BI} = T_{IV} - T_{IH}$$

$$\Delta T_{BW} = T_{BWV} - T_{BWH}$$

$$\Delta e_W = e_{WV} - e_{WH}$$

$$\Delta e_I = e_{IV} - e_{IH}$$

The general form of (5-8) can be written as

$$C = A\Delta TB + B \quad (5\text{-}9)$$

Coefficients A and B can be obtained from (5-8). When the assumption is made that the difference between vertical and horizontal sea-ice brightness temperatures is independent of ice type, (5-8) is known as the Hughes algorithm. From this equation, one can determine the dependence of the total ice-cover estimate on the variability in atmospheric conditions. The variability of ΔT_{BI} with the ice type present within the field of view can also influence the accuracy of the calculations. This type of algorithm can be tuned to be very sensitive to presence of sea ice within the pixel if the weather conditions are used as input to calculate A and B. In the absence of meteorological information, it can lead to false classification of an ice-free pixel for a wind-roughened ocean or an overcast sky (Rubinstein [65]; Rubinstein and Ramseier [67]). Errors in estimating the ice concentration also can occur if an inappropriate value of ΔT_{BI} is selected as a threshold value for ice presence within a pixel.

Weather Effects

During the validation of retrieval accuracies for these algorithms it became evident that weather-related effects on brightness temperature should be evaluated and included in the retrieval procedures. Weather effects on the microwave emissive properties on the ice-free ocean surface were described by Walters et al. [91], Wilheit et al. [93], and others. The changes in observed brightness temperatures induced by heavy clouds and surface wind action (Stogryn [78]) on the ocean can lead to brightness temperatures similar to those of an ice-covered ocean (Fig. 5.28; Fig. 5.29).

Several different techniques were devised to reduce weather effects on calculation of sea-ice concentrations. One of the methods (NASA Team) was based on the observations that the spectral gradient ratio, $GR = (T_{V37} - T_{V18})/(T_{V37} + T_{V18})$, calculated for pixels containing ice, is below 0.08 (for SMMR

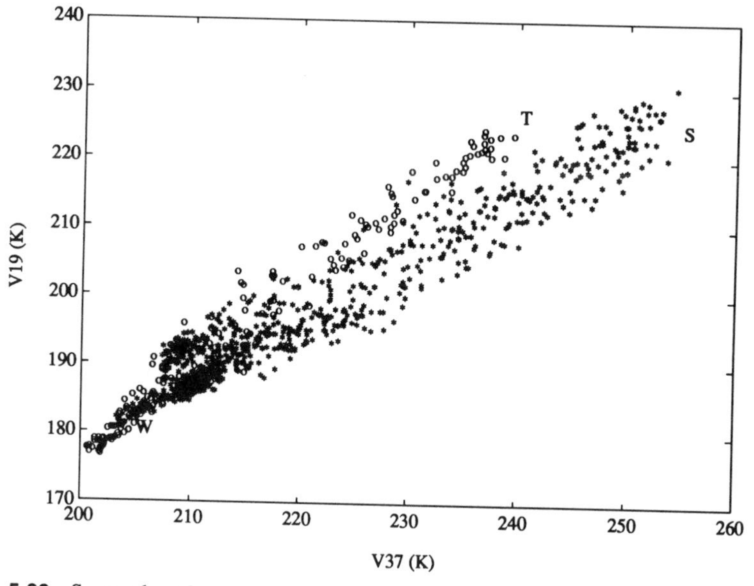

Figure 5.28 Scatterplot of vertical 19 and 37 GHz brightness temperatures showing the atmospheric effects on the observed brightness temperatures. W indicates location of the ice free calm ocean data. The brightness temperatures (along WT) for new/young ice type are represented with a symbol ○. Symbol * is used for the brightness temperatures (along WS) extracted from ice free waters under storm conditions data for 17 February 1992, Newfoundland waters.

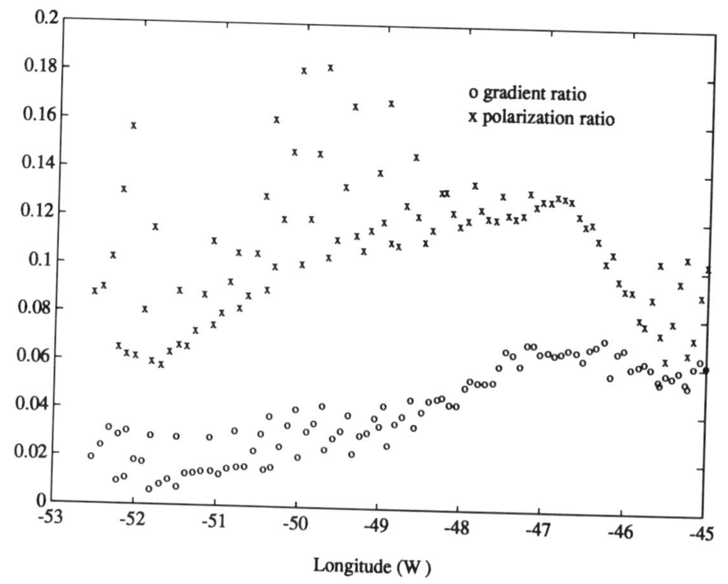

Figure 5.29 Gradient and polarization ratios calculated for 17 February 1992 SSM/I brightness temperatures observed along 48N, Newfoundland Waters. Ice edge was located at 48°N, 48.5°W, strong winds and precipitating cloud (freezing rain/snow) east of 48°W.

data). Only data points with spectral gradient values below the selected cutoff were classified as containing ice. In addition, introduction of a cutoff value for a polarization ratio for discriminating ice-containing from ice-free pixels reduces errors. This method has reduced the number of erroneous classifications but, at the same time, has limited accuracy for retrieving ice concentrations of less than 12%. A more complex method was developed by Walters et al. [91] and is used by scientists at the University of Amherst (Carsey [10]). In addition to 18- and 37-GHz data, the 21-GHz channel (which is sensitive to water vapor presence) is used to reduce weather effects on the estimates of total ice concentration.

Figures 5.30, 5.31, and 5.32 illustrate errors in ice concentrations calculated without correction for weather effects and the results of using the D function to restrict retrieval by the $D \geq 0$ criterium.

Algorithm Errors

The uncertainty experienced in retrieving sea-ice parameters from passive microwave sensors is due to errors induced by (a) instrument noise and calibration errors; (b) uncertainty in selected signature brightness temperatures (i.e., the variability in the brightness temperature of old ice and new (thin) ice types); and (c) the use of statistically derived equations relating brightness temperatures and ice parameters.

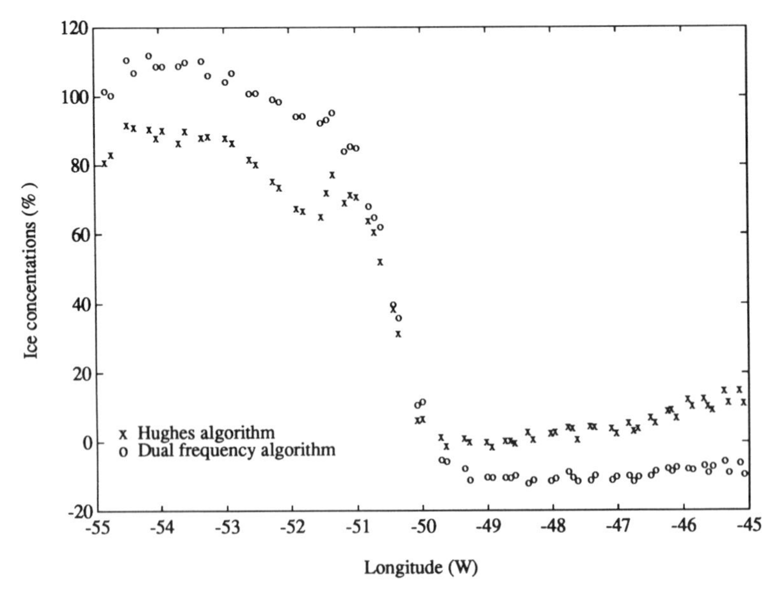

Figure 5.30 Single and dual frequency algorithm calculation of ice concentrations along 51°N transect (16 February 1992, orbit #1135), without corrections for weather effects. Symbols used: × Hughes (SSM/I prelaunch) algorithm calculations; ○ dual frequency (AES/York without weather correcting routine) algorithm calculations.

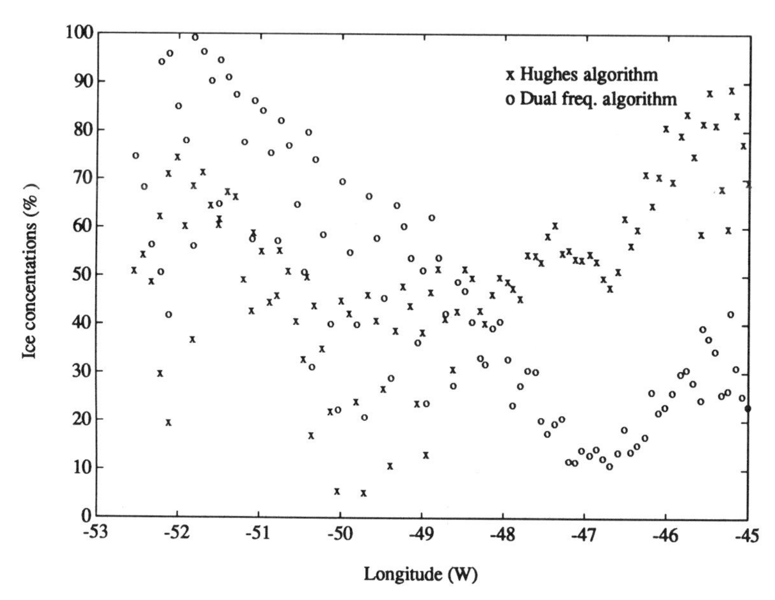

Figure 5.31 Single- and dual-frequency algorithm calculation of ice concentrations along 48°N (Newfoundland waters) on 17 February 1992, SSM/I orbit #1149 without corrections for weather effects.

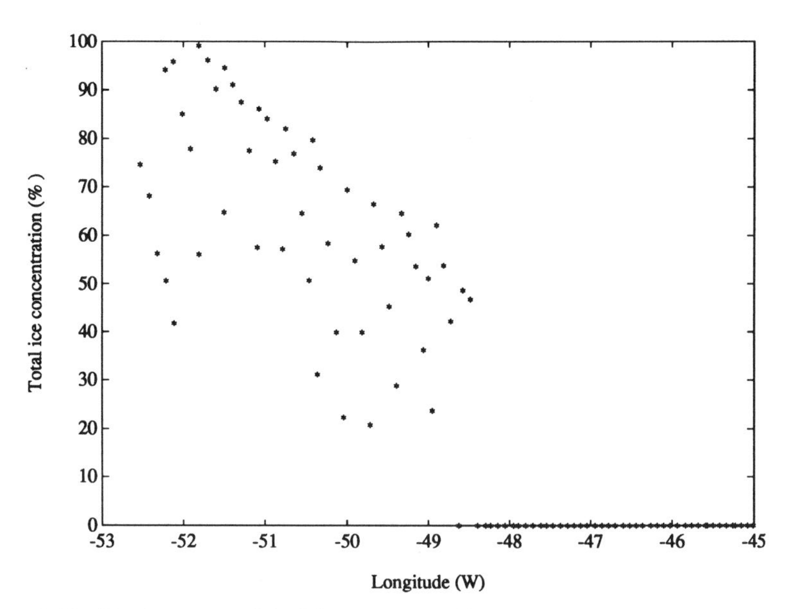

Figure 5.32 Weather corrected dual frequency algorithm calculations along 48°N (Newfoundland waters, SSM/I orbit #1149).

All existing algorithms are subject to errors from these sources. The effect of radiometer noise can be calculated using the instrument status reports (it is about 1 K). The absolute radiometer calibration should not cause problems since algorithms retrieve ice parameters by comparing brightness temperatures with signature values (tie points). Monitoring regions selected as containing pure surface classes will provide users with information about the changes in calibration stability.

The development of faster computers has removed some of the obstacles, for example computation-time limitations, that were imposed on earlier satellite algorithms for data retrieval in near-real time. The algorithm described in Subsection 5.4.2, although containing more logical decision steps than other existing methods, was developed for use in near-real time.

5.4.2 AES/York Algorithm

To accommodate variabilities in sea-ice cover, such as differences in size and distribution of floes in the marginal ice zone compared to pack ice, an algorithm was developed that calls on different retrieval equations for different ice conditions. Separate modules were devised for different tasks. Logical decisions, based on empirical and/or theoretical results, invoke different calculations within the retrieval procedure. This algorithm is known as the AES/York algorithm as it was developed at the Institute for Space and Terrestrial Science (ISTS), York University. It is used at both ISTS and the Canadian Meteorological Centre.

An important goal in developing the AES/York algorithm was to simulate the thinking process used by an analyst for spectroscopic classification. The retrieval procedure does not use all four channels at the same time but under preselected conditions, it evaluates the calculated values for consistency with the brightness temperatures from channels that were not used in calculating the total ice concentration and/or ice fractions. The AES/York algorithm was designed to determine old, first-year, and new ice classes. Only two ice types and the percentage of the ice-free area are estimated for each pixel. The decision on which two ice types are present in the pixel is made within the algorithm.

For any two channels (frequencies f_1 and f_2) with different sensitivities to atmospheric conditions and with different sea-ice (T_{I1} and T_{I2}) and open-ocean (T_{BW1} and T_{BW2}) signatures, Equation 5-4 can be rewritten as follows:

$$T_{B1} = \exp(-\tau_1)[CT_{I1} + (1 - C)T_{BW1} + (Ce_{W1} - Ce_{I1})T_{12}] + T_{11} + (Ce_{W1} - Ce_{I1})T_{sp1} \exp(-2\tau_1) \quad (5\text{-}10)$$

$$T_{B2} = \exp(-\tau_2)[CT_{I2} + (1 - C)T_{BW2} + (Ce_{W2} - Ce_{I2})T_{22}] + T_{21} + (Ce_{W2} - Ce_{I2})T_{sp2} \exp(-2\tau_2) \quad (5\text{-}11)$$

Assuming that only two ice types are present within the pixel (C_{F1} is concentration of ice type one, C_{F2} is concentration of the second ice type), and C_W is the fraction of open water, (5-11) and (5-12) can be solved for the ice fractions under the following constraint:

$$C_{F1} + C_{F2} + C_W = 1$$

Replacing $(1 - C)$ by $1 - C_{F1} - C_{F2}$, one can show that solutions for C_{F1} and C_{F2} have the form

$$C_{F1} = A_1 T_{B1} + B_1 T_{B2} + C_1 \quad (5\text{-}12)$$

$$C_{F2} = A_2 T_{B1} + B_2 T_{B2} + C_2 \quad (5\text{-}13)$$

$$C = C_{F1} + C_{F2} = A_C T_{B1} + B_C T_{B2} + C_C \quad (5\text{-}14)$$

Coefficients A_1, B_1, C_1, A_2, B_2, C_2, A_C, B_C, and C_C are obtained from (5-12) and (5-13).

Current knowledge of microwave brightness temperatures for sea-ice and open-ocean signatures allows retrieval of open-water, first-year, and old-ice fractions for pixels identified as containing old ice (Ramseier et al. [63]). First-year and new ice fractions are calculated in the absence of old ice. One of the largest sources of error in using signature brightness temperatures to calculate ice fractions is the spatial and temporal variability in the old- and new-ice signatures.

The spatial and seasonal variability of ice signatures also is considered in the retrieval procedure. Analysis of several years of 18- and 37-GHz SMMR brightness temperature data, supplemented with 19-GHz SSM/I data, generated an ice-classification technique that includes information on the range of signature variability. In addition, correction of spurious ice concentration calculations (labeled the weather-filtering routine, Rubinstein [66]; Rubinstein and Ramseier [70]), caused by the effects of wind and cloud or precipitation on brightness temperatures, is included in the algorithm. Severe weather effects can be corrected by imposing threshold values on $T_{B1} - T_{B2}$, but this method results in eliminating ice concentrations of less than 15% (this cutoff is probably higher for new-ice areas). Using the channel with the best resolution to correct the severity of the cut-off procedure allows reclassification of pixels identified as ice free. One of the shortcomings of a dual-frequency algorithm is that the use of the 19-GHz frequency reduces the spatial accuracy since the lower frequency channel has a lower resolution. This loss of resolution is not noticeable if the pixels within the field of view are located in dielectrically homogeneous areas. For pixels with ice concentrations of less than 70%, only the 37-GHz channel is used. The variability of old ice signatures is less severe at 19 GHz; therefore, this channel is used as an indicator for the presence of old ice. The brightness temperature of the ice area within the sample is calculated housing the 19-GHz V-polarized channel as

$$T_{BI} = (T_{BV19} - (1 - C)T_{WV19})/C$$

where

T_{BV19} = observed brightness temperature
T_{WV19} = open-ocean 19-GHz V-polarized signature brightness temperature
C = calculated total ice concentration

If T_{BI} is below $(0.35TO_{V19} + 0.65TF_{V19})$, the pixel is identified as containing old ice (i.e., a pixel containing 35% or more of old ice), where TO_{V19} is the vertical component of the signature brightness temperature at 19 GHz for old ice and TF_{V19} is the vertical component of the signature brightness temperature at 19 GHz for first-year ice. The procedure for exact ice-fraction calculations is more complex (Hollinger [42]) and involves several logical tests before arriving at a final value. Tests on the observed brightness temperatures will indicate if calculations for old-ice retrieval must be done.

To accommodate regional and seasonal variabilities of ice signatures, the algorithm contains several testing routines for ice brightness temperature. Clustering of observations for data representative of different regions was used to define the range of pure signature classes, as shown in Fig. 5.33, for open water, new, first-year, and old ice. Pixels containing mixtures of pure classes can be identified as such if they are located within the limits of the following lines: points along a line drawn from O to F contain old and first-year ice;

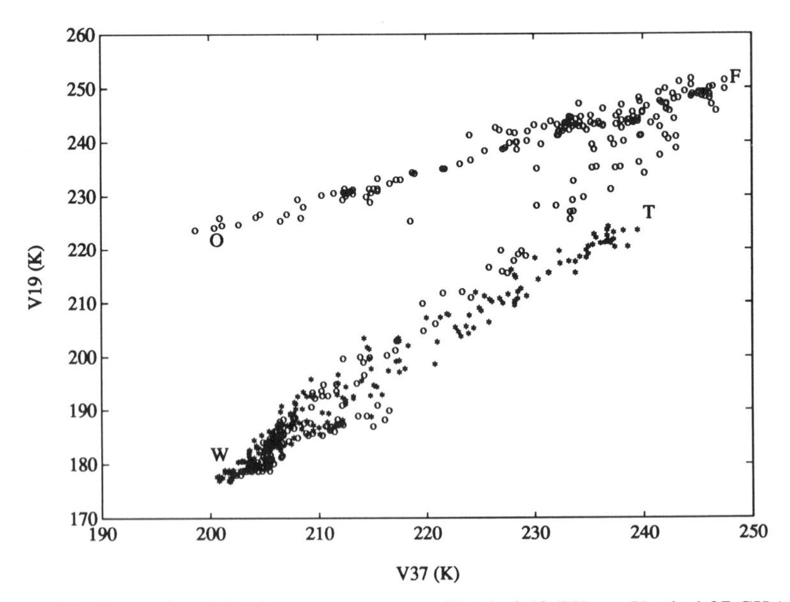

Figure 5.33 Scatterplot of brightness temperatures (Vertical 19 GHz vs Vertical 37 GHz) used for dual-frequency ice classification routine. Letter W identifies ice free, calm ocean brightness temperatures (200, 180), O marks old ice signature (198, 225), and F indicates first-year ice signature (248, 250) brightness temperature. Brightness temperatures for new ice/open ocean mixtures are assumed to be located along WT.

points along a line drawn from W to F contain open water and first-year ice; and points along a line from W to T contain water and new ice.

As the calculated ice fractions should be real positive numbers, special tests are performed on the incoming brightness temperatures if any of the calculated fractions are negative. One of the reasons for fractions to be negative is an underestimation of the atmospheric opacity. New calculations of the ice fraction are then performed with optical opacity set to a different value.

The weather effects, causing false identification of ice-free areas as containing ice, were reduced using discriminating functions D and R. Function

$$D = a_D + b_D[T_B(37V) - T_B(19V)]$$

uses the difference between $T_B(19V)$ and $T_B(37V)$. It becomes negative when $DV = T_B(37V) - T_B(19V)$ exceeds a threshold value determined statistically. The coefficients a_D and b_D were calculated using brightness temperatures for ice-free and ice-covered areas. In addition, for pixels with total ice concentrations less than 70% (algorithm estimate), reduction of false retrieval is achieved using the "roughness" function $R = a_R T_B(37V) + b_R T_B(37H) + c_R$. This function reduces false retrievals that could be caused by wind effects on the ocean surface. Figures 5.34 and 5.35 illustrate the behavior of the discriminating functions calculated along 51°N (16 February 1992, clear sky, calm ocean for the ice-free area) and 48°N (17 February 1992, winter storm east of 48°W).

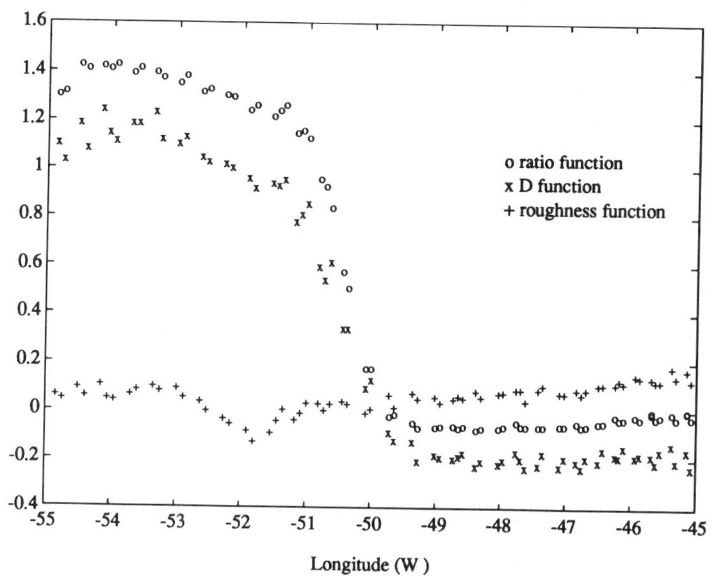

Figure 5.34 Roughness and D functions for a transect along 51°N, in the Labrador Sea, with clear sky and calm ocean (ice edge at 51°N, 50°W).

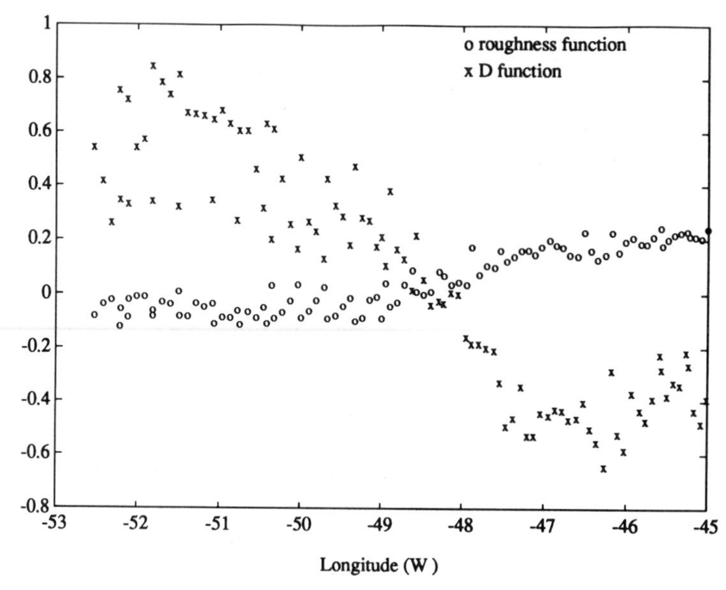

Figure 5.35 Roughness and D functions for a transect along 51°N, Labrador Sea, during a winter storm (ice edge location at 48°N, 48.5°W).

The operational usefulness of the algorithm is enhanced by the following additional functions. The brightness temperatures for pixels identified as ice free are sent through an oceanic parameters algorithm. Ocean-surface winds and cloud cover are calculated for these points (Rubinstein, [67]; Rubinstein [69]; Rubinstein and Ramseier [70]). If any of the pixels contain ice and are not classified correctly, the calculated surface wind speed and/or cloud cover will look out of place when the processed data are plotted. This synergistic use of atmospheric and ocean-surface wind speed algorithms provides additional information to the user. For pixels near to the ice edge, comparison of oceanic retrievals with weather information can be used to further reduce false-ice (or ice-free) classification. Pixels identified as ice free but actually containing very low amounts of ice or very thin ice may be identified as cloud covered and/or with ocean-surface wind speeds much higher than other pixels in the area (see Fig. 5-36). The modular structure of the AES/York algorithm allows the use of each module as an individual algorithm. This retrieval technique, although it appears to be more complex than other methods, can be used for global and regional monitoring of sea-ice parameters. Equations (5-11) and (5-12) can be used for channels other than 19 and 37 GHz. The structure of the algorithm allows switching to different channels because most of the constants are calculated from the signature brightness temperatures.

The uncertainty in retrieval is evaluated using the δT_B values induced by (a) instrument noise and calibration errors; (b) uncertainty in selected signature brightness temperatures for ocean and ice-covered surfaces; and (c) the use of

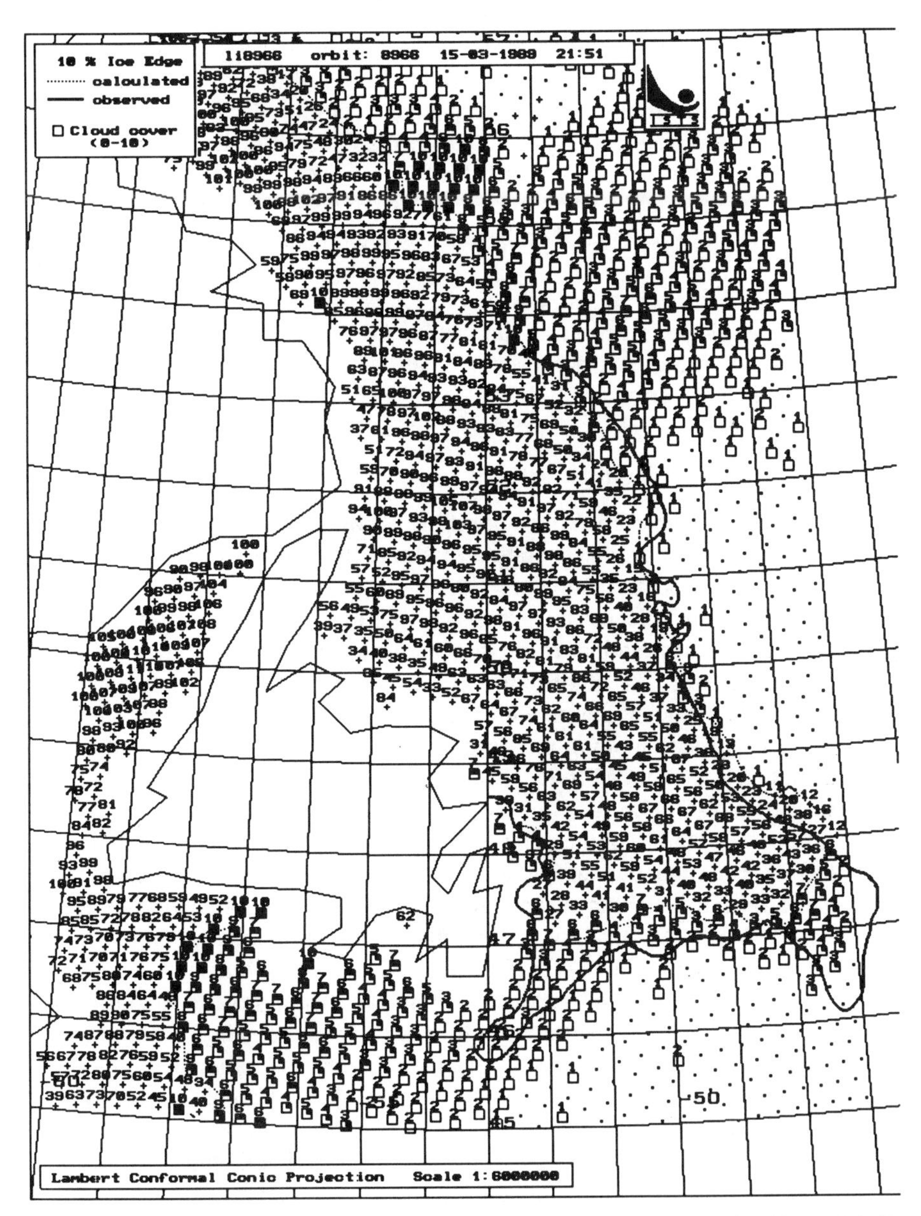

Figure 5.36 Comparison of the retrieved ice concentrations from AES/York algorithm and AES information 15 March 1989, Labrador Sea. Note areas in the southern part of the chart classified as ice free are identified (by the AES/York algorithm) as cloud covered.

empirical equations relating brightness temperatures and ice parameters. Calculations for the AES/York algorithm yielded the following: instrument noise leads to 1.65% uncertainty in first-year ice-concentration retrieval, and 3.7% uncertainty in old-ice–concentration retrieval. The errors in selecting atmospheric and signature parameters can lead to as much as 9.5% uncertainty in calculations of total ice concentration. This source of error can be reduced to 2%–3% by using a regional data base for sea-ice signatures and atmospheric parameters.

The variability in the emissivity of old ice is the largest source of errors in estimating the old-ice fraction: it can cause errors of more than 50%. Input of previously known regional parameters is one of the recommended procedures for correcting these errors.

5.4.3 Algorithm Evaluation

Prior to accepting the AES/York algorithm for operational retrievel of SSM/I data in near-real time or for generation of climatological sea-ice data sets, AES contracted two Canadian companies (Ph.D. Associates Inc., North York, Ontario, and Norland Science and Engineering Ltd., Ottawa) to determine the accuracy of ice information from that algorithm and the prelaunch SSM/I (Hughes) algorithm. The algorithm was evaluated by comparing the maps of retrieved sea-ice concentrations with nearly time-coincident synthetic aperture radar (SAR) and side-looking airborne radar (SLAR) imagery and with daily charts produced by the AES Ice Branch. The results of this evaluation are summarized in Table 5.3.

Additional evaluation of the performance of the AES/York algorithm in near-real time was carried out during numerous sea-ice research campaigns. For example, airborne SAR imagery acquired during the Labrador Ice Margin Experiment (LIMEX'89, January to March 1989) was compared with SSM/I estimates (Rubinstein et al. [71]). Figures 5.37 and 5.38 illustrate some of the results. Gray tone values and variances within each scene were used to translate SAR imagery into ice-concentration information. Sea-ice charts produced with the AES/York algorithm were used for planning and navigational support during other sea-ice research campaigns in the Greenland, Barents, Weddell, Labrador, and Beaufort Seas. The SSM/I data (received via high-speed modem from the Civilian Navy Ocean Data Distribution Service Computer after processing at ISTS) were made available to the ship captains within 15 min of data-acquisition time (Ramseier et al. [63]).

The AES/York algorithm total ice-retrieval module was adapted for use at FNOC (Monterey, CA) and it is used as one of the sea-ice information sources at the Naval Polar Oceanographic Center.

The Canadian Meteorological Centre (Dorval, Quebec) has used the AES/York algorithm since 1990 to produce sea-ice information for input in weather-forecasting models. Validation of the existing algorithms (AES/York and NASA Team) has been carried out (Ramseier et al. [63]; Cavalieri et al. [12]) and in

TABLE 5.3 Final Conclusions on the AES/YORK Algorithm Performance

Group category	Ice concentration			Ice Edge results		
	Mean diff (%)	Meets criteria	95% C.I.*	Displacement (km)	Meets criteria	95% C.I.
I-Combined Area:						
Pooled	−8.9	yes	0.9	−0.2	yes	1.9
I-A, class A	−4.2	yes	3.1	—	—	—
I-B, class B	−6.8	yes	1.6	—	—	—
I-C, class C	−10.3	no	1.1	—	—	—
I-AB, class A+B	−6.3	yes	1.4	—	—	—
I-1, ice formation	−9.1	marginal	1.3	−0.7	yes	2.6
I-2, winter	−5.5	yes	1.0	same as III-2	—	—
I-3, initial ice melt	same as II-3	—	—	same as II-3	—	—
I-4, ice melt	same as II-4	—	—	same as II-4	—	—
II-Artic:						
Pooled	−8.6	yes	0.9	−3.1	yes	2.1
II-A, class A	+0.4	yes	2.3	—	—	—
II-B, class B	−7.1	yes	1.8	—	—	—
II-C, class C	−10.0	no	1.2	—	—	—
II-AB, class A+B	−5.8	yes	1.6	—	—	—
II-1, ice formation	−9.1	marginal	1.5	−5.1	yes	2.9

TABLE 5.3 (*Continued*)

Group category	Ice concentration			Ice Edge results		
	Mean diff (%)	Meets criteria	95% C.I.*	Displacement (km)	Meets criteria	95% C.I.
II-2, winter	−1.5	yes	0.5	—	—	—
II-3, initial ice melt	−5.1	yes	2.7	−4.0	marginal	9.5
II-4, ice melt	−11.5	no	2.7	−4.3	yes	3.6
III-Gulf of St. Lawrence:						
Pooled	−10.1	marginal	1.6	+7.6	yes	4.2
III-A, class A	−12.1	no	4.5	—	—	—
III-B, class B	−5.8	yes	2.5	—	—	—
III-C, class C	−12.1	no	2.2	—	—	—
III-AB, class A+B	−7.8	yes	2.2	—	—	—
III-1, ice formation	−8.9	marginal	2.3	+6.2	yes	4.5
III-2, winter	−11.6	no	2.9	+21.1	no	9.6
Ice Fraction:						
First-year	−8.0	marginal	2.7	—	—	—
Old ice	+6.5	yes	2.6	—	—	—

*C.I. = confidence interval

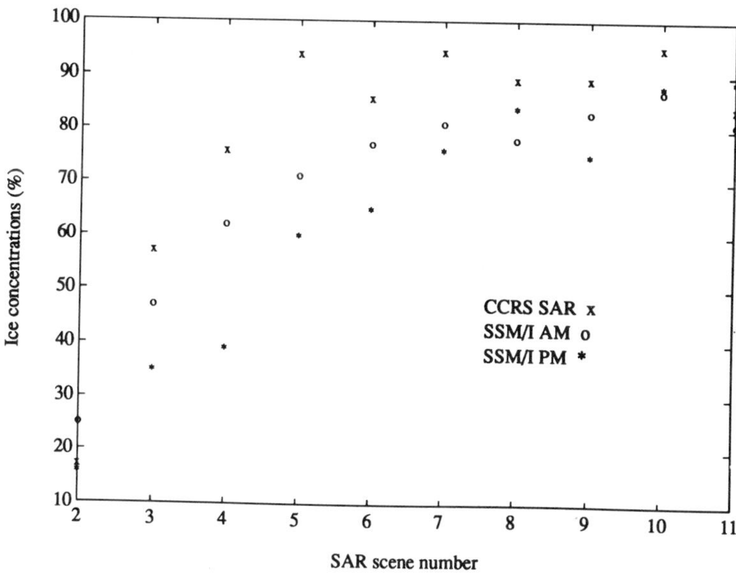

Figure 5.37 Comparison of calculated ice concentrations (SSM/I data AES/York algorithm) with SAR imagery interpretation, 14 March 1989. Two different methods (stdv-thr = standard deviation minimum threshold; stdv ent-thr = standard deviation of entropy threshold) were used to interpret SAR imagery (Rubinstein and Shokr, [71]).

addition, comparisons have been made between seven known algorithms (Steffen et al. [76]). The algorithms were tested for their sensitivity to weather effects and to regional ice signatures. The AES/York algorithm and an algorithm developed by Comiso showed better performance in identifying ice-free pixels, compared to the NASA Team and University of Massachusetts algorithms (Steffen et al. [76]). All the algorithms showed good agreement for cold ice conditions in the Arctic Ocean. The AES/York algorithm showed no sensitivity to cloud presence, while other algorithms indicated lower ice concentrations for the cloud-covered areas. Large areas covered with young ice were classified incorrectly as areas with reduced ice concentrations by all the algorithms except the AES/York algorithm. For areas with 65% to 85% total ice concentrations, the concentrations were overestimated by the AES/York and Comiso algorithms as compared with estimates of ice cover from coincident Landsat imagery.

5.5 PASSIVE MICROWAVE APPLICATIONS

5.5.1 Ice Branch Requirements

The Atmospheric Environment Service of Environment Canada has a mandate to provide timely and accurate ice information for Canadian waters on an

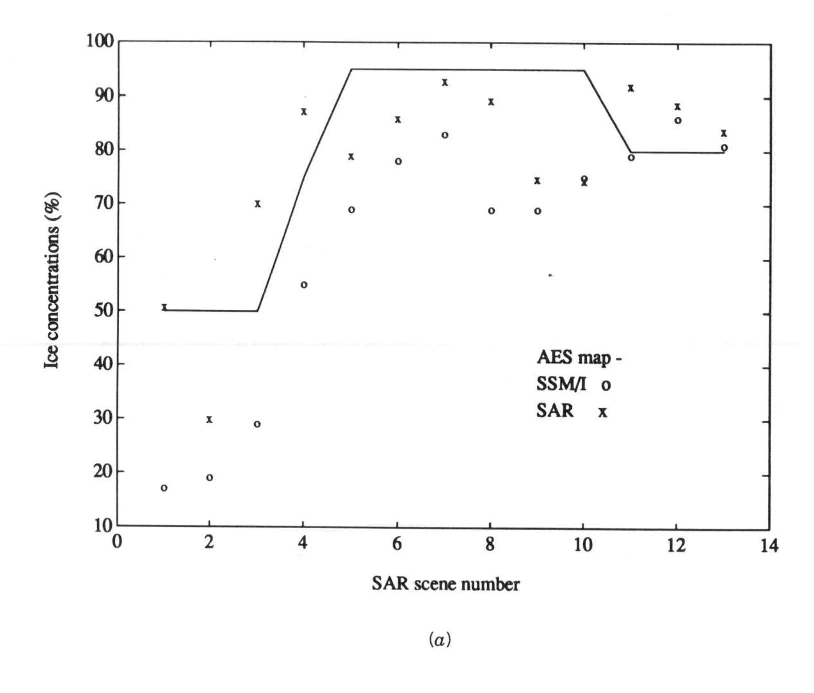

(*a*)

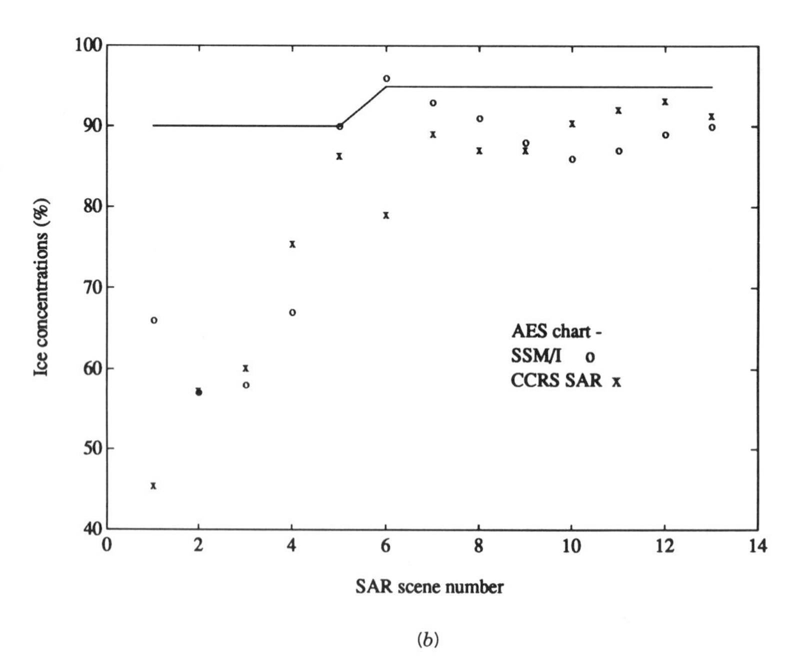

(*b*)

Figure 5.38 Comparison of calculated ice concentrations (SSM/I data AES/York algorithm) with SAR imagery interpretation, 15 March 1989.

operational basis and to provide the national archives for the data. The AES Ice Branch undertakes this responsibility: it provides routine ice reconnaissance and provides ice forecasting and advisory services to support Canadian marine activities. Other groups within the Ice Branch concern themselves with ice climatology and research and development (see also Chapter 12).

The utilization of any type of ice information by the Ice Branch is contingent on the data meeting accuracy requirements based on operational needs. Numerous Canadian and international studies have been conducted to evaluate the capabilities of passive microwave techniques for monitoring sea-ice extent and ice-type classification (Cavalieri and Swift [16]; Ramseier et al. [62]; Steffen and Schweiger [77]).

Within Canada, several investigations have compared passive microwave-derived ice information with operational ice charts produced by Ice Branch analysts (Nazarenko et al. [55]). These charts are compiled from a variety of data sources, including visual reconnaissance, SAR, SLAR, satellite imagery, and reports from ships and weather stations.

SMMR Evaluations

Two studies, by Henderson [39] and Moreau et al. [52], evaluated the accuracy of SMMR-derived ice-edge positioning with the AES/PhD algorithm (the latter evolved into the AES/York algorithm after its adaptation to process SSM/I data with the addition of modules for new ice calculations and correction of mathematical deficiencies). Henderson determined that there were large mean absolute deviations between the positioning of SMMR and AES ice-chart ice edges, with the ice charts more accurate in positioning the ice edge and the SMMR product more accurate at positioning the 60% contour. The AES ice chart 40% and the SMMR 35% concentration contours were consistently located at comparable positions.

Henderson also evaluated the SMMR and AES chart information against SLAR and satellite multispectral imagery (NOAA 7 advanced very high resolution radiometer [AVHRR]) of the same area. The results of his evaluation point out important constraints to the validation of interpreted ice products. He found that neither the AES ice charts nor the SMMR data yielded ice-concentration results statistically comparable to the information available from the higher resolution imagery. Furthermore, the acknowledged problems in georectification of either reference data set (SLAR or NOAA AVHRR imagery) meant that an absolute standard against which to compare the SMMR-derived ice concentrations was not available, prohibiting the absolute definition of SMMR ice information reliability.

Moreau et al. [52] compared the SMMR 10% concentration limit with AES ice charts in the Beaufort Sea and east coast waters, for several dates in 1984 when there was concurrent SMMR and SLAR or visual data. Their findings indicate that the mean difference in position between the SMMR and charted ice edges (the two sources were not quite concurrent) in the Beaufort Sea was 7.2 km, with a root mean square distance (RMSD) of 17.8 km. The SMMR

TABLE 5.4 Deviation Between SMMR-Derived Ice Edge and Ice Reconnaissance Observations (km) [39, 52]

Contour Level	Comparison Source	Location	Mean	RMSD	Max	Min.	No. of Obs.
10%	AES Ice charts	Beaufort	7.2	17.8	50	−4	149
10%	SLAR/Visual observation	East coast	−4.4	31.2	93	−144	223
35%	SLAR/Visual	East coast	3.6	21.5	50	−85	103

ice edge generally overestimated the ice margin compared to its ice chart location. On the east coast, where the SMMR ice information was evaluated against same-day SLAR imagery or visual observations, the mean deviation of the SMMR 10% contour was −4.4 km, with an RMSD of 31.2 km. This range of values was much larger than that of the Beaufort Sea data, although comparison of the 35% contour produces better agreement between the SMMR ice limit and the ice-reconnaissance observations. Table 5.4 summarizes the statistical results of the evaluations carried out by Henderson [39] and Moreau et al. [52].

Since the early interest in SMMR, further evaluation of the SMMR data against AES ice chart information has been carried out for Hudson Bay (Etkin [29]). Etkin compared SMMR and AES chart data using monthly averages, except for the freeze-up and breakup period. Total concentration differences were calculated on a monthly basis for a 3-year period from 1979 to 1981, in an area of exclusive first-year ice coverage. In addition, eight smaller subareas and a larger rectangular region in central Hudson Bay were studied in a similar fashion to identify potential variability within the larger Hudson Bay ice regime. Overall, the two data sources were found to be in good agreement for 10 of the 12 months of the year. Figs. 5.39 and 5.40 show the monthly concentration differences, in tenths, for the large box in central Hudson Bay. During the melt period, the SMMR ice information tended to underestimate the total ice concentrations, while during freeze-up the SMMR data overestimated concentrations, compared to the AES ice charts. The results were consistent for all subareas. The deviations were attributed to surface melt water during the breakup period which affected the ice emissivity. During freeze-up, the positive difference between SMMR and AES chart concentrations was attributed to the more sensitive detection of new ice formation by the passive microwave sensor.

SSM/I Evaluation

Since the launch of the DMSP SSM/I radiometer and fine tuning of the AES/York algorithm, passive microwave data from SSM/I is processed by AES for operational use. The coarse resolution of SSM/I data does not meet the Ice Branch requirements for areas of shipping activity (Carrieres and Crosbie [9]); however, the sensor's wide coverage and frequent repeat cycle provide a com-

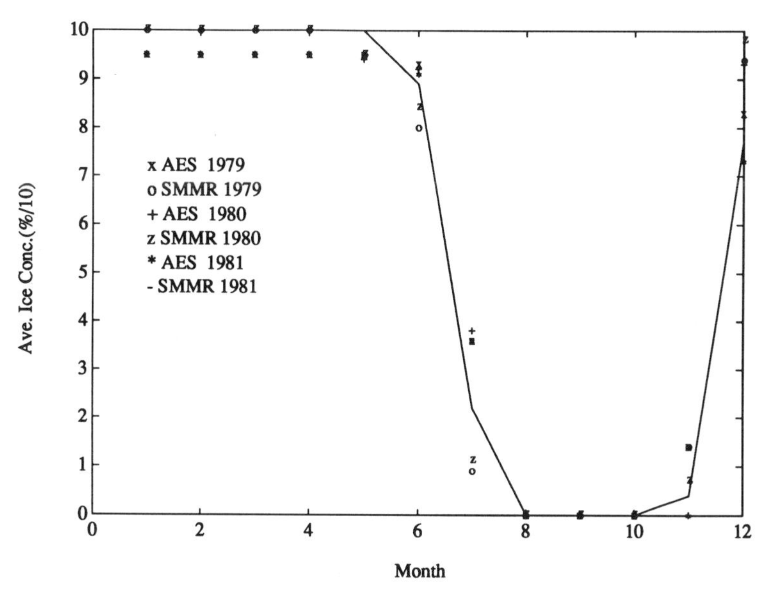

Figure 5.39 Comparison of SMMR and AES ice charts for an area of Hudson Bay (Etkin [29]). Symbols used: × (1979), + (1980), and * (1981) AES data setes; ○, z and solid line are used for 1979, 1980 and 1981 SMMR data correspondingly.

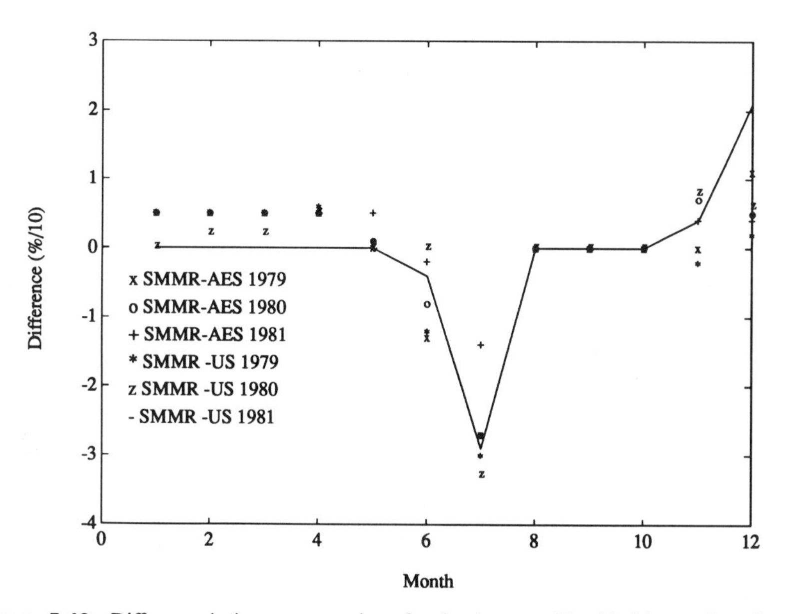

Figure 5.40 Difference in ice concentrations for the data sets identified in previous figure.

prehensive, relatively inexpensive data source that is available consistently. It can be used effectively in monitoring ice conditions where accuracy requirements are less stringent.

The Ice Branch recently conducted a comparison of ice information obtained from airborne SAR, NOAA AVHRR, and SSM/I data along the east coast of Canada. The data sets were roughly concurrent and provided an opportunity to evaluate the relative utility of each information type in meeting Ice Branch objectives for ice information, including delineation of the ice edge and evaluation of features within the ice pack. Although a quantitative evaluation was not made, the results indicate adequate agreement between the three sensors in the general trend of the ice edge. It was found that the SSM/I-derived ice edge was consistent with the SAR- and AVHRR-derived ice edges, within the resolution limitations of the sensor (Carrieres and Crosbie [9]; Rubinstein et al. [71]). This comparison demonstrates the value of SSM/I information in providing broader regional representation of ice conditions. Whereas the coarser resolution of the SSM/I precludes its use for identification of small-scale ice features, the same Ice Branch evaluators noted that the SSM/I ice-edge information could be used to guide the interpretation of AVHRR imagery in areas obscured (or partially obscured) by cloud.

5.5.2 Incorporation into Operational Ice Forecasts

Pack ice, formed in the polar basin, is in almost constant motion. This motion, coupled with deformation of the ice cover, affects the heat, salt, and momentum fluxes (water stress) between the atmosphere and ocean. The use of numerical models and satellite data to study these complex interactions is gaining an increased acceptance. Satellite sensors provide sea-ice information over large areas; then numerical models can be used to infer and forecast parameters, for example ice drift, and to "guesstimate" ice thickness that cannot be retrieved from the available observations.

Lemke et al. [47] were some of the first investigators to use passive microwave imagery in a stochastic–dynamic model. Since then, satellite-derived sea-ice extent has been used in several numerical models (Preller et al. [61]). One of the best examples of ice-model–satellite-data synergetic uses is documented by Serreze et al. [73]. The cause of a large area of reduced ice concentrations within the Arctic pack ice in September 1988 was diagnosed with sea-ice concentrations derived from passive microwave data, augmented with visible-band imagery, and combined with an ice model. In another example, the performance and initialization of a prediction model for ice-edge location (5% to 10% ice-concentration contour lines) predicted the advance and retreat of the ice edge near Newfoundland (El-Tahan and Rubinstein [27]). It was tested using SSM/I data during the Canadian Atlantic Storm Project experiment in March 1992.

In addition to an improved understanding of the physical processes governing polar seas, passive microwave data can be used to develop detection techniques for extreme climatic conditions, such as significant reduction in the old-ice fraction or change in total ice-covered surface area (Le Drew et al. [48]). Although the passive microwave records span more than 20 years, it is still premature to assess significant climatic trends in sea-ice concentration (Parkinson [56]). The incorporation of remotely sensed data into atmospheric and oceanographic models is recognized as a requirement for improving forecast capabilities. This objective has also been identified within the ice-modeling community (Department of Fisheries and Oceans Report, 1992). Only recently have significant advances been made in the area of remote sensing data assimilation into operational sea-ice models. The U.S. Navy uses three Hiber-styled sea-ice models (Preller et al. [60]) to forecast ice conditions in the polar regions and the East Greenland and Barents seas. The availability of SSM/I-derived ice information, at a comparable resolution to the ice models, has led the Navy Ocean and Atmospheric Research Laboratory (NOARL) to experiment with integrating SSM/I data into their models.

Currently, each model is initialized once per week using U.S. Naval Polar Oceanography Center (NPOC) ice analyses. Between NPOC analyses, real-time SSM/I data are interpolated to the model grids and used as a data source for interim initialization. Although they are preliminary, forecast results are encouraging and plans are in place to continue evaluation and refinement of the assimilation process (Preller et al. [60]). The techniques offer potential for other regional models, such as those used by the AES Ice Branch as input to its ice-analysis program. Also in Canada, the Canadian Meteorological Centre (CMC) uses numerical prediction techniques to provide weather forecasts for North America, including the Arctic and Canadian marine areas. Past operational procedures have dealt with sea-ice cover using climatological ice information or weekly ice-analysis charts, depending on the weather model being used. Recently, CMC has begun to evaluate the utility of real-time SSM/I ice information to provide input to their models. Rather than climatological or weekly composite inputs, the SSM/I data are used to produce a 24-h average total ice concentration for each model grid cell (about 100 km $\times$ 100 km). Where no SSM/I data are available, the procedure gradually returns to the climatological ice values. Preliminary results have shown that SSM/I data can have a significant impact on CMC's ability to forecast finer scale weather features such as polar lows (Halle [38]).

5.5.3 Support for Marine Operations

Remotely sensed ice information plays an important role in assisting the marine transportation industry to navigate in ice. SAR is used routinely by the Ice Branch to produce ice analyses which are transmitted to ships in Canadian waters. Some vessels also have the capability of receiving SAR imagery directly, and have been using the imagery itself for tactical navigation. The utility of passive microwave ice information has also been tested for marine navigation applications and has proven useful as a tool for strategic planning and navigation. SSM/I has been used as a source of ice information for several research voyages in the Weddell, Barents, and East Greenland seas (for example, Lovas et al. [50]).

SSM/I data processed at ISTS were used by the Canadian Coast Guard for navigational and planning support for fishing vessels. The timeliness of the ice information and its accuracy resulted in substantial time and fuel savings. The ice concentration charts from SSM/I data, processed using the AES/York algorithm by Ph.D. Associates Inc., are provided to the ship masters navigating in Antarctic waters (for example, the U.S. research ship *N. Palmer* is one of the users).

Voyage of Rossia

In August 1990, the Soviet icebreaker *Rossia* made a voyage from Murmansk to the north pole supported by SSM/I-derived ice information provided by ISTS. The ship while it was en route, received SSM/I information and visual observations of ice conditions (made by a Canadian observer from Norland Science and Engineering Ltd.) which were used to assess the SSM/I ice information. Figures 5.41 and 5.42 show total ice concentrations for the area on 3 and 10 August, respectively, and the locations where *Rossia* entered and exited the ice pack show good agreement with the SSM/I-derived ice edge (Wells [92]).

The change in ice concentration north of 80°N, between 3 and 10 August, corresponded to observed differences in ice conditions: northbound *Rossia* encountered heavy ice conditions, while on the southbound transit, somewhat lighter ice conditions were encountered. Shipboard observations near the pole recorded 9 to 9+ tenths total ice concentrations which were lower than SSM/I estimates of 99% in the area. This difference was attributed to successful tactical navigation which allowed the ship to take advantage of more favorable ice conditions (such as leads around the edges of large ice floes), leading to localized observations of lower ice concentrations.

5.5.4 Synoptic-Scale Variability

The application of passive microwave satellite imagery to studies of climate has been identified as an important source of information concerning sea ice and climate interactions (Crane et al. [24]; Gloersen and Campbell [33]; Parkinson [52]; Davies et al. [26]) because of its global coverage on a frequent basis for the past 20 years. The SMMR-derived data from 1978 to 1987 were the main source of passive microwave data used in climatological research. The merging of SMMR and SSM/I data into a climatological data base requires development of a common spatial grid. Our discussion of synoptic-scale variability applications will mainly consider the use of ESMR and SMMR data. Research on climatological uses of SSM/I data is in progress at numerous research institutions (LeDrew et al. [48]).

Numerous studies have used passive microwave data to evaluate variability

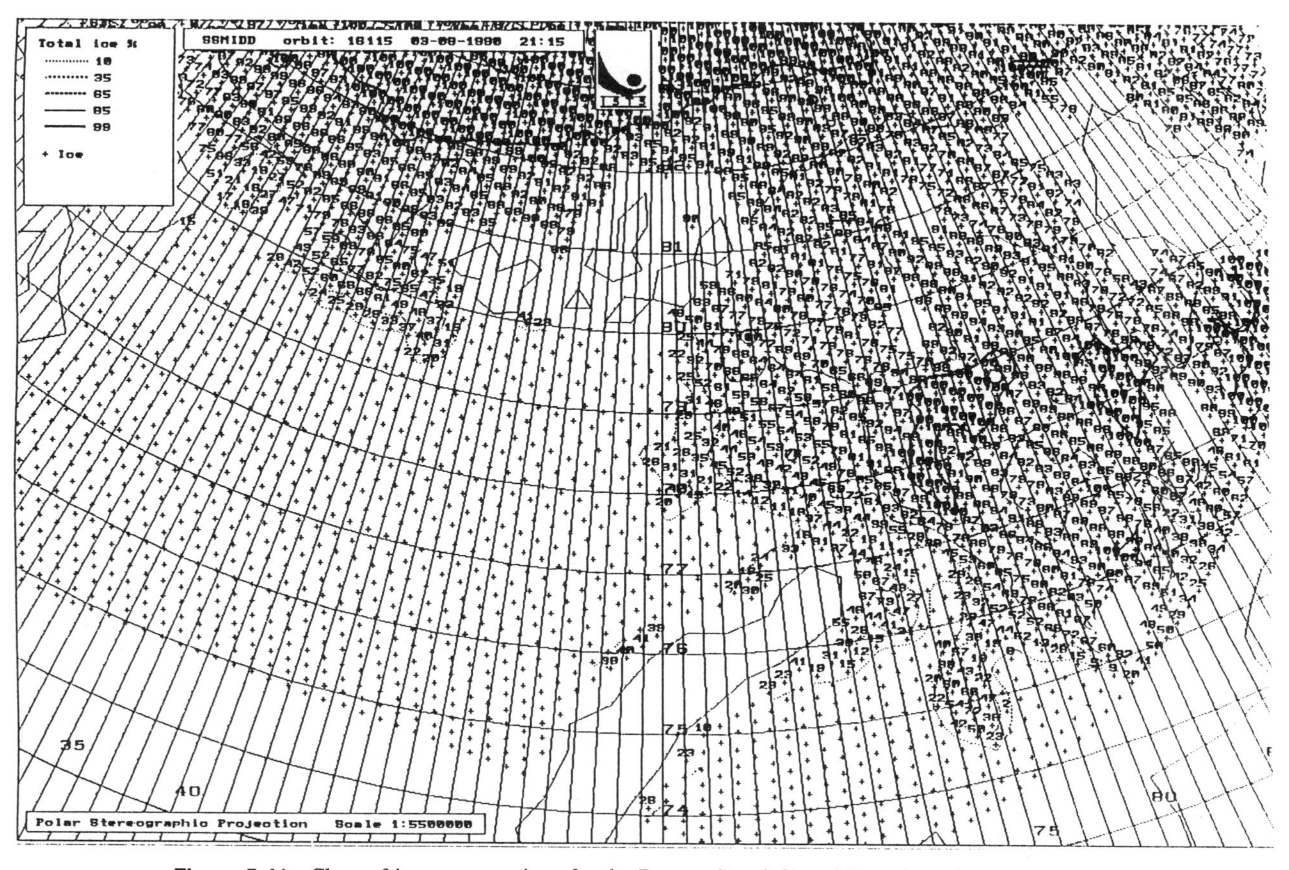

Figure 5.41 Chart of ice concentrations for the Barents Sea (left) and Kara Sea (right), separated by Novaya Zemlya, produced at ISTS using SSM/I data on 3 August 1990.

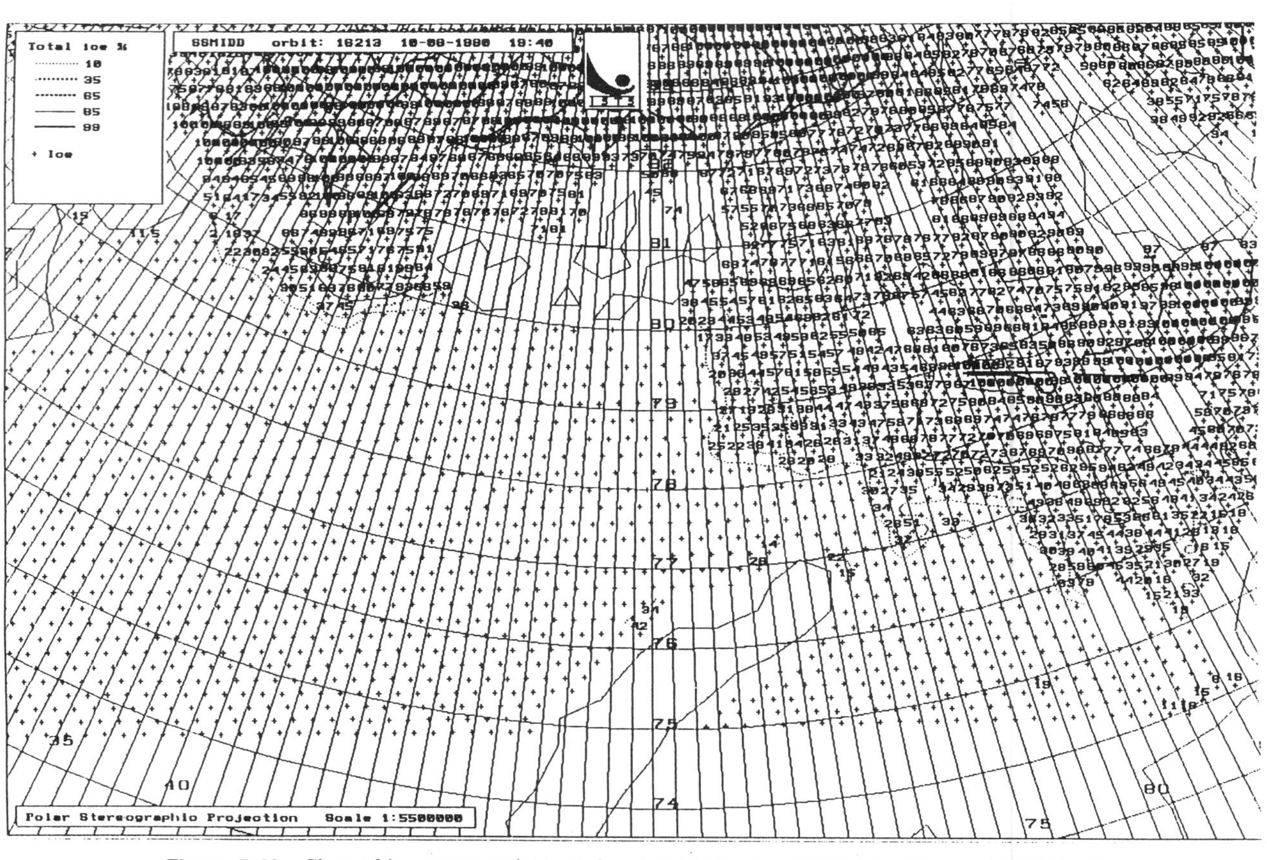

Figure 5.42 Chart of ice concentrations produced at ISTS using SSM/I data 10 August 1990 for the same locations as in Figure 5.41.

of ice conditions over monthly, seasonal, and annual time scales. Applications have been evaluated on regional to hemispheric scales. Campbell et al. [8] used a time series of early satellite passive microwave imagery to study regional variations in Arctic sea-ice cover. Cavalieri and Parkinson [15] used passive microwave information to correlate Antarctic sea-ice extent with associated (hemispheric) atmospheric circulation patterns. Parkinson and Gratz [58] used 3-day–averaged ESMR data to monitor the regional growth and decay of the ice cover in the Sea of Okhotsk between 1973 and 1976. Their qualitative evaluation focused on the interaction between the ice cover and the local oceanography in a protected basin and demonstrated the utility of satellite microwave-derived ice information in time-series analysis.

Parkinson et al. [57] have compiled a time series of Arctic sea-ice conditions for the 1973–1976 period using satellite passive microwave information. A similar time series has been compiled for the Antarctic (Zwally et al. [96]). Cahalan and Chiu [7] were concerned with synoptic-scale atmospheric forcing on sea ice and successfully employed ESMR data to evaluate space–time variability of the Antarctic marginal ice zone. Their findings indicated strong correlation between the spatial pattern and advection of sea-ice anomalies and atmospheric sea-level pressure. Crane [23] showed that brightness temperature variability can be related to sea-level atmospheric pressure and temperature at synoptic scales in a study of ice conditions in the Beaufort and Chukchi seas.

Walker [89] utilized ESMR and SMMR data from 1973–1985 to study interannual variability of ice concentration and extent along the coast of Canada. Following computation of annual ice area and extent, attempts were made to account for the interannual variability by comparing ice conditions to atmospheric and oceanographic conditions. Table 5.4 shows correlation coefficients for ice area and extent versus several indicators of atmospheric and oceanographic forcing. Further comparisons with the mean position of the Icelandic Low and prevailing wind direction and speed provided low correlations. However, from the comparisons with temperature information, Walker [89] noted that annual ice variability showed a significant correlation with seasonal air and water temperatures (see Table 5.5).

Variability of sea-ice cover in transition regions of the marginal ice zone has been the subject of numerous research studies. The SMMR was the satellite microwave sensor during several marginal ice-zone experiments (the Marginal

TABLE 5.5 Correlation Between Ice Conditions and Indicators

	Correlation Coefficient	
Indicator	Ice Area (km^2)	Ice Extent (km^2)
Mean February T_{air}	0.02	−0.10
Mean March T_{air}	−0.53	−0.45
Seasonal mean (Dec.–Feb.) T_{air}	−0.82	−0.82
FDD accumulation (to end of Feb.)	0.79	0.78
Mean annual SST anomaly	−0.79	−0.78

Ice Zone Experiment [MIZEX] series in the Chukchi Sea and LIMEX series off the Canadian east coast) providing multichannel data for large-scale ice observations during MIZEX'84, MIZEX'87, and LIMEX'87.

A study of Nazarenko [53] looked at synoptic-scale ice variability in the Canadian east coast seasonal ice zone using SMMR data for the 1980–1981 and 1984–1985 ice seasons. The study area (Davis Strait) represents an area of widely varying environmental conditions. The dynamics of this environment are experienced over a broad range of spatial and temporal scales, one product of which is a highly variable ice regime.

In response to large-scale circulation patterns, synoptic activity throughout this region exhibits a noticeable summer–winter variation with more intense activity during the winter months. During winter, an average of about 7 lows per month affect the Grand Banks area, with progressively lower numbers of storms to the north. Many of the lows entering the Grand Banks–Labrador Sea region slow down and can persist in the area for up to 2 weeks (Bursey et al. [6]). The typical extent of these lows is on the order of 1000 km.

In the summer, the storm tracks tend to pass much further north with an average of 5 to 6 cyclones per month per 10° square over the Labrador Sea. At this time, the area of cyclogenesis is much larger, extending from the eastern seaboard through the Labrador Sea as far as the southern edge of Greenland. Despite a wider extent, these storms generally are weaker than those occurring during winter and also tend to exhibit less persistence in the area (Bursey et al. [6]).

In modeling ice extent in the Labrador Sea, Ikeda et al. [44] suggested that the critical factors influencing ice extent were air temperature and wind velocity. This is supported by empirical results which indicate a strong correlation between ice-velocity variability and wind conditions (Nazarenko and Miller [54]; Fissel and Tang [30]). Winds can influence both areal extent and near-edge ice-pack concentrations, producing a highly concentrated pack with a sharply defined edge when winds are toward the shore. The opposite is true when winds move over the ice towards open water. The dynamic variability of the ice near the pack edge is important. Frequent cyclonic disturbances passing through the area can produce significant ice displacements over short periods of time.

Ice concentrations derived from brightness temperatures recorded by the Nimbus 7 SMMR provided the basis for regular evaluation of ice conditions through the study area for the 1980–1981 and 1984–1985 ice seasons, corresponding to "light" and "heavy" ice seasons, respectively. Total ice concentration data for individual footprints were used to interpolate a regularly spaced ice concentration grid from which ice distributions were mapped. These were then used for evaluation against individual storm events during the two seasons.

Using storms which tracked through the study area, basic storm information was documented, including the minimum observed pressure, the pressure range while in the study area, storm velocity, the latitude and longitude at which the storm entered the study region, the latitude at which it left, and the net storm bearing through the study area. For each storm, a tally was kept of the number of synoptic observations within the study area, providing an indication of the storm duration.

Two factors were of interest in assessing storm influence on ice conditions: ice-extent variation and concentration-change within the ice pack. The former was considered by examining the changes in the ice limit through the storm period by contouring the limits of detected ice. Ice-concentration variations were then investigated by plotting the difference in concentration between the first and last appropriate SMMR ice charts.

Analysis of a selected storm serves to illustrate the dynamic behavior of the ice cover. Between the 20 and 26 of January 1985, a storm of severe intensity passed through the southwest quadrant of the study area, tracking north-northeast over Newfoundland. Figure 5.43 shows the change in SMMR-derived ice concentration over the study area from the first date (20 January) to the last (26 January). The map shows that as the storm passed, little change in ice concentration occurred off the Newfoundland coast, whereas concentrations increased off central Labrador and decreased in the northern part of the study area.

Similar analyses were made for the other 25 storms. Consistently, passing storms resulted in parts of the ice edge advancing while other parts retreated or remained stable. Within the ice pack, changes in concentration, as displayed by the SMMR data, varied from increases to no change to decreases. For a given storm, areas of both convergence and divergence were observed, especially during midseason when the ice cover was near its maximum extent. With expansion of the ice edge, ice divergence frequently occurred inshore, and the SMMR charts showed ice concentrations correspondingly reduced.

A stepwise multiple regression was used to correlate grid cells showing concentration increase and decrease and along the ice edge, cells showing edge advance and retreat, against the various storm parameters. In all cases, the highest correlations were achieved with only one or two of the storm variables. The highest correlation was found for concentration increase where R^2 was 38% using storm-start longitude and the westerly component of the storm velocity. Storm parameters were also able to explain over 30% of the edge advance variability ($R^2 = 0.31$). Several authors have noted similar correlations between meteorological conditions and ice variability (Thorndike and Colony [84]; Ikeda et al. [44]; Nazarenko and Miller [54]).

Correlations were considerably lower for concentration decrease and ice-edge retreat (18% and 19%, respectively), suggesting that the dynamic storm parameters used in this study may be less important in their influence on this ice behavior during periods of ice retreat. The role of thermodynamic effects on marginal ice-cover variability has been noted previously at both seasonal (Ikeda et al. [44]) and synoptic (Andreas [2]) scales, and may have some importance here as well. Despite these limitations, the study demonstrated that passive microwave-derived ice information could be used to monitor high-frequency variability in the marginal ice zone.

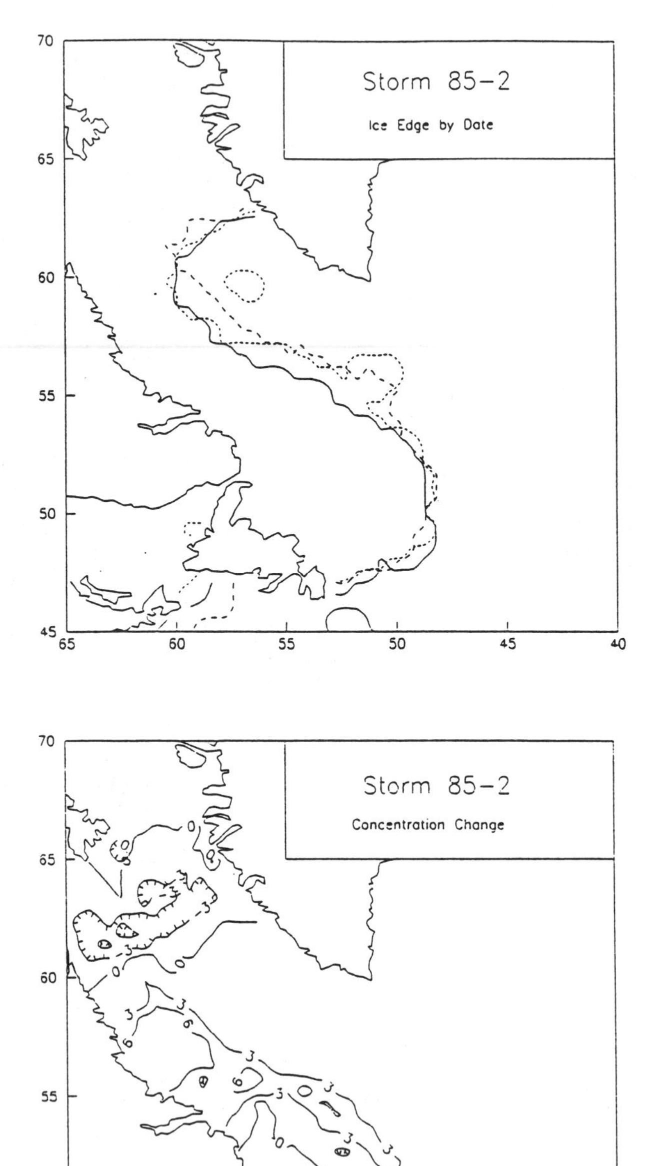

Figure 5.43 Example of ice dynamics during mid-season. Ice concentrations obtained from SMMR data on 20 to 26 January 1985. Line patterns for the ice edge plots are: solid for 20 January, dashes for 23 January, and fine dashes for 26 January 1985.

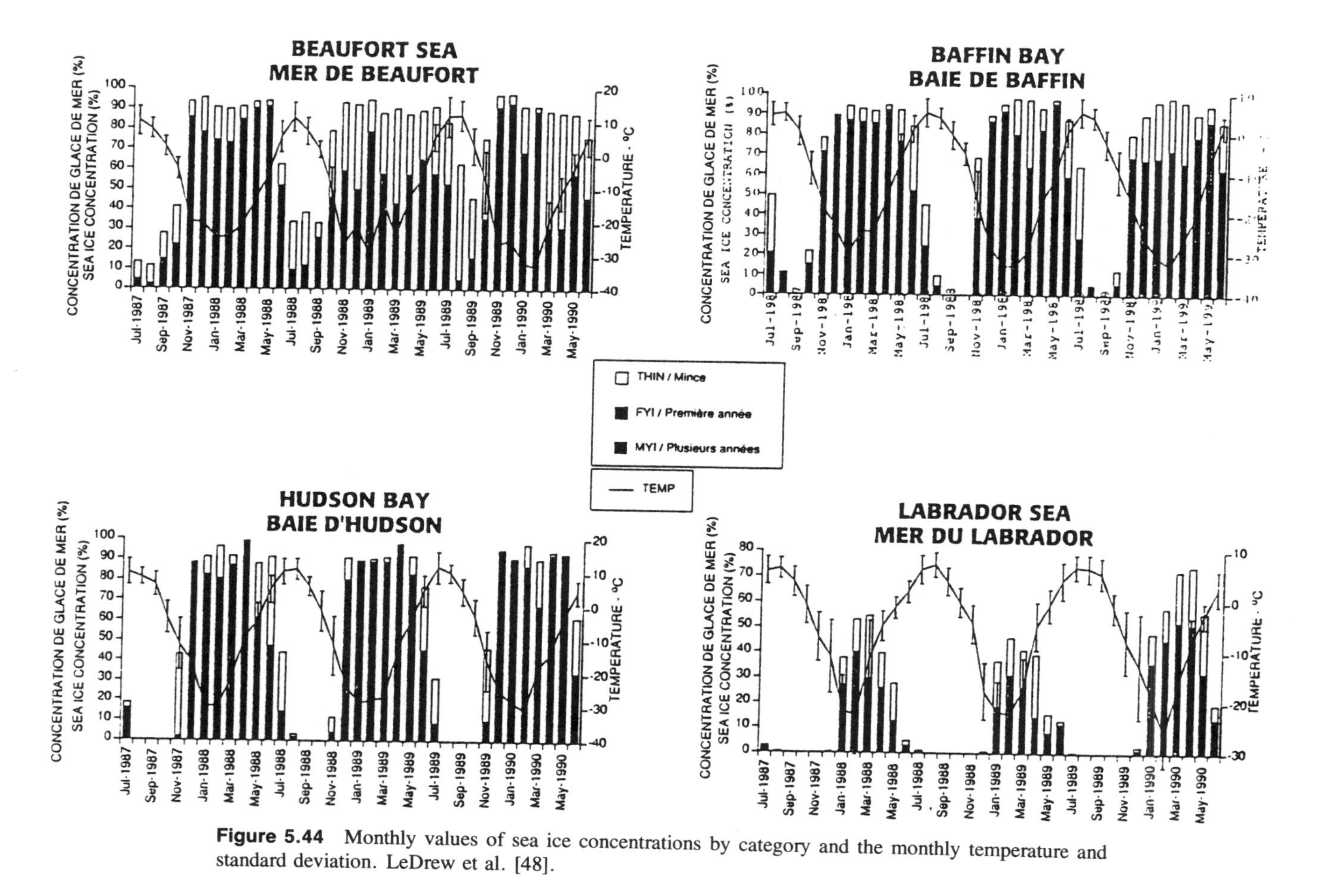

Figure 5.44 Monthly values of sea ice concentrations by category and the monthly temperature and standard deviation. LeDrew et al. [48].

The results of the use of SSM/I data for similar studies are beginning to appear in publications. The Canadian Sea Ice Atlas (LeDrew et al. [48]), produced with SSM/I data and the AES/York algorithm, from July 1987 to June 1990 illustrates the use of satellite data in analysis of the relationship between the sea-ice concentration and surface air temperature (see Fig. 5.44).

5.6 SUMMARY

The importance of developing long-term data on ice parameters is well recognized. Prior to relying on any new source of derived sea-ice information, the capabilities and the limitations of retrieval techniques must be subjected to rigorous validation exercises. In this case, quantitative relationships between sea-ice parameters derived from algorithms for passive microwave measurements and the same parameters derived from other sources have been (and will continue to be) investigated, with comparisons made for different seasons and for as many geographical areas as possible. Sea-ice parameters under scrutiny were (and are): location of the ice edge, total ice concentrations, and ice-fraction (new, first-year, and old ice) estimates. In addition, algorithm accuracies were (and are) evaluated for compatibility with operational ice information, produced by the AES Ice Branch and the U.S. NPOC ice centers. The near-real time operational application of the satellite data was tested during sea-ice research campaigns and through transfer of the derived sea-ice information to users (e.g., ship captains, Canadian Coast Guard). In the course of this evaluation process, passive microwave remote sensing data became an accepted source of sea-ice information for large-scale ice research, operational ice forecasts and navigational support.

The spatial and temporal scale of the data from satellite sensors lends itself to studies of the earth's cryosphere on a global scale. In addition, on a synoptic scale, these data are used to understand atmosphere–ocean–ice interactions. The numerous algorithm-validation campaigns not only allowed critical evaluation of the retrieval techniques used, but also expanded our understanding of how sea-ice information obtained from different sensors can be used synergistically. One of the unresolved problems in passive microwave remote sensing is the coarseness of the resolution and because of this, the integration of satellite data with SAR or AVHRR imagery of much higher resolution is a challenging task. The performance of existing algorithms can be improved by assimilating data from other sensors. For example, the use of atmospheric information deduced from AVHRR imagery could improve the retrieval accuracy of the AES/York old-ice fraction which would reduce some of the anomalies caused by precipitation.

Most of the recommendations for improving algorithm performance imply the use of sources of information in addition to passive microwave data. The techniques for incorporation of ancillary data are at the exploratory stage. The Kalman filter technique (Thomas and Rothrock [83]) is one of the methods that could be used for combining different data types when rapid processing is not required. Knowledge-based procedures are becoming very popular (see Chapter 13). Their ability to produce groups of possible outputs reduces some of the errors inherent in other classification methods. For example, neural-network results were found to be more accurate than the supervised maximum likelihood method (Key et al. [46]).

The important advantages of passive microwave remote sensing, such as

1. day- and night-sensing capabilities (important at polar latitudes)
2. relatively unimpeded by clouds
3. ability to provide multispectral data from a single platform
4. repetitive, regional coverage providing daily updates of information over broad areas
5. data base of global observations covering a time span of 25 years
6. maturity of the retrieval techniques used to interpret satellite data

have resulted in considerable effort in Canada to use spaceborne passive microwave data for environmental research and in support of the sea-ice and weather-forecasting activities. Research at the present time is directed towards developing techniques for improving spatial resolution.

The assimilation of auxiliary information and higher frequency microwave data into the retrieval algorithms is another challenge for the research community. Our understanding of seasonal–regional variabilities of microwave sea-ice signatures is far from complete. The use of in-situ microwave observations, i.e., basic research of microwave emission and scattering properties, and theoretical models will enhance the accuracy of the retrieval procedures.

ACKNOWLEDGMENTS

This work was made possible with funding from the following:

Ice Branch, AES, Canada: Translation of SMMR algorithm to SSM/I application (AES/PhD algorithm); Cal./Validation 1987–1989; operational testing for Canadian waters (1989–1990); partial support for data acquisition (navigational application evaluation during *Polarstern* cruises).

Canadian Meteorological Centre: Development of the Satellite Image Display Software tools; validation.

Aerojet Electronic System GenCorp.: Partial support (1989–1991) for: validation and development of the algorithm modules; data acquisition during research and operational testing campaigns.

Institute for Space and Terrestrial Science: Project #9-97605, 1988–1992.

One of the authors (IGR) would like to thank Drs. F. W. Thirkettle and F. E. Bunn (Ph.D. Associates Inc.) for support and encouragement during the development of the earlier AES/PhD version of the algorithm. Their involvement in the passive microwave sensor and data-application projects went beyond their contractual commitments.

The assistance of Mr. A. Davies (meteorology) and the Atmospheric Research and Modeling consulting staff (programming) is gratefully acknowledged. In addition, editorial assistance from Dr. D. Barber (University of Manitoba) and Mr. H. McLaughlin (ISTS) is gratefully acknowledged.

We thank Dr. R. O. Ramseier (Ice Branch, AES, Environment Canada) for providing the opportunity to become involved in this research.

LIST OF SYMBOLS

B_f	Blackbody spectral brightness ($Wm^{-2}sr^{-1}Hz^{-1}$)
I_f	Radiation energy flux ($Wm^{-2}Hz^{-1}$)
f	Frequency (GHz or Hz)
h	Planck's constant
k	Boltzmann's constant
T_B	Brightness temperature (K)
P_f	Spectral power
T_A	Antenna radiometric temperature
η	Antenna radiation efficiency
α_m	Antenna main beam efficiency
G	Receiving power gain
T_{sys}	System noise temperature
τ	Integration time
τ	Optical opacity
B	Bandwidth
L	Loss factor
e	Emissivity
C	Total ice concentration
T_{Bsp}	Brightness temperature of the surface target, "p" polarization
PR	Polarization ratio
GR	Gradient ratio

LIST OF ACRONYMS

AES	Atmospheric Environment Service (Canada)
AIMR	Airborne Imagining Microwave Radiometer
AVHRR	Advanced Very High Resolution Radiometer
CMC	Canadian Meteorological Center
CME	Cabin-Mounted Equipment
DMSP	Defense Meteorological Satellite Program
EPA	External Payload Assembly

ESMR	Electronically Scanning Microwave Radiometer
FDD	Freezing Degree Days
FNOC	Fleet Numeric Oceanographic Center
GOLA	Gaussian OpticsLens Antenna
IFOV	Instantaneous Field of View
LIMEX	Labrador Ice Margin Experiment
MIZEX	Marginal Ice Zone Experiment
NIMBUS	U.S. satellite
NOAA	National Oceanic and Atmospheric Administration
NOARL	Replaced by NRL
NPOC	Naval Polar Oceanographic Center
NRL	Naval Research Laboratories
NTSC	National Television Standards Commission
SAR	Synthetic Aperture Radar
SEASAT	U.S. satellite
SLAR	Sideways-Looking Airborne Radar
SMMR	Scanning Multichannel Microwave Radiometer
SSM/I	Special Scanning Microwave Imager
SST	Sea Surface Temperature
VME	Versa Modular Eurocard

REFERENCES

[1] M. R. Anderson (1987) ''The onset of spring melt in first-year ice regions of the Arctic as determined from Scanning Multichannel Microwave Radiometer data for 1979 and 1980,'' *J. Geophys. Res.*, **92**, 13,153–13,163.

[2] E. L. Andreas (1985) ''Heat and moisture advection over Antarctic sea ice,'' *Monthly Weather Rev.*, **113**, 736–746.

[3] A. H. Barrett and E. Lilley (1963) ''Mariner-2 microwave observations of Venus,'' *Sky and Telescope*, 192–195.

[4] A. E. Basharinov, A. S. Gurvich, S. T. Yegorov, A. A. Kurskaya, D. T. Matvyev, and A. M. Shutko (1971) ''The results of microwave sounding of the earth's surface according to experimental data from satellite Cosmos 243,'' *Space Res.*, **11**, Akademie-Verlag, Berlin.

[5] C. Bjerklund, R. O. Ramseier, and I. G. Rubinstein (1990) ''Validation of the SSM/I and AES/York algorithms for sea ice parameters,'' S. F. Ackley and W. F. Weeks (Eds.), *Sea Ice Properties and Processes*, Monog. 90-1, Cold Region Res. and Eng. Lab., Hanover, NH, 206–208.

[6] J. O. Bursey, W. J. Snowden, A. D. Gates and C. L. Blackwood (1977) ''The climate of the Labrador Sea,'' POAC-77, St. John's, Newfoundland, 938–951.

[7] R. F. Cahalan and L. G. Chiu (1986) ''Large-scale short-period sea-ice interaction,'' *J. Geophys. Res.*, **91** (C9), 10,709–10,717.

[8] W. J. Campbell, P. Gloersen, and J. J. Zwally (1984) ''Aspects of Arctic sea ice observable by sequential passive microwave observations from the Nimbus-5 sat-

ellite,'' in I. Dyer and C. Chryssostomidis (Eds.), *Arctic Technology and Policy*, Hemisphere Publishing, Washington, DC, 197–222.

[9] T. Carrieres and D. Crosbie (1991) ''A comparison of remote sensing systems for operational ice analysis,'' in *Proc. of 2nd WMO Operational Ice Remote Sensing Workshop*, Ottawa, 51–61.

[10] F. D. Carsey (Ed.) (1992) *Microwave Remote Sensing of Sea Ice*, AGU Monog. 68, American Geophysical Union, Washington, DC.

[11] D. J. Cavalieri, B. Burns, and R. Onstott (1990) ''Investigation of effects of summer melt on the calculation of sea ice concentration using active and passive microwave data,'' *J. Geophys. Res.*, **95** (C4), 5359–5369.

[12] D. J. Cavalieri, J. P. Crawford, M. R. Drinkwater, D. T. Eppler, L. D. Farmer, R. R. Jentz, and C. C. Wackerman (1991) ''Aircraft active and passive validation of sea-ice concentration from the DMSP SSM/I,'' *J. Geophys. Res.*, **96**, 21,989–22,008.

[13] D. J. Cavalieri, P. Gloersen, and W. J. Campbell (1984) ''Determination of sea-ice parameters with the NIMBUS-7 SMMR,'' *J. Geophys. Res.*, **89**, 5355–5369.

[14] D. J. Cavalieri, P. Gloersen, and T. T. Wilheit (1986) ''Aircraft and satellite passive microwave observations of the Bering Sea ice cover during MIZEX West,'' *IEEE Trans. Geosci. and Remote Sens.*, **Ge-24** (3) 268–377.

[15] D. J. Cavalieri and C. L. Parkinson (1981) ''Large-scale variations in observed Antarctic sea-ice extent and associated atmospheric circulation,'' *Monthly Weather Rev.*, **109**, 2323–2336.

[16] D. J. Cavalieri and C. T. Swift (1987) ''NASA sea-ice and snow validation plan for the DMSP, SSM/I,'' NASA Tech. Memo. 100683.

[17] D. J. Cavalieri and H. J. Zwally (1985) ''Satellite observations of sea-ice,'' *Advances in Space Res.*, **5**, 247–255.

[18] S. Charton (1965) ''Radar systems sensitivity in modern radar,'' in R. S. Berkowitz (Ed.), *Modern Radar*, John Wiley & Sons, New York.

[19] J. C. Comiso (1983) ''Sea ice effective emissivities from satellite passive microwave and infrared observations,'' *J. Geophys. Res.*, **88**, 7686–7704.

[20] J. C. Comiso (1990) ''Multiyear ice classification and summer ice cover using Arctic passive microwave data,'' *J. Geophys. Res.*, **95**, 975–994.

[21] J. C. Comiso, S. F. Ackley, and A. L. Gordon (1984) ''Antarctic sea ice microwave signature and their correlation with in-situ ice observations,'' *J. Geophys. Res.*, **89**, 662–672.

[22] J. C. Comiso, T. C. Grenfell, D. L. Bell, M. A. Lange, and S. F. Ackley (1989) ''Passive microwave in situ observations of winter Weddell Sea ice,'' *J. Geophys. Res.*, **94** (C8), 10,891–10,905.

[23] R. G. Crane (1983) ''Atmosphere–sea-ice interactions in the Beaufort/Chukchi Sea and in the European sector of the Arctic,'' *J. Geophys. Res.*, **88**, 4505–4523.

[24] R. G. Crane, R. G. Barry, and H. J. Zwally (1982) ''Analysis of atmosphere-sea ice interactions in the Arctic Basin using ESMR microwave data,'' *Int. J. of Remote Sens.*, **3**, 259–276.

[25] R. W. Crawford and A. Madej (1992) ''CASP II Meteorology field and data summary,'' *Environment Canada, Atmospheric Environment Service, Cloud Physics Research Division Report*.

[26] A. F. Davies, I. G. Rubinstein, and R. O. Ramseier (1990) "Climatological Oceanic Information from the special sensor microwave imager," in *Can. Meteorol. and Oceanogr. Soc. Symp.*, Victoria, BC.

[27] M. El-Tahan and I. G. Rubinstein (1993) "The use of SSM/I data in ice edge prediction models during CASP II," Work in progress.

[28] D. T. Eppler, L. D. Farmer, and A. W. Lohanick (1990) "On the relationship between ice thickness and 33.6 GHz brightness temperature observed for first-season sea-ice," in S. F. Ackley and W. W. Weeks (Eds.), *Sea Ice Properties and Processes*, Cold Region Res. and Eng. Lab. Monog. 90-1, Hanover, NH, 229–232.

[29] D. A. Etkin (1990) "A comparison of conventional and passive microwave sea ice data sets for Hudson Bay," M.Sc. Thesis, York University, Centre for Research in Experimental Space Science, North York, Ontario.

[30] D. B. Fissel and C. L. Tang (1991) "Response of sea-ice drift to wind forcing on the northeastern Newfoundland shelf," *J. Geophys. Res.*, **96**, 18,397–18,409.

[31] G. R. Fowles (1975) *Introduction to Modern Optics*, Holt, Rinehart and Winston, New York.

[32] J. A. Gagliano (1987) "MMW Radiometry," in N. Currie and C. Brown (Eds.), *Principles and Applications of Millimetre-Wave Radar*, Artech House, Norwood, MA.

[33] P. Gloersen and W. J. Campbell (1988) "Variations in the Arctic, Antarctic, and global sea ice cover during 1978–1987 as observed with the NIMBUS-7 SMMR," *J. Geophys. Res.*, **93** (C9), 10,669–10,674.

[34] T. C. Grenfell (1992) "Surface-based passive microwave studies of multiyear sea ice," *J. Geophys. Res.*, **97**, 3485–3501.

[35] T. C. Grenfell and J. C. Comiso (1986) "Multifrequency passive microwave observations of first-year sea ice grown in tank," *IEEE Trans. Geosci. and Remote Sens.*, **GE-24** (6), 826–831.

[36] T. C. Grenfell, D. Bell, A. W. Lohanick, C. T. Swift, and K. St. Germain (1988) "Multifrequency passive microwave observations of saline ice grown in tank," in *Proc. IGARSS'88 Symp.*, ESA Spec. Pub. SP-284, 1687–1690.

[37] T. C. Grenfell and A. W. Lohanick (1985) "Temporal variations of the microwave signatures of sea-ice during the late spring and early summer near Mould Bay, NWT," *J. Geophys. Res.*, **90**, 5063–5074.

[38] J. Halle (1981) "Objective analysis of ice cover at CMC and its use in numerical weather prediction," in *Proc. 2nd WMO Operational Ice Remote Sensing Workshop*, Ottawa, 75–85.

[39] D. Henderson (1985) "Evaluation of accuracy of SMMR derived ice concentrations produced for the Labrador Coast during the operational demonstration period: Jan. 15–March 31, 1984," Environment Canada, Atmospheric Environment Service, Ice Product Development Division Report.

[40] J. Hollinger, R. Lo, and G. Poe (1987) "Special sensor microwave imager user's guide," Naval Research Laboratory Report, Washington, DC.

[41] J. Hollinger (1989) "DMSP special microwave/imager calibration/validation, Final Report: Vol. I," NRL, Washington, DC.

[42] J. Hollinger (1991) "DMSP special sensor microwave/imager calibration/validation, Final Report: Vol. II," NRL, Washington, DC.

[43] Hughes Aircraft Company (1980) "Special sensor microwave imager (SSM/I) critical design review," Contract No. FO4701-79-C-0061 Report.

[44] M. Ikeda, G. Symonds, and T. Yao (1988) "Simulated fluctuations in annual Labrador sea ice cover," *Atmosphere-Ocean*, **26** (1), 16–39.

[45] R. C. Johnson (1984) "Introduction to antennas," in R. C. Johnson and H. Jasek (Eds.), *Antenna Engineering Handbook*, 2nd ed., McGraw-Hill, New York.

[46] J. R. Key, J. A. Maslanik, and A. J. Schweiger (1989) "Classification of merged AVHRR and SMMR Arctic data with neural networks," *Photogram. Eng. and Remote Sens.*, **55**, 1331–1338.

[47] P. E. Lemke, W. Trinkl, and K. Hasselman (1980) "Stochastic dynamic analysis of polar sea-ice variability," *J. Phys. Oceanog.*, **10**, 2100–2120.

[48] E. LeDrew, D. Barber, T. Agnew, and D. Dunlop (1992) "Canadian sea ice atlas from microwave remotely sensed imagery: July 1987 to June 1990," Atmospheric Environmental Service, Environment Canada, Climatological Study No. 44, p. 80.

[49] C. E. Livingstone, R. G. Onstott, L. D. Arsenault, A. L. Gray, and K. P. Singh (1987) "Microwave sea-ice signatures near the onset of melt," *IEEE Trans. Geosci. and Remote Sens.*, **GE-25** (2), 174–187.

[50] S. M. Lovas, S. Vefsnmo, and R. O. Ramseier (1991) "Barents sea ice conditions as observed by passive microwave and other Techniques," in *Proc. 2nd WMO Operational Ice Remote Sensing Workshop*, Ottawa, 101–115.

[51] J. Maslanik, J. Key, and A. J. Schweiger (1990) "Neural network identification of sea-ice seasons in passive microwave data," in *Proc. IGARSS'90*, 1281–1284.

[52] T. A. Moreau, E. L. Sudeikis, I. G. Rubinstein, and F. W. Thirkettle (1985) "A study and report on operational ice mapping, Nimbus-7, SMMR," AES Contr. 01SE.KM147-4-0501, Ph.D. Associates Ltd., North York, Ontario.

[53] D. M. Nazarenko (1990) *Synoptic scale ice-atmosphere interaction off the east coast of Canada*, Centre for Climate and Global Change, McGill University, Montreal.

[54] D. M. Nazarenko and J. D. Miller (1986) "Sea ice drift measurements using satellite tracking beacons," *Proc. Canadian Sea-Ice Workshop*, Halifax.

[55] D. M. Nazarenko, I. G. Rubinstein, and T. Carrieres (1991) "The operational utility of satellite passive microwave imagery in supporting Canadian regional sea ice monitoring," in *Proc. WMO Sea Ice Workshop*, Ottawa, 47–53.

[56] C. L. Parkinson (1989) "On the value of long-term satellite passive microwave data sets for sea-ice/climate studies," *Geophys. J.*, **18** (1), 9–20.

[57] C. L. Parkinson, J. C. Comiso, H. J. Zwally, D. J. Cavalieri, P. Gloersen, and W. J. Campbell (1987) "Arctic sea-ice, 1973–1976, satellite passive microwave observations," NASA Scientific and Technical Branch Rept. NASA SP-489.

[58] C. L. Parkinson and A. J. Gratz (1983) "On the seasonal sea ice cover of the Sea of Okhotsk," *J. Geophys. Res.*, **88** (C5), 2793–2802.

[59] L. T. Pedersen (1991) "Retrieval of sea-ice concentration by means of microwave radiometry," Ph.D. Thesis, Technical University of Denmark, Lyngby, Denmark.

[60] R. H. Preller, P. G. Posey, and A. Cheng (1991) "Existing and planned uses of remotely sensed data in the U.S. Navy's sea ice forecast models," in *Proc. 2nd WMO Operational Ice Remote Sensing Workshop*, Ottawa, 235–255.

[61] R. H. Preller, J. E. Walsh, and J. A. Maslanik (1992) ''The use of satellite observations in ice cover simulations,'' in F. D Carsey (Ed.), *Microwave Remote Sensing of Sea Ice*, AGU Monog. 68, American Geophysical Union, Washington, DC, 385–404.

[62] R. O. Ramseier, I. G. Rubinstein, and A. F. Davies (1988) ''Operational evaluation of special sensor microwave imager,'' Environment Canada, Atmospheric Environment Service, Ice Branch Report, ISTS/Microwave Group Report.

[63] R. O. Ramseier, D. Lapp, I. G. Rubinstein, and K. Asmus (1989) ''Canadian validation of the SSM/I and AES/York algorithms for sea-ice parameters,'' ISTS/ Microwave Group Report, North York, Ontario.

[64] D. A. Rothrock, D. R. Thomas, and A. S. Thorndike (1988) ''Principal component analysis of satellite passive microwave data over sea ice,'' *J. Geophys. Res.*, **93** (C3), 2321–2332.

[65] I. G. Rubinstein (1983) ''Retrieval of sea ice properties from Nimbus-7 SMMR data,'' in *Proc. of Can. Oceanogr. and Meteorol. Soc.*, Banff, Alberta.

[66] I. G. Rubinstein (1984) ''Sea state and weather correction to sea-ice Algorithms,'' in P. Gloersen (Ed.), *Proc. of Nimbus-7 Experiment Team Meeting*, NASA Goddard Space Center, Greenbelt, MD.

[67] I. G. Rubinstein and R. O. Ramseier (1985) ''Retrieval of sea-ice properties and ocean surface wind speeds from satellite borne radiometer,'' in Ph.D. Assoc. Inc. (Ed.), *Proc. of Intl. Sea Ice Workshop*, York University, North York, Ontario.

[68] I. G. Rubinstein and R. O. Ramseier (1985) ''Scientific application of passive microwave satellite data for ice monitoring and research,'' in *Proc. of ESA Conf. on the Use of Satellite Data in Climate Models*, ESA SP-244, Alpach, Austria, 117–124.

[69] I. G. Rubinstein (1987) ''Retrieval of sea-ice properties and ocean surface wind speeds from satellite borne radiometer,'' in V. V. Varadan and V. K. Varadan (Eds.), *Multiple Scattering of Waves in Random Media and Random Rough Surfaces*, The Pennsylvania State University Press, University Park, PA, 851–866.

[70] I. G. Rubinstein and R. O. Ramseier (1987) ''Optimum use of dual frequency passive microwave measurements for ice/ocean interactions,'' in *Proc. IGARSS'87 Symp.*, Ann Arbor, MI, 1147–1150.

[71] I. G. Rubinstein, M. Shokr, and P. Ostiguy (1993) ''Comparison of ice concentration retrieval from passive and active microwave sensors,'' Proc. WMO Sea Ice Workshop, Ottawa.

[72] T. V. Seling and D. K. Nance (1968) ''Sensitive microwave radiometer detects small icebergs,'' unpublished.

[73] M. C. Serreze, J. A. Maslanik, R. H. Preller, and R. G. Barry (1990) ''Sea ice concentrations in the Canada Basin during 1988; Comparison with other years and evidence of multiple forcing mechanisms,'' *J. Geophys. Res.*, **95** (C12), 22,253–22,267.

[74] N. Skou (1989) *Microwave Radiometer Systems*, Artech House, Norwood, NH.

[75] D. H. Staelin, A. H. Barett, J. W. Waters, F. T. Barath, E. J. Johnston, P. W. Rosenkrantz, N. E. Grant, and W. R. Lenoir (1973) ''Microwave spectrometer on the NIMBUS-5 satellite, meteorological and geophysical data,'' *Science.*, **182**, 1339–1341.

[76] K. Steffen, J. Key, D. J. Cavalieri, J. Comiso, P. Gloersen, K. St. Germain, and

I. Rubinstein (1992) "The estimation of geophysical parameters using passive microwave algorithms," in F. D. Carsey (Ed.), *Microwave Remote Sensing of Sea-Ice*, AGU Monog. 68, American Geophysical Union, Washington, DC, 201–232.

[77] K. Steffen and A. Schweiger (1991) "NASA Team algorithm for sea ice concentration retrieval from Defence Meteorological Satellite program special sensor microwave imager, comparison with LANDSAT satellite imagery," *J. Geophys. Res.*, **96** (C12), 21,971–21,987.

[78] A. Stogryn (1972) "The emissivity of sea foam at microwave frequencies," *J. Geophys. Res.*, **77**, 1658–1666.

[79] A. Stogryn and G. J. Desargent (1985) "The dielectric properties of brine in sea-ice at microwave frequencies," *IEEE Trans. on Antennas and Propag.*, **AP-33**, 523–532.

[80] A. Stogryn (1985) "A study of some microwave properties of sea-ice and snow," Rept. 7788, Aerojet Electrosystems Co., Azusa, CA.

[81] R. Stone (1991) "Keeping tabs on a big berg," *Science*, **254**, 1290.

[82] A. W. Straiton, C. W. Tolbert, and C. O. Britt (1958) "Apparent temperature distribution of some terrestrial materials and the sun at 4.3 mm wavelength," *J. Appl. Phys.*, **29**, 776–782.

[83] D. R. Thomas and D. A. Rothrock (1989) "Blending sequential scanning multichannel microwave radiometer data into a sea ice model," *J. Geophys. Res.*, **94**, 10,907–10,920.

[84] A. S. Thorndike and R. Colony (1982) "Sea ice motion in response to geostrophic winds," *J. Geophys. Res.*, **87** (C8), 5845–5852.

[85] W. B. Tucker, III, T. C. Grenfell, R. G. Onstott, D. K. Perovich, A. J. Gow, R. A. Schuchman, and L. L. Sutherland (1991) "Microwave and physical properties of sea-ice in the winter marginal ice zone," *J. Geophys. Res.*, **96**, 4573–4587.

[86] F. T. Ulaby, R. K. Moore, and A. K. Fung (1981) *Microwave Remote Sensing: Active and Passive*, Vol. I, Addison-Wesley, Reading, MA.

[87] F. T. Ulaby, R. K. Moore, and A. K. Fung (1982) *Microwave Remote Sensing: Active and Passive*, Vol. II, Addison-Wesley, Reading, MA.

[88] F. T. Ulaby, R. K. Moore, and A. K. Fung (1986) *Microwave Remote Sensing: Active and Passive*, Vol. III, Artech House, Norwood, MA.

[89] A. E. Walker (1986) "A climatological analysis of eastern Canadian seaboard maximum ice extent using passive microwave Data (1973–1985)," M. A. Thesis, Carleton University.

[90] J. E. Walsh and R. G. Crane (1992) "A comparison of GCM simulations of arctic climate," *Geophys. Res. Letters*, **19** (1), 29–32.

[91] J. M. Walters, C. Ruf, and C. T. Swift (1988) "A microwave radiometer weather-correcting sea ice algorithm," *J. Geophys. Res.*, **92** (C6), 6521–6534.

[92] D. G. Wells (1990) "Trafficability study for the Soviet nuclear icebreaker *Rossia* Voyage to the North Pole, July 31 to August 15, 1990," Report, Norland Science and Engineering, Ottawa.

[93] T. Wilheit, A. T. C. Chang, and A. S. Milman (1980) "Atmospheric corrections to passive microwave observations of the ocean," *Boundary-Layer Meteorol.* 65–77.

[94] D. R. P. Williams (1986) "Evaluation and applications of sea ice models to Nimbus-7 satellite passive microwave observations," M.Sc. Thesis, York University, Centre for Research in Experimental Space Science, North York, Ontario.

[95] H. J. Zwally (1984) "Observing polar ice variability," *Ann. of Glaciology*, **5**, 191–198.

[96] H. J. Zwally, J. Comiso, C. L. Parkinson, W. J. Campbell, F. D. Carsey, and P. Gloersen (1983) "Antarctic sea ice, 1973–1976, satellite passive microwave observations," NASA SP-459, p. 206.

[97] H. J. Zwally and P. Gloersen (1977) "Passive microwave images of the polar regions and research applications," *Polar Rec.*, **18**, 431–450.

6

ACTIVE MICROWAVE SYSTEMS

R. KEITH RANEY
Canada Centre for Remote Sensing
Energy, Mines and Resources
Ottawa, Ontario, Canada

6.1 INTRODUCTION

The following chapters in this book deal extensively with specific types of active microwave systems. Those radars have much in common with each other at fundamental levels. In principle, the signal received by any radar contains information on the range (time delay), amplitude, phase, azimuth position, and polarization for each reflection. The main objective of this chapter is to describe the ability of remote sensing radars to achieve such observations, and to highlight the essential similarities as well as the significant differences between them.

The traditional introduction to radar system performance starts with the radar equation for a single-pulse observation of an isolated, discrete scatterer (e.g., Ridenour [19]; Skolnik [21]; or Ulaby et al. [23]). The radar equation is usually derived, or more aptly explained, in the *power* domain. Such an approach depends on the concept of radar cross section, which is a definition of scattering in terms of a power ratio. The traditional radar equation is helpful in understanding the first-order parametric dependencies of radar systems and historically has been an essential tool for system design. Unfortunately, the traditional approach is not appropriate for quantitative consideration of multipulse signal-processing radars.

Virtually no remote sensing radar uses only a single pulse, nor is the prime

Remote Sensing of Sea Ice and Icebergs, Edited by Simon Haykin, Edward O. Lewis, R. Keith Raney, and James R. Rossiter.
ISBN 0-471-55494-4

objective the observation of only one discrete scatterer. Radar signals combine according to the properties of their (complex) *amplitude*. The effective power of a set of signals depends not only on the individual powers but also on the relative phases of the constituents. A disadvantage of the traditional radar equation approach is that subtle and often more important aspects, such as details of reflectivity from area scatterers or extension of the single pulse case to the coherent multiple-pulse situation of a synthetic aperture radar (SAR), are less than obvious if each signal element is described only by its power.

The approach taken in this chapter differs from traditional treatments. The starting point is through the *amplitude* of the signal received from an individual pulse. The resulting model is particularly suitable for remote sensing radars which rely on two-dimensional signal processing in the complex voltage domain. The result is a description of the salient features of a variety of systems through a unified model.

First, the single-pulse response, or *complex range line*, is described, of a radar observing an extended scattering surface, of which ice is an excellent and interesting example. The detected range line is introduced, and multiple scatterer phenomena are described. The discussion then develops an expression for the two-dimensional response of a generalized radar system, based upon suitable combinations of a sequence of complex range lines. Specific cases of this model include the principal radars of interest in this book. The radar equation in each case may be found from the ratio of signal and noise power levels in the output expressions. The concept of coherence is explored, and is used to differentiate between various radars used in remote sensing. Finally, the model is extended to include measurement of the polarimetric scattering matrix of the scene.

6.2 ANALYTIC RADAR MODEL, SINGLE PULSE

The model is an analytic statement which is equivalent to a simple verbal description of the operation of a radar. In words, for each transmission, a radar generates an electromagnetic pulse which is radiated by an antenna and propagates to the scene, of which a fraction is reflected back towards the radar where it is gathered by the receiving antenna. The received signal, being weak, suffers from additive noise which accompanies the signal through the rest of the system. Both signal and noise are amplified in the receiver, demodulated from the carrier (microwave) frequency, filtered to match the original pulse characteristics, and then detected. A block diagram of this sequence is shown in Fig. 6.1. Comments on each stage of this model are noted below, from which follow the desired analytic expressions.

All radars of relevance to remote sensing transmit a sequence of pulses at a rate known as the pulse repetition frequency. Most remote sensing radars improve their performance by processing over a group of received signals. In

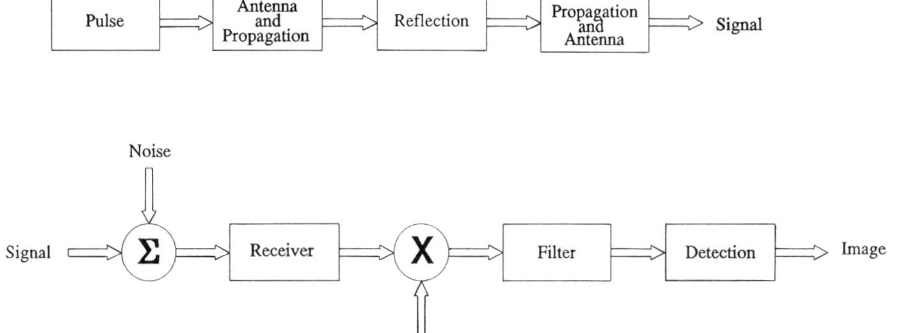

Figure 6.1 Model of signal flow in a radar. Functional steps, parallel to those discussed in the text, followed the history of a single pulse from generation and transmission, through reflection, reception, and receiver functions, to form the detected "image" for a single range line.

anticipation of the need to handle such cases quantitatively later in the discussion, the one signal of interest at this point is that indexed by the letter i.

Similarly, a remote sensing radar usually illuminates a scene that is comprised of many small scatterers. The development of this section starts with the *voltage* response for one elemental scatterer, then extends coverage to the illuminated scene by complex summation over the individual scattering elements. The classical *point-target radar cross section* in the power domain follows as a special case of the detected voltage response. Note that this order is backwards when compared to the traditional method, but arguably adheres more closely to the physics involved. Furthermore, this approach allows direct extension of the single-pulse results to multiple-pulse data-processing radars that observe extended scattering scenes, and to complex scattering phenomena such as quadrature polarimetry.

The purpose of the model is to enable insight into the principles governing radar performance rather than to determine design of the electronic components of the system. In this spirit, the details of physical implementation are stripped away in the discussion to reveal the central signal theoretic issues. Specific system issues are highlighted as required in subsequent chapters.

6.2.1 Pulse

In this model, the hardware comprising the reference oscillator, pulse generation, high-power amplifier, and so forth, are all included in one compact expression. One needs only to know the characteristics of the pulse as delivered to the antenna. The ith pulse may be written as

$$a_T p_t(t) \sin(\omega_0 t + \varphi_i^T) \tag{6-1}$$

where

a_T = amplitude of the transmitted signal pulse
$p_t(t)$ = shape of the pulse, including envelope and phase modulation
ω_0 = radar carrier frequency (radians) = $2\pi f_0$
φ_i^T = reference phase of the transmitted signal

Note that the reference phase, in general, is different for each pulse unless the radar is designed otherwise. Time t is assumed to start from zero at the instant that each pulse leaves the radar. The *peak power* of the radiated pulse is a_T^2, where the envelope $p_t(t)$ has maximum value of unity, by definition. The *energy* of the pulse is given by the product of its power and its length τ. The sequence of pulses is transmitted at the *pulse repetition frequency* f_p, having uniform *interpulse period* T_p. *Average power* $\overline{P_T}$ is defined as pulse energy divided by the interpulse period, which is equivalent to the peak power multiplied by the *duty cycle* τ/T_p, or, equivalently, $P_T\tau f_p$.

The signal is emitted as an electromagnetic (EM) field, elegantly described by Maxwell's equations (Maxwell [11]; Kerr [9]; Ramo and Whinnery [16]; Born and Wolf [1]). The electric vector $\boldsymbol{E}$ of the EM field is essentially equivalent to a potential disturbance with units of *voltage*. Note that an EM field is transverse: its vector components are orthogonal with respect to the direction of field propagation. It follows that the spatial orientation of the $\boldsymbol{E}$ component, which determines the polarization of the field, must be included in a complete description. The vector nature of the propagation and reflection of the wave fields is the focus of the polarization section which closes this chapter. For the moment, the discussion is in terms of a scalar theory, for which the field orientation is not an issue.

Phase is an essential property of $\boldsymbol{E}$. This fact has far-reaching consequences penetrating to the heart of radar operation, and is revisited several times in the discussions that follow. Phase is not present in the *power* domain; power is a real (and positive) number. Power is described in EM theory as the magnitude of the Poynting vector, which in turn may be shown to be proportional to the square of the magnitude of the $\boldsymbol{E}$ vector, hence to voltage squared. Phase preservation through the radar when using many pulses leads to the concept of coherence, discussed as needed below.

6.2.2 Antenna and Propagation

The task of the transmitting antenna is to direct illumination towards the scene of interest and to minimize coverage of all other regions. For this purpose, the antenna is usually built to have a *pattern*, with fan and pencil shapes (narrow in one and two angular directions, respectively) being examples commonly met in remote sensing radar systems. Signals radiated by a directive antenna benefit from gain within the beam. For the moment, let the one-way voltage gain of the transmitting antenna, in the direction of the scatterer, be g_T. Antenna losses are included in the definition of *gain*, which accordingly is distinguished from the *directivity* of the antennal pattern.

As the radiated signal propagates to range distance R, two things happen of first-order importance: the strength of the signal decreases and there is a time delay. The *voltage* of the radiating pulse decreases in proportion to R. This follows from the principle of conservation of energy, applicable for the non-dispersive media surrounding most remote sensing radars. The total power of an EM wave observed on a spherical surface centered at the radiating source must be inversely proportional to the surface area. The surface area of a sphere with radius R is $4\pi R^2$. Thus, the radiating pulse voltage is decreased in amplitude by $2R\sqrt{\pi}$ at range R. This is known as spreading loss.

As the pulse propagates, it travels at the speed of light, $c = 3 \times 10^8$ m s^{-1}. The time delay at range R is R/c, leading to

$$t \rightarrow t - \frac{R}{c} \tag{6-2}$$

as the appropriate time-delayed argument in (6-1) as it appears at range R.

Putting these results together, the operator corresponding to the antenna and propagation box of the model becomes

$$\frac{g_T}{2\sqrt{\pi}R}, t \rightarrow t - \frac{R}{c} \tag{6-3}$$

Equation (6-3) is to be used as a multiplier and time delay when applied to the pulse expression of (6-1).

Note, from the known relationships of the wave equation

$$\frac{\omega_0}{c} = \frac{2\pi f_0}{c} = \frac{2\pi}{\lambda} \doteq k \tag{6-4}$$

that the phase shift corresponding to the one-way time delay of (6-3) may be written as $-kR$, where k is the wave number of the radar.

6.2.3 Reflection

The purpose of (remote sensing) radars is to observe properties of the scene illuminated by the system. The only way in which sensible observations arise is through the mechanism of reflection. Hence, description of the reflection process needs to be handled with care. In this chapter, reflection is modeled in sufficient detail to support discussion of the various radars of interest to the book. Specialized aspects of observable scattering are covered elsewhere, such as dielectric material properties (Chapter 2), Bragg scattering (Chapter 7), and spatial (speckle) interference noise (Chapter 11).

It is assumed in this discussion that the so-called monostatic radar configuration is under study. A *monostatic* radar is one for which the transmit antenna and the receive antenna are essentially in the same location, in contrast to a *bistatic* configuration for which the two antennas may be separated by a substantial angular difference as viewed from the reflecting element in the scene. The model could be extended to include this case, but it would complicate the development unnecessarily and is not required for radars of primary interest in this book. In the following, only the reflection back towards the radar, the *backscatter*, is considered.

Reflection as formally defined is intuitively satisfying in principle: for each small element illuminated by the radar, the field re-radiated back towards the radar is a simple fraction Γ of the incident field. If the field incident on a scattering element is $\boldsymbol{E}_i$, then the field $\boldsymbol{E}_r$ reflected back towards the radar from that element is

$$\boldsymbol{E}_r = \Gamma \boldsymbol{E}_i. \tag{6-5}$$

The quantity Γ, the (voltage) reflectivity of the element, is a complex number as the field may suffer a phase change upon reflection. The magnitude of Γ represents the efficiency of field reflection, so it has magnitude less than unity. The coefficient of reflectivity as used here is closely related to the Fresnel scattering coefficient Γ (Stratton [22]) which generally is defined only for a smooth plain surface. Extension of this discussion to account for polarization is considered in Section 6.4.

The total amount of signal reflected will increase as the illuminated area increases. There are many subtleties involved here. For the moment, it is sufficient to restrict the elemental area, Δ, contributing to one reflection to be very small. As the reflected field has phase as well as amplitude, the way in which it may combine with fields reflected from other elemental scatterers needs to be treated with care. The consequences of this observation are developed in greater detail below. At this point, all that is required is the reflectivity operation in the model which appears as a (complex) multiplier $\Gamma\Delta$ for the field reflected from one small element. The conventional geometric and radiometric parameters used to quantify reflectivity in remote sensing applications are discussed in Subsection 6.2.9.

6.2.4 Propagation and Antenna

Once reflected, the field must propagate back to the radar. On the way, the returning field undergoes spreading loss and time delay, just as occurred in the transmission direction. The numbers that describe these effects are the same as those discussed in Subsection 6.2.2.

Back in the vicinity of the radar, the portion of the reflected (voltage) field that is collected is determined by the square root of the effective area of the receive antenna, which follows from reapplication of the conservation of energy argument of Subsection 6.2.2. For convenience, it is assumed that the same antenna is used for reception as is used for transmission, which often is true. The effective receiving antenna area A_R is related to the one-way voltage gain g_T of the antenna according to $\sqrt{A_R} = g_T\lambda/2\sqrt{\pi}$. As was the case for the transmit antenna, further discussion of antenna gain is postponed until a later section.

Putting these results together, the operator corresponding to the propagation and antenna box in the receive path of the model becomes

$$\frac{g_T\lambda}{4\pi R}, t - \frac{R}{c} \rightarrow t - \frac{2R}{c} \tag{6-6}$$

which, as before, is to be applied as a multiplier and a time delay on the preceding expressions.

The sequence above completes derivation of the signal increment $ds_{t,i}^{\omega_0}(t, R)$ that is presented to the radar from the ith transmission reflected from the scattering element Δ. From (6-1) through (6-4) and (6-6), we have

$$ds_{t,i}^{\omega_0}(t, R) = \frac{a_T g_T^2 \lambda}{(4\pi)^{3/2} R^2} \Gamma \Delta p_t \left(t - \frac{2R}{c}\right) \sin\left(\omega_0 t + \varphi_i^T - 2kR\right) \tag{6-7}$$

based on the development of the model up to the receiver. In this and the following, the superscript ω_0 denotes a signal expressed at the carrier frequency.

For any scene typical of remote sensing applications, there will be many scattering elements distributed over the illuminated area. For all those at range R, their signals will appear simultaneously at the receiver, and thus will combine in some way to build the observed signal from the ensemble of elemental reflections. This important issue is discussed following the description of the signal path through the remainder of the system.

6.2.5 Receiver

When it arrives back at the radar, the signal is extremely weak. The first task of the receiver is to amplify the signal to a more useful level. Unfortunately, the signal must compete with background noise. Any physical device generates internal noise, which in effect is added to the signal, particularly where the signal is weakest. In the case of a radar, this is at the input to the receiver. Amplification gain derived through the receiver and following components affects the signal and the noise equally. The overall receiver voltage gain is modeled as unity, a convention that is quite satisfactory for the purposes of this discussion (although it certainly would not meet with the approval of a hardware system designer).

The desired signal output of the radar in comparison to output noise is a featured aspect of this discussion. From a signal-to-noise ratio point of view, the new parameters required are those that describe a receiver's noise characteristics. The receiver noise model is considered in two parts. The first step is to obtain an expression for the noise at the input of the receiver. The second step is to quantify the noise level as it competes with the desired signal, which is deferred to later sections.

There are many sources of additive noise that may occur in the receiver or enter the system by being received at the antenna (e.g., Skolnik [21]). For systems of interest in terrestrial remote sensing, the dominant additive noise is thermal noise in the "front end" of the system. Such noise is well modeled as a stationary Gaussian random process modulating a carrier at the radar frequency (Davenport and Root [5]). The noise n encumbering the ith signal in the passband of the receiver may be expressed as

$$n_i^{\omega_0}(t) = n_{c,i}(t) \cos \omega_0 t - n_{s,i} \sin \omega_0 t \tag{6-8}$$

where $n_{c,i}$ and $n_{s,i}$ are statistically independent random processes. Although narrow band with respect to the radar frequency ω_0, the noise essentially has a flat spectrum across the receiver bandwidth used in remote sensing radars.

An alternative representation of (6-8) is in terms of envelope and phase random processes according to

$$n_i^{\omega_0}(t) = \nu_{N,i}(t) \sin [\omega_0 t + \varphi_i^N(t)] \tag{6-9}$$

in which the envelope

$$\nu_{N,i}(t) = [n_{c,i}^2(t) + n_{s,i}^2(t)]^{1/2} \tag{6-10}$$

has a Rayleigh probability density function, and the phase of (6-9) (given by the inverse tangent of the ratio of the Gaussian random variables) is uniformly distributed between zero and 2π. The Rayleigh and associated probability distribution functions are discussed in Subsection 6-2-10.

6.2.6 Demodulation

The information of interest in the received signal is in the envelope and phase of the radar frequency, not in the carrier frequency itself. The carrier may be removed by a process known as demodulation, through which the output of the receiver is multiplied by a sinusoidal signal in a device known as a mixer, followed by a lowpass filter. The result of the multiplication is the creation of two versions of the receiver output, one centered at the sum of the radar carrier and demodulator frequencies, and one centered at their difference frequency. The filter selects the signal components at the difference frequency.

Demodulation may take place in two or more stages in any given radar. In the event that two demodulations are used, the first operation leads to a so-called intermediate frequency, fondly known as the IF for those in the trade. For the purposes of this discussion, however, it is sufficient to group all demodulation stages into one stage.

The mean frequency after demodulation is always much less than that of the original radar frequency. The model discussed here assumes that the demodulation frequency is matched to the radar frequency, so that the output of the demodulator is at zero average frequency, hence at baseband. To do this without creating problems, complex or *quadrature* demodulation (e.g., Brown [2]) must be used. This method simplifies the mathematics, conforms exactly to certain radars of importance in remote sensing, and includes offset baseband systems as a special case.

Let the demodulator use $\sin(\omega_0 t + \varphi_i^{LO})$ upon reception of the ith pulse, where φ_i^{LO} is the reference (local oscillator) phase, and ω_0 is the radar carrier (radian) frequency. The signal plus noise input to the demodulator may be written

$$a \sin(\omega_0 t + \varphi) + \nu_{N,i}(t) \sin[\omega_0 t + \varphi_i^N(t)] \tag{6-11}$$

where a and φ are placeholders for the more cumbersome expressions [(6-7)] of amplitude and phase of the received signal. Then the lowpass output of the demodulator for the sinusoidal reference oscillator is

$$a \cos(\varphi - \varphi_i^{LO}) + \nu_{N,i}(t) \cos(\varphi_i^N - \varphi_i^{LO})$$

Alternatively, $\cos(\omega_0 t + \varphi_i^{LO})$ could be used instead of the sinusoidal reference. Applying the cosine to the mixer input of (6-11), the lowpass output of the demodulator for the cosinusoidal reference oscillator is

$$a \sin(\varphi - \varphi_i^{LO}) + \nu_{N,i}(t) \sin(\varphi_i^N - \varphi_i^{LO})$$

When both sine and cosine demodulation references are used in the same system, then there are two output signals available. These are known as the in-phase (I) and the quadrature (Q) components. When treated as an ordered pair, $(I + jQ)$, in subsequent processing, complex representation of the demodulated signal results, where $j = \sqrt{-1}$. Quadrature demodulation allows more effective estimation of the scene reflectivity and is commonly employed in certain remote sensing radars.

Based on these observations, the remaining steps of the radar model are represented with complex notation. The local oscillator signal for demodulation may be written as $\exp\{j(\omega_0 t + \varphi_i^{LO})\}$. Using this as the mixer reference, and the signal [(6-7)] plus noise as input, the output of the demodulator becomes

$$ds_{t,i}(t, R) = \frac{a_T g_T^2 \lambda}{(4\pi)^{3/2} R^2} \Gamma \Delta p_t \left(t - \frac{2R}{c}\right) \exp\{-j(\varphi_i^{LO} - \varphi_i^T + 2kR)\} + n_i(t) \tag{6-12}$$

The noise, after being demodulated to baseband, appears as a complex Gaussian random process $\{(n_i(t)\}$, the lowpass counterpart to (6-8) or (6-9).

Equation (6-12) represents the elemental signal after demodulation but before filtering operations. In a digital implementation, the analog I and Q signals are digitized and stored in a memory device for subsequent processing. The discussion is continued in the analog domain, however, as it is simpler in notation and sufficient to convey the main ideas.

The relative gain terms of (6-12) are the subject of traditional signal-to-noise ratio developments. For signal-processing radars, however, the phase structure takes on pivotal significance, since combination of two or more signals at the same radar carrier frequency depends intimately on their relative phase properties.

6.2.7 Filter

In many radars the transmitted pulse is phase modulated. The task of the filter is to compress the pulse to a simple short impulse response and to establish the receiver bandwidth together with possible additional weighting. Under the constraint that the output signal-to-noise ratio is to be maximized, it may be shown that the optimum filter transfer function, known as the *matched filter*, is the complex conjugate of the Fourier transform of the original pulse (North [13]). The main reason for using a modulated pulse is that the radar is better able to transmit a relatively large amount of energy in each pulse under the constraint that the peak power cannot exceed a given value. Radar performance is determined by the energy (or average power) in each pulse rather than the peak power, so this is an excellent scheme and often used.

The operation of pulse compression is illustrated in Fig. 6.2. The principle feature is that for an input pulse $p(t)$ of unity amplitude, time duration τ, and bandwidth B_p, the amplitude of the output pulse is $\sqrt{(\tau B_p)}$ and its width in time τ_0 is $1/B_p$. The quantity τB_p is known as the time-bandwidth product (*TBP*), a parameter of fundamental importance for pulse compression systems. Typically, $TBP > 100$ in remote sensing radars. Note that the *TBP* of the *output* pulse is unity. In practice, weighting may be used on both the original pulse shape and the filter envelope to reduce sidelobes of the compressed pulse. Weighting causes the output pulse to have a peak level less than optimum, and a width greater than optimum. If the actual input pulse width τ and output pulse width τ_0 are measured, then the more general expression of the filter action is in terms of the range *pulse compression ratio*, $C_R = \tau/\tau_0$.

The role of pulse compression may be incorporated explicitly into the model through the filter convolution integral

$$\sqrt{C_R}p_0(t) = \int p^*(t')p(t - t')\, dt'$$

which expresses the matched filter transformation of the modulated pulse $p(t)$ into the compressed pulse, $\sqrt{C_R}p_0(t)$. The peak value of $p_0(t)$ is normalized to one, by definition. In the event that a simple unmodulated pulse is transmitted,

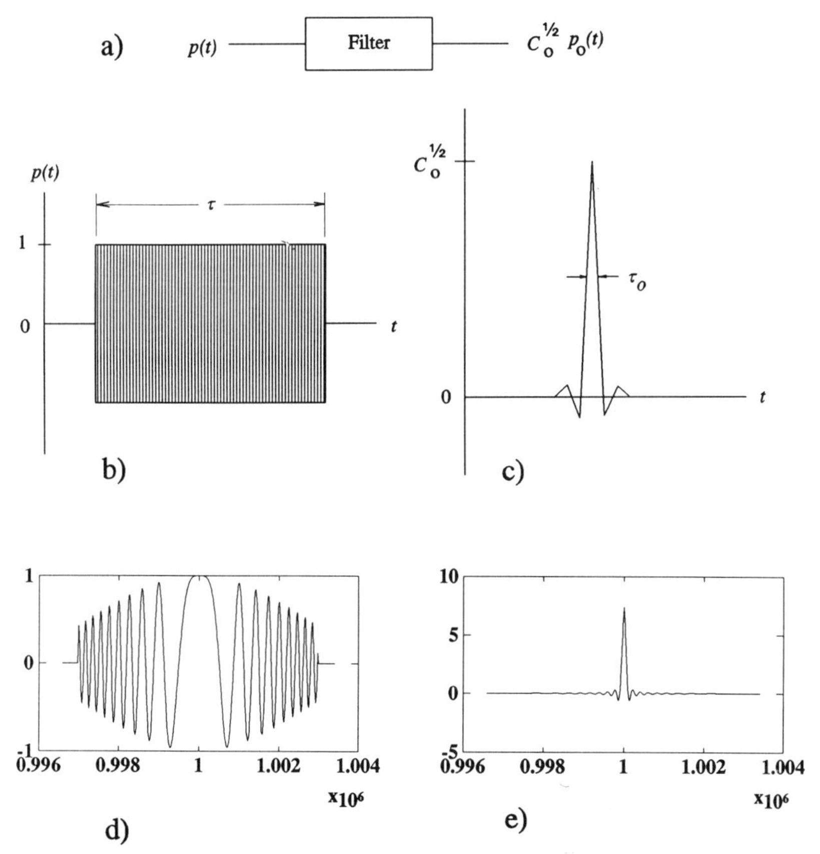

Figure 6.2 Action of a pulse compression filter: (a) Filter model for pulse compression ratio of $C_0^{1/2}$; (b) Large time-bandwidth product pulse, normalized to unity amplitude, and pulse duration τ; (c) Compressed pulse of enhanced amplitude and width τ_0; (d) Modulated pulse typical of a spacecraft radar, length 6-km and TBP = 100, from reflector at 1000-km range; and (e) Compressed pulse at output of filter. Amplitude of the compressed pulse is less than $100^{1/2} = 10$ due to amplitude weighting on modulated phase.

this same approach may be used noting that the compression gain in this case would be unity.

In short, the benefit of a matched filter is to sharpen the impulse response and to increase the effective gain for individual scatterers. For coherent radars, these benefits may be realized also in the azimuth channel in a manner analogous to the range channel studied in this section. These points are developed in Section 6.3.

6.2.8 The Complex Range Line

The *complex range line* is the set of data resulting from each pulse, considered after filtering and other processing steps, before detection. The complex range line is easily manipulated as a digital file. It cannot be displayed as a visual product: to do so requires a detection operation. The idea of a complex range line is helpful in the present development and is an essential concept for signal-processing radars which exploit attributes of phase structure in a set of received signals, with benefits such as improvement of resolution or Doppler measurement. The fundamental building block in signal-processing radars is the complex range line.

Finding the value at each sample position in the complex range line requires a suitable combination of the response from all illuminated elemental scatterers. It is reasonable to assume that the individual returns combine linearly. Linearity is true for the implicit summation that occurs at any instant in the received signal, because the fields reflected from all scatterers at a given range are present simultaneously and superimposed at the receive antenna. The assumption of linearity is reasonably true for the explicit summations done in the radar and signal processor, with possible limitations imposed by the hardware or software linearity.

To perform the summations analytically, use must be made of a specific illumination geometry relating the radar and the scene. In general, azimuth integration at any instant is over the angular variable θ, having a spatial increment $dx = R d\theta$ over the wavefront at range R. This representation is adequate for radars fixed in one location, but is awkward for radars that do quasi-linear scanning, as is the case for airborne or spaceborne imaging systems. The geometric issue is simplified for side-looking scanning systems if it is assumed that the radar has an azimuth antenna pattern that is very narrow. In this context, "very narrow" implies that the difference between the actual wavefront of radius R and a plane wave approximation to the wavefront is less than the range resolution of the radar, within the azimuth beamwidth, a requirement that is reasonably well satisfied for all radars of interest at the level of this discussion. For wide beam surface-based radars, such as those used for over-the-horizon work (Chapter 7), the azimuth integration should be left in the azimuth angular variable. The formalities of the following development apply to both cases.

Under the narrow antenna assumption, the illumination geometry in the scene may be described by a rectangular coordinate system (x, r) as in Fig. 6.3. The coordinate system is in the slant range plane, thus closely resembling the ground plane for earth-based radars. Although the line of sight radar–scatterer distance, or "slant range," is the natural dimension proportional to time delay of pulse propagation and reflection, for airborne or satellite radars, a planimetric surface map is usually demanded for applications which must be expressed in terms of "ground range." A geometric transformation from the slant range plane to the ground plane is required. The transformation applies both to the map scale and to pertinent local scales such as range resolution and range pixel spacing.

The scattering element has area Δ orthogonal to the slant range vector, and hence facing the radar, denoted the *incident area* plane by Cosgriff et al. [3],

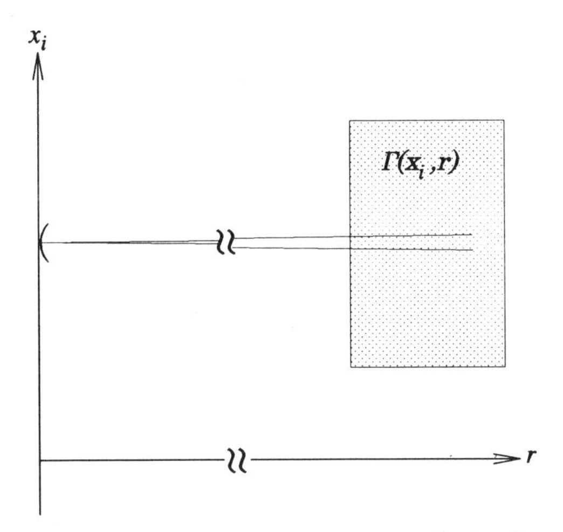

Figure 6.3 Rectangular, orthogonal slant range scene coordinates for a generalized remote sensing radar.

as in Fig. 6.4. The natural coordinate system for the radar, however, is in azimuth and range, which for any element $dx\, dr$ implicitly defines an element of the incident area plane. Let

$$\Delta = dx\, dz(r)$$

in which $z(r)$ is introduced as an auxiliary spatial variable to assist in the organization of the following development.

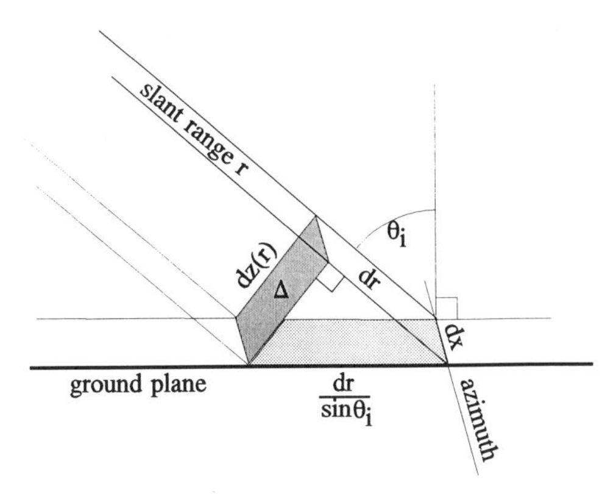

Figure 6.4 Slant range viewing geometry of the illuminated scene, assuming locally level features (detail). Incremental area Δ is determined by the azimuth width dx and the orthogonal elevation dimension $dz(r)$, which is set by the range increment dr and the (local) incidence angle θ_i. The geometry is applicable to calculation of the range resolution in slant range, ground range, or projected onto the local slope (if any), and for integrals over signal increments.

Let the radar be at $(x_i, 0)$ for the ith transmission. Note, for radars on a moving platform, x_i will be different for each transmission, whereas for ground-based radars, x_i usually is constant for all pulses. The scattering from each element may be written

$$\Gamma(x, r)\Delta \rightarrow \Gamma(x, r) \exp\left\{-jk \frac{(x - x_i)^2}{r}\right\} dx\, dz(r) \tag{6-13}$$

where the exponential term (for the ith sample) represents the required correction in phase, relative to the beam center at x_i, to move from the tangent plane wavefront aligned with the (x, r) coordinate grid to the scatterer illuminated by the spherical wavefront of radius r (Fig. 6.5).

To complete this section, two more intermediate steps are required. Previously, the voltage gain of the antenna in the direction of the elemental scatterer was abbreviated as g_T. The gain of the antenna applicable to any given reflection is determined by the relative position of the scatterer within the antenna pattern. Normally, the radiated pattern of an antenna is described in terms of angular coordinates, elevation ϕ and azimuth θ. It is more useful in the present context to describe the pattern in spatial coordinates. For the radar located at $(x_i, 0)$ let the one-way *voltage pattern*, towards the scatterer position (x, r), be

$$g_T \rightarrow \sqrt{G_0}\, g_V(x_i - x, r)$$

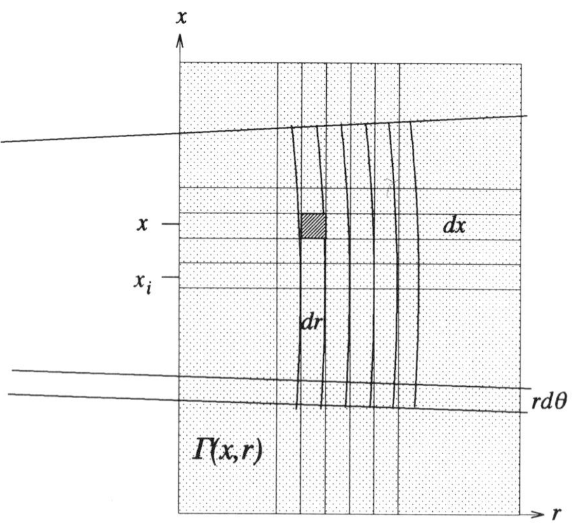

Figure 6.5 Coordinate system used for narrow beam radars requiring rectangular scene description. The only substantive difference between a cell in (x, r) space and the same cell in $(r\theta, r)$ space is a phase shift $k(x - x_i)^2/r$, as discussed in the text. For geometries in which the angular bandwidth is less than about 1°, the approach is excellent. For high performance systems, however, higher order effects must be included, as noted in Chapter 11.

where G_0 is the one-way peak *power* gain of the antenna, and $g_V(x, r)$ describes the one-way voltage-gain pattern in azimuth and range.

Anticipating that the radar data products eventually are to be expressed as a function of spatial position, it is convenient to convert the impulse-response independent variable from time, t, to slant range, r. This conversion is easily done, noting that the change is a scaling in the slant range domain by $r = ct/2$, where c is the speed of light. Thus, the slant range spatial impulse response may be written as $p_0(r)$. The independent variable of all other functions may be treated in the spatial domain in similar fashion.

Based on the foregoing developments, the single-pulse radar response may be written directly. At each range, the value of the complex range line is the linear sum over all illuminated scattering elements. Passing to the limit, the sums become integrals. Then the ith range line complex response is

$$i_i^c(x_i, r) = n_i(r) + g_1 \exp\{-j(\varphi_i^{LO} - \varphi_i^T)\} \iint \Gamma(x', r') g_V^2(x_i - x', r) \cdot p_0(r - r')\Phi_{s,i}(x_i - x'; r)\, dx\, dz(r') \tag{6-14}$$

where

$$g_1 \doteq \frac{a_T G_0 \lambda \sqrt{C_R}}{(4\pi)^{3/2} r^2} \quad \text{and} \quad \Phi_{s,i}(x; r) = \exp\left\{-jk\frac{x^2}{r}\right\}$$

The single-pulse (amplitude) gain term g_1 shows up repeatedly; it includes parameters that are determined primarily by the radar hardware. The range dependence of this gain is a slowly varying function (expressed in terms of r rather than r') and thus it may be brought out of the integral. This is reasonable because the gain change is negligible from this term over the width of the range impulse response within the integral.

The factor $\Phi_{s,i}$ ( ) is the phase of the scattering element with response to x_i, and is determined by the range-dependent phases of (6-12) and (6-13), in turn determined by the particular imaging geometry of a given system.

Several aspects of the complex range line should be noted. First and foremost, phase is present both inside and outside of the integrals. These phase terms, together with the implied phase of the complex reflectivity term, are fundamental to the operation of all radars and are central to the discussion of Section 6.3.

The azimuth integral is, in effect, a summation over all reflecting material within each range resolution cell, weighted by the azimuth antenna pattern. The range integral is a convolution of the voltage range impulse response over the scene reflectivity, and being a convolution, has the same functional form which occurs in the analytical description of any linear imaging system.

6.2.9 The One-Dimensional Detected Range Line

The output of a radar is expected to appear as a visual image product. This requires that the complex range line be detected. There are several options available for detection, but the discussion here is restricted only to the magnitude-squared operator. It has cleaner mathematical representation than other forms and is commonly used in practice.

Although it is a simple operation to implement, detection has subtle and deep consequences. In addition to stripping away the phase terms outside of the integrals inherent to the complex data, detection may lead to surprising results in the output, an example of which is noted in Subsection 6.2.10. There are random phases inherent in each reflection, so that the scattering term needs to be handled as a sample function of a random process. An expression for the detected image may be found only by taking an average over the output of the detector.

Let the detected range line corresponding to the ith pulse be $I_i(x_i, r)$ where, as before, this represents the single-pulse slant range response gathered by the radar at position x_i. Then

$$I_i(x, r) = \boldsymbol{E}[|i_i^C(x_i, r)|^2] \tag{6-15}$$

where $\boldsymbol{E}$ is the expectation operator (e.g., Davenport and Root [5]). Looking at (6-14), there is the potential for four terms when evaluating (6-15). However, the signal is statistically independent of the noise, and the noise voltage has zero mean value. It follows that the expected value of both the signal and noise cross terms is zero. The expected value of $|n_i|^2$ is the mean noise power $\overline{P_{n,i}}$.

Evaluation of the remaining term of the expansion implied by (6-15) leads to an integral of the form

$$\iiiint \boldsymbol{E}[\Gamma(x, r)\Gamma^*(x', r')]f(x, r)f^*(x', r')\, dx\, dx'\, dz(r)\, dz(r') \tag{6-16}$$

where the asterisk denotes complex conjugate, and $f(\ )$ is shorthand representation for the remaining factors in the integrand. The essential feature of this equation is determined by the average operation $\boldsymbol{E}[\ ]$ over the scattering process.

The discussion now considers the most important part of the radar remote sensing issue: the way in which elemental reflections combine. For each pulse, the wave incident upon the scene is essentially at one frequency, or monochromatic to use the optical analogical term. For any radar, this implies that the interaction between all illuminated elements within a range resolution shell is coherent, so that their phases interact causing either constructive or destructive interference in the net signal. For a distributed scene such as ice, the elemental scatterers each have phases that are statistically independent. Furthermore, the scattering elements for such a scene are always much more densely spaced then the radar is able to resolve. This set of scatterers may be modeled by mathematics equivalent to that which describe additive noise in the receiver of a radar. It follows that the reflection from a distributed scene is a complex Gaussian random process for which the cross correlation between any two elements has value zero for all scatterer positions that do not coincide. The mathematical expression of this is that

$$E[\Gamma(x, r)\Gamma^*(x', r')] = |\Gamma(x', r')|^2 \delta(x - x')\delta(r - r') \qquad (6\text{-}17)$$

where the $\delta(\ )$ represents the Dirac delta distribution (e.g., Papoulis [14]). The impact of the Dirac distribution is to cause the integral to exist only for the values for which their argument is zero. The effect of this is to reduce (6-16) to the form

$$\iint |\Gamma(x', r')|^2 |f(x', r')|^2 \, dx' \, dz(r')$$

This equation is a consequence of the random phases for each of the scattering elements: the total reflected power is the sum over the power of each of the elemental reflections. (N.B.: If there is non-zero phase correlation between the scattering elements, then the total power observed by the radar is not equal to the sum of the individual powers.)

The quantity $|\Gamma|^2$ is the average reflectivity of the scene, normalized with respect to the incidence area of Fig. 6.4. From the geometry in the figure,

$$|\Gamma(x', r')|^2 \, dx' \, dz(r') = \frac{\gamma(x', r')}{\tan \theta_i(r')} \, dx' \, dr'$$

In references that prefer this normalization (e.g., Cosgriff et al. [3]), $|\Gamma|^2$ is denoted by γ. It is interesting to note that for Lambertian scattering (Born and Wolf [1]) which is typical of reflection from a rough planar surface, γ is constant as a function of incidence angle.

For remote sensing radars, the more conventional way to describe reflectivity is σ^0, so-called *sigma nought*, which is the (average) reflectivity per unit area in the nominal ground plane. Ground-plane normalization is preferred by most people in the remote sensing community, as it has a more direct intuitive interpretation. In this case, the geometric conversion leads to transformation of the $|\Gamma|^2$ representation according to

$$|\Gamma(x', r')|^2 \, dx' \, dz(r') = \frac{\sigma^0(x', r')}{\sin \theta_i(r')} \, dx' \, dr' \qquad (6\text{-}18)$$

which is a simple scaling relationship determined by the incident angle θ_i at range r'. When used in an integral expression, generally the incidence angle changes insignificantly over the width of the range resolution cell. This allows the incidence angle factor to be brought out of the integrand in expressions used to model the received signal.

Using the σ^0 convention, the expression for the detected response to distributed scatterers for a single transmitted pulse is

$$I_i(x_i, r) = \overline{P_{n,i}} + \frac{g_1^2}{\sin \theta_i(r)} \iint \sigma^0(x', r') g_V^4(x_i - x', r) p_0^2(r - r')\, dx'\, dr' \tag{6-19}$$

The output is given by the convolution of the (power) impulse response over the average scattering function of the scene. This result is generalizable to two-dimensional images, as shown in Section 6-3. It is worth noting that the detected range line has no phase information remaining. The noise is indexed by the pulse number for use in subsequent discussions.

6.2.10 Distributed Scatterers

For a given scene, σ^0 represents the mean power ratio of scattering per unit area on the surface and as such is a dimensionless quantity. It is expressed in terms of deciBels (dB), and typically ranges from more than -10 dB (e.g., very rough second-year ice) to less than -25 dB (e.g., smooth new ice). In general, σ^0 is a function of incidence angle and radar frequency as well as physical characteristics of the medium being imaged. There are extensive tabulated data on the σ^0 properties of ice and other natural scenes. For incidence angles approaching vertical, σ^0 includes scattering enhanced by the so-called coherent component because of coincidence of the illuminating wavefront and many elemental scatterers in the scene.

The discussion to this point has been concerned with the *average* scattering strength as observed in a (magnitude-squared) detected form. It is of interest to examine the variability of each estimate, for which the probability distribution of σ^0 is required. Consider the common situation in which the reflectivity Γ may be described as a complex, Gaussian random process. The voltage envelope of such a process is the Rayleigh distribution [Fig. 6.6(b)]. The probability distribution of the corresponding power, and hence of σ^0, is exponential in form [Fig. 6.6(a)] (e.g., Cramér [4]), which has two consequences worth comment.

Consider the observed reflectivity at point (x_i, r_0). At any point in the detected image, the random process $\{\sigma^0\}$ has an exponential probability distribution. Neglecting the effect of additive noise, at any given image location, the estimate of mean reflectivity is, say, $\sigma^0(x_i, r_0)$, for which the most likely observed value is zero. There will be a variety of values for the estimate at nearby points, even for nominally constant σ^0. The variation in reflectivity from point to point, arising from a scene characterized by a given σ^0, is known

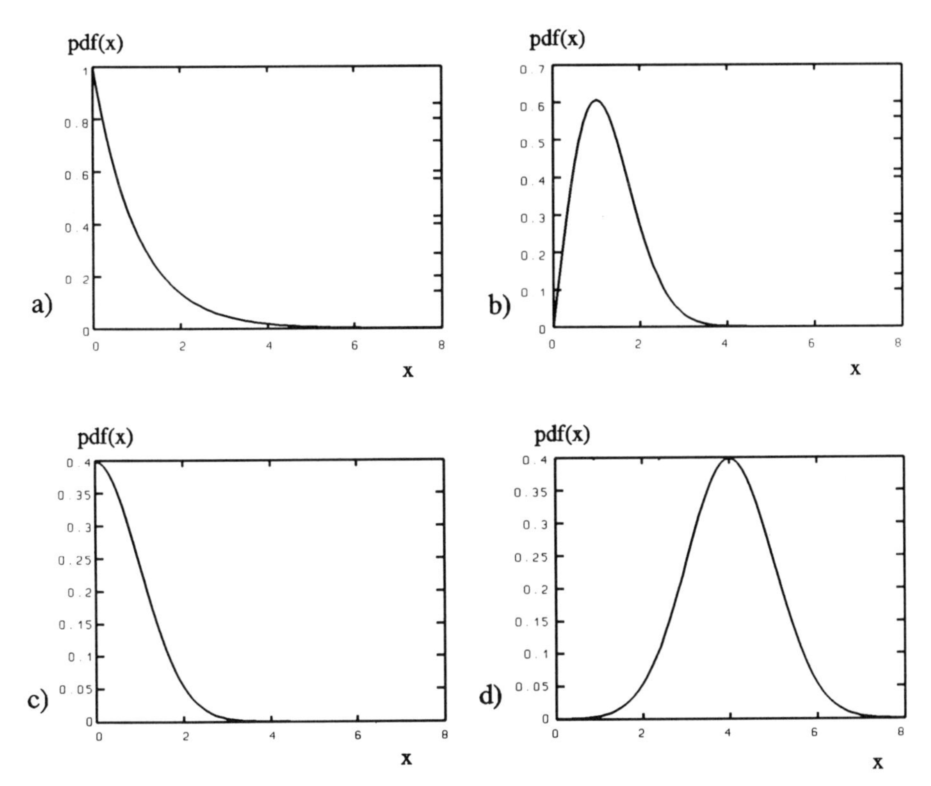

Figure 6.6 Probability distribution functions (*pdf*) of importance in remote sensing radar applications: (a) Exponential, $\langle x \rangle = 1$; (b) Rayleigh, $\langle x \rangle = 1$; (c) Gaussian, $\langle x \rangle = 1$ and $\sigma_x = 1$ (symmetrical, plotted only for $x > 0$); and (d) Gaussian, $\langle x \rangle = 4$ and $\sigma_x = 1$. Analytic expressions for these functions are listed in Table 6.1. In these figures, the independent variable x is a placeholder for scene reflectivity, σ^0, target reflectivity σ, or any other random variable encountered in radar applications.

as fading. The same phenomenon occurs for synthetic aperture radars and is known as speckle, discussed in Chapter 11.

The standard deviation σ_x is one measure of the width of a distribution of a random variable $\{x\}$ with respect to its mean value $\langle x \rangle$. For an exponential distribution, the standard deviation equals the mean value. Clearly, one single observation by a radar of a point in such a Gaussian scene does not lead to a reliable estimate of mean reflectivity.

The estimate may be improved by averaging more than one statistically independent observation, which implies integration after formation of the detected image. This leads to the chi-squared distribution, examples of which are plotted in Fig. 6.7. When N independent samples are used for an estimate, the standard deviation of the result is $1/\sqrt{N}$ of that of one sample. Virtually all remote sensing radars use some form of post-detection averaging for this reason. The distribution of sample values converges towards the ensemble mean

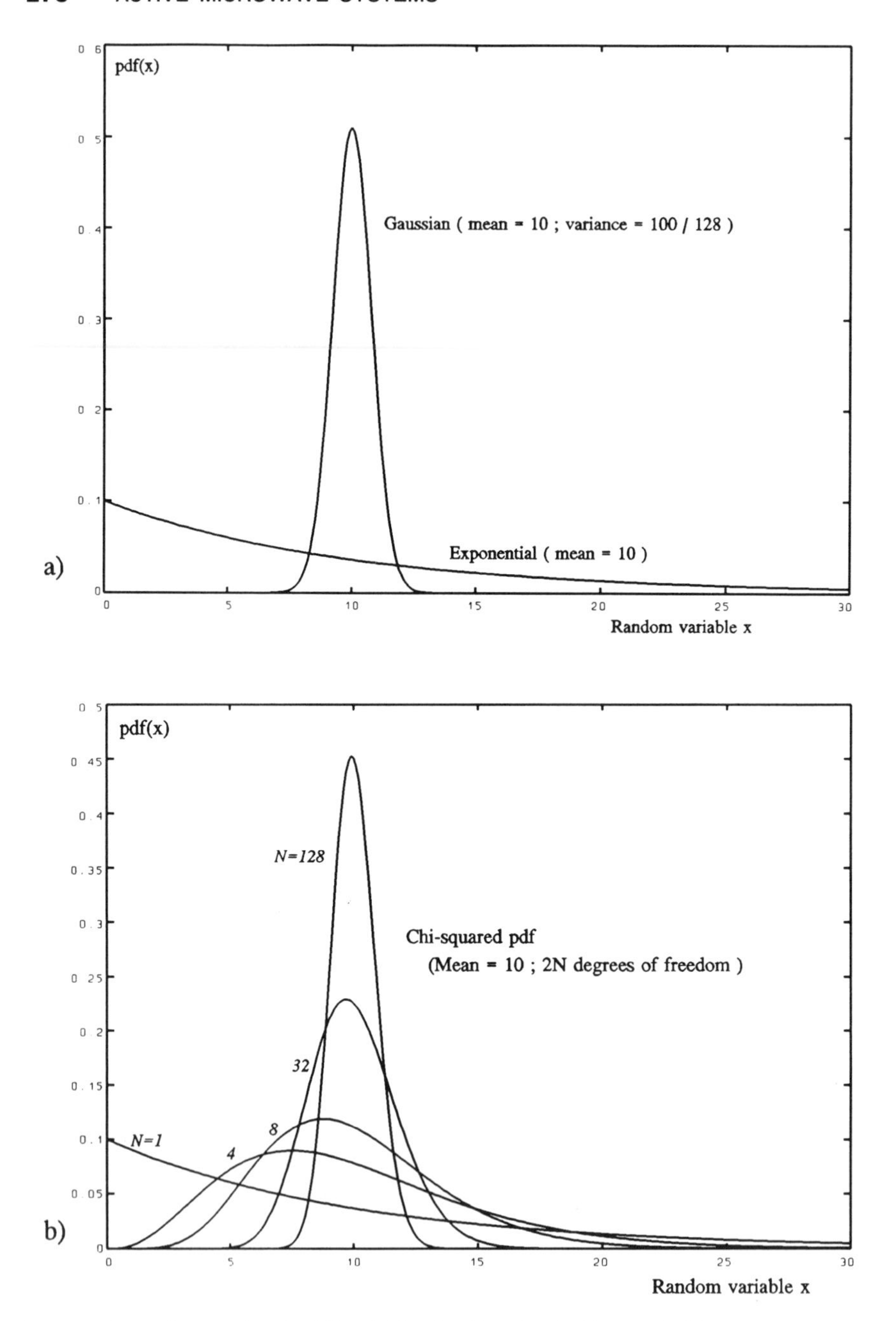

Figure 6.7 Chi-squared probability distribution function (*pdf*), and limiting forms: (a) Exponential *pdf*, $\langle x \rangle = 10$, limiting form of chi-squared *pdf* for $N = 1$, and Gaussian *pdf*, $\langle x \rangle = 10$, limiting form of chi-squared *pdf* for very large N; and (b) Chi-squared *pdf*, $\langle x \rangle = 10$, for several N, where $2N$ is known as degrees of freedom. Note the relatively slow convergence of the distribution towards its mean value for increasing N. (See caption on Figure 6.6 for comments on the role of x in these *pdfs*.)

TABLE 6.1 Probability Distribution Functions

Mean of $x = \langle x \rangle = m$; *Variance of* $x = \langle (x - m)^2 \rangle = \sigma_x^2$

(Note potential confusion between the parameter σ_x, which is the *standard deviation* of the random variable x, and the independent variable $x \to \sigma^0$, which is the random variable of distributed scene reflectivity in radar applications. Both usages of "σ" are well established. They may be distinguished only through context.)

Exponential, *defined for* $x \geq 0$

$$p_e(x) = \frac{1}{m} \exp\left\{-\frac{x}{m}\right\}; \; m = \langle x \rangle; \; \sigma_x = m$$

Rayleigh; *defined for* $x \geq 0$

$$p_R(x) = \frac{x}{\sigma_R^2} \exp\left\{-\frac{x^2}{2\sigma_R^2}\right\}; \; m = \sqrt{\frac{\pi}{2}}\,\sigma_R^2; \; \sigma_x^2 = \left(2 - \frac{\pi}{2}\right)\sigma_R^2$$

Gaussian, *defined for* $-\infty \leq x \leq \infty$

$$p_G(x) = \frac{1}{\sqrt{2\pi}\,\sigma_G} \exp\left\{-\frac{(x-m)^2}{2\sigma_G^2}\right\}; \; m = \langle x \rangle; \; \sigma_x = \sigma_G$$

Chi-squared with $2N$ degrees of freedom, *defined for* $x \geq 0$

$$p_{2N}(x) = \frac{x^{N-1}}{2^N \sigma_X^{2N} \Gamma(N)} \exp\left\{-\frac{x}{2\sigma_X^2}\right\}; \; m = 2N\sigma_X^2; \; \sigma_X^2 = 4N\sigma_X^4.$$

Note that $m^2/\sigma_x^2 = N$. The complete gamma function $\Gamma(N)$ is available in tables.

with increasing N, although not rapidly, as is evident from examination of the figure. The values of N selected are representative of a single-look SAR ($N = 1$), Seasat SAR ($N = 4$), typical airborne SAR systems ($N = 8$), certain altimeters ($N = 32$), and scatterometers or real aperture imaging systems ($N > 100$).

Probability distribution functions that occur frequently in radar work are listed in Table 6.1.

6.2.11 Discrete Scatterers

If the reflected field arises from an isolated discrete scatterer, a special case of the range line expression occurs. Such scatterers are often called *point targets*.

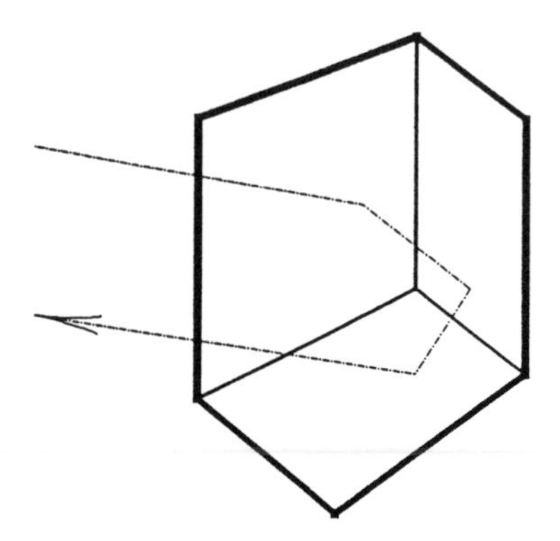

Figure 6.8 Example of a corner reflector, an object which redirects most illumination back in the source direction.

In this context, *point* means that the physical size of the reflector is much smaller than the impulse response width of the radar. A good example of a point target is a corner reflector, as shown in Fig. 6.8. The backscattered signal strength from a corner reflector is relatively large as all the reflections tend to be directed back in the direction of the radar's illumination. Carefully constructed corner reflectors are useful as calibration references. Corner reflectors also occur in a natural situation, such as the intersection between the side of an iceberg and the sea surface. The properties of corner reflectors are reviewed extensively in Ruck et al. [20].

Radar response for the specular reflection from a discrete scatterer is a special case of (6-19). For a corner reflector or other point target located at (x_i, R), noting the reflectivity normalization of (6-18), let

$$\frac{\sigma^0(x', r')}{\sin \theta_i(r')} \rightarrow \sigma\delta(x_i - x')\delta(R - r')$$

The Dirac distributions reduce the double integral of (6-19) to evaluation of the integrand at the scatterer position. It follows that

$$I_{CR,i}(x_i, r; R) = \frac{P_T G^2(x_t - x_i, R)\lambda^2 C_R}{(4\pi)^3 R^4} \sigma p_0^2(r - R) + \overline{P_{n,i}} \qquad (6\text{-}20)$$

is the detected single pulse returned from the corner reflector. In this expression, the more customary transmitted power P_T has been used, rather than the square of the transmitted amplitude. The antenna terms have been combined into the one-way power antenna gain $G(x_t - x_i, R)$ in the direction of the target, including an offset of $x_t - x_i$ from the peak gain position in azimuth.

Equation (6-20) is the *single-pulse range-impulse response*, in the power domain, of a radar. Note the distinction between the nominal range position R of the scatterer and the slant range independent variable r. The impulse response shape is determined by $p_0^2(r - R)$, generally a sharply peaked function centered on the target position R, where its value is one. In the event that range pulse modulation is used, range compression gain is expressed by C_R.

The new and interesting number is σ (*sigma*) of the target, with units of area, which is a measure of the strength of scattering from a discrete reflecting object. It may be shown, for a given range R, that the area, σ, is the *radar cross section* (RCS) of a hypothetical, isotropically scattering sphere that is large enough to cause the same reflected power observed at the radar as that seen from the target. Radar cross section, σ, bears rather little resemblance to the size of an object: it is much more dependent on backscatter directivity caused by the shape of the object. For example, at X band, a human being and an F-18 aircraft both have (average) cross sections of about 1 m^2 or less, whereas a triangular corner reflector with sides 1-m long at their intersection, has a cross section of about 4500 m^2.

The reflection from a point scatterer in general is stable in phase, in contrast to diffuse scattering. If two or more point scatterers are present in the same resolution cell of a radar, then signals reflected simultaneously from them interfere, and fading occurs. (N.B.: When there is a nonrandom phase relationship between scattering elements, then the total radar cross section may be calculated only by linear combination over the complex amplitudes of individual scatterers, followed by detection.)

6.2.12 The Radar Equation

The ratio of the signal and noise gain coefficients of the detected response function is the most fundamental expression of the radar equation for any radar. The most commonly cited version of the radar equation is the peak signal-to-noise ratio for the single-pulse output of a radar when a point target is observed. The required expression follows directly from (6-20) and related discussions.

It is customary to characterize the average noise power $\overline{P_n}$ that competes with the output detected signal in terms of radar system parameters (e.g., Ridenour [19]). One may show that

$$\overline{P_n} = KTB_nF_n \tag{6-21}$$

where

K = Boltzmann's Constant (1.38×10^{-23} Joule/Kelvin)
T = receiver temperature (Kelvin)
B_n = receiver noise bandwidth (Hz)
F_n = receiver noise figure

The single-pulse signal-to-noise ratio may be found directly from (6-14), (6-20), and (6-21), and is

$$\text{SNR}_1 = \frac{g_1^2}{\overline{P_n}L}\sigma = \frac{P_T G(R)^2 \lambda^2 C_R \sigma}{(4\pi)^3 R^4 KTB_nF_nL} \tag{6-22}$$

where $G(R)$ is the one-way power gain of the antenna in the direction of the target at range R. Note the inclusion of the range compression ratio C_R. System losses not otherwise included often are represented as a loss term L in the denominator. Propagation losses or multipath effects, which may be significant for certain instances, have been neglected for the purposes of this development.

The single-pulse radar equation (6-22) has an inverse fourth power dependency on range R. Usually better performance is available for multipulse radars used in remote sensing, as will be seen below.

Although the single-pulse SNR is the most frequently cited performance parameter of a remote sensing radar, it should be noted that this equation has been derived using only additive Gaussian noise as the nonsignal part of the receiver output. In general, there are many potential sources of signal contamination, including additive noise that is proportional to signal level (such as digital quantization noise), multiplicative noise (such as fading), and ambiguities (due to sampling in a multipulse system). Further consideration of these issues may be found in Skolnik [21].

6.3 RADAR MODEL FOR REMOTE SENSING

The development to this point has considered the observed return from only one pulse. All radars of interest in remote sensing depend on the extensive use of many pulses. The single-pulse model may be extended to handle multipulse radars. The results of this section embrace surface-based systems, such as pulse/Doppler radars, and airborne or satellite side-looking imaging systems, such as synthetic aperture radars.

6.3.1 The Generalized Model

Multipulse radars are important because they may be designed to take advantage of the phase structure in the azimuth (or pulse number) dimension as well as in the range dimension of the signal ensemble. To do so requires that the data from each pulse be stored in memory and processed as a two-dimensional complex data set. The two dimensions are range, r, and pulse number, i. Signal processing over the ensemble of signals is designed to take advantage of the interesting structure that evolves during the time that the signals are received. The several types of radars used in remote sensing are distinguished by what *interesting* implies in each instance.

Any systematic movement between the radar and the scene leads to phase shifts from each elemental reflector that are a function of pulse number. If the relative imaging geometry is known, then the phase shifts may be matched by a filter acting in the azimuth direction. Let the azimuth filter operator be $\Phi(M, i; r)$, which appears as a multiplier on the ith received signal. Like the range operator $p(\ )$ (Subsection 6.2.7), $|\Phi|$ is less than or equal to one. It is assumed that the received signal ensemble is to be combined linearly, indexed by pulse number. Then the two-dimensional complex image of the generalized radar model is a sum over the set of single-pulse complex range lines $i_i^C(\ )$ [6-14] each weighted by the appropriate operator $\Phi(\ )$. Thus,

$$i^C(M, r) = \sum_{i=-N_T/2+1}^{N_T/2} \Phi(M, i; r) i_i^C(x_i, r) \tag{6-23}$$

where M is the pulse number that locates the group being processed, and N_T is the number of pulses summed. Equation (6-23) may be modeled as a filter on the azimuth index, analogous to the range filter of Subsection 6.2.7.

The detected two-dimensional generalized image is

$$I(M, r) = \mathbf{E}[|i^C(M, r)|^2] \tag{6-24}$$

which generally is subjected to additional post-detection filtering. A conceptual block diagram of the model derived from (6-23) and (6-24) is shown in Fig. 6.9. All radars used for remote sensing of sea ice are represented by this model. Each specific type of radar is determined by applicable constraints on the phase and summation properties of the complex data set.

The approach of this section is based on linear processing. Nonlinear processing may be employed where it is advantageous. Examples of such processing include knowledge-based feature enhancement, or use of context-sensitive inclusion or exclusion rules. Nonlinear processing is considered further in Chapter 9.

Most radars use smoothing, integration, or other linear processing stages after image detection. For completeness, these are incorporated into the general

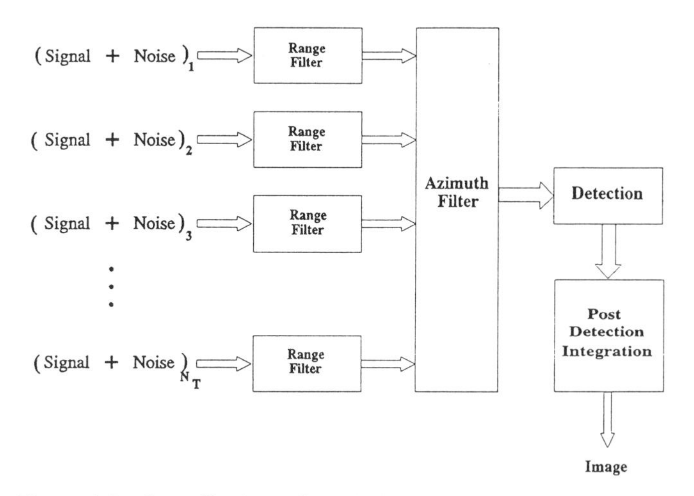

Figure 6.9 Generalized two-dimensional data processing radar model.

model, although discussion of the techniques and benefits for each type of remote sensing radar goes beyond the intent of this discussion. The reader is directed in particular to the multilook discussions of Chapters 10 and 11 for additional development of this point.

6.3.2 Comments on Coherence

A coherent system is one for which signal phase structure is sensible and important in signal combination. Extra effort is required to design and build a coherent radar.

For typical radar hardware, both the transmitter and the local oscillator phase references are random values, uniformly distributed between zero and 2π. It is clear from (6-14) that if the reference phase $(\varphi_i^{LO} - \varphi_i^T)$ were different for every i of the N_T pulses in the sequence, then the signal phase would be randomized across the sequence and phase processing in azimuth would not be possible. Pulse-to-pulse phase randomness is the situation characteristic of so-called *noncoherent* radars, whether surface based, airborne, or on spacecraft. In the event that the radar is noncoherent, then the image expression of (6-24) becomes

$$I_{Non\text{-}Coh}(M, r) = \sum_{i=-N_T/2+1}^{N_T/2} |\Phi(M, i; r)|^2 |n_i(r)|^2 + g_1^2 \iint |\Gamma(x', r')|^2 \cdot p_0^2(r - r') \sum_{i=-N_T/2+1}^{N_T/2} |\Phi(M, i; r)|^2 g_V^4(x_i - x', r)\, dx'\, dr' \tag{6-25}$$

where the gain definition introduced after (6-14) has been used. Equation (6-25) is the starting point for discussion of the noncoherent radars in Subsections 6.3.3 and 6.3.4.

The radar is *coherent* if, and only if, the phase difference $(\varphi_i^{LO} - \varphi_i^T)$ in the complex image has the same value for all received pulses. To achieve this, the radar may be designed to lock the phase of both the transmitter and the receiver at a constant value. Alternatively, the local oscillator phase for each transmission may be reset to match the phase of the transmitter, in which case the system is said to be *coherent on receive*. In either case, the interpulse phase structure in the signal ensemble is stable and, in principle, is available for processing. For the model of a coherent remote sensing radar in this discussion, the reference phase difference is assumed to be zero for all received pulses.

The benefits of a coherent radar may be quantified through analysis of the effects of the azimuth operator. In the case of a coherent imaging system using linear processing, the azimuth filter Φ, in combination with the known azimuth phase and weighting $\Phi_{s,i}$ of the signal, may be written as the compressed azimuth impulse response, Φ_0, as

$$\sum_{i=-N_T/2+1}^{N_T/2} g_\nu^2(x' - x_i, r)\Phi(x_M - x_i; r)\Phi_{s,i}(x' - x_i; r) \doteq \sqrt{C_A}\Phi_0(x_M - x'; r) \tag{6-26}$$

which is analogous to the range compressed pulse. The peak magnitude of Φ_0 is normalized to unity. In general, the phase structure is a parametric function of range r, so that the azimuth-matched filter also must be range dependent. Like the action of the coherent range filter (Subsection 6.2.7), the azimuth processing filter exploits known signal phase properties to achieve an improvement in the impulse response width and the azimuth compression gain C_A. From the impulse response of (6-26) and the imaging model of (6-24),

$$I_{Coh}(M, r) = \sum_{i=-N_Y/2+1}^{N_T/2} |\Phi(x_M - x_i; r)|^2 P_{n,i} + \frac{g_1^2 C_A}{\sin\theta_i(r)} \iint \sigma^0(x', r') \cdot p_0^2(r - r')\Phi_0^2(x_M - x'; r)\, dx'\, dr' \tag{6-27}$$

Note that the detected image I_{Coh} is proportional to the convolution of a two-dimensional system function $p_0^2(\ )\Phi_0^2(\ )$ over the average scene reflectivity. Imaging geometry determines the specific structure of the system function, which is known as the radar's *ambiguity function* (e.g., Evans and Hagfors [7]) when the image coordinates are range and range rate (Doppler). Equation (6-27) is the starting point for discussion of coherent radars in Subsections 6.3.5 and 6.3.6.

6.3.3 Noncoherent Surface-Based Radars

A noncoherent radar simply adds the detected returns from N_T pulses, known as post-detection integration. Usually the noise is statistically independent pulse to pulse. Hence the total noise power is proportional to the number of pulses summed. For a noncoherent radar, the same argument is true for the combination of signals. In the event that the beam is not scanning, using (6-19) and (6-25) for distributed scatterers, the detected output from a noncoherent surface-based radar is

$$I_{NC,G}(M, r) = N_T\overline{P}_n + \frac{N_T g_1^2}{\sin\theta_i(r)} \iint \sigma^0(x', r') \cdot g_V^4(x_M - x', r)p_0^2(r - r')\, dx'\, dr' \tag{6-28}$$

An example of a scanning noncoherent radar is found in Chapter 8.

The signal-to-noise ratio of the sum of pulses in this case would appear to be the same as that of a single pulse, but this is a bit misleading. A sum of N_T

pulses reduces the standard deviation of the summed estimates of scene reflectivity by $\sqrt{N_T}$, so that for small contrast changes, as between the backscatter from the sea and floating ice, post-detection integration is helpful. In such a situation, the desired distributed scene content is *signal*, and the unwanted content is *clutter*.

If the mean reflectivity, σ^0, is nominally constant across the illuminated area at range r, then the integrations may be completed formally in terms of known system constants. Under this assumption, the integrated radar output is

$$I_{NC,G:\sigma 0}(M, r) = N_T\overline{P_n} + N_T g_1^2 \frac{\sigma^2 r\beta_h r_{sr}}{\sin\theta_i(r)} \tag{6-29}$$

where β_h is the equivalent rectangle angular width of the azimuth (two-way power) antenna pattern, g_ν^2. As g_ν is normalized, β_h is simply the value of the x' integral. Similarly, r_{sr} is slant range resolution (on r) which, for the equivalent rectangle norm, is the value of the integral on r'. Note that the azimuth integration brings a range (r) proportionally to the signal gain which combines with the r^{-4} in g_1^2 [(6-14)] to yield a r^{-3} range dependency in output signal level. This range dependency occurs for most radars in which there is integration across the azimuth extent of the antenna pattern.

6.3.4 Noncoherent Side-Looking Radars

Consider a side-looking noncoherent radar operating from an aircraft, having forward velocity V_{ac}. Side-looking airborne radar (SLAR) is the name often used for such systems, but RAR (real aperture radar) is more descriptive and less ambiguous (Ulaby et al. [23]). For an interpulse period $T_p = 1/f_p$, and assuming rectangular imaging geometry, the along-track dimension is incremented by $\Delta x = V_{ac}T_P$ for each transmission. (Constraints on the size of the interpulse spacing Δx are considered in Subsection 6.3.6). Assuming that the weighting is uniform, $|\Phi| = 1$, and that N_T pulses are summed, the expected detected RAR image at position (x_M, r) is

$$I_{\text{RAR}}(x_M, r) = N_T\overline{P_n} + \frac{r\beta_h}{\Delta x}\frac{g_1^2}{\sin\theta_i(r)} \iint \sigma^0(x', r')g_V^4(x_M - x', r) \cdot p_0^2(r - r')\, dx'\, dr' \tag{6-30}$$

which has subtle differences when compared to the nonscanning model of Subsection 6.3.3. Equation (6-30) represents the imaging process for a scanning, airborne, side-looking noncoherent radar as a two-dimensional convolution of the impulse response function over the scene reflectivity. As such, it is based on the summation of $r\beta_h/\Delta x$ pulses from one area of the scene as the radar scans by, rather than the adding up of all areas of the scene illuminated by the antenna at one transmission. The reader should be cautioned that most references use a description similar to (6-29) to describe such a radar, rather than the more suitable scanning description (6-30). Hence, claims that may be found in the literature for RAR imagery signal-to-noise ratios should be treated with care.

Most RAR systems operate according to (6-25), such that the number of pulses summed, N_T, is constant with range. When this is the case, the image signal-to-noise ratio of a RAR could be improved by making the number of pulses proportional to range, thus reducing the noise added to the image at all ranges less than the maximum.

6.3.5 Coherent Surface-Based Radars

Coherent, surface-based radars may be used to measure small frequency changes in each reflection, a shift so small that it is observable only by comparing the phases of signals in the received sequence. The shift appears in the signal ensemble as a Doppler frequency in the pulse index, and such systems are known as *Range-Doppler* radars. In this case, the appropriate operator is a reference frequency, transforming (6-23) to the form

$$i_{RD}^{C}(M, r) = \sum_{i=-N_T/2+1}^{N_T/2} \exp\left\{-j2\pi \frac{iM}{N_T}\right\} i_i^C(x_i, r)$$

This equation may be recognized as the discrete Fourier transform across the set of complex range lines at constant slant range r. The detected image is a map of the reflected power as a function of range and Doppler shift, which leads to a two-dimensional spatial map when there is differential Doppler over the illuminated scene. The technique is used in radar astronomy to image planets based on their rotation, and in certain surface-based radars to distinguish moving objects from those that are not moving. Coherent surface-based radars are considered further in Chapters 7 and 9. Range-Doppler signal processing also is a natural antecedent to synthetic aperture radars, the subject of Chapter 11.

6.3.6 Coherent Side-Looking Radars

The most important example of a radar of this type is a synthetic aperture radar (SAR). The name derives from coherent integration over line number, which is the mathematical equivalent to the coherent signal combination achieved by a very large aperture.

For an SAR, the detected image corresponding to (6-27) may be shown to be

$$I_{\text{SAR}}(x_M, r) = N_T\overline{P_n} + \frac{g_1^2 N_T^2 G_p}{\sin\theta_i(r)} \iint \sigma^0(x', r') \cdot p_0^2(r - r')\Phi_0^2(x_M - x'; r)\, dx'\, dr' \qquad (6\text{-}31)$$

The width of the range impulse response $p_0^2(\ )$ is slant range resolution (r_{sr}) as before, and in like manner, the azimuth resolution (r_a) is the equivalent rectangle width of $\Phi_0^2(\ )$. The SAR performance of (6-31) should be compared with the RAR image of (6-30). Both expressions show that the respective images are formed from a two-dimensional convolution of the system impulse response functions over the illuminated scene. They differ only with respect to the effective gain derived through coherent azimuth processing and, as a corollary, they have different azimuth impulse response functions. Although azimuth resolution is discussed at greater length in Chapter 11 of this book, it is important to observe at this point that SAR azimuth resolution (r_a) is constant with range and has a (minimum) magnitude on the order of one-half of the azimuth aperture size of the radar antenna. For airborne systems usually this is less than a meter, and for spacecraft radars, several meters. These small numbers correspond to *high-resolution* performance.

The factor G_p represents the processing gain enjoyed by distributed scatterers when the radar sampling rate is greater than the Doppler bandwidth of the system (Raney [18]). The pulse repetition frequency f_p must always be at least a little larger than the Doppler bandwidth in order to unambiguously sample the received signal. This source of processing gain, which occurs only for coherent radars, arises from correlation between adjacent pulses in the processor and is (approximately) equal to the f_p-to-Doppler bandwidth ratio. The f_p sets the interpulse spacing Δx along the path of the radar. To satisfy the Nyquist unambiguous sampling requirement, interpulse spacing must be less than about half of the azimuth antenna aperture dimension along the line of flight of the radar, and proportionally less than that at the illuminated surface, determined by the viewing geometry (Raney [18]).

The number of azimuth pulses integrated, N_T, is selected at the time of data processing and, for (6-31) to be valid, should be no greater than the number of pulses available to be integrated from each range r. Many SAR processors are operated with the pulse integration set to maximize the signal-to-noise ratio so that N_T is proportional to range. Under this condition, the signal-to-noise ratio with an SAR from a distributed scatterer of mean reflectivity, σ^0, at slant range R, becomes

$$\mathrm{SNR}_{\sigma^0,\mathrm{SAR}} = \frac{\overline{P_T} G^2(R) \lambda^2 B_p G_p \beta_h \sigma^0 r_a r_{sr}}{(4\pi)^3 R^3 K T B_n F_n L V_{S/C} \sin \theta_i(R)} \tag{6-32}$$

where $G(R)$ is the antenna elevation gain parameter, $\overline{P_T}$ is the average transmitted power in the radar frame of reference, and B_p is the processing bandwidth in azimuth which, in general, is different from the nominal Doppler bandwidth $\beta_h V_{S/C}$. Equation (6-32) may be compared to the single-pulse, point-target case of (6-22); but perhaps of greater interest, the comparison should be with the response of a real-aperture side-looking radar. In this case, when confronted with an extended scene of nominal reflectivity σ^0, the signal-to-noise ratio response of the two radars is the same, except for the processing gain G_p. The

SAR processing gain follows directly from azimuth coherence in the signal ensemble.

There are many different forms of the SAR–SNR equation to be found in the literature. In general, there is ambiguous treatment of the velocity term, which leads to discrepancies when extending conventional expressions to the satellite SAR case. Processing gain, noise and processing bandwidths, and resolution approximations are often unsatisfactory. The subtleties arise, in a situation that should be straightforward, through the coherent summation of pulses in azimuth. For orbital geometries, more pulses are available to be summed than predicted from the rectilinear geometry usually applied to the aircraft geometry. For detailed quantitative work, the SNR expression in terms of discrete digital sample spacing (pixels) is preferred over the analog expression in terms of spatial resolution conventionally used.

Just as in the single-pulse case, the reflectivity estimate at each image location is subject to fading, known as speckle in the SAR community. Speckle may be reduced by averaging statistically independent estimates, a process known as multilooking. More on these and related topics may be found in Chapter 11.

6.4 POLARIZATION

The complete set of information available in a signal obtained by a monostatic radar from the observed scene embraces amplitude, range delay, frequency, phase, and time, all considered above, and *polarization*, the subject of this section. Polarization describes the spatial orientation of the $\boldsymbol{E}$ vector of an electromagnetic wave. Most radars operate with a single antenna, which determines the polarization of the transmitted wave and selectively receives the same polarization component of the backscattered waves. In effect, such a radar may be viewed as providing only a scalar estimate of the scene's reflectivity for a given polarization, which served as the basis for (6-5) and related discussions.

To capture the complete polarization characteristics of the scene, a *vector* measurement is required. In most situations, additional information about the desired scene features may be deduced from the vector measurement. For any given polarization of the transmitted wave, the reflectivity process in general will give rise to a diversity of polarizations in the backscattered wave. To observe these, the radar must be built such that more than one polarization can be received. Similarly, the reflectivity is a function of the polarization of the transmitted wave. Thus, more than one polarization must be transmitted as well.

A necessary and sufficient condition for representation of an arbitrary wave requires the simultaneous knowledge of only two orthogonal polarization components (e.g., Nespor et al. [12]; Giuli [8]). Thus, if the radar itself satisfies certain conditions, the complete polarization characteristics of a scene may be observed by a radar with two orthogonal polarizations in the transmitted signal, and also two orthogonal polarizations in the received signal, leading to a set of four estimates of the polarimetric reflectivity for each element in the scene.

The polarimetric signature can vary over the observation interval of the radar. In general, polarization properties of a wave may be decomposed into two parts: *polarized*, in which the polarization does not change; and *unpolarized*, in which the polarization changes randomly with time. The ratio between the power of the polarized component and the total power of the wave is defined as the *degree of polarization*.

Knowledge of preferred states of polarization of certain reflectors has been used to advantage to design radars tailored for a particular purpose, such as aircraft detection in the presence of rain (e.g., Long [10]). An important aspect of relevance to this book, however, is to explore the dimension of polarization *adaptability* in the context of radar remote sensing. Adaptability requires that a set of scattering estimates having specific polarization properties be gathered for each point in the scene. Such radars are known as multipolarization, or polarization-diversity, systems which are of concern for the remainder of this section.

Estimation of a scene's polarimetric scattering properties is implemented usually by transmitting with one polarization and receiving with both polarizations simultaneously (requiring two nominally identical receive and record channels in the system), followed by the same procedure using the orthogonally polarized state on the subsequent transmission. The polarization channels of the radar must be well isolated from each other and in general there must be a careful calibration of the system in the polarimetric mode (Zebker et al. [25]). Design of a polarimetric radar system goes beyond the scope of this discussion. In the following, it is assumed that the radar has uncoupled polarization channels. Although this is not possible in practice, calibration procedures are sufficient to compensate for realistic levels of inter-channel contamination.

The completeness with which the scene's polarimetric signature may be estimated depends on whether the radar is coherent or noncoherent. These two situations are described in corresponding sections below, following the basic analytic formulation.

6.4.1 Analytic Representation

There is a well-established literature that describes the polarization of an electromagnetic (EM) wave (e.g., Maxwell [11]; Poincaré [15]; Born and Wolf [1]; and Giuli [8]). Key points from these references are summarized here as they bear on issues encountered in active microwave remote sensing. The radar in question is assumed to be monostatic and satisfies the condition of linearity, reciprocity, and quasi-monochromaticity as employed previously in this chapter.

Assume for the moment a Cartesian coordinate system in which r is the direction of wave propagation. Then, an arbitrary monochromatic EM wave traveling in the positive r direction may be representated as a vector

$$\begin{bmatrix} E_H(r, t) \\ E_V(r, t) \end{bmatrix} = \begin{bmatrix} a_H \cos [(\omega t - kr) + \delta_H] \\ a_V \cos [(\omega t - kr) + \delta_V] \end{bmatrix} \tag{6-33}$$

where the orthogonal components of the polarized wave have been chosen in the horizontal (subscript H) and vertical (subscript V) directions in the plane orthogonal to the direction of propagation, the representation that is most frequently used in remote sensing radars. (Notation in this and the following subsections has been adapted from Born and Wolf [1] and conforms to the standard in most regards.) The *polarization state* of this wave is determined by the relative amplitude, a, and phase, δ, of the wave components at the H and V polarizations.

A geometrical interpretation of polarization follows from solution of (6-33) to eliminate the wave propagation term $(\omega t - kr)$. The result is

$$\left(\frac{E_H}{a_H}\right)^2 + \left(\frac{E_V}{a_V}\right)^2 - 2\frac{E_H}{a_H}\frac{E_V}{a_V}\cos\delta = \sin^2\delta \tag{6-34}$$

where

$$\delta = \delta_V - \delta_H$$

The locus of points (E_H, E_V) describes an ellipse as shown in Fig. 6.10. In this formulation, the polarization of any wave is described by two parameters: the *orientation* ψ, and the *ellipticity* χ. These may be derived from (6-34). After some trigonometry, the following may be obtained. Defining

$$\tan\alpha = \frac{a_V}{a_H}, \qquad 0 \le \alpha \le \pi/2$$

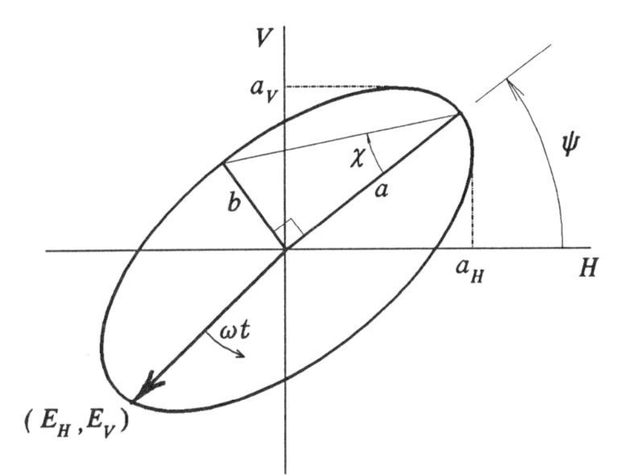

Figure 6.10 The Polarization Ellipse. Any state of polarization of a transverse wave may be represented by only two parameters: ellipticity χ, and orientation ψ.

then the principal semiaxes a and b and the angle ψ, which describe an arbitrary polarization ellipse with respect to the reference coordinate system, are given by the relations

$$a^2 + b^2 = a_H^2 + a_V^2$$

$$\tan 2\psi = (\tan 2\alpha) \cos \delta, \qquad 0 \leq \psi \leq \pi$$

$$\sin 2\chi = (\sin 2\alpha) \sin \delta, \qquad -\pi/4 \leq \chi \leq \pi/4$$

and

$$\tan \chi = \pm \frac{b}{a}$$

All possible states of polarization of a wave may be represented by appropriate association of the two canonical parameters. In general, a wave has elliptical polarization. Limiting forms include *linear polarization*, which by definition corresponds to $\chi = 0$. Orientation of an elliptically or linearly polarized wave is described by ψ, although in remote sensing applications, the more familiar terms *vertical* V or *horizontal* H are used, or an angle with respect to one of these two directions. *Circular polarization* corresponds to $\chi = |45°|$, with *right circular* and *left circular* polarizations determined by the sign of χ.

The set of possible states of polarization has been described elegantly by Poincaré [15] as the ensemble of points on a unit sphere whose polar angular coordinates are 2χ and 2ψ. In the *Poincaré sphere*, right circular and left circular polarizations constitute an orthogonal pair in the sense of the discussion above.

From this development, it follows that the components of the $\boldsymbol{E}$-vector of a wave may be written (to within a complex multiplicative phase constant) in terms of the polarization state variables. Although any orthogonal pair could be used, the following discussion is with reference to the customary H and V components of polarization as summarized by (6-32).

Primary interest in polarimetric radars derives from replacement of the scalar form of reflectivity [see (6-5)] by its vector counterpart. Thus, when either H or V polarizations are incident upon a scattering element, Δ, in general both polarizations are reflected according to

$$\begin{bmatrix} E_H^r \\ E_V^r \end{bmatrix} = \begin{bmatrix} \Gamma_{HH} & \Gamma_{VH} \\ \Gamma_{HV} & \Gamma_{VV} \end{bmatrix} \begin{bmatrix} E_H^i \\ E_V^i \end{bmatrix} \tag{6-35}$$

The new terms of interest represent the 2×2 *scattering matrix* of the scene. This is required to describe a scene that is illuminated by an incident field with both horizontal and vertical polarization components, giving rise to both components after reflection. The scattering matrix is a quantitative description of the transformation of the polarization state upon reflection, as well as the reflection within each state.

In the literature, one often encounters S used in place of Γ as elements of the scattering matrix. The variety in notation arises from the practitioners coming to polarimetric radar from a variety of fields, including classical optics and physics as well as from electrical engineering. The Γ notation has been used here since it is consistent with the treatment of reflectivity summarized in earlier sections.

A multichannel radar system model could be developed based on the polarimetric vector formulation by following through in parallel to the development in Sections 6.2 and 6.3 for each of the four reflectivity components identified in (6-35). However, it is adequate for this discussion, and much more efficient, to consider the consequences from a different point of view. In the following, the scattering matrix is developed as seen first with a noncoherent radar, and then with a coherent radar. It is left to the reader to work through the gains and radar equations appropriate for each type of radar, and to each combination of polarizations.

6.4.2 Noncoherent Radars

With a noncoherent radar, signal phase is random, and is not available for exploitation in processing and interpretation. It follows that the output of a noncoherent radar with polarization diversity is restricted to only the mean reflected power corresponding to each pair of radar polarization states. With a radar using (V, H) polarizations, for each position in the output data set, an estimate is available of σ^0 for the four possible combinations of transmit and receive orientations, leading to

$$\begin{bmatrix} \sigma^0_{HH} & \sigma^0_{VH} \\ \sigma^0_{HV} & \sigma^0_{VV} \end{bmatrix} \tag{6-36}$$

as the data available for subsequent processing. In this notation, the first subscript is taken to mean the polarization state of the transmitted wave, and the second subscript describes the state of the received wave. Although this is the standard in remote sensing literature, it is by no means standard for all of the radar (or optical) literature of relevance to the field. The matrix of (6-36) is sometimes referred to as the "scattering matrix," although more correctly it is the *reflectivity* or *sigma nought* matrix. The term *scattering matrix* properly refers only to the coherent representation of scattering as given in (6-35).

There are several additional observations that might be made at this point. First, as is satisfied in most situations by almost all radars, reciprocity applies, such that the transmit and receive conditions may be interchanged. Thus, in general $\sigma^0_{HV} = \sigma^0_{VH}$, so that there are three nominally independent estimates of reflectivity for each scene element available from a polarization diversity radar that is not coherent. This equivalence may be turned to advantage to increase the effective signal power available in the cross-polarized component (Raney [17]).

Second, transformation of the transmitted polarization state by reflection requires two or more actual reflections within the scene before the wave assumes its backscattered direction. Upon each reflection, the wave amplitude is reduced. Except for very special circumstances, therefore, the cross-polarized component is always weaker than the like-polarized component. On average, the difference is about 6 dB when observing linearly polarized signals, with a range of 2 dB to 10 dB commonly encountered (Long [10]). If one is interested in the cross-polarized signal in applications, the polarization loss must be included in the relevant signal-to-noise ratio or other form of the radar equation during system-design stages, from which follows a requirement for higher transmitter power than would otherwise be needed.

Third, as an illustration of the incompleteness of the noncoherent representation, note the following. Although H and V are orthogonal, the magnitude of the reflectivity with respect to these two orientations is *not sufficient* to allow transformation into any other polarization state. To do so requires phase information. For example, it is not possible to estimate the expected reflectivity matrix of a scene in terms of circular polarization components based solely on the noncoherent reflectivity elements of the σ^0 matrix.

Fourth, the strength of reflectivity in one polarization state relative to another polarization state has an operational advantage: such a ratio is determined primarily by the physical characteristics of the reflecting element, and rather little by radar system characteristics. The *copolarization ratio* and the *depolarization ratio*, for example, have quantitative significance even for a radar that is not absolutely calibrated. Interchannel gain verification is sufficient. Interchannel signature comparisons are important in applications as is apparent from the extensive discussion evident in the literature.

There is extensive tabulated σ^0 data for ice and other materials as a function of polarization (HH, VV, and HV), as well as for frequency and incidence angle, the two other geometric variables of first-order importance. These are referenced as needed in other chapters of this book.

6.4.3 Coherent Radars

With a coherent radar, by definition, phase is available in the complex image for signal processing. In this case, operation of a polarimetric radar would seem to lead to an estimate of the (complex) scattering matrix $[\Gamma]$ of (6-35), but only for the one polarization basis that corresponds to the particular configuration of the radar. However, a powerful principle comes to bear at this point. With both phase and amplitude information present, transformation from one polarization basis to any other may be accomplished by a simple matrix multiply (Nespor et al. [12]). Thus, with data available from a coherent polarimetric radar, all possible polarimetric possibilities may be synthesized from that data set. The technique is of importance in remote sensing and has become one of the most of the active radar remote sensing research areas (e.g., Evans et al. [6] or Zebker et al. [22]). When four channels are used, the radar is said to be *quadrature-polarimetric*, and "quad-pol" for short, thus distinguishing them from polarity diversity radars usually having only two noncoherently related channels. Examples of quad-pol radars include a P-, L-, and C-band SAR operated by the pioneer in this specialty, the Jet Propulsion Laboratory (Pasadena, California), an L-, C-, and X-band SAR operated by the Environmental Research Institute of Michigan (Ann Arbor, Michigan), a K_a-band SAR operated by Lincoln Laboratories (Lexington, Massachusetts), and a C-band SAR operated by the Canada Centre for Remote Sensing (Ottawa, Ontario). (See Chapter 11 for more on this radar.)

The importance of polarimetric synthesis may be illustrated easily. For certain reflecting geometries, one may predict the expected response of reflection over all possible states of transmit and receive polarization, which leads to the concept of *polarization signature* (van Zyl et al. [24]) whereby the response of each element of a scene may be plotted over the full span of the polarization variables. Examples of polarization signatures are plotted in Fig. 6.11. Both Bragg scattering and 2-bounce scattering play an important role in remote sensing of sea ice. Reflection for the sea is dominated by Bragg scattering, whereas reflection from many ice features, such as the near edge of ice floes or bergs, is dominated by double reflection from the sea surface and the rise of the ice above the sea. By using polarimetric discrimination, these two classes of reflection may be differentiated based on their polarimetric signatures alone.

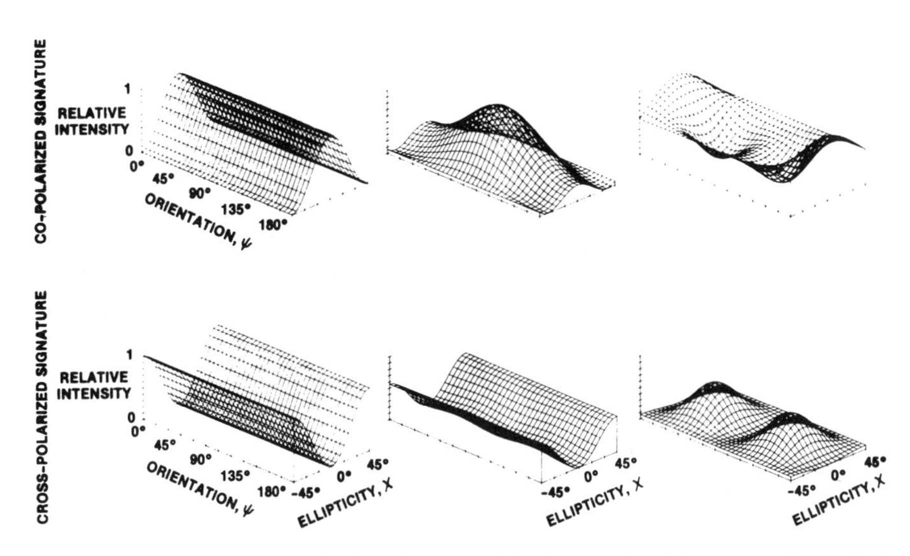

Figure 6.11 Examples of Polarization Signatures: (a) sphere model; (b) Bragg model; and (c) 2-bounce model. (Courtesy J. van Zyl, Jet Propulsion Laboratory)

6.5 CONCLUSIONS

This chapter has addressed the issue of active microwave remote sensing, with a consideration for application of the technique to the observation of floating ice. The underlying philosophy of the discussion has been to downplay superficial differences and to concentrate on common themes central to the performance of these systems. The signal received by an active microwave system at a given frequency carries information about the scene primarily in the parameters of amplitude, phase, and polarization. A generalized radar model has been developed to handle the first two of these parameters quantitatively, at least for all radar systems of interest to this book. Polarization has been approached within the framework of the model, with coherence of the radar identified as the main distinguishing feature for achieving polarization diversity. Although theoretical in appearance, the method of this chapter is designed to link all the radar chapters of this book and to provide a basis for quantitative intercomparison of different types of systems.

Both active microwave systems and their applications to sea ice and icebergs are continuing to develop. It is hoped that the perspective offered by this chapter, reinforced by related discussions in this book, might help to focus future work, as well as to put the present status and past accomplishments into sharper focus.

REFERENCES

[1] M. Born and E. Wolf (1959) *Principles of Optics*, Pergamon Press, Macmillan, New York.

[2] W. M. Brown (1963) *Analysis of Linear Time-Invariant Systems*, McGraw-Hill, New York.

[3] R. L. Cosgriff, W. H. Peake, and R. C. Taylor (1960) *Terrain Scattering Properties for Sensor System Design*, Terrain Handbook II, Engineering Experiment Station, The Ohio State University, Columbus, OH.

[4] H. Cramér (1955) *The Elements of Probability Theory*, Wiley, New York.

[5] W. B. Davenport and W. L. Root (1958) *An Introduction to the Theory of Random Signals and Noise*, McGraw-Hill, New York.

[6] D. L. Evans, T. G. Farr, J. J. van Zyl, and H. A. Zebker (1988) "Radar polarimetry: Analysis tools and applications," *IEEE Trans. Geosc. and Remote Sens.*, **26**(6), 774–789.

[7] J. V. Evans and T. Hagfors (1968) *Radar Astronomy*, McGraw-Hill, New York.

[8] D. Giuli (1986) Polarization diversity in radars," *Proc. IEEE*, **74**(2), 245–269.

[9] D. E. Kerr (1951) *Propagation of Short Radio Waves*, Vol. 13, MIT Radiation Lab. Ser., McGraw-Hill, New York.

[10] M. W. Long (1975) *Radar Reflectivity of Land and Sea*, Lexington Books, D.C. Heath, Toronto/London.

[11] J. C. Maxwell (1873) *A Treatise on Electricity and Magnetism*, Oxford.

[12] J. D. Nespor, A. P. Agrawal, and W. M. Boerner (1985) "Theory and design of a dual polarization radar for clutter analysis," in W. M. Boerner et al. (Eds.), *Inverse Methods in Electromagnetic Imaging*, Pt 1., Reidel, Hingham, MA, 643-659.

[13] D. O. North (1943) "An analysis of the factors which determine signal/noise discrimination in pulsed-carrier systems," *RCA Tech. Rep. PTR-6C* (ATI 14009).

[14] A. Papoulis (1965) *Probability, Random Variables, and Stochastic Processes*, McGraw-Hill, New York.

[15] H. Poincaré (1892) *Théorie Mathématique de la Lumière*, Georges Carre, Paris.

[16] S. Ramo and J. R. Whinnery (1953) *Fields and Waves in Modern Radio*, Wiley, New York.

[17] R. K. Raney (1988) "A "Free" 3-dB cross-polarized SAR data," *IEEE Trans. Geosci. and Remote Sens.*, **26**(5), 700–702.

[18] R. K. Raney (1991) "Considerations for SAR image quantification unique to orbital systems," *IEEE Trans. Geosci. and Remote Sens.*, **29**(5), 754–760.

[19] L. N. Ridenour (1947) *Radar System Engineering*, Vol. 1, MIT Radiation Lab. Ser., McGraw-Hill, New York.

[20] G. T. Ruck (Ed.) (1970) *Radar Cross Section Handbook*, Vols. 1 and 2, Plenum Press, New York.

[21] M. I. Skolnik (1962) *Introduction to Radar Systems*, McGraw-Hill, New York.

[22] J. A. Stratton (1941) *Electromagnetic Theory*, McGraw-Hill, New York.

[23] F. T. Ulaby, R. K. Moore, and A. K. Fung (1982) *Microwave Remote Sensing: Active and Passive*, Vol. II, Addison-Wesley, Reading, MA.

[24] J. J. van Zyl, H. A. Zebker, and C. Elachi (1987) "Imaging radar polarization signatures: Theory and observation," *Radio Sci.*, **22**, 529–543.

[25] H. A. Zebker, J. J. van Zyl, S. L. Durden, and L. Norikane (1991) "Calibrated imaging radar polarimetry: Technique, examples, and applications," *IEEE Trans. Geosci. and Remote Sens.*, **29**(6), 942–961.

7

OVER-THE-HORIZON RADAR

SATISH K. SRIVASTAVA
RADARSAT Program Office
Canadian Space Agency
Ottawa, Ontario, Canada

JOHN WALSH
Faculty of Engineering
Memorial University
St. John's, Newfoundland

7.1 INTRODUCTION

With resource exploration and development off the east coast of Canada, more emphasis is being placed on solving problems associated with navigation and exploration in ice-infested waters. Sea ice and icebergs pose major problems on the Grand Banks of Newfoundland and off the Labrador coast. Although precautions are being taken to build ships and drilling platforms which will withstand large ice forces, early detection of advancing ice hazards allows better planning of navigation routes and drilling schedules, thereby avoiding expensive delays and possible disaster. Microwave marine radars are used extensively for ice-hazard detection (Chapters 8 and 9), but their detection range is limited generally to that of the visible horizon, in the range of a few tens of kilometers. Airborne radars provide wide-area ice surveillance but maintaining continuous or frequent observation is expensive (Chapters 10 and 11). Moreover, weather often poses severe limitations on conducting such surveillance. Satellite sensors are much more expensive and may not provide the desired resolution or continuous coverage (Chapters 5 and 12). Some of these limitations can be overcome by the use of the over-the-horizon (OTH) radar.

Remote Sensing of Sea Ice and Icebergs, Edited by Simon Haykin, Edward O. Lewis, R. Keith Raney, and James R. Rossiter.
ISBN 0-471-55494-4

In the last two decades, significant advances have been made in exploiting the potential of OTH radars. The capability of this type of radar has been demonstrated for measuring ocean surface current, measuring sea state, and in detection of hard targets (including icebergs, ships, and low-flying aircraft). OTH radars operate in the HF band (2 to 30 MHz) and follow the ground-wave mode of radio propagation, in which the radiated wave essentially follows the curvature of the earth by diffraction. This mode can provide a detection range far in excess of that of conventional microwave radar, to hundreds of kilometers. Radars operating in the HF band but following the sky-wave mode of propagation, usually called sky-wave radars, can provide detection ranges out to thousands of kilometers. In Canada, however, research on OTH radars is concentrated on the ground-wave–mode type only. Therefore, the discussion here is limited to the OTH radars operating in the ground-wave mode.

The range of an OTH radar is determined primarily by its operating frequency and the electrical conductivity of the earth. Because of the high conductivity of seawater, the radars are very effective over the ocean surface. Of course, other factors such as peak transmitted power, duty cycle, antenna gains, and size of the target also have to be considered in determining the maximum range. A range of detection out to 400 km would require a radar operating at the lower end of the HF band (2 to 6 MHz), where the transmission loss is low. In addition to the target, the ocean surface also acts as a scatterer and results in sea-clutter return. The level of this sea-clutter return is determined by the prevailing sea conditions and increases with radar frequency. The sea clutter also is dependent upon the size of the radar resolution cell, which in turn depends on transmitted bandwidth and beamwidth of the receive antenna.

As ocean waves move with different phase velocities, they induce different Doppler frequencies in the incident radio waves. As a result, the return from the ocean surface consists of a band of frequencies in the form of a Doppler spectrum. Depending upon its velocity, a moving hard target also produces a Doppler shift. Thus, by making an OTH radar a coherent system, a target's return is distinguished from sea clutter by spectral analysis of the backscattered signal. Sea clutter contains a wealth of information concerning the marine environment. For example, from analysis of this clutter we can determine: surface water current (Barrick et al. [5]), surface wind conditions (Long and Trizna [18]), and directional ocean wave height spectrum (Barrick [4]; Lipa and Barrick [17]; Wyatt [57]).

A major task in distinguishing a target's return from that of sea clutter, or in using sea clutter itself for deducing information about the marine environment, is interpretation of the sea echo. This task requires an understanding of the interaction process occurring in the scattering of radio waves from the ocean surface. The subject becomes complicated further due to the randomness of the ocean surface. A theoretical analysis of this sea clutter was given first by Barrick [2, 3].

In this chapter, the material covered about the OTH radar consists of: Canadian research and development; its limitations; theoretical discussions on ground-wave propagation; backscattered cross-section models for the ocean surface, ice edge and icebergs; radar systems; experimental results; comparisons with model predictions; and conclusions.

7.1.1 Research and Development in Canada

In 1978, exploitation of potential uses of OTH radars became a focus for research at the Memorial University of Newfoundland. Research began with the development of a more general and advanced analytical model of sea clutter, using a new technique (Walsh [48]; Srivastava [35]; Walsh and Srivastava [54, 55]). Investigation into HF propagation for the case of a mixed path with discontinuities was also conducted by Walsh [49], Ryan [32], and Donnelley [12]. As an application of the results in remote sensing of sea ice, cross-section estimates for different types of sea ice were presented by Ryan and Walsh [33]. Walsh [50] also described a general analytical technique for scattering from layered media. This work led to the development of a detection model for icebergs by OTH radars (Walsh and Srivastava [53]). A generalization of the analytical technique to deal with problems of propagation and scattering for mixed paths with discontinuity was developed later by Walsh and Donnelley [51]. Further, a consolidated approach to two-body scattering problems was described by Walsh and Donnelley [52], which can be used for a variety of scattering problems, including rough surfaces, mixed paths, and layered media. Dawe [11] studied ground-wave propagation over a rough spherical earth and developed computationally efficient software for calculating the transmission loss for varying sea states.

Iceberg detection by an OTH radar was demonstrated in 1984 (Walsh et al. [56]). The detection trials were conducted at Byron Bay, Labrador, using a portable commercial radar (CODAR) operating at 25.4 MHz. This radar has a limited range capability ($\approx$40 km) and was built primarily for measuring surface currents. However, for the iceberg experiment, its receive antenna was replaced by an array forming a narrow beam (3.33°). The experiment also verified the software model developed previously for iceberg detection in the HF region.

Enthused by the iceberg-detection experiment and other results, including ship detection and ocean surface current and sea-state measurements, it was decided within the Memorial University research group in 1986 that the technology was ready to be transferred to industry for commercial development. This transfer was accomplished through Northern Radar Systems Limited. Sufficient funding was obtained through private investment and the Atlantic Canada Opportunities Agency to develop and build a full-scale monostatic OTH radar system. Construction began in 1988 and the radar became operational in 1990. The strategic location of the facility, at Cape Race, Newfoundland, permits coverage of the Hibernia and Terra Nova oil fields as well as a major portion of the Grand Banks fishing area.

The radar has selectable frequency and uses a frequency modulated, interrupted continuous wave form, allowing a range resolution down to 400 m. The receive antenna can be steered over a 120° sector with a nominal beam width of 3.5°. This steering is accomplished, at present, with a combination of hardware and software beam-forming techniques. The system is fully automated and can be controlled and operated via a modem and telephone line. The results may be displayed in real time on a personal computer.

The radar has been in operation in excess of 10,000 hours, routinely producing a variety of data products out to 200 km from shore. These products include ship detection and tracking, sea-ice detection and tracking, and environmental measurements, such as ocean surface currents, waves, and winds. There are plans to upgrade the system to an operational range in excess of 370 km (200 nautical miles) within the next 2 years. A further description of this radar, as well as examples of its data products, are given in Section 7.4.

About 1986, NORDCO Limited of St. John's, Newfoundland, also started research and development work in OTH radars. The main funding agencies were the Department of National Defence (DND), Government of Canada and the Department of Development (DOD), Government of Newfoundland. It developed a generalized computer-simulation package, using the analytical models developed by Memorial University (Bryant et al. [8]; Ponsford et al. [27]). The simulation package is used to predict the performance of OTH radars for differing tasks and environmental conditions. An experimental monostatic radar system has been developed by NORDCO under DND contract at a site on the Cape Bonavista Peninsula, Newfoundland. Similar to the Northern Radar's site, this promontory also offers excellent coverage of the Hibernia and Terra Nova oil fields, as well as a part of the Grands Banks fishing ground. The radar system is described later in Section 7.3, where experimental results are discussed. However, it is worthwhile to mention here that using this radar facility, icebergs have been detected out to ranges in excess of 280 km (Srivastava and Ponsford [37]).

7.1.2 Limitations of OTH Radar

OTH radar has several limitations. The biggest drawback is the requirement for a large ground area for the antenna system. In order to form a directional beam, an array of antenna elements is required which can be large depending upon the frequency and beamwidth requirements: for example, NORDCO's receive array operating at 1.95 MHz consists of 11 doublets, with an aperture length of 900 m. External noise and spectral congestion in the HF band are lesser drawbacks to the operation of an OTH radar. Both of these limitations are location dependent. The maximum range of detection is limited by the external noise. The radar has to be compatible with other uses of the HF spectrum. Interference both from and to other users limits the usable bandwidth and, therefore, the desired resolution in range may not be achieved. This is a severe limitation when compared to that of which is achieved by microwave radars.

7.2 OTH RADAR THEORY

7.2.1 Radar Range Equation

For a pulsed OTH radar, with ground-wave mode of propagation, the radar range equation for the received power spectrum, S_r, at the receiver input for the backscattered signal from a target or range cell area of the ocean surface, is given by (Barrick [2]; Walsh and Srivastava [55]);

$$S_r(\omega_d) = \frac{P_t g_t g_r \lambda^2 |W|^4}{64\pi^3 \rho_o^4} \sigma_s(\omega_d) \tag{7-1}$$

where

P_t = transmit pulse power (W)
g_t, g_r = free space transmit, receive antenna maximum gain or directivity
λ = free space radar wavelength (m)
W = ground-wave attenuation factor including surface roughness effect
ρ_o = range of the target or ocean range cell (m)
$\sigma_s(\omega_d)$ = spectral radar cross section of the target or ocean range cell (m^2/rad/s)
ω_d = Doppler frequency (rad/s)

It is assumed that targets lie along the antenna boresight. Angular variation of antenna gain in the case of the ocean surface is accounted for in the definition of the spectral cross section of the ocean range cell.

From the received power spectrum, the received power, P_r, may be calculated as

$$P_r = \frac{1}{2\pi} \int S_r(\omega_d)\, d\omega_d \tag{7-2}$$

Equations (7-1) and (7-2) imply

$$\sigma = \frac{1}{2\pi} \int \sigma_s(\omega_d)\, d\omega_d \tag{7-3}$$

where σ is the radar cross section of the target or ocean range cell in m^2. Note that this is "radar cross section," not the normalized reflectivity per unit area. This radar cross section terminology is frequently found in the literature on the subject.

7.2.2 Ground-Wave Attenuation Factor

Radio propagation over the earth has been studied by many investigators. In 1909, Sommerfeld [34] investigated radio propagation over a planar earth model for a dipole source and presented the field solution in the form of a complex integral equation. An asymptotic series expansion of the Sommerfeld integral equation was provided much later by Norton [22], from which the field solution may be computed easily. Norton represented the propagation loss due to the finite conductivity of the earth's surface by a function, now commonly referred to as the ground-wave attenuation factor for the planar earth model. The same result was derived by Wait [44, 46], using a concept of normalized surface impedance (surface impedance normalized to impedance of free space, which is 120π ohms).

The attenuation factor depends on propagation distance, frequency, and normalized surface impedance. The normalized surface impedance is a function of frequency and of the permittivity and conductivity of the surface. It describes the effectiveness of the surface in supporting the ground-wave mode of propagation. The lower the normalized surface impedance, the better the propagation can be. In the limiting case of a perfectly conducting surface the attenuation factor attains a constant value of unity. For a highly conducting surface, such as the ocean, in the HF band this impedance is low, giving long ground-wave propagation. However, for sea ice, because the conductivity is relatively much lower, propagation over the ice is very limited.

The phase of the normalized surface impedance also affects the attenuation factor. When the phase exceeds 45°, the propagation follows a trapped surface wave, as discussed by Wait [46] for the case of layered media. In this phenomenon the attenuation factor reverses to an enhancement up to a certain propagation distance, i.e., its value exceeds unity and increases with distance to a point where the trapped surface-wave enhancement balances the normal attenuation. Theoretical analysis has shown this effect also may occur for HF propagation over ocean waves, due to surface roughness for certain wind and transmit frequency combinations (Barrick [1]).

Radio propagation over a smooth spherical earth model has been investigated by van der Pol and Bremmer [41, 42], giving field solutions in the form of a residue series. In an independent analysis, Fock [13] has presented a similar residue series. Compared to the flat earth model, both analysis and results for the spherical earth model are much more complex. However, as in the planar earth model, a smooth spherical earth attenuation factor is given in terms of a residue series. There are limitations on applicability of the residue series in terms of frequency, normalized surface impedance, and propagation distance (Bremmer [7]). However, the series can be used for calculating the attenuation factor for OTH radars if the frequency of operation is in the HF band and if propagation is over the ocean surface (Dawe [11]). A computer program for calculating this attenuation factor has been developed by Berry and Chrisman [6]. A similar but computationally efficient program has been developed by Dawe [11].

The effect of surface roughness, for example waves, on ground-wave propagation results in additional loss. This effect can be included in the propagation model by modifying the normalized surface impedance. Using a perturbation approach, Barrick [1] has provided an expression for the modified surface impedance for the ocean surface. By treating ocean waves as a random rough surface, the roughness is represented by the measurable, directional, wave-height spectrum. Thus, the expression for the modified surface impedance involves the directional wave-height spectrum and hence, it is dependent on the wind condition, in addition to the transmit frequency and permittivity and conductivity of the seawater.

Attenuation increases with wind speed due to an increase in the ocean surface roughness, as expected. Also, for a given wind speed, the added loss increases with frequency since the modified surface impedance increases. It should be mentioned again that a trapped surface-wave phenomenon may exist, depending upon the radar frequency and wind condition. Using a different approach, Srivastava [35] derived an expression for a modified surface impedance (see also Walsh and Srivastava [55]). The two expressions given in algebraic form do not agree completely; when computed, however, the results agree very closely (Dawe [11]).

7.2.3 Cross Section of the Ocean Surface

Perhaps the most challenging task in using OTH radars, whether for target detection or for the measurement of sea-state parameters, is the interpretation of the signal returned by the ocean surface. The return varies with changes in radar frequency, antenna system (wide beam or narrow beam), and mode of operation (monostatic or bistatic) for any given set of ocean conditions. These complications demand an understanding of the interaction processes occurring in the scattering of radio waves from the ocean surface. This understanding would be simpler if the ocean surface was deterministic. Unfortunately, it falls under the category of a time-varying, random rough surface.

Electromagnetic scattering from rough surfaces has been treated by many investigators, using classical techniques. These techniques are physical and geometrical optics (tangent plane method) and perturbation. These methods may be applied to a random rough surface in a statistical average sense. An excellent review of these methods as applied to the ocean surface is presented by Valenzuela [43]. The physical and geometrical optics technique is limited to those surfaces whose principal radii of curvature are much greater than the incident radio wavelength. Hence, in the case of the ocean surface, the technique is not suited for the long radio wavelengths comprising ground-wave frequencies.

In 1896, a perturbation technique was introduced by Rayleigh [29] to study the acoustic reflection from a sinusoidal surface. The first application of this technique to electromagnetic problems was presented by Rice [30] to find the scattered field of a plane wave from a nontime-varying, random rough surface. Rice obtained first- and second-order scattered field solutions for both vertical and horizontal polarizations but he assumed the surface was perfectly conducting. Furthermore, he derived the first-order solution for horizontal polarization when the surface was taken to be finitely conducting. Of course, the complexity of the analysis increases with the order of perturbation and for the finitely conducting case. The order of perturbation is assumed to be the same as the order of the smallness of the surface parameters (height and slope). The analysis is limited to the surfaces having small height and slope variations. The perturbation technique has been extended by other investigators for finitely conducting surfaces and vertical polarization. Moreover, the analysis can be simplified by using a surface impedance boundary condition. Some notable papers are by Wait [47], Barrick [1], and Rosich and Wait [31].

From Crombie's [9] pioneering experiment on radar returns from the ocean at 13.56 MHz, it is evident that HF radars are suitable for detecting long ocean waves (gravity waves). Based on measurements of the backscattered Doppler spectrum, Crombie found that two peaks in the spectrum were caused by two ocean waves, one moving toward and the other moving away from the radar, each having a wavelength equal to one-half the radar wavelength. Further, at ground-wave frequencies, the restrictions imposed by the perturbation analysis are met in the case of the ocean surface under moderate sea conditions. These positive results led Barrick [2, 3] to extend the perturbation technique for deriving the Doppler spectrum of the ocean surface. For this he included the third-dimensional variation in time in the rough surface model of Rice; i.e., modeling the ocean surface as a three-dimensional periodic surface in space and time with random Fourier coefficients. Thus, he derived first- and second-orders of the spectral radar cross section of the ocean surface (discussed later in this section) in terms of the directional wave-height spectrum of the ocean surface for vertical polarization and he confirmed the experimental observations made by Crombie.

In a different approach, Walsh [48] presented a detailed general formulation for rough surface propagation and scattering problems based on a concept of generalized functions. The surface is assumed to be time invariant. The technique uses a three-dimensional (spatial and temporal) Fourier transform representation for the field quantities and requires, in the most general case, the solution of a pair of coupled vector, integral equations for surface field quantities. However, under certain assumptions (reasonable for the ocean at ground-wave frequencies) for the refractive index and spatial frequency spectrum of surface and electrical fields, these equations may be reduced to a single vector, integral equation in one unknown, which happens to be the surface field intensity. This unknown is the prime quantity of interest for ground-wave propagation and scattering. The formulation is open to any finite source, as opposed to the Rice perturbation technique where a plane-wave incidence generally is used.

Srivastava [35] applied Walsh's formulation to a two-dimensional periodic surface having a good conductivity. To obtain a solution for the surface electric field in the transform domain, the Neumann series expansion was used as the formal inverse of an operator. It was shown how, in the limit, Rice's perturbation result for a perfectly conducting periodic surface with plane-wave incidence may be derived from the Neumann series solution. The series solution for a surface electric field is partially summed to form another series which is in a more tractable form. Although the solution is in the spatial and temporal Fourier transform domain and is relatively complex, its advantage lies in the fact that any finite source can be used. The solution is general in the sense that it can be used for any good conductivity periodic surface but which is not necessarily perfectly conducting. Moreover, the solution provides a better physical interpretation of the scattering mechanisms. The above formulation and solution are available in Walsh and Srivastava [54] in a condensed form.

To study the Doppler spectrum of sea echo, an application of the series solution to the ocean surface was reported by Srivastava [35]. Like Barrick, he assumed the ocean surface can be described as a three-dimensional periodic surface with random Fourier coefficients. He chose the source to be an omnidirectional antenna, located close to the surface and excited by a pulsed sinusoidal current, which models closely a pulsed OTH radar system. To facilitate the analysis, the surface slopes are assumed to be small compared to unity. Assuming a narrow-beam receive antenna, expressions for the first and second orders of the backscattered power spectrum were derived. Consequently, two orders of backscattered spectral cross section are derived. A summary of the analytical method can be found in Srivastava and Walsh [38]. The extension of this application to deal with wide-beam or omnidirectional reception followed in Srivastava and Walsh [39] and the resulting backscattered spectral cross-section expressions for the two orders are discussed in Walsh and Srivastava [55].

Based on the theoretical model developed, a physical interpretation of the scattering mechanism was described by Srivastava and Walsh [38, 39]. The first-order backscattered signal represents a single scattering of the transmitted signal from the ocean surface. The scattering occurs in an angular section of a circular annulus of the surface located at ρ_o from the antenna, where $2\rho_o$ is the two-way propagation path distance corresponding to the time delay between transmitted and received signals. The radial width of the annulus or the resolution cell is $2\Delta\rho$, where $2\Delta\rho = c\tau/2$ for a pulsed radar: c is the velocity of light and τ is the transmitted pulse duration. The angular width depends upon the receiving antenna beamwidth. For omnidirectional reception, the angular width is 2π. For narrow-beam reception, the scattering area reduces to a patch as shown in Fig. 7.1.

The second-order signal represents a double scattering of the transmitted signal. However, the scattering points are not restricted to lie in the annulus or in the patch defined by the first order. The two scattering points may be anywhere on the surface, provided the total travel time for the signal is the same as for the first order. The second-order results consist of three parts. The first part represents the case where both first and second scatterings occur in the annulus or patch. The second part represents the case where a first scattering occurs at the transmit point and the second scattering occurs in the annulus or patch. The third part represents the case where the two scatterings do not occur

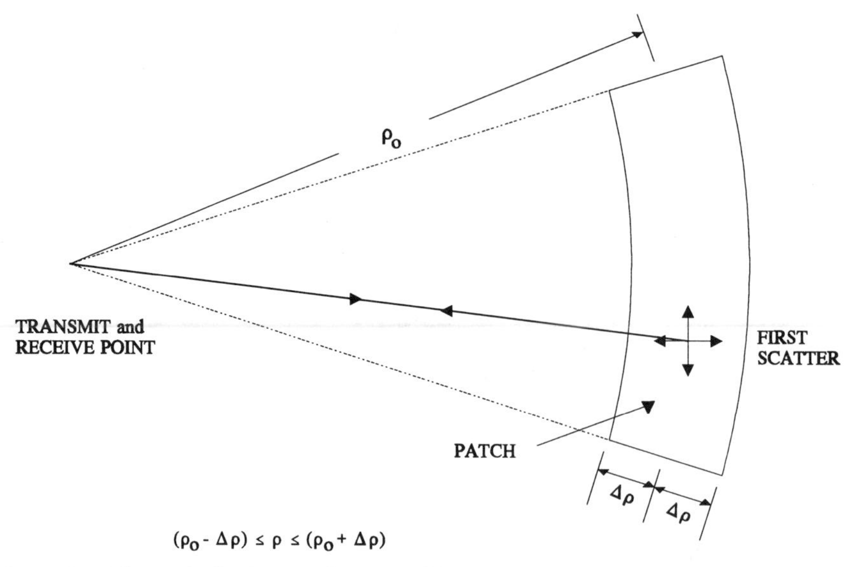

Figure 7.1 First-order backscatter from a surface patch for omnidirectional transmission and narrow-beam reception (ρ_0 = patch distance, $2\Delta\rho$ = radial width of patch).

in the annulus or at the transmit point. Of course, the last two parts imply that the transmit antenna is surrounded by the ocean, such as on a ship or offshore platform. Numerical examples show that the contribution to the total second-order Doppler spectrum from the third part is insignificant compared to that of the first and second parts, at least in the Doppler region used for measuring the ocean wave-height spectrum (Howell et al. [15]; Walsh and Srivastava [55]) and hence this part is not considered here. An illustration of the first and second parts of the second-order scattering for narrow-beam reception is given in Fig. 7.2.

Based on the Srivastava and Walsh sea-clutter model, a few plots of the backscattered spectral cross section normalized to the patch area for narrow beam reception are presented. A model for the directional ocean wave-height spectrum is required for computation of the spectral cross section. Several semi-empirical oceanographic models are available for the directional ocean wave-height spectrum for a fully developed sea (Kinsman [16]; Longuet-Higgins et al. [19]). The Pierson-Moskowitz frequency spectrum, with a cardioid directional distribution as described in Srivastava [35], has been used in the figures. A plot of the first-order spectral cross section for a patch of the ocean surface is shown in Fig. 7.3. It contains two dominant peaks, called the resonant or Bragg peaks. The expression for the first-order spectral cross section shows that these peaks are caused by two ocean gravity waves whose wavelengths are equal to one-half the radar wavelength, one moving toward the radar (positive Doppler) while the other is moving away from it (negative Doppler). The

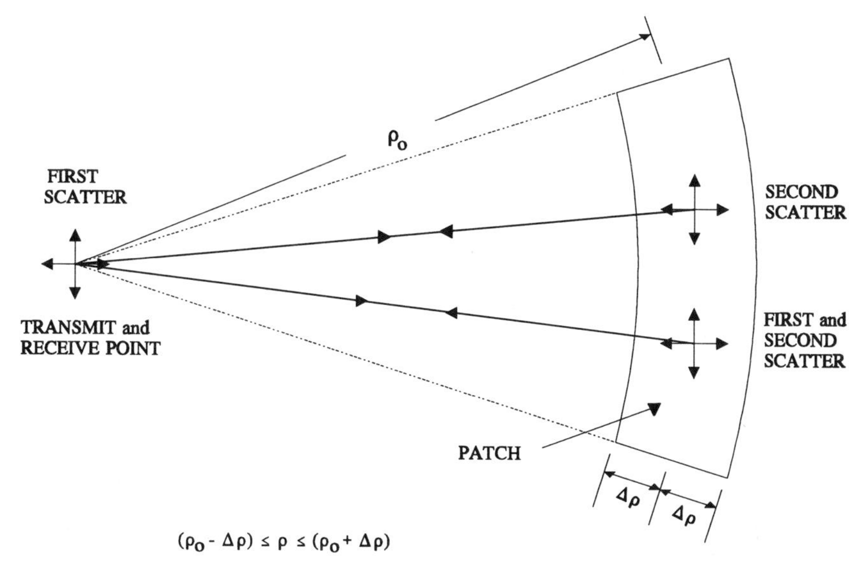

Figure 7.2 Two parts of the second-order backscatter from a surface patch for omnidirectional transmission and narrow beam reception for the open sea condition (ρ_0 = patch distance, $2\Delta\rho$ = radial width of patch).

corresponding Doppler frequencies are called the Bragg frequencies. Between the two peaks, it is a continuum, as shown in Fig. 7.3.

In a comparison, Barrick's first-order result contains only two impulses, one located at each Bragg frequency. However, in the limit when the radial width of the patch is taken to be infinity, the result does go to that derived by Barrick [2, 3], as shown by Srivastava [35]. The Doppler shift in the first-order continuum increases with the increasing frequency of the sinusoidal components of the ocean wave spectrum causing the backscatter. Further, except for the two Bragg peaks, it does not follow the principle of conventional Doppler shift based on velocity. This theoretical behavior of the first-order Doppler shift has also been observed by Wait [45] and Crombie [10].

In the model assumed for the ocean surface it has been considered that the water mass is not moving, only the waves are moving. If the water also is physically moving in the form of a surface water current, the two first-order peaks are shifted equally from the Bragg position by a small amount in the same direction. The shift towards positive or negative Doppler and the size of the shift depend upon the radial component of the surface current vector. The presence of a current modifies the phase velocity of ocean waves. Thus, by measuring this shift in the Doppler spectrum, one can obtain the radial component of the mean surface current vector (Barrick et al. [52]).

An illustration of the first part of the second-order spectral cross section, normalized to the patch area, is given in Fig. 7.4. This part is equivalent to

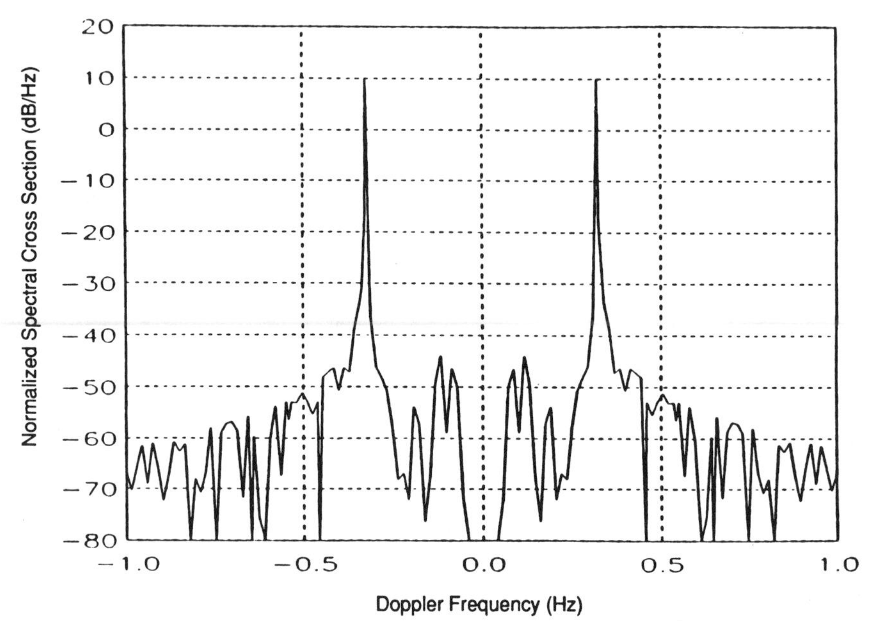

Figure 7.3 First-order sea clutter at a frequency of 10 MHz, wind speed of 15 m/s at 90°.

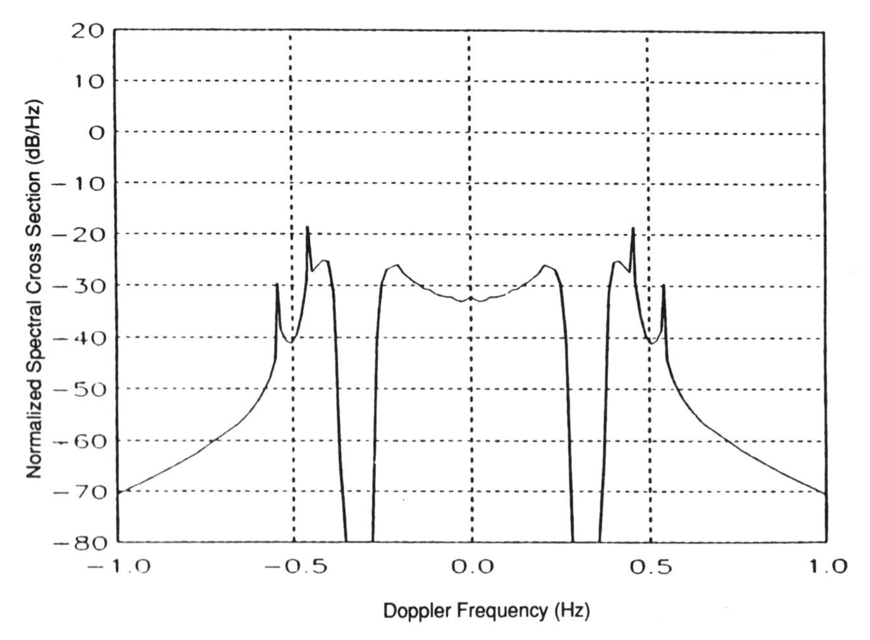

Figure 7.4 First part of second-order sea clutter at a frequency of 10 MHz, wind speed of 15 m/s at 90°.

Barrick's complete second-order result. It contains well-known peaks at $\pm\sqrt{2}$ and $\pm 2^{0.75}$ times the Bragg frequency, later known as the corner reflector effect (Barrick [3]). The second part of the second-order spectral cross section is shown in Fig. 7.5. This part produces peaks at zero Doppler and at + and − twice the Bragg frequency, depending on the sea condition. These peaks may be viewed as a repeated first-order phenomenon, first at the transmit point (from behind) then in the patch. This part also produces a significant clutter level with a much slower rolloff at high Doppler frequencies when compared to either the first-order (Fig. 7.3) or the first part of the second-order (Fig. 7.4). The Doppler shift increases with the frequency of the sinusoidal ocean wave component, similar to the first-order continuum. It may be seen that the Doppler regions near the two Bragg frequencies, which are commonly used for estimating the ocean wave parameters using the first part (Lipa and Barrick [17]; Howell et al. [15]), are relatively unaffected by the second part. Therefore in these regions the second-order cross section may be described adequately by the first part alone. On the other hand, in target detection, the second part may be quite significant when the target's Doppler frequency is around zero (icebergs, sea ice, and ships) or beyond $\pm 2^{0.75}$ times the Bragg frequency (high speed targets such as low-flying aircraft). Of course, when the transmit antenna is located on a beach or near a shore, this part may not be wholly present. The total (first and second order) spectral cross section with the second part is shown in Fig. 7.6 and without the second part in Fig. 7.7.

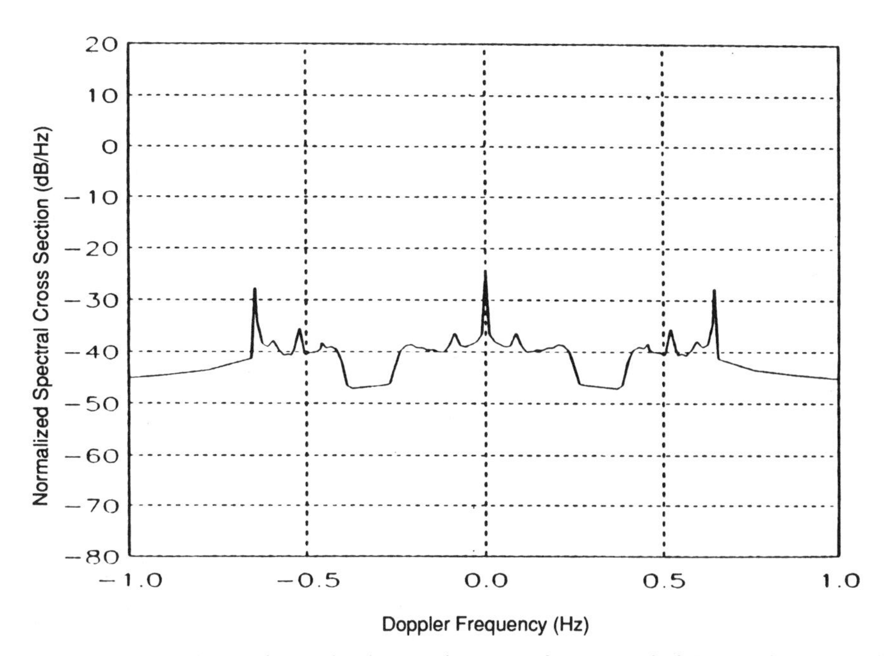

Figure 7.5 Second part of second-order sea clutter at a frequency of 10 MHz, wind speed of 15 m/s at 90°.

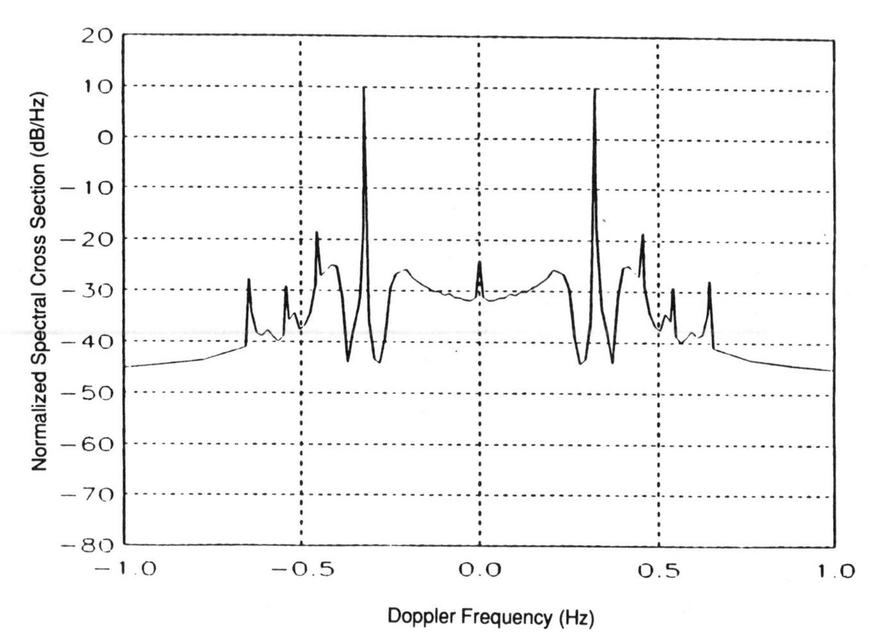

Figure 7.6 Total sea clutter (with second part) at a frequency of 10 MHz, wind speed of 15 m/s at 90°.

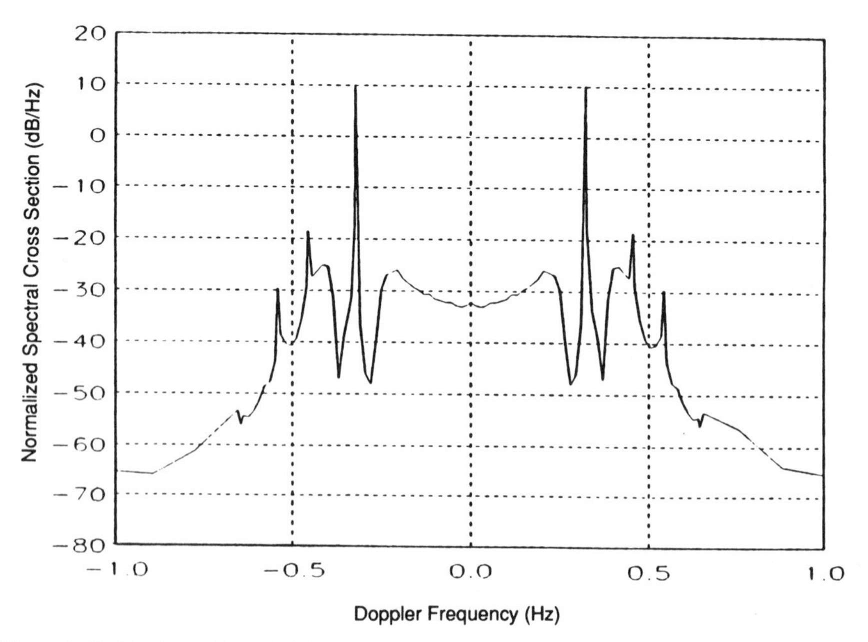

Figure 7.7 Total sea clutter (without second part) at a frequency of 10 MHz, wind speed of 15 m/s at 90°.

7.2.4 Ice-Edge Detection

Remote sensing of sea ice in open water by an OTH radar is based on detection of the ice edge. Therefore, it calls for an estimate of the backscattered radar cross section of the ice edge. Electromagnetic scattering from an ice edge is essentially a "mixed path" propagation problem. Several investigators have treated the mixed path problem using different analytical techniques. In a semi-empirical work, Millington [20] predicted an enhancement in the electrical field when the field was propagating from a low-conducting medium to a higher conducting medium. This behavior of the field is commonly referred to as the "recovery effect." General methods for treating the problem by Wait [46] and Furutsu [14] emphasize calculating the field propagating past the boundary or the edge. However, for the radar application in remote detection of an ice edge, an estimate for the field far from the boundary is required.

Based on the technique of generalized functions developed by Walsh [48] for rough surface scattering problems, Ryan and Walsh [33] presented a method for analyzing the mixed path case (see also Walsh [49]; Ryan [32]). The model assumes the complete space has an upper and lower half consisting of three media: the upper half is free space and the lower half below has two media separated by a vertical plane, each medium having different electrical properties, e.g., seawater and sea ice. The source is a vertical electric dipole located in the upper half. For this model, they derived an expression for the forward (past the boundary) propagated field which was in agreement with the results of other investigators. They also derived an expression for the field backscattered from the boundary. From this expression, a result was derived for the backscattered cross section of the boundary in the HF band.

As an application of the result to remote sensing of sea ice, Ryan and Walsh calculated the backscattered cross section for different types of sea ice at different radar frequencies, assuming a length of 300 m for the ice edge. For first-year ice with salinity of 10%, cross sections were computed to be 29.1, 31.5, and 33.4 dBm^2 at 5, 10, and 25 MHz, respectively. The cross-section values increase to 34.0, 34.7, and 34.8 dBm^2 for the same frequencies for old ice with a salinity of 1%. Also, these values are range independent when the ice edge is far (> 10 km) from the radar. These are valuable results because a high cross-section prediction combined with a long-range propagation benefit over seawater at HF indicates that OTH radars are suitable for ice detection at long ranges.

7.2.5 Iceberg Detection

An iceberg is an object floating in the ocean with very different electrical properties compared to those of seawater. Typical values for relative permittivity and conductivity for icebergs are 5 and 10^{-5} s/m, respectively, whereas for seawater these values are 80 and 4 s/m. Thus, electromagnetic scattering problems for icebergs fall in the category of layered media combined with mixed paths and discontinuities. A general analytical technique for layered media for an arbitrary source has been given by Walsh [49, 50]. He presented a similar technique for analyzing mixed paths with discontinuities. By using these results, Walsh and Srivastava [53] derived an expression for the backscattered radar cross section for icebergs having arbitrary shape and size. The source is assumed to be an elementary vertical dipole, an appropriate model for ground-wave OTH excitation.

The cross-section expression requires an integral evaluation over the contour of a given iceberg at the waterline. It is difficult to model the actual shape of the contour as it differs from one iceberg to another. However, from the remote sensing point of view, one is interested in an average cross section, independent of shape for a given size of iceberg at a given frequency. For this purpose, Walsh et al. [56] proposed an averaging technique. As an approximation to common iceberg shapes, they considered a rectangular contour shape with different aspect ratios. The aspect ratio is defined as a/b, where a is the dimension of the berg along the radar boresight and b is its dimension in the perpendicular direction. The cross-section plot as a function of contour area for a given aspect ratio and frequency, shown by Walsh et al. [56], contains peaks and nulls. These peaks and nulls occur due to phase addition and phase cancellation, respectively, of signal returns from the leading and trailing edges of the iceberg. For example, for a square contour (aspect ratio = 1) peaks in the cross section occur when the perimeter of the contour corresponds to odd integer multiples of the radar wavelength. Similarly, nulls occur when the perimeter corresponds to even multiples of the wavelength. In reality, icebergs have highly irregular shapes, so that these extreme variations in cross section do not often occur naturally. Therefore, a smoothing technique is used for the cross section and this process is repeated for different aspect ratios. Finally, smoothed cross section values are averaged amongst different ratios for the same contour area, to yield an average estimate of the cross section independent of iceberg shape.

To study the dependence of iceberg cross section on radar frequency, Ponsford and Srivastava [24] computed the average cross section normalized to the contour area at different frequencies covering the HF band, using the technique suggested by Walsh et al. [56]. The result is presented in Fig. 7.8. The iceberg size classification, as given by Murray [21], is based on height (above water) and length (at waterline), as shown in Table 7.1 (this classification differs, slightly, from the WMO classification given in Table 2.1).

The size classification has been incorporated into Fig. 7.8 by using the corresponding waterline contour area and assuming a square shape. Considering sizes commonly found in the North Atlantic, cross sections have been computed up to large icebergs. It is interesting to note that the normalized cross section at any frequency in the HF band remains almost constant (variations within 3 dB only) as the size of the iceberg increases from small to large. Therefore, in this size range, the average cross section is directly proportional to the contour area. In addition, the normalized cross section remains fairly constant (variations within 4 dB only) with respect to the frequency. As the frequency increases, this behavior extends to both bergy bits and growlers. Based on this

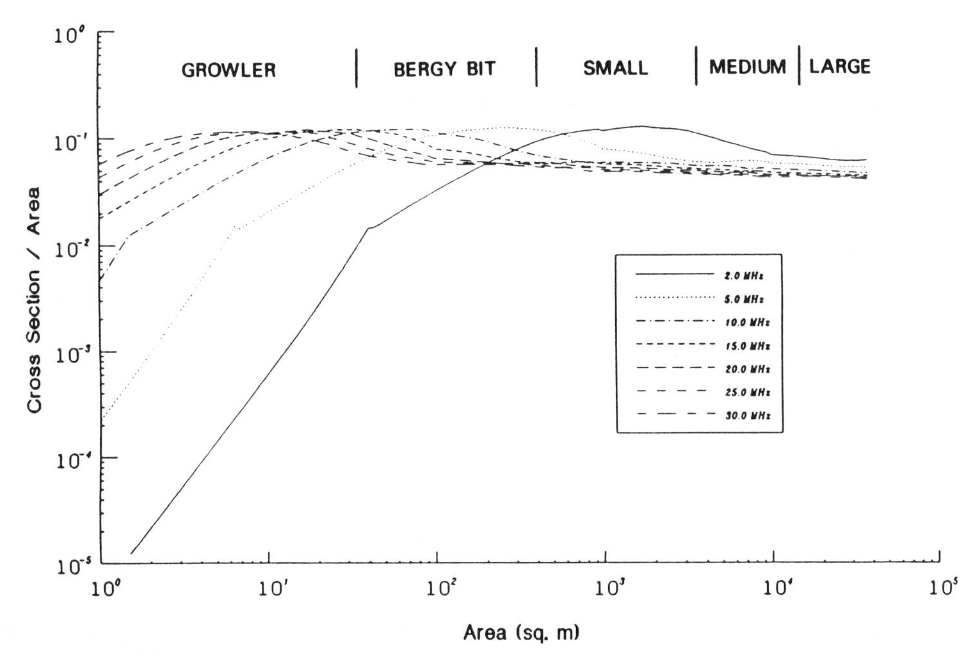

Figure 7.8 Average normalized iceberg cross section at different frequencies.

TABLE 7.1 Size Classification of Icebergs (after Murray [21])

Classification	Height (m)	Length (m)
Growler	<2.5	<6
Bergy bit	2.5–5	6–21
Small iceberg	5–15	21–61
Medium iceberg	15–46	61–122
Large iceberg	46–78	122–213
Very large iceberg	>78	>213

model, the average cross section may be estimated (Ponsford and Srivastava [24]) according to Table 7.2.

The cross sections for bergy bits and growlers in the lower HF band are significantly smaller than given in Table 7.2. Also, the normalized cross section varies significantly with size and frequency, as shown in Fig. 7.8.

7.3 CAPE BONAVISTA SYSTEM

To study the capabilities of OTH radar in detection and tracking of icebergs, trials were conducted by NORDCO Limited, using the Department of National

TABLE 7.2 Variation of Average Cross Section of Icebergs with Frequency

Classification	Average Backscattered Cross Section (m^2)	Frequency Range (MHz)
Small–Large	0.06 × contour area	2–30
Bergy bit	0.07 × contour area	5–30
Growler	0.08 × contour area	20–30

Defence (DND) OTH radar facility at Cape Bonavista, Newfoundland. A summary of the experiment and detection results are presented here; for more details, reference is made to a report by Srivastava et al. [40].

7.3.1 Radar Description

The transmitter consists of a modified LORAN-A transmitter, operating at 1.95 MHz with a peak power of 1 MW. The transmitted pulse is shaped to a raised cosine function of 50-μs duration between the 3-dB points, with a choice of pulse repetition frequencies (PRF) of either 25 or 50 Hz. The transmit antenna is a quarter-wave monopole at 1.95 MHz. The receiver site is closely located to the transmitter site and covers an area of about 40 hectares of enclosed land. The receive antenna consists of a linear array of 11 doublets. The array elements are 18-m whip antennas with an interelement spacing, along the line of the array, of one-half of a wavelength at 1.95 MHz and one-quarter of a wavelength between the elements of the doublet.

The radar receiver and processing unit consists of three main sections: timing control unit, radar receiver, and VME-based processing unit. The timing control unit provides the necessary control signals to the radar receiver and processing unit. These control signals are synchronized with the radar's transmitted pulse and the master reference (5 MHz). To maximize the operational range of the radar, it is necessary that the radar receiver has atmospheric noise-level sensitivity. In addition, the receiver must have sufficient selectivity to be able to reject strong radio transmissions that may typically be 120 dB greater than the desired backscattered signal. The radar receiver used in the study covers the entire HF band (1.9 to 30.0 MHz). The receiver output is a 16-bit baseband signal, centered at 25 kHz, with a bandwidth of 30 kHz, and sampled at a rate of 125 kHz. A spurious free dynamic range of greater than 120 dB is achieved in the analogue stages. The 16-bit digital output is transferred to VME-based signal processor via a data buffer.

7.3.2 Radar Signal Processing Software

Functions for extracting Doppler spectra are implemented in software on the VME signal-processing system (Ponsford et al. [27]; Srivastava et al. [40].

The incoming data are stored initially on a hard disk. A radar (whose primary function is target detection) is operated usually with a matched filter receiver, such that the signal-to-noise ratio is maximized. However, when the radar is required to track a target, the matched filter approach may not be optimum. In the case of an OTH radar with range resolution on the order of kilometers, it is necessary to refine the target's range, normally by oversampling. This technique implies that the transmitted pulse shape has been preserved and necessitates using a receiver employing an HF filter with a time–bandwidth product greater than unity. To maintain a linear phase response in the passband, a finite impulse response filter was employed.

The received bandpass signal is centered at one-fifth of the sampling frequency. To extract the Doppler information, it is required to shift this frequency to baseband. It is also necessary to resolve both positive and negative Doppler shifts in the baseband signal, which is achieved by mixing the bandpass signal with in-phase cosine and quadrature sine waves, whose frequencies are equal to one-fifth of the sampling frequency. To remove the unwanted products of mixing, it is necessary to lowpass filter the data, which is achieved with the same finite impulse response (FIR) filter subroutine as for the bandpass filter; however, in this case, filter coefficients for a lowpass filter are used. This process is repeated for each transmitted pulse that occurs during the coherent integration period, or dwell time.

The time-accumulated and range-gated data are processed next using a fast Fourier transform (FFT) subroutine, on a range-gate by range-gate basis. The frequency resolution of the Doppler spectrum is determined by the reciprocal of the dwell time. Therefore, the number of time samples taken for spectral analysis is dependent upon the required Doppler resolution. A limit is set on the achievable Doppler resolution by the velocity of the target, as it is required that the target remain within the range cell for the dwell time. The complex output from the FFT was converted into a power spectrum. The Doppler frequency, range, and spectral power are stored in a data file for further processing, e.g., automatic detection of targets above a given threshold using a peak detection routine.

Use of the complete dwell time for increased resolution, without a corresponding increase in the data storage requirements, can be made by decimating, or undersampling, the data before the FFT, but this will cause an increase in noise due to aliasing and thus, an increased risk of aliased interference. For each range gate, baseband samples differ in time by the pulse-repetition interval. To reduce the aliased noise, it is possible to block-average small data sets, of length equal to the under-sampling factor, and to output a single value for further analysis.

During the spring 1990 trials, primary interest was in detecting icebergs. At 1.95 MHz, the Bragg resonant frequencies occur at ± 0.142 Hz; this represents an approaching and receding seawave at 10.92 m/s. As targets with radial velocities greater than 15 m/s were not expected, the Doppler frequency range of interest was restricted to ± 0.2 Hz. Icebergs travel very slowly (typically

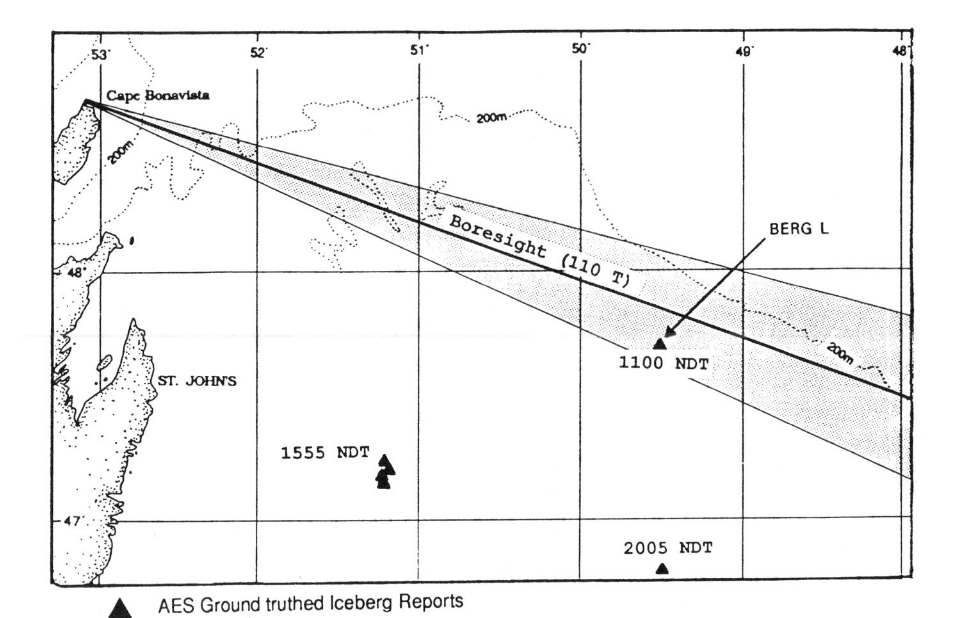

Figure 7.9 Positions of Berg L and other iceberg targets within and around the coverage area of the Cape Bonavista OTH radar on 23 May 1990, as reported by the Atmospheric Environment Service (NDT = Newfoundland Daylight Time).

less than 0.5 m/s) and thus their Doppler frequencies fall within the two Bragg resonances. Considering a dwell period close to 18 min and restricting data for FFT to 4096 samples for a PRF of 50 Hz, it was permissible to undersample the baseband data by a factor of 13. For the situation described herein, the Doppler resolution, therefore, will be 9.4×10^{-4} Hz and the corresponding resolvable Doppler bandwidth would be 3.8 Hz.

7.3.3 Detection of a Known Iceberg

The field trials took place in May 1990, with cooperation from the Atmospheric Environment Service (AES) Ice Branch and an observing vessel, MV *Clipper Crusader*. About 20 icebergs were reported in the area (see Fig. 7.9) and the OTH radar detected and tracked several of these (Srivastava et al. [40]). Of particular interest was an iceberg (labelled Berg L) which the MV *Clipper Crusader* documented in detail. It was a medium iceberg: 151 m in length, 56 m in width, and 30 m above the waterline; thence, the waterline area was about 8500 m^2 and the iceberg cross section was estimated as 27 dBm^2. From Fig. 7.9, it can be seen that Berg L was in the radar boresight at a range of 283 km (153 nautical miles).

Berg L was detected and tracked through four consecutive data files. The first detection occurred at a range of 282 km from the radar, with the iceberg

TABLE 7.3 Detection Summary for Berg L on 23 May 1990

Time (NDT)	Doppler (Hz)	Speed (m/s)	Spectral Power (dBu/Hz)	Range Sample No.	Range (km)
1106	+0.002817	−0.2	9.1	227	282
1434	+0.004069	−0.3	14.9	225	280
1533	+0.004695	−0.4	14.5	223	278
1934	+0.005634	−0.4	19.6	219	273

travelling at a radial velocity of 0.2 m/s toward the radar at 1106 NDT (Newfoundland Daylight Time). At 1434 NDT, the iceberg was detected at 280 km, travelling towards the radar with a velocity of 0.3 m/s. The third detection occurred at 1533 NDT, with Berg L 278 km away, and travelling towards the radar with a radial velocity of 0.4 m/s. The final detection occurred at 1934 NDT, at a range of 273 km from the radar. Again the iceberg was travelling towards the radar with a radial velocity of 0.4 m/s. The detection data for this iceberg are summarized in Table 7.3. The spectral power given here is on a common relative scale and not on an absolute power scale. Therefore, the spectral power levels shown in Table 7.3 and in subsequent three-dimensional iceberg-detection plots are in dB units per Hertz (dBu/Hz).

The three-dimensional plots that were computed for radar data collection during these four time periods are shown in Figs. 7.10 to 7.15. Figure 7.10 is a plot of the initial detection with the return from Berg L in range sample 227. A return from a larger target, M, possibly an iceberg also, can be seen peaking

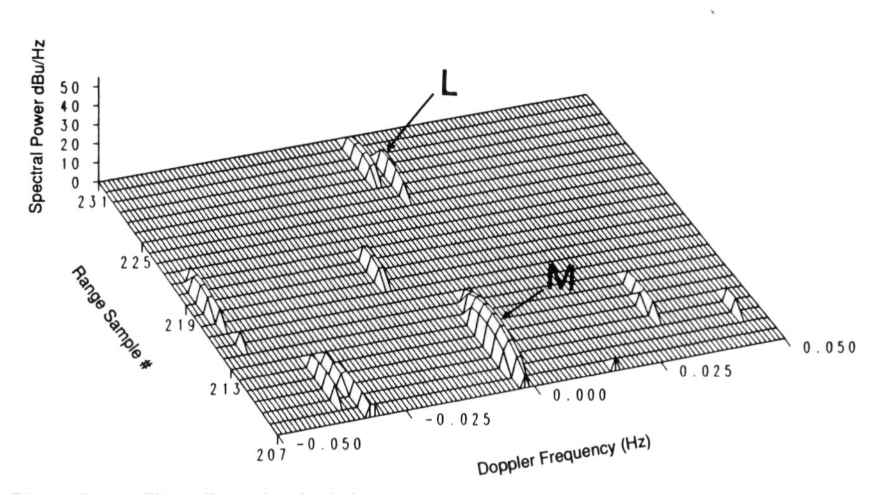

Figure 7.10 Three-dimensional relative spectral power plot showing Bergs L and M, as detected by the Cape Bonavista radar at 1106 NDT on 23 May 1990. Signal threshold for the plot is taken as 6 dBu/Hz.

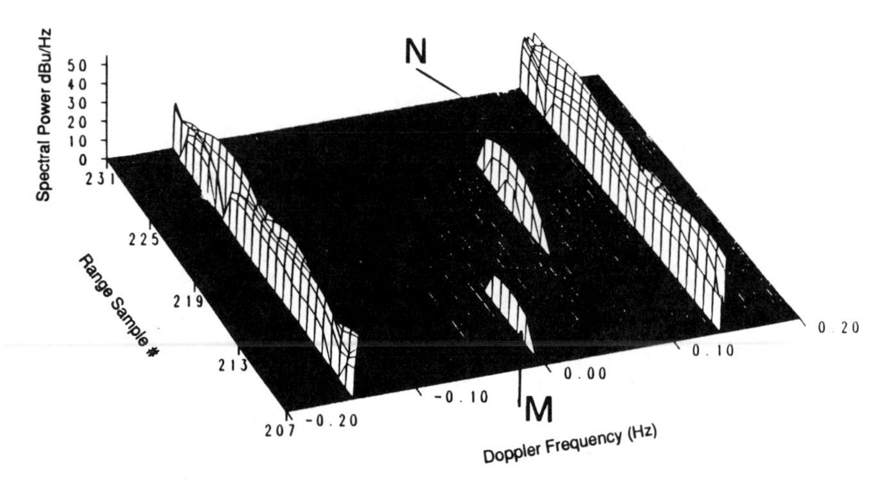

Figure 7.11 Three-dimensional relative spectral power plot showing targets N and M detected at 1106 NDT on 23 May 1990. The plot also shows advance and recede Braggs, as the Doppler range is extended. Signal threshold for the plot is taken as 12 dBu/Hz.

in range sample 211, with a negative Doppler shift of −0.00188 Hz. This corresponds to a return from a target moving away from the radar with a radial velocity of 0.1 m/s. It is interesting to note that by expanding the Doppler range on the plot, as presented in Fig. 7.11, the observing vessel, the MV *Clipper Crusader*, can be seen. This has been identified as target N in the plot. The vessel is approaching the radar at a range of 275 km, with a radial velocity of 4.8 m/s. In this plot, the baseline threshold level has been raised and Berg L is below the baseline, thus it is not visible. The two dominant peaks at about ±0.142 Hz are the Bragg peaks as discussed previously.

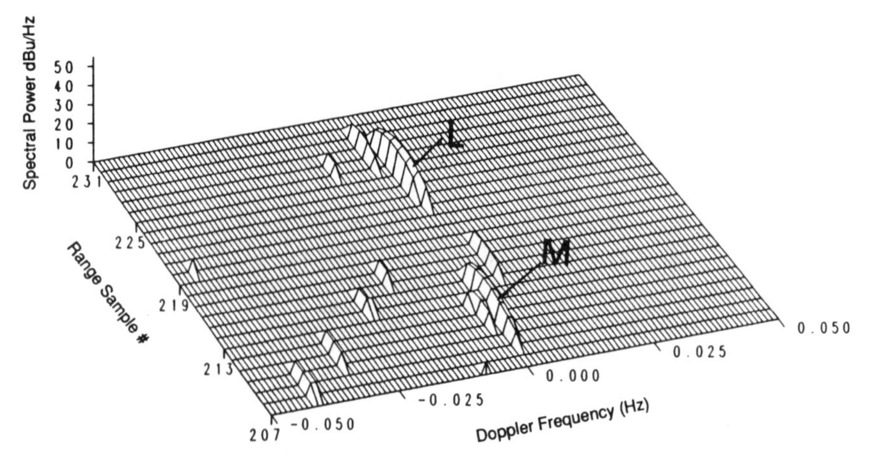

Figure 7.12 Three-dimensional relative spectral power plot showing Bergs L and M detected at 1434 NDT on 23 May 1990. Signal threshold for the plot is taken as 6 dBu/Hz.

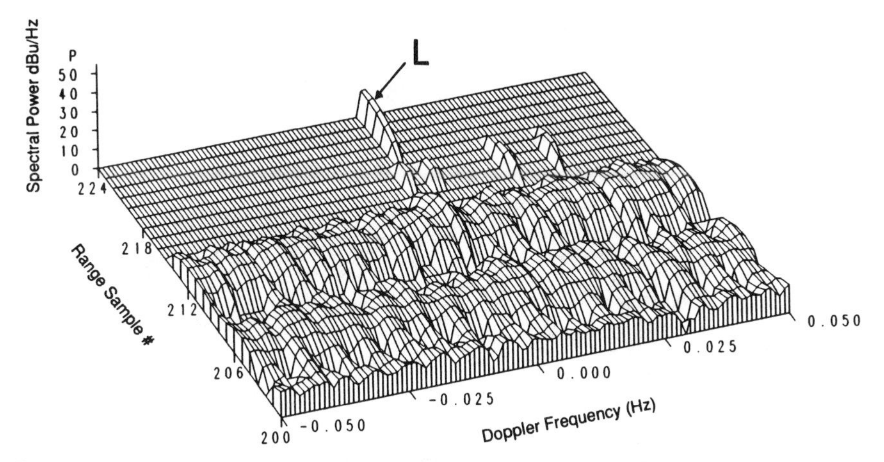

Figure 7.13 Three-dimensional relative spectral power plot showing Berg L detected at 1533 NDT on 23 May 1990. Signal threshold for the plot is taken as 6 dBu/Hz. Berg M is lost in high-level ionosphere return from the F layer.

The second detection is shown in Fig. 7.12. The return from Berg L is clearly visible, peaking in range sample 225. Target M was detected again in range sample 213; however, it was moving towards the radar with a radial velocity of 0.1 m/s.

Figure 7.13 is a plot of the third detection. The return from Berg L is clearly visible, but target M is lost in high-level clutter returns caused by ionospheric interference from the F layer. An expanded version of this plot is presented in Fig. 7.14, from which it can be observed that the ionospheric interference is

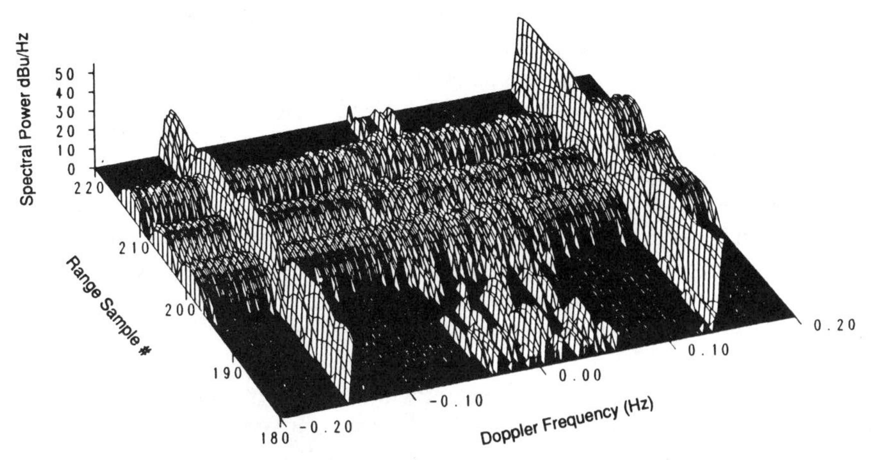

Figure 7.14 An extended three-dimensional relative spectral power plot shows ionospheric return is localized between range samples 195 and 215. Advance and recede Braggs are also visible. Signal threshold for the plot is taken as 6 dBu/Hz.

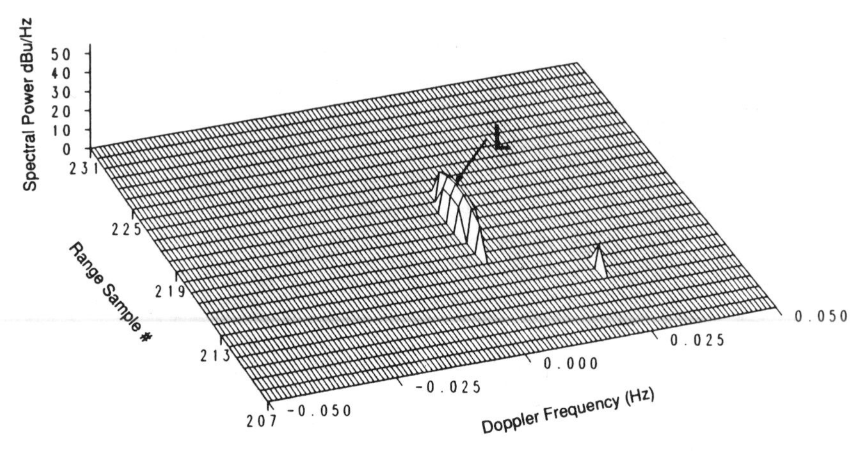

Figure 7.15 Three-dimensional relative spectral power plot showing Berg L detected at 1934 NDT on 23 May 1990. Signal threshold for the plot is taken as 6 dBu/Hz.

localized between range samples 195 and 215. It should be noted that ionospheric interference is dependent upon both the operating frequency and time of day; it is the result of sky-wave propagation from the radar transmit antenna.

The final detection is shown in Fig. 7.15. The return from Berg L peaks in range sample 219. It is interesting to note that the ionospheric interference is no longer noticeable in this range block which is consistent with the behavior of the F layer over this time period.

Comparison with Model Prediction

The radar data obtained from the measured iceberg, Berg L, were compared with the model prediction using NORDCO's generalized computer-simulation package (Ponsford et al. [27]; Srivastava and Ponsford [36]). The radar cross section for this berg has been estimated to be 27 dBm^2, as given above.

Ocean data were provided by a DATAWELL directional waverider buoy deployed 50 km from the radar along the line of boresight (110° True). The significant wave height at 1100 NDT on 23 May was 283 cm, with a mean wave direction of 064° magnetic (042° True) which corresponds to sea state 5 with a wind speed of 12 m/s blowing from 042° True. For the simulation purpose, this is equivalent to a wind blowing to 248° with respect to boresight. It was assumed that the buoy measurements represented the general wind condition in the radar coverage area, including the Berg L location. Such a wind direction would act to drive icebergs towards the radar.

In addition to wind, the surface current will influence the motion of icebergs. The extraction of surface current from the location of Bragg peaks is well established (Barrick et al. [5]). A two-dimensional plot from the iceberg-detection data collected at 1106 NDT is presented in Fig. 7.16. The plot shows Doppler frequency versus received power spectrum for range sample 227 (282 km): it can be seen that the Bragg lines are shifted to the left, no longer

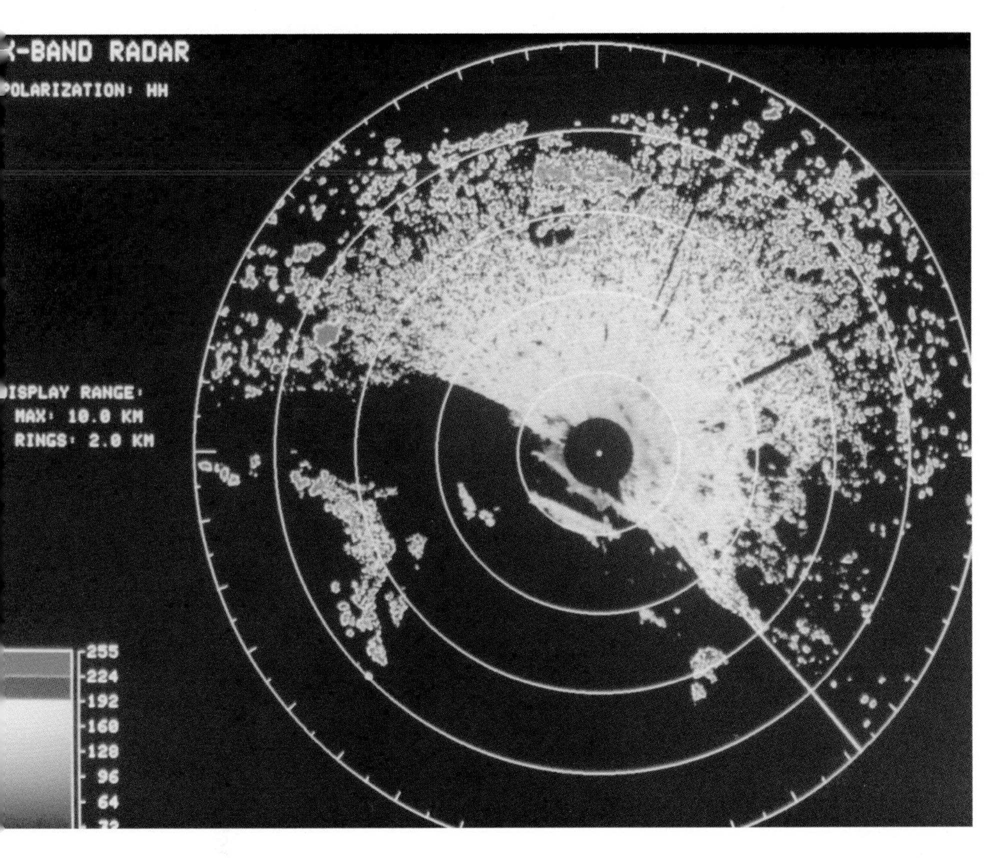

Figure 8.62 Color-coded image. Red and blue represent strong signal returns while the yellow highlights detected edges.

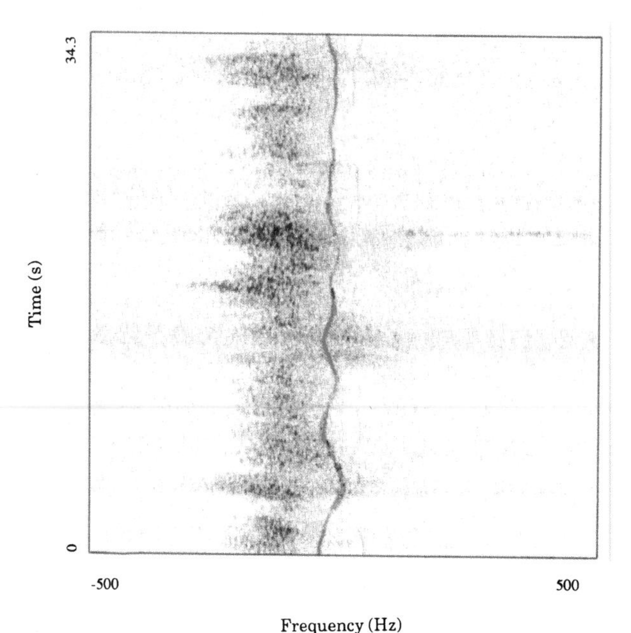

Figure 9.19 *HH/VV* power spectral ratio, in time–frequency format, for the range cell containing a growler in sea clutter, at a bearing of 119° and a range of 6.5 km. Data are the same as those in Fig. 9.14. The image brightness varies as the sum of the powers, whereas the color shows the ratio of the powers. Note how the growler's ratio is near unity, whereas for the sea clutter, *VV* is more powerful at smaller Doppler shifts and *HH* is more powerful at larger Doppler shifts.

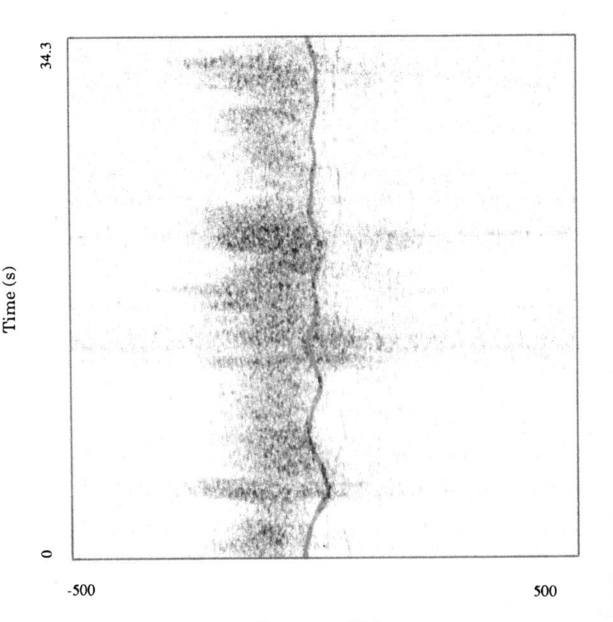

Figure 9.20 *HH/VV* spectral coefficient phase difference, in time–frequency format, for the range cell containing a growler in sea clutter, at a bearing of 119° and a range of 6.5 km. Data are the same as those in Fig. 9.19. The image brightness varies as the product of the powers, whereas the color shows the phase difference. Note how the growler's phase difference is consistent, whereas for the sea clutter, the phase difference is much more variable.

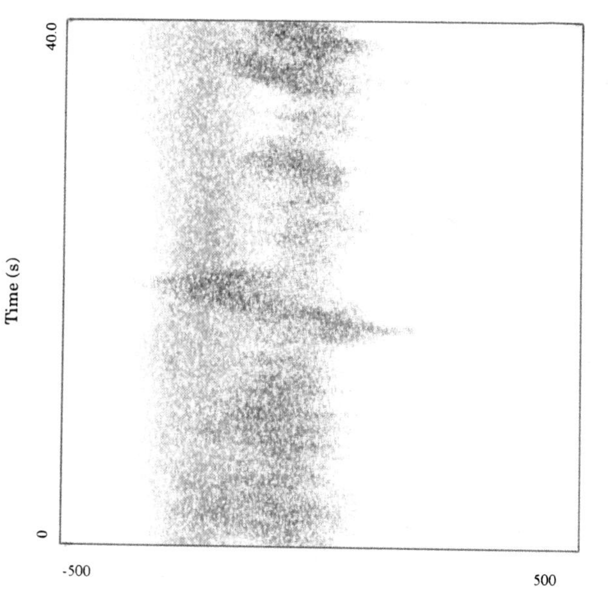

Figure 9.22 *HH/VV* spectral coefficient phase, in time–frequency format, for the range cell containing both rain and sea clutter, at a bearing of 30° and a range of 5.8 km. Data are from a data set similar to that used for Fig. 9.11. The image brightness varies as the product of the powers, whereas the color shows the phase difference. Note how the phase difference for rain clutter is consistent but for the sea clutter, the phase difference is much more variable.

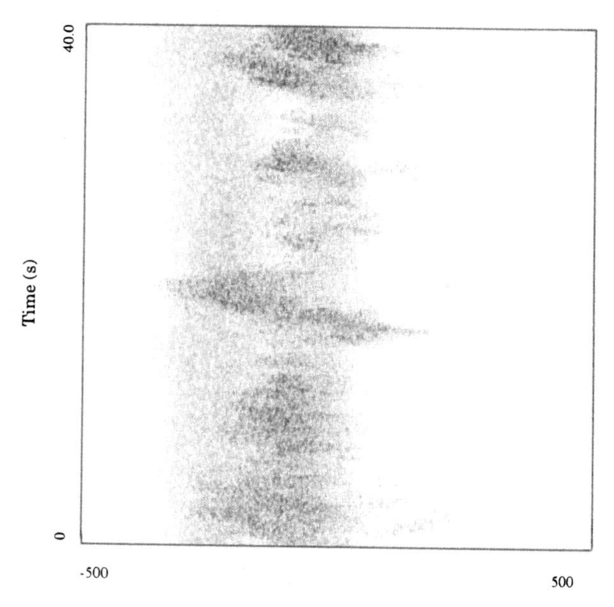

HH/VV = -10 dB HH/VV = 0 dB HH/VV = +10 dB

Figure 9.25 *HH/VV* power spectral ratio, in time–frequency format, for the range cell containing both rain and sea clutter, at a bearing of 30° and a range of 5.8 km. Data are the same as those used for Fig. 9.22. Note how the rain clutter's ratio is consistently near unity but for the sea clutter, the ratio is much more variable.

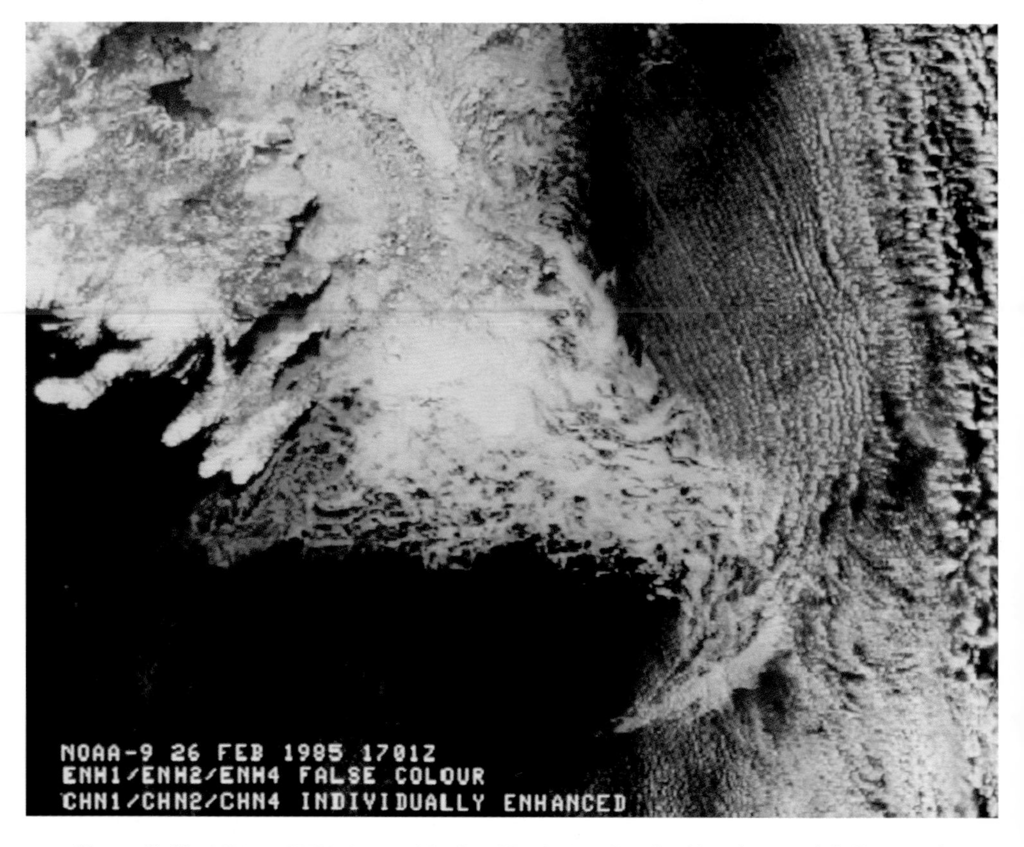

Figure 12.16 NOAA AVHRR image of the Grand Banks. Newfoundland is at the upper left. Because of the enhancement, the Labrador pack ice in the center of the image is clearly distinguishable from the clouds at the right, even where the clouds partially obscure the ice.

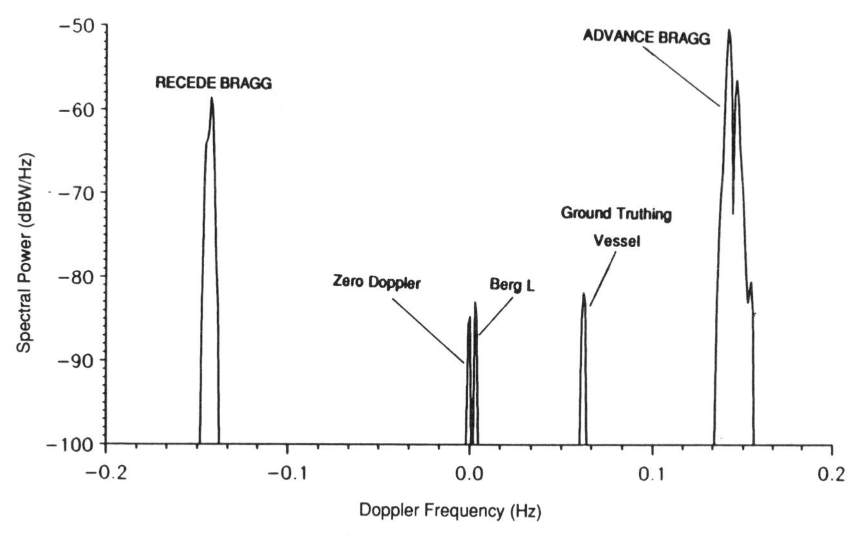

Figue 7.16 Absolute spectral power plot for Berg L detected at 1106 NDT on 23 May 1990 at range sample # 227 (282 km).

displaced symmetrically about zero. This effect is caused by the surface current having a radial component away from the radar. In this case, the radial component of the current is deduced to be 0.1 m/s.

The effect of this wind condition is to move the iceberg towards the radar, whereas the deduced surface current is moving it away. The general movement of icebergs is influenced by other factors in addition to surface winds and surface currents. The net effect in this case was to move the icebergs towards the radar and was generally observed for all icebergs tracked during the experiment.

The result of running the simulation software is presented in Fig. 7.17. This can be compared to the measured data previously presented in Fig. 7.16. To ease the comparison, the modeled spectra have been replotted, at the same absolute scale, in Fig. 7.18. Excellent agreement exists between the measured and predicted results, other than a small and constant difference in absolute levels of various peaks. This may be attributed to a small uncertainty in the estimation of system gain. In the model simulation, the ocean waves producing the Bragg scatter were assumed to be coherent over the entire dwell period. This assumption may not be valid for the long dwell period of 18 min used in this iceberg-detection exercise, as the ocean waves producing the Bragg scatter may lose coherency. The loss of coherency would result in a smearing of the Bragg peaks. Also, because the ocean surface current may not remain constant during the dwell period as well as over the complete resolution cell, a broadening of the Bragg peaks may occur. These effects have not been incorporated in the simulation software.

From the target visibility profile presented in Fig. 7.17, it can be seen that the maximum detection range for a medium-size iceberg such as Berg L is on

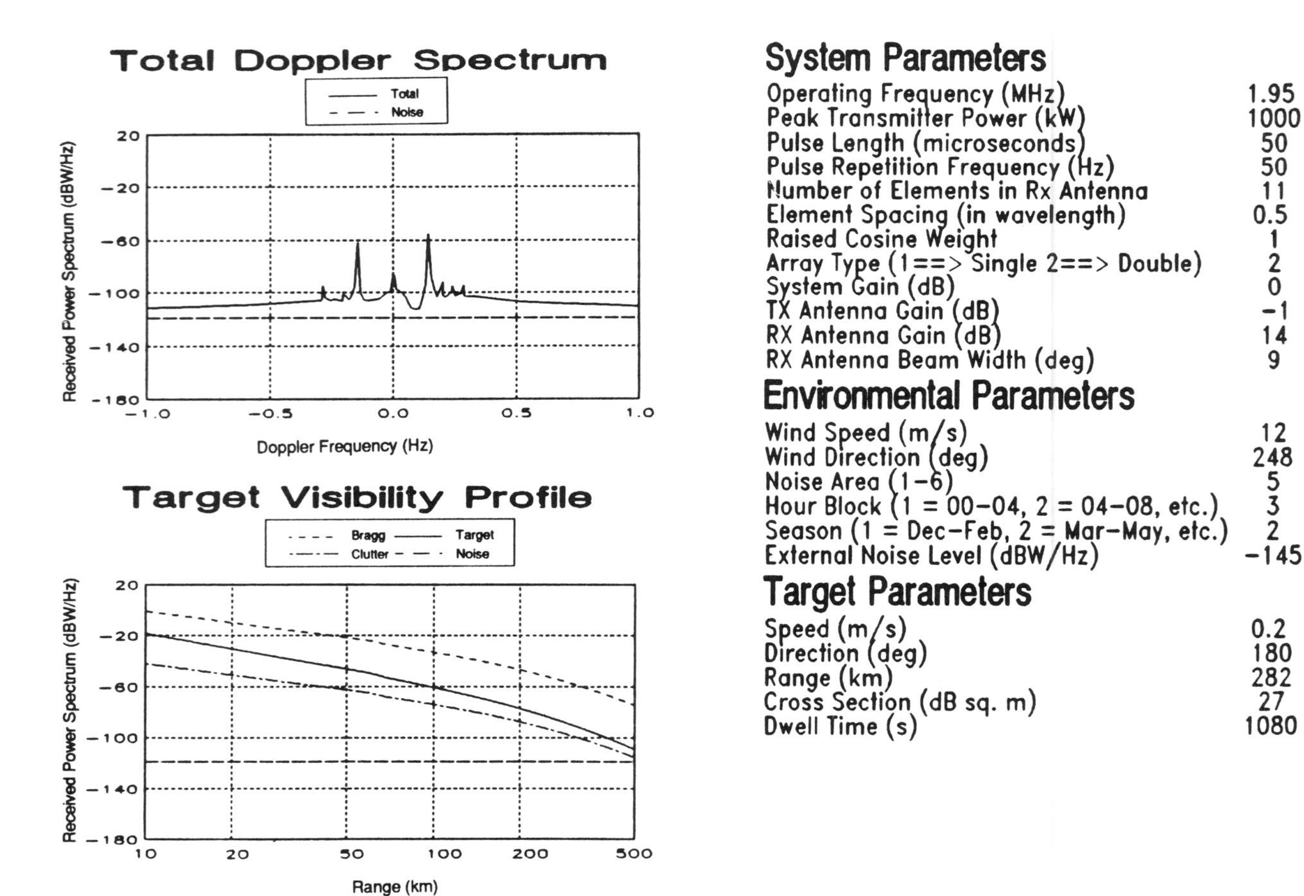

Figure 7.17 Detection performance as predicted by the simulation software for Berg L, located at a range of 282 km.

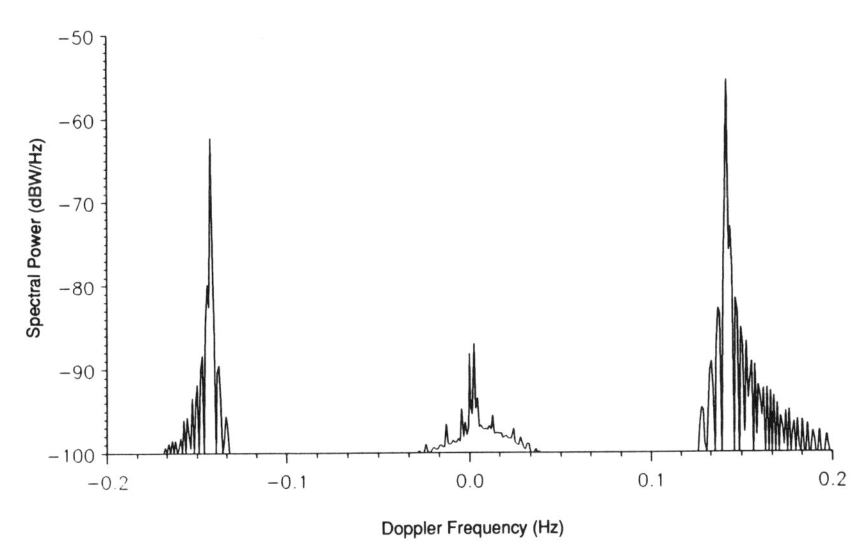

Figure 7.18 Simulated absolute spectral power plot for Berg L, located at a range of 282 km.

the order of 500 km. The detection result for this iceberg and its comparison with the software simulation has been reported previously by Srivastava and Ponsford [37].

7.3.4 Vessel and Aircraft Detection and Tracking

Trials have been conducted from the Cape Bonavista radar site that have demonstrated the ability of the OTH radar to reliably detect both airborne and surface targets, well beyond the horizon. In the 1989 trials for the Canadian Department of National Defence, the 47-m Fleet Auxiliary vessel *Bluethroat* was successfully detected and tracked over a 5-day period. Excellent agreement existed between measured and predicted radar performance (Ponsford and Srivastava [25]). Aircraft were also flown and detected (Ponsford and Srivastava [26]). During the trials, targets of opportunity, including ships and aircraft, were detected and tracked to 500 km and 300 km, respectively (Ponsford and Chan [23]).

7.4 CAPE RACE SYSTEM

7.4.1 Radar Description

This radar system, built and operated by Northern Radar Systems Limited, is located on a 2.5-km strip of land along the coast at Cape Race, Newfoundland. Figure 7.19 shows the radar location and the coverage area. The system has an effective range out to 200 km (110 nautical miles) of the sector shown,

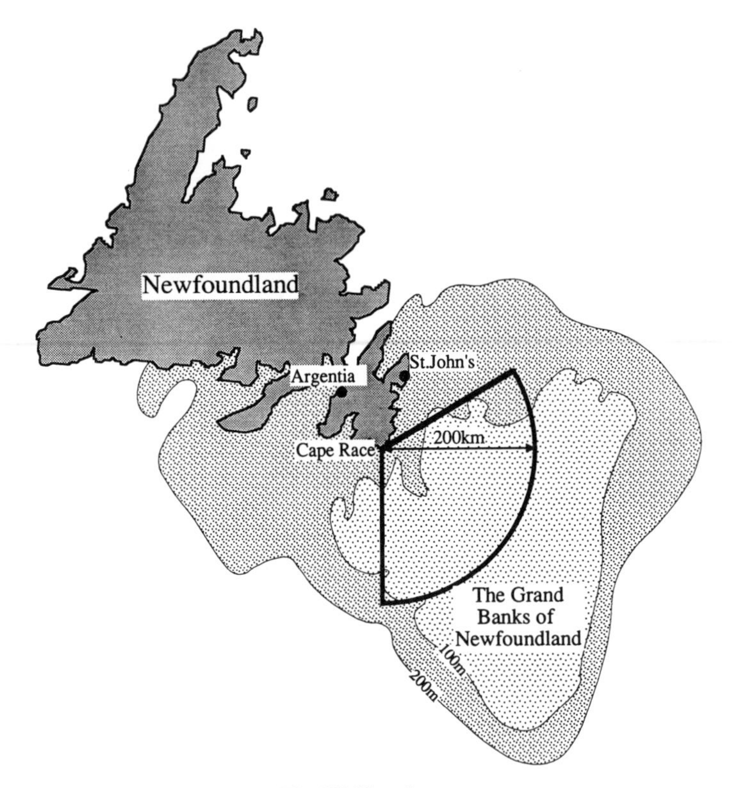

Figure 7.19 The NRSL radar coverage area.

while an upgraded radar (to become operational within two years) will have a range in excess of that shown in Fig. 7.19. The site consists of receive and transmit antennas, a building for receive, control, and accommodations, and several outbuildings. Figure 7.20 is an artist's drawing of the site, which is difficult to photograph effectively. The receive antenna is a 40-element–phase array, custom designed for the Cape Race system, and is nearly 1 km in length. The current transmit antenna is a standard, commercial log-periodic array. It will be replaced by a new antenna, custom designed to optimize ground-wave propagation for the upgraded system. The radar hardware system consists of analog receivers, transmitter, control section, and a digital processing section. The analog section provides basic discrimination and detection of the received radio frequency signals and effects a digital conversion; the digital section performs the detailed target discrimination, detection, and tracking functions. The radar hardware is capable of operation anywhere in the HF band (3 to 30 MHz); however, the existing version is restricted to a frequency range of about 5.4 to 10.0 MHz because of transmit antenna and filter limitations. The upgraded system will operate in a frequency band from 3 to 14 MHz in order to achieve the required range and to accommodate the various necessary remote

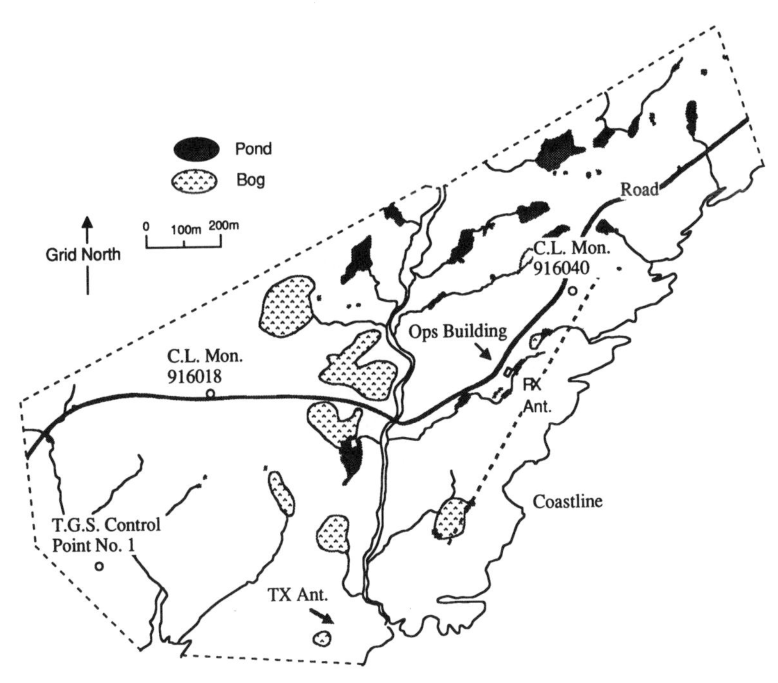

Figure 7.20 An artist's drawing of the radar site showing the transmit (TX) and receive (RX) antennas, the operations (Ops) building, and several survey points (numbered circles).

sensing tasks. The radar is capable of 400-m range resolution requiring 375-kHz bandwidth; however, this range resolution and bandwidth, as well as the operating frequency, may be varied to accommodate different target types and for frequency management purposes. In addition, specific frequencies within the radar bandwidth may be suppressed, again for frequency management purposes and interference suppression.

All the operations of the Cape Race radar system can be controlled remotely by a modem link over a standard telephone line which allows real-time monitoring of the radar data from the premises of any user. The salient features of the system may be summarized as follows:

1. Operation in the HF band to achieve OTH detection.
2. Employment of advanced "swept carrier" (frequency modulated, pulsed continuous wave) radar technology.
3. Continuous all-weather operation capability.
4. Automatic detection of targets or features with near-real time operation.
5. Cost-effective coverage of over 160,000 sq km of the Grand Banks, to the east and southeast of Newfoundland (with greater than 400-km range, 120° sector, for the upgraded system).

6. Detection and tracking of vessels and icebergs; measurement of ocean waves and surface currents.
7. Resolution capability of 400 m in range and 3° in azimuth.

7.4.2 Iceberg-Detection Experiments

During the spring of 1991, the Cape Race facility was used for iceberg surveillance. Two government agencies which provide operational ice information assisted by providing surface information. The AES Ice Centre, Ottawa, provided daily maps of ice coverage and iceberg sightings in the coastal areas. The International Ice Patrol (IIP) of the U.S. Coast Guard flew regular patrols during the iceberg season and provided Northern Radar with data on icebergs within the radar-coverage sector. Unfortunately, the priorities of the IIP excluded this area from their flight patterns during most of the 1990–1991 season, which limited the quality of surface verification data.

As an example of iceberg tracking, one segment of radar data collection was from 0802 NDT, 24 July 1991 to 1544 NDT, on 31 July 1991. There was no data collection on 29 July 1991. The only surface verification data available was from Ice Centre reports. Although their reports contained hundreds of iceberg positions, only a dozen or so were within the radar coverage areas. Of

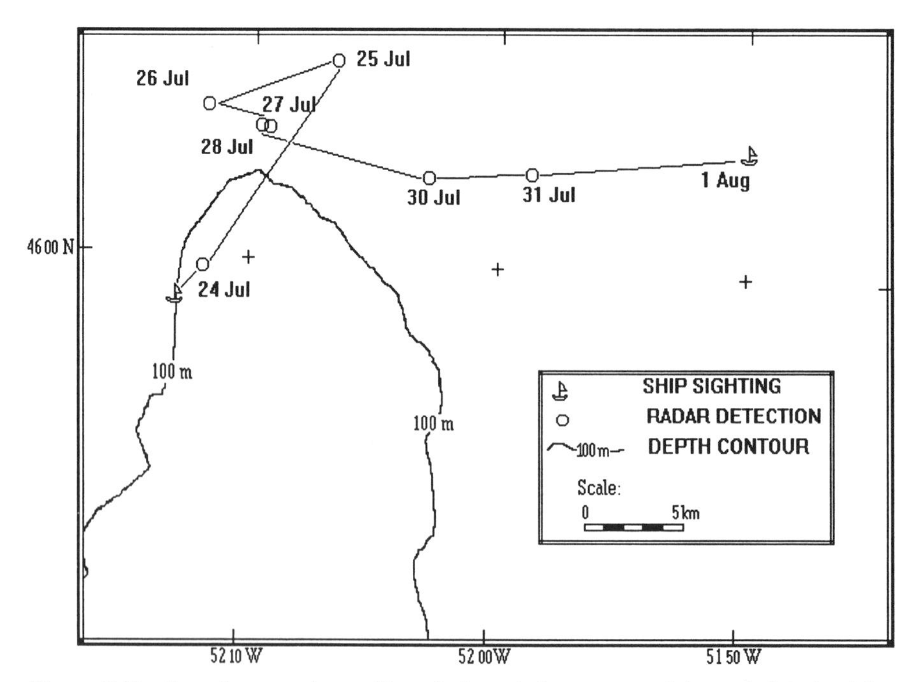

Figure 7.21 Ground-wave radar tracking of a large iceberg over an 8-day period during July–August 1991. Available ship sightings of the same berg are indicated also.

these, 10 were in close proximity to radar-detected targets. Figure 7.21 shows an iceberg track within the observation period. There were 14 iceberg detections, two for each day, and two ship observations, on 24 July and 1 August. The iceberg appeared to be interacting with the bottom in the period of 26 to 28 July because it slowed down considerably and appeared to backtrack from 27 to 28 July. After 30 July, the iceberg began to drift in an eastward direction which is in agreement with the bathymetry in the region, with regard to depth and probable current patterns. The first ship report on 24 July indicated, by both radar and visual contact, that the iceberg was large in size (height: 46 to 75 m; length: 121 to 200 m). The second ship report on 1 August was by radar contact only. Both ship sightings are in agreement with the OTH radar target track information.

7.4.3 Sea-Ice Detection

During early spring 1991, a significant amount of sea ice drifted through the radar coverage area and was detected by the OTH radar. This subsection presents an example of sea-ice cover, recorded by the OTH system, compared to sea-ice information provided by AES Ice Centre charts. Figure 7.22 follows the change in sea ice over a 2-day period in early April. The AES charts show a segment of 9+ sea ice (greater than 90% ice concentration), which diminished in size as it moved eastward. There is good correlation between the radar-generated maps and the AES ice charts.

7.4.4 Vessel Detection and Tracking

During the summer and autumn of 1990, five vessel-detection experiments were performed using the Northern Radar OTH system. These experiments were critical in the preliminary calibration and testing of the various components of the radar system. The most comprehensive vessel detection and tracking trial was carried out, in October 1991, under contract to the Department of National Defence. The Canadian Navy vessel, HMCS *Margaree*, was assisting in fishery patrol duties on the Grand Banks. Operational capabilities of the OTH system, for daytime as well as nighttime operations, were evaluated. Multiple-target tracking and detection were demonstrated with, in one instance, the observation of a vessel intercepting and escorting a second vessel. During this trial, targets were tracked consistently for extended periods, at ranges beyond 200 km.

Figure 7.23 demonstrates the ship-tracking capability of the Cape Race radar system.

7.4.5 Ocean Surface Parameters

In addition to hard-target detection and tracking such as depicted in Subsections 7.4.2 through 7.4.4, the radar is capable of providing environmental infor-

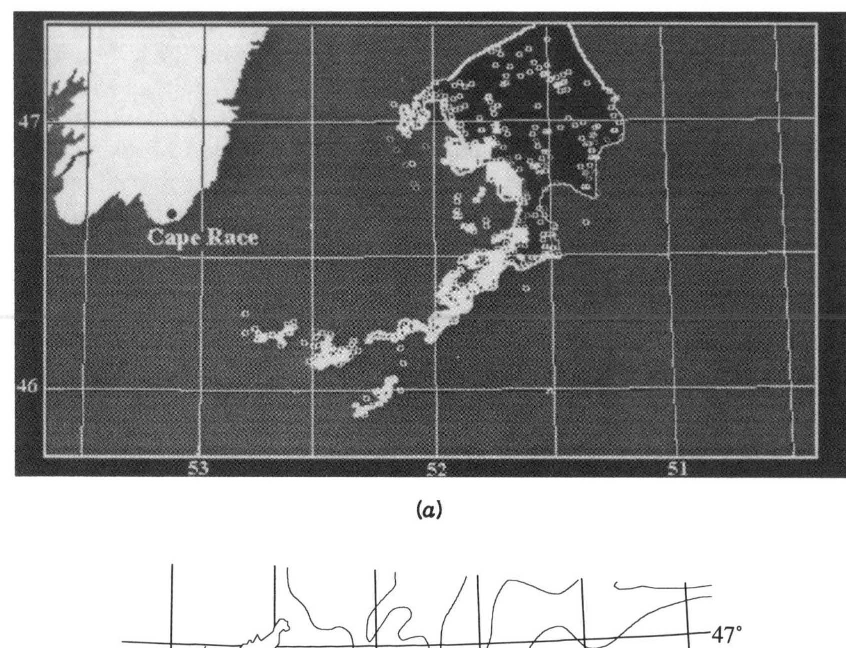

(*a*)

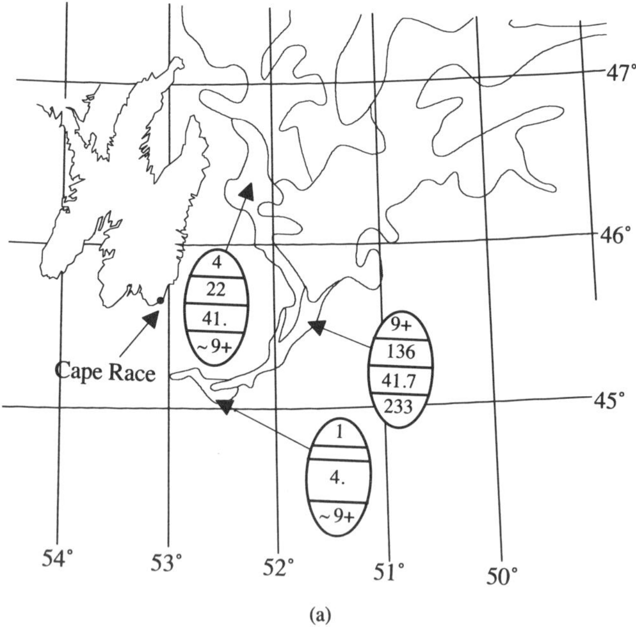

(a)

Figure 7.22 Detection of sea ice with ground-wave radar and comparison with AES ice maps for (a) April 3 and (b) April 4, 1991. The top numbers in the ovals indicate the percentage of ice cover, while the remaining numbers (reading downward by column) are AES Ice Centre classifications which give details of the cover (see AES Ice Centre report, 1991).

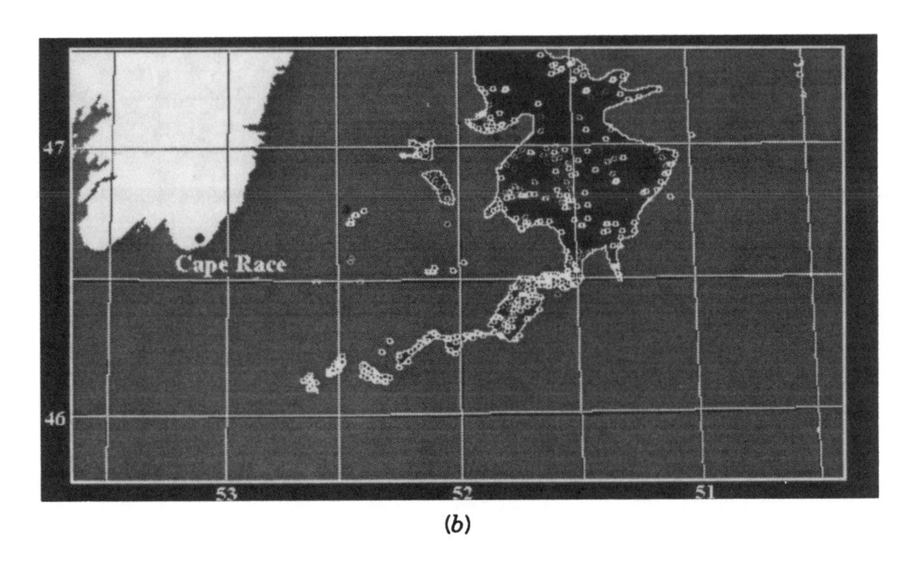

(b)

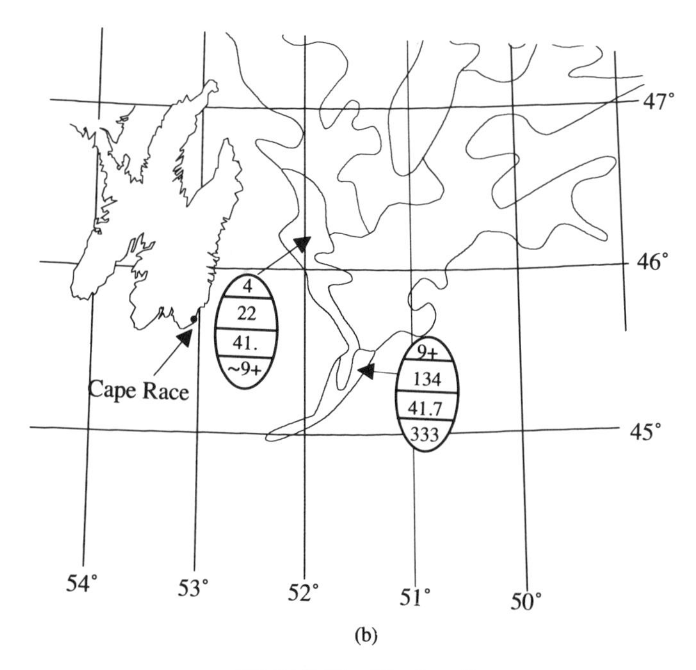

(b)

Figure 7.22 (*Continued*)

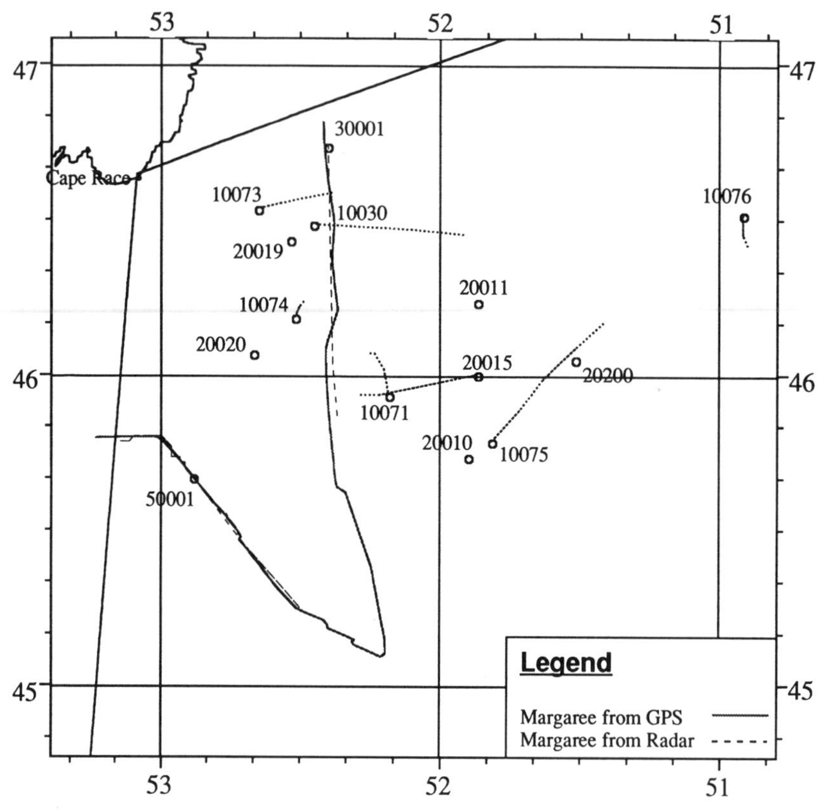

Figure 7.23 Multiple-target tracking with the NRSL radar during night-time operation. Included is the ground-truthed track of the HMCS *Margaree* between positions 50001 and 30001.

mation. This includes surface current maps, nondirectional wave spectra with their associated parameters (e.g., significant waveheight), and surface wind speeds. During low interference periods, the radar range extends from 25 km to 200 km for wave and wind measurements and to about 330 km for surface currents.

In the fall of 1992, the Northern Radar system, under contract to the Department of Fisheries and Oceans (DFO), was used to determine the radial surface current regime—i.e., the projection of the currents along the radar look direction—across the entire coverage area within a maximum range of 200 km. Surface current measurements were also obtained from the tracks of drifter buoys deployed by DFO. Figure 7.24 shows a scattergram of the radial currents provided by the radar and one of the drifters. Clearly, there is very good agreement between radar-produced values and those obtained from the drifter motion.

Figure 7.25 depicts wave information gathered in November 1991 during

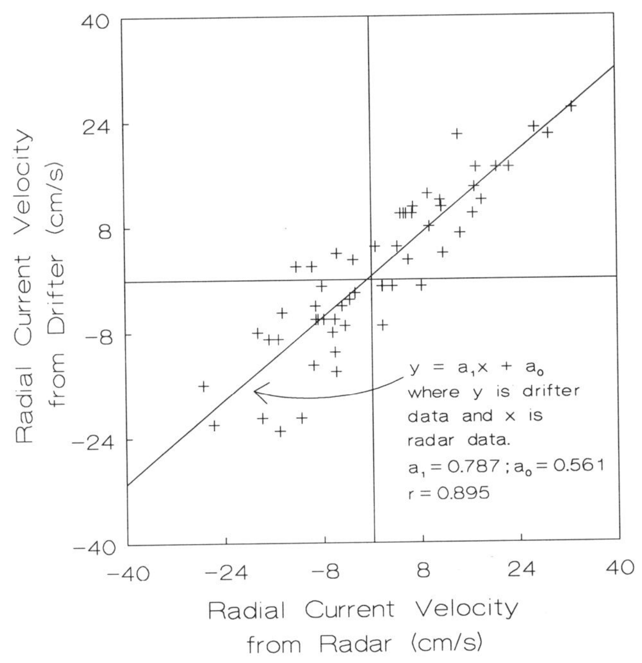

Figure 7.24 Scattergram of radial surface currents provided by the radar and an Accurate Surface Tracker drifter deployed by DFO in the fall of 1992. The radar values are accurate to within ± 4 cm/s.

ERS-1 SAR Wave Spectra Validation Experiment. The Northern Radar facility was one of several sensors used at that time in conjunction with the overflights of the European Remote Sensing Satellite (ERS-1). Part (a) of the figure shows radar-deduced significant waveheights—i.e., the average height of the 1/3 highest waves compared with waveheight contours provided by the Canadian Forces Meteorology and Oceanography Centre (METOC). In part (b), a nondirectional ocean wave spectrum obtained from the radar data is compared to that provided by a Waverider buoy. To get this result, radar spectra were averaged over 11 range cells, with the spectrum for each cell being obtained using a 256-point FFT. The buoy estimate represents an average of six FFT data segments over a 20-min period. Considering that the range is at the limit of the radar wave-processing capability, the comparison is very reasonable.

In addition to the above environmental data products, wind speeds are determined from the radar data. Table 7.4 contains comparisons with values obtained from AES during the wave experiment referred to above. The AES values are considered accurate to within about 4.6 km/h. The mean difference between the AES and radar wind speeds is 16.4%, with a standard deviation of 21.0%. Removal of the outlier on 23 November causes an improvement in these values to 8.3% and 7.2%, respectively.

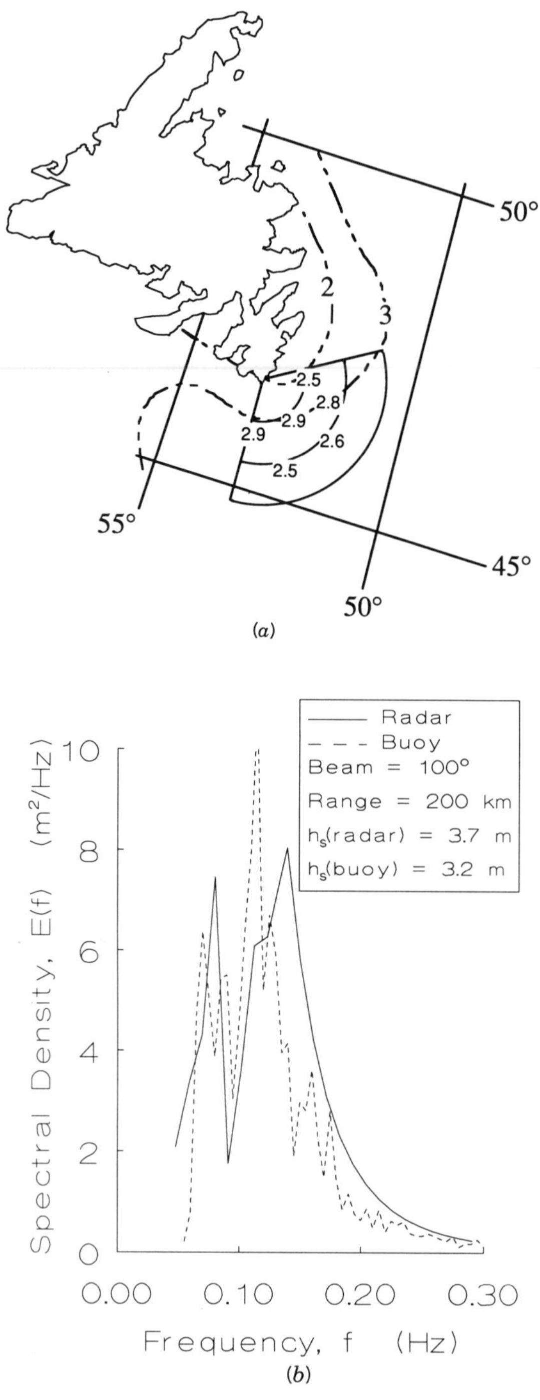

Figure 7.25 (a) Comparison between radar-deduced significant waveheights in meters (sector numbers) and METOC values (on contour lines). (b) A typical comparison between radar and Waverider buoy nondirectional wave spectra obtained during the *ERS-1* SAR Wave Spectra Validation Experiment, November 1991.

TABLE 7.4 Comparison of Radar and AES Wind Speed

		Wind Speed (km/h)	
Date	Time	AES	Radar
11/11/91	08:30	19	19
14/11/91	08:30	46	44
14/11/91	20:30	28	30
20/11/91	20:30	37	44
23/11/91	08:30	28	44
26/11/91	08:30	46	41

7.5 CONCLUSIONS

This chapter provides a summary of significant advancements made in research and development of OTH radars in Canada. It describes ground-wave propagation and the development of theoretical models for the cross sections of ocean surface, ice edge, and iceberg, which are key to understanding the principle of the OTH radar as well as to realizing its full potential. It also describes the two OTH radar systems developed in Newfoundland, namely, the Cape Bonavista radar and the Cape Race radar. A number of trials have been conducted at both radar facilities on ice-edge detection, iceberg detection and tracking, vessel detection and tracking, and on ocean surface parameters measurements. Radar data collected at these facilities have consistently shown excellent agreements with model predictions. These results have demonstrated that the OTH radar is well suited for the remote sensing of targets on or above the ocean surface and ocean surface parameters at long ranges, over wide areas, and for continued time.

7.6 FUTURE DEVELOPMENTS

Work led by Raytheon Canada Limited is currently in progress to develop an Integrated Coastal Surveillance System with coverage to beyond 500 km on the east coast of Canada. The system, as described by Ponsford et al. [28], consists of a number of remote OTH radar stations with track data brought back to a central location for fusion and "mosaicing" to form seamless coverage of the exclusive economic zone. Data obtained from other sensors will also be fused together to identify and to form a real-time display of targets on and above the ocean surface. The information will be made available to the user community to assist in such areas as marine safety, pollution prevention, search and rescue, fisheries and vessel traffic, etc. A prototype system is expected to be completed by the spring of 1995.

ACKNOWLEDGMENTS

The authors gratefully acknowledge the assistance of Dr. A. M. (Tony) Ponsford in providing the simulation software results and technical data regarding the operation and performance of the Cape Bonavista ground-wave OTH radar. The original radar system and simulation software were developed by NORDCO Limited under contract from the Department of National Defence, Canada. The radar has progressively evolved, and since 1990 Raytheon Canada Limited, under contract from National Defence, Canada, has been responsible for the development and operation of the system.

The Cape Bonavista iceberg detection trials were undertaken by NORDCO Limited and jointly funded by the Hibernia Management and Development Company (HMDC) and the Canada Newfoundland Offshore Development Fund. Ground truth data was provided by Atmospheric Environment Service. The authors acknowledge the permission of HMDC and Raytheon Canada Limited to publish the data.

The authors also gratefully acknowledge the support of Northern Radar Systems Limited in providing the experimental results from their Cape Race prototype radar. In particular the efforts of Mr. Eric Gill in preparation of the text and figures is very much appreciated.

The research, construction of, and development of the Cape Race system has been funded by various agencies, the Natural Sciences and Engineering Research Council (NSERC), the Department of National Defence, the Petroleum Industry, the Atlantic Canada Opportunities Agency, the Department of Fisheries and Oceans, and private investment. The Cape Race radar has been operational since 1990.

REFERENCES

[1] D. E. Barrick (1971) "Theory of HF and VHF propagation across the rough sea, Parts 1 and 2," *Radio Sci.*, **6** (5), 517–534.

[2] D. E. Barrick (1972) "First-order theory and analysis of MF/HF/VHF scatter from the sea," *IEEE Trans. Antennas and Propagat.*, **AP-20** (1) 1, 2–10.

[3] D. E. Barrick (1972) "Remote sensing of sea state by radar," in *Remote Sensing of the Troposphere*, V. E. Derr (Ed.), U.S. Govt. Printing, Washington, DC.

[4] D. E. Barrick, (1977) "The ocean waveheight nondirectional spectrum from inversion of the HF sea echo Doppler spectrum," *Remote Sensing Environ.*, **6**, 201–227.

[5] D. E. Barrick, J. M. Headrick, R. W. Bogle, and D. D. Crombie (1974) "Sea backscatter at HF: Interpretation and utilization of the echo," *Proc. IEEE*, **62**, 673–680.

[6] L. A. Berry and M. E. Christman (1966) "A fortran program for ground wave propagation over homogeneous spherical earth for dipole antennas," U.S. Dept. of Commerce, Boulder, CO, NBS Tech. Rep. 9178.

[7] H. Bremmer (1949) *Terrestrial Radio Waves*, Elsevier, New York.

[8] D. S. Bryant, A. M. Ponsford, and S. K. Srivastava (1988) "A computer package for the parameter optimization of groundwave radar," *Proc. Oceans'88 Conf.*, **2**, 485–490.

[9] D. D. Crombie (1955) "Doppler spectrum of sea echo at 13.56 Mc/s," *Nature*, **175**, 681–682.

[10] D. D. Crombie (1971) "Backscatter of HF radio waves from the sea," in *Electromagnetic Probing in Geophysics*, J. R. Wait (Ed.), Golem Press, Boulder, CO, 131–162.

[11] B. J. Dawe (1988) "Radio wave propagation over earth: Field calculations and an implementation of roughness effect," M.Eng. Thesis, Memorial Univ. of Newfoundland, St. John's, Canada.

[12] R. Donnelly (1983) "Electromagnetic scattering from a vertical half-space discontinuity: Operator decomposition approach," M.Eng. Thesis, Memorial Univ. of Newfoundland, St. John's, Canada.

[13] V. A. Fock (1965) *Electromagnetic Diffraction and Propagation Problems*, Pergamon Press, New York.

[14] K. Furutsu (1982) "A systematic theory of wave propagation over irregular terrain," *Radio Sci.*, **17** (5), 1037–1050.

[15] R. Howell, R. S. Srivastava, S. K. Srivastava, and J. Walsh (1987) "Remote sensing of sea state using ground wave radar," *Proc. Oceans'87 Conf.*, **3**, 877–882.

[16] B. Kinsman (1965) *Wind Waves*, Prentice-Hall, New Jersey.

[17] B. J. Lipa and D. E. Barrick (1982) "Analysis methods for narrow-beam high-frequency radar sea echo," U.S. Dept. of Commerce, Boulder, CO, NOAA Tech. Rep. ERL420-WPL56.

[18] A. E. Long and D. B. Trizna (1973) "Mapping of North Atlantic winds by HF radar sea backscatter interpretation," *IEEE Trans. Antennas and Propagat.*, **AP-21**, 680–685.

[19] M. S. Longuet-Higgins, D. E. Cartwright, and N. D. Smith (1963) "Observations of the directional spectrum of sea waves using the motions of a floating buoy," in *Ocean Wave Spectra*, Prentice-Hall, New Jersey, 111–136.

[20] G. Millington (1949) "Ground wave propagation over an inhomogeneous smooth earth," *Proc. Instn. Elect. Engrs.*, **96**, 53–70.

[21] J. E. Murray (1969) "The drift, deterioration, and distribution of icebergs in the North Atlantic ocean," *Proc. Ice Seminar, Pet. Soc. of Can. Inst. Min. Metall.*, **10**, 3–18.

[22] K. A. Norton (1937) "The propagation of radio waves over the surface of the earth and in the upper atmosphere, Part II," *Proc. IRE*, **25** (9), 1203–1236.

[23] A. M. Ponsford and H. C. Chan (1992) "Some experimental results from the Cape Bonavista surface wave radar facility," *Proc. 5th Annu. CRL/TRIO/ISTS Radar Workshop*, ISTS.

[24] A. M. Ponsford and S. K. Srivastava (1991) "Ground wave radar development at NORDCO Limited—Phase 1, Part 2," Final Rep. prepared for Atlantic Accord Offshore Development, Govt. of Newfoundland, St. John's, Canada, 230–239.

[25] A. M. Ponsford and S. K. Srivastava (1991) "A comparison between predicted

and measured vessel cross-sections for ground wave radars,'' in *Proc. 5th IEE Int. Conf.* HF Radio, **339**, 61–65.

[26] A. M. Ponsford and S. K. Srivastava (1991) ''Long range detection of icebergs, ships and aircraft using ground wave radar,'' in *Proc. Newfoundland Ocean Industries Assoc. (NOIA) Conf.*, 186–198.

[27] A. M. Ponsford, S. K. Srivastava, and T. N. R. Coyne (1989) ''Groundwave, over-the-horizon, radar development at NORDCO,'' *Proc. IGARSS'89*, **5**, 2953–2956.

[28] A. M. Ponsford, R. E. Moutray, and R. White (1993) ''Surveillance of the exclusive economic zone,'' in *Proc. Canadian Maritime Ind. Assoc. 45th Annu. Tech. Conf.*, section A7.

[29] Lord Rayleigh (J. W. Strutt) (1945) *The Theory of Sound*, Vol. II, Dover, New York.

[30] S. O. Rice (1951) ''Reflection of electromagnetic waves from slightly rough surfaces,'' in *Theory of Electromagnetic Waves*, M. Kline (Ed.), Interscience, New York, 351–378.

[31] R. K. Rosich and J. R. Wait (1977) ''A general perturbation solution for reflection from two-dimensional periodic surfaces,'' *Radio Sci.*, **12**, 719–729.

[32] J. Ryan (1983) ''The electromagnetic scattering from a vertical discontinuity with application to ice hazard detection: An operator expansion approach,'' M.Eng. Thesis, Memorial Univ. of Newfoundland, St. John's, Canada.

[33] J. Ryan and J. Walsh (1985) ''Electric dipole fields over a quarter space earth inhomogeneity and application to ice hazard detection,'' *Radio Sci.*, **20**, 1518–1528.

[34] A. Sommerfeld (1949) *Partial Differential Equations*, Academic Press, New York.

[35] S. K. Srivastava (1984) ''Scattering of high-frequency electromagnetic waves from an ocean surface: An alternative approach incorporating a dipole source,'' Ph.D. Thesis, Memorial Univ. of Newfoundland, St. John's, Canada.

[36] S. K. Srivastava and A. M. Ponsford (1989) ''Verification of additional sea clutter from a groundwave radar,'' in *Proc. 2nd Ann. TRIO/CRL Radar Symp.*, 2.1.1–2.1.7.

[37] S. K. Srivastava and A. M. Ponsford (1991) ''Long range detection of icebergs using ground wave radar,'' Proc. 1st (1991) Int. Offshore and Polar Eng. Conf., **II**, 471–476.

[38] S. K. Srivastava and J. Walsh (1985) ''An analysis of the second-order Doppler return from the ocean surface,'' *IEEE J. Oceanic Eng.* **OE-10** (4), 443–445.

[39] S. K. Srivastava and J. Walsh (1986) ''An analytical model for the HF backscattered Doppler spectrum for the ocean surface,'' *IEEE J. Oceanic Eng.*, **OE-11** (2), 293–295.

[40] S. K. Srivastava, A. M. Ponsford, and L. D. Cuff (1990) ''Groundwave radar iceberg detection exercises Spring 1990,'' Final Rep. prepared for Mobil Oil Canada Properties, St. John's, Canada.

[41] B. van der Pol and H. Bremmer (1938) ''The propagation of radio waves over a finitely conducting spherical earth,'' *Phil. Mag.*, **7** (25), 817–834.

[42] B. van der Pol and H. Bremmer (1939) ''Further note on the propagation of radio waves over a finitely conducting spherical earth,'' *Phil. Mag.*, **7** (27), 261–275.

[43] G. R. Valenzuela (1978) "Theories for the interaction of electromagnetic and oceanic waves—A review," *Boundary-Layer Meteorology*, **13**, 61–85.

[44] J. R. Wait (1964) "Electromagnetic surface waves," in *Advances in Radio Research*, J. A. Saxton (Ed.), Academic Press, New York, 157–217.

[45] J. R. Wait (1969) "Concerning the theory of scatter of HF radio ground waves from periodic sea waves," GPO, Washington, DC, ESSA Tech. Rep. ERL 145-OD-3.

[46] J. R. Wait (1970) *Electromagnetic Waves in Stratified Media*, Pergamon Press, New York.

[47] J. R. Wait (1971) "Perturbation analysis for reflection from two-dimensional periodic sea waves," *Radio Sci.*, **6**, 387–391.

[48] J. Walsh (1980) "On the theory of electromagnetic propagation across a rough surface and calculations in the VHF region," Memorial Univ. of Newfoundland, St. John's, Canada, OEIC Tech. Rep. N00232.

[49] J. Walsh (1983) "Propagation and scatter for mixed paths with discontinuities and application to remote sensing of sea ice with HF radar," Memorial Univ. of Newfoundland, St. John's, Canada, C-CORE Tech. Rep. 83-16.

[50] J. Walsh (1982) "A general theory of the interaction of electromagnetic waves with isotropic, horizontally layered media, and applications to propagation over sea ice," Memorial Univ. of Newfoundland, St. John's, Canada, C-CORE Tech. Rep. 82-9.

[51] J. Walsh and R. Donnelly (1987) "A new technique for studying propagation and scattering for mixed paths with discontinuities," *Proc. Roy. Soc. Lond.*, A 412, 125–167.

[52] J. Walsh and R. Donnelly (1987) "Consolidated approach to two-body electromagnetic scattering," *Phys. Rev. A*, **36** (9), 4474–4485.

[53] J. Walsh and S. K. Srivastava (1984) "Model development for feasibility studies of HF radars as ice hazard remote sensors," Memorial Univ. of Newfoundland, St. John's, Canada, OEIC Tech. Rep. N00397.

[54] J. Walsh and S. K. Srivastava (1987) "Rough surface propagation and scatter 1, general formulation and solution for periodic surfaces," *Radio Sci.*, **22** (2), 193–208.

[55] J. Walsh and S. K. Srivastava (1988) "Rough surface propagation and scatter with applications to ground wave remote sensing in an ocean environment," in *Proc. AGARD (NATO) EPP Specialists' Meeting on Scattering and Propagation in Random Media*, no. 419, 23.1–23.15.

[56] J. Walsh, B. J. Dawe, and S. K. Srivastava (1986) "Remote sensing of icebergs by ground-wave Doppler radar," *IEEE J. Oceanic Eng.*, **OE-11** (2), 276–284.

[57] L. Wyatt (1986) "The measurement of the ocean wave directional spectrum from HF radar Doppler spectra," *Radio Sci.*, **21** (3), 473–485.

8

SURFACE-BASED RADAR: NONCOHERENT

EDWARD O. LEWIS
Bayfield Institute
Fisheries and Oceans Canada
Burlington, Ontario, Canada

BRIAN W. CURRIE and SIMON HAYKIN
Communications Research Laboratory
McMaster University
Hamilton, Ontario, Canada

8.1 INTRODUCTION

8.1.1 Ice Information Requirements

Knowledge of both the presence and type of ice in their surrounding area is required by vessels and drilling platforms. In a heavy ice cover, as exists for most of the year in the Beaufort Sea, Arctic archipelago, or Northwest Passage, classification of ice type is of prime importance. In predominantly open-water conditions, such as off Newfoundland, or Lancaster Sound/Baffin Bay in summer, detection of sea ice and glacial-origin ice is crucial.

The operational demands for ice information are generally broken down into categories based on the temporal and geographical limits of the ice information requirement. The terms used are: strategic, tactical, and close-tactical.

Strategic information is required in the time frame of several days to several weeks or longer before a planned activity and covers large geographical areas, such as the complete route of a ship navigating the Northwest Passage, or the complete upstream ocean environment for a drilling rig. Strategic information may be available from historic sources, from satellite imagery, and from long-range patrol aircraft.

Remote Sensing of Sea Ice and Icebergs, Edited by Simon Haykin, Edward O. Lewis, R. Keith Raney, and James R. Rossiter.
ISBN 0-471-55494-4

Ice information required for tactical purposes spans the period from a few hours to 1 or 2 days ahead and covers a geographical area equivalent to the temporal requirement. In the case of a ship in transit, the geographical coverage is the distance the ship can expect to cover in the following 12 to 48 hours (250 to 1000 km). This information is used for route planning for the tactical period. For an offshore facility (drill rig, production platform), the geographical coverage for tactical purposes is that area from which ice may reach the facility within the tactical period. The ice information is used for icebreaking, ice deflecting, iceberg towing, or for preparing to move the facility itself. Tactical information may be provided by other vessels upstream or from overflying aircraft.

For the close-tactical case, the information is required continuously for the immediate period and up to 3 to 6 hours ahead (50 to 150 km), covering the contiguous vicinity of the operation. This information may be provided by aircraft, if the aircraft is immediately and continuously in the vicinity, or if the ice conditions are stable enough that little or no change takes place between the time the ice is sensed by the aircraft and the time that the operation encounters the ice. Airborne systems usually use side-looking airborne radar (SLAR) or synthetic aperture radar (SAR) as discussed in Chapters 10 and 11. Severe weather and realistic financial considerations make continuous airborne coverage impractical. Alternately, tactical ice information may be acquired from systems onboard the vessel or drilling platform. Surface-based radar is still the main and most effective of these surface-based sensors.

8.1.2 Chapter Outline

This chapter deals with the use of a surface-based radar for the classification and detection of ice. The radar systems considered in this chapter use noncoherent detection, extracting only the magnitude of the returned signal. Coherent detection is addressed in Chapter 9.

A surface-based radar, using a rotating real-aperture antenna operating near the earth's surface, and beaming essentially horizontally, has a number of related factors which must be taken into account in establishing the performance of such a system. These factors are discussed in Section 8.2.

Section 8.3 presents a brief history of the use of a surface-based radar (usually shipboard) for ice detection, summarizes the results of some early studies, and describes the shortcomings of current systems.

Recognizing the need for improved systems, the authors have undertaken a multi-year experimental research program with the aim of producing an improved system. The program is described in Section 8.4. Two distinct but complementary applications were recognized: classification of ice type in a full ice cover situation, and detection of ice pieces floating in open water. The results of the research program and the work of others, related to ice classification, are presented in Section 8.4.3. Section 8.4.4 presents the results for the ice-detection studies.

The experience and knowledge gained from the execution of the various experiments led to two parallel development programs. The first resulted in the development of a prototype ice-classification radar, described in Section 8.5, which was field tested in the Arctic. The second program, still ongoing, involves the development of an improved radar signal processor and display system, described in Section 8.6, and intended for use with the ice-classification radar system.

Section 8.7 concludes the chapter by summarizing the major points related to the use of a noncoherent surface-based radar and suggests research topic areas remaining to be studied.

8.2 FACTORS AFFECTING SURFACE-BASED RADAR

8.2.1 Basic Radar Equation

The generalized radar equation has been shown in Chapter 6, (6-22). It is repeated here in a form more customary for a single pulse for surface-based applications:

$$P_r = \frac{P_t G^2 \lambda^2 \sigma_t F^4}{(4\pi)^3 R^4 L} \tag{8-1}$$

where

P_r = power received
P_t = power transmitted
G = antenna gain
λ = radar wavelength
F = pattern propagation factor ($F = 1$ for free space)
L = system losses
σ_t = radar cross section of target
R = range to target

From this equation, it can be seen that for the free space case, the power received is inversely proportional to the fourth power of range, directly proportional to the square of the wavelength of the transmitted signal, and directly proportional to the radar cross section of the target.

The value of σ_t from (8-1) is a measure of the portion of the incident energy reflected back to the radar. By definition, the radar cross section is the strength of a radar target expressed in terms of the cross section of an ideal spherical reflector which would give the same returned signal strength. The radar cross section has units of area. The signals returned from a target vary with wavelength and polarization of the transmitted signal, target geometry, aspect, and reflectivity. The observed radar cross section for a given target can vary significantly. Any cross-section value given for a target is therefore usually the mean of the associated distribution of measured values.

In the application considered here, ice targets with flat faces (new pressure ridges, rafted faces, and icebergs with flat surfaces) may appear to have substantially larger radar cross sections than ice features with spherical or rounded surfaces (weathered pressure ridges, hummocked ice, etc.), even if the flat surfaces are at an angle to the radar beam. However, it is very difficult to predict the radar cross section of ice targets because of their random shape, height, and surface roughness. Empirical data remain the most important source of predicted ice-scattering strength.

8.2.2 Incidence and Grazing Angles

In describing the geometry of the interaction of a radar signal and a surface target, three terms are used: incidence angle, depression angle, and grazing angle. These are shown in Fig. 8.1. Depression angle is defined as the angle of radiation with respect to the local horizon at the transmitter. For surface-based work, the incidence angle is defined as the angle between the impinging radar beam and a perpendicular to the surface at the target. The grazing angle is defined as the angle with respect to the local horizon at the target and is the complement of the incidence angle.

For surface-based radar applications, an even more important definition is the term "local" grazing angle, which accounts for the effect of the local surface topography. As indicated earlier, the grazing angle is the angle the radar wave makes with respect to the horizon at the target. As shown in Fig. 8.2, for a smooth surface, the grazing angle for a surface-based radar is small. For a rough surface, the grazing angle is also small; however, the "local" grazing angle between the tangent to the surface at the target and the radar wave can be considerably larger. This larger "local" grazing angle results in significantly more energy being returned to the radar.

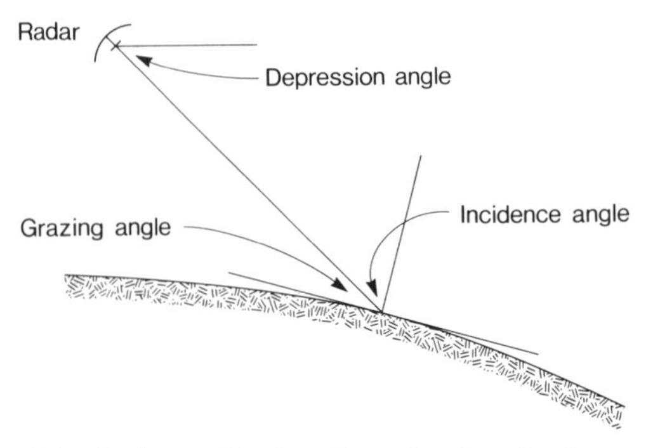

Figure 8.1 Angles used to describe surface-based radar operation.

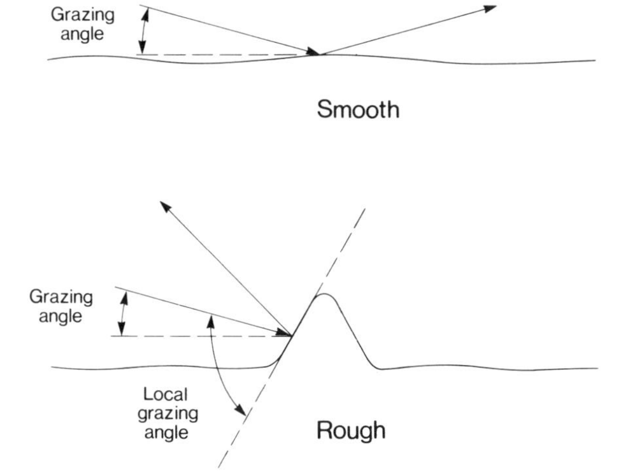

Figure 8.2 Designation of local grazing angle.

8.2.3 Radar Resolution

Range. The minimum separation required between two targets for the radar to resolve them individually is termed the radar resolution. In range, the returned signals from two targets must be separated in time, for a simple pulse system, by at least the pulse length (τ_0 in seconds) of the transmitted signal, that is,

$$\text{Range resolution (temporal)} = \tau_0 \tag{8-2}$$

For a pulse-compressed system, the targets must be separated by the effective compressed pulse length in order to be individually resolvable.

In terms of the physical separation in range that two targets must have, this resolution is expressed as

$$\text{Range resolution (spatial)} = \frac{c\tau_0}{2} \tag{8-3}$$

where c = speed of light. For surface radar, the range resolution is independent of the range of the target from the radar and is influenced only by the effective length of the pulse. For a 1-μsec pulse, the spatial range resolution is 150.0 m; for a 250-nsec pulse, the resolution is 37.5 m.

Azimuth. As in the case of range resolution, the azimuthal resolution can be expressed in two ways. The angular azimuthal resolution is uniquely specified by the horizontal beamwidth (β_h) of the radar antenna in radians. Thus,

$$\text{Azimuthal resolution (angular)} = \beta_h \tag{8-4}$$

The actual spatial azimuthal resolution (the distance by which two targets must be separated in bearing to be resolvable individually) is not only a function of the horizontal beamwidth of the antenna, but also is directly proportional to the range to the target:

$$\text{Azimuthal resolution (spatial)} = R\beta_h \tag{8-5}$$

Hence spatial azimuthal resolution degrades as range to the target increases. Figure 8.3 shows the azimuthal resolution for several horizontal beamwidths versus range.

8.2.4 Area-Clutter Radar Equation

In the case where the radar illuminates the surface of the earth or the ocean, not only is the target illuminated, but also the surface itself. In this case, the power returned is from both the target of interest and from the surface.

In the situation where the target is large compared to the radar resolution cell, or is made up of a large number of individual scatterers, σ_t in (8-1) must be modified to take into account the distributed nature of the target. Signal

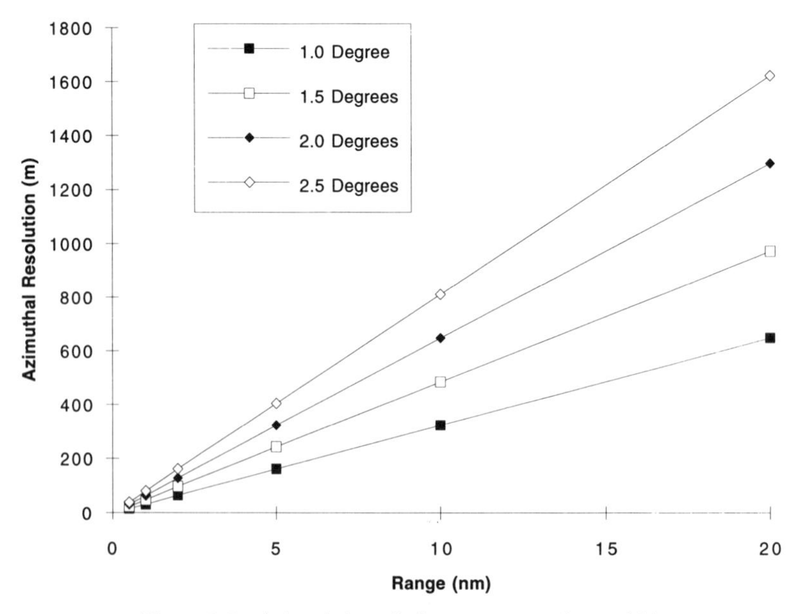

Figure 8.3 Azimuthal resolution vs. antenna beamwidth.

returns from a distributed target of this nature are often referred to as clutter, since they mask point target echoes. In some instances the clutter is actually the target of interest, as in the classification of ice features. In other instances (e.g., the detection of a growler in the sea), the clutter from the sea is an impediment to detection. Hence, the role of clutter varies with the application of interest.

The surface area illuminated at a given instant, referred to as the resolution cell, is determined by the product of the range and spatial resolutions and takes into account the grazing angle, ψ, as shown in Fig. 8.4. The factor of 2 accounts for the two-way signal path. The area illuminated then becomes

$$A_c = \frac{c\tau_0 R\beta_h \sec \psi}{2} \tag{8-6}$$

where ψ is the grazing angle. The area illuminated (resolution cell) as a function of range is shown in Table 8.1.

The average clutter cross section, σ_c^0, is equal to the areal extent of the radar illumination (8-6) times the normalized radar cross section, σ^0, of the illuminated area:

$$\sigma_c^0 = A_c\sigma^0 \tag{8-7}$$

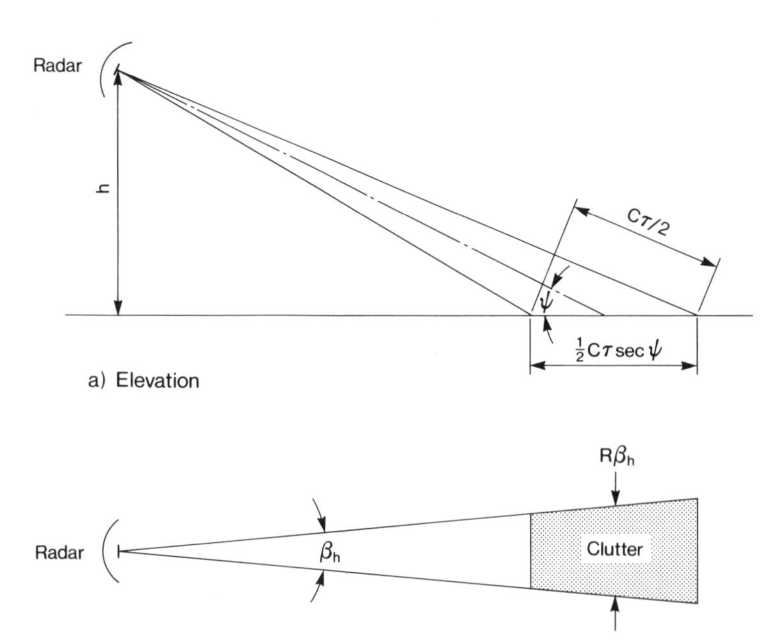

Figure 8.4 Geometry of radar clutter cell.

TABLE 8.1 Grazing Angle and Resolution Cell Size versus Range for a Typical Marine Radar. Resolution Cell for a 1-μsec Pulse with 1.9° Horizontal Beamwidth, at an Antenna Height of 35 m.

Range (nm)	Grazing Angle (°)	Resolution Cell (m^2)
0.5	2.30	4,605
1.0	1.10	9,210
2.0	0.57	18,420
3.0	0.38	27,630
4.0	0.28	36,841
5.0	0.23	46,051
6.0	0.19	55,261

where the normalized radar cross section is usually defined with reference to 1 m^2.

For the low-grazing angles associated with surface-based radar, sec ψ in (8-6) is approximately equal to 1, such that the equation for σ_c^0 reduces to

$$\sigma_c^0 = \frac{c\tau_0 R\beta_h\sigma^0}{2} \tag{8-8}$$

Using the expression for σ_c^0 in place of σ_t, (8-1) becomes

$$P_r = \frac{P_t G^2\lambda^2 F^4 c\tau_0 R\beta_h\sigma^0}{2(4\pi)^3 R^4 L} \tag{8-9}$$

Rearranging this equation to solve for σ^0 yields

$$\sigma^0 = \frac{2P_r(4\pi)^3 R^3 L}{P_t G^2\lambda^2 F^4 c\tau_0\beta_h} \tag{8-10}$$

Comparison of (8-1) and (8-10) shows that the typical R^4 relationship between σ_t and range [as shown in (8-1) for a point target] implies an R^3 relationship for a distributed target filling the antenna pattern.

8.2.5 Radar Horizon

For a ''normally'' refractive atmosphere as discussed in Section 8.2.9 (one in which the refractive-index gradient decreases linearly with altitude), the radar horizon (in km) is defined by

TABLE 8.2 Range-to-Radar Horizon for Targets of Different Heights at Various Antenna Heights

Antenna Height (m)	Range-to-Radar Horizon (km)			
	10-m High Target	5-m High Target	1-m High Target	0-m High (Surface)
5	22.2	18.4	13.3	9.2
10	26.0	22.2	17.1	13.0
20	31.4	27.6	22.5	18.4
50	42.1	38.3	33.2	29.1
100	54.1	50.3	45.2	41.1
200	71.1	67.3	62.2	58.1

$$R_h = 4.11(\sqrt{h_a} + \sqrt{h_t}) \tag{8-11}$$

where h_a is the height of the antenna (m) and h_t is the height of the target (m).

Table 8.2 shows distance to the radar horizon (in km) for various antenna heights above the surface, and for target heights of 0 (surface), 1, 5, and 10 m.

8.2.6 Skin Depth of Ice

The penetration of the radar wave into ice is dictated by the dielectric characteristics of the ice, as discussed in Chapter 2, Section 2.4.1. The loss tangent determines how quickly the transmitted signal will attenuate as it propagates into the ice. From Evans [12], the field strength of the electromagnetic field decreases with exponential attenuation, α:

$$\alpha = \frac{2\pi}{\lambda}\left\{\frac{\epsilon_r'}{2}\left(\sqrt{1 + \tan^2 \delta} - 1\right)\right\}^{1/2} \tag{8-12}$$

where ϵ_r' is the dielectric constant and $\tan \delta$ is the loss tangent.

The skin depth, $1/\alpha$ (at which the incident electromagnetic energy has been reduced to $1/e$) defines the depth of penetration of an electromagnetic wave into ice. Figure 8.5 shows the skin depth as a function of loss tangent and frequency. Typical ranges of loss tangent for first-year and old ice and glacier ice are indicated. From (8-12) and Fig. 8.5, it can be seen that the higher losses in first-year ice significantly reduce the penetration into the ice, whereas the lower losses in old ice will allow significant penetration. This effect is noted by Weeks and Ackley [65] who define young and first-year as a "high-loss" and old ice as a "low-loss" dielectric material.

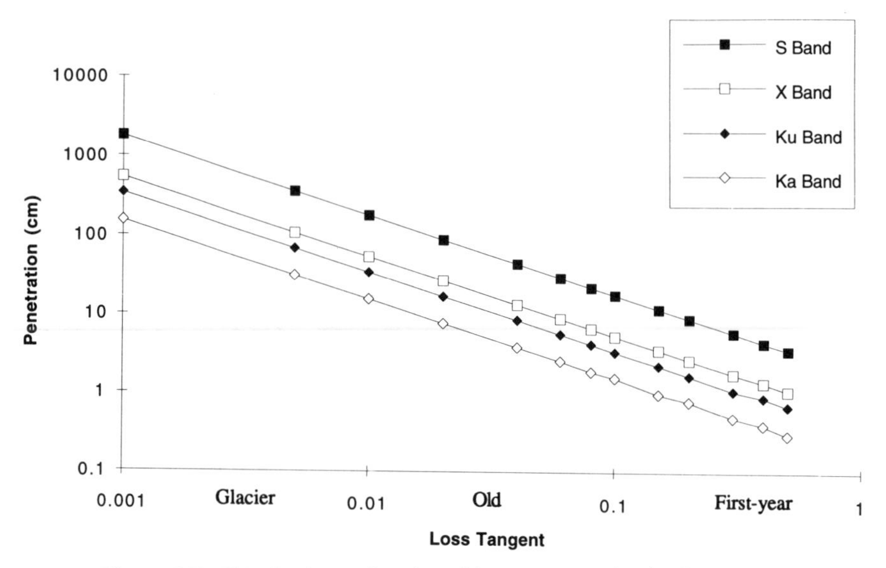

Figure 8.5 Skin depth as a function of loss tangent and radar frequency.

8.2.7 Rayleigh Roughness

The energy incident on a surface will be reflected in one of two distinct fashions. If the surface is "smooth," the energy will be reflected specularly in a single direction away from the radar, but if the surface is "rough," the energy will be diffused and scattered in many directions. This effect can be used to define rough and smooth surfaces. A surface that reflects the incident energy specularly will be considered smooth, whereas a surface that scatters the energy in various directions will be considered rough. From the perspective of the incident energy, the smoothness or roughness of a surface is dependent on three variables: the wavelength of the incident energy, the angle of incidence, and the physical relief of the surface.

The Rayleigh roughness criterion relates these variables and provides a mathematical determination of surface roughness. If the surface is considered as shown in Fig. 8.6, the path difference between a ray striking the top of the

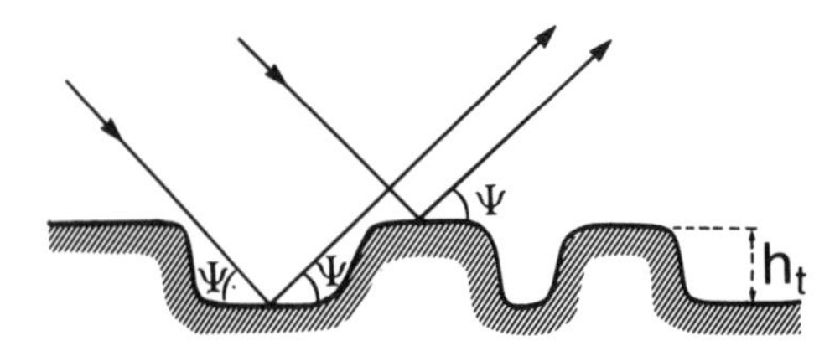

Figure 8.6 Derivation of the Rayleigh roughness criterion (reprinted with permission from Beckmann and Spizzichino [3]).

surface and another striking the bottom of the surface can be defined as

$$\Delta r = 2h_t \sin \psi \tag{8-13}$$

where h_t is the height of the surface feature and ψ is the grazing angle. This path difference can be converted to a phase difference between the two rays as

$$\Delta\phi = \frac{2\pi}{\lambda}\,\Delta r = \frac{4\pi h}{\lambda}\sin \psi \tag{8-14}$$

If the surface is smooth, the two rays will be in phase. However, as the phase difference increases, the two rays will begin to interfere until at $\Delta\phi = \pi$ they will be out of phase and will cancel each other, and no energy will be reflected in the forward direction. By the law of conservation of energy, however, the energy has not been lost: consequently, it must have been scattered in other directions. It can, therefore, be said that on a surface where the phase difference between the ray striking the top of the surface and the one striking the bottom is π, the surface is definitely rough. Conversely, where the difference is zero, the surface is smooth.

A division between rough and smooth can be established arbitrarily by choosing a point halfway between the two points, for example $\pi/2$, which then yields the normal Rayleigh criterion where a surface is considered smooth for

$$h < \frac{\lambda}{8 \sin \psi} \tag{8-15}$$

Table 8.3 shows the relationship between the grazing angle, the frequency (or wavelength), and the surface height (in cm) for a surface which is considered rough. See Table 8.1 for the relationship between grazing angle, antenna height,

TABLE 8.3 Rayleigh Roughness Criterion for Radars of Different Wavelength

	Roughness (cm)			
Grazing Angle (deg)	*S* Band 10.0 cm	*X* Band 3.0 cm	K_u Band 1.9 cm	K_a Band 0.86 cm
5.00	14.0	4.3	2.7	1.2
2.00	36.0	11.0	7.0	3.1
1.00	71.0	21.0	14.0	6.0
0.75	95.0	29.0	18.0	7.0
0.50	143.0	43.0	27.0	12.0
0.25	286.0	86.0	54.0	25.0

and range. Note that for a 1° grazing angle, at *S*-band the surface is not rough until the surface features vary by 71 cm, but at K_a-band the surface only need vary by 6 cm to be considered rough. The practical implications of this roughness criterion are shown later in this chapter.

8.2.8 Multipath Interference

The pattern propagation factor, F, of (8-1) takes into account nonfree-space propagation conditions caused by multipath interference.

As shown in Fig. 8.7, multipath interference (also known as the Lloyd's Mirror Effect) occurs when a direct signal transmitted from a radar near the earth's surface is interfered with by a signal reflected off the surface. This is a phenomena well known over the sea surface and will be shown to occur over a smooth ice surface as well. Multipath interference results in destructive or constructive interference, depending on the phase between the direct and indirect signals.

The radar equation (8-1) with $F = 1$ is for a signal propagating in free space. However, for propagation over a plane reflecting surface (earth, sea, ice, etc.), the propagation factor F has to be modified to take into account multipath interference. This modified radar equation (Skolnik [58])

$$P_r = \frac{P_t G^2 \lambda^2 \sigma_t}{(4\pi)^3 R^4 L} \, 16 \sin^4 \left(\frac{2\pi h_a h_t}{\lambda R} \right) \tag{8-16}$$

shows that the returned power is modulated by the term

$$F^4 = 16 \sin^4 \left(\frac{2\pi h_a h_t}{\lambda R} \right) \tag{8-17}$$

which cycles from a maximum of 16 (in decibels, 12 dB) to a minimum of zero.

For small angles where $\sin x \cong x$, (8-16) can be reduced to

$$P_r = \frac{4\pi P_t G^2 \sigma_t (h_a h_t)^4}{\lambda^2 R^8 L} \tag{8-18}$$

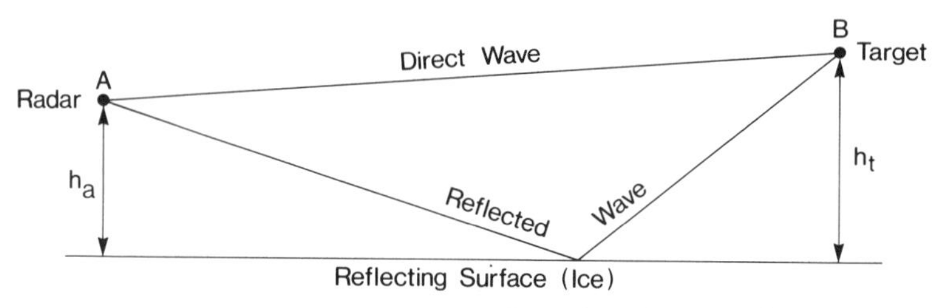

Figure 8.7 Diagrammatic representation of multipath interference.

For a point target, the power returned falls off as $1/R^4$ when the effect of multipath interference is ignored. However, when the modulating term as shown in (8-17) is added, the return power drops off at the $1/R^4$ rate only out to what is known as the transition range, defined as

$$R_T = \frac{4h_a h_t}{\lambda} \tag{8-19}$$

after which the average power of a single pulse drops off as $1/R^8$.

Within the interference region, that is the area from the transmitter out to the transition range, the return power is modulated by (8-17) such that the power experiences successive peaks and nulls, as shown in Fig. 8.8. The locations of the peaks and nulls are determined by the height of the antenna and of the target, and by the wavelength of the transmitted signal.

The effect of multipath interference at a particular range can be visualized by plotting the modulating lobes as a function of height of the target above the surface. Figure 8.9 is a plot of the first three interference lobes for an *X*-band (3-cm) radar operating at a height of 30 m and a range of 5600 m. It can be seen that lobe maxima occur at 1.4, 4.2, and 7.0 m, and lobe minima occur at 2.8 and 5.6 m. Equations (8-16) and (8-18) are based on the assumption of a perfectly reflecting surface. As this is seldom the case, the theoretical maxima will not be realized and the nulls (minima) will be partially filled in.

Equation (8-18) shows that by increasing the height of the antenna and

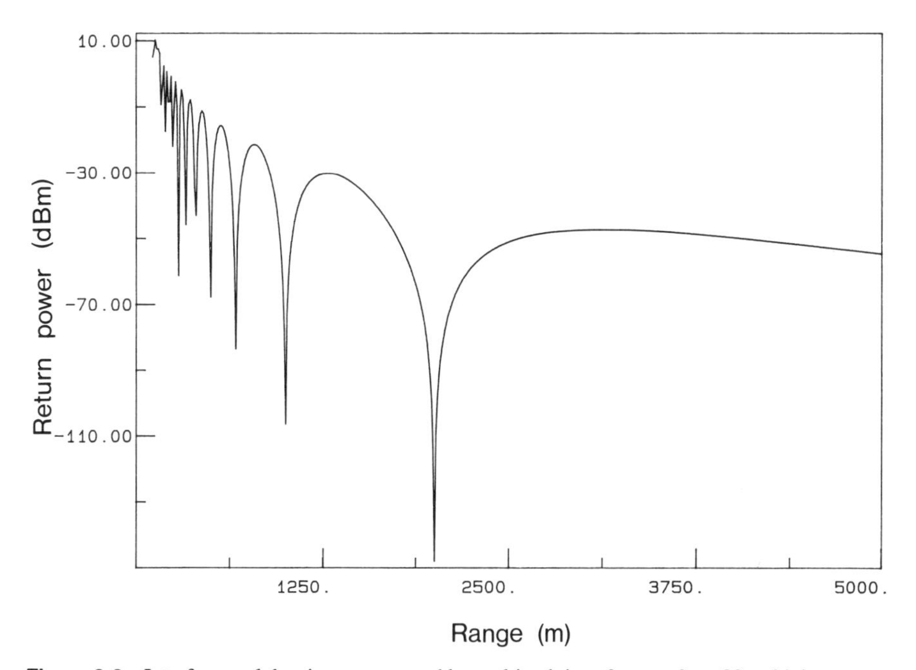

Figure 8.8 Interference lobes in range caused by multipath interference for a 30-m high antenna.

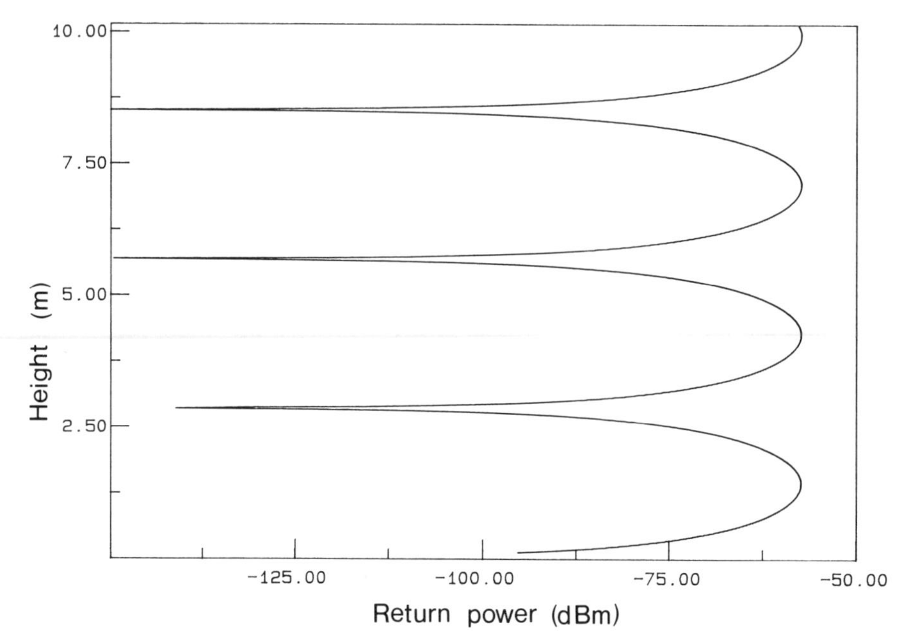

Figure 8.9 Multipath interference lobes versus height at range of 5600 m, for *X*-band radar with antenna height of 30 m.

decreasing the wavelength of the transmitted signal (increasing its frequency), the power returned from a given low-lying target can be increased. These effects are illustrated in Fig. 8.10(a) and (b). Results of experimental measurements of multipath over an ice cover, showing excellent correlation with theoretical predictions, can be found in Lewis et al. [36]. These results highlight the extreme care which must be taken when deploying radar reflectors of known cross section for in-field calibration of a radar system. Depending upon the location of the reflector and the smoothness of the intervening surface, multipath may alter the radar cross section significantly from the value which is expected.

8.2.9 Anomalous Propagation Conditions

In addition to problems caused by multipath interference, radar returns from targets near the surface can be significantly affected by varying atmospheric conditions.

Variations in the refractive index of the atmosphere can have significant effects on the propagation of radar signals near the earth's surface. The refractive index (n) is defined as the ratio of the velocity of an electromagnetic wave in vacuum to its velocity in the atmosphere. For a standard atmosphere, the refractive index decreases exponentially with increasing height, with the result that at low altitudes it decreases nearly linearly with height. The refractive

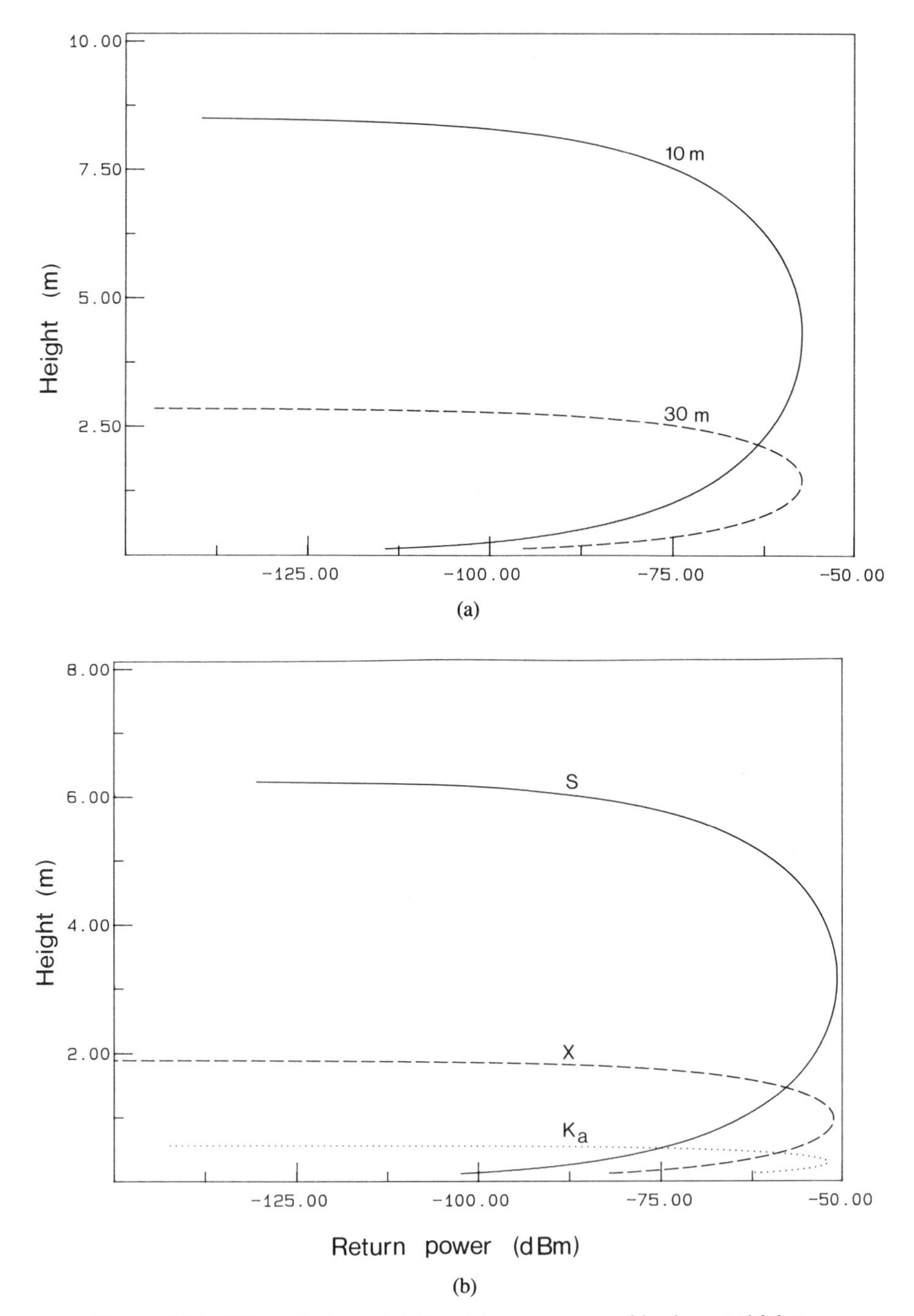

Figure 8.10 Effect of antenna height and frequency on multipath-created lobes.

TABLE 8.4 Refractive Index for Selected Propagation Conditions

Condition	N_r Gradient (N/km)
Trapping	$dN_r/dh \leq -157$
Superrefractive	$-157 < dN_r/dh \leq -79$
Standard	$-79 \leq dN_r/dh \leq 0$
Subrefractive	$dN_r/dh > 0$

index varies from an earth's surface value of approximately 1.0003 to 1 at high altitude.

A more common method of expressing the refractive nature of the atmosphere is in N-units (normally referred to as refractivity) where

$$N_r = (n - 1) \times 10^6 \tag{8-20}$$

which yields a value of approximately 300 at the earth's surface.

With a normally refractive atmosphere (where N_r decreases at 39 N_r-units/km near the surface), radio waves refract towards the earth's surface but with a radius less than that of the earth. Standard conditions are considered to exist when the N_r gradient is between -79 and 0 N_r-units/km.

Anomalous propagation conditions exist when there are abnormal vertical distributions in the refractive index. If the N_r gradient is between -156 and -79, the radio waves will be bent downward at a rate greater than standard but less than the earth's curvature. This is known as *superrefraction* and results in extended propagation range. If the gradient is less than -157 N_r-units/km, the radio waves are refracted downward with a radius exceeding that of the earth, and the waves can become trapped in a channel or duct. This ducting phenomenon has the effect usually of greatly extending the range of propagation. If the N_r gradient is greater than zero, *subrefraction* occurs and the radio waves bend less, thus reducing propagation range. The ranges of N_r values and the associated propagation conditions are shown in Table 8.4 (Hitney et al. [21]).

8.2.10 Target Detection

The ability of a radar system to detect a target of interest is influenced by the ratio of the average power of desired signal to the average power of interfering signal received. The interfering signals take one of two forms: noise and clutter. Noise is composed of thermal noise associated with the receiver, as well as the many sources of man-made noise. Clutter, in general, is an unwanted signal returned from the area illuminated by the radar signal.

The next subsection presents expressions to determine the average signal-to-noise ratio and the signal-to-clutter ratio. Detection performance depends both upon these ratios of average power and upon the statistics of the target and interference powers.

Noise-Limited Case. In the noise-limited case, the objective is to choose radar parameters such that sufficient signal is received from the target to exceed system noise. Ice-related radar situations which are noise-limited include the detection of small pieces of ice in low sea-states and targets at longer ranges.

The receiver noise power is given by

$$P_n = F_n KTB_n \tag{8-21}$$

where

K = Boltzmann's Constant
= 1.38×10^{-23} W/(Hz°K)
T = receiver temperature (°K)
B_n = receiver noise bandwidth (Hz)
F_n = receiver noise figure

The term KTB_n is the thermal noise power generated at the receiver input. For a typical marine radar receiver with a bandwidth of 10 MHz, the thermal noise power is about −104 dBm. Thus for a practical receiver with a noise figure of 10 dB, the minimum detectable signal level is −94 dBm. This power level is termed the minimum detectable signal (MDS) of the radar, and determines the minimum target power required for detection. Using (8-1) for the target power and (8-21) for the noise power, the expression for the signal-to-noise ratio (S/N) is

$$S/N = \frac{P_{\text{target}}}{P_{\text{noise}}} = \frac{P_t G^2 \lambda^2 \sigma_t F^4}{(4\pi)^3 R^4 LKTB_n F_n} \tag{8-22}$$

This equation describes the mean power ratio. The ratio, hence target detection, can be improved by changing a number of radar parameters, including increasing the transmitter power, P_t, or the antenna gain, G.

The signals are characterized statistically by probability density functions (pdf). The distribution of noise at the front end of a radar receiver is typically Gaussian. However, after rectification by a video detector, the noise amplitude becomes Rayleigh distributed. Figure 8.11 shows plots of typical pdfs for noise and signal-plus-noise. Target detection is accomplished by setting a threshold between the means of the pdfs, and a target is declared whenever the threshold is exceeded.

The integral of the area under the signal-plus-noise curve above the threshold is defined as the probability of detection (P_d), whereas the integral of the area under the noise curve above the threshold is termed the probability of false alarm (P_{fa}). All noise returns above the threshold will be declared incorrectly as targets. Likewise, the integral of the area under the signal-plus-noise curve below the threshold presents the probability of missed targets (P_m) as a target below the threshold level will not be declared. The setting of the threshold

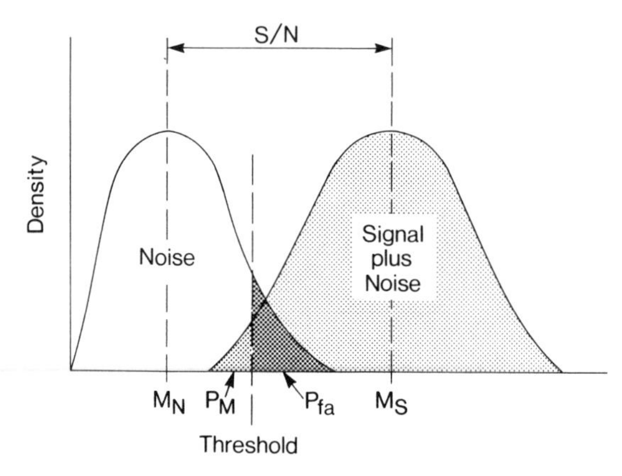

Figure 8.11 Relationship between P_{fa}, P_d, and S/N.

determines the trade-off between P_d and P_{fa}. Varying the threshold and plotting the resulting P_d versus P_{fa} produces what is termed the receiver operating curve (ROC). In a radar detection system, the threshold is commonly chosen so as to provide a constant P_{fa}. This procedure is referred to as the constant false alarm rate (CFAR) operation.

The interrelationship between P_{fa}, P_d, and signal-to-noise ratio (S/N) can be seen in Fig. 8.12, which has been constructed for the single-pulse detection of a nonfluctuating target signal return in Gaussian noise. This figure can be used in a number of ways to determine the parameters in the detection process. If the specifications were given that a P_d of 50% were required with a P_{fa} of 10^{-6}, then the required S/N of 11.3 dB can be determined. Conversely, if the S/N were known to be 10 dB, and a P_{fa} of 10^{-6} were required, the P_d can be immediately determined to be about 22%.

Clutter-Limited Case. The more general situation in the marine environment is the clutter-limited case. The main source of clutter is the sea itself, with additional sources being weather (precipitation), or in the case of navigation in ice, ice that is obscuring the target of interest (such as icebergs or another ship). Considering only sea clutter, and using (8-1) and (8-9) for the average target and clutter powers, respectively, the expression for signal-to-clutter ratio (S/C) is

$$S/C = \frac{P_{\text{target}}}{P_{\text{clutter}}} = \frac{2\sigma_t}{c\tau_0 R\beta_n\sigma^0} \tag{8-23}$$

The ratio, for a given target and clutter reflectivity, can only be increased by improving the radar resolution. Increasing radar parameters such as transmitter power or antenna gain strengthens both the target and clutter returns equally.

In the case of sea or weather clutter, the returns vary with time and may be

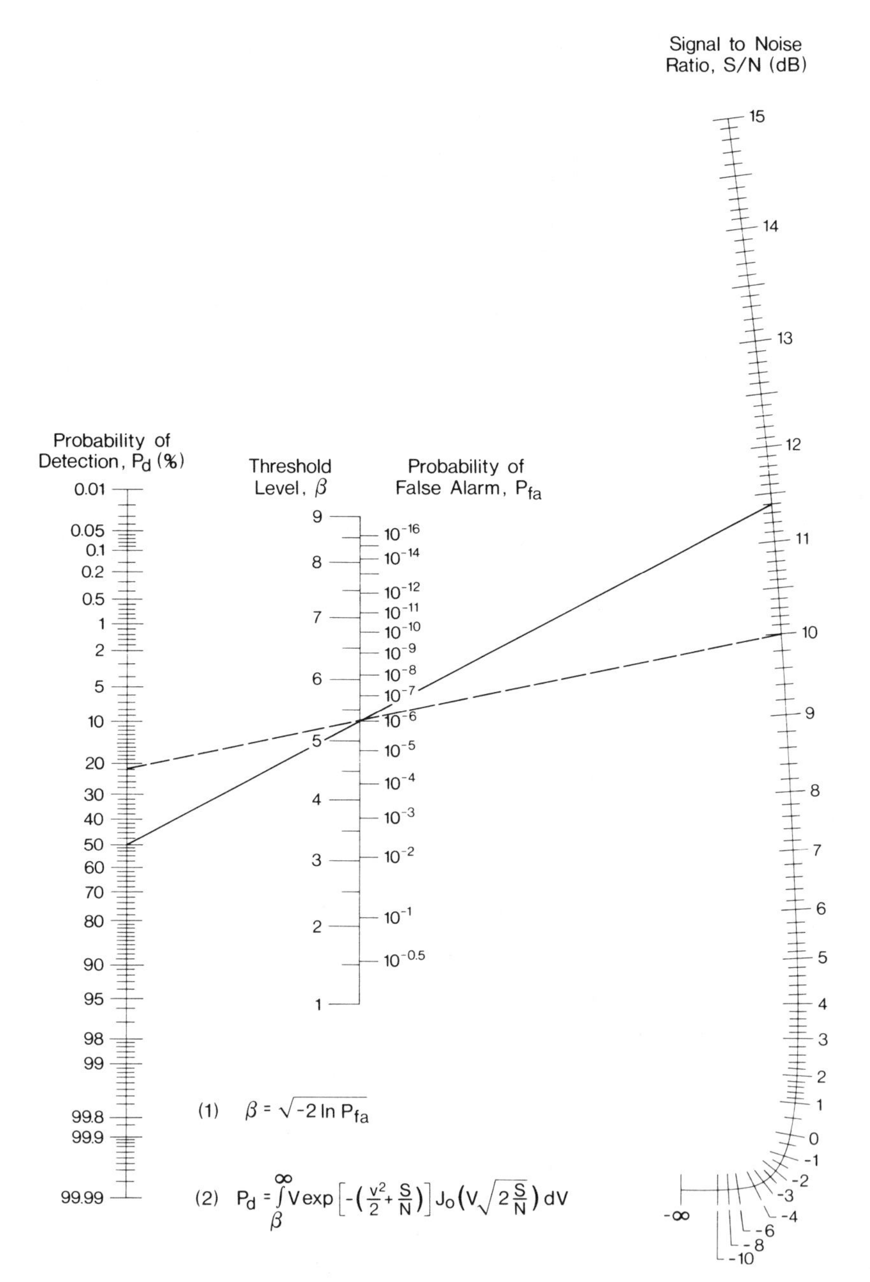

Figure 8.12 Probability of detection for a single pulse for a non-fluctuating target (Bailey et al. [1]).

described statistically. If it is assumed that the signal return from any single scatterer in the clutter is not large compared to the total return from the radar resolution cell, then the set of returns can be considered as conforming with the Central Limit Theorem, yielding a probability distribution that is Gaussian. As in the case of the noise discussed earlier, after video detection this results in a Rayleigh distribution. This is a reasonable assumption for low- to medium-intensity sea clutter, using low-resolution radar systems. As the clutter intensity increases, or the radar resolution is improved, the statistics of the clutter begin to depart substantially from the Rayleigh distribution. Non-Rayleigh sea clutter statistics are discussed in Section 8.4.4.

Pulse-to-Pulse Integration. The true S/N or S/C is determined by the ratio of the average powers of the signal and the noise or clutter. In practice, however, the radar provides only a limited number of samples with which to make a target detection. From this limited number of samples, the noise and clutter average powers must be estimated, then used to set appropriate detection thresholds. Incorrect setting of the threshold based on poor power estimates will result in increased false alarms or missed detections. The aim, therefore, is to include as many samples as practical to improve the power estimates and the threshold setting, hence improving detection performance.

In a scanning radar environment, there are two time scales on which data from a given range and bearing location are collected: within one scan and from multiple scans. (The latter is addressed in the next section.) Within each scan, the radar illuminates the target for a number of successive pulses as the antenna beam scans past it. This group of pulses can be integrated to produce an improved estimate of the average power. The effectiveness of integration depends upon the samples being statistically independent. This is the case for noise. For clutter, however, the successive pulses are correlated, reducing the effectiveness of the integration.

The values in Fig. 8.12 are appropriate for only a single pulse for a nonfluctuating target return. Using integration to improve the estimate of the average, which results in a more accurate setting of the threshold, can be equated to increasing the apparent S/N or S/C for the single-pulse case.

For noise, if a number, N, of successive pulses are available, and the target echo is correlated from pulse-to-pulse whereas the noise is statistically independent, integration of the pulses will increase the S/N by a factor of approximately $\sqrt{N}$. Thus, the required single pulse S/N is reduced by $\sqrt{N}$. Curves are available (Nathanson [43]) which show this smaller S/N required for a given P_d and P_{fa} for a variety of number of pulses integrated. For a typical scanning marine radar, 10 to 20 successive pulses are available for integration. The integration may be performed on a pulse-to-pulse and/or a scan-to-scan integration basis. For the situation where the returns from the target are fluctuating, the values in Fig. 8.12 are not appropriate. Swerling (Marcum and Swerling [38]) defined four target fluctuation models (cases) and developed curves for signal-to-noise ratios required for given P_d and P_{fa}. Swerling Cases

1 and 2 apply for targets that can be represented as a number of independently fluctuating reflectors of about equal echoing area; Cases 3 and 4 apply for targets that can be represented as one large reflector together with a number of small reflectors or as one large reflector subject to small changes in orientation. Cases 2 and 4 apply when fluctuations occur from pulse to pulse. Cases 1 and 3 apply when echo fluctuations are statistically independent from scan to scan but are perfectly correlated pulse to pulse. For the application of radar sensing and detection of ice, Swerling Cases 1 and 3 are the ones that apply, as the returns from the slowly moving ice targets are normally correlated on a pulse-to-pulse basis.

For clutter, if it is assumed to be Rayleigh distributed, the same method as used in the noise-limited case can be employed to determine the single-pulse P_d or P_{fa}. Unlike the noise, however, the clutter is correlated pulse to pulse, reducing the effectiveness of integration. To determine the integration gain as above, the number of samples, N, is replaced with the equivalent number of independent samples, which is dependent upon the normalized autocorrelation function of the clutter (Nathanson [43]). As the correlation of the clutter increases, the equivalent number of independent samples decreases, reducing the integration gain.

Scan-to-Scan Integration. A marine radar antenna typically scans at 30 rpm, thereby illuminating a given range and bearing location every 2 s. The slowly drifting ice targets of interest here typically remain in a given radar resolution cell for perhaps tens of seconds, permitting the integration of 5–10 scans. It is presumed that within each scan, pulse-to-pulse integration will have been used. Scan-to-scan integration can prove to be very effective in improving target detection since most of the sea-clutter signal is decorrelated over the period of tens of seconds.

In Section 8.4.4, the effectiveness of both pulse-to-pulse and scan-to-scan integration is considered in modeling radar performance.

Constant-False-Alarm-Rate Processing. In practice, the detection threshold in Fig. 8.11 is set at a value equal to the estimated average power of the noise (clutter) plus a certain factor times the estimated standard deviation of the noise (clutter). Assuming that the shape of the noise (clutter) distribution is known (e.g., Rayleigh distributed), the factor can be chosen theoretically to provide a desired probability of false alarm. However, a difficulty with clutter is that its true average power may vary with range and bearing. If this varying average power can be properly estimated, then the detection threshold may be adjusted accordingly, thereby permitting CFAR operation. The basic method behind CFAR processing is to adaptively make an estimate of the local average power of the clutter, based on radar resolution cells neighboring in range and bearing, called reference cells. Figure 8.13 shows the form of a basic cell–averaging CFAR processor operating in range. Since a target echo may be contained in the cell-under-test and its return may spill into adjacent range cells, the cell-

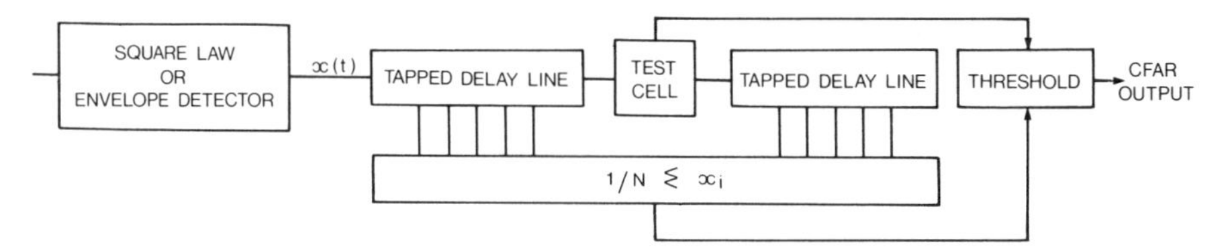

Figure 8.13 Conventional constant false alarm rate (CFAR) processor.

under-test and its neighboring range cells (called guard cells) should be excluded from the estimation process so as to prevent contamination of the clutter power estimate by the target return. The performance of this type of CFAR processor will degrade when the statistical nature of the clutter in the reference cells is not uniform, such as might occur at a clutter edge, yielding an inaccurate estimate of the average clutter power and an incorrect setting of the detection threshold. Performance also suffers as the clutter statistics become more non-Rayleigh. Various techniques have been applied to mitigate these difficulties (Moore and Lawrence [40]; Rohling [51]).

CFAR processing is an application of one form of an adaptive filtering technique. There are many alternate approaches to adaptive processing. One method, involving the use of chaos theory and neural networks, is described in the next chapter, Section 9.4.2.

8.3 HISTORY OF SURFACE-BASED ICE REMOTE SENSING

8.3.1 Using Radar to Observe Ice

The first documented use of a marine radar in an ice environment is reported by Luse [37]. The incident occurred in the spring of 1943 when the U.S. Coast Guard icebreaker *Storis* was escorting a troop carrier through ice fields off Greenland. Luse reported that with the radar gain turned down, they could see the escorted vessel as well as a fjord entrance, whereas with the gain turned up, they could see the leads through the ice field.

By the end of the Second World War, *X*-band (3-cm) marine radar employing plan position indicator (PPI) displays were in common use. Following the war, surplus military radars were installed on many commercial ships and many of these operated in ice-infested waters.

During 1945 and 1946, the U.S. Coast Guard carried out a program in the Grand Banks area of Newfoundland to investigate the detection of discrete pieces of ice. The program employed American service radars at *X* and *S* band with both PPI displays with the antennas scanning, and A-scope displays with the antenna stationary on a target. This experiment resulted in the first quantitative analysis of radar returns from ice. The results from this experiment were classified and were not released until they were consolidated with work reported by Budinger et al. [7], discussed later in this section.

The National Research Council of Canada released a report in 1947 entitled "Canadian Marine Radar," which in the final section discussed radar in Arctic waters. The report briefly discusses trials with the Type 268 *X*-band radar (developed by the National Research Council and produced in quantities exceeding 1600 during the war) in detecting icebergs and growlers in Arctic waters.

One of the first detailed reports of ice detection by shipborne radar was that of Larsson [30]. Larsson reported tests of *X*- and *S*-band radars aboard the Swedish icebreaker *Ymer* in 1947 and 1948. Some of the observations from these experiments were:

1. Smooth ice surfaces did not show up on the screen, but reflections appeared from areas where the surface had been broken. Broken, rugged ice surfaces always gave clear reflections up to 2 to 3 nm (nautical miles) away, with the 3-cm (*X*-band) system giving a clearer representation.
2. The relative strength of ice returns could be determined by adjusting the gain control to compensate for the limited dynamic range of the PPI.
3. Old channels in the ice, covered over with snow and invisible to the naked eye, were often visible on the screen because of the radar signal's ability to penetrate the snow.
4. A direct relationship was noted between the intensity of the returns on the radar screen and the resistance to navigation through the ice, with the brighter the return, the heavier the resistance.
5. Firm ice could be distinguished from open water by the lack of sea clutter.

Le Page and Milwright [31] reported results of tests carried out on board the Canadian icebreaker *N.B. McLean* to investigate the detection of ice hazards in Hudson Bay and its approaches. As a result of experiments carried out during the voyage, they concluded that "radar is an invaluable aid to ships navigating in bad visibility in ice-infested waters, provided it is used wisely and the limitations of the radar are known." Other observations cited include:

1. In calm seas, icebergs were detected out to 15 to 20 nm, small growlers were detected only out to 2 nm.
2. Under rough sea conditions when sea clutter extends beyond 1 nm, growlers large enough to cause damage to the ship can go undetected.
3. Fields of pack ice should be detectable under all sea conditions out to 3 nm.

Perry [48] reported similar investigations aboard the SS *North Anglia* on a voyage to Port Churchill. Perry's results were similar to those of Le Page and Milwright. In smooth seas, growlers could be detected at 2–3 nm, but in rougher seas, growlers sufficiently large to cause disaster could not be detected.

On one occasion, close pack ice looked like sea clutter on the 1- and 3-mile ranges. Hummocks in the ice could only be isolated using the anti-clutter control.

Hood [22, 23] analyzed reports submitted by commercial shipping entering Hudson Bay which described the performance of marine radar in detecting ice. He concluded that for icebergs (large "point" targets), the ranges of detection were reasonably close to theoretical and experimental data using a $1/R^4$ range relationship.

8.3.2 Previous Studies

With the possible exception of the work done in 1945 and 1946 off Newfoundland, the first major, systematic, and quantitative experiment conducted to quantify the capability of marine radar to detect ice of potential hazard to shipping was carried out by the U.S. Coast Guard between March and October 1959. The results of these experiments are reported by Budinger et al. [7] and were based on detailed analysis of 128 ice targets for the 1959 experiment and 24 targets from the 1945 experiment. The maximum range of detection for the 152 targets conformed very closely to the anticipated $1/R^4$ relationship. However, on several occasions it was not possible to detect growlers in sea clutter. The reflection coefficient for icebergs on the Grand Banks was found to be about 0.33, and icebergs were found to reflect radar energy 60 times less than a ship with an equal-sized physical cross section. During the experiment, subnormal propagation conditions were encountered and Budinger noted that (during the spring) such conditions would be the rule rather than the exception. He commented that this would lead to slightly reduced ranges of detection for small targets and greatly reduced ranges of detection for large targets. Budinger concluded with the following warning:

> "All shipmasters, mates, and owners are warned that safe passage through iceberg areas of the North Atlantic Ocean cannot be assured by the use of radar."

During 1966, Holden of Decca Radar, as reported by Williams [66], carried out controlled trials on the detection of growlers in various sea states. Contrary to expected results, it was found that for the detection of growlers in high sea states, a wider beamwidth, hence lower azimuthal resolution antenna, gave superior results to a narrow beamwidth antenna. It is anticipated that the wider smear of the ice target using the wider beamwidth antenna made the target easier to see on the PPI display.

Tabata et al. [61–63] and Tabata [60] reported on work carried out on the detection of pack ice off Hokkaido, Japan, using three coastal C-band (5-cm) radar stations. During these experiments, it was found that pack ice could be tracked over extended periods. By tracking the ice, drift velocities could be determined. Ice-free patches (polynyas) within the ice pack could be determined by their low returns, whereas pressure ridges could be identified by their bright returns. With the heights of the radars between 220 and 400 m above sea level, pack ice could be observed 60 to 85 km in range.

Hagman et al. [17] reported on results of marine radar studies carried out as part of the Sea Ice-75 program. During this program, *X*- and *S*-band radars were operated from the icebreaker *Tor* lodged in an area of ice. Some of the conclusions they reached were:

1. The ship's radar could distinguish the difference between open water and ice, open water and the ice edge, ice ridges and level ice, and vast thick floes refrozen in level ice.
2. The *X*-band radar gave a clearer representation than the *S*-band radar.
3. With the heights of the antennas on the *Tor* at 22 to 24 m, good information was provided on the ice situation out to 1.5 nm. Larger ridges (>2 to 3 m) gave radar echoes out to 3.5 nm.
4. The ship's radar could not display the difference between:
   - snow-covered ice and snow-free ice unless the snow-covered ice was very rough;
   - the height of the ice ridges;
   - ice of different thickness; and
   - a narrow channel and a small ridge.

During the late 1970s and 1980s, considerable work was carried out in support of the Canadian offshore-resource industry on the detection, classification, and tracking of ice. Benedict and Hall [4] prepared a report examining the benefits of a multisensor system, including radar, for the detection of icebergs. Subsequent to this report, the Canadian Government, in conjunction with the offshore-resource industry, funded field trials to evaluate the various sensors. Results of some of this work is found in Ryan et al. [55], Ryan [54], and Harvey and Ryan [18]. Other work in Atlantic Canada included monitoring the interaction of (first-year) pack ice against and around an island, using an inexpensive marine radar together with time-lapse movie cameras (Klein and Butt [28]). At periodic intervals the cameras recorded the radar display. Upon playback, the changes and movement in the ice cover were readily appreciated.

During the same period, Canadian oil companies were working on improving radar detection and tracking of ice around drillships in the Beaufort Sea. The RIDS (radar image display system) was developed, capable of processing, displaying, and recording data (Raman et al. [50]; Mercer et al. [39]). The system incorporated pulse-to-pulse and scan-to-scan averaging and was used to track the movement of ice around stationary platforms.

Other interesting work continues, including the integration of shipboard sensing and display systems (Sneyd et al. [59]), the monitoring of ice encroaching on hydroelectric facilities in the Niagara River (Crissman and Lalumiere [8]), and work on new display technologies.

8.3.3 Shortcomings of Existing Systems

During the period from the first documented use of radar in an ice environment in 1943, as reported by Luse [37], until the late 1970s, all navigation in ice (at least for civilian applications) was accomplished using conventional, commercial marine radar equipment. These systems were not designed for the detection and classification of ice targets; rather, they were optimized for detecting and tracking large targets, often man-made.

All conventional systems use horizontal polarization only. However, a significant advantage has been demonstrated for employing dual polarization in both classification and detection. Although these existing systems have relatively narrow azimuthal beamwidths, their pulse lengths are long (in order to achieve useful average power on target), resulting in large resolution cells. In the presence of sea clutter, large resolution cell size results in poor signal-to-clutter ratios. Superior signal-to-clutter performance can be achieved by employing shorter effective pulses through the use of pulse compression.

A major deficiency of existing marine radar systems is the radar display itself. For a shipborne radar illuminating an ice cover, the range of signal intensities for targets of interest spans 30 to 40 dB or more, but the dynamic range of the conventional PPI display is limited to less than 12 dB. When the display is adjusted to show small targets, larger targets cause saturation of the display. This, then, conceals the difference between smaller, less hazardous targets and larger, dangerous targets (including icebergs), as they are all displayed at essentially the same intensity.

Another problem results from the spreading of the radar energy with range. Targets near the radar reflect proportionally more energy than similar-strength targets further away, because more incident energy falls on them. For a surface-based radar illuminating the ice surface, this decrease in average radar energy reflected back to the radar varies as $1/R^3$, and after the transition range as $1/R^7$, where R is the range to the target. If the display is adjusted to show targets close to the radar, similar-sized targets further away may not be displayed because of this nonuniform-range dependence. Conversely, if the display is adjusted to show long-range targets, strong blooming near the display center results in the loss of individual short-range targets.

The image on the PPI is maintained by employing tube persistence which results in limited display intensity, and the display cannot be viewed in direct light. In daylight hours, the display must be observed through a viewing hood, which restricts its operation to one person at a time. Finally, to make a range measurement on a target, the variable range marker must be made to pass through the target, which can take several scans of the radar sweep to accomplish, thus limiting the speed at which range measurements can be made.

These and other shortcomings of current radar displays can be overcome with suitable signal processing and raster scan display technology. Research in this area is described in Section 8.6.

It was only in the early 1980s that radar equipment began to be modified or constructed to operate in an ice environment. These modifications, however, have been on a limited scale and, with the exception of combining *X*- and *S*-band radars, have concentrated mainly on signal-processing and radar-display improvements. In addition, until the early 1980s, the experimental programs undertaken have been confined to using conventional marine radar hardware. It was not until the commencement of the research discussed in this chapter in 1981 that a serious evaluation and examination of many radar parameters was undertaken. This research had led to a better understanding of the interaction between a surface-transmitted radar signal and ice, and has led to the installation and use of sophisticated third-generation radar systems optimized for sensing of ice. Work is continuing to define the radar parameters and associated processing for optimum ice-detection performance.

8.4 RESEARCH PROGRAM

In 1981, the Canadian Department of Fisheries and Oceans, in collaboration with the Communications Research Laboratory of McMaster University, began a systematic study of surface-based radar classification and detection of ice. The results are the subject of this chapter and some of the results are reported in Lewis and Currie [33, 34], Lewis et al. [35, 36], Currie and Lewis [10], and Haykin et al. [20].

8.4.1 Objectives

One of the serious deficiencies in the use of surface-based radar systems in an ice environment was a lack of understanding of the interaction between surface-transmitted radar signals and ice, together with the experimental data through which to develop this understanding.

The research program had the objective of developing a better understanding of the interaction of a surface-based radar signal with ice of many types: first-year and old sea ice, pressure ridges, icebergs, bergy bits, and growlers. The philosophy throughout the research has been to conduct tightly controlled experiments with good calibration and detailed surface verification. The program focused on two general scenarios: classification of ice type in an ice cover, and detection of ice floating in open water.

8.4.2 Description

To undertake the research, it was necessary to choose an experimental site which gave exposure to the ice types of interest. To provide good experimental control and surface verification for classification studies, it was essential that the ice cover remain stationary during the desired duration of the experiment. In anticipation of future open-water experiments, it was important that the landfast ice clear away from the selected location for some period during the summer season.

Bearing these requirements in mind, a site was selected on the north end of Borden Peninsula on northern Baffin Island (Fig. 8.14), overlooking Lancaster Sound. The site selected was just over 30 m above sea level (approximately at a ship's mast height). The site overlooked the transition zone between Navy Board Inlet where the ice is usually landfast and the more mobile ice in Lancaster Sound. The area has a good selection of icebergs; in most years, a significant number run aground or are frozen into the ice at this location.

Major field experiments were carried out at the Borden Radar Station in 1981, 1982, 1983, and 1984, with a subsequent shipboard program in 1986. The 1981 experiment, carried out under fast-ice conditions in March and April, was the first attempt to systematically evaluate the effects of radar parameters on the ability of the radar to classify and detect ice types and features. To evaluate frequency effects, radars were operated at *X* and *S* bands. The effects of antenna beamwidth, transmit pulse length, and antenna elevation were also evaluated. A limited attempt (using a bistatic radar system) was made to evaluate polarization effects.

To provide surface information for later analysis, aerial photographs were taken with a Wild RC-10 aerial camera and were mosaiced to provide a detailed record of the ice conditions in the survey area. The photo mosaic was used extensively to correlate radar returns with ice features.

As a result of findings from the 1981 experiment and a general interest in confirming the presence of multipath interference over an ice surface, a dedi-

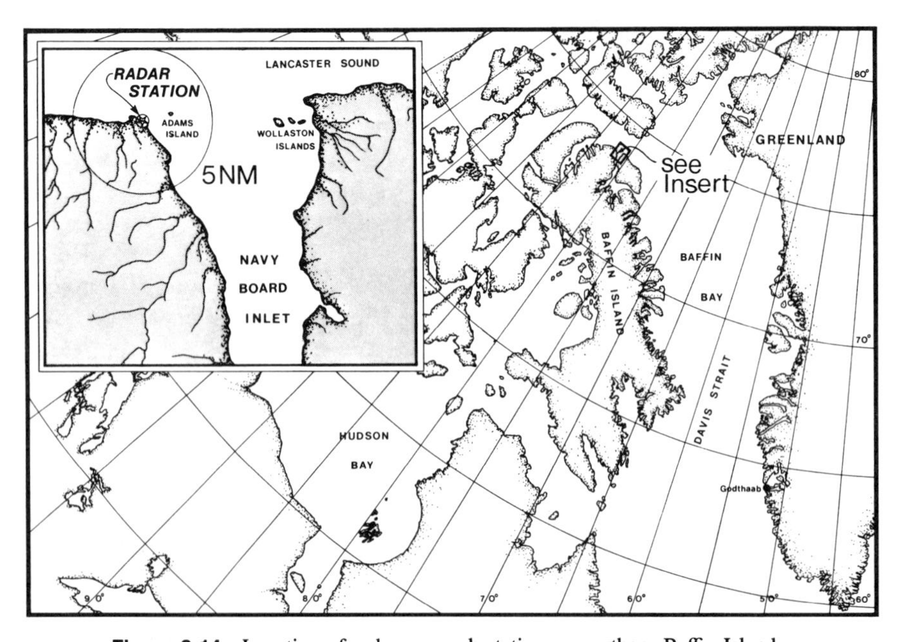

Figure 8.14 Location of radar research station on northern Baffin Island.

cated multipath experiment was performed in 1982. This experiment was carried out under fast-ice conditions in May.

The 1983 experiment, also conducted under fast-ice conditions, was a multifaceted program to investigate the effect of radar frequency (over as wide a range as possible) on the ability of the radar to differentiate ice types; to investigate the use of polarization to differentiate ice types; and to conduct further investigations on multipath interference over ice. Photographs were again taken with a high-resolution aerial camera to document ice conditions in the survey area.

Radars were operated at the S, X, K_u, and K_a bands. In addition, the radars at X, K_u, and K_a bands were operated in a dual-polarized configuration. Table 8.5 gives the major parameters for the radar systems used in the experimental program.

The three experiments were all conducted during the late spring months when a full and fixed ice cover was present. The 1984 experiment, however, was conducted in the fall during the open-water period. This experiment used a dual-polarized X-band radar to evaluate the effects of polarization on the ability to detect ice targets in open water. The detailed objectives of this experiment were to investigate the depolarizing characteristics of icebergs, bergy bits, and growlers in open water, and to investigate the depolarizing characteristics of sea clutter.

The results of the research based on these experiments are presented in the following sections. Section 8.4.3 deals with the ice-type classification scenario, beginning by showing the effect of changing radar frequency and polarization. Scattering theory is then introduced to explain the observed results. The effect of changing other parameters, such as antenna height and resolution is shown. The section closes with some ice-classification image-processing software studies.

Section 8.4.4 deals with the floating ice detection scenario. First, the results based on the 1984 experiment conducted in the Arctic are presented. Experiments by McMaster and others on the Canadian east coast are summarized. The section closes with a description of radar performance modeling studies.

The 1986 shipboard experiment is discussed in Section 8.5.

In the subsequent discussion, the following notation will be employed: *HH*

TABLE 8.5 Radar Parameters for 1983 Experiment

	Frequency (GHz)	Polarization	Transmit Power (kW)	Transmit Pulse (ns)	Antenna Beamwidth (°)
S Band	3.04	H	30	25, 250, 1000	2.0
X Band	9.4	H&V	100	510, 2100	3.5
K_u Band	16.5	H&V	28	400	1.5
K_a Band	35.0	H&V	4.5	140	1.0

will represent horizontal transmit/horizontal receive, and *HV* will represent horizontal transmit/vertical receive; *VV* will represent vertical transmit/vertical receive, and *VH* will represent vertical transmit/horizontal receive.

8.4.3 Ice Type Classification

Frequency Results

Analysis of the data collected during the 1981 experiment, and subsequently in more detail in the 1983 experiment, shows substantial differences in the radar returns from the ice surface as a function of frequency. The radars were operated with the antenna stationary, illuminating a radial line containing several different types of ice targets. The lines were numbered according to compass bearing. Figure 8.15 shows the photo mosaic for a section of ice corresponding to Line 247 and the corresponding radar returns for several frequencies.

The transmitter is located at X on the left side of the image, and the range from the transmitter to the right side of the mosaic is about 9.0 km. The ice surface consisted of relatively smooth, first-year ice in the area A to B, rougher first-year ice from B to C, smooth first-year ice from C to D, medium-sized first-year pressure ridges at D and E, rough first-year ice in the area E to F, and a large old-ice floe in the area F to G.

The normalized radar cross sections, σ^0_{HH}, for S, X, K_u, and K_a bands are shown below the mosaic. The X- and K_u-band plots are very "busy," with substantial returns from all along the line. From a review of the aerial mosaic and extensive surface verification, it was concluded that many of these strong returns were caused by targets that would not be significant from a navigational point of view; that is, they would not represent a threat to a ship or be a major impediment to navigation.

The S-band plot, on the other hand, is considerably less busy. The strong signal returns in the region B to C correspond to significant roughness in this area, visible on the aerial photograph and substantiated by surface investigation, and the low signal returns from C to D correspond to the smooth area on the mosaic. The ice ridges at D and E are well represented by strong peaks. The roughness immediately behind the ridges is also displayed, as well as the smoothness of the old ice floe. A ridge in the center of this floe is also evident.

There are two frequency effects that contribute to the generally lower returns at lower frequencies. First, the Rayleigh roughness criterion specifies that targets small with respect to the wavelength will not appear rough to the radar wave and thus will return little energy. As the wavelength of the radar signal decreases, targets will appear increasingly rough and will reflect more energy. Second, as the frequency increases, the lower side of the first multipath interference lobe will move closer to the surface, putting more energy on the surface.

The K_a-band results were more complex. The K_a-band radar suffered from low output power and high waveguide losses, resulting in weak returns. This effect is evident in Fig. 8.15 and makes the interpretation of the K_a-band returns

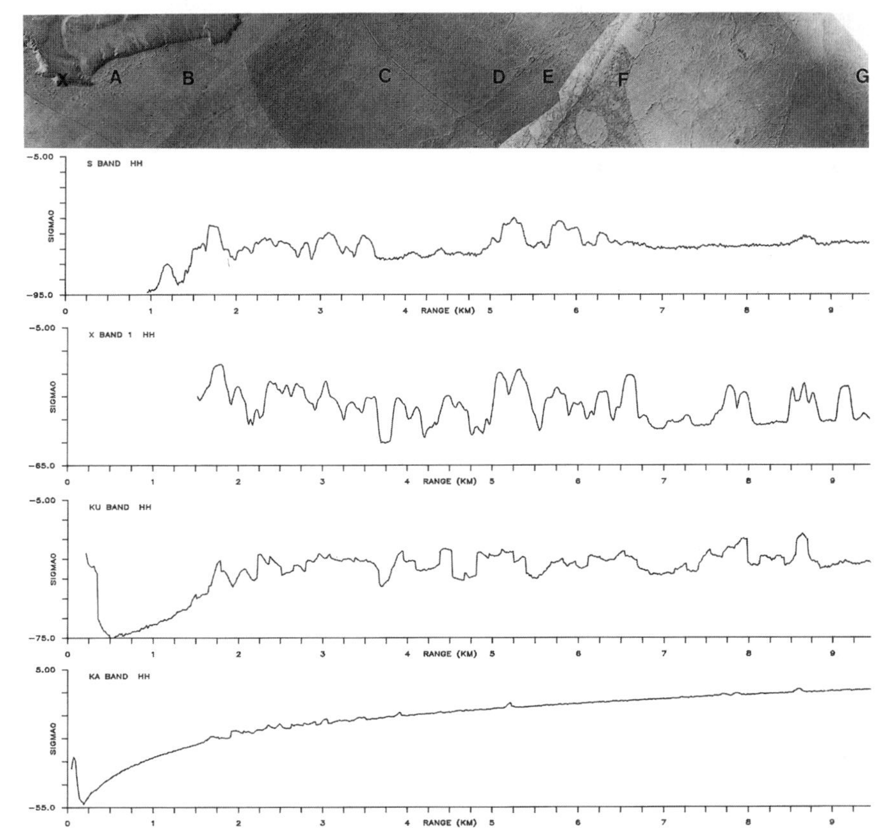

Figure 8.15 Photomosaic and radar returns for S, X, K_u, and K_a bands along Line 247. Radar is located at X on left of figure. Area from A to B is smooth first-year ice; B to C, rough FY ice; C to D, smooth FY ice; medium-sized pressure ridges at D and E; rough FY ice from E to F; and a large old-ice floe from F to G.

difficult. However, the returns, although at low levels, are significant. The targets associated with the returns in the area F to G were old-ice pressure ridges on the old-ice floe. The return at D was caused by a substantial first-year pressure ridge which contained a number of old-ice fragments.

An analysis of the returns from a second line, Line 341, yields similar results. From a navigational point of view, the targets of interest (Fig. 8.16) were an iceberg at B and old-ice floes at A and C. Again it was evident that the signal returns became more cluttered in going from S, to X, to K_u band because the surface appeared increasingly rough as the frequency increased, and as the frequency increased, the first interference lobe moved closer to the ice. The K_a-band return was reduced significantly by the low power of the transmitter and the high waveguide losses.

The S-band radar returns reflected the general roughness of the ice: the

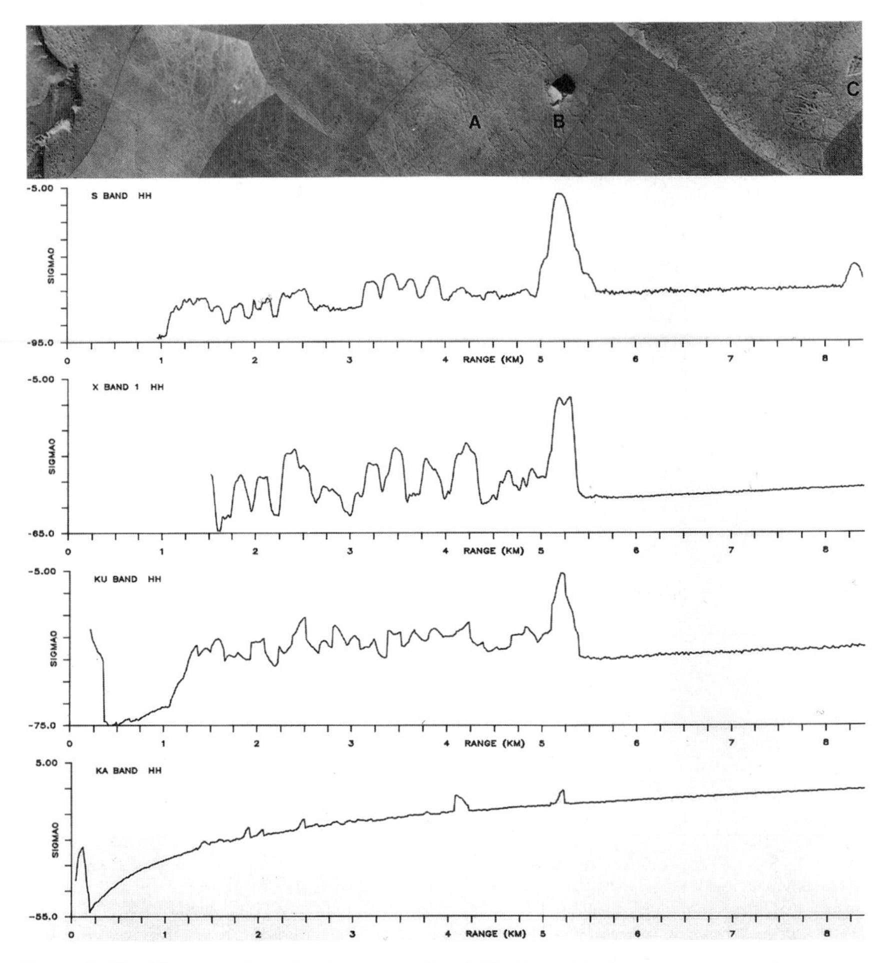

Figure 8.16 Photomosaic and radar returns for S, X, K_u, and K_a bands along Line 341. Targets of interest are old-ice floes at A and C, and iceberg at B.

iceberg at B was clearly detected but the old-ice floe at A was missed. The S-band radar detected the ice floe at C because of its wider beamwidth, whereas the other systems did not illuminate the floe. The X- and K_u-band systems gave good returns from the iceberg but did not detect other targets. The K_a-band radar, although suffering from low-level returns, detected the iceberg at B and the old-ice floe at A.

The analysis of data sets from several other lines confirmed the above findings. The S-band system gave returns that were strongly related to the target's significant vertical extent, thus displaying targets which would be a threat to navigation. However, the S-band system did not detect old-ice floes that did not have significant surface relief. The X- and K_u-band radars showed responses to surface features that were much smaller in vertical extent (many of which would not have been a threat from a navigational point of view) than those detected by the *S*-band radar. The *X*- and K_u-band radars presented a cluttered response that was not easy to interpret. The K_a-band returns, on the other hand, although lower in level, correlated well with targets of interest, including icebergs, old-ice floes, and some first-year pressure ridges.

The normalized, like-polarized radar cross section values (σ^0) for first-year, old, and glacier ice are shown in detail in Table 8.6. As expected, the mean σ^0 values for each ice type increased with frequency. Although the values themselves are interesting, it is more informative at this point to examine the change in σ^0 for each ice type as the frequency increases. These results are shown in Table 8.7, and although they will be examined in some detail in the subsection on frequency effects, it is worth noting here that the change in the σ^0 for glacier ice between *S* and K_a band, *X* and K_a band, and K_u and K_a band is considerably less than that for first-year and old ice. In addition, there are, in general, systematic differences in the change of σ^0 for old and first-year ice in the three frequency pairs.

TABLE 8.6 Normalized Radar Cross Section ($\sigma°$) of Ice versus Frequency

	Normalized Radar Cross Section $\sigma°$ (dBm^2)			
Feature	*S* Band 3 GHz	*X* Band 10 GHz	K_u Band 16 GHz	K_a Band 35 GHz
ICEBERG				
Line 292	−41	−20	−6	−13
Line 307 (1)	−21	−22	−11	−6
(2)	−54	−35	−36	−17
Line 341	−9	−13	−6	−6
Mean	−31.3	−22.5	−14.8	−10.5
OLD ICE				
Line 247 (1)	−56	−30	−22	−4
(2)	. . .	−30	−25	−6
Line 277 (1)	−60	−26	−27	−12
(2)	−53	−33	−23	−5
Line 292	−56	−30	−26	−8
Line 341	−63	−30	−28	−8
Mean	−57.6	−29.8	−25.3	−7.2
FIRST-YEAR ICE				
Line 0	−61	−38	−30	−18
Line 247 (Ridge)	−45	−23	−32	−10
Line 277 (Ridge)	−39	−23	−30	−13
Mean	−48.3	−28.0	−30.7	−13.7
ISLAND	−35	−22	−21	−5

TABLE 8.7 Change in Normalized Radar Cross Section of Ice as a Function of Change in Radar Frequency

	Change in σ^0 (dBm^2)		
Feature	from S to K_a	from X to K_a	from K_u to K_a
First-year ice	35	14	16
Old ice	50	22	18
Glacier ice	20	12	4

With the exception of the change between K_u and K_a bands, the change in the σ^0 for old ice is more substantial than for first-year ice.

Analysis of the line data gives a quantitative measure of the effect of frequency on the detection and classification of ice features, and a visual examination of images for each of the frequencies supports the line data. Figure 8.17 shows corresponding segments of images from the 1983 experiment for each of the S, X, K_u, and K_a bands for horizontal polarization.

Surface verification of the area established that the targets of interest from a navigational perspective are the first-year pressure ridge at A; icebergs at B, C, D, and E; an island at F; and a large number of old-ice floes distributed throughout the ice field. Examination of the radar images shows that the S-band image best defines the first-year presure ridge, as well as adequately displaying the four icebergs and the island. The smallest of the icebergs, D, however, is somewhat difficult to detect in the surrounding ice clutter. The S-band image also highlights several of the old-ice floes north of (above) the pressure ridge. The X- and K_u-band images show returns from a large number of targets, many of which are not significant navigational threats. These returns in many cases conceal returns from targets of interest. Neither the X- nor the K_u-band image clearly displays the first-year pressure ridge. In addition, the icebergs and island are more difficult to discern, particularly in the X-band image. Also, the small iceberg at D is not detectable in either the X- or K_u-band images.

Finally, although the signal levels received with the K_a-band system were low, all the icebergs (including the small one at D) as well as the island at F are identifiable on the K_a-band image. Although the returns from far range targets are low, the targets that are displayed in the area of G correspond to known old-ice floes.

In conclusion, examination of radar images of the ice field at each of the frequencies showed that the S and K_a bands provide information about the ice not apparent in the other frequency images. The K_a-band radar in particular was able to enhance the detectability of the icebergs and some of the old-ice floes.

Polarization

The second major parameter investigated was polarization. Analysis of cross-polarized data at X band collected in 1981 showed that there were distinct

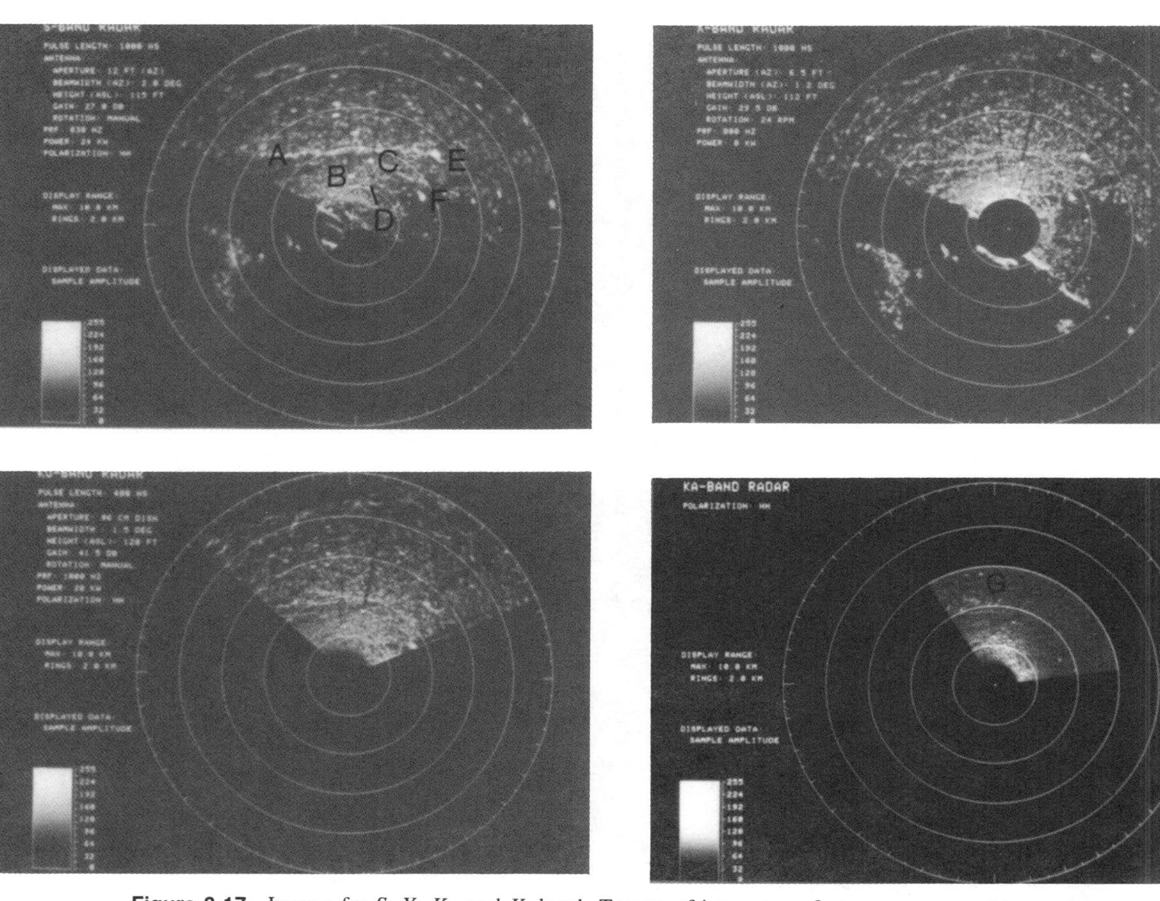

Figure 8.17 Images for S, X, K_u, and K_a band. Targets of interest are first-year pressure ridge at A; Icebergs at B, C, D, and E; and Island at F.

differences between the like- and cross-polarized echoes. These findings prompted the major frequency/polarization experiment conducted in 1983.

Figure 8.18 shows a strip of the aerial photomosaic corresponding to Line 307 from 1983, with the X-, K_u-, and K_a-band returns for like- and cross-polarization shown below. Two targets of interest, which demonstrate the use of polarization to differentiate and detect ice features, are shown. The target at A was a large iceberg with a height of 45 m and width of 75 m, yielding a physical cross section of about 3400 m^2. The low signal levels to the right of the iceberg in the K_u- and K_a-band returns were the result of the radar shadow cast by the iceberg. The wider beamwidth of the X-band antenna illuminated targets which were not directly behind the iceberg. Target B to the left of the large iceberg was another small iceberg (300 m^2) which would have been vitally important to detect as a hazard to shipping.

All three frequencies using like-polarization detected the iceberg at A. However, by comparing the like- and cross-polarized returns it can be seen that, with the exception of the return from the iceberg, the cross-polarized signals are considerably lower than the like-polarized returns and this makes the iceberg more readily observable. The improvement in target-to-clutter ratio (where the clutter is considered to be all those signals from ice targets that are not of interest from a navigational point of view) means that the target was more easily differentiated from the surrounding signals. In this situation the clutter was predominantly caused by returns from first-year ice. The improvement in target-to-clutter ratio is particularly apparent in the K_u-band returns where the improvement exceeds 10 dB.

The second target, the small iceberg at B, was not detectable in the like-polarized returns with any of the frequencies, without prior knowledge of its location. The X-band like-polarized return had a sizable return at the correct range, but the return was distributed over a considerable range extent and thus did not identify a unique target. The X- and K_u-band cross-polarized returns, on the other hand, distinctly detected the small iceberg at B. The low K_a-band return (caused by low power transmission and lossy waveguide) did not provide enough cross-polarized energy to allow detection. Other examples of K_a-band returns, however, confirmed that the cross-polarized returns gave similar results to the X- and K_u-band radars.

For the subsequent analysis of polarization from the 1983 data, only the K_u-band results will be considered. There are two reasons for this. First, the K_a-band returns were too weak and in many cases did not provide adequate signals for analysis. Second, the X-band radar employed a wide (3.5°) azimuthal beam width antenna which illuminated a large area of ice, particularly at far ranges, so that returns from targets well off the line of interest contaminated the returns under analysis. In addition, the X-band antenna had poor cross-polarized isolation which may have resulted in a leak of the like-polarized energy into the cross-polarized channel. The higher cross-channel isolation and the narrower beamwidth (1.5°) of the K_u-band antenna avoided these difficulties.

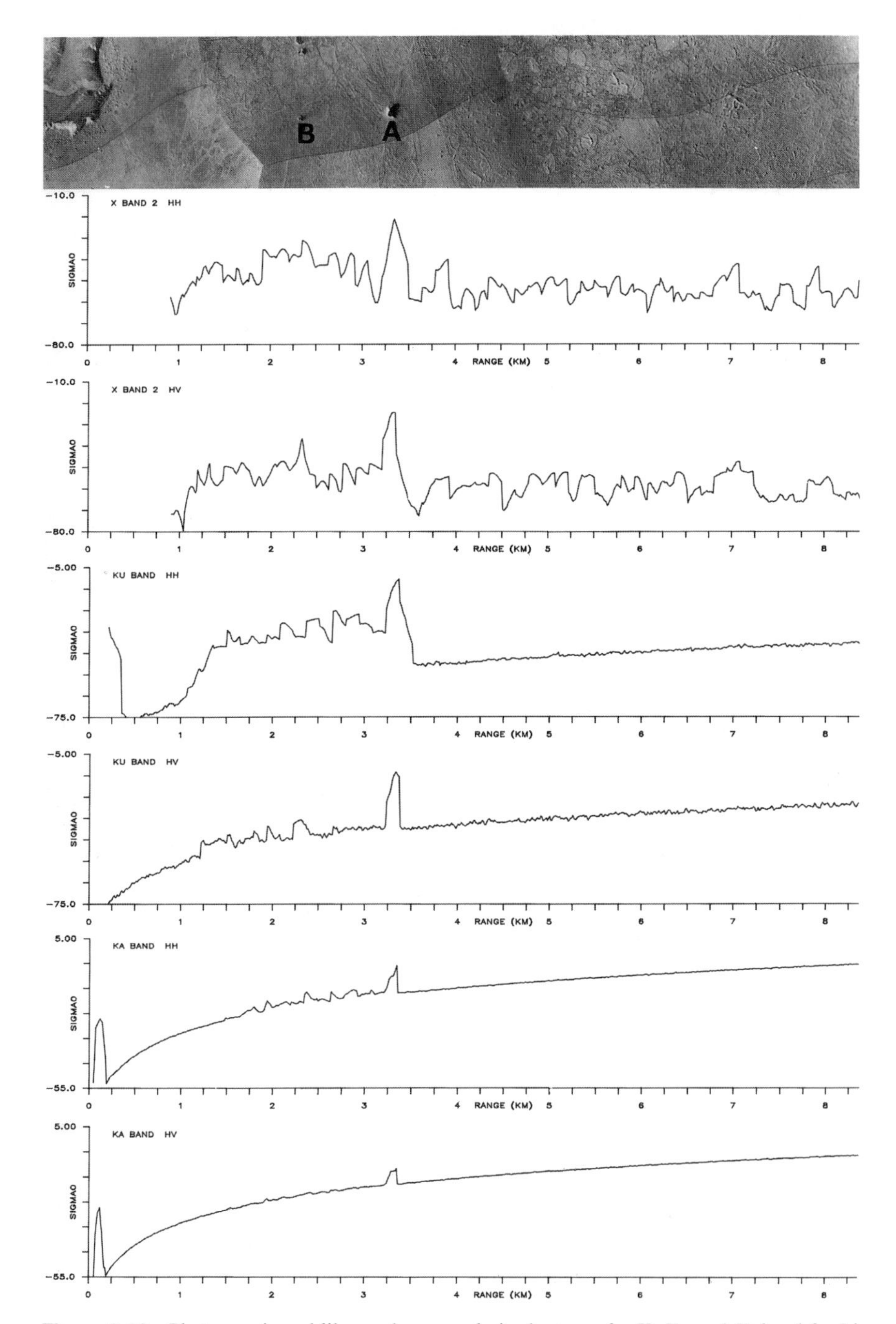

Figure 8.18 Photomosaic and like- and cross-polarized returns for X, K_u, and K_a band for Line 307. Targets of interest include the large iceberg at A and a smaller iceberg at B.

For the analysis of the 1983 data, only *HH* and *HV* returns are analyzed since the values of *HV* and *VH* were essentially identical for the same targets. The horizontal transmit polarization was chosen because of its general usage.

Closer examination of the traces for K_u-band σ^0_{HH} and σ^0_{HV} in Fig. 8.18 shows the improvement in the "visibility" of the icebergs against the sea ice background afforded through the use of cross-polarization. The target at A was a large iceberg, with a height of 45 m and a width of 75 m, yielding a physical cross section of about 3400 m^2. The very low signal levels behind the iceberg were a result of the radar shadow cast by the iceberg. Comparison of the like- and cross-polarized returns shows that the cross-polarized signals, with the exception of the return from the iceberg, were considerably lower than the like-polarized returns so that the iceberg is more readily noticed in the cross-polarized plot. The target-to-clutter ratio for the large iceberg at A was 16 dB in the *HH* case and increased to 26 dB in the *HV* case, an improvement of 10 dB.

Of even more significance, however, is the return for the target at B. Target B was a small iceberg (300 m^2) in front of the large iceberg and although this target is not identifiable in the like-polarized plot, it is well defined in the cross-polarized plot. For the small iceberg at B, the target was 8 dB below the surrounding clutter level in *HH* polarization, but in *HV* polarization the target was about 7 dB above the surrounding clutter level, an improvement in target-to-clutter ratio of 15 dB. In this case, the cross-polarized return resulted in the detection of a hazardous target that was undetectable with the conventional *HH* polarization.

Another example of the use of cross-polarization in detecting targets of navigational interest is shown in Fig. 8.19. The photomosaic and radar returns are for Line 341 from the 1983 data. The K_u-band σ^0_{HH} is at the top, and the σ^0_{HV} is below the photomosaic. The targets of concern are the iceberg at A and the old-ice floe at B. The large iceberg (15,200 m^2) was well represented in both the like- and cross-polarized returns, although the iceberg was more clearly detectable in the cross-polarized response. There was, however, no target-to-clutter advantage for this target in the *HV* return.

The target at B, which was not distinguishable in the like-polarized return but was easily distinguishable in the cross-polarized trace, was an old-ice floe that would have posed significant resistance to navigation. In the like-polarized return, the old-ice floe gave the same σ^0 level as the surrounding clutter, whereas in the cross-polarized return, the floe was 10 dB above the clutter, resulting in a 10-dB target-to-clutter improvement.

Analysis of a large number of targets yielded target-to-clutter improvements of 10 to 17 dB for icebergs and of 7 to 10 dB for old-ice floes.

Although *X*-band polarization results have not been shown because of difficulties with the *X*-band antenna, it is important to emphasize that the improvements resulting from using the cross-polarized returns also occur at *X* band, although possibly to a somewhat lesser extent. Revisiting Fig. 8.18, the *X*-band results for Line 307 may be compared to those for K_u band. The cross-

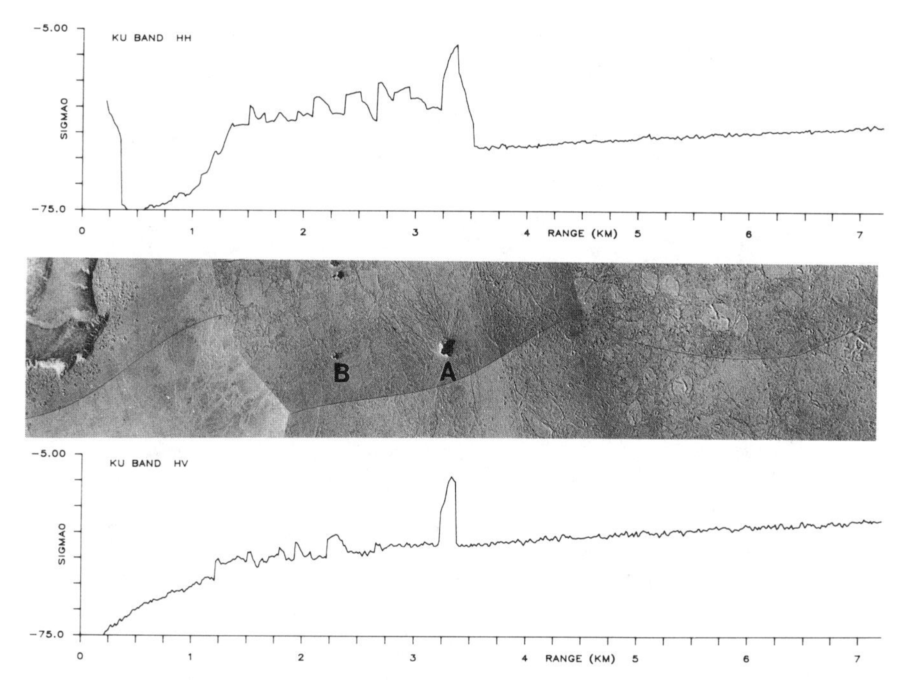

Figure 8.19 Photomosaic and K_u-band like- and cross-polarized returns for Line 341. Targets of interest include old-ice Floe at A, iceberg at B.

polarized return provided a 6-dB target-to-clutter improvement for the large iceberg, and a 7-dB target-to-clutter improvement for the smaller iceberg at B. These improvements would be even more substantial if a better antenna were employed.

The improvements in detectability of desired ice targets discussed in the previous examples can be observed clearly in image format, as shown in Fig. 8.20. The images were generated on a high-resolution raster scan display system and photographed in the laboratory. The image at the top of the figure shows the K_u-band like-polarized returns and the image below shows the cross-polarized returns.

The large iceberg (1) on Line 307 was detectable on both images, whereas the smaller iceberg (2) closer to the radar was not visible in the like-polarized image. The iceberg (3) on Line 341 was identifiable on both the like- and cross-polarized image, although it was more clearly discernible on the cross-polarized image. The old ice floe (4) on Line 341 was not identifiable on the like-polarized image but was identifiable on the cross-polarized image. The old-ice floe (5) on Line 277 was not identifiable or separable from the first-year pressure ridge on the like-polarized image but again was clearly identifiable on the cross-polarized image. Although the island (6) on Line 0 was discernible on the like-polarized return, it was more clearly detectable on the cross-polarized image.

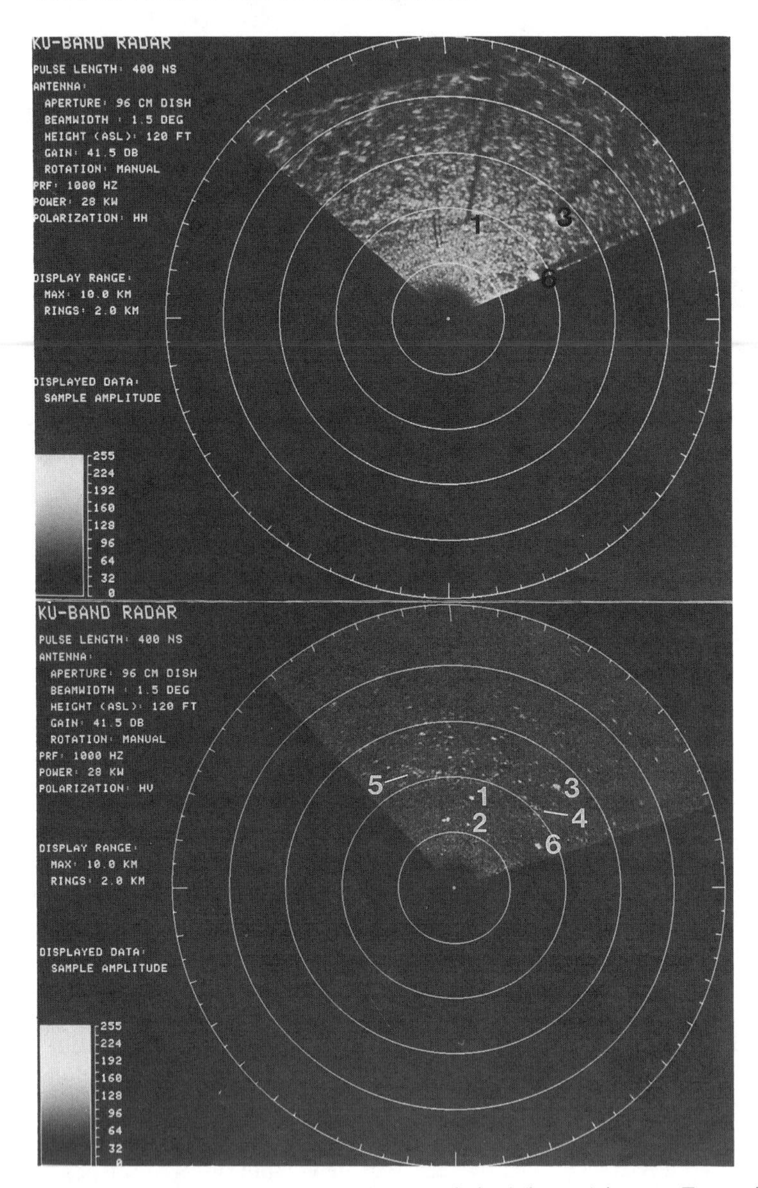

Figure 8.20 K_u-band like-polarized (top) and cross-polarized (bottom) images. Targets include (1) Large iceberg, Line 307; (2) Small iceberg, Line 307; (3) Iceberg, Line 341; (4) Old-ice floe, Line 341; (5) Old-ice floe, Line 277; (6) Island.

In general, there were significantly fewer targets on the cross-polarized image, and of the ones that were displayed, the majority were known to be either iceberg or old ice. No known targets of interest (icebergs or old ice) were absent from the cross-polarized image.

Although the K_a-band results have not, in general, been displayed because of the low signal levels associated with that frequency, it is worthwhile to examine the cross-polarized image for K_a-band operation (Fig. 8.21) and to compare this to the cross-polarized image for the K_u-band operation shown in Fig. 8.20. The K_a-band image clearly identified the icebergs at 1, 2, and 7 and the island at 6. The iceberg at 3 was poorly displayed because of the drop-off in returned power with range. Also, as the antenna was scanned manually from northwest to southeast, the main vertical lobe of the antenna was focused closer and closer in range, resulting in a further decrease in illumination for distant targets in the northeast and east sections of the image. However, several old-ice floes, particularly at 8, 9, and 10, were displayed more clearly on the K_a-band cross-polarized image than in the K_u-band cross-polarized image.

Table 8.8 summarizes the σ^0 values, the ratios between *HH* and *VV*, and the ratio between the like- and cross-polarized components for the targets. The table shows that icebergs have the highest like-polarized and cross-polarized cross sections, which result from their greater vertical relief. The target with the next highest σ^0 value both in like- and cross-polarization is the island. Old ice has the third strongest like- and cross-polarized σ^0 values, rough first-year ice and moderately rough first-year ice have the lowest σ^0 values. The mean values of horizontal like-polarized σ^0 for the various targets are −7.8 dB for iceberg ice, −21.0 dB for the island, −27.5 dB for old ice, −24.0 dB for

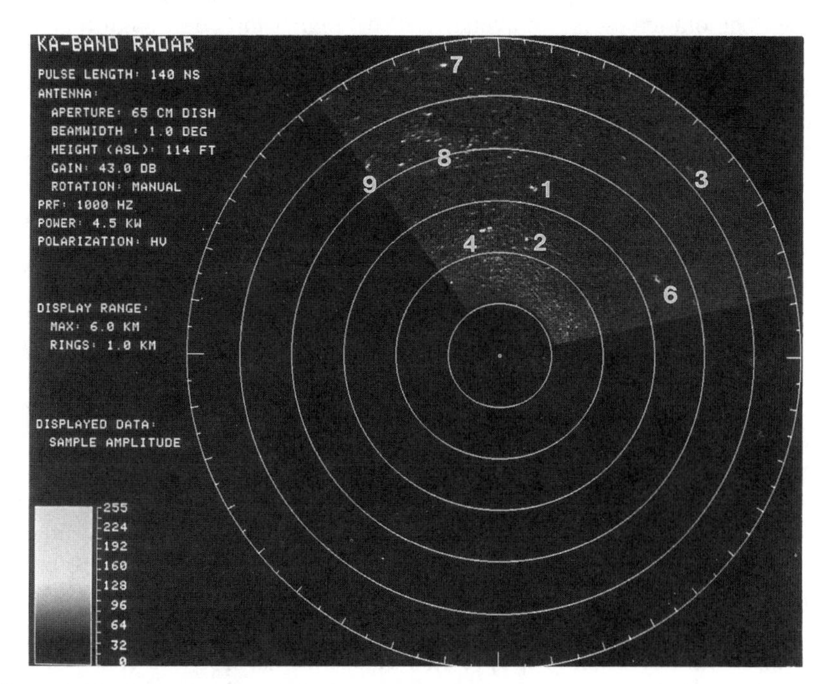

Figure 8.21 K_a-band cross-polarized image. Targets include: (1) Large iceberg, Line 307; (2) Small iceberg, Line 307; (3) Iceberg, Line 341; (4) Icebergs; (6) Island; (7), (8), and (9) Old-ice floes.

TABLE 8.8 Normalized Radar Cross Sections at K_u-Band for Various Ice Features

	Radar Cross Section (dBm^2)							
Features	σ^0_{HH}	σ^0_{VV}	σ^0_{HV}	σ^0_{VH}	$\sigma^0_{HH}/\sigma^0_{VV}$	$\sigma^0_{HH}/\sigma^0_{HV}$	$\sigma^0_{VV}/\sigma^0_{VH}$	Average Like/ Cross
Icebergs	−7.8	−11.6	−16.2	−16.2	3.8	8.4	4.6	6.5
Old ice	−27.5	−25.9	−27.0	−27.2	−1.7	−0.5	1.3	0.4
First-year ridges	−28.4	−29.5	−37.1	−37.8	1.3	8.7	7.9	8.3
First-year rough ice	−28.7	−28.7	−42.0	−42.0	0	13.3	13.3	13.3
Island	−21.0	−14.0	−18.0	−20.0	−7.0	−3.0	6.0	1.5

rough first-year ice, and −27.0 dB for moderately rough first-year ice. The average σ^0 values for the last three categories span only 1.2 dB, which supports the difficulty, as shown in the figures earlier, of differentiating old ice from first-year ice. In contrast, the cross-polarized σ^0 values give average values of −16.2 dB for the icebergs, −18.0 dB for the island, −27.0 dB for old ice, −37.1 dB for rough first-year ice, and −42.0 dB for moderately rough first-year ice. The old ice σ^0 is, on average, 10 dB higher than the rough first-year ice which is, in turn, 5 dB stronger than the moderately rough first-year ice. As shown in the figures discussed previously, this substantially aids in the differentiation of these ice types.

Another interesting observation to be made from the data in Table 8.8 is the degree of depolarization caused by each target type. Columns *HH*/*HV* and *VV*/*VH* show the ratio of the like- to cross-polarized σ^0 for horizontal and vertical transmit, respectively. The average of the cross-polarized returns shows that the old ice was the strongest depolarizer with the like-polarization on average having a σ^0 of only 0.4 dB above the cross-polarization. The island was the next strongest depolarizer, with the like-polarized σ^0 being 1.5 dB above the cross-polarized σ^0. The icebergs were third, with a like-to-cross ratio of 6.5 dB, and the rough and moderately rough first-year ice had ratios of 8.3 dB and 13.3 dB, respectively.

The final point to make with respect to Table 8.8 concerns the ratio $\sigma^0_{HH}/\sigma^0_{VV}$. Considering only the ice, σ^0_{HH} is essentially equal to σ^0_{VV}, although for the iceberg σ^0_{HH} is higher by 3.8 dB. This result suggests that the conventional choice of horizontal transmit is also the most appropriate for surface-based sensing of ice.

The results tabulated in Table 8.8 can be demonstrated more graphically as shown in Fig. 8.22, where σ^0_{HV} is plotted versus ice hazard. It is clear that the lowest values (and the lesser hazards) are associated with moderately rough first-year ice, whereas the highest values (and greatest hazards) are associated

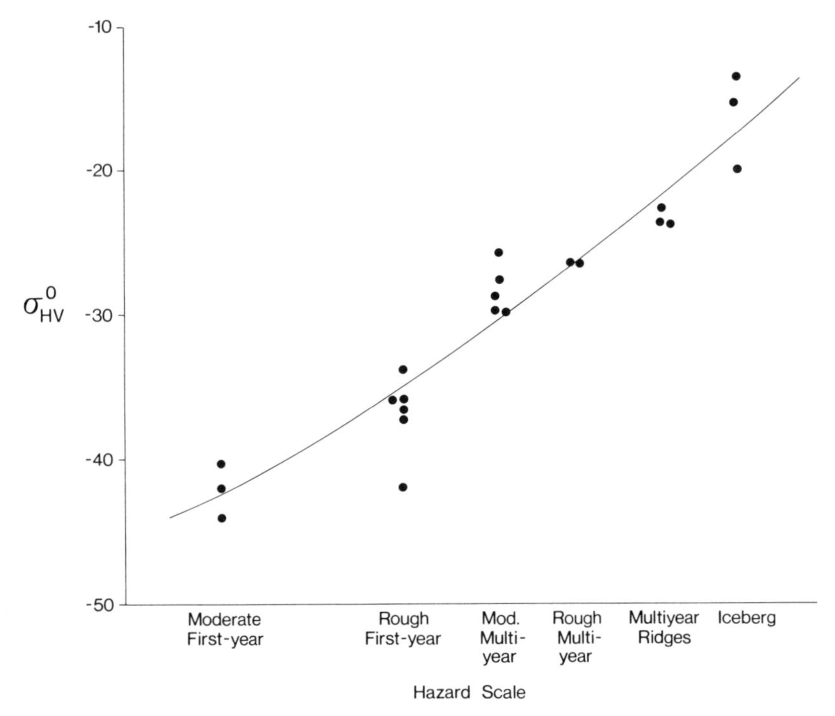

Figure 8.22 Plot of K_u-band cross-polarized σ^0 versus hazard scale.

with icebergs. In addition, there is a linear increase in σ^0_{HV} from moderately rough first-year, to rough first-year, to moderately rough old ice, up to the highest values associated with the icebergs.

Finally, a plot of the like- versus cross-polarized σ^0 values can be constructed, as shown in Fig. 8.23. It shows that a classification system can be established which would distinguish between icebergs and old or first-year ice, and between old ice and smooth or rough first-year ice. The separation between rough first-year ice and old ice is not as substantial. In classifying ice as old or first-year based on the σ^0_{HV} value, it would appear better to err on the side of caution and to declare any target with σ^0_{HV} greater than about -35 dB as old ice.

Scattering Theory

An understanding of the mechanisms that control the interaction of a surface-transmitted radar signal and various types of ice can now be formulated with reference to the experimental results previously discussed and a review of current theoretical models for radar backscatter.

It is first appropriate to reiterate the pertinent properties of the ice types under consideration. First-year sea ice is a high-loss dielectric medium that strongly attenuates radar energy. As shown in Table 8.4, the skin depth for

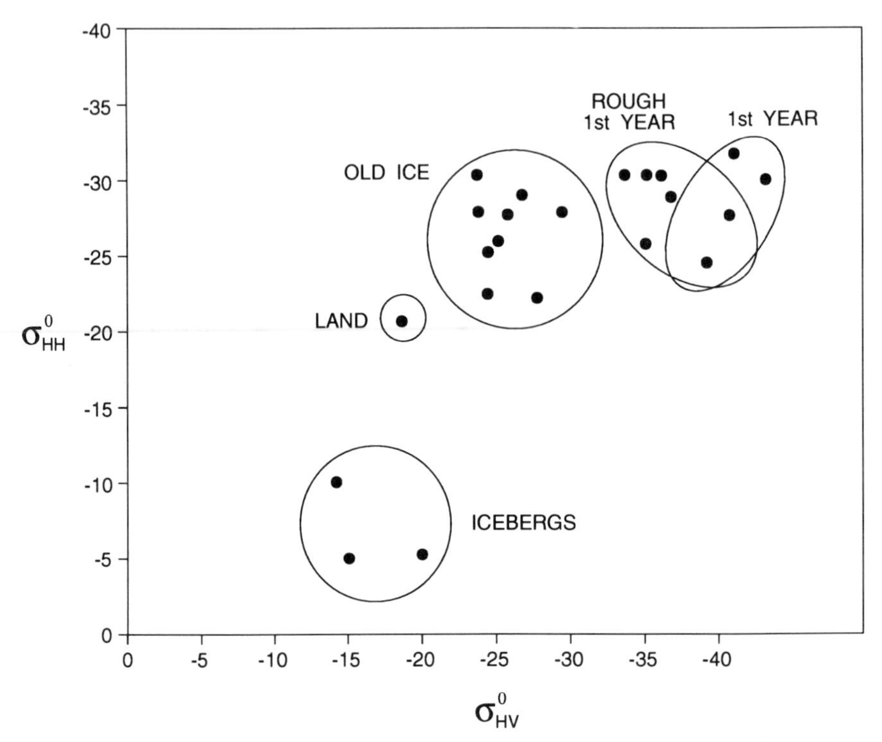

Figure 8.23 Plot of K_u-band like-polarized σ^0 versus cross-polarized σ^0.

first-year ice at *X* band is about 5 cm. Therefore, there will be little penetration of the radar wave into the ice and the signal returns will result mainly from surface scattering. The surface of first-year ice can be heavily deformed, producing significant vertical relief. With the high-loss and therefore reflective surface, considerable energy can be expected to return to the radar.

Old sea ice has low salinity (particularly in high-relief areas) and is a low-loss dielectric. Table 8.4 shows that the skin depth for old ice is considerable, resulting in significant penetration of the radar wave into the ice. This penetration will permit interaction between the radar wave and the inhomogeneities in the ice. The top layer of old ice contains a concentration of relatively large (1 to 2 mm, with as large as 5 mm) air bubbles or voids. Old hummocks and ridges have significant vertical relief, thus providing large local grazing angles for the microwave energy.

Ice of glacial origin (icebergs and their fragments) is salt-free and is a very low-loss dielectric. As Table 8.4 shows, the skin depth for glacier ice is also considerable, with low attenuation of radar energy permitting deep penetration of the signal. This results in interaction between the radar wave and any inhomogeneities in the ice. The inhomogeneities usually consist of small air bubbles (0.1 to 0.5 mm) entrapped in the ice during its conversion from snow, and cracks caused by pressure. Icebergs and bergy bits exhibit significant vertical relief; hence, they have appreciable local grazing angles for penetration by the radar wave. Growlers, on the other hand, have a low profile and can be well smoothed, providing local grazing angles equal to the radar grazing angle. In open water, the detection problem is further compounded by the effect of wave wash and sea spray which wet the surface of the growler and may be expected to reduce signal penetration into the ice.

Over the past 15 to 20 years considerable effort has been directed toward the modeling of radar-scattering phenomena. However, it is only in recent years that models have been developed that agree reasonably well with experimental results. Early work by Fung [15], and Beckmann and Spizzichino [3], amongst others, concluded that polarized backscatter was related exclusively to surface statistics and that depolarization resulted from edge effects only; thus early models could not account for any significant depolarization. Leader [32] and Rouse [53] showed that the depolarization was not exclusively a surface phenomena but was related to a subsurface scattering effect. Rouse [53], in fact, showed that for a single surface, polarization power was, to a first order, completely dependent upon the volume scattering. He concluded that the polarized return also contains a contribution from the subsurface and that for highly reflective surfaces, multiple scattering at the surface is enhanced and the subsurface contributions decrease.

Blanchard and Rouse [5], following up on the work of Rouse [53], used two formulations to account for each of surface and subsurface scattering. They concluded that the like-polarized cross section consists of contributions from both the surface and the subsurface, and that as the volume reflection coefficient increases, the contribution from the subsurface begins to dominate. They also concluded that the depolarized energy was independent of surface conditions and was only a function of volumetric scattering geometry and subsurface permittivity. Sheives [57] noted that a low density of scatterers in the volume caused only single scattering and did not contribute to depolarization, whereas a high particle density resulted in significant depolarization.

Fung and Eom [16], using an intensity (rather than a field) approach with a combined rough surface and volume scattering theory, modeled backscatter from sea ice. Their results, which agree well with observed data, suggest that:

1. The cross-polarized return results from multiple scattering in the volume that is enhanced by the rough surface.
2. For near-vertical incidence angles, the backscatter is dominated by surface scattering.
3. Beyond 25° incidence angle, the like-polarized return is dominated by volume scattering, which in turn is controlled by the permittivity of the medium and the rough surface boundary.
4. The absolute value of the like- and depolarized-returns and their separation is controlled by the scattering albedo of the volume and the boundary roughness.

Ellis [11] used the radiative transfer theory (accounting for volume and surface scattering) to model the scattering from icebergs. He modeled his results at X and K_u bands, and showed that K_u-band returns should be stronger than X-band returns and that there is little difference between horizontal (HH) or vertical (VV) returns. He also showed that cross-polarized returns should be quite high, being 2 to 8 dB below the like polarization. Ellis also looked at signal-to-clutter improvements in detecting icebergs in the ocean and in pack ice, and concluded that cross-polarization provides the best signal-to-clutter situation.

Kim et al. [26] used two models for backscatter returns from sea ice. They employed a surface scattering model for first-year ice that ignores volume scattering contributions, and a combined volume and surface scattering model for old ice. Using a physical optics model as an approximation to the Kirchhoff model, and with an exponential correlation function, they showed results that are in good agreement with experimental data for scattering from lossy first-year ice and wet (summer) old ice.

Using the radiative transfer method (intensity rather than field solution), Kim et al. [26] showed that the volume scattering characteristics are controlled by the dielectric constant of the ice and the size of the air bubbles. For old ice, air bubbles in the recrystallized ice layer at the top were assumed to be the primary volume scatterers. They noted that surface scattering is always present and that at frequencies below X band the volume scattering contribution for old ice can be lower than the surface scattering contribution. It is concluded that depolarization is a secondary effect for the surface scattering associated with first-year ice whereas volume scattering causes considerable depolarization, and this should facilitate discriminating old from first-year ice.

All the modeling results discussed have assumed that any scatterers within the medium have been Rayleigh scatterers (that is, the radar wavelength is large compared to the circumference of the scattering element). If the scatterers begin to become non-Rayleigh, the model results cease to apply.

From the modeling results, it is reasonable to assume that for high-loss first-year ice, the major scattering mechanism is surface scattering. As the modeling efforts discussed do not account for any depolarization caused by surface scattering, the depolarization that does occur must result from multiple scattering at the surface, and (possibly) from a small amount of volume scattering in the near-surface layer. The low average values of cross-polarized σ^0_{HV}, discussed earlier in this subsection and the high like- to cross-polarized ratios are consistent with this explanation. Radar returns from highly saline first-year ice can be assumed to be predominantly the result of surface scattering, and as such will be related directly to surface roughness.

From the modeling results discussed, the returned energy from low-loss dielectrics, such as glacial and old ice, will be assumed to be the result of three scattering mechanisms: surface scattering, multiple scattering at the surface, and volume scattering. The contribution of each will depend on the roughness of the surface and the size of the scatterers with respect to the wavelength within the volume.

Frequency Effects. Weeks and Ackley [65] stated that if scatterers in the volume are less than about 0.1 wavelengths of the incident radiation, then the contribution of the scatterers can be neglected. Conversely, as particle sizes approach the wavelength of the radiation, then scattering by individual particles becomes significant.

For typical old ice with near-surface scatterers of about 2.0-mm diameter, according to Weeks and Ackley [65], the wavelength should be no longer than 2.0 cm for the individual scatterers to have an effect on the total return. Therefore, for old ice it is anticipated that significant scattering from the inhomogeneities will not result at X band but will result at K_u and K_a bands. For glacial ice with bubble sizes assumed to be 0.5 mm, the radar wavelength should be no longer than 0.5 cm for scattering from inhomogeneities to be significant, so that it is not anticipated that even K_a band will show strong evidence of scattering from the inhomogeneities.

The situation can be appreciated in another fashion by looking at Fig. 8.24 (based on Skolnik [58]), which plots the radar cross section of a sphere as a function of the ratio of circumference of the sphere to the radar wavelength. Targets that are small compared to the wavelength are Rayleigh scatterers. They will have small radar cross sections, and therefore scatter little energy. As the size of the particle increases relative to the wavelength, the cross section will increase, resulting in the scattering of more energy. Superimposed on this figure are points representing scatterers in old and glacier ice at X and K_a bands.

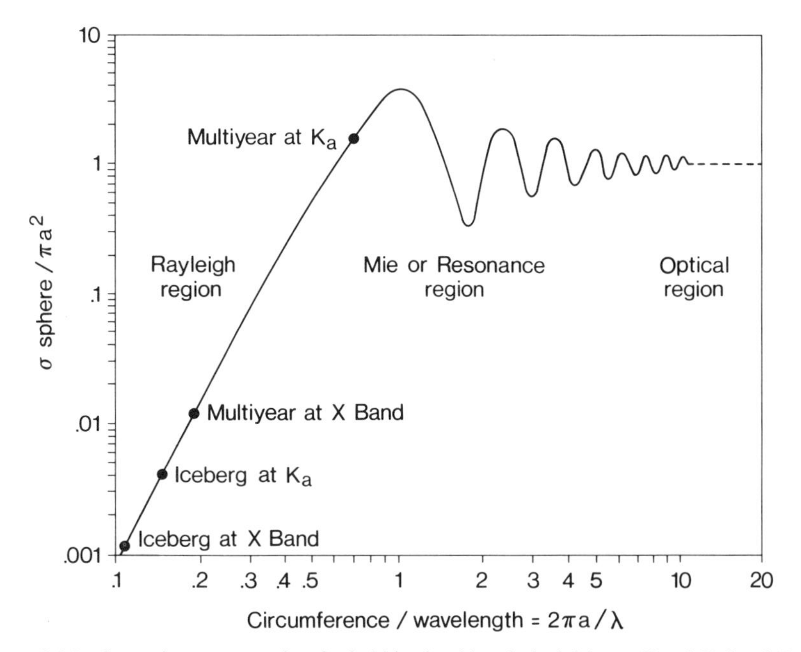

Figure 8.24 Scattering cross-section for bubbles in old and glacial ice at X and K_a band (Based on Skolnik [58]).

The scatterers in the old ice are again assumed to be 2.0 mm in diameter and the scatterers in the glacier ice are assumed to be 0.5 mm in diameter. It is apparent that at the same frequency the scatterers in old ice have a larger cross section than the scatterers in glacier ice, and as the frequency increases the cross section of the scatterers in a particular ice type increases towards the optical limit.

Kim et al. [26] modeled the effect of air bubble size on the scattering albedo of ice. Figure 8.25, after Kim et al., shows the albedo as a function of frequency for a number of bubble sizes. Curves for old ice (large bubble diameter 5.0 mm) and iceberg ice (bubble diameter 0.5 mm) have been added to Kim's plot. Again, it is clear that although the 2-mm (or larger) bubbles in old ice will scatter some energy at X band, the albedo increases significantly at K_u band. It is equally clear that the 0.5-mm bubbles in iceberg ice have a low albedo even at K_u band.

The validity of these assumptions can be judged by reviewing the results discussed in Section 8.4.3 and shown in Table 8.6: the average σ^0 values for first-year ice, old ice, and glacier ice increased by 35 dB, 50 dB, and 20 dB, respectively, from S band to K_a band. Weathering of old ice during the summer melt makes the surface smoother, in a large-scale sense, than that of first-year ice, although it may have the same vertical relief. Assuming that the surface scattering is related directly to surface roughness, as the frequency increases from S to K_a band (causing the ice to appear rougher to the radar), the rougher first-year ice would be expected to show a larger increase in backscatter than the smoother old ice. However, the σ^0 of old ice increased 15 dB more than the first-year ice. Therefore, the increase in the σ^0 value of old ice over and above the increase for first-year ice must be caused by volume scattering.

Another effect of increasing the radar frequency is to direct more radar energy onto the ice surface, as discussed in the subsection on multipath inter-

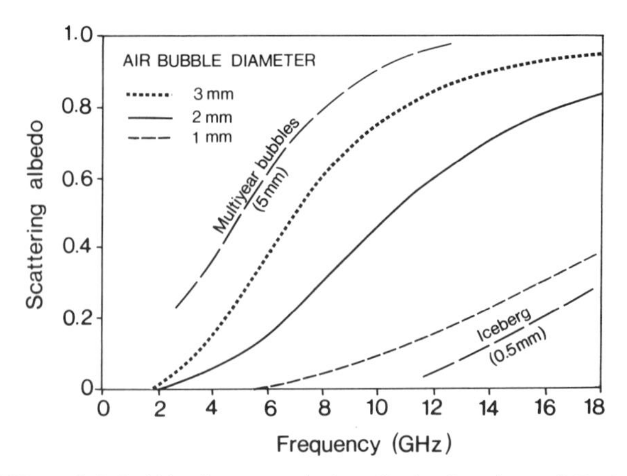

Figure 8.25 Effect of air bubble size on scattering albedo. Density = 0.7 g/cm^3, e = 3.15 − j0.01 (Kim et al. [26]).

ference. As a result, both the first-year and old ice will exhibit an additional increase in backscatter as the frequency increases because of this improved surface illumination. The iceberg, with larger vertical extent, will not benefit from this effect to the same degree. For example, the σ^0 value for glacier ice increased by only 20 dB as compared to the significantly larger increases for first-year and old ice. The larger part of the 20-dB increase for glacier ice is attributed to volume scattering. However, the volume scattering effect is significantly less than in the old ice because of the smaller sizes of the scatterers in the glacier ice.

Table 8.7 shows the change in σ^0 values from K_u to K_a band: the average first-year ice σ_0 increased by 16 dB, the old ice σ^0 increased by 18 dB, and the glacier ice increased by 4 dB. The increase in σ^0 for first-year ice is caused mainly by the increase in roughness of the first-year ice at K_a band. However, it is apparent that there may be, in fact, some volume scattering from the first-year ice as well. First-year ice contains air bubbles of about 0.5 mm as well as brine drainage pockets. As the first-year σ^0 values were collected in the late spring, it is probable that some of these pockets contained air voids. Although the radar signals would not penetrate far into such saline ice, these voids near the surface could contribute to some volume scattering.

The increase in old ice σ^0 value from K_u to K_a band (Table 8.7) was partially a result of the increased surface roughness at K_a band, but to a greater extent resulted from increased volume scattering from the inhomogeneities in the ice. The σ^0 for glacier ice, on the other hand, has on average only increased 4 dB from K_u to K_a band. It is concluded that, as the surface roughness would have increased at K_a band, there could have been little contribution to scattering from small air pockets in the glacier ice.

It is anticipated that volume scattering for glacier and old ice is made up of two components. There is scattering from surfaces within the volume as well as scattering from air bubbles within the volume. For old ice, there is substantial scattering from air bubbles as well as a contribution from the internal surfaces. For glacier ice, the volume scattering is largely from the internal surfaces because the air bubbles are considerably smaller, which results in considerably less contribution from volume scattering for glacier ice.

In summary, as the wavelength of the radar signal approaches the physical size of inhomogeneities in the ice, significant increase in returned energy will be experienced because of enhanced volume scattering. For old ice, this occurs at frequencies above X band and for glacier ice it is anticipated to occur at frequencies above K_a band. The experimental results also suggest that there may be some volume scattering from late-season first-year ice.

Polarization Effects. As discussed in the review of theoretical modeling of scattering, surface scattering does not depolarize the incident energy. Multiple scattering at the surface results in some depolarization, although the majority of the depolarized energy is expected to result from volume scattering.

Table 8.8 shows that for K_u band, the ratio of like- to cross-polarized σ^0 for old ice is 0.4 dB on average, indicating that old ice is a strong depolarizer. As indicated in the subsection on polarization, the K_u-band results were examined because of low return levels from the K_a-band radar, particularly for cross-polarized signals. Table 8.8 shows that glacier ice is another reasonably strong depolarizer, with an average like- to cross-polarized σ^0 ratio of 6.5 dB. Although both ice types are penetrated easily by radar waves, the old ice depolarized the radar signal by about 6 dB more than the icebergs. This larger depolarizing effect in the old ice was caused by a larger contribution to volume scattering from inhomogeneities within the ice. Again referring to Table 8.8, rough first-year ice had an average like- to cross-polarized ratio of 8.3 dB, whereas moderately rough first-year ice had a ratio of 13.3 dB. It is, therefore, proposed that much of this depolarizing effect resulted from surface roughness, that is, multiple scattering at the surface, although there may be some small contribution from volume scattering caused by air bubbles in the volume.

The relatively small difference between the depolarizing ratio for glacier ice and that for rough first-year ice is accounted for by the depolarization in both cases coming mostly from multiple scattering at the surface. The 2-dB higher ratio for icebergs probably results from the multiple scattering from surfaces within the volume of the icebergs. In contrast, the strong depolarization occurring in old ice is caused mainly by volume scattering.

In the analysis of the 1984 open-water data (described in Section 8.4.4), it is shown that icebergs in open water also depolarize the radar energy to a significant extent. Unfortunately, the sea also depolarizes the radar wave such that there is little or no target-to-clutter improvement in using the depolarized returns.

Summary of Scattering Results

The foregoing sections have formulated an explanation for the radar backscatter from ice which agrees with the experimental results discussed and agrees with the modeling. The theoretical models and experimental data suggest that for first-year ice, the main scattering mechanism is surface scattering, with multiple scattering at the surface contributing to and causing a small amount of depolarization. There is a possibility that there may be a small contribution from volume scattering in late-season first-year ice.

The scattering from old ice is a result of surface scattering, multiple scattering at the surface, and predominantly, volume scattering. The depolarization of the radar wave is caused to a large extent by volume scattering.

For glacier ice there is extensive penetration of the radar wave into the ice. The scattering mechanisms controlling the return of energy are surface and multiple scattering. The relatively smaller depolarization, when compared to that of old ice, is mainly a result of multiple scattering at the surface and scattering from internal surfaces within the volume.

These conclusions are consistent with the smaller X-band depolarization noted in the 1983 data, whereas K_u- and K_a-band radars produced more substantial depolarizing effects. They also explain the stronger depolarizing effect in old ice than glacier ice and the strong like-polarized returns from old ice at K_a band, as there are larger scatterers in old ice.

Other Parameters

There are several other radar parameters that can have an effect on the ability of the radar to classify (and detect) ice targets. Three of the parameters examined are height of radar antenna, azimuthal beamwidth of radar antenna, and transmitted pulse length. The last two define the resolution of the radar.

Antenna Height. The height of the radar antenna above the sea surface has a major effect on the performance of the radar system. It is intuitively obvious that for longer range detection, a higher antenna is required. This has been the conventional wisdom and is demonstrated in shipborne installations, where the radar antenna is located as high as possible. For most conditions this wisdom is well founded, although there are some exceptions.

The radar horizon, for a normally refractive atmosphere, was given in (8-11). The range increases as a function of the square root of the height of the antenna. For an antenna height of 15 m, the radar range to the horizon is 8.6 nm, whereas for a 30-m antenna it is 12.2 nm. The higher antenna results in a significant improvement in the maximum range of detection, given sufficient signal return. In addition to affecting the range to the radar horizon and maximum range of detection, the antenna height affects the propagation factor, F, as (8-17) shows. Figure 8.10 showed a higher antenna results in the first antenna lobe maximum occurring closer to the surface, putting more energy on low-lying targets and improving the probability of their detection.

As a further consequence of the multipath phenomenon, the target average single-pulse power return drops off as $1/R^4$ out to the transition range, and thereafter drops off at a $1/R^8$ rate. This effect results in targets beyond the transition range being illuminated with significantly less power. As (8-19) shows, the transition range is directly proportional to the height of the antenna and thus by increasing the height of the antenna the transition range can be extended, which will result in improved probability of detection out to farther ranges.

As shown in Fig. 8.26, as the height of the antenna increases, the grazing angle for a particular range also increases. This increased grazing angle will allow increased penetration of the radar signal into the ice target; in the case of classification of ice targets, it will allow more volume scattering and will improve the ability to differentiate ice types. In the case of the detection of an ice target in open water, the increased penetration will also result in more volume scattering, and improve its detectability by separating the ice target from the sea clutter.

One possible drawback of increasing the height of the antenna is that as the height of the antenna increases, the range to which sea clutter is a problem also increases. For larger targets, the extended range of the sea clutter and its increased intensity may visually mask the returns from the target. To detect

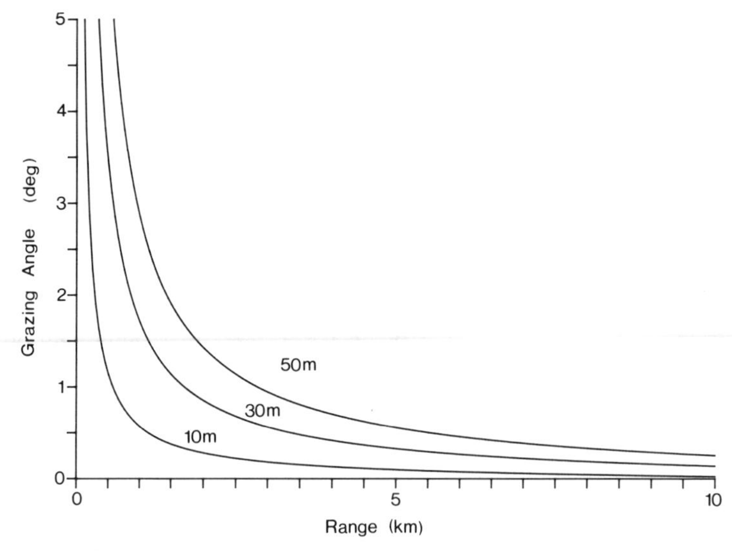

Figure 8.26 Grazing angle versus range for antenna heights of 10, 30, and 50 m.

small targets, however, it is necessary to have the radar energy on the surface, and for a fixed radar frequency, this is achieved by raising the antenna. As noted by Williams [68], to detect small targets it is first necessary to display the clutter. Once the surface has been illuminated as evidenced by the appearance of clutter, clutter-suppression techniques can be employed to make the target detectable.

Another potential difficulty with raising the antenna height is related to the ducting phenomena. The propagation ducts anticipated to be encountered in the Canadian offshore area are generally shallow, near-surface ducts, and the optimum placement of the radar antenna to take advantage of such ducting conditions is within the duct; therefore, lower, near-surface placement of the antenna is suggested. However, although the optimum placement of the antenna is within the duct, antennas placed above the duct can experience extended propagation ranges because of the leaky nature of the duct, although not to the same extent as a lower antenna.

Several points related to antenna height are illustrated in Figs. 8.27 and 8.28. These figures show images of radar returns from a full ice cover with an *X*-band radar. The radar and ice conditions for both images are identical, except that Fig. 8.27 is for the radar antenna mounted at 37 m, whereas Fig. 8.28 is for the antenna mounted at 29 m. Although the difference in elevation between the two antennas is minimal, there are significant differences between the two images. Figure 8.27, from the higher antenna, has stronger signal returns and more texture, particularly at far range, resulting from more energy reaching the ice because of the multipath factor. Figure 8.27 also shows detection of targets at greater ranges, particularly in the area of A.

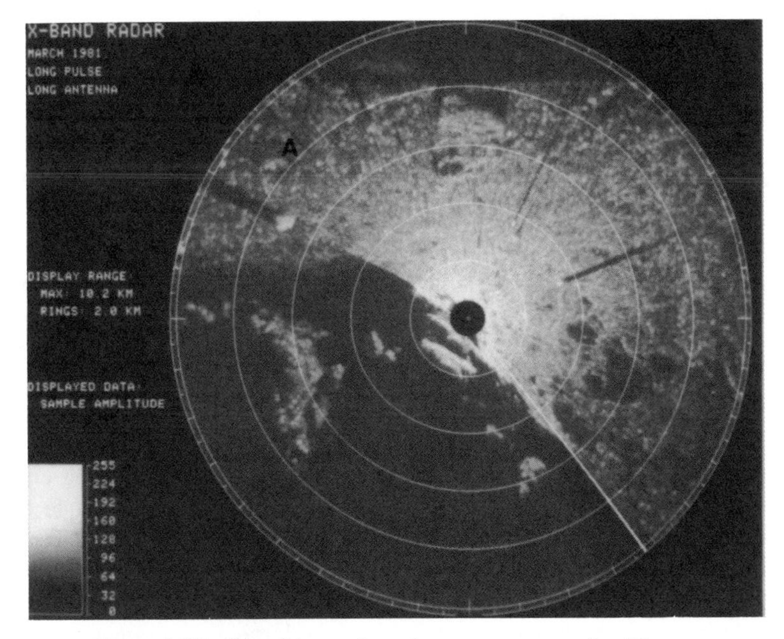

Figure 8.27 *X*-band image for radar antenna mounted at 37 m.

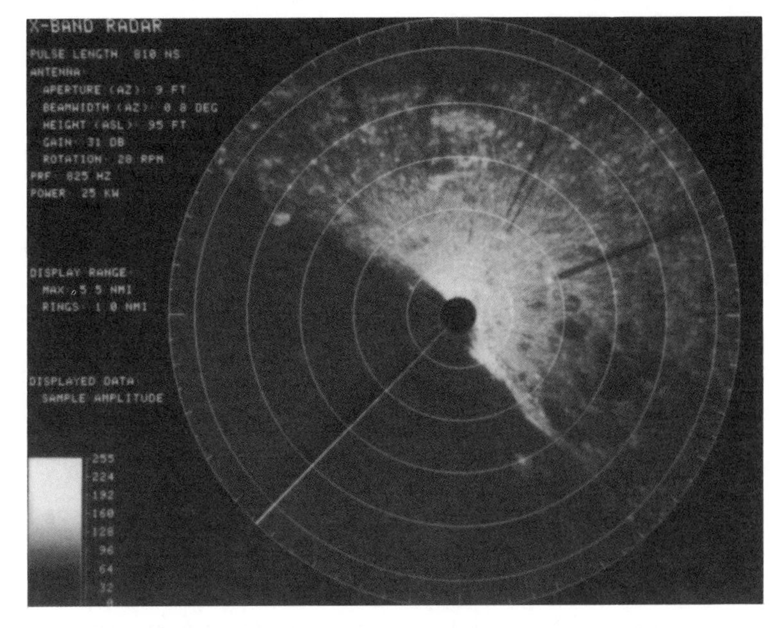

Figure 8.28 *X*-band image for radar antenna mounted at 29 m.

Figure 8.28 shows longer shadows behind the large targets as a result of the lower height of the antenna. In addition, the image shows more discrete targets in the intermediate ranges, because there is less energy near the surface, caused by the multipath factor, and only targets with some significant vertical extent can be illuminated. These effects lead to a reduction in the indication of area roughness.

Radar Resolution. Radar resolution is defined as the ability to separate two targets. It has both a range and an azimuthal component. Resolution in range is determined by the pulse length of the radar and azimuthal resolution is determined by the beamwidth (3 dB) of the antenna. The effects of varying pulse length and azimuthal resolution are examined separately below.

Azimuthal Beamwidth. Azimuthal beamwidth of the antenna, as described in Section 8.2.3, controls the azimuthal resolution of the radar. As the antenna beamwidth increases from 1 to 2°, the resolution is degraded by a factor of two; that is, for two targets to be resolved independently they must be twice as far apart in bearing with a 2° antenna as with a 1° antenna. As the spatial azimuthal resolution is also a function of range, the spatial resolution in azimuth for a wide beamwidth antenna can become very large as the range increases.

In target detection mode in sea clutter, the wide beamwidth antenna results in large clutter cells. The illuminated area is equal to the azimuthal resolution times the range resolution, so that a doubling of azimuthal beamwidth results in a doubling of the clutter cell, and subsequently causes a reduction in the target-to-clutter signal ratio and a reduced detection probability. However, both Croney et al. [9] and Williams [67] show examples of increased target detection with wider azimuthal beamwidth antennas. The explanation of this is twofold. First, the wider azimuthal beamwidth antenna smears or broadens the return from the target and thus makes it more recognizable on a PPI display. Second, as the resolution of the radar system is increased, the statistics of the returns depart from Rayleigh and develop a higher probability of large returns. It is these large returns that make the target more difficult to see.

As just noted, the wider beamwidth antenna results in a smearing of target returns in azimuth. In many cases, as in the detection of a ship at sea, this smearing results in easier identification of the target. However, for ice classification this can be a disadvantage, as shown by Figs. 8.29 and 8.30. The figures show images of the radar returns with an *X*-band radar for a full ice cover. The radar parameters for the two images are identical, except that the image in Fig. 8.29 was collected with a 2.8-m antenna with an azimuthal beamwidth of 0.8°, while the image in Fig. 8.30 was collected with a 1.2-m antenna with an azimuthal beamwidth of 1.9°. Figure 8.29 is a sharper image than Fig. 8.30 which is blurred or smeared. In particular, the shadows behind the large ice targets at A and B were filled in completely by the larger beamwidth in Fig. 8.30, thus significantly decreasing their visibility. The shadow behind target C was also reduced, but was still visible. Shadows are a valuable

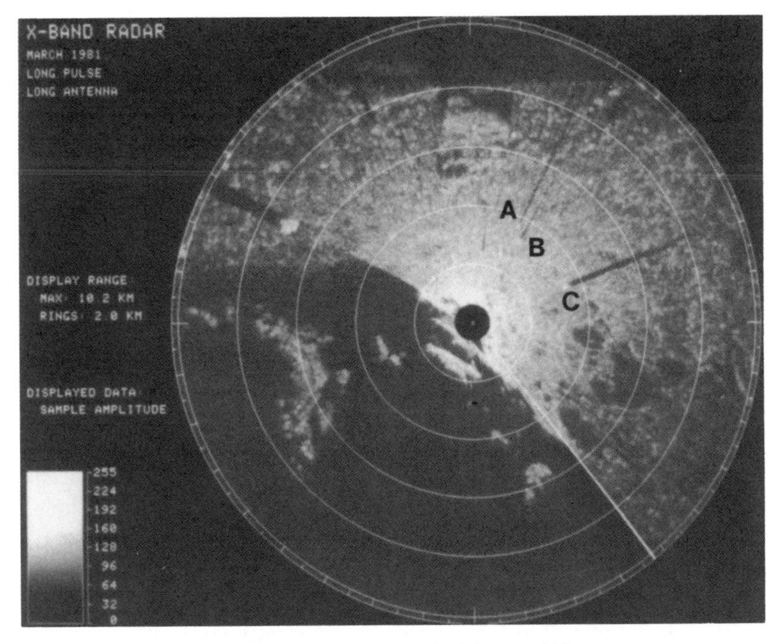

Figure 8.29 *X*-band image for 0.8° beamwidth antenna.

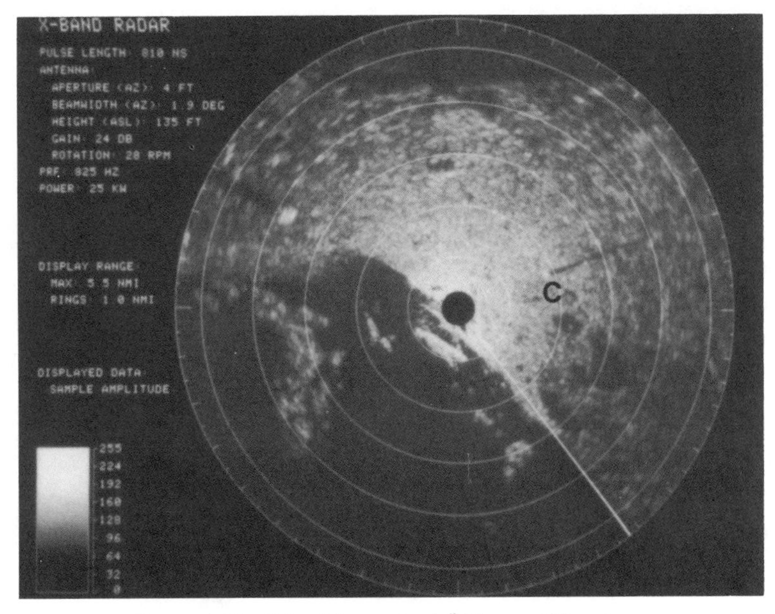

Figure 8.30 *X*-band image for 1.9° beamwidth antenna.

indication that a target has significant vertical relief. Target C was less visible because of the smearing of its return, as were other targets, particularly at intermediate and far ranges.

Transmitted Pulse Length. The resolution of the radar in range was defined in (8-3), where the resolution in range is directly proportional to the output pulse length. If the pulse length is doubled, the distance in range between two targets must also double for them to be resolved individually.

Although it is desirable to have the maximum resolution possible, a serious problem can develop as the pulse length is shortened. The energy reflected from a target is related directly to the average power incident upon it, and as the average power is determined by the peak power times the PRF times the duration of the pulse, shortening the pulse reduces the energy returned from the target. Figure 8.31 shows images from a radar with the same radar parameters, with the exception of pulse length. As the pulse length decreases from 810, to 220, to 70 ns, the image becomes sharper because of the increasing range resolution. However, also noticeable in Fig. 8.31 is the decrease in range of coverage resulting from the reduction in average power (about 3 dB less power per pulse reduction). In the 70-ns image, targets at far range are not detected and the returns from closer targets are significantly reduced.

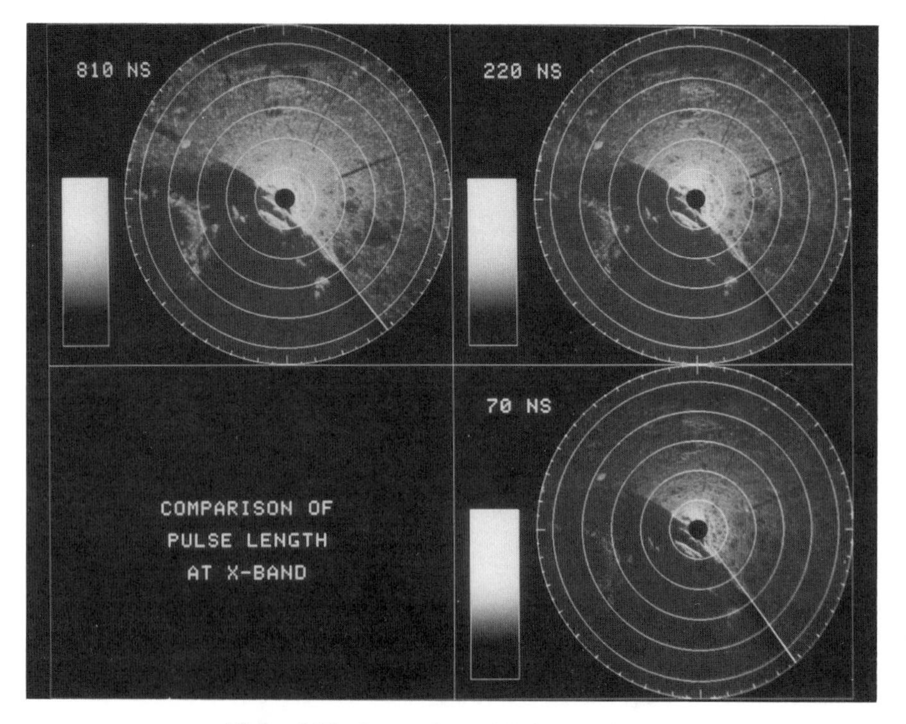

Figure 8.31 Images for various pulse lengths.

This problem of reduced coverage with shorter pulses can be overcome with pulse compression techniques (see Chapter 6, Section 6.2.7). Pulse compression uses long coded transmission pulses which, upon reception, are decoded to yield an effective pulse length much shorter than that of the transmitted pulse. The longer transmitted pulse yields higher average transmitter power and thus longer range of coverage. Compression ratios, that is the ratio of the length of the actual transmitted pulse to the length of the effective compressed pulse, are typically in the range of 50 to 200. For a compression ratio of 50, for example, a real pulse of 1 μs with resulting high average power could be transmitted, but range resolution corresponding to a 20-ns pulse could be achieved. The difficulty with pulse compression systems is that the coded pulse is usually generated at the intermediate frequency of the radar, using surface acoustic wave (SAW) devices for example. The pulse must then be upconverted to the radar's transmit frequency, and amplified for transmission. The inexpensive magnetron transmitter usually used in noncoherent radar is not suitable for pulse compressed operation.

A further difficulty with short pulses, either natural or compressed, is their effect on sea clutter characteristics as the pulses become progressively shorter. This phenomenon is addressed in Section 8.4.4.

Statistical and Neural Classifiers

Researchers at McMaster University have investigated the use of various advanced techniques for classifying the ice types contained within the radar images collected during the 1983 field program.

Bayesian Classification. The initial investigation (Murthy and Haykin [41]) involved the use of the Bayes classification procedure to discriminate among four classes: first-year ice, old ice, icebergs, and radar shadows. The data consisted of K_u-band like- and cross-polarized images, and a like-polarized S-band image. Images of a single polarization were analyzed in B-scan form.

If classes A_i, $i = 1, M$, define a sample space, and the probability $P(A_i) > 0$ for all i, then the conditional probability of A_i given B is defined by

$$P(A_i|B) = \frac{P(B|A_i)P(A_i)}{\sum_{i=1}^{M} P(B|A_i)P(A_i)} \tag{8-24}$$

For the ice classification problem, A_i corresponds to the four image classes, and B corresponds to the feature vector of an unknown pixel x. $P(B|A_i)$ simply describes the probability of an event B belonging to class A_i. $P(A_i)$ represents the probability of occurrence of class A_i in the field of interest. The $P(B|A_i)$ are obtained by approximating the observed histograms for different ice types by continuous density functions. $P(A_i)$ was assumed to be equal for all cases, that is, $P(A_i) = 1/M$ where M is the number of classes (in this case, 4). Classification is performed by maximizing the a posteriori probability obtained in the above equation which is the Bayes rule for optimal classification. For the ice classification problem considered, the feature vector was one dimensional, with the feature that was measured being reflectivity.

Before using the images for classification, preliminary processing was performed in four steps:

1. A 3 × 3 median filter was used to remove noise spikes.
2. The digitized gray levels were converted to return power using the radar calibration data.
3. The data was normalized in range to remove the range R^{-3} dependence.
4. Thresholding was performed to remove noise-only samples which were within two standard deviations of the noise mean.

The first step in the classification procedure was to accumulate reflectivity histograms for each of the four classes. Using the Pearson rule (Johnson and Kotz [25]) for unimodal distributions, the distribution which most closely fit the histograms was determined to be the beta distribution. Each of the histograms was then fitted with the appropriate beta distribution, whose parameters were estimated from the training data. These distributions were then used to train the classifier.

The results of the classification are presented in color images in Murthy and Haykin [41]. To provide a numerical indication of performance, the classification accuracy was determined as the percentage of correctly classified sample points out of the total number of sample points classified. Because of the limited amount of data available, it was necessary to use the training data in this evaluation, thereby biasing the result. The classification performance measurements, based on single-polarization images, are given in Table 8.9, opposite "Beta." For comparison, results using a Gaussian classifier are also given. Murthy indicated that the misclassification could be reduced by logically combining the like- and cross-polarized results, and by taking advantage of the observation that all the icebergs had sufficient vertical height to cast radar shadows. Therefore, correlating the occurrence of shadows with the location of suspected icebergs should improve the classification performance.

Gaussian and Neural Network Classifiers. Later research at McMaster University (Orlando et al. [46]) addressed the simultaneous use of both like- and cross-polarized data, examining four approaches in all. The initial approach was to combine the like- and cross-polarized data into one composite image, yielding a one-dimensional vector, using principal component analysis. The composite image was then classified using a one-dimensional Gaussian classifier, which was found to offer essentially the same performance as a Beta classifier. This method succeeded in increasing the separation of old ice and iceberg classes as compared to using only the like-polarized image, and increased the separation between shadows and first-year ice classes compared to

TABLE 8.9 Summary of Classifier Results (% Correctly Classified)

	Shadows	First-Year Ice	Old Ice	Icebergs	Average
SINGLE-IMAGE INPUT					
Like-Polarized Image					
Gaussian	93.3	83.1	57.1	73.6	76.8
Beta	96.9	82.9	75.7	*	85.2
Cross-Polarized Image					
Gaussian	58.9	43.6	43.3	68.7	53.6
Beta	**	**	52.1	67.5	59.8
TWO-IMAGE INPUT					
Principal Component Analysis +					
1D Gaussian	86.9	84.2	58.7	76.9	76.7
2D Gaussian	93.3	84.2	72.7	77.8	82.0
Multilayer Perceptron (6 hidden units)	91.8	86.2	73.6	78.7	82.6
Self-Organizing Feature Map (40 neurons with LVQ)	90.7	88.7	74.2	74.9	82.1

*For the like-polarized image, the results for the beta classifier are available only when old ice and icebergs are treated as one class.
**For the cross-polarized image, the results for the beta classifier are available only for the old and iceberg classes.

the cross-polarized image. The classification results are given in Table 8.9, under Two-Image Input.

A second approach was to use a two-dimensional Gaussian classifier. The sample vector was two-dimensional, with samples from the like- and cross-polarized images at corresponding spatial coordinates. The use of the combined image data improved the separation of old ice and iceberg classes. Table 8.9 shows the results, opposite 2D Gaussian.

Two more classifiers were tested, each based on artificial neural network designs. The first was a multilayer perceptron, trained with the back-propagation learning algorithm. The tested network consisted of three layers of neurons: an input layer with like- and cross-polarization inputs, a hidden layer with a variable number of neurons, and an output layer with four neurons, one for each of the four output classes. The three layers were connected by variable weights, designated by the lines in Fig. 8.32. A given neuron computes the weighted sum of its inputs from the previous layer, deducts a threshold term from this value, and sets its output according to the following rule:

$$f(net) = \frac{1}{1 + \exp(-net)} \tag{8-25}$$

where *net* is the result of the weighted sum/subtraction operation.

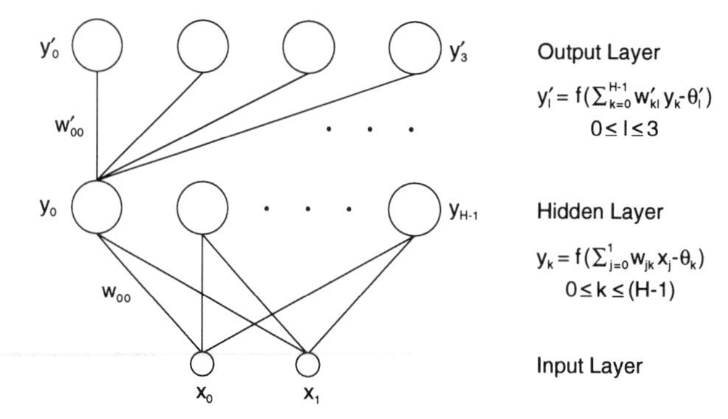

Figure 8.32 Multilayer perceptron classifier. Each circle represents one neuron. There are two input neurons (one for the like-polarized channel, and the second for the cross-polarized channel) and four output neurons (one for each of shadows, first-year ice, old ice, and icebergs). There are H hidden units between the input and output layers. The connecting lines are variable connection strengths (or weights) that describe the network behaviour.

The network is trained so that the weights model the input distribution. The training phase consists of applying a series of training patterns (sets of like- and cross-polarized data) to the network, propagating the results forward to observe the output, and correcting the weights if the output is incorrect. The weights are corrected by feeding the error backwards to apply a small correction to the weight values. The back-propagation process is an iterative algorithm, with the learning speed controlled by two parameters: the learning rate and a momentum constant. For the ice data used for testing, the classification performance did not improve significantly for networks with more than six neurons in the hidden layer. Table 8.9 gives the classification results using a multilayer perceptron with six hidden units.

The second neural network approach involved the Kohonen self-organizing feature map (Kohonen [29]), which forms an approximation to the input distribution, used to classify the data. A Kohonen network consists of a linear array of neurons, arranged as shown in Fig. 8.33. The input to the network, an M-dimensional feature vector, is connected to every neuron. Each neuron has a weight vector associated with it, of the same dimension as the input vector. When the input vector is presented to the network, it is broadcast to all cells, which compute the Euclidean distance between the input vector and each weight vector. The neuron with the weight vector which gives the minimum distance is termed the winning neuron. The weight vectors in a neighborhood around the winning neuron are then updated iteratively to be nearer the input vector, with the size of the neighborhood gradually decreasing until only the winning neuron is included.

To improve the resulting decision boundaries of the classifier, an additional procedure called learning vector quantization (LVQ) was applied. This procedure involves making small changes to the weight vectors, so as to smooth the decision regions of the classifier. Table 8.9 summarizes the classification

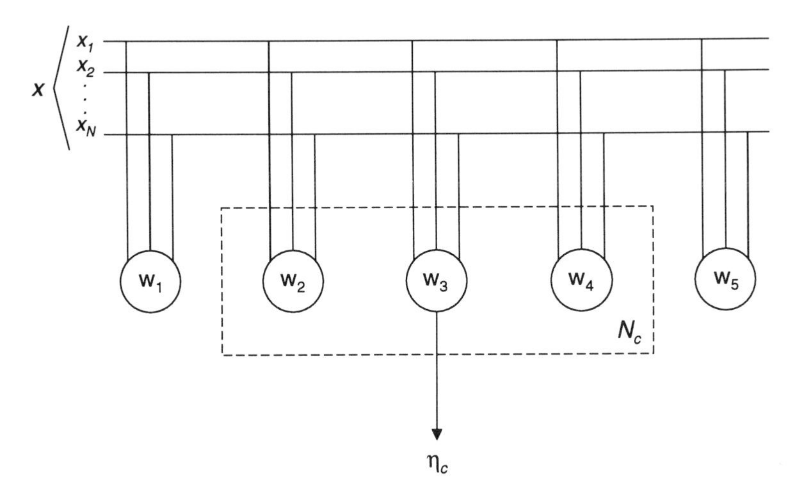

Figure 8.33 A one-dimensional Kohonen feature-map neural network. See description in text.

performance of a self-organizing feature map containing 40 neurons, combined with LVQ.

Conclusion. For classification based on a single polarization, the beta classifier worked well for like-polarized images. The major drawback was its inability to distinguish between old ice and icebergs. By using both the like- and cross-polarized images simultaneously, this latter problem was overcome, and good classification performance was achieved. Encouraging results were obtained for the use of neural network classifiers, whose performance equaled or surpassed that of the traditional Gaussian technique. In addition, neural networks have a number of properties (Haykin [19]), such as the ability to generalize, which make their use very attractive for a practical system. Their application warrants further study. Overall, the above studies showed that it is possible to construct a classifier which will automatically separate radar images into four classes: shadows, first-year ice, old ice, and icebergs. This image-classification capability has significant application potential for ice navigation requirements.

8.4.4 Ice Detection

Although significant progress has been made on the classification of large ice forms from a surfaced-based sensor, much remains to be done with respect to the detection of small ice targets in the sea. A research project carried out by the authors investigated target detection in low sea states using a noncoherent radar. Further experimental measurements made on the east coast by McMaster University, and earlier field evaluations conducted under the auspices of the Environmental Studies Revolving Fund (ESRF), contribute additional understanding. Models for predicting the detection performance of a noncoherent radar have been developed.

DFO/McMaster Arctic Experiment

Whereas the ice targets examined in the previous sections were locked in a stationary ice cover during the field programs, the ice targets examined in the 1984 experiment were in open-water conditions and many experienced significant motion and aspect change during the data collection. All the measurements during the 1984 program were made with a dual-polarized *X*-band radar system. It was expected that the targets whose aspect changed significantly with respect to the radar would experience large changes in returned power, and that smaller and lower targets would fluctuate because of multipath interference. However, all the targets, including several grounded icebergs which experienced little or no aspect change with respect to the radar, experienced large variations in return power. In addition, the variability occurred with a periodicity such that the changes could not be ascribed to the actual physical motion of the target.

These changes in returned power profoundly affect the probability of detection of ice targets of all sizes. However, as the following analysis shows, there are offsetting phenomena which may permit improved probability of detection.

First a look at target variability. Target B22, Fig. 8.34 (a small, grounded iceberg with about 400 m^2 of surface exposed to the radar) had a variability of up to 14 dB in the *HH* return and 10 dB in the *HV* return (Fig. 8.35) within a 30-s sampling period. The 128 samples shown were collected at equally spaced intervals over the 30-s sample period. Over a 4-day period, the *HH* return for this target varied 14 dB and the *HV* return varied 24 dB. For vertical transmit, the short-term (30-s) variation was 9 dB for *VV* returns and 8 dB for *VH* returns and the 4-day variability was 10 dB for *VV* returns and 9 dB for *VH* returns. This target exhibited no noticeable change in aspect with respect to the radar

Figure 8.34 Photo of ice target B22 (with vessel to the right).

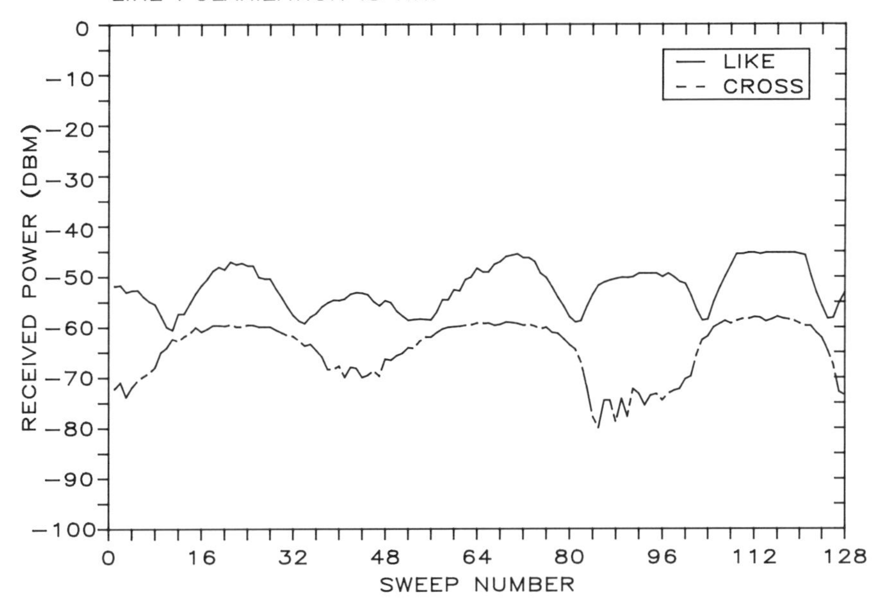

Figure 8.35 Variability of like- (*HH*) and cross-polarized (*HV*) return from target B22 over a 30-s period.

during the experiment and was high enough (15 m) to minimize the expected effects of multipath interference. The target was at a range of only 2500 m, and should not have been affected by any propagation effects from changing atmospheric conditions. In fact, the weather and the sea state were essentially stable during the experimental program.

The difference between the like- and cross-polarized returns for a specific target was just as variable as the returns themselves, with the like-to-cross ratio changing from 7 to 25 dB within one data set (30 s) to a range of −1 (stronger cross return) to +8 dB within another set on another day.

Another important observation from the experiment was that all targets significantly depolarized the radar signal, although with high variability. This depolarizing effect was also shown in the previous results for the stationary ice environment. As the dual-polarized dish antenna used for this experiment had a cross-polarized isolation of better than 25 dB, like-to-cross ratios smaller than 25 dB can be attributed to depolarization caused by reflection in the illuminated area.

For the larger, stationary targets, the like-to-cross ratio varied from −5.5 dB (a larger cross-polarized return) to +37.0 dB. An analysis of returns from these targets shows that the average like-to-cross ratio was 4 dB. There was some dependence on transmitter polarization, as the average *HH*/*HV* ratio was 2.0 dB, and the average *VV*/*VH* ratio was 9.7 dB.

To acquire statistics on returns from the larger stationary targets, all data from four large targets (B22, B23, B26, B27) were accumulated over 5 days. The radar equation (8-1) was used to convert the return power to radar cross section of the targets. This cross section was then normalized by the target's physical area, as seen by the radar, to yield the normalized radar backscatter coefficient, σ^0, expressed in dB with respect to 1 m^2. The cross section was normalized with only the target area (not the entire radar cell) as the surrounding ocean did not contribute any significant return. Figure 8.36 shows the histogram of σ^0 values for like- and cross-polarizations for target B22. The range of like-polarized σ^0 values covered 27 dB with an average value of -20.0 dB.

The like-polarized histogram for another target (B23) had a range of 26 dB, with a higher average σ^0 of -10.3 dB. The other two larger targets (B26 and B27) had similar spreads of σ^0 values and yielded like-polarized averages of -7.0 dB (B26) and -11.5 dB (B27). The lower σ^0 value associated with B22 probably resulted from the flat smooth face of the target sloping away from the radar.

To determine the ice target sensitivity to transmitter polarization, separate *HH* and *VV* histograms were accumulated. Targets B23 and B27 had average σ^0_{VV} of 5.1 dB and 3.4 dB, respectively, above σ^0_{HH}; conversely, target B22 had an average σ^0_{HH} that was 2.6 dB above σ^0_{VV}, whereas target B26 exhibited no difference.

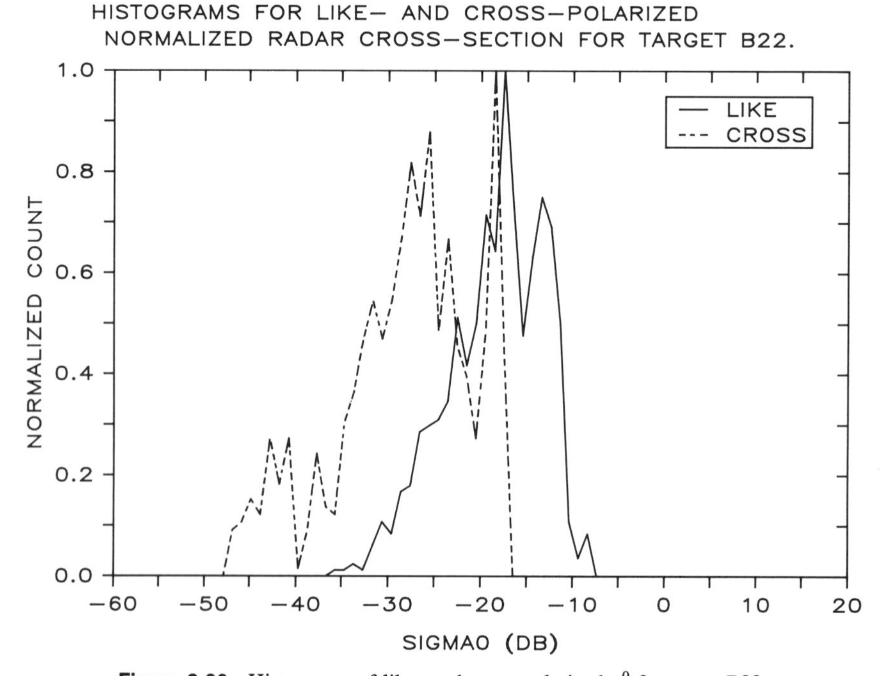

Figure 8.36 Histograms of like- and cross-polarized σ^0 for target B22.

Regardless of polarization, the like-to-cross ratio of σ^0 ranged between 6.4 and 10.2 dB for the four targets, with an average of 5.0 dB. A representative sample of the smaller targets showed an $\sigma^0_{HH}/\sigma^0_{HV}$ ratio of about 5 dB, and the $\sigma^0_{VV}/\sigma^0_{VH}$ ratio was about 7 dB.

One of the smaller targets, B33 (with physical cross section of about 11 m^2) for which a large number of data samples were collected, was analyzed to provide statistics on returns from small nonstationary targets. Figure 8.37 shows the appearance of the target and Fig. 8.38 shows the histograms of like- and cross-polarized σ^0 values for all data collected for this target. The average like-polarized σ^0 value for all returns was -15.1 dB, while the range of σ^0 values was from -37 to $+2$ dB. There was little difference between the average σ^0_{HH} (-15.2 dB) and σ^0_{VV} (-14.9 dB). In addition, σ^0_{HV} and σ^0_{VH} were within 0.8 dB, whereas the average like-to-cross ratio was about 7.5 dB. Table 8.10 summarizes σ^0 values for all of the targets.

The average σ^0_{HH} value of -13.9 dBm^2 from this open-water experiment is somewhat larger than the average *X*-band value (-22.5 dBm^2) for the frozen ice situation (Table 8.6). However, a more exhaustive analysis (Rossiter et al., [52]) of the data used to generate Table 8.2 using carefully derived physical cross sections for two of the icebergs yielded an average σ_{HH} value of -12.1 dBm^2 which agrees favorably with the like σ_0 value from Table 8.10.

In reviewing the open water data, it became apparent that although all targets occasionally returned a cross-polarized signal stronger than the like-polarized return, this phenomenon happened more frequently with the smaller targets. Target B28, a growler of about 26 m^2, gave cross-polarized returns greater than 13 dB above the like-polarized returns on several sample sets. Figure 8.39

Figure 8.37 Photo of target B33 (with 0.5-m buoy in ice at right).

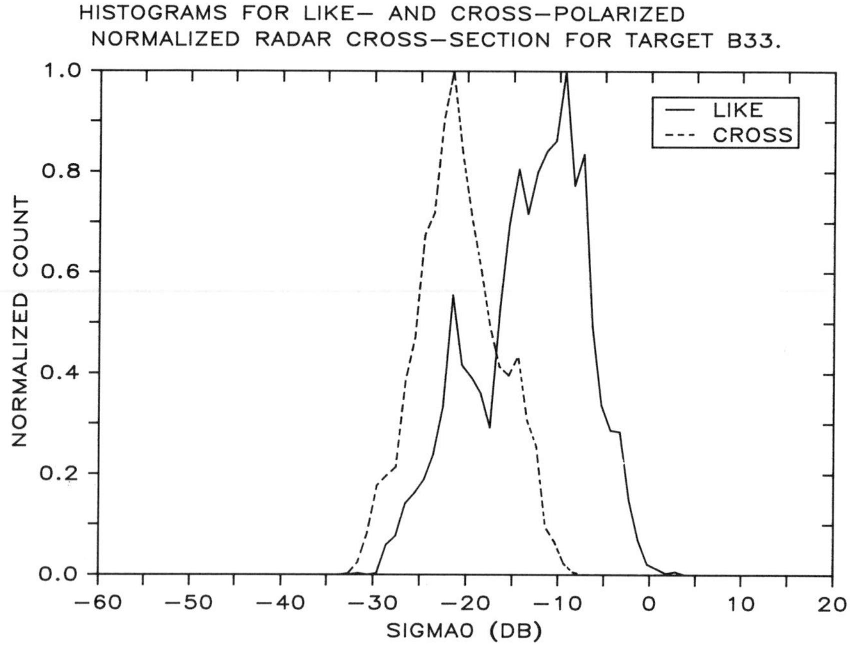

HISTOGRAMS FOR LIKE- AND CROSS-POLARIZED NORMALIZED RADAR CROSS-SECTION FOR TARGET B33.

Figure 8.38 Histograms of like- and cross-polarized σ^0 for target B33.

plots the received power for 128 samples collected at uniform intervals over a 30-s period. The like-polarized return is plotted with a solid line, and the cross-polarized return is plotted with a dotted line. This situation of substantial signal advantage for the cross-polarized return lasted for periods of up to several minutes. Similarly, target B31 (about 1.5 m^2) gave cross-polarized returns up to 10 dB stronger than the like-polarized returns. This higher cross-polarized return often permitted target detection, even though the like-polarized return was lost in clutter or noise.

For the smaller ice targets, it is interesting to note that there was a strong correlation between the frequency of variation of the returns from the target and from the sea clutter surrounding it. Figure 8.40 shows a situation where the period of the target oscillations in (B) is the same as the sea-clutter oscillations in (A). This situation occurred because the target was not only being vertically displaced by the waves at the wave frequency (and thus at the sea-clutter frequency), but also was being obscured partially by the waves, causing target return scintillation.

Finally, from a detection point of view, of particular significance is the observation that although the like- and cross-polarized returns tended to vary considerably with time, these variations were often not in phase. When the like-polarized return was reaching a minimum, the cross-polarized return was often high, and vice versa. Examination of many other sets of data confirms

TABLE 8.10 Normalized Radar Cross Sections for *X* Band for Selected Ice Targets

Target Sizes (m^2)	Average (m^2)	σ^0 Like	σ^0 Cross	σ^0_{HH}	σ^0_{HV}	σ^0_{VV}	σ^0_{VH}	Like/Cross	$\sigma^0_{HH}/\sigma^0_{HV}$	$\sigma^0_{VV}/\sigma^0_{VH}$
< 100	20.4	−13.3	−22.9	−13.6	−22.9	−12.5	−22.0	8.7	8.1	9.1
> 100 < 1000	328.0	−17.9	−24.8	−18.9	−25.4	−15.1	−24.0	7.0	6.6	8.4
> 1000	2766.7	−9.0	−19.1	−9.2	−18.5	−8.7	−22.0	9.7	9.3	11.9
Average		−13.4	−22.3	−13.9	−22.3	−12.1	−22.7	8.5	8.0	9.8

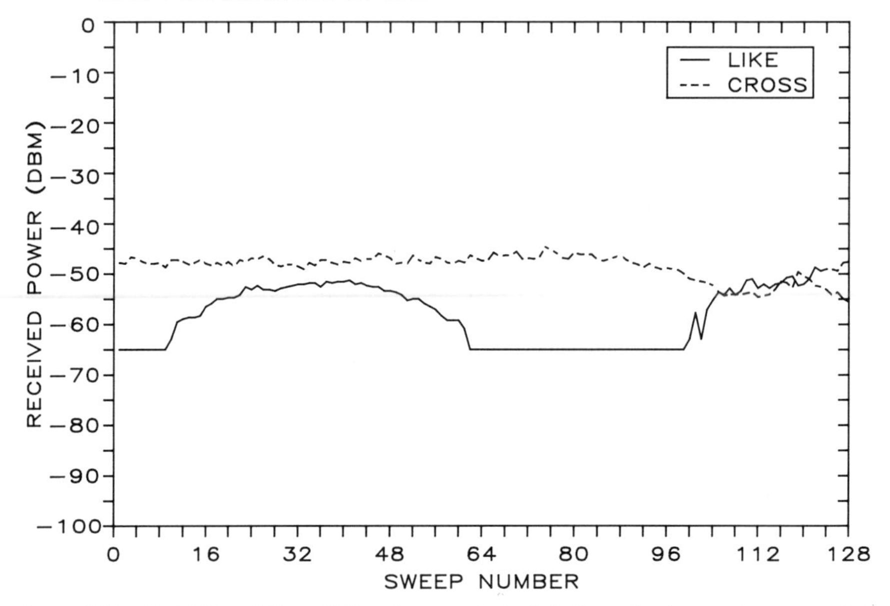

Figure 8.39 Variability of like- (*HH*) and cross-polarized (*HV*) received power from target B28 over a 30-s period.

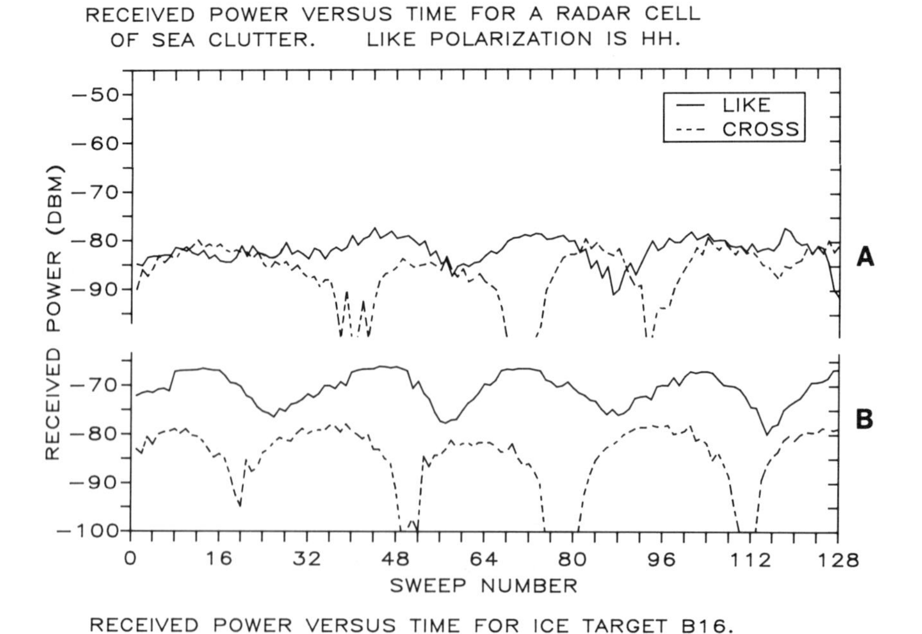

Figure 8.40 Correlation between variations in sea clutter return (top traces) and small ice target returns (bottom traces).

the frequent independence of like- and cross-polarized returns and this implies the possibility of improved target detection through the combined use of like- and cross-polarized signals.

This improved detection can be visualized by reference to Fig. 8.41, which plots the like- and cross-polarized returned power from a small ice target for 128 consecutive sweeps. With an arbitrary threshold level of −60 dB (horizontal line in the figure), a target would be declared 59% of the time by the like-polarized channel. Using only the cross-polarized channel, a target would be declared 45% of the time, but if both channels were employed, a target would be declared 81% of the time. Use of both channels improved the detection by 22%.

Figure 8.42 combines the results of like- and cross-polarized σ^0 for icebergs from the open-water experiment with the values obtained for icebergs from the fast-ice experiment. It must be remembered, however, that the open-water data were collected at X band, whereas the fast-ice data (with the exception of the two asterisked points) were collected at K_u band. Therefore, it is not surprising that the X-band values are lower, because the radar cross section of ice targets increases with frequency.

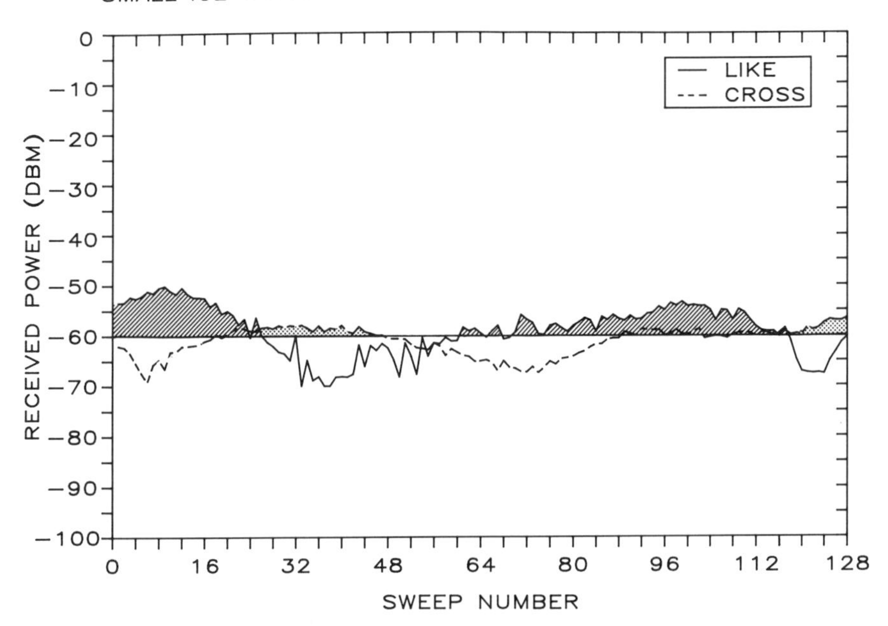

Figure 8.41 Demonstration of improved probability of detection using both like- (HH) and cross-polarized (HV) channels. With threshold of −60 dBm, detection occurs 59% of the time with only like-polarized channel, 45% of the time with cross-polarized channel, but 81% of the time by using both channels.

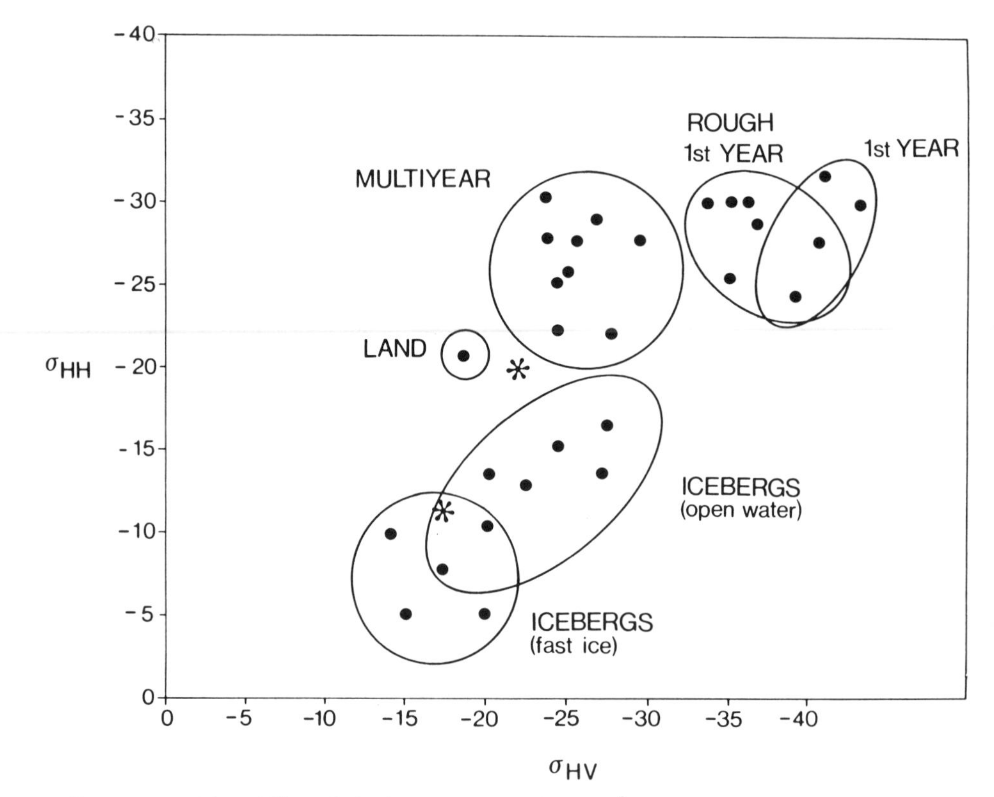

Figure 8.42 Plot of like-polarized versus cross-polarized σ^0. The values for icebergs in open water were collected at X band, all other values are for K_u band.

The two asterisks shown represent data points at X band collected during the 1983 fast-ice experiment, and although these values are of poorer quality (because of antenna problems), they show that the open-water values are in the appropriate range. These two values suggest that the cross sections for the open-water targets are probably not significantly lower than for the fast-ice situation.

East Coast Experiments

McMaster. To design suitable detection algorithms, it is necessary to understand the statistics of the interfering signals, the main one of which is sea clutter. As part of its coherent radar program, described in the next chapter, McMaster University conducted an experiment on the east coast of Canada, at Cape Bonavista. One of the investigations used only the amplitude of the returned signal, as a noncoherent radar would measure, to study the statistics of the sea-clutter signal. Any divergence from the normally assumed Rayleigh statistics, as is increasingly the case as the radar resolution is improved, quickly degrades the normal radar's performance.

The experiment collected sufficient sea-clutter data to permit determination of the form of the distribution. To provide the required dwell time and uncorrelated samples, the radar was operated at a PRF of 200 Hz, and dwell times of about 7 min per file. The antenna was stationary. Fixed polarizations were used, with changes on sequential files. Since the important part of the distribution was in its upper "tail" region, the normalized amplitude histograms were plotted on a log scale. The data were normalized with respect to its mean so that comparisons between different data sets could be made easily.

The data were plotted and the statistics were found to be well represented by the K-distribution (Nohara et al. [44]). The K-distribution is defined as

$$p(\sigma;\, r) = \frac{b}{\sqrt{\sigma}\,\Gamma(v)}\,(b\sqrt{\sigma}/2)^{v} K_{v-1}(b\sqrt{\sigma}) \qquad (8\text{-}26)$$

where Γ is the Gamma function, and $K_{v-1}(\cdot)$ is the modified Bessel function. This form of the equation describes the statistics of the radar cross section of clutter, σ. In dealing with the amplitude statistics, the form used by Watt [64] is

$$p(\alpha;\, r) = \frac{4c}{\Gamma(v)}\,(c\alpha)^{v} K_{v-1}(2c\alpha) \qquad (8\text{-}27)$$

which is related to (8-26) by a simple transformation of variables, and describes the statistics of the clutter amplitude α. In both cases, v is the so-called shape parameter, and b and c are related scale parameters. As the parameter v reduces in value, the distribution in the tail becomes more pronounced. With the increasing occurrence of larger clutter amplitudes, the clutter is said to be more "spiky."

Figure 8.43 shows a plot of the amplitude histogram for like- and cross-polarized sea-clutter data for the 32 ns compressed pulse, H transmit. Theoretical curves are shown for the K-distribution with the associated v parameter, which best fits the data. Shown in the solid line is the theoretical Rayleigh distribution. The sea-clutter data is clearly non-Rayleigh, and is well fitted by the K-distribution for both like- and cross-polarized data. Similar results hold for the other transmit polarization.

The value of v, that is, the spikiness of the clutter, seems to be related to two factors. As the angle between the radar's look direction and the direction of the wind and waves varies, so does the spikiness. The spikiness was found to be maximum in the case of the radar looking upwind. The second factor was the pulse length, that is, the radar resolution. As the pulse length increased, the spikiness of the clutter decreased. This latter effect is evident in examining Figs. 8.44 and 8.45.

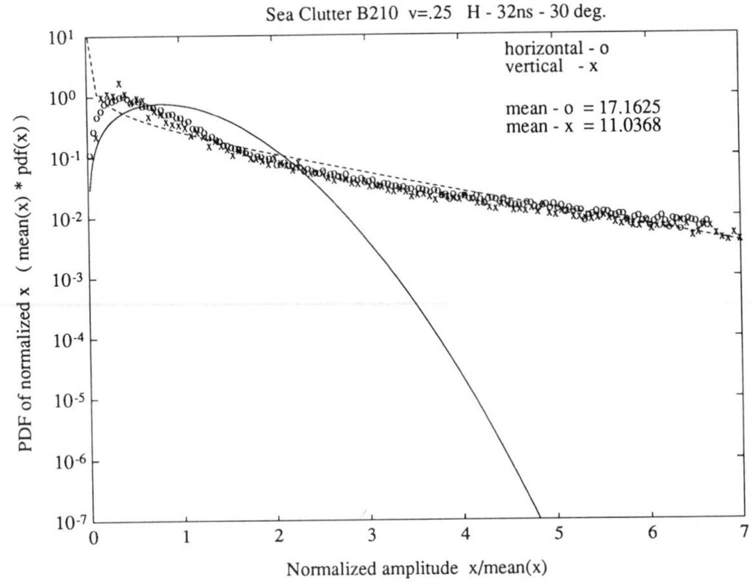

Figure 8.43 Probability distribution function of the normalized amplitude of the sea clutter return signal; transmitted polarization horizontal, pulse length 32 ns (compressed) at bearing 30°. The *HH* return is indicated by 0, the *HV* return by x. The solid line indicates the Rayleigh curve [$v = \infty$ in (8.27)] and the dashed line corresponds to $v = 0.25$ (approaching the log-normal limit). The latter curve fits both the *HH* and *HV* curves well.

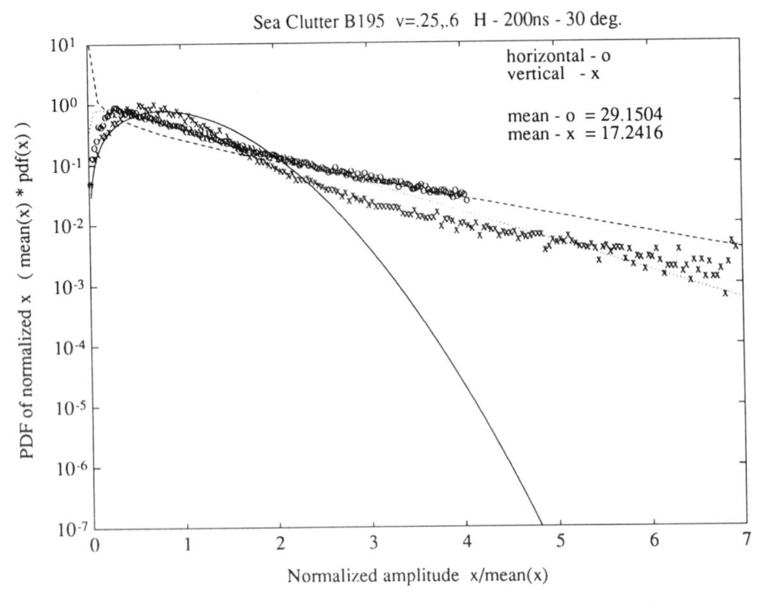

Figure 8.44 Same as Figure 8.43, but for a pulse length of 200 ns. The *HH* curve is still well represented by the dashed curve ($v = 0.25$), but the *HV* curve has a larger v value; the dotted curve is a *K* distribution with $v = 0.6$.

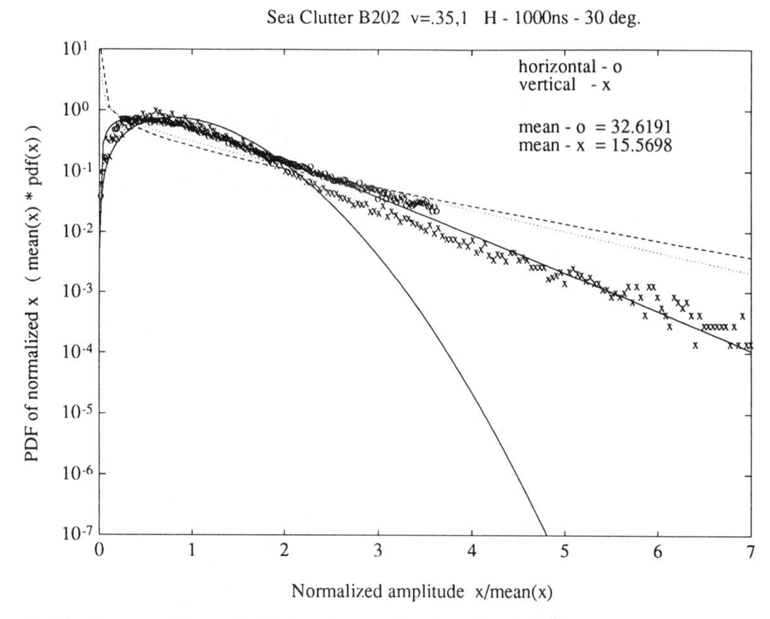

Figure 8.45 Same as Figure 8.43, but for a pulse length of 1000 ns. The *HH* return is shown with a *K*-distributed curve with $v = 0.35$, and the *HV* curve with $v = 1.0$.

The sea-clutter data were also used to measure the temporal and spatial coherence of the sea surface (Nohara [45]). Figure 8.46 shows an amplitude-versus-time image for sea-clutter data collected at the 200-Hz PRF. The antenna was stationary, and the radar signal was sampled in the range interval shown. For each horizontal line in the image, approximately 500 successive sweeps have been averaged. Thus, the capillary wave component of the sea clutter has been averaged out; what remains is the underlying undulating mean power, in response to the large gravity waves. This is the sort of model to which the *K*-distribution is well matched.

Figure 8.47 shows spatial and temporal correlation measurements made on a sea-clutter data file, using 10 s of data. Two distinct correlation times are evident. The fast speckle component is determined to have a correlation time of about 10 ms. Furthermore, the much longer correlation time of the second component is observed. For a lag of 1 s, this component is still very correlated due to the fact that the mean signal level of the sea is strongly correlated over this period. Using 60 s of data to estimate temporal lags out to 10 s, the correlation of the mean level component is shown to be very strong, persisting for several seconds.

The range correlation shows that over the distance of 30 m (the pulse length), the spatial correlation has dropped to less than 60%. This indicates the mean level component varies much more quickly in resolved range than it does in time.

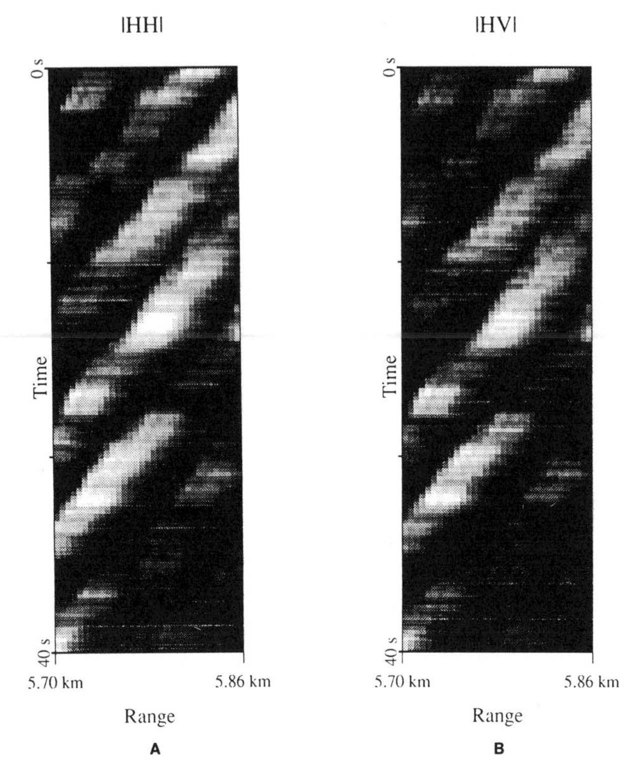

Figure 8.46 The underlying, mean sea clutter backscatter of file B347. The figures show the well-developed swell structure of the sea. (a) The *HH* polarization returns; (b) the *HV* polarization returns.

ESRF Studies. As part of resource exploration activities on the east coast of Canada, the resource development companies, through the program known as the Environmental Studies Revolving Fund, supported a number of experiments to examine the ice-detection performance of commercially available, noncoherent marine radars.

One study (Ryan et al. [55]) was conducted from a semisubmersible drilling platform. There were four radar systems on the rig: a collocated *X*- and *S*-band system which used a back-to-back antenna arrangement mounted on the derrick top at 75 m above sea level, and separate *X*- and *S*-band systems at heights of 45 m and 35 m, respectively. The systems were operated from March 1 to May 20, 1984, with intensive collection during the final month. Data was collected on 12 well-documented icebergs, and four additional icebergs, bergy bits, and growlers.

The radar video was digitized using a computer-based sampling system. The radars were calibrated by sampling known injected power levels, so that sampled data could be converted to absolute received power, and in turn, into radar cross-section estimates.

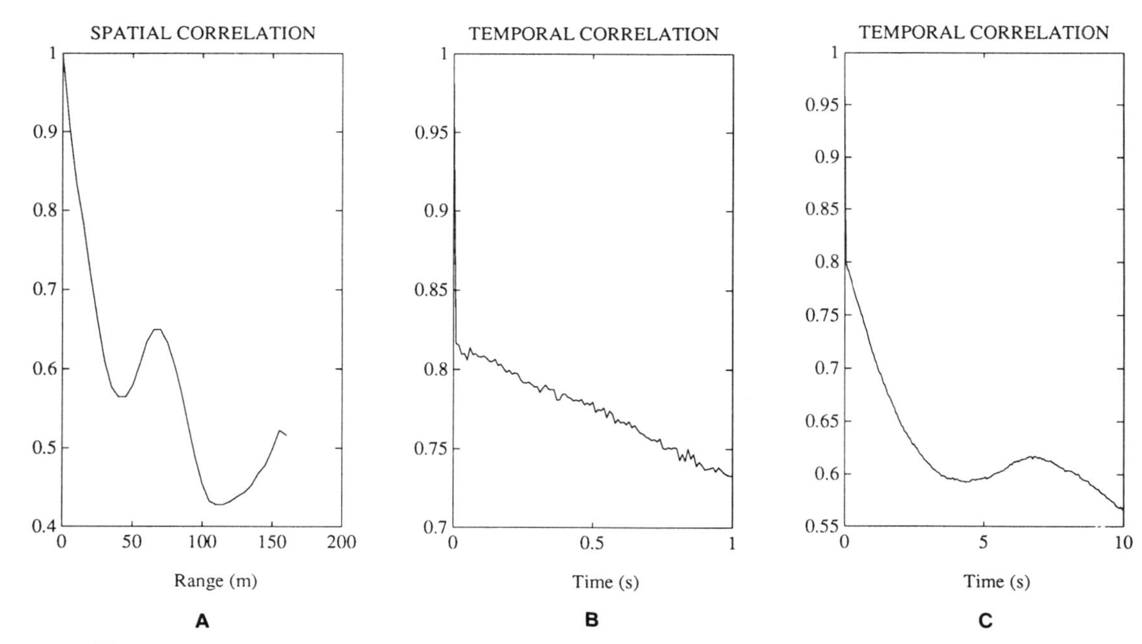

Figure 8.47 Estimates of the temporal and spatial correlation properties of sea clutter (B195). In (a), the range correlation of the sea clutter is shown; in (b), the temporal correlation for lags up to 1 s; while (c) shows lags up to 10 s.

The purpose of the experiment was twofold: to investigate the environmental effects on propagation and detection, and to consider the relationship between the physical above-water size of the iceberg and its radar cross section.

In addition to the ice targets, there was a neighboring rig and several supply boats. The strength of the return from the neighboring rig was monitored daily, and significant changes were used to indicate changes in radar signal propagation. These propagation changes were correlated to atmospheric conditions, and indicated the presence of ducting from refraction changes. Superrefractive conditions led to longer range detection of ice and vessel targets, as compared to standard and subrefractive days.

The calculated iceberg radar cross sections were plotted versus the visible physical cross section, as shown in Fig. 8.48. In addition, the relationship determined by Budinger et al. [7] of 0.056 (−12.5 dB) between radar and physical cross sections was plotted, and the data showed reasonable agreement, with 67% of the points being within 5 dB of Budinger's line. The exceptions tended to occur during periods of ducting.

In spring 1985, a second field experiment was conducted (Harvey and Ryan [18]). The combination X- and S-band system was mounted on a vessel, at a height of 15 m, and data were collected for 38 measured icebergs. Again the radar system was calibrated. The radar video was sampled every nautical mile as the vessel approached and receded from the ice target being studied.

For a set of five icebergs, of small (height less than 16 m), medium (height 16 to 48 m), and large (height greater than 48 m) size, the calculated average normalized radar cross section σ^0, was −10 dB for X band and −15 dB for S band. The iceberg's radar cross section was calculated from return power using the point target, free-space form of the radar equation (8-1), and then normalized by dividing by the physical cross sectional area of the iceberg. The X-band value of −10 dB agrees favorably with the average value of −13 shown in Table 8.10 for open water icebergs in the Arctic.

The radar cross section of the iceberg versus the initial detection range (defined as hits on two successive scans) is shown in Fig. 8.49. The resulting fitted curve led to the rule of thumb that all icebergs with radar cross sections greater than 100 m^2 will be detected at a range of 4 nm or more, and 1000 m^2 icebergs at 8 nm or more, for an antenna height of 15 m.

Radar Performance Modeling

There was recognition that experimental studies would never be able to address all ice target types and sizes of interest, at a variety of ranges, and in a variety of sea conditions. Instead, the approach was to develop a model to predict the performance of an ice-detection radar, and to use experimental measurements as control points to validate the model.

Consequently, during the period of the above experiments, and continuing to the present, a radar performance prediction model has been developed (Johnson and Ryan [24]). There are a number of elements to the total model.

In modeling the iceberg and sea clutter signals, two parameters need to be

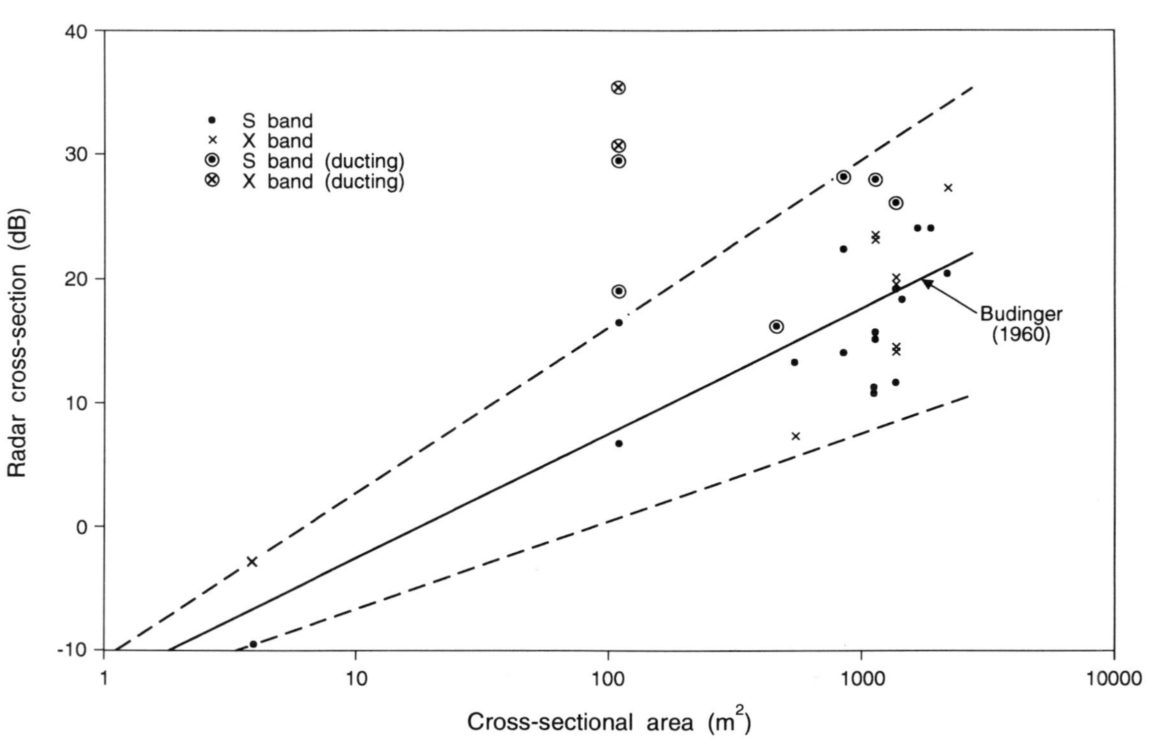

Figure 8.48 Radar cross section of icebergs, as a function of physical cross-sectional area.

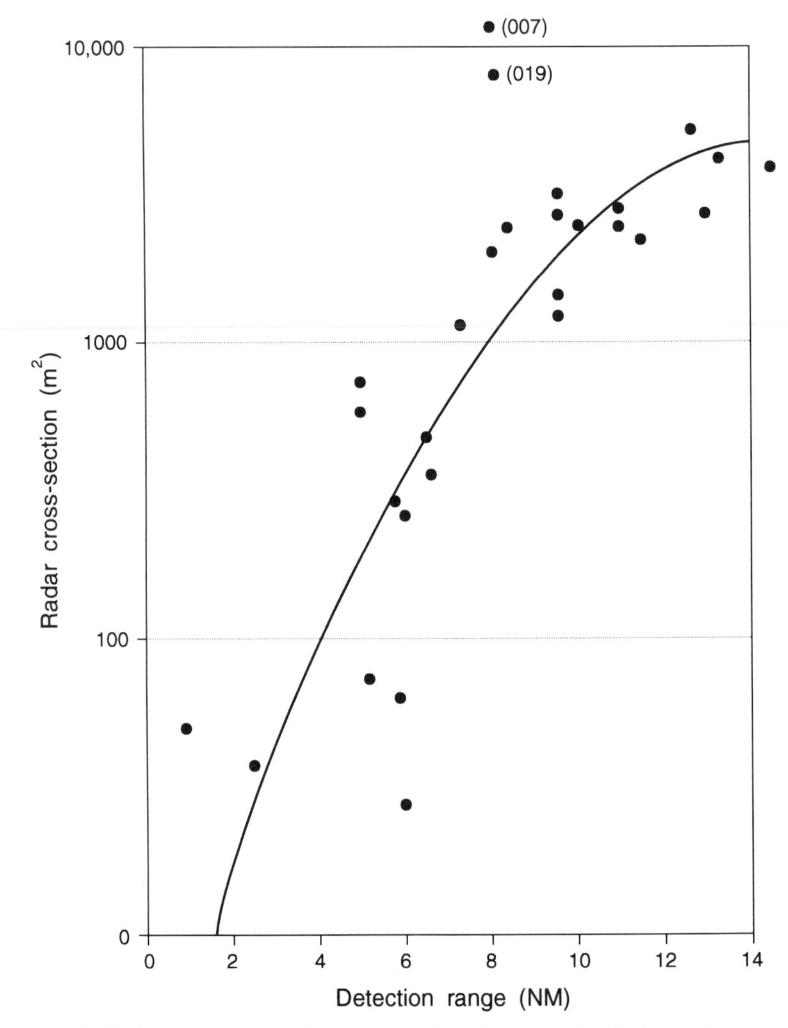

Figure 8.49 Radar cross section (area) versus detection range for icebergs, from a typical workboat with antenna at 15 m.

specified: the average return power expected, and the distribution (or statistics) of the return power. Both of these parameters affect attainable detection performance.

For the iceberg, analysis of the experimental data led to an empirical model relating physical iceberg size to expected radar cross section at X and S bands. As well, the iceberg statistics were found to be best represented with a Chi-square distribution, with 2 degrees of freedom which is equivalent to a Swerling Case 1, and leads to a Rayleigh-distributed cross section.

For the sea clutter, a few models are available to predict the expected average sea clutter cross section. The Georgia Institute of Technology model (Ewell et al. [13]) was found to fit the experimental data most closely. For the statistics of the sea clutter, the *K*-distribution model was chosen. The *K* distribution is a two-parameter model: the scale parameter is related to the second moment of the data, that is, the mean power level, and the shape parameter is a measure of the spikiness of the signal.

Knowing the distributions for the iceberg and sea clutter for a given set of conditions, the model first calculates the single-pulse probabilities of false alarm and detection. The model can then accommodate different types of signal processing, such as pulse-to-pulse integration, frequency agility, and scan-to-scan integration.

Figures 8.50 to 8.53 show some predictions generated by the model. The output is based on a bergy bit with a 1-m^2 radar cross section and an *X*-band radar with a 30-ns pulse length (compressed). The bergy bit is assumed to be a Swerling 1 target. The probability of false alarm is 10^{-6} and the mean sea-clutter level is generated using the Georgia Institute of Technology model. The drop in detection performance at medium range is due to the presence of sea clutter, whereas at longer range, detection becomes noise-limited. Figure 8.50 shows how strongly the increasing spikiness of the sea clutter (smaller v) adversely affects detection. Figure 8.51 shows the effect of an increased sea state. Figure 8.52 shows the pulse-to-pulse integration gain, which for a fixed frequency radar, is only effective against noise. Use of frequency agility decorrelates the short-term sea clutter component, yielding improvement in the clutter-limited ranges. Finally, Fig. 8.53 shows the scan-to-scan integration

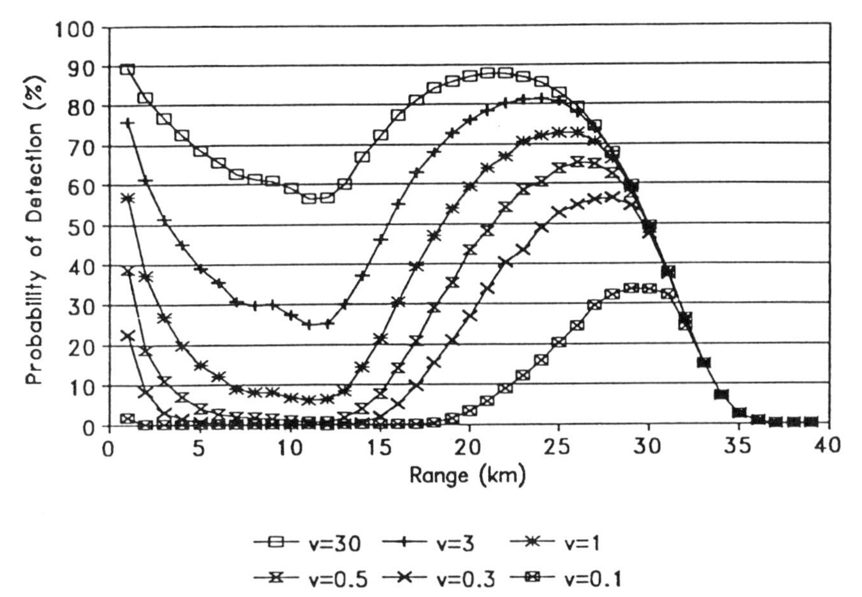

Figure 8.50 Effect of shape parameter v (frequency agile radar, 10 pulses integrated, sea state 3).

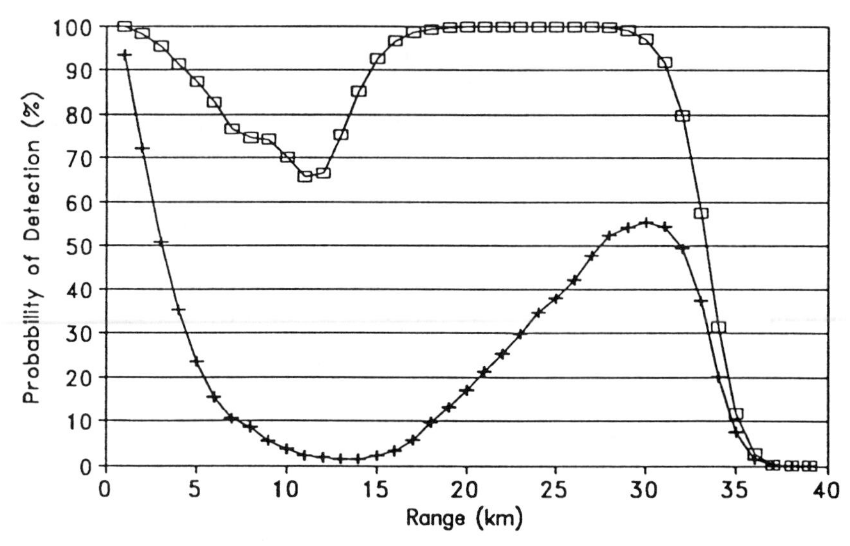

Figure 8.51 Effect of clutter mean—sea state 3 versus sea state 4 (v = 1, frequency agile radar, 10 pulses and 5 scans integrated).

gain. Here much of the clutter is assumed to decorrelate from scan to scan, yielding integration improvement even in the clutter-limited ranges.

Work is continuing, both on the east coast (Ryan, Sigma Engineering, personal communication) and at McMaster University, to extend this modeling approach to include the use of additional radar features such as coherence and

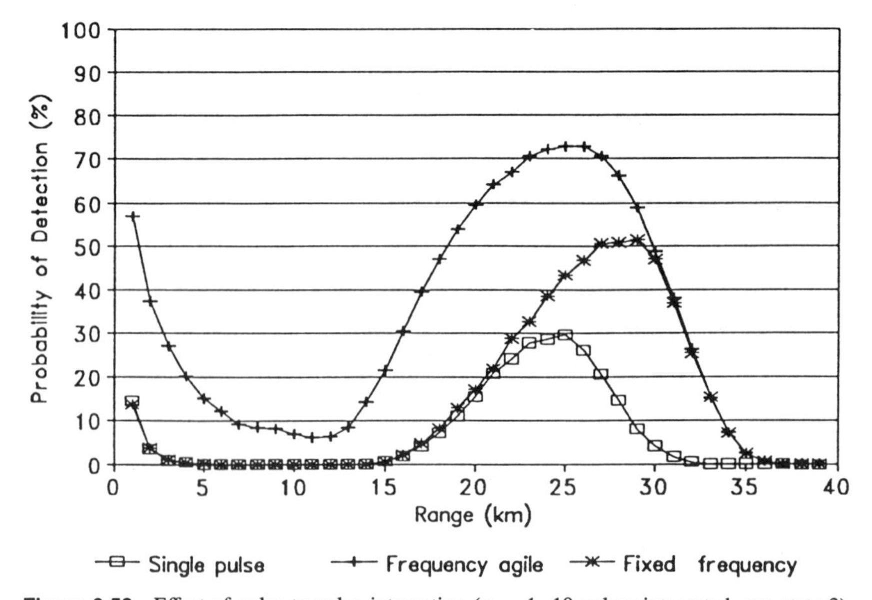

Figure 8.52 Effect of pulse-to-pulse integration (v = 1, 10 pulses integrated, sea state 3).

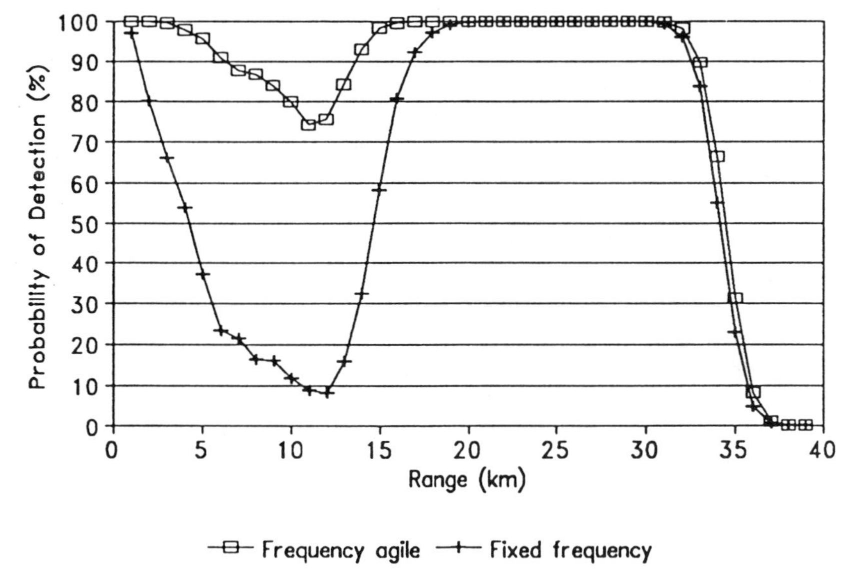

Figure 8.53 Effect of scan-to-scan integration (v = 0.3, 10 pulses and 10 scans integrated, sea state 3).

dual-polarization. Some qualitative results from McMaster's studies are given in the next chapter, Section 9.4.

8.5 PROTOTYPE RADAR

Based on the research work previously discussed, the elements of a dual-polarized *X*-band radar system were integrated into an operational system for evaluation at sea. In April of 1986, DFO installed the radar system on the Canadian ice-strengthened carrier, MV *Arctic*.

8.5.1 System Description

The radar system was comprised of two conventional 25-kW marine-band radar systems and a custom-designed turning unit and antenna system. The antenna system consisted of separate collocated horizontally and vertically polarized antennas. One of the radar systems was connected to the horizontally polarized antenna and was configured to transmit and receive horizontally polarized energy. The second radar, configured for receive-only, was connected to the vertically polarized antenna and displayed vertically polarized returns. The radars fed identical conventional PPI displays which were mounted side-by-side on the forward bridge of the ship. This arrangement allowed comparison of the like- and cross-polarized signals and gave easy access for the officers of the ship.

The antenna system, Fig. 8.54, consisted of a conventional 9-ft (2.7 m) *X*-band, horizontally polarized marine antenna (top of the figure) and a specially designed and constructed vertically polarized antenna (bottom of the figure). The latter antenna was designed and constructed by the Department of Electrical Engineering at the University of Manitoba, Winnipeg, Manitoba. The horizontally polarized antenna had 0.8° azimuthal beamwidth and 20° vertical beamwidth, whereas the vertically polarized antenna had 1.0° azimuthal beamwidth and 15° vertical beamwidth. The two antennas were rotated as one by a modified *S*-band turning unit which contained a dual-channel *X*-band rotary joint.

During the course of research voyages carried out in June and November–December of 1986, the radar equipment was augmented with data acquisition equipment and a radar digitizer/scan converter and digital display. The scan-converted digital images were displayed on monochrome and color monitors.

8.5.2 Test Program

The radar system was tested successfully during late November and early December of 1986. The MV *Arctic*, an ice-strengthened bulk carrier, departed Antwerp, Belgium, bound for Nanisivik on Baffin Island for late-season ice trials. The ship encountered a complete cover of first-year ice north of Disko Island on the west coast of Greenland. Mixed with the first-year ice were a large number of icebergs, bergy bits, and growlers. Near the entrance to Lancaster Sound, the ship encountered a high concentration of old ice contained in a full cover of first-year ice.

During the course of the voyage a large quantity of dual-polarized radar data

Figure 8.54 Dual-polarized *X*-band radar antenna with horizontally polarized element on top and vertically polarized element below.

was collected from sea clutter, icebergs in open water, icebergs in full ice cover, and old ice in full ice cover.

In the early portion of the experiment, data were collected from several targets that were sighted visually, described, and photographed. However, during a major part of the experiment the ship was operating in 24-h darkness, and only targets that were close to the ship and within range of the ship's searchlights could be documented.

8.5.3 Operational Use

After a short introductory period, the radar system was used extensively by the officers of the ship to navigate in difficult ice situations. While navigating through a high concentration of icebergs and bergy bits in a full ice cover, the cross-polarized display provided a superior indication of icebergs and bergy bits compared to conventional marine radar. This improvement is shown in Fig. 8.55. The image at the top is the output of the digital display system of the like-polarized channel, and the image at the bottom is the cross-polarized image from the same display. The range from the center to the edge of the

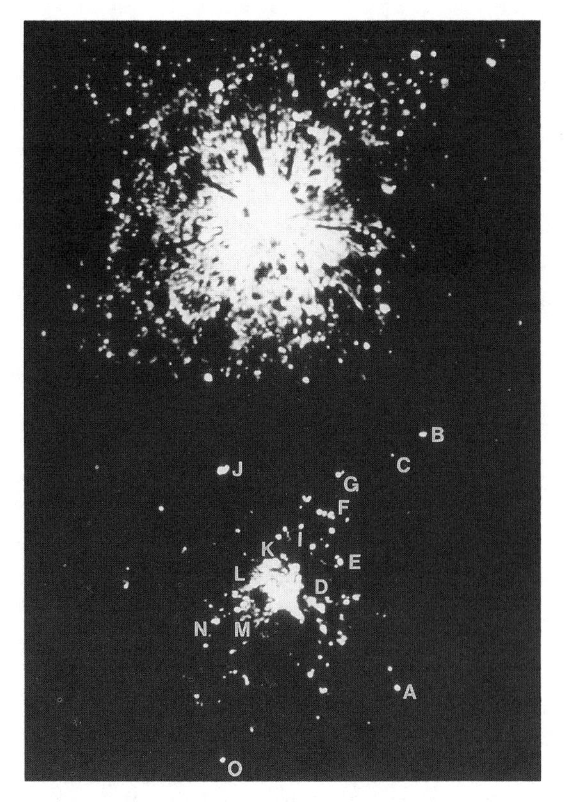

Figure 8.55 Like- (top) and cross-polarized (bottom) images of icebergs in a full first-year ice cover: icebergs at A through O.

image is 3 nm. Note the substantial improvement in display of iceberg targets at points A through O.

The smallest target detected with the cross-polarized display during the voyage was a growler of 1.5 m by 15 m, which was detected initially at a range of 2 nm. The target was never distinguishable on the like-polarized display and was only visually detected at about 0.75 nm.

As the cross-polarized image of Fig. 8.55 shows, there is still a substantial amount of return energy out to about 0.5 nm from the ship, and which is not necessarily associated with hazardous targets. These high returns result from a different scattering mechanism, caused by steeper grazing angles in this range and saturation of the display at close ranges because of the strong range-dependency of the returned power. The operational procedure used during the voyage was to set the gain of the cross-polarized display to give clutter returns to about 0.5 nm in range and then to declare hazardous any target that was detected at a range greater than 1.0 nm. This technique resulted in no false target declarations and no known target misses.

The second scenario for which the radar provided invaluable information was in navigating in a field of old-ice floes frozen into a full first-year ice cover. In this situation, the radar system performance was even more impressive. The cross-polarized display was able to display and delineate old-ice floes clearly, as shown in Fig. 8.56. The image at the top is from the like-polarized radar, the image at the bottom is from the cross-polarized radar. These images are photographs of the PPIs of the two radars. The range from the center to the edge of the images is 6 nm. In addition to showing the old ice floes at points A through E clearly, the cross-polarized display also showed the icebergs at F through O. Using the cross-polarized display, the navigator was able to maneuver the ship through the gap in the old ice between C and D, as shown in the cross-polarized image. This route would not have been identified with the conventional like-polarized image at the top. The operational procedure for the radar in this environment of first-year ice with old-ice floes was to set the cross-polarized radar gain to display clutter out to about 1.0 nm, and to declare any target displayed at a range of greater than 1.5 nm to be hazardous. Again, no targets were falsely declared as hazardous, nor were any known hazardous targets missed.

Once the cross-polarized radar settings were established for the two ice regimes, the settings were left constant for operational use. It should be emphasized, however, that no matter how the settings of the like-polarized display were manipulated, the performance of the cross-polarized display could not be achieved.

During the experimental program it was discovered that there was still a slight pattern misalignment between the horizontally polarized and the vertically polarized antennas. This resulted in reduced returns in the cross-polarized channel and reduced performance of the cross-polarized display. In addition, the gain of the cross-polarized antenna was below its specified value due to the damage caused to the antenna by transmitting into it during the 1986 spring

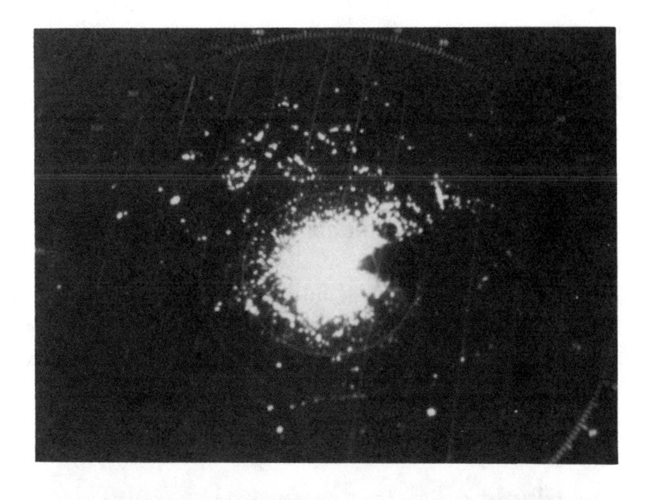

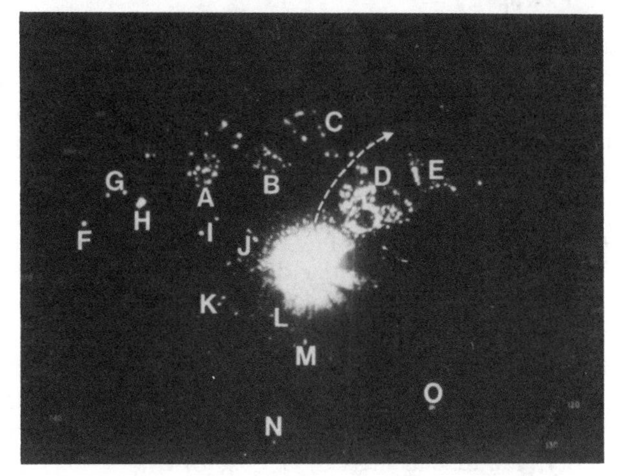

Figure 8.56 Like- (top) and cross-polarized (bottom) images of old-ice floes and icebergs in a full first-year ice cover: old ice at A through E; icebergs at F through O.

experiment. It is anticipated that with these problems resolved, a 5- to 8-dB improvement in the sensitivity of the cross-polarized radar can be achieved.

Canarctic Shipping Company Ltd., operators of the MV *Arctic*, have purchased and installed new higher power dual *X*-band radars and a commercial dual-polarized antenna as a permanent ice navigation system on the MV *Arctic*.

Use of the dual-polarized radar, and in particular, the cross-polarized display, provided substantial improvements in the ability to navigate in ice fields containing icebergs and old ice. The higher power (50-kW) radars have made the cross-polarized image useful out to a greater range, which significantly enhances its operational effectiveness (Bob Gorman, Canarctic Shipping, personal communication, 1991).

8.6 SIGNAL PROCESSING AND DISPLAY IMPROVEMENTS

Based on the experience gained from the use of the prototype radar system, it became apparent that an improved display system, supported by appropriate signal processing, was required to maximize the effectiveness of the dual-polarized radar system. For example, the display should produce a clear, sharp, bright image, and permit various combinations of the like- and cross-polarized images as required by the operating situation. The following sections discuss the use of raster scan technology, and introduce a number of signal and image processing techniques to improve the display of the radar image. Several example images illustrating the improvement obtainable are included. The illustrated techniques are chosen based on their effectiveness and their suitability for real-time implementation.

8.6.1 Raster Display

The conventional PPI marine radar display has several shortcomings, as outlined in Section 8.3.3. One way to overcome these problems is through the use of a raster scan (TV-type) display. Because of its rapid update rate (1/60 of a second), the display appears to be free of flicker. The entire image is displayed continuously and cursor position measurements can be made at any time. In addition, because of the much brighter display, the device can be viewed under normal light. Most important, however, is the much wider dynamic range of this device. The wider dynamic range and the use of color allows full advantage of image processing to be taken. By using a black-and-white display with resolution of 8 bits per pixel, a dynamic range of 48 dB can be accommodated, although the discernible dynamic range is expected to be in the 20- to 30-dB range. By using a color display, an even wider dynamic range can be accommodated, and signals of particular interest, or in particular ranges of intensity, can be enhanced by the use of color.

A major precondition of employing this type of display technology is the requirement to digitize the incoming radar signal. However, with modern digital circuitry, the analogue-to-digital conversion is readily accomplished. To accurately represent the analogue signal in digital format, it is necessary to sample and digitize the signal at least twice per range resolution cell, where the range resolution is the minimum range separation for two targets to be resolved individually, and is determined by the pulse length of the transmitted radar signal. For a typical marine radar with a pulse length of 200 ns (30-m resolution), the signal must be digitized at about 10 MHz.

A major consideration in using a raster scan display is that the incoming radar data are sampled in a range-bearing (polar) format, whereas the raster scan system displays in an *X–Y* (rectangular) format. The required polar-to-rectangular mapping is known as scan conversion.

Generally, there are more sampled data points in a polar image than can be displayed in the X–Y system, and this is particularly true near the center of the image. The resulting reduction of data points in converting to the X–Y display must be managed carefully to avoid loss of information. If not properly implemented, the scan-conversion routine will retain the final range–bearing value appropriate for a particular X–Y coordinate which can easily result in the loss of a point target and could have serious implications in the case of detection of ice targets in open water. To overcome this problem, peak detect-and-hold circuitry could be implemented to assign the highest value detected to the X–Y coordinate, or some averaged or median value could be calculated and used.

Conversely, near the outside of the image there will be fewer polar data points because of the diverging nature of the polar scan lines. In fact, some X–Y locations (pixels) will not have a data value assigned. This problem results in the formation of moiré patterns, particularly at the 45° points of each quadrant of the display. These patterns are the result of missed pixels all being displayed at the same intensity.

As the location of these pixels can be determined for a given radar geometry, one solution would be to fill these pixels with random, but relatively low-level, values. Some commercial scan-conversion routines duplicate the portion of the sweep data near the display edge to fill the missed pixels. Conversely, an interpolation between two adjacent X–Y pixel values could fill the missing pixels. There are a number of scan-conversion procedures that address these and other problems. It is necessary to select the procedure that is most suitable to the image being converted, and to the nature of the targets that are to be displayed.

Figure 8.57 shows a photograph of the original radar data on a typical PPI display. (To permit fair comparison of the images, the contrast of the images has been adjusted to use the full available gray scale.)

Figure 8.58 shows a photograph of the digitized data displayed on a high-resolution display system with 1280 × 1024 pixels. (Because of the digitizing hardware available, the data were digitized to 8 bits at 11 MHz.) To produce this image, the scan-conversion routine retained the last polar value assigned to each pixel on the X–Y display. This technique has the disadvantage that small isolated targets can be lost as they are replaced by a subsequent lower level signal. For the application of displaying imagery of an ice-covered region in which the targets are spatially large, this would not be a major problem. However, for the detection of singular targets in ice-infested waters, a more elaborate technique, possibly using a peak detector, would be required.

Examination of the differences in Figs. 8.57 and 8.58 shows the image improvement afforded by the wider dynamic range of the raster scan display. The individual targets, such as the icebergs at B and C, are lost in the saturated video of the PPI display, but are clearly shown on the raster scan display. If the PPI display intensity is turned down, small but significant targets at far range are lost.

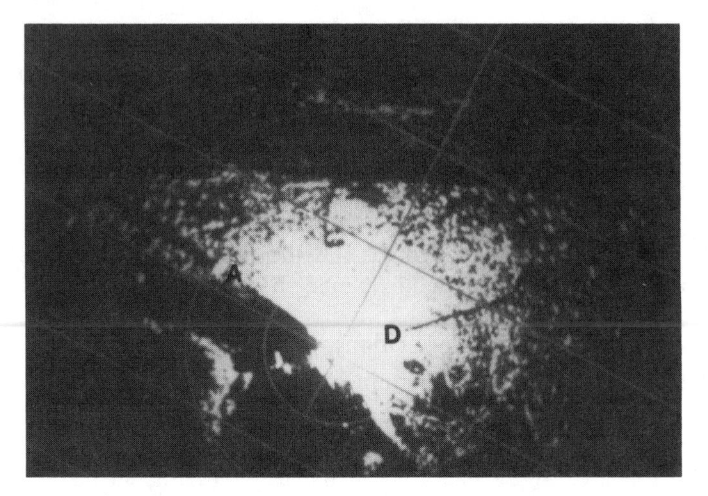

Figure 8.57 Original radar image on PPI display.

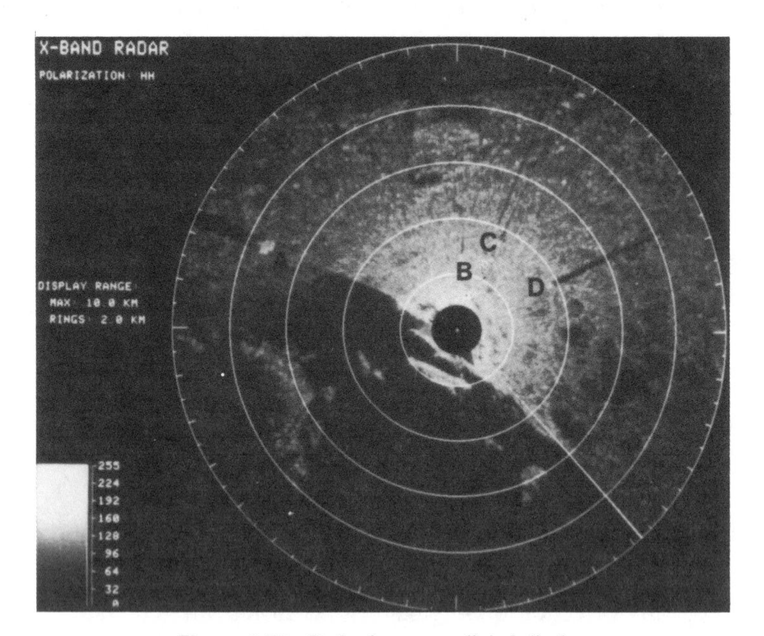

Figure 8.58 Radar image on digital display.

8.6.2 Image Processing

Range Normalization

Once the radar signal has been digitized, the last of the major problems associated with the standard PPI display can be resolved. By normalizing the radar returns as a function of range, the blooming problem described earlier can be eliminated. All targets of equal radar cross section will then be displayed with the same intensity, regardless of their range from the radar. The normalizing process is carried out by multiplying the return signal by an appropriate factor to compensate for the range dependence. Although this normalization can be achieved with most radars by employing the sensitivity time control (STC) function, the resulting signals are usually uncalibrated, thus making further processing difficult.

Figure 8.59 shows the result of applying range normalization, preceded by thresholding. The image is discussed further under the noise reduction section.

Image Enhancement

Enhancement can include any or all of the following techniques: smoothing, high- or low-pass filtering, histogram operations, edge sharpening, bit slicing, and color coding.

With the exception of noise reduction, which might be considered a restoration process, all processing of ice radar images discussed below can be considered as image enhancement.

Noise Reduction. Most noise in the radar image is introduced by one of two sources: noise (both thermal and other) associated with the radar system, and noise associated with the digitizing process. Both noise sources are white and additive, and they are usually a single sample value in duration; successive noise samples are uncorrelated.

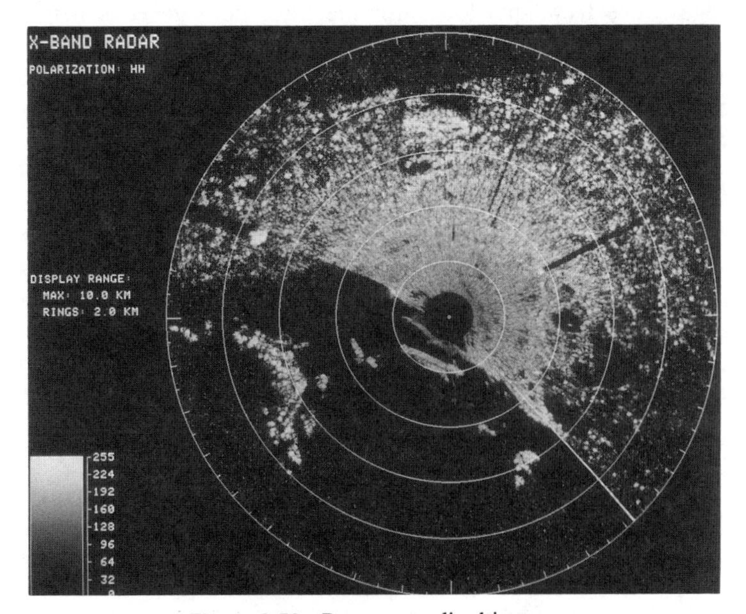

Figure 8.59 Range normalized image.

There are a number of methods by which noise reduction can be achieved. One simple method of reducing low-level noise signals is to set a threshold for incoming samples at some selectable level, such that all values near the minimum detectable signal (MDS) level of the radar are set to zero, whereas higher level samples are passed unaltered.

However, noise reduction has traditionally involved averaging, either temporally or spatially. One temporal method averages a number of images to reduce the uncorrelated noise (random signals). This method, known as scan-to-scan integration, is a viable method of noise or clutter reduction. A spatial method of noise reduction involves low-pass filtering of the image, thus reducing high-frequency, spiky noise. If the loss-pass filtering is implemented by spatial averaging, it has the disadvantage of smoothing and blurring the image, particularly around the edges of features. As edges represent a major source of information for the remote sensing of ice, this technique is undesirable.

A spatial filtering method that reduces this smearing effect is median filtering. In this technique the image is sampled with a moving window. At each pixel, the values within the window are arranged in order and the median value is selected and substituted for the original pixel value. Yang and Huang [69] found that median filtering preserved step-(sharp) edge locations better than averaging, and for impulse noise improved the accuracy of edge-location estimation. Blom and Daily [6] found that "median value filtering of radar images removes speckle with minimum edge effects and resolution degradation."

Another smoothing technique, edge-preserving smoothing, developed by Nagao and Matsuyama [42] removes noise in flat regions and does not blur sharp edges. The technique looks for the most homogeneous neighborhood around each point and then assigns to the point the average gray-level value of the selected neighborhood. By repeated application of this technique, the noise is reduced while the edges remain sharp. In actual operation, a bar mask is rotated around each point. The algorithm detects which mask has the smallest variance of gray-scale levels within the mask and assigns that mask's mean value to the point. This process is repeated until "almost no points in the picture change." To reduce noise, most other methods usually take an average over a relatively large region, resulting in the smoothing of small regions. This technique, by iterating many times, is able to achieve noise reduction and look at only small regions (3×3 or 5×5 pixels). It not only smooths without blurring but, in fact, sharpens blurred edges. The technique, however, being iterative in nature, is not well suited to real-time display applications with limited real-time computing capability.

The following paragraphs illustrate the results of some of the above techniques. To reduce background noise, the incoming data were subjected to a threshold level. The threshold level was determined by performing a histogram on the system noise in a nonsignal area of the image and selecting the level that rejected 95% of the noise samples. This process substantially reduced system noise. Care must be taken in choosing the threshold value so that small targets are not lost. If the thresholding is not performed prior to range normalization, the R^3 normalization weighting causes the noise level to increase with range, and this is distracting to the observer.

The next operation carried out was range normalization, which eliminated the effect of the range factor in the radar equation and displayed targets of equal radar cross section at equal intensity regardless of range. Figure 8.59 shows the results of applying a $1/R^3$ normalization. In comparing Fig. 8.59 with the raw radar image, it can be seen that the normalized image has a superior appearance, simply because there is a uniform intensity across the image. An examination of the figure, however, shows a considerable amount of noise, particularly at far range.

The choice of a smoothing (filtering) routine to remove the noise was more difficult. The routine that yielded the best results, edge-preserving smoothing, requires multiple passes to be effective and, as such, could not be implemented in real or near-real time. The median filter, however, yielded acceptable results and can be implemented relatively easily in hardware to operate in real time.

As a median filter removes ''spikes'' that are shorter than one-half of the filter length, the filter length must be chosen so that it will not eliminate single targets. At the 11-MHz digitizing rate, there are two samples per pulse length and a minimum of two returns from a single-point target, therefore a filter cannot be used which is larger than 3×3 pixels. Figure 8.60 shows the results of applying the median filter. Note the reduction in noise at far range.

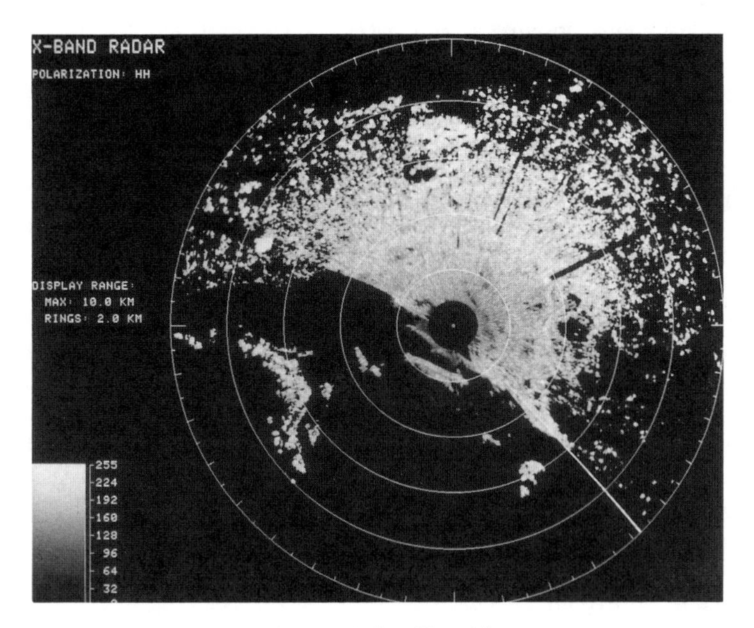

Figure 8.60 Median filtered image.

Histogram Modification. Another group of methods that can provide improvements in the visual quality of an image is histogram modification. A histogram of the pixel values of an image can provide valuable descriptive information regarding enhancement methods that may be applied effectively. Many images suffer from limited use of available gray-scale values, resulting in low contrast across the image. A histogram will show the frequency of occurrence of each intensity value and that few of the available gray-scale values are being used.

The problem of limited gray-scale use (limited contrast) can be resolved to a certain extent by a technique known as contrast stretching. With this technique, the gray-scale values from the original image are mapped into new values (often by a look-up table) which use the full range of available gray-scale values more effectively. Using this technique, it is common to ignore a small percentage of the extreme pixel values by assigning them to the black or white end of the gray-scale to allow for maximum stretching. The basic shape of the histogram, however, is unchanged.

A more effective method of histogram modification is a technique known as histogram equalization. This method improves the visual contrast of an image by adjusting gray-scale values so that each is represented in relatively equal numbers in the image. That is, the histogram of the resulting image is made quite flat. This technique is based on the assumption of a linear human brightness response. It can be accomplished by calculating the cumulative distribution function (the integral of the original histogram) and then mapping the gray-scale values of the cumulative distribution function onto the desired cumulative distribution function. In the case of simple histogram equalization, the desired distribution function is a linearly increasing function. The mapped values are then put into a look-up table and by using the remapped gray-scale values, the image is equalized.

Another method of histogram modification is a technique called histogram hyperbolization, described by Frei [14]. The basis of this approach is that a model of human brightness perception suggests that the distribution of displayed brightness levels should be hyperbolic, not linear. The processing technique should aim for an image with a uniform distribution of perceived brightness levels, rather than an actual uniform distribution of gray levels.

Edge Detection and Enhancement. Another powerful method of image improvement is edge detection and enhancement. As noted by Ballard and Brown [2], edges are "extremely important; often an object can be recognized from only a crude outline." For the purposes of this study, it has been assumed that edge detection will result in display of just the edges, whereas edge enhancement will include some method of adding the detected edge information to the original image.

As edges represent abrupt changes in gray-scale level, they have a significant high-frequency contact. As such, edge information can be extracted through the use of high-frequency filtering.

High-frequency filtering can be accomplished in the frequency domain through the use of Fourier transforms. However, this method requires that the whole image be present before processing, whereas other techniques allow operations to take place as the data are being loaded into memory. For real-time applications, such as image enhancement for marine radars, it is preferable to operate on image data as they are received. For example, high-pass filtering can be accomplished in the spatial domain where a small, square, edge operator is convolved with the image. This technique requires that only a small portion of the image be stored at any one time. Some of these edge operators perform a simple differential function at each pixel of the image, others provide a measure of the gradient at each point. Because edge operator response varies with edge orientation, operators which are horizontally and vertically sensitive are usually combined to ensure that all edge information is extracted.

Edge detection involves the application of an edge operator, a comparison of the operator output to a threshold, and a resulting image which contains only edge information. Edge enhancement combines this extracted edge information with the original image to produce a sharpened result.

There are two common classes of edge detectors: local and regional. According to Shaw [56], "regional operators can be applied on windows several times the width of the edge and can be expected to give responses when the entire region is well described by two constant levels." And again, according to Shaw "regional operators need to find only one edge in a region . . . that one may expect such operators to fail if clutter or substantial textural variation is expected."

As the images to be processed for ice and iceberg applications are made up of regions with substantial variation and clutter, regional operators were not expected to give useful results and thus were not investigated further.

Local edge-detection operators are usually applied on small windows (of the order of 3×3 or 5×5 pixels), although they can be larger. By applying two directional masks rotated by 90°, the response and an angle of maximum response for the directional local operators can be calculated. For edge detection, the edge operators, which consist of masks of varying weights, are convolved with the image. The masks normally are 2×2, 3×3, or 5×5 pixels, although there are several larger masks referenced in the literature. The output image can consist of the actual operator output values which are related to edge strength, or the operator output can be compared to a threshold and a binary edge/no-edge image produced. The threshold can be selected interactively or set to pass a fixed percentage of the strongest edges. Pratt [49] described an edge-crispening technique, known as discrete convolution filtering, where a high-pass mask was convolved with the image.

A method of edge sharpening, without regard to direction, employing a Laplacian mask, was shown in Pratt [49]. This method also convolved the mask with the image. One of the most often-cited gradient operators (Peli and Malah [47]) is the Roberts gradient operator. The application of this equation for the enhancement of icebergs from SLAR images is described by Kirby [27]. Another local operator referenced in the literature (Shaw [56]) is the Sobel operator which also is an approximation to the gradient. The magnitude of the Sobel operator is evaluated by taking the square root of the sum of the squares of the horizontal and vertical responses.

Another simple operator is the Mero-Vassey operator (Shaw [56]) which is measured as the sum of the absolute values of the vertical and horizontal responses. Figure 8.61 shows the edges detected with the Mero-Vassey operator after performing a thresholding process. An improvement in image quality by adding the edges back in color can be achieved and is shown in a later image.

There are a number of other operators suggested in the literature and there are several papers which compare these operators. However, they seem to perform differently with different types of images, such that no operator appears to be universally superior.

Gray-Scale or Color Table Manipulations. There are several techniques, involving some manipulation of the gray-scale or color assignment table, that can be used to enhance information in the image. A technique known as "bit slicing" can be used to highlight certain gray-scale levels within the image. By manipulating the video look-up table of the display, a small range of gray-scale levels can be assigned a unique gray-scale or color level, so that features with those gray-scale values will be highlighted in the image. It is possible to have the gray-level range of the bit slice controlled interactively so that the highlighted window can be moved through the gray-scale table and the resulting highlighting seen on the image.

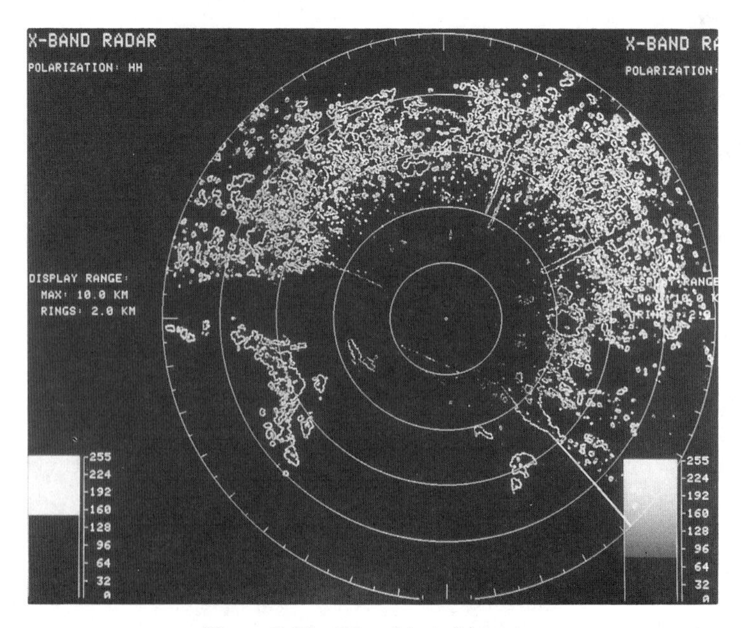

Figure 8.61 Edge-detected image.

As an extension to this technique, density slicing gives all values over a certain gray-scale level a single assigned gray-scale (or color) level and all values below the cut-off level are assigned to a second gray-scale (or color) value, resulting in a binary image.

Finally, the introduction of pseudocolor processing can further enhance images. Pseudocolor is taken to mean the assignment of a color, based on an intensity level value, where no color was present before. The maximum usable number of colors is limited by the spatially diverse nature of the radar image. Adjacent pixels may have substantially different intensities, and to be separable by a human operator they must have color assignments that are readily distinguishable. The application of a few well-chosen colors makes it easier to distinguish between adjacent signal levels; the assignment of darker colors for low signals and brighter colors for higher signals provides a recognizable progression of intensities. Figure 8.62 (see color insert) shows the use of two colors to highlight strong returns and a third to display detected edges.

8.6.3 Hardware Considerations

This section discusses some necessary considerations involved in implementing (in hardware) the techniques illustrated in the previous section.

Although all the techniques discussed in the previous sections improve image quality, some are more suitable for continuous real-time processing of radar images with high data rates. It is useful to review the processing time con-

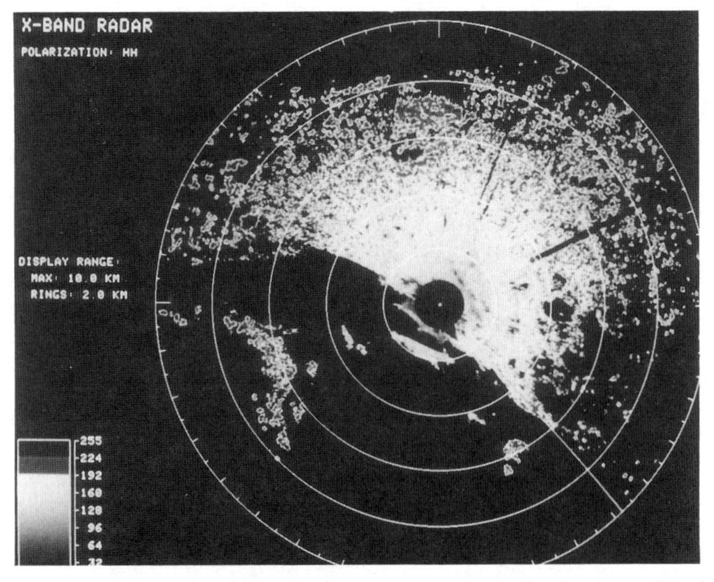

Figure 8-62 Color-coded image. Red and blue represent strong signal returns while the yellow highlights detected edges (see color insert).

straints. The data must be digitized at about 10 MHz, which allows less than 100 ns to perform processing operations between digitized samples. However, because the pulse repetition frequency (PRF) of the radar is typically 1000 Hz or less, giving a new set of signal returns every 1 ms, and because in this case only data out to 200 μs (or 30 km, the typical maximum radar horizon) are of concern, there are about 800 μs available at the end of each sweep to perform additional processing between pulses.

It is also important to define real time in the context of this research. Real time is understood to be the ability to handle the data out to a 30-km range on a continuous basis without having to miss incoming data. Image enhancement is accomplished by taking advantage of the dead time at the end of each sweep.

The analogue radar video is digitized using a readily available single-chip A/D converter. The converter output is passed through a comparator circuit whose output is zero for input values below the operator-controlled threshold. Values above the threshold are passed unchanged. Using a look-up table which is based on radar calibration data, the A/D converter value is mapped to represent received power. The data are then normalized in range through multiplication by a range-dependent constant taken from a read-only memory (ROM). Each range position has a different multiplier value in ROM.

To minimize the hardware, the 3 × 3 pixel median filter could be implemented sequentially in range then bearing, using two three-point, one-dimensional filters. The range filtering would be done along each sweep as the data are received. The range filter would then be fed into a three-sweep buffer. The buffer is then accessed to provide the output for the three-point–bearing median filter.

The bearing filter output is passed on to the edge detection and enhancement circuitry. The use of a 3 × 3 pixel edge operator requires another three-sweep buffer. As the Mero-Vassey operator coefficients are only zeroes and ones, multiplication reduces to addition and subtraction. The horizontal and vertical operator magnitudes are combined to produce the edge operator output. For an image of detected edges, this output is subjected to a threshold and passed to the scan converter for display. For edge-enhanced images, the edge operator output is combined in a weighted sum with the buffered median-filter output. This sum is passed on to the scan converter for display. With more complex circuitry providing programmable multipliers, any 3 × 3 convolution operator can be implemented. Alternatively, the edge image can be treated as a color overlay through manipulation of the color look-up table.

For scan conversion, the circuitry has the basic purpose of storing the range-bearing sample in the appropriate X–Y display location (pixel). The mapping from range-bearing sample coordinates to X–Y location can be implemented through look-up table techniques using ROM. To reduce storage requirements, the look-up table, because of symmetry, need only contain the mapping for a single half quadrant.

Assuming the use of a 1024 line by 1024 pixels per line display area, there is a maximum of 512 pixels (range positions) per sweep. To allow one look-up table to suffice for various range settings on the display, the number of incoming data points per sweep fed to the scan converter must be maintained at 512. For the 10-km range coverage at the 10-MHz sampling rate, there are about 1024 samples per sweep. To reduce this to the required 512 displayable points, with minimum loss of information, successive pairs of samples are averaged. When the maximum range being displayed is reduced to 5 km, the first 512 sample points are used. As the maximum range being displayed is reduced further by factors of two, the appropriate data points are replicated to maintain an input of 512 data points to the scan converter.

Raster scan displays are currently available in 1280 pixel by 1024 line format. These provide a square 1024 $\times$ 1024 area for the radar image and a side area for annotation. Each screen pixel typically has an 8-bit storage location associated with it. This 8-bit pixel value is used to index a video look-up table whose output determines the color of the pixel, by controlling the intensity of each of the red, green, and blue guns. Most colors can be produced by an appropriate mixture of red, green, and blue, and for the gray-scale, the red, green, and blue intensities at each level are set equal. The video look-up table is a powerful tool in that the appearance of the image can be changed without altering the original data. For example, such techniques as bit slicing and pseudocolor mapping can be implemented in this manner.

These processes were initially investigated off-line in the laboratory because of the significant computational power needed to perform them. However, a new generation of hardware radar signal processing systems has been developed which is now capable of performing these tasks in real time aboard ship. The radar signal processor incorporates a digitally controlled analogue gain (STC), analogue signal processing, pulse-to-pulse and scan-to-scan integration, two-radar channel mixing, range normalization, and color table manipulation. At the time of this writing, the hardware is undergoing testing.

8.7 CONCLUSIONS AND FUTURE WORK

This chapter has examined the use of noncoherent surface-based radars in the detection and classification of ice. It has been shown that different radar frequencies enhance different ice types and features. Lower frequency systems (S band) emphasize larger features and surface roughness, while higher frequency systems (K_a band) enhance the detection of icebergs and old ice.

The chapter shows the significant improvement in identification of old ice and icebergs through the use of a classification system based on a combination of σ_{HH} and σ_{HV}. A scattering process was discussed which explained the experimental results, consistent with accepted models and theories.

A number of other radar parameters, including antenna beamwidth, transmitted pulse length, and antenna height, were examined and their effect on sensing of ice discussed.

Experimental data showed that the high variability in the radar returns from large, seemingly stable targets could adversely affect detection of these targets. For some targets, the use of a dual-polarized system provided substantial improvement in detection.

It was shown that significant improvements in the interpretability of ice images could be achieved through the use of relatively simple signal and image processing techniques. The conversion itself to a digital display provided considerable improvements.

Although the radar problems involved in detecting ice and icebergs have received considerable attention, there are several areas requiring additional work. In particular, much remains to be done to optimize the display of the radar data. Some additional experimental work with higher frequency systems (particularly K_a band), and addressing target detection in open water, is required.

Further research into the use of neural network techniques for the real-time classification of ice types based on like- and cross-polarized radar returns is warranted. More data on the detection of ice targets in open water are required in order to validate and improve detection performance modeling.

A dual-polarized, coherent *X*-band radar has been developed at McMaster University and is being used to conduct tests at a coastal site in Newfoundland. Data are being collected for sea clutter and for icebergs. Results from this research are described in Chapter 9.

The experimental work reported in this chapter, culminating in the successful demonstration of a dual-polarized ice radar, indicates that significant improvements in the use of surface-based radar for ice surveillance are indeed possible. Further work should realize even greater gains in ice-radar performance.

ACKNOWLEDGMENTS

The authors wish to acknowledge the dedicated effort of many supporting staff both at the Department of Fisheries and Oceans and at the Communications Research Laboratory at McMaster University. In particular, Trevor Dyas and Bill Montgomery, at the Department of Fisheries and Oceans, are acknowledged for their excellent technical contribution to the program through many difficult field and shipborne trials. Terry Greenlay at McMaster is acknowledged for his contributions to the first experimental field study in 1981. The authors acknowledge the contribution of figures and data to Section 8.4.4 by Joe Ryan and Tim Nohara.

The funding for the research program undertaken by the authors was provided by the Canadian Energy Research and Development Fund, the Natural Sciences and Engineering Research Council, and the Department of Fisheries and Oceans. Without the generous support of these agencies, this research would not have been possible.

We also wish to acknowledge Canarctic Shipping for the provision of shiptime during the seaborne tests, and the evaluation of the prototype system.

Finally, the lead author wishes to acknowledge the support and encouragement of his late friend and colleague Dr. Nelson Freeman.

REFERENCES

[1] E. B. Bailey and N. C. Randall (1961) "Detecting signals in noise," Electronics, Vol. 34, No. 11, 66.

[2] D. H. Ballard and C. M. Brown (1982) *Computer Vision*, Prentice Hall, Englewood Cliffs, NJ.

[3] P. Beckmann and A. Spizzichino (1963) *The Scattering of Electromagnetic Waves from Rough Surfaces*, Pergamon Press, Oxford.

[4] P. Benedict and J. Hall (1979) "A study of iceberg detection for marine navigation," Transport Canada, Ottawa, Rept. TP 2409.

[5] A. J. Blanchard and J. W. Rouse, Jr. (1980) "Depolarization of electromagnetic waves scattered from an inhomogeneous half space bounded by a rough surface," *Radio Sci.*, **15** (4), 773–779.

[6] R. G. Blom and M. Daily (1982) "Radar image processing for rock type discrimination," *IEEE Trans. Geosci. and Remote Sens.*, **GE-20** (3), 343–351.

[7] T. F. Budinger, R. P. Dinsmore, P. A. Morill, and F. M. Soule (1960) "Iceberg detection by radar," Int. Ice Patrol Bull. No. 45, U.S. Coast Guard, 49–97.

[8] R. D. Crissman and L. A. Lalumiere (1990) "Radar monitoring of ice on the upper Niagara River," in *Proc. Sixth Int. Cold Regions Eng. Specialty Conf.*, 406–415.

[9] J. A. Croney, A. Woroncow, and B. R. Gladman (1975) "Further observations on the detection of small targets in sea clutter," *The Radio and Electron. Eng.*, **45** (3), 105–115.

[10] B. W. Currie and E. O. Lewis (1985) "Evaluation of dual-polarization in the surface-based radar detection of icebergs," in *Proc. 8th Int. Conf. on Ports and Oceans Eng. under Arctic Conditions*, 757–766.

[11] R. N. Ellis (1984) "Theoretical study of radar scattering from an iceberg," M.Sc. Thesis, McMaster University, Hamilton, Ontario.

[12] S. Evans (1965) "Dielectric properties of ice and snow—a review." *J. Glaciology*, **5** (42), 773–791.

[13] G. W. Ewell, M. M. Horst, and M. T. Tuley (1979) "Predicting the performance of low-angle microwave search radars—targets, sea clutter, and the detection process," *IEEE Oceans '79*, 373–378.

[14] W. Frei (1977) "Image processing by histogram hyperbolization," *Comput. Graphics and Image Processing*, **6**, 286–294.

[15] A. K. Fung (1967) "Theory of cross polarized power returned from a random surface," *Appl. Sci. Res.*, **18**, 50–60.

[16] A. K. Fung and H. Eom (1982) "Application of a combined rough surface and volume scattering theory to sea ice and snow backscatter," *IEEE Trans. Geosci. and Remote Sens.*, **GE-20** (4), 528–536.

[17] T. Hagman, J. Nilsson, and Y. Nilsson (1976) "Sea Ice-75: *Flar*, *Odar*, ship's radar," Swedish/Finnish Winter Navigation Research Board Rept. No. 16:5.

[18] M. Harvey and J. Ryan (1986) ''Further studies on the assessment of marine radars for the detection of icebergs,'' Environmental Studies Revolving Fund Rept. No. 035, Ottawa.

[19] S. Haykin (1994) *Neural Networks: A Comprehensive Foundation*, Macmillan, New York.

[20] S. Haykin, B. W. Currie, E. O. Lewis, and K. Nickerson (1985) ''Surface-based radar imaging of sea ice,'' *Proc. IEEE,* **73** (2), 233–251.

[21] H. V. Hitney, J. H. Richter, R. A. Pappert, K. D. Anderson, and G. B. Baumgartner, Jr. (1985) ''Tropospheric radio propagation assessment,'' *Proc. IEEE,* **73** (2), 265–283.

[22] A. D. Hood (1954) ''An analysis of radar ice reports submitted by Hudson Bay Shipping (1953),'' National Research Council of Canada, Radio and Electrical Engineering Division, Ottawa, No. 3301.

[23] A. D. Hood (1954) ''An analysis of radar ice reports submitted by Hudson Bay Shipping (1954),'' National Research Council of Canada, Radio and Electrical Engineering Division, Ottawa, No. 3560.

[24] M. Johnson and J. Ryan (1991) ''A radar performance prediction model for iceberg detection,'' in *Proc. 11th Int. Conf. on Port and Ocean Eng. under Arctic Conditions, POAC'91.*

[25] N. L. Johnson and S. Kotz (1970) *Continuous Univariable Distributions*, Vols. 1 and 2, Houghton Mifflin, New York.

[26] Y.-S. Kim, R. K. Moore, and R. G. Onstott (1984) ''Theoretical and experimental study of radar backscatter from sea ice,'' Remote Sensing Lab., Center for Research, Inc., The University of Kansas, Lawrence, Tech. Rept. 331–37.

[27] M. E. Kirby (1981) ''Applications of the Roberts gradient operator for the digital enhancement of icebergs from SLAR imagery,'' in *7th Can. Remote Sensing Symp.*

[28] K. Klein and A. Butt (1986) ''Improving the display of marine radar returns from first-year sea ice,'' C-CORE Tech. Rept. 86-1, St. John's, Newfoundland.

[29] T. Kohonen (1988) *Self-Organization and Associative Memory*, 2nd Ed., Springer, Berlin.

[30] H. Larsson (1949) ''The use of radar in the ice-breaker service,'' *J. Inst. Navigation*, **2**, 315–323.

[31] L. S. Le Page and A. L. P. Milwright (1953) ''Radar and ice,'' *J. Inst. Navigation*, **4** (2), 113–130.

[32] J. C. Leader (1970) ''Multiple scattering of electromagnetic waves from rough surfaces,'' *J. Optical Soc. Am.*, **60**, 1552.

[33] E. O. Lewis and B. W. Currie (1984) ''Effects of frequency and polarization on marine radar detection of ice targets in an ice cover,'' in *Int. Assoc. Hydraulic Res.—Ice Symp.* 375–384.

[34] E. O. Lewis and B. W. Currie (1984) ''Improved marine radar display for navigation in ice-covered water,'' in *Int. Assoc. Hydraulic Res.—Ice Symp.* 363–374.

[35] E. O. Lewis, S. Haykin, and B. W. Currie (1985) ''Detection of sea-ice features using *Ka*-band radar,'' *Electron. Letters*, **21** (11), 499–500.

[36] E. O. Lewis, B. W. Currie, and S. Haykin (1987) *Detection and Classification of Ice*, Research Studies Press, Letchworth, England.

[37] J. D. Luse (1981) "A brief history of the use of marine radar," *J. Inst. Navigation*, **28** (3), 199–205.

[38] J. I. Marcum and P. Swerling (1960) "Studies of target detection by pulsed radar," *IRE Trans.*, **IT-6** (2), 59–267.

[39] J. B. Mercer, R. T. Lowry, and E. D. Leavitt (1985) "An integrated ice monitoring system for offshore drilling," in *Proc. of Arctic 1985*.

[40] J. D. Moore and N. B. Lawrence (1980) "Comparison of two CFAR methods used with square law detection of Swerling I targets," in *IEEE Intl. Radar Conf.*, 403–409.

[41] H. Murthy and S. Haykin (1987) "Bayesian classification of surface-based ice-radar images," *IEEE J. Oceanic Eng.*, **OE-12** (3).

[42] M. Nagao and T. Matsuyama (1979) "Edge preserving smoothing," *Comput. Graphics and Image Processing*, **9**, 394–407.

[43] F. E. Nathanson (1969) *Radar Design Principles*, McGraw-Hill, New York.

[44] T. J. Nohara, S. Haykin, B. W. Currie, and C. Krasnor (1989) "Towards the improved detection of small ice targets in K-distributed sea clutter," in *Int. Symp. on Noise and Clutter Rejection in Radars and Image Sensors*.

[45] T. J. Nohara (1991) "Detection of growlers in sea clutter using an *X*-band pulse-doppler radar," Ph.D. Thesis, McMaster University, Hamilton, Ontario.

[46] J. Orlando, R. Mann, and S. Haykin (1990) "Classification of sea-ice images using a dual-polarized radar," *IEEE J. Oceanic Eng.*, **15** (3).

[47] T. Peli and D. Malah (1982) "A study of edge detection algorithms," *Comput. Graphics and Image Processing*, **20**, 1–21.

[48] R. E. Perry (1953) "A record of radar performance in ice conditions," *J. Inst. Navigation*, **V1** (1), 74–85.

[49] W. K. Pratt (1978) *Digital Image Processing*, Wiley, New York.

[50] B. Raman, R. T. Lowry, and J. B. Mercer (1983) "Improved marine radar performance for ice-hazard detection," in *7th Int. Conf. on Port and Ocean Eng. under Arctic Conditions*, Vol. 4, 835–844.

[51] H. Rohling (1983) "Radar CFAR thresholding in clutter multiple target situations," *IEEE Trans. Aerosp. Electron. Syst.*, **AES-19** (4), 608–619.

[52] J. R. Rossiter, B. W. Currie and E. O. Lewis (1985) "Radar cross sections of two cold icebergs," Iceberg Research, No. 11, 3–9.

[53] J. W. Rouse, Jr. (1972) "The effect of the subsurface on the depolarization of rough-surface backscatter," *Radio Sci.*, **7** (10), 889–895.

[54] J. Ryan (1986) "Enhancement of the radar detectability of icebergs," Environmental Studies Revolving Fund Rept. No. 022, Ottawa.

[55] J. Ryan, M. Harvey, and A. Kent (1985) "Assessment of marine radars for the detection of ice and icebergs," Environmental Studies Revolving Fund Rept. No. 008, Ottawa.

[56] G. B. Shaw (1979) "Local and regional edge detectors: some comparisons," *Comput. Graphics and Image Processing*, **9**, 135–149.

[57] T. C. Sheives (1974) "A study of a dual polarization laser backscatter system for remote identification and measurement of water pollution," Remote Sensing Center, Texas A&M University, College Station, TX, Tech. Rept. RSC-52.

[58] M. I. Skolnik (1980) *Introduction to Radar Systems*, McGraw-Hill, New York.

[59] A. R. Sneyd, M. P. Luce, R. W. Gorman, et al. (1986) "Operational use of remote sensing for commercial arctic class vessel navigation," in *10th Canadian Symp. on Remote Sensing*.

[60] T. Tabata (1975) "Sea-ice reconnaissance by radar," *J. Glaciology*, **15** (73), 215–224.

[61] T. Tabata, M. Aota, and M. Oi (1967) "On the observation of sea ice distribution with the sea ice radar (preliminary report)," *Low Temp. Sci.*, Ser. A (Physical Sci.), No. 25, 233–239. Translated by E. R. Hope, 1970, Defense Research Board of Canada, Ottawa, T102J.

[62] T. Tabata, M. Aota, M. Oi, and M. Ishikawa (1969) "Observation of drift ice movements with the sea ice radar network," *Low Temp. Sci.*, Ser. A (Physical Sci.), No. 17, 295–315. Translated by E. R. Hope, 1970, Defense Research Board of Canada, Ottawa, T105J.

[63] T. Tabata, M. Aota, M. Oi, and M. Ishikawa (1969) "Distribution of pack ice field off Hokaido observed with sea ice radar network," *Low Temp. Sci.*, Ser. A (Physical Sci.), No. 27, 23–38, Data Rept. Suppl. Translated by E. R. Hope, Defense Research Board of Canada, Ottawa, T103J.

[64] S. Watt (1985) "Radar detection prediction in sea clutter using the compound *K*-distribution model," *IEEE Proc.*, **132**, Pt. F, No. 7.

[65] W. F. Weeks and S. F. Ackley (1982) "The growth, structure and properties of sea ice," U.S. Army Corps of Engineers, Cold Regions Research and Engineering Laboratory, Hanover, NH, Monog. 82-1.

[66] P. D. L. Williams (1973) "Detection of sea ice growlers by radar," in *IEEE Radar Conf.*, 239–244.

[67] P. D. L. Williams (1975) "Limitations of radar techniques for the detection of small surface agents in clutter," *The Radio and Electron. Eng.*, **45** (8), 379–389.

[68] P. D. L. Williams (1979) "The detection of ice at sea by radar," *The Radio and Electron. Eng.*, **49** (6), 275–287.

[69] G. J. Yang and T. S. Huang (1985) "The effect of median filtering on edge location estimation," *Comput. Graph. and Image Processing*, **15**, 224–245.

9

SURFACE-BASED RADAR: COHERENT

SIMON HAYKIN, BRIAN W. CURRIE,
AND VITAS KEZYS

Communications Research Laboratory
McMaster University
Hamilton, Ontario, Canada

9.1 INTRODUCTION

The noncoherent marine radars described in Chapter 8 process only the amplitude of the returned radar echo. Often there is insufficient separation, in the amplitude domain, between the signal of interest and the interfering noise and clutter to permit satisfactory ice detection. This chapter explores the use of other signal domains which offer greater separation between the desired signal and the unwanted interference.

In addition to amplitude, the electromagnetic signal has the properties of frequency, phase, and polarization, which can be combined with temporal and spatial processing to provide improved performance. The experimental program using noncoherent radars (Chapter 8) showed significant improvements in performance afforded by using various radar frequencies and dual polarization. This chapter extends the research through the use of a coherent radar and presents experimental results, advanced processing techniques, and examples of improved performance. The experimental coherent radar system, known as the IPIX radar, is described in Subsection 9.2.1.

The following paragraphs are intended to indicate the major capabilities afforded through the use of such a system. It is assumed that the radar is monostatic, that is, it employs the same antenna for both transmit and receive.

In a coherent radar, the transmitted and received signals are processed using phase-locked oscillators, such that the phase of the received signal relative to

Remote Sensing of Sea Ice and Icebergs, Edited by Simon Haykin, Edward O. Lewis, R. Keith Raney, and James R. Rossiter.
ISBN 0-471-55494-4

the transmitted signal can be measured, in addition to measuring the return signal amplitude. This ability to measure the phase on each pulse provides significantly increased information content, which can be exploited in three signal domains: Doppler, polarization, and frequency.

9.1.1 Doppler

The phase actually provides a very accurate measure of the range of the target, but as the phase angle is modulo 2π, the distance is ambiguous on the scale of the radar wavelength. However, for targets which move less than a radar wavelength between radar transmissions, the change in the phase measurement from pulse to pulse yields a measure of the target's velocity component in the radial direction relative to the radar. This phase change versus time gives rise to an apparent shift of returned frequency, termed the Doppler frequency shift, often referred to simply as the Doppler frequency. A target moving with radial velocity, v, will produce a Doppler frequency shift, f, given by

$$f = \frac{2v}{\lambda}$$

where λ is the radar wavelength. Therefore, calculation of the Doppler frequency spectrum provides a description of the velocity profile of the target. The usual approach is to collect a time series of samples of the target's return, taken at the pulse repetition frequency (PRF), then transform the time series into the Doppler frequency domain. As detailed in Subsection 9.2.3, the Doppler frequency domain reveals differences in the target and interference behavior, which are not accessible with amplitude-only processing.

9.1.2 Polarization

As discussed in Section 6.4.1, the polarization scattering matrix of a target defines its scattering response to incident illumination, in terms of its electric field vector, $\boldsymbol{E}$. For illumination using a horizontally (H) and vertically (V) dual-polarized wave, the complex coefficients, Γ, in (6-35) can be expanded to give the coherent scattering matrix, written in matrix form as

$$\begin{bmatrix} \mathbf{E}_h \\ \mathbf{E}_v \end{bmatrix}_{\text{receive}} = A \begin{bmatrix} |S_{\text{HH}}| e^{j\theta_{HH}} & |S_{HV}| e^{j\theta_{HV}} \\ |S_{VH}| e^{j\theta_{VH}} & |S_{VV}| e^{j\theta_{VV}} \end{bmatrix} \cdot \begin{bmatrix} \mathbf{E}_h \\ \mathbf{E}_v \end{bmatrix}_{\text{transmit}} \tag{9-1}$$

Each Γ coefficient has been expanded into a scattering magnitude, $|S|$, and a phase angle, θ. A is a multiplicative scaling factor.

For a monostatic radar, some simplifications may be made. The absolute phase is a function of range and is not a target-related parameter; therefore, there is no loss in generality if the cross-polarized phase is set to zero, and the other phases are measured relative to it. Further, if the intervening medium and the two channels of the radar are assumed to be linear and isotropic, reciprocity permits the two cross-polarized terms to be set equal. Therefore, (9-1) becomes

$$\begin{bmatrix} \mathbf{E}_h \\ \mathbf{E}_v \end{bmatrix}_{\text{receive}} = A \begin{bmatrix} |S_{\text{HH}}| e^{j\theta_{HH} - \theta_{VH}} & |S_{HV}| \\ |S_{VH}| & |S_{VV}| e^{j\theta_{VV} - \theta_{VH}} \end{bmatrix} \cdot \begin{bmatrix} \mathbf{E}_h \\ \mathbf{E}_v \end{bmatrix}_{\text{transmit}}$$

The scattering matrix is determined by the measurement of five parameters: three amplitudes and two relative phase angles.

The IPIX radar transmits only one polarization at a time. However, it is equipped with a high-speed waveguide switch, which allows alternation of the transmitted polarization on a pulse-to-pulse basis. At a PRF of 2 kHz, the delay between subsequent transmissions is 0.5 ms. For slowly moving ice targets, dual-polarized measurements made with this spacing can be considered essentially instantaneous, and approximately so for sea clutter.

As detailed in Subsection 9.2.4, the differing properties of the targets and interference in the polarization domain permit improved detection.

9.1.3 Frequency

The illumination of a target by a number of radar frequencies can yield additional information compared to single-frequency operation. Changing the transmit frequency from one pulse to the next, by more than the pulse bandwidth, provides an uncorrelated measurement. This frequency agility improves the effectiveness of pulse-to-pulse integration for increasing the target-to-clutter ratio, hence providing an improvement in detection capability.

Illuminating a slowly moving target with a series of pulses, each at a different but phase-related frequency, permits synthesis of a wide-bandwidth pulse, yielding improved range resolution. This capability is related to pulse compression techniques.

Simultaneous transmission of two radar frequencies (Palmer [18]) permits the use of interferometric processing techniques, which can provide information on scatterers with spacing on the order of the wavelength of the difference frequency. This method provides some measure of the spatial coherence or rigidity of the scattering source, and is expected to show considerable contrast between the fluid, dynamic sea surface and the more solid ice. This technique has been extended to the use of more than two frequencies (Schuler et al. [21]). Such multifrequency capability is incorporated into the IPIX radar.

9.1.4 Chapter Outline

The remainder of the chapter is organized in four sections. Section 9.2 presents results of the experimental program conducted with the IPIX radar system. The radar system is described as is the field experiment during which the data on which most of the results are based were collected. Doppler processing of the radar data is presented, followed by polarization results.

Section 9.3 presents some qualitative results of the theoretical studies, which involve modeling the physical dynamics of the ocean and floating targets, combined with radar scattering models, to simulate the expected nature of the radar returns.

Section 9.4 addresses target detection techniques. The initial approaches involve processing in the Doppler frequency domain, with additional information obtained from the two received polarizations. The section then discusses the improvements in performance made possible with nonlinear processing techniques, including the chaos theory and its application using neural networks.

Section 9.5 summarizes the results.

9.2 EXPERIMENTAL PROGRAM

Based on experience gained in the experiments with noncoherent radars, as reported in Chapter 8, it became apparent that to advance, in an effective manner, the research for a radar which would have enhanced ice-detection capability, a ground-based radar system specifically designed for such research was required. At the same time, there were a number of radar parameters remaining to be studied which were not available in a single existing radar. As a result, the Communications Research Laboratory of McMaster University, beginning in 1984, designed and constructed an experimental radar system, known as the IPIX radar (Haykin et al. [10] and Currie et al. [4]).

9.2.1 The IPIX Radar

The IPIX (*i*ntelligent *pix*el-processing) radar was designed to include as many radar measurement domains as practical, and to permit as much flexibility in its operation as possible. However, once selected, parameters needed to remain accurate and stable. Built into the radar was the capability to document the parameter settings and operational configuration.

The radar underwent a major revision in 1991–1992, including incorporation of a multifrequency capability, improved sampling system, and installation in a trailer, making the facility portable. Table 9.1 lists the parameters of the radar as it currently exists. As the analysis results presented below are based on data collected in 1989, differences in design specific to that time will be indicated where appropriate.

9.2.2 Field Experiment

A major field experiment, using an early version of the IPIX radar, was conducted from Cape Bonavista, Newfoundland, between 30 May and 13 June

TABLE 9.1 Major Parameters of the IPIX Radar System

Parameter	Value
Transmitter	
Type	travelling-wave tube amplifier
Peak power	8 kW
Duty cycle	2% maximum
Pulse width	
uncoded	20 to 5000 ns
coded (nonlinear FM)	5000 ns compressed to 32 ns
Frequency	9.39 GHz (fixed)} individually or 8.9–9.4 GHz (agile)} simultaneously
Polarization	H or V; switchable pulse-to-pulse
Pulse repetition interval	arbitrary; 100 μs minimum
Receivers	
Number of units	2
Type	linear
Polarization	H or V (each receiver)
Tuning	fixed or agile transmit frequency
Sensitivity time control	0 to 80 dB attenuation, programmable
Dynamic range	>50 dB instantaneous
Antenna	
Type	parabolic dish
Diameter	2.4 m
Beamwidth	0.9° (pencil beam)
Polarization	H and V
Gain	44 dB
Sidelobes	< −30 dB
Cross-polarization isolation	>33 dB
Rotation rate	0 to 30 rpm
Angular coverage:	(rotating or stationary)
in azimuth	0° to 360°
in elevation	−5° to +90°
repeatability	<0.1°
control	computer
Data Acquisition	
Number of channels	4
A/D word size	8, 10, or 10 mapped onto 8 bits
Signal source	in-phase (I) and quadrature (Q) for each of 2 receivers
A/D sampling rate	50 MHz maximum, matched to pulse length
Sampling window	defined by start–stop values in angle and range
Sampling capacity (per real-time event)	256-Mb random access memory
Archival devices	8 mm tape; >2 Gb per tape optical disk; 500 Mb per side

TABLE 9.1 (*Continued*)

Parameter	Value
Calibration	
Internal (receivers and data acquisition)	inject internally generated signals into the receivers; develop tables for injected power versus A/D values
External (adds transmitter and antenna)	custom-built transponder: returns the transmitted signal with known amplification and phase shift (to calibrate Doppler)
Data Verification	
Capabilities available in the field with the radar system	color workstation, data visualization software, array processor board

1989. Figure 9.1 shows the location of the radar site. The antenna was about 22 m above sea level, and the land sloped downward from the front of the antenna to the sea. The radar had an unobstructed view of, and could collect useful data from, the ocean and targets at compass bearings from 20° to 140°. For this experiment, a 1.2-m, dual-polarized dish antenna (beamwidth 2.2°) was used.

Atmospheric data were taken from log sheets recorded by the Atmospheric Environment Service weather station in Bonavista, within 5 km of the radar site. These measurements included wind speed and direction, gust speeds, temperature, pressure, precipitation, and visibility.

During the period of the experiment, wind speeds were typically 15–20 knots, with some periods of 30-knot winds with 40-knot gusts. Temperatures varied, from 8°C to 18°C for 10 of the days, and from 2°C to 8°C for 4 of the days. There were 2 days with rain. Barometric pressure varied from 100.5 to 102.5 kPa.

To measure ocean conditions, a nondirectional Datawell Waverider buoy was deployed throughout the experiment, at a range of 6.8 km and bearing of 72°. Data were transmitted in real time to the shore station, and results were printed every hour for significant wave height, maximum wave height, and peak period. These values were also continuously displayed on the data-logging computer screen. Wave periods of 6 to 11 s were encountered, with significant wave heights of 1 to 3 m, and maximum wave heights of 5 m.

Ice targets included three growlers and one large iceberg. To act as a reference target, a Luneberg lens reflector, with nominal radar cross section of 2 m^2, was mounted atop a fiberglass spar buoy, which was moored at a range of 8.1 km and a bearing of 67°. The reflector was about 3 m above the water.

Sampled radar data were collected from the targets and the sea surface with the following typical radar parameters. The transmit pulse length was 200 or 1000 ns for the uncoded pulse. Pulse compression was used, but a hardware

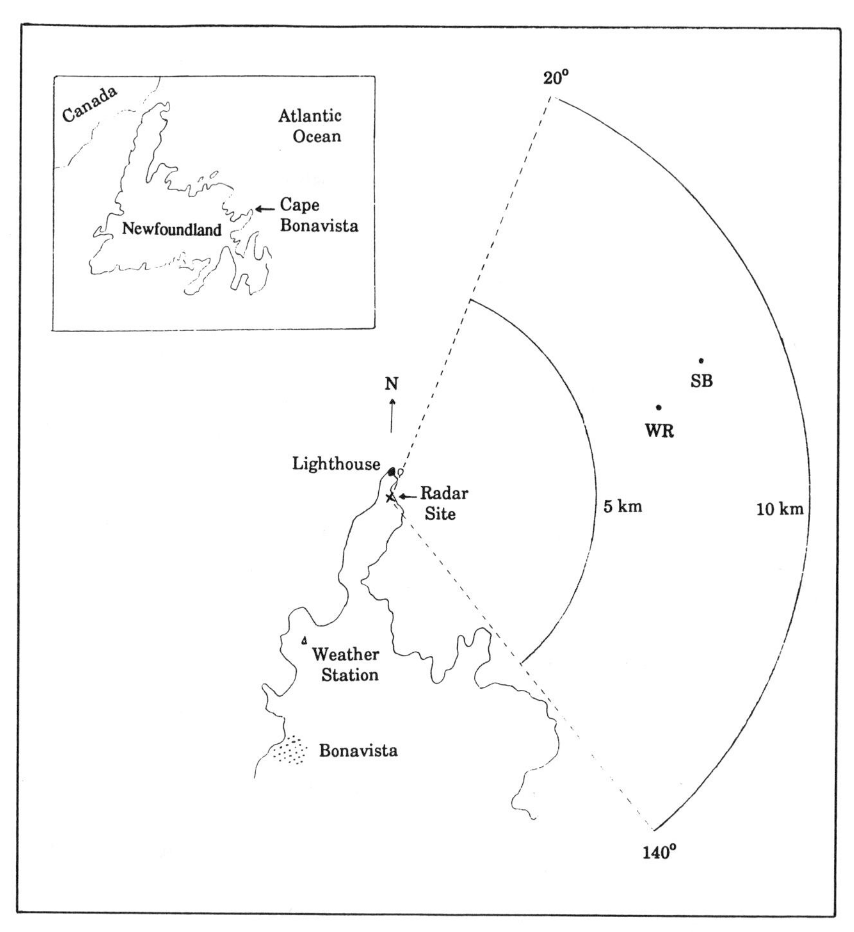

Figure 9.1 Location of the field experiment conducted using the IPIX radar system in May/June 1989. The radar had an unobstructed view of the ocean from about 20° to 140°, relative to true north. The location of the moored Waverider is designated by WR, and the spar buoy with radar reflector by SB.

fault made the data less useful. For long dwell time and uncorrelated amplitude data, a PRF of 200 Hz was used. Most of the time the PRF was 1000 Hz for fixed polarization, and 2000 Hz otherwise. Data were often collected in sets of three, one for each of the fixed polarizations on transmit, then the third one for alternating polarization. Most of the data were collected with the antenna stationary; the scanning antenna data were taken with the antenna rotating at 3, 15, or 30 rpm. The data were taken within a sampling window, defined by limits set in range and bearing (time), and which usually was set to provide tens of seconds of dwell time with the antenna stationary.

The early version of the IPIX radar system used in the 1989 experiment, had a sampling memory capacity of 16 Mb; about 400 files of 16 Mb each were collected in the field.

9.2.3 Doppler Results

The nature of the spectrum of the Doppler frequency shift, and its evolution with time, shows a number of interesting properties of the target and the interference. Time histories of in-phase (I) and quadrature (Q) samples were constructed for given range cells. For each data stream, successive groups of 512 points, overlapping by 64 points, were used to calculate the Doppler spectrum with a fast Fourier transform (FFT) algorithm. Figure 9.2 shows an example of the amplitude of the returned signal from sea clutter at a range of 6.4 km, and the associated Doppler spectrum. For the results presented, the convention is such that negative frequencies in the spectrum represent targets approaching the radar, and positive frequencies represent receding targets. As an indication of the power of Doppler processing to reveal a target, Fig. 9.3 shows a similar amplitude plot and the corresponding Doppler spectrum for the same data file, but taken at a range of 6.5 km, for a cell containing a growler in sea clutter. Although there is no significant increase in the strength of the amplitude plot, the growler's peak in the Doppler spectrum is identified easily as separate from the sea clutter. The presence of the growler was confirmed visually from the shore.

To permit examination of the time evolution of these Doppler spectra, the log-magnitudes of each spectrum were gray-scale coded, and shown on a raster display as one line. Successive spectra were shown as successive lines. This technique created time-frequency plots, which are presented in the following subsections. There are three factors which should be kept in mind when examining these time-frequency plots, all related to data digitization issues:

1. If there is any residual DC offset in the data after subtraction of the estimated mean, a spike will appear at 0 Hz in the Doppler spectrum. However, this factor is generally inconsequential.

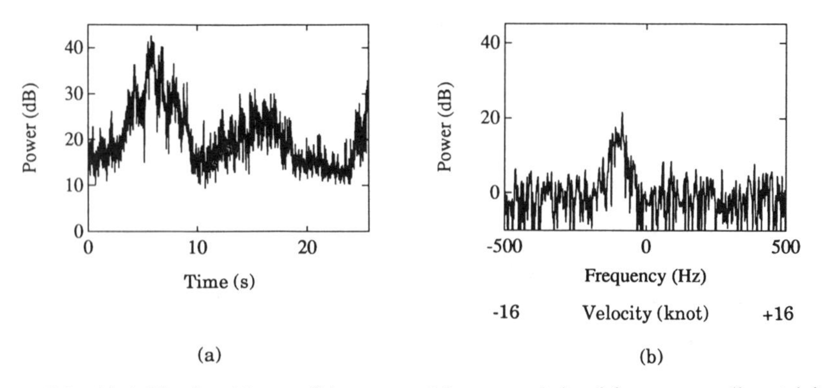

Figure 9.2 (a) A 25-s time history of the power of the returned signal for a range cell containing only sea clutter, at a range of 6.4 km. (b) Doppler spectrum of a 0.5-s segment of I, Q data taken from the time history shown in (a).

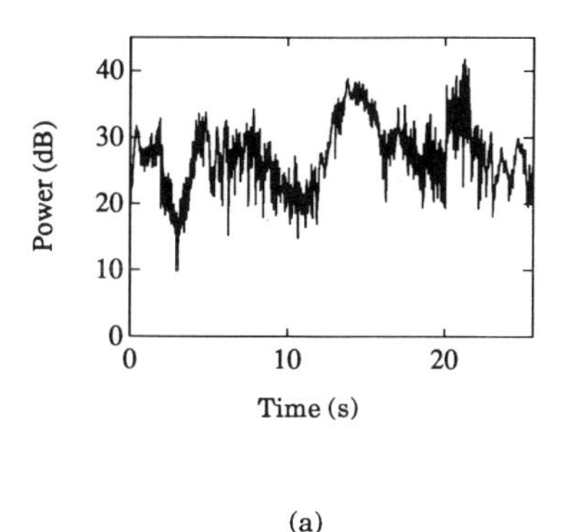

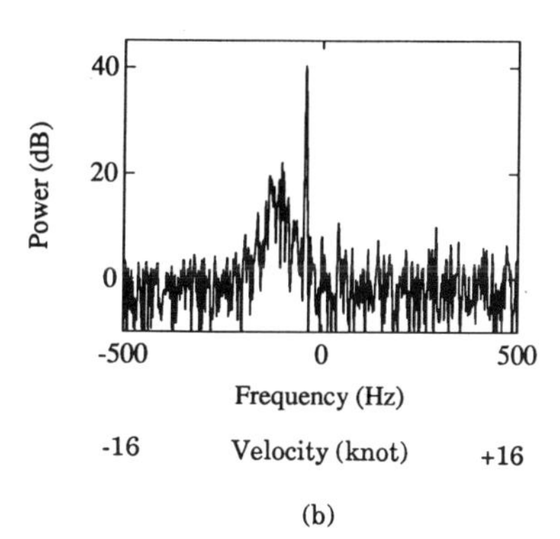

Figure 9.3 (a) A 25-s time history of the returned power for a range cell containing sea clutter and a growler, at a range of 6.5 km. These data and those for Fig. 9.2 were taken from the same data file, that is, they were collected simultaneously. Note that the presence of the growler does not significantly increase the signal power as compared to the sea clutter alone in Fig. 9.2(a). (b) Doppler spectrum of a 0.5-s segment of I, Q data taken from the time history shown in (a). The presence of the growler, which was confirmed visually, is clearly evident.

2. If the gains of the I and Q channels are not equal, or if the I and Q outputs are not exactly orthogonal, a real nonzero Doppler frequency peak will produce a mirror image at the same frequency but with opposite sign. The strength of the image peak will vary in relation to the amount of imbalance and nonorthogonality. This I–Q balance and orthogonality, which originates in the radar system hardware, can be estimated from the data itself, but may be only partially corrected.
3. If either the I or Q data are clipped, spurious frequency peaks will be generated. The severity of this "noise" will be proportional to the severity of the clipping. To keep aware of possible clipping, at the right side of each Doppler spectrum, a black dot is entered at each line for which one or more of the input data samples have been clipped. If the amount of clipping is slight, the spectrum will contain merely an increased background level at a number of frequencies, but the main trend of the data still will be evident. Data which have been clipped are nonrecoverable, although if the clipping is not too severe, interpolation or prediction techniques may be used to estimate the clipped data.

Some results of the Doppler analysis follow (Currie et al. [2]), based on the source of the reflections: sea clutter, rain clutter, growlers, and iceberg.

Sea Clutter

A simplistic view of the scattering mechanism for radar returns of sea clutter is the following. The actual reflectors are the small-scale waves with short wavelengths which are initially generated by wind, and which may be sustained only with wind, or may develop into slightly larger scale waves, which persist even after the wind ceases. The effective position of these small-scale wave reflectors is driven by the underlying swell. As the swell wave propagates through the radar cell, the actual scatterers—the small waves—are thrust up in to the radar beam and also undergo horizontal displacement because of the orbital motion induced by the swell. This orbital motion moves the scatterers in the direction of the swell at the top of the swell wave, and then backward as the trough moves through. The other physical factor of consequence is that as the swell increases in amplitude, more and more of the sea surface (i.e., the troughs) is shadowed by the intervening swell peaks because of the very low grazing angle (less than 1°) at which the radar views the waves.

To study sea clutter alone, data were collected on 12 June 1989 from 12:00 to 20:00, at three bearings, 30°, 75°, and 120°. The ocean had a significant wave height of 2.5 m, a period of 8 s, with the swell direction (not directly measured) from about 120°. The winds were about 20 knots, gusting as high as 30 knots, and coming initially from a bearing of 120°, then gradually changing to 150°. The samples were taken at a range of about 5.8 km.

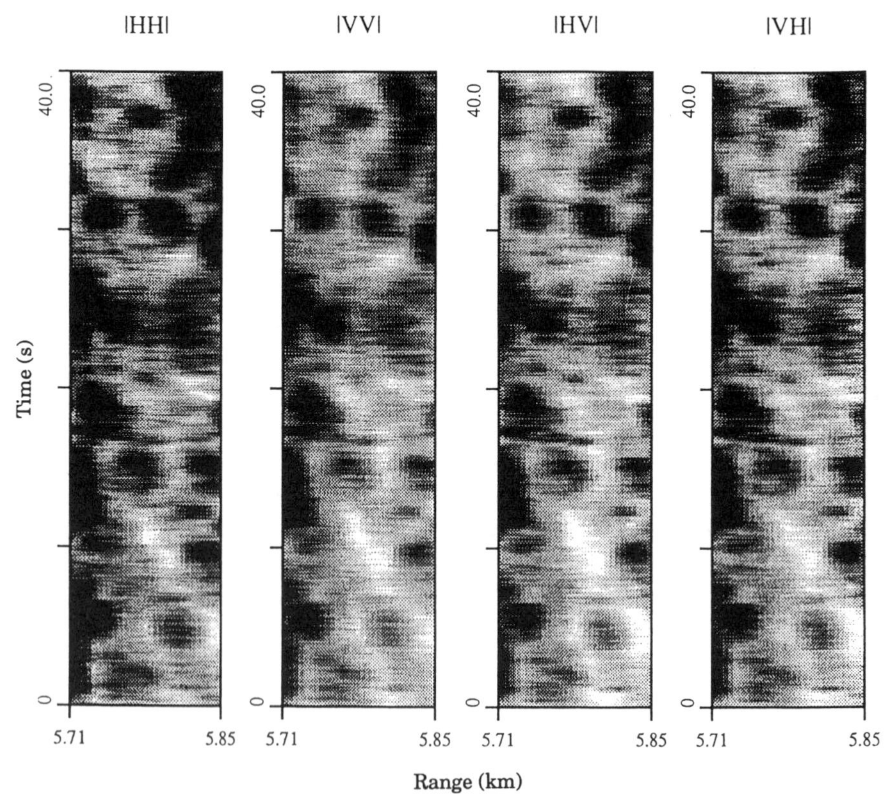

Figure 9.4 Time history of the amplitude versus range for sea clutter, at a bearing of 30° and a range of 5.8 km. The Waverider measured a significant wave height of 2.5 m and a period of 8 s. The swell was approaching the radar from a bearing of about 120°. Winds were about 20 knots from 120°.

The resulting data are presented in amplitude plots, one for each channel, in which each line shows the gray-scale–coded amplitude collected in range for one (or the average of several) sweep; successive lines reveal the time history.

The amplitude plot (Fig. 9.4) for bearing 30°, shows no particular wave pattern: an expected result as the radar is looking more or less crosswind and cross-swell. This situation is in contrast to the amplitude plot for bearing 75° (Fig. 9.5), which begins to show some indication of the swell waves. Finally, the amplitude plot for bearing 120° (Fig. 9.6) shows the swell waves to be well defined: again an expected result as the radar was looking into the swell and the wind direction. (Note that because about 6 dB of radio frequency attenuation was used on the like-polarized receivers, but was not used on the cross-polarized receivers, the like- and cross-polarized powers look similar.)

These three data sets were taken with alternating polarization, which permits comparison of both Doppler and polarization properties. Consider the Doppler plot for bearing 30°, at range 5.75 km, as shown in Fig. 9.7. Because there

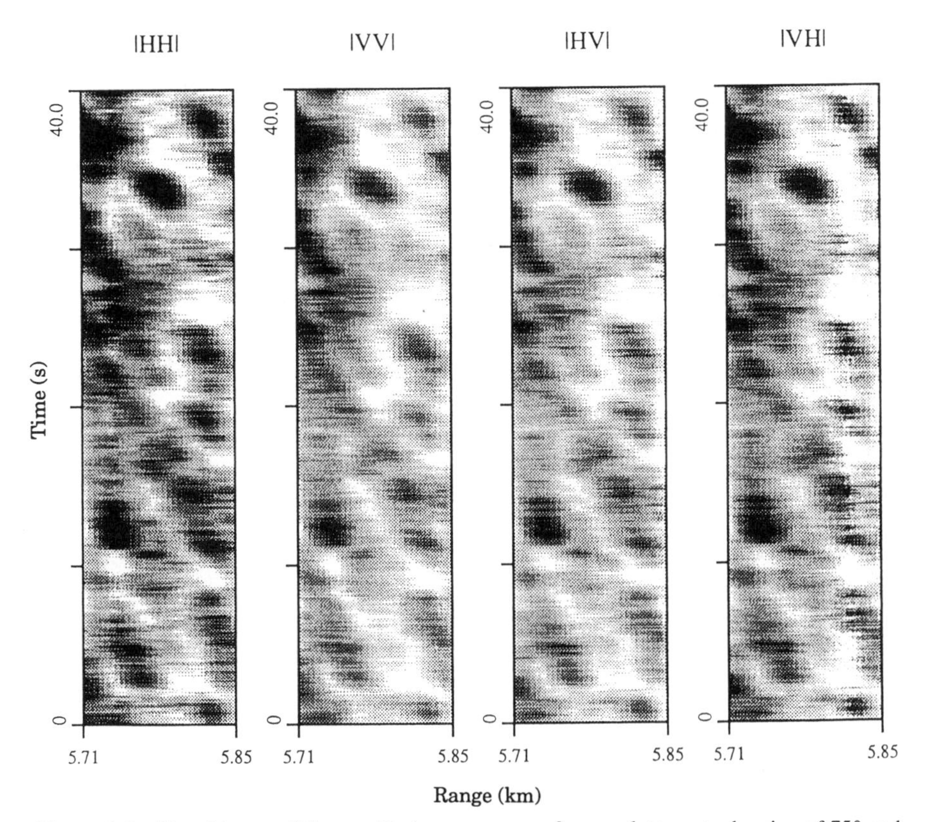

Figure 9.5 Time history of the amplitude versus range for sea clutter, at a bearing of 75° and a range of 5.8 km. The Waverider measured a significant wave height of 2.5 m and a period of 8 s. The swell was approaching the radar from a bearing of about 120°. Winds were about 20 knots from 120°.

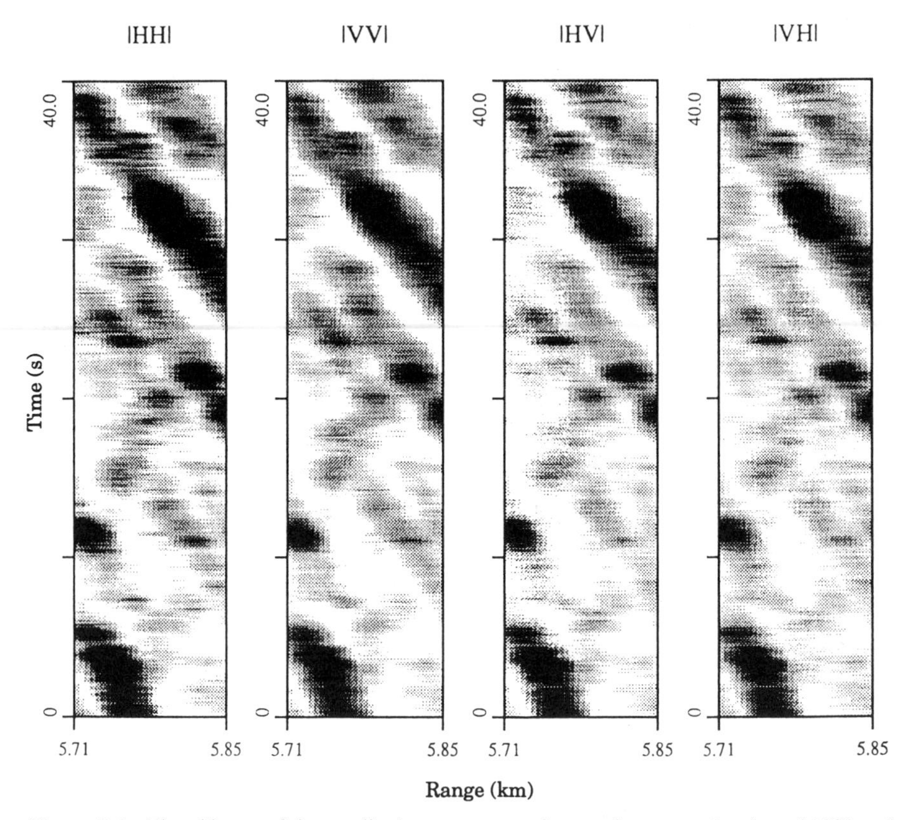

Figure 9.6 Time history of the amplitude versus range for sea clutter, at a bearing of 120° and a range of 5.8 km. The Waverider measured a significant wave height of 2.5 m and a period of 8 s. The swell was approaching the radar from a bearing of about 120°. Winds were about 20 knots from 120°. Note the increasingly pronounced wave crest pattern in progressing through Figs. 9.4 and 9.5 to this figure, a sequence in which the radar looked more upwind and upwave.

also happens to be rain clutter (discussed below) in the like-polarized plots, the HV Doppler plot is examined here to study the sea clutter alone, as the rain clutter is too weak to be detectable. From the plot, a number of observations can be made:

1. The strength of the Doppler content varies periodically, with about five cycles in the 40-s record, or 8 s per cycle. This correlates to the period of the swell. The appearance of swell peaks and troughs, and related shadowing, modulates the strength of the return.
2. Most of the clutter spectrum power is in negative frequencies, indicating swell approaching the radar. However, there is also content at small positive frequencies, indicating components of the waves moving away from the radar. This effect appears to be consistent with the cross-swell observation angle, for the action on the rear of the swell waves is visible, and it can move away from the radar as the wave propagates.

3. The maximum frequency shift (i.e., furthest away from 0 Hz) occurs when the return strength is the maximum. This result is conjectured to be associated with the passing of the peak of the swell wave. The wave exposes the largest number of scatterers, and gives those scatterers near the peak their maximum horizontal movement (toward the radar) because of the orbital action. By similar reasoning, the horizontal movement away from the radar occurs in the troughs. However, shadowing reduces the return power, making this effect less pronounced. Thus the positive frequency side of the spectrum shows less variation with time.
4. The average Doppler shift is about 75 Hz, with an average spectral width of about 250 Hz.
5. Comparison of the HV plot (Fig. 9.7) and the VH plot (not shown) reveals no visible difference in the Doppler behavior.
6. Comparison of the HH plot (Fig. 9.8) and the VV plot (not shown) reveals some small differences in detail, but the overall behavior is quite similar.

Consider next the HH plot (Fig. 9.9) for bearing 75° (at about 45° to the wind and the swell). The observations are:

1. The most obvious change is in the average Doppler shift, which appears to be about 200 Hz. As the radar was looking closer to upswell and upwave than in the previous data set, the radial velocity components are larger.

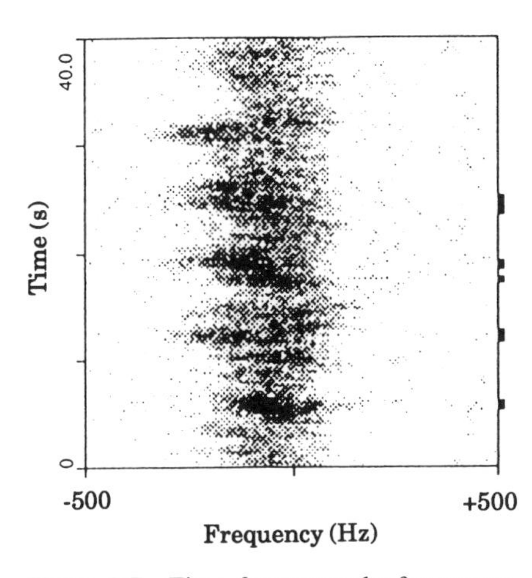

Figure 9.7 Time-frequency plot for a range cell containing sea clutter, at a bearing of 30° and a range of 5.75 km. The data is taken from the same file as for Fig. 9.4. The HV polarization is presented as the rain clutter component is not significant.

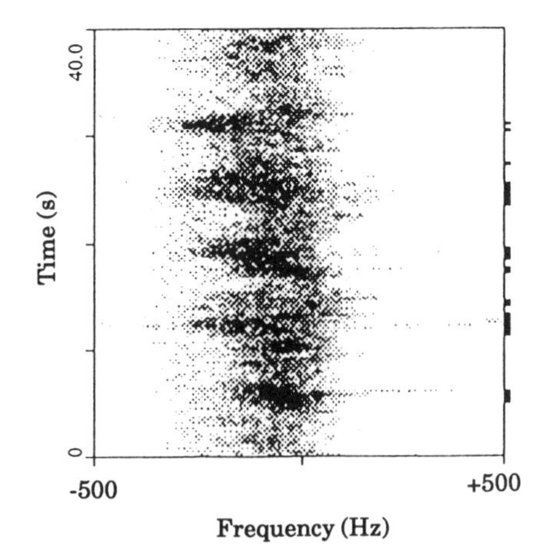

Figure 9.8 Time-frequency plot for a range cell containing sea clutter, at a bearing of 30° and a range of 5.75 km. The data are taken from the same file as those for Fig. 9.4, HH polarization. Comparison with Fig. 9.7 reveals the larger rain clutter component.

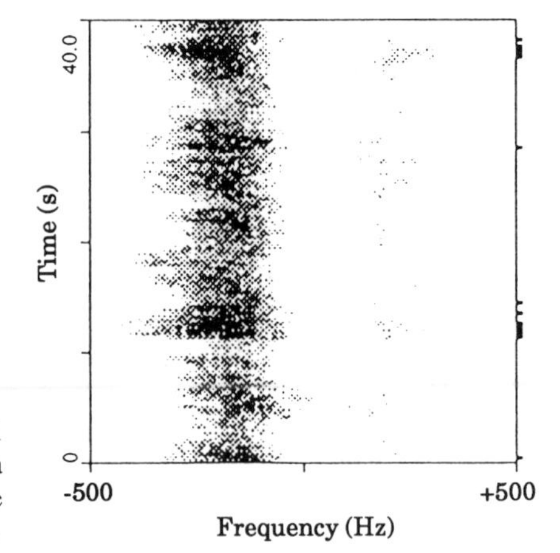

Figure 9.9 Time–frequency plot for a range cell containing sea clutter, at a bearing of 75° and a range of 5.75 km. The data are taken from the same file as those for Fig. 9.5, HH polarization.

2. The average spectral width is somewhat narrower, at about 200 Hz.
3. There is no longer any positive frequency content. (The light sprinkling of positive frequency dots is caused by spectral images resulting from strong returns and small I–Q imbalance and nonorthogonality.) The explanation is that the troughs in which the outward horizontal displacement occurs are shadowed by the swell peaks nearer to the radar. The radar cannot sample the full spectral content of the sea surface. Therefore, the side of the spectrum nearer to 0 Hz is much more uniform with time. This inability to see the complete, true Doppler spectral content also will bias the estimate of the mean Doppler shift and the spectral width.

The HH plot (Fig. 9.10) for bearing 120°, presumed to be looking upwind and upswell, shows a Doppler shift and spread similar to that for bearing 75°.

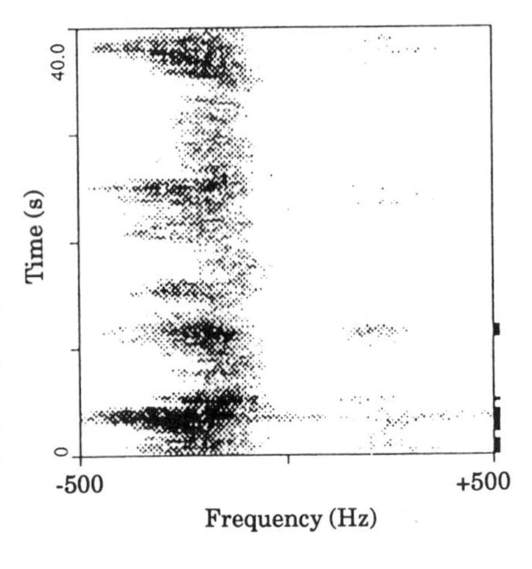

Figure 9.10 Time–frequency plot for a range cell containing sea clutter, at a bearing of 120° and a range of 5.75 km. The data are taken from the same file as those for Fig. 9.6, HH polarization. Note the increasing shift in the average Doppler frequency from Figs. 9.8 to 9.9, to this figure, as the radar faces more upwind and upwave.

However, the occurrence of the wave peaks is better defined, with less spectral power between the peaks. This result may imply more wave shadowing than at bearing 75°; a reasonable assumption as the radar is now looking into the swell.

Rain Clutter

Five data sets captured rain clutter, in addition to sea clutter. The rain was being driven by a strong wind, 25–30 knots from 110°–120°. The data were collected over a period of about 30 min, apparently as a cell of rain showers moved through the radar sampling area. When present, the rain occurred at all ranges. No estimate of rainfall rate could be made.

The like-polarized Doppler plot, for example, the HH data (Fig. 9.11) at a range of 5.80 km and bearing of 30°, clearly shows the presence of the rain clutter: the band of uniform speckles in the −200 Hz to −325 Hz portion of the spectrum. On the other hand, the cross-polarized plot, for example, the HV data (Fig. 9.12), shows essentially no rain clutter, only sea clutter.

The rain clutter is the backscatter from a large number of raindrops of a variety of sizes. These scatterers, which in this circumstance appear to the radar to be nearly symmetrical, induce very little depolarization of the radar signal. The movement of the scatterers is dependent upon the effects of gravity and of the wind which, averaged over the volume of the radar cell, produces a more uniform spectrum than does the sea clutter.

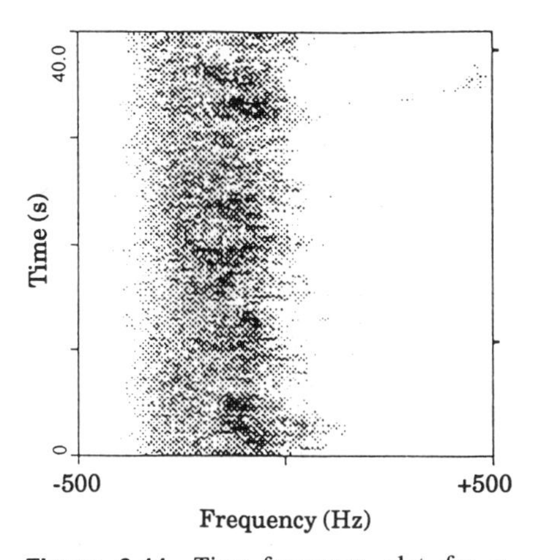

Figure 9.11 Time–frequency plot for a range cell containing sea and rain clutter, at a bearing of 30° and a range of 5.80 km, HH polarization. The rain clutter shows as the more uniform speckled component between −200 and −325 Hz.

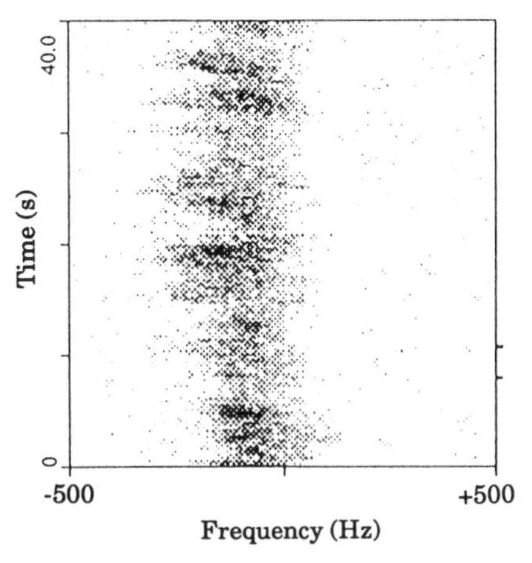

Figure 9.12 Time–frequency plot for a range cell containing sea and rain clutter, at a bearing of 30° and a range of 5.80 km, from the same data file as Fig. 9.11, but for HV polarization. The rain clutter component is essentially nonexistent in comparison to Fig. 9.11; only the sea clutter is detected.

Growlers

One data set included a growler in sea clutter, at a range of 6.5 km and a bearing 119°, on 7 June 1989 at 12:50 (the same data set used to generate Fig. 9.2 and 9.3). The winds were light and variable. The growler is visible in the amplitude plot (Fig. 9.13) because of the ability of the eye to detect and integrate the growler's return throughout the 34.5 s. However, if only short portions of the total data set are considered, the growler's return is often weaker than that of the neighboring clutter, and would not be identifiable. Note, as well, that the strength of the growler's return, compared to the sea clutter, is less in the cross-polarized than in the like-polarized data. This indicates that the sea depolarized the radar signal more than the ice did in this particular experiment.

The HH Doppler plot (Fig. 9.14) shows the greater detectability of the target in the Doppler domain. The growler's return is the nearly black undulating line. Even over short portions of data, the growler is detectable.

The target's return has been integrated through the FFT into a narrow Dopp-

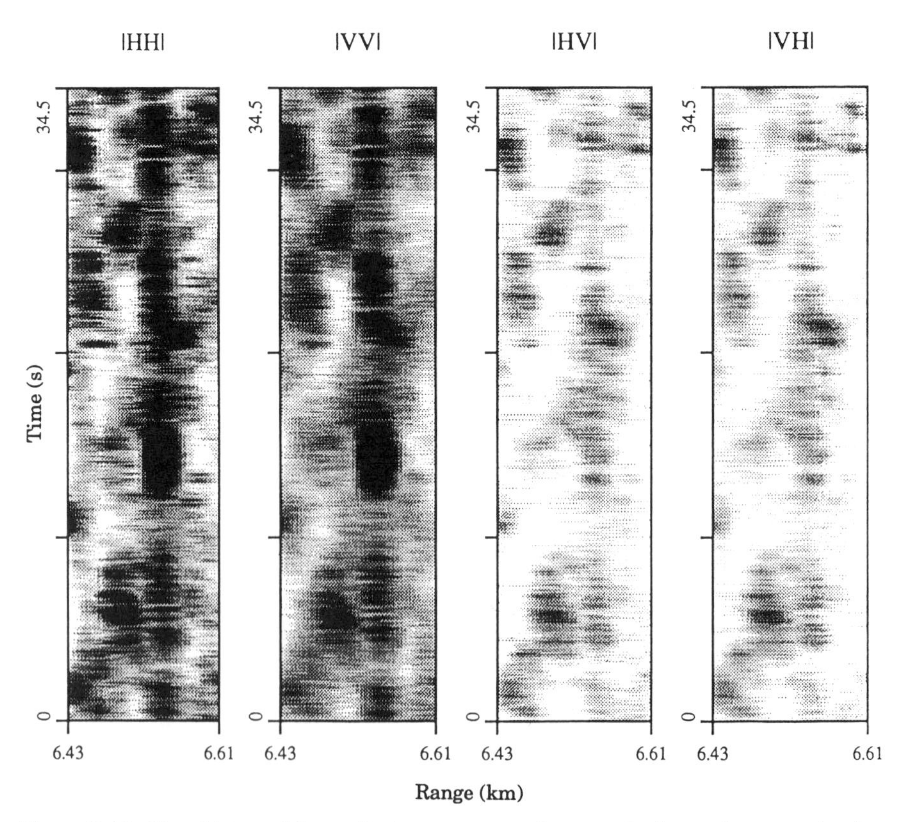

Figure 9.13 Time history of the amplitude versus range for a growler in sea clutter, at a bearing of 119° and a range of about 6.5 km. Note that there are many times when the growler's amplitude is only equal to, or less than, that of the sea clutter.

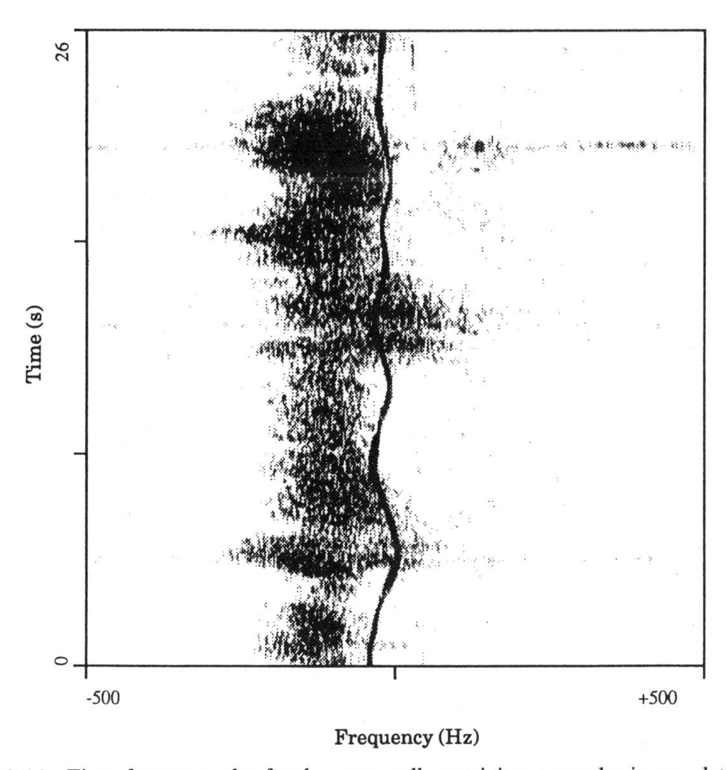

Figure 9.14 Time–frequency plot for the range cell containing a growler in sea clutter, at a bearing of 119° and a range of 6.5 km (data is the same as in Fig. 9.13). Note how the growler's narrow spectral component permits its clear detection in the sea-clutter background.

ler peak, whereas the sea-clutter power has spread over a wider spectral range, and in this case, tends to occupy a different portion of the spectrum. The target was a reasonably rigid object, which must move as a whole. As a consequence, at any given moment, it will have a narrow distribution of relative velocities since all its scattering centers are connected, and because of its relatively small size, the growler will contain a small number of scatterers, closely spaced. By contrast, the radar beam illuminates a larger area of sea surface, containing a larger number of scatterers, which can have more independent relative velocities, hence wider Doppler spreads.

The growler was a low-lying target, with 80–90% of its mass below the surface. Therefore, its overall translational movement was due primarily to current, with speeds on the order of 0.5 to 1.0 knot. However, the short-term velocity of the growler can be influenced (modulated) by the local wave action, such as the swell. Indeed, Fig. 9.14 shows the growler attained a maximum velocity of about 1.5 knots for a few seconds.

The amplitude plot for a different data set (Fig. 9.15) shows the presence of two growlers, at a range of about 4.60 km, and bearing of 57°. In Chapter

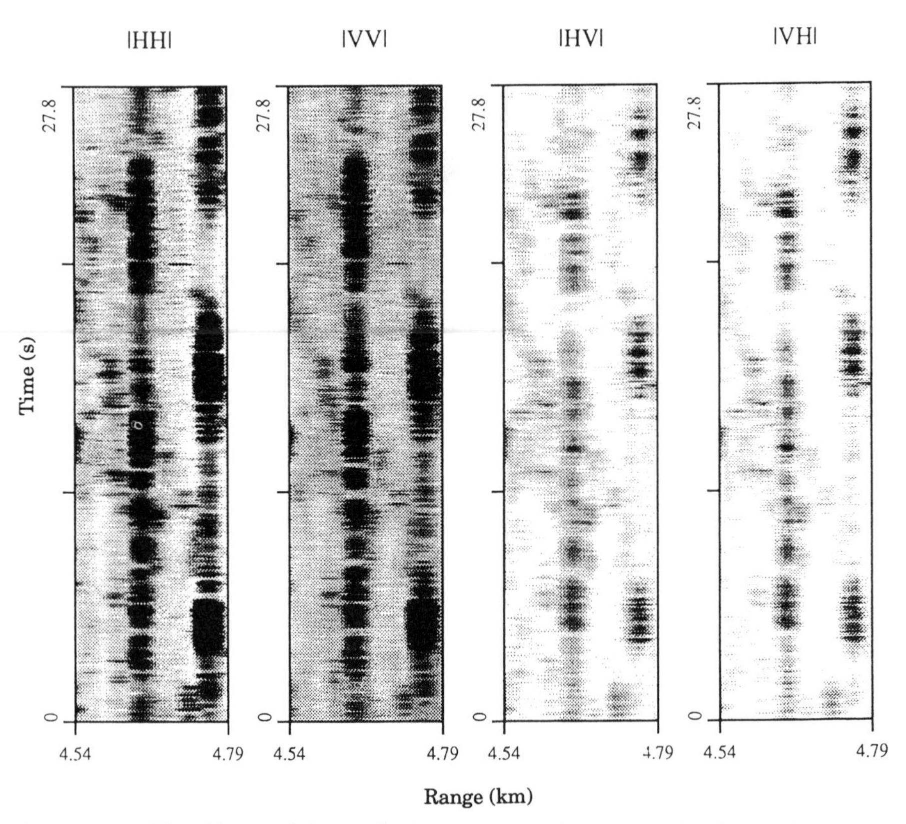

Figure 9.15 Time history of the amplitude versus range for two growlers in sea clutter, at a bearing of 57° and a range of about 4.6 km.

8, it was observed that targets may exhibit significant variation in their amplitude over periods of only a few seconds, and this data set supports the observation. The HH Doppler plots (Fig. 9.16) show the two targets: (a) at range 4.66 km, and (b) at range 4.77 km. The differing nature of their Doppler peak variation gives further confirmation that they were indeed separate targets. Figure 9.16(b) also shows what is assumed to be circling bird activity. The presence of floating objects disturbs the sea surface and creates upwelling of deeper water because of the temperature differential between the ice and the ocean. This upwelling brings nutrients to the surface, attracts fish, which in turn attract the fish-feeding birds. The presence of birds near growlers and icebergs was frequently observed, although it was not specifically confirmed for any particular data set during this experiment.

Iceberg

Several data sets were collected for an iceberg, at a range of 14.8 km with bearing of 40°. At that range, there was no neighboring sea clutter. Six se-

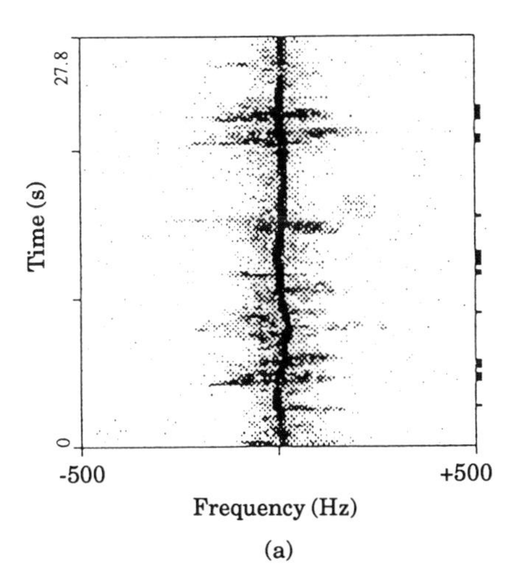

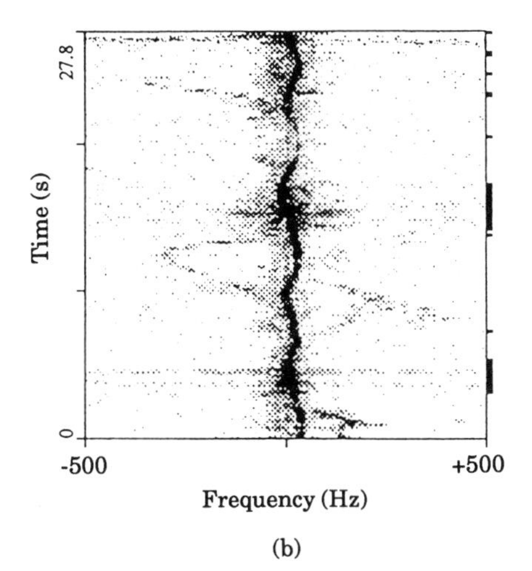

Figure 9.16 Time–frequency plots for the two range cells, each containing a different growler in sea clutter, at a bearing of 57° and a range of (a) 4.66 km and (b) 4.77 km, for the same data set as shown in Fig. 9.15. Again the growlers are clearly detectable. Plot (b) also shows a Doppler component which was varying quickly, presumably caused by circling bird activity.

quential data sets showed the iceberg to have a small, positive Doppler shift, constant over the 30-s record.

Later data sets show more interesting behavior. One HH plot (Fig. 9.17), with the iceberg at a range of 14.9 km, shows some diffuse spectral content in conjunction with the iceberg's return peak which does not occur for sea areas away from the iceberg. It is thought to be the result of wave interaction with the iceberg, which caused splashing. In addition, the position of the Doppler peak from the iceberg varies slightly, indicating some rocking of the iceberg. This motion, in turn, can create wave action around the iceberg. Another factor which may contribute to the Doppler behavior is the multiple-bounce reflection between the iceberg and the sea, which will be modulated by the sea motion.

To examine the longer term variation in the Doppler peak resulting from

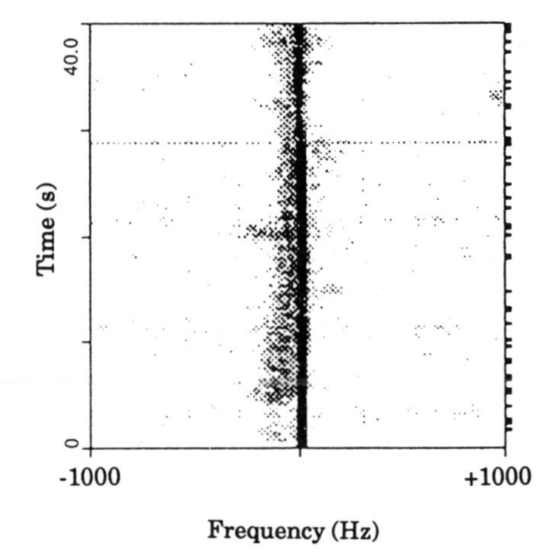

Figure 9.17 Time–frequency plot for the range cell containing an iceberg, at a bearing of 40° and a range of 14.8 km. The spectral component that appears to be sea clutter is presumed to be caused by interaction (splashing and rocking) of the iceberg and the ocean, as it does not appear at comparable ranges nearby with returns only from the sea surface.

the iceberg, another HH data set (Fig. 9.18) was studied. This data set was collected using pulse compression at a PRF of 200 Hz, and yielded a time dwell of 588 s. The iceberg exhibited a two-component variation in its Doppler peak location. There was an underlying long-term variation of about 12 Hz with a period of about 100 s. This Doppler shift was modulated in the shorter term by a smaller variation of about 5 Hz and a period of about 8 s, approximating the period of the swell. Thus, the iceberg appeared to be influenced by the swell, and also to be rolling, rocking, or bobbing in harmonic resonance with the ocean but at a much longer period.

9.2.4 Polarization Results

To investigate the polarization properties of scatterers (Currie et al. [3]), it is desirable to collect data for both transmit polarizations within the decorrelation time of the scatterers, so that in both polarizations the radar observes the same target. For the scatterers considered here, the sea clutter decorrelates the fastest. With maximum Doppler spectral widths of about 400 Hz, the decorrelation time is on the order of 2.5 ms. The only way to collect data for two transmit polarizations in this time frame is to use alternating polarization.

By visual inspection, the amplitude, and particularly the Doppler plots, appear to be quite similar for HH and VV, and for HV and VH. The only way to detect subtle differences is to superimpose the plots, or combine the data on a point-by-point basis into one plot. The latter method was chosen, with two formats being used. Both formats start with two Doppler plots from two different polarization combinations, then the corresponding Doppler coefficients are combined through some operation, and the resulting coefficients are plotted on one graph. To provide an extra plotting dimension, one of the parameters is coded in color. The color system used is brightness, hue, and saturation. The last is kept at maximum.

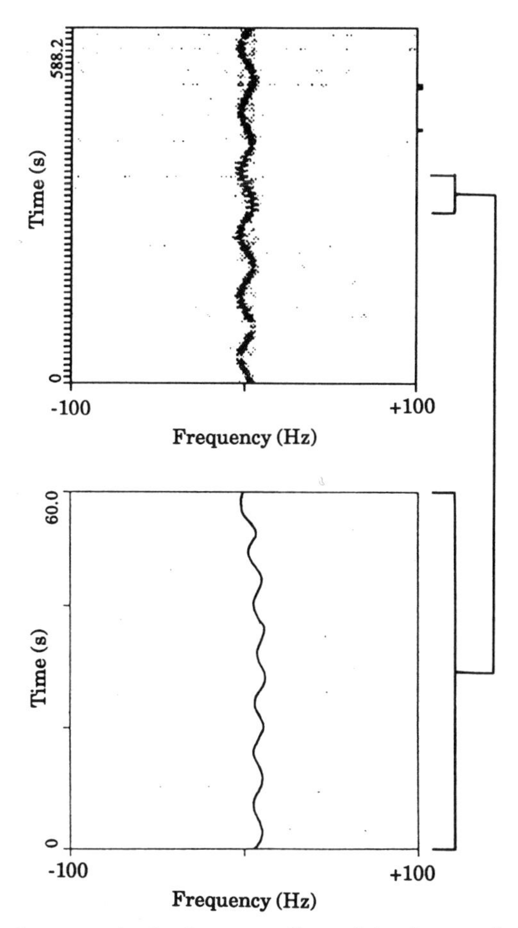

Figure 9.18 Time–frequency plot for the range cell containing the same iceberg shown in Fig. 9.17, at a bearing of 40° and a range of 14.8 km. The PRF has been reduced to 200 Hz, yielding a dwell time of nearly 10 min. Note the two components of the iceberg's Doppler variation with time.

The first format indicates the relative powers of the two input spectra. The brightness of the output image represents the sum of the input powers, whereas the color represents the ratio of the powers. For example, for input spectra A and B, red might be used to indicate the power of A is so many dB less than B; blue might indicate A is so many dB greater than B; and green might indicate that A and B are about equal in power. The presence of nongreen areas indicates polarization-related power differences.

The second format examines the phase difference between the two input spectra at each Doppler frequency. The brightness of the image represents the product of the input powers. This technique ensures that only significant signals are displayed. The phase difference, which varies from 0 to 2 π, is represented by hue.

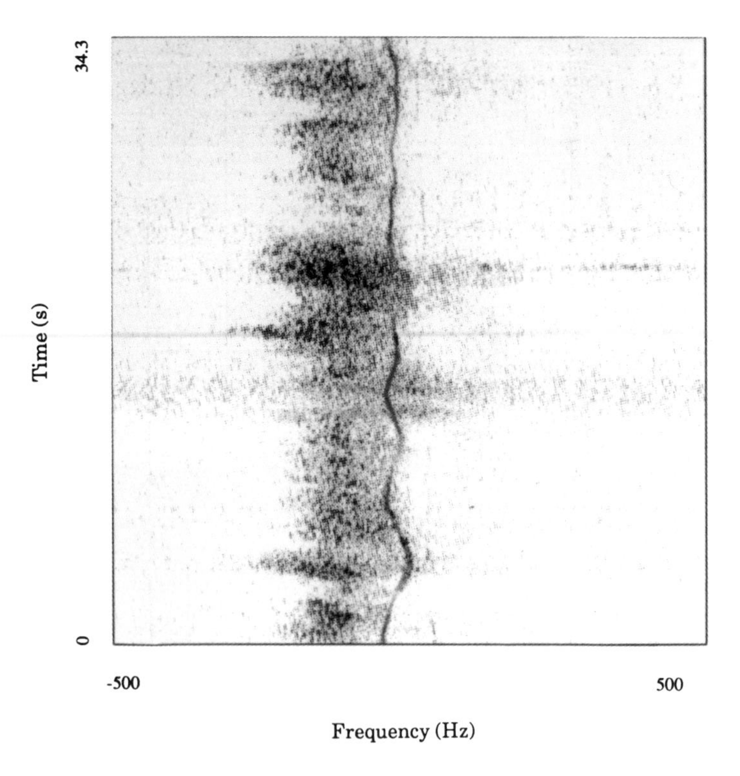

Figure 9.19 HH/VV power spectral ratio, in time–frequency format, for the range cell containing a growler in sea clutter, at a bearing of 119° and a range of 6.5 km. Data are the same as those in Fig. 9.14. The image brightness varies as the sum of the powers, whereas the color shows the ratio of the powers. Note how the growler's ratio is near unity, whereas for the sea clutter, VV is more powerful at smaller Doppler shifts and HH is more powerful at large Doppler shifts (see color insert).

Figure 9.19 (see color insert) shows the power comparison of the HH and VV channels for the growler in sea clutter that was used to produce Fig. 9.14. Consider only the sea clutter for a moment. An immediate observation is that for the larger absolute Doppler shifts, the HH polarized data are more powerful, whereas for the smaller shifts, the VV data are stronger. Thus it appears that the nature of the Doppler spectrum for sea clutter depends upon the transmit polarization. Further analysis on this point is given below. On the other hand, the growler, for the most part, shows similar power in HH and VV.

Figure 9.20 (see color insert) shows the same data set but analyzed for phase difference. The nearly constant hue of the Doppler peak from the growler shows that it maintains essentially the same phase difference between HH and VV, whereas the sea clutter exhibits a wide variation in hue, hence phase difference. It is conjectured that the ice target, being a rigid body with an irregular shape

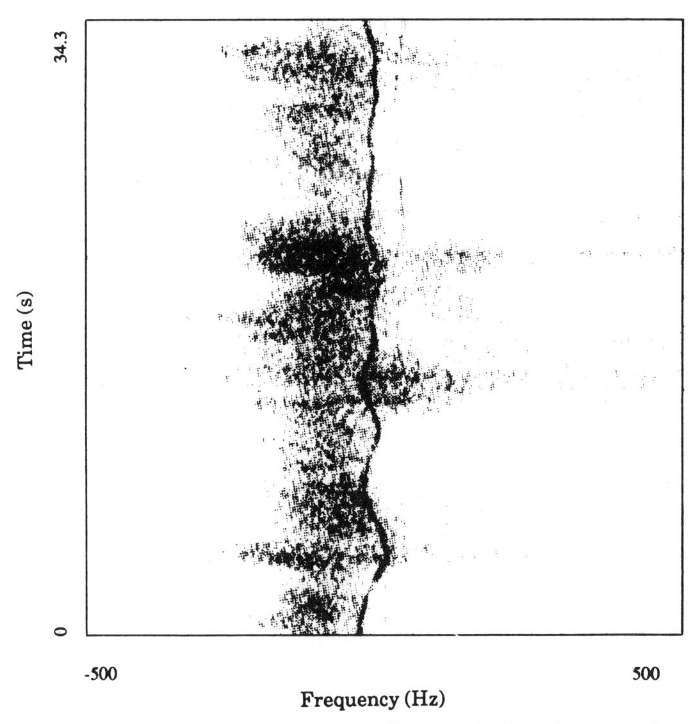

Figure 9.20 HH/VV spectral coefficient phase difference, in time–frequency format, for the range cell containing a growler in sea clutter, at a bearing of 119° and a range of 6.5 km. Data are the same as those in Fig. 9.19. The image brightness varies as the product of the powers, whereas the color shows the phase difference. Note how the growler's phase difference is consistent, whereas for the sea clutter, the phase difference is much more variable (see color insert).

and surface, offers surface scatterers (with no significant signal penetration) and produces the same response to the two polarizations. The power comparison of Fig. 9.19 showed HH and VV signals for the target to be of equal power. On the other hand, the sea clutter looked different to the two polarizations, with its echo generated as the sum of many scatterers, but with different scatterers for different polarizations. Consequently, the phase difference is much more variable for sea clutter. Figure 9.21 illustrates this further by comparing the histograms of the phase difference (HH and VV) for the growler's Doppler component and the sea-clutter (only) Doppler component.

Another interesting result is shown in Fig. 9.22, the phase difference plot (HH and VV) for a data set containing both sea and rain clutter (see color insert). The Doppler components attributable to sea and rain clutter are separated easily. The sea clutter again shows its wide distribution of phase differences. The rain clutter, however, seems to have a much narrower phase difference distribution, which is attributed to the uniformity in the shape (but not size) of the rain particles, such that over time, they produce a similar response to the radar.

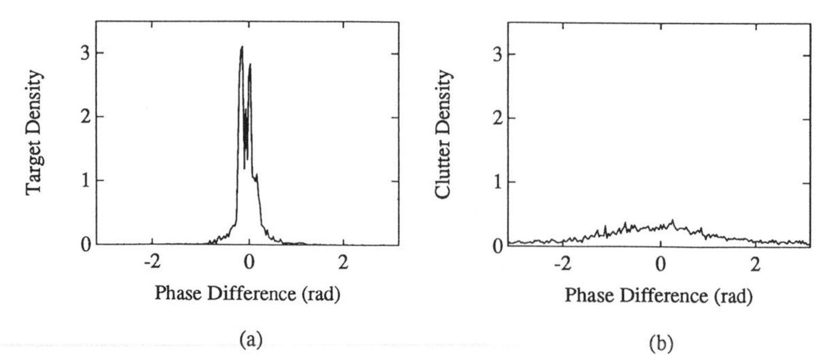

Figure 9.21 Distribution of the HH/VV spectral coefficient phase difference, taken from the data of Fig. 9.20, for (a) the growler, and (b) sea clutter alone. Note the much wider distribution for the sea clutter as compared to the growler.

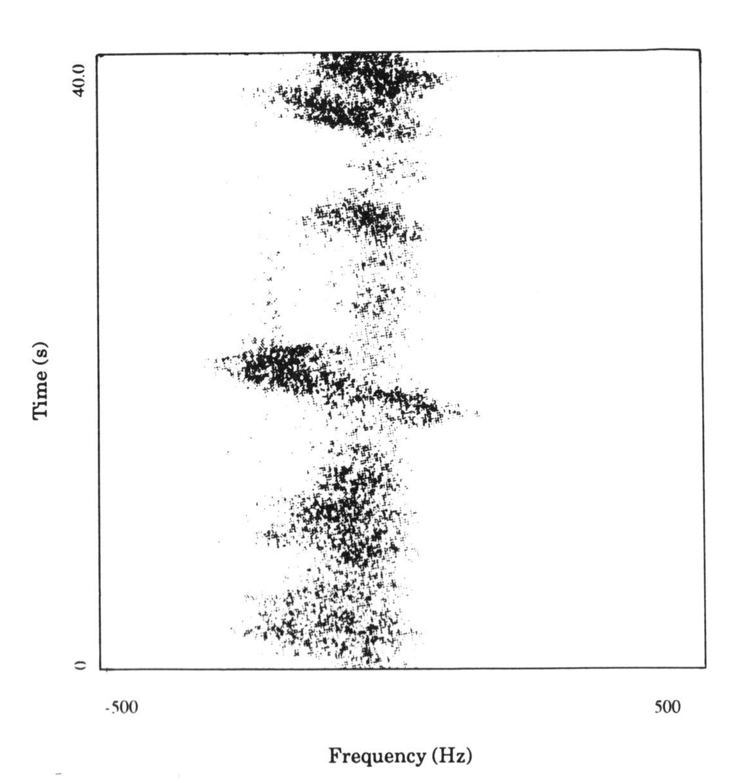

Figure 9.22 HH/VV spectral coefficient phase difference in time–frequency format, for the range cell containing both rain and sea clutter, at a bearing of 30° and a range of 5.8 km. Data are from a data set similar to that used for Fig. 9.11. The image brightness varies as the product of the powers, whereas the color shows the phase difference. Note how the phase difference for rain clutter is consistent but for the sea clutter, the phase difference is much more variable (see color insert).

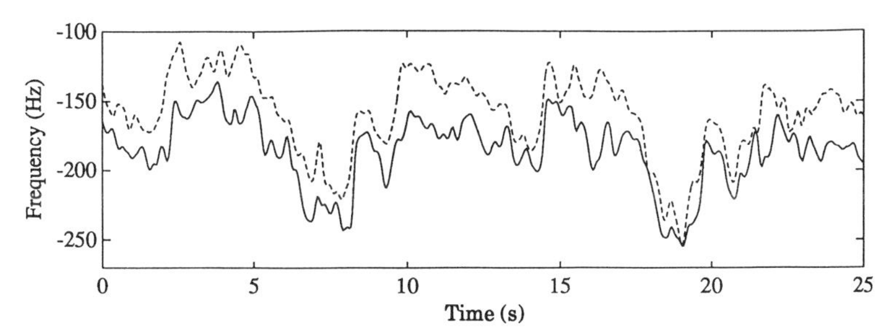

Figure 9.23 Mean Doppler frequency shift versus time, for HH (solid) and VV (dashed) polarizations, for an upwind sea clutter data set, similar to that of Fig. 9.10. The HH shift is consistently greater than the VV shift. The large swings in the center frequency are correlated to large changes in the power of the sea clutter return.

Revisiting the differing Doppler shifts of the HH and VV sea clutter, two measures were extracted from each input Doppler spectrum: the mean Doppler shift and the Doppler spread. These measures are used to indicate data trends. As discussed earlier, wave obscuration by swells tends to bias such spectral estimates because the full, true spectrum cannot be measured. Figure 9.23 shows the mean Doppler shift versus time for the HH and VV data sets of sea clutter. It is apparent that the HH channel has a consistently larger Doppler shift than the VV channel, which agrees with results reported by Trizna [23]. To examine the Doppler spreads, histograms were accumulated. Figure 9.24 shows the results for HH and VV. The VV channel, on the whole, shows a wider spread than the HH channel, and confirms that differing scattering mechanisms are taking place for the two polarizations.

Finally, Fig. 9.25 shows the spectral ratio plot for the data set containing both sea and rain clutter (see color insert). Whereas the sea clutter shows differing HH and VV powers at various Doppler frequencies, the rain clutter shows essentially equal HH and VV powers throughout the frequencies in its Doppler spectrum.

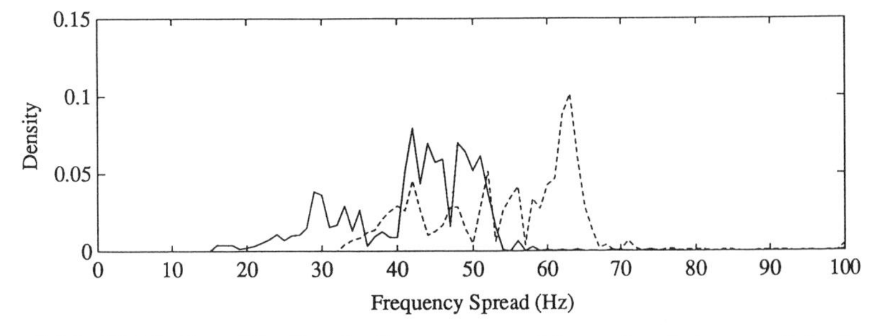

Figure 9.24 Distribution of the Doppler frequency spread (second central moment) for the same data as those used for Fig. 9.23. The spread of the HH spectra is less than that for VV spectra. The contribution of each width estimate was weighted by the SNR of the spectra on which it was based.

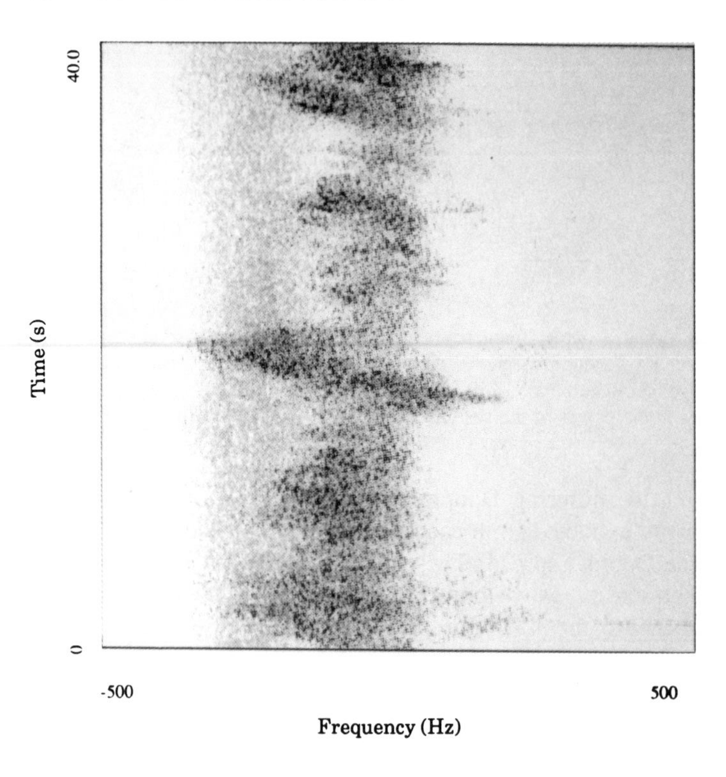

HH/VV = -10 dB HH/VV = 0 dB HH/VV = +10 dB

Figure 9.25 HH/VV power spectral ratio, in time–frequency format, for the range cell containing both rain and sea clutter, at a bearing of 30° and a range of 5.8 km. Data are the same as those used for Fig. 9.22. Note how the rain clutter's ratio is consistently near unity but for the sea clutter, the ratio is much more variable (see color insert).

9.3 THEORETICAL PROGRAM

In considering the measured experimental data, the question arises as to what physical mechanisms are responsible for the observations reported in Section 9.2. Indeed, a better understanding of the relationships between the physical processes involved and the radar backscatter signal should permit the development of better detection algorithms. Therefore, a program of theoretical studies was begun to address these issues.

It is not difficult to see that the motion of a floating target and the dynamics of the sea surface must be in some way related. The dynamics of the surface are responsible, in large measure, for the dynamics of the target. Therefore, the radar signal returned from the sea surface (i.e., the clutter) may provide information about the motion of the target.

Conventionally, the aim is to suppress the clutter signal before performing detection algorithms. Even in most constant-false-alarm-rate systems (as discussed in Subsection 8.2.10), the clutter signal is used only to provide some estimated measure of its power and thus to set detection thresholds. However, the sea clutter could be viewed as a source of information about the surface dynamics, hence target motion, rather than just an interference effect to be minimized. This source, in turn, could provide *a priori* information in detection algorithms to improve their performance.

To gain such an understanding of the processes involved, a computer simulation was developed. This software embodied models of the physical processes involved to compute simulated radar backscatter data in both Doppler and polarization domains. As the purpose of the effort was primarily to gain insight, and hopefully to obtain qualitative agreement with the measurements, fairly basic models were chosen and many simplifying assumptions were made.

9.3.1 Overview of Models

In formulating a simulated backscatter signal, several processes must be modeled. These include:

- sea surface dynamics
- target dynamics
- sea surface scatter
- target scatter
- radar system characteristics

The characterization of the sea surface, its dynamics, and its scattering are used to compute the radar cross section and Doppler spectrum of the backscattered signal. It is also required to compute the motion of the target and its shadowing by the sea. The quantities of interest are computed over a grid of discrete points in two dimensions of space, representing the surface, and in time. The models require the input of various environmental, radar, target, and geometric parameters.

9.3.2 Sea-Surface Dynamics

The sea surface is characterized by evaluating its height, velocity, and slope over the two-dimensional surface and over time. The model takes as input the wind speed and direction and water depth. As an intermediate step, the temporal and spatial correlation of the above quantities are computed in the form of frequency and wave-number spectra. From this statistical description, sample functions of the above quantities are computed over the grid.

The following derivation is based on a one-dimensional surface to present the concept in a simplified notation. We start with the standard assumptions of incompressible and irrotational flow and define the potential function, Φ_w, such that

$$\mathbf{V} = \nabla \Phi_w$$

where **V** is the velocity field of the fluid and ∇ is the gradient operator. Using the Bernoulli and continuity equations with suitable boundary conditions, the potential can be expressed as (Phillips [19])

$$\Phi_w = \sum_{n=1}^{\infty} \xi^n (\Phi_w)_n$$

where ξ is a series expansion coefficient in units of surface slope.

Keeping only the term linear in ξ (i.e., the first-order Stokes wave), the solution for the potential, as a function of position and time, is

$$\Phi_{w1} = \frac{gA}{\omega} \frac{\cosh k_w s}{\cosh k_w d} \sin k_w (x - Ct)$$

where

d = depth of water
y = wave travel in the positive (up) direction
$s = d + y$
x = horizontal direction of wave travel
k_w = wave number of the fluid
A = wave amplitude
ω = wave frequency
C = phase velocity, ω/k_w
g = gravitational constant
cosh = hyperbolic cosine

From this potential, one can determine the velocity of a fluid element

$$\mathbf{V}_x = A\omega \frac{\cosh k_w s}{\sinh k_w d} \cos k_w (x - Ct)$$

$$\mathbf{V}_y = A\omega \frac{\sinh k_w s}{\sinh k_w d} \sin k_w (x - Ct)$$

where sinh is the hyperbolic sine.

Thus it can be seen that when water is deep, the motion is circular. The radius of this circular motion is equal to the wave amplitude, A. The surface height, ζ, is given by

$$\zeta = -\frac{1}{g}\frac{\partial \Phi_w}{\partial t}\Bigg\|_{y=0} = A \cos k_w(x - Ct)$$

The relation between the wave number and frequency of a wave is known as the dispersion relation. In its first-order form, it can be written as

$$\omega^2 = gk_w\left(1 + \frac{\gamma k_w^2}{g}\right) \tanh k_w d$$

where γ is the ratio of surface tension to water density, and tanh is the hyperbolic tangent. When the above equations are generalized to a two-dimensional surface, as in the simulation implemented, the wave numbers become vector quantities.

The surface, in general, does not consist of just a single wave but can be represented by the superposition of many waves, covering a continuum of frequencies. The amplitude and phase of these component waves are random quantities, fixed for one realization of the process. For a Gaussian surface, which is appropriate for a first-order Stokes approximation, the phases and amplitudes are drawn from uniform and Rayleigh distributions, respectively. A frequency spectrum (or wave-height spectrum), $S(\omega)$, is required to determine the distribution of wave height versus frequency.

There exist a number of models that relate the environmental parameters to the frequency spectrum. One such relation is the Pierson-Moskowitz frequency spectrum (Pierson and Moskowitz [20]). It represents the condition of a fully developed sea, with unlimited fetch.

$$S(\omega) = 0.0081g^2\omega^{-5} \exp\left[-0.74\left(\frac{\omega U_\omega}{g}\right)^{-4}\right]$$

where U_ω represents the wind speed. The wave frequency of the spectral peak (using MKS units) is

$$\omega_p = 0.877\frac{g}{U_W}$$

The frequency spectrum can be extended to the directional frequency spectrum, Ψ_0, by considering a directional distribution (Donelan and Pierson [5]) such as

$$D(\omega,\theta) = \text{sech}^2\left(\beta(\omega)\theta\right)$$

where sech is the hyperbolic secant, and

$$\beta(\omega) = \begin{cases} 2.44(\omega/0.95\ \omega_p)^{+1.3} & \text{for } 0.56 < \omega/\omega_p < 0.95 \\ 2.44(\omega/0.95\ \omega_p)^{-1.3} & \text{for } 0.95 < \omega/\omega_p < 1.6 \\ 1.24 \text{ otherwise} & \end{cases}$$

and

$$\Psi_0(\omega, \theta) = S(\omega)D(\omega, \theta)$$

The direction of the wind is taken as $\theta = 0$.

The directional wave number spectrum is derived from the directional frequency spectrum by using the relation:

$$k_w dk_w = 2 \frac{\omega^3}{g^2} d\omega$$

Given the wave number spectrum and the dispersion relation, a sample function of the surface is computed. This sample function is three-dimensional: two spatial dimensions and time. The spectrum is made discrete over wave number, randomized, and converted to a surface by using an FFT subroutine. The procedure is repeated to compute the velocities and surface slopes. Figure 9.26 depicts a surface, at one instant of time, over a region of 1 km × 1 km computed with a grid point spacing of 8 m × 8 m. The vertical dimension is exaggerated to show the waves better.

The implementation can be extended to include higher order Stokes waves resulting in a more realistic non-Gaussian surface. In addition, more complex wave spectra and measured directional wave spectra can be used.

9.3.3 Target Dynamics

There are several rather involved methods for computing the response of an object (e.g., ship on the sea, offshore drilling platform, moored structures) when it is subjected to hydrodynamic forces (Chakrabarti [1]). However, there is particular interest in growlers and bergy bits which are difficult targets to detect, even visually, because of their small size, low freeboard, and (frequently) rough ocean surroundings. Using wave tank studies, it has been shown that a floating object smaller than 1/13th of the wavelength of the waves behaves as a particle on the surface (Lever et al. [12]). Thus for a surface with one wave frequency predominating, the motion of the floating object will be approximately circular, as it was for a water particle. For a wave approaching the radar, the object will have maximum speed toward the radar when it is at the crest of the wave, and maximum speed away from the radar when in the trough.

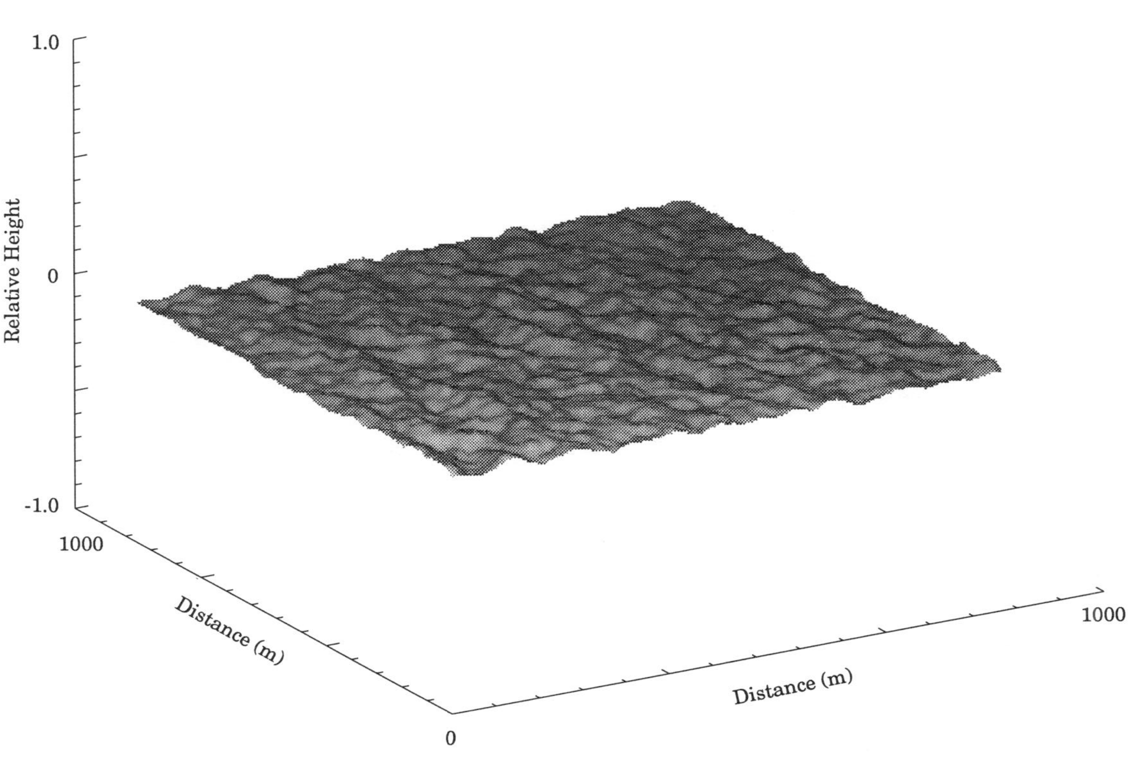

Figure 9.26 Example of a simulated ocean surface at one instant of time, based on a theoretical model. Shown is the surface in a 1 km × 1 km region, using a grid point spacing of 8 m by 8 m. The vertical dimension has been exaggerated to show the waves better. Calculation of a series of such surfaces can model the temporal behavior of the ocean surface.

In considering the forces on the floating object from progressively shorter waves, there should be a point at which the forces diminish. For the purpose of this study, the response of the floating object follows a filtered version of the particle displacement computed from the surface dynamics. The filter is a second-order, low-pass filter with a cutoff at a wave number corresponding to twice the diameter of the object. This method is certainly an empirical approach and could be replaced by a more rigorous one.

9.3.4 Sea-Surface Scatter

For the purpose of the surface scatter model, two types of waves are identified. First, there are gravity waves, with wavelengths ranging from 200 m to a fraction of a meter. The dominant restoring force for these waves is the force of gravity. Second, there are smaller capillary waves, with wavelengths on the order of centimeters or less. The dominant restoring force for these waves is surface tension.

The radar backscatter from the surface at each patch, defined by the gridded surface, is determined by using a two-scale composite model (Wright [25]). At microwave frequencies and low-to-medium grazing angles, the mechanism is Bragg scattering. The features of the surface with the same scale as half the radar wavelength are predominant scatterers. At microwave frequencies, the Bragg scatter is from capillary waves. The two-scale composite scattering model considers the Bragg return as it is modulated by the larger scale gravity waves. Gravity waves influence the backscatter by

- surface tilting
- advection of small scales from orbital velocity
- straining of small scales caused by acceleration

The surface, considered first to be frozen in time, and the backscatter cross section, at points defined by the surface grid, are computed. Each surface grid point may be viewed as a plane tilted by the large-scale gravity waves. This plane is perturbed by capillary waves acting as Bragg scatterers. The backscatter cross section of such a plane is a function of the radar incidence angle, as well as the local surface tilt both in the plane of propagation and perpendicular to it. The backscatter also depends on the transmit and receive polarization. The exact relationship between the sea-surface wave number and scattering geometry can be found in Donelan and Pierson [5].

However, at low grazing angles, not every patch on the surface may be visible. Two different shadowing mechanisms are considered. One is to ignore those patches that are tilted away from the radar by more than 90°. The other is a statistical technique (Wetzel [24]) that considers the probability of a point on the surface being obscured by a wave closer to the observer. For low grazing angles, one can compute a threshold height above which points on the surface are almost certain to be visible, whereas those below it are not. A more realistic, but much more difficult, approach would be to include the effects of diffraction and multiple scattering.

Using the above techniques, the backscatter cross section is computed for each patch on a surface with

- grazing angle = 10°
- area represented = 1 km × 1 km on 8 m × 8 m grid points
- wind speed = 10 m/s
- radar looking downwind

The spatial grid resolution effectively determines the cutoff between the two scales. The radar backscatter is represented by shaded surfaces in Figs. 9.27, 9.28, and 9.29 for the HH, VV, and HV polarizations, respectively. As noted in measured results, the HH polarization is more spiky than the VV polarization.

These figures represent the spatial distribution of expected backscattered power at one instant of time. Because of its nonlinear relation to surface slope, the backscattered power appears spiky. Whereas a radar with low resolution, which illuminates a larger surface area, would average out such spikes, one with higher resolution would not, leading to spiky temporal behavior. This spikiness also creates a change in the statistics of the clutter signal, which in turn may degrade the performance of statistically based detection algorithms. The transition in designation from low- to high-resolution radar is not well defined, but it begins to occur as the antenna beamwidth is reduced below 1° and the transmitted pulse width is shortened to less than 0.5 μs.

Next, the effect of the surface motion is considered by looking at the Doppler spectrum of the backscattered signal. The Doppler shift from simple Bragg scattering is equal to the frequency of the surface wave with a wave number k_B:

$$k_B = \frac{2\pi}{\lambda} \sin \theta_i$$

where λ is the radar wavelength and θ_i is the incidence angle at the tilted patch.

In addition, there is a contribution to the Doppler shift resulting from the translation of the patch resulting from the orbital velocity of the gravity waves. In the case considered here, it is modeled by the instantaneous velocity of the grid point. The straining of the capillary waves was not considered in this simulation.

9.3.5 Target Scatter

The floating target was modeled as a simple point scatterer with a cross section that varied according to shadowing.

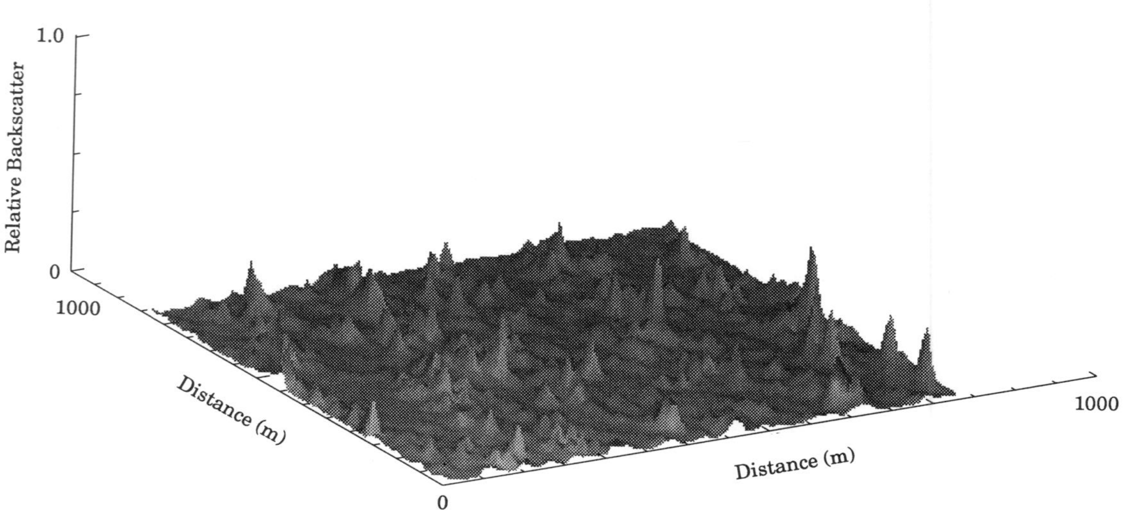

Figure 9.27 Example of simulated radar backscatter from the ocean surface at one instant of time, for HH polarization, based on the theoretical model. Backscatter is shown for a surface in a 1 km × 1 km region, using a grid point spacing of 8 m by 8 m. The vertical dimension has been exaggerated.

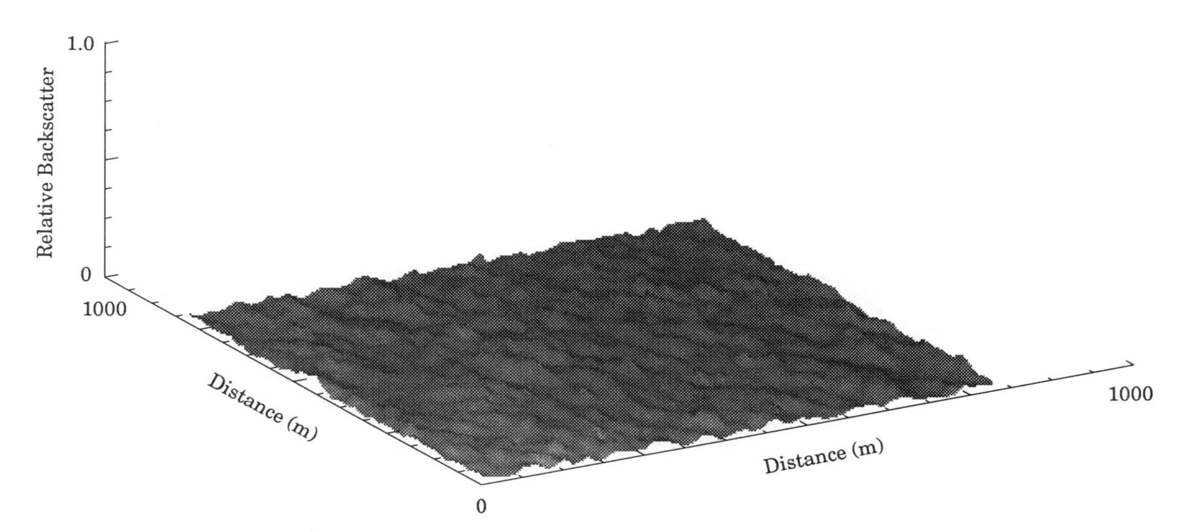

Figure 9.28 Example of simulated radar backscatter from the ocean surface at one instant of time, for VV polarization, based on the theoretical model. Backscatter is shown for a surface in a 1 km × 1 km region, using a grid point spacing of 8 m by 8 m. The vertical dimension has been exaggerated. The backscatter is not as spiky as the HH polarization of Fig. 9.27.

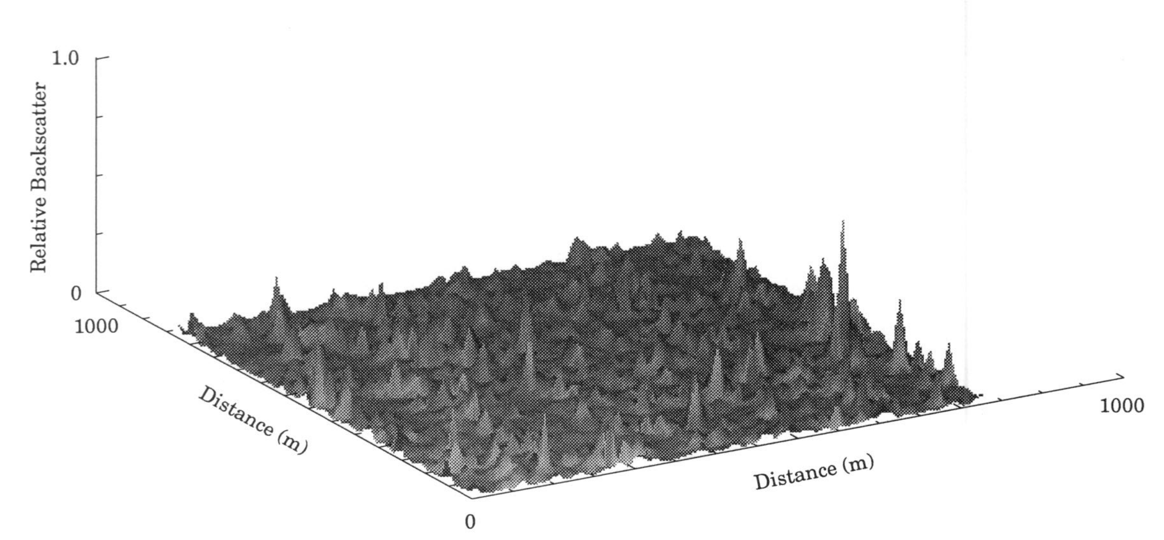

Figure 9.29 Example of simulated radar backscatter from the ocean surface at one instant of time, for HV polarization, based on the theoretical model. Backscatter is shown for a surface in a 1 km × 1 km region, using a grid point spacing of 8 m by 8 m. The vertical dimension has been exaggerated. The backscatter exhibits some spiky behavior caused by the H component of the polarization.

9.3.6 Radar Model

The backscatter signal is represented in terms of "instantaneous" Doppler spectra computed at each grid time point. It is calculated by fitting a Gaussian Doppler spectral shape to the power-weighted Doppler shifts of each patch that makes up the radar resolution cell.

9.3.7 Surface-Scatter, Doppler Simulation Results

The models were used to generate the time-Doppler plot shown in Fig. 9.30 under the following conditions:

- grazing angle of 1°
- radar azimuth upwind
- range resolution of 40 m
- cross-range resolution of 256 m
- grid size of 8 m × 8 m

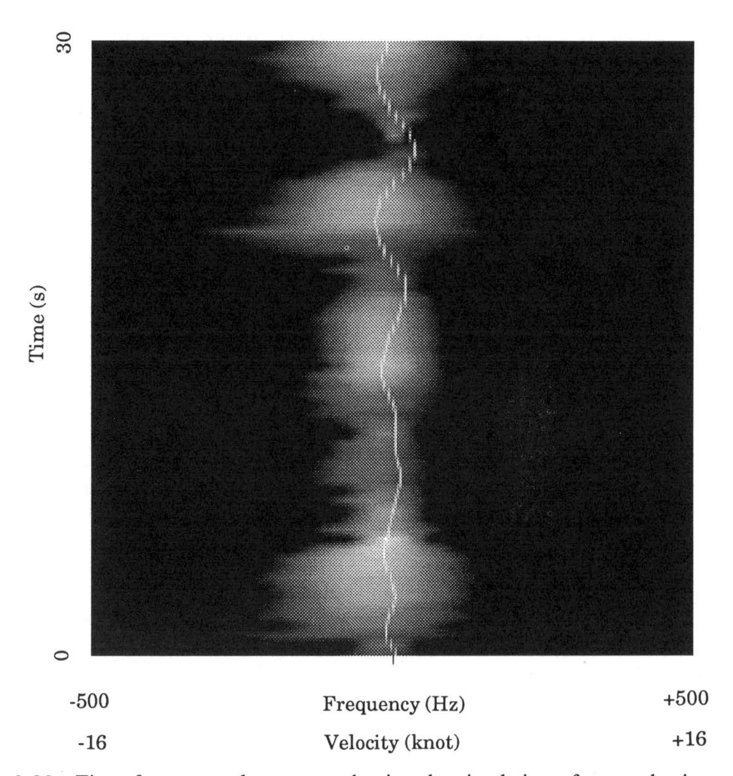

Figure 9.30 Time–frequency plot generated using the simulation of a growler in sea clutter. The model parameters are given in the text. Note the strong resemblance to the time–frequency plot in Fig. 9.14, obtained experimentally for a real growler in sea clutter.

- wind speed of 15 m/s
- fully developed wave spectrum
- target radius of 10 m

This plot bears a strong resemblance, in a qualitative sense, to the experimental spectrum for the growler in sea clutter, displayed in Fig. 9.14.

9.3.8 Effect of Results on Detection Algorithms

Having gained some understanding of the physical processes involved, a detection method that exploits the target dynamics and the spatial characteristics of the clutter is proposed.

The model indicates that the shape of the clutter Doppler spectrum does not vary abruptly with range. This property permits the technique of whitening the clutter component of the radar signal before detection. If the radar has sufficient range resolution, data from range cells adjacent to the cell under test can be used to estimate the Doppler spectrum of the clutter only. These data in turn can be used to whiten the time series of the signal from the cell under test. If a parametric spectral model (based on the observed Gaussian shape of the Doppler spectrum) is used, one should ensure that the technique is robust to the presence of other phenomena, such as rain or birds.

The target's dynamic behavior, as indicated in the time-frequency plot, could be used to improve detection performance. The target exhibits a Doppler shift (i.e., radial velocity) that is highly correlated in time. In fact, if the directional wave spectrum is known (or can be estimated from the clutter return), this correlation function can be determined.

Consider the scenario where the radar gathers data from the cell under test over a number of short (e.g., 0.5 s) epochs, not necessarily contiguous, spanning a total time of up to several minutes. For each epoch of data, the clutter component can be whitened, and the frequency of the Doppler peak can be estimated. If no target exists in the cell under test, one would expect that the frequency of the peak would vary randomly from epoch to epoch. However, if a target was present, this peak would tend to vary according to the target's radial velocity correlation function. Using the sequence of frequency peaks, one could perform a statistical test as to whether the sequence does indeed conform to this correlation function, and use this result as the detection decision.

As the processing of each epoch of data involves extracting the peak of the Doppler spectrum, the epoch must be sufficiently short so that the peak is not smeared by changes in target radial velocity. The method can be extended by noting that the target's acceleration correlation function can be computed from the velocity correlation function. By taking the acceleration into account, the epoch can be extended to an interval over which the acceleration needs only to remain relatively constant.

In cases where the signal-to-clutter ratio is low, the peak of the spectrum, when a target is present, may not correspond to that of the target. To account for this, a small number of Doppler peaks could be extracted from the Doppler spectrum for each epoch. Combinations of these peaks, one from each epoch, can be used to form all possible sequences. Each of these sequences then can be tested against the correlation function.

The detection approach outlined above has the advantage of not performing hard decisions at each epoch. To be practical, the method outlined should account for shadowing of the target and should provide some mechanism for allowing for target drift. Algorithms based on the above techniques are currently under development.

9.4 DETECTION TECHNIQUES

This section discusses various techniques for the detection of an ice target in the presence of sea clutter, and presents results using the IPIX radar data. As a frame of reference, the coherent data are first converted to amplitude-only and detection performance is evaluated using noncoherent processing, including integration.

Coherent processing techniques are then examined, operating mainly in the Doppler frequency domain. The Doppler properties for each of the two orthogonal linear polarizations are compared.

Finally, the section discusses promising new techniques based on the use of nonlinear concepts, including the use of the chaos theory and neural networks.

9.4.1 Noncoherent Detection

Detection consists of comparing some statistic of the received signal to a threshold. If the statistic exceeds the threshold, a target is declared; otherwise, a decision is made that there is no target present in the received signal. In the simplest case, the statistic can be the measured power (square of the amplitude) of the signal. To estimate the probability of false alarm (P_{fa}) when using the IPIX data base, a histogram is made of the signal power for a set of clutter-only files. Normalizing the histogram gives the estimated probability density function. Integrating the density gives the power distribution, which when subtracted from unity, gives the P_{fa} as a function of threshold. To estimate the probability of detection (P_d), the above steps (histogram, normalization, and integration) are repeated on a set of target files, but only at ranges at which targets are present. Using a threshold based on signal power, P_{fa} and P_d versus threshold curves for the selected IPIX radar files can be computed. By using threshold as a parameter, the curve for P_{fa} versus P_d can be plotted, giving the receiver operating curve (ROC) for the particular detection algorithm. The ROC for the simple power detection algorithm is shown as curve A in Fig. 9.31.

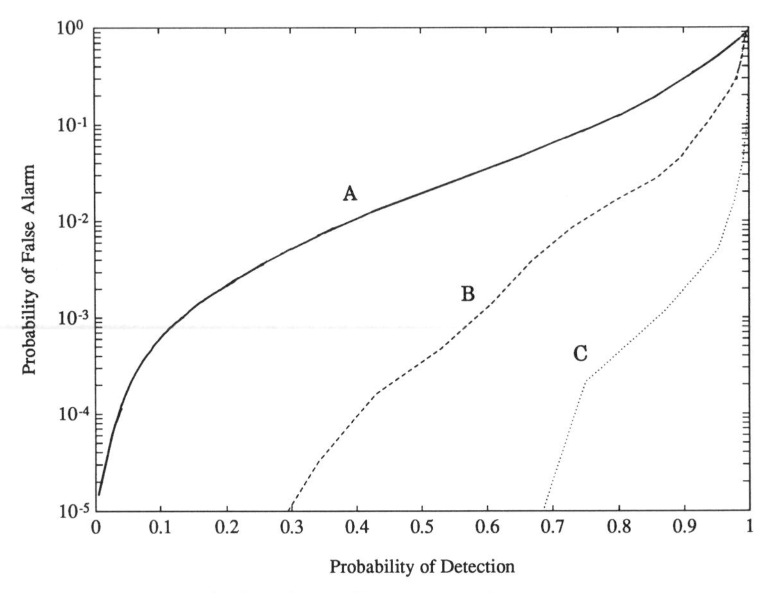

Figure 9.31 Improvement in detection performance provided by scan-to-scan integration, as shown by the receiver operating curves. Curve A is for simple threshold detection. Using data taken with the antenna stationary, curve B is for simulation of four scans at an antenna rotation rate of 5 rpm. Curve C simulates 32 scans at 42 rpm.

Noncoherent Integration

A noncoherent radar can improve detection by integrating (averaging) a number of return pulses. Strictly speaking, noncoherent integration does not improve the signal-to-noise ratio (SNR) or the signal-to-clutter ratio (SCR), because noise or clutter power is integrated just as well as target power. Rather, noncoherent integration potentially improves detection in a statistical manner, as the average of a number of pulses has (possibly) less variance than the individual pulses themselves. Bringing the returns closer to their average values allows a detection threshold to be set nearer to average interference power. Noncoherent integration works quite well for noise, where the mean power does not vary rapidly with time (and thus can be estimated accurately) and where consecutive samples are uncorrelated. The absence of both of these factors makes noncoherent integration more difficult for sea clutter.

Figure 9.32 shows sea clutter power, on a pulse-to-pulse basis, from an IPIX file at range 6.46 km. The data file is the same one used for Fig. 9.2(a). Figure 9.33 shows the same data after a sliding-window integration of 100 pulses. Note the dynamic range of the clutter power. The diagrams show the two orders of time on which the sea varies: a short-term (pulse-to-pulse) fluctuation about a long-term, highly varying mean; the long-term variance is caused by the large-scale motion of the waves. The short-term fluctuations can be removed easily by noncoherent integration (compare the two figures), but

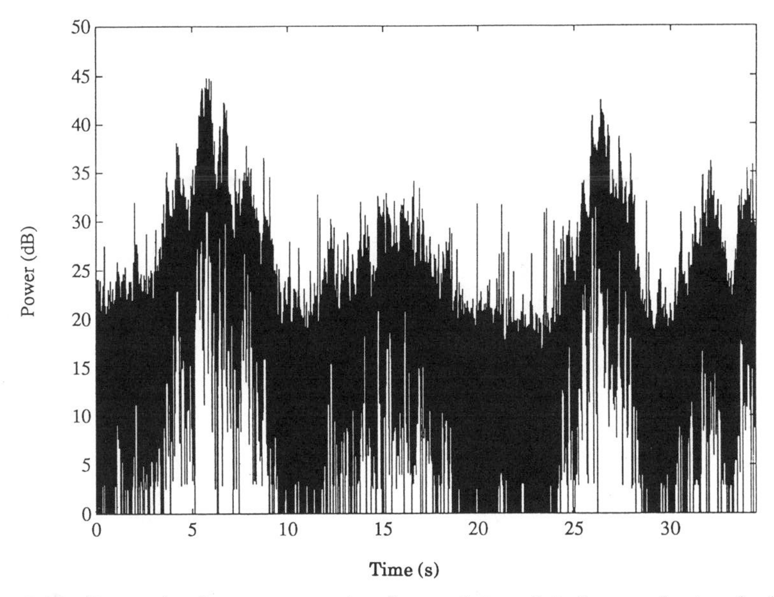

Figure 9.32 Return signal power versus time for sea clutter, plotted on a pulse-to-pulse basis. Note the significant variation in the return power, both on the short pulse-to-pulse time scale, and on the longer period of seconds.

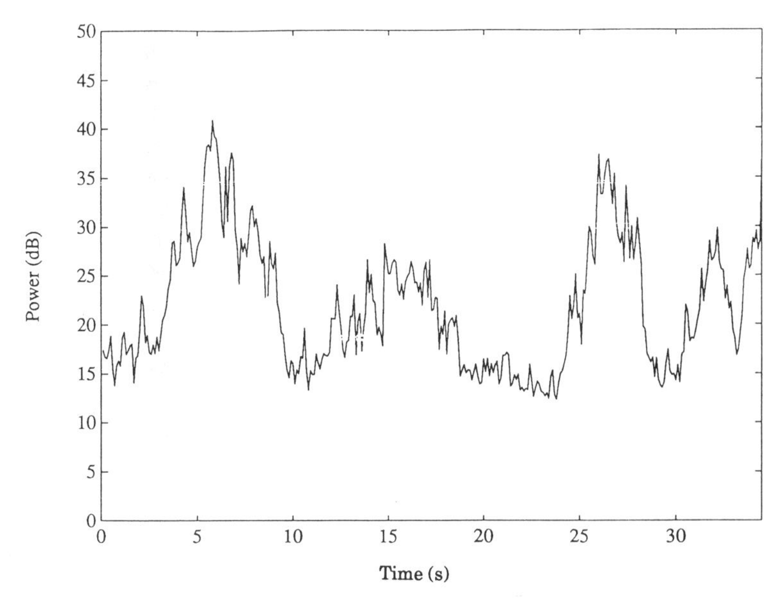

Figure 9.33 The data of Fig. 9.32 after a sliding-window averaging of 100 points. Note that the pulse-to-pulse fluctuation is removed, but the long-term variation still remains. These data are the same as those used in Fig. 9.2(a), but with more smoothing and for a slightly longer duration.

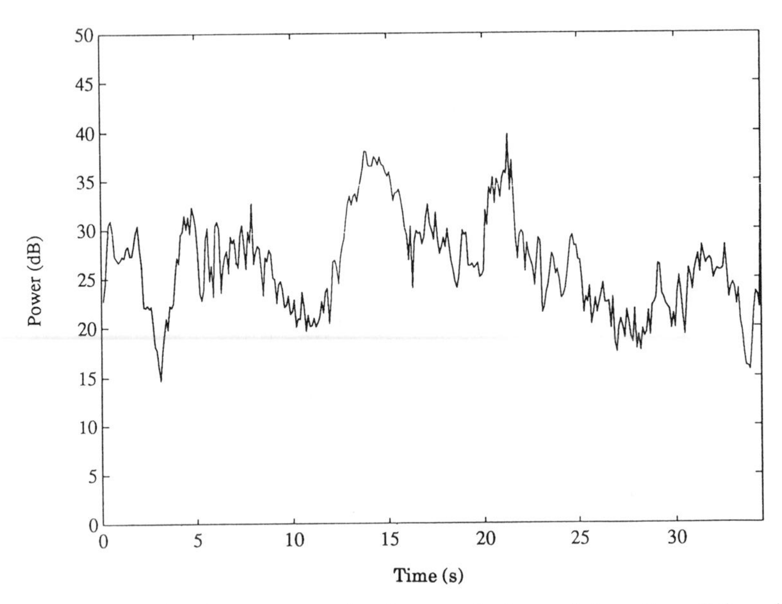

Figure 9.34 Return signal power versus time for a growler in sea clutter, after a sliding-window averaging of 100 points, using the same data as those in Fig. 9.3(a), but with more smoothing and for a slightly longer duration. Comparison with Fig. 9.33 shows that the growler cannot be detected reliably using simple threshold detection.

the long-term variance remains as a major problem. Figure 9.34 shows the integrated plot of the received power from a growler in sea clutter (same file, at range 6.52 km). The data file is the same one used for Fig. 9.3(a). Two points to note are: (1) the growler return is stronger for a greater percentage of the time, but (2) over short periods, the clutter power can be greater than the growler power. A detection threshold that ensures a low false-alarm rate for sea clutter, by being set higher than the expected strongest sea-clutter powers, would rarely be exceeded by the growler returns. The first point could be exploited by a detector that integrates (possibly nonlinearly) the signal over a long enough period that the difference in statistics is exploited: by observing the process over a time at least on the order of the wave period. Operationally, in a surveillance mode, this method usually involves the use of scan-to-scan integration.

Scan-to-Scan Integration

Scan-to-scan integration is a method of obtaining a long-term look at a region of (range-azimuth) space in surveillance mode, when many such regions must be monitored. The concept is to rotate the antenna and to integrate the return signal from the same spatial region over many scans (and thus many seconds). This time scale permits the averaging out of the longer term variations in the sea-clutter power. Because the targets of interest are moving, the antenna is usually rotated at a rapid rate (30–300 rpm) to acquire as many scans as possible within a fixed time period. A potential problem with scan-to-scan integration is the motion of the radar platform. If the radar moves a significant fraction of the width of the spatial regions, then compensation for the motion must be made.

Because of the limited data base of scanning antenna files, the scan-to-scan integration was simulated using staring (nonrotating antenna) IPIX files. The data were integrated as they would be if received from a rotating antenna. An antenna beamwidth of 2° was assumed, as used in the previous version of the IPIX radar. Also, with the limited amount of data, false-alarm performance is statistically accurate down to about 2×10^{-4}; the lowest parts of the curves may be inaccurate.

Two antenna rotation rates were simulated: (1) a dwell time of 8 ms to pass through the 2°, equivalent to 42 rpm, and (2) a dwell-time of 32 ms, equivalent to about 5 rpm. With a PRF of 2 kHz and the IPIX files of about 45 s in duration, the number of successive scans simulated was 32 for the 42-rpm antenna rotation rate, but only 4 scans for the 5-rpm rate.

Figure 9.31 shows the two ROC curves resulting from scan-to-scan integration, with curve B for the 5-rpm antenna rate, and curve C for the 42-rpm rate. At the lower rotation rate, the radar receives fewer scans in a given period, and within each scan the data are strongly correlated. At the higher rotation rate, the radar obtains a larger number of scans, providing more low-correlation looks at the sea clutter. Consequently, noncoherent integration is more effective at the higher scan rates, as shown in the improved detection performance.

9.4.2 Coherent Detection

A coherent radar allows, through Doppler processing, the extraction of spectral information from the return signal. Figures 9.2(b) and 9.3(b) showed the short-term (0.5 s) power spectra of sea clutter alone and a growler target in sea clutter, respectively. The target shows up as a narrow, powerful spike in frequency, which indicates that coherence, followed by appropriate Doppler processing, can improve detection. By repeating this process for a number of periods (while the radar is staring at the same direction in azimuth), time-frequency plots of the processes can be formed.

Figure 9.14 showed the time-frequency plot for a growler in sea clutter. The target spectral content forms a narrow track in time, modulated in Doppler as it is being tossed by the waves. The wider, diffuse spectral component is from the ocean surface. This component varies in intensity and in width as a function of the moving waves. The target-to-clutter ratio in this case is about 3 dB, which is not enough for a simple noncoherent radar to detect the target reliably. By the coherent integration of forming the spectrum, the target-to-clutter ratio has been improved by about 7 dB because the clutter power is spread over a wider band of frequencies. Thus, the description of the received signal as a function of frequency allows better separation of target and clutter on a signal power basis.

The plot shows a few reasons why detection can be improved by such coherent processing:

1. The growler is consistently narrower in spectral width compared to the clutter.
2. The growler is stronger than the clutter more often than before the processing (because of the previous point).
3. The behavior of the target (slowly varying in frequency) allows it to be tracked, if revisited often enough.

Because of these properties, detection can be enhanced by processing strategies that employ banks of filters with pass bands equal to the target spectral width, so that the signal-to-clutter ratio is maximized in the filter containing the target. A detection in any filter is considered valid. However, the time and spatial variation of clutter power still prevents the use of fixed thresholds following the filter bank. The next section discusses techniques used to adjust the threshold adaptively in response to the data to improve false-alarm performance.

General CFAR Principles

A cell-averaging, constant-false-alarm-rate (CFAR) algorithm (Minkler and Minkler [16]) estimates clutter power at a given position and time (the cell under test), divides the actual received power by the estimated power, and compares it to a detection threshold. The division by the estimated power is an attempt to maintain a low CFAR. The estimate is usually some type of average over a number of (spatially) nearby samples, known as reference cells. This type of CFAR processor assumes that the interference has a distribution, over the reference cells and the cell under test, such that only its mean power is unknown (e.g., the clutter power is locally, exponentially distributed, with unknown mean). To avoid having the target corrupt the estimate and potentially prevent its own detection, the estimated clutter power is based on nearby regions in space and time, far enough away that the same target is not present. Using past time samples from the same region does not work in this application, because the target will have been present. The CFAR detector depends on the presence of a target in the cell under test raising the ratio above the threshold. The threshold determines the false-alarm rate. By taking statistics of the ratio, the threshold can be determined experimentally. For example, if a false-alarm probability of 10^{-4} is required, the threshold has the value such that only one clutter sample in 10^4 exceeds it.

To perform well, the cell-averaging CFAR processor requires spatially homogeneous interference, where the mean power does not vary greatly over the width of the reference region. It performs well when the interference is noise, extended weather clutter, or sea clutter from a low-resolution radar (not IPIX). The cell-averaging CFAR processor does not perform well for spatially nonhomogeneous interference, such as high-resolution sea clutter. Spatial homogeneity of sea clutter can be increased in two ways:

1. By integrating spatially over a region on the order of the sea's wavelength;
2. By integrating temporally over a time on the order of the sea's wave period.

Option 1 reduces the threshold but does not help in detection performance, since target power is diluted spatially (it is the equivalent of having a lower resolution system). Option 2, scan-to-scan integration, was shown in Subsection 9.4.1 to offer significant detection improvement for noncoherent processing.

Doppler CFAR

The separation between an ice target and sea clutter is improved by transforming coherent radar data into the Doppler frequency domain. The next requirement is to determine a suitable method for setting the decision threshold for detection. Because of the larger spectral width of the clutter, the Fourier transformation improved the SCR, by 7 dB or so. Despite this improvement, algorithms with fixed detection thresholds following the Doppler filters still have poor ROCs. However, the clutter power, because of its wider spectrum, is divided among a number of consecutive filters. This allows adjacent filters to be used as CFAR reference cells. A Doppler CFAR detection algorithm declares a detection (at any range) if the power in any Doppler filter exceeds the average power of nearby filters by a multiplicative threshold (typically 20 dB). Sea-clutter spectra are wider and more homogeneous than target spectra (a requirement for CFAR to be effective). In the time or spatial domains, this is not the case; the clutter is nonhomogeneous, making CFAR in the time or spatial dimensions not as effective. Reference filters are selected close in frequency to the filter under test, with a guard region equal to the target's spectral width.

The following summarizes the Doppler CFAR algorithm, as it would be implemented (continuously) in an operational processor:

1. Extract a time-series of samples from each range gate.
2. Form a sampled power spectrum estimate by:
   - windowing the time series (Hamming window is fine for the IPIX radar dynamic range)
   - discrete Fourier transform (DFT)
   - magnitude squared of each spectral sample.
3. Divide each spectral sample by its estimated mean interference power and compare the values to a threshold.

The interference power, as a function of Doppler frequency, is estimated by averaging the power spectrum samples from nearby frequencies. If the ratio at any frequency exceeds the threshold, a detection is declared. Detection decisions are a function of range, time, and frequency; the combination of these parameters would be used in postdetection processing (tracking).

Using a technique similar to the CFAR processor discussed in Chapter 8, the interference power estimate for a given test cell at frequency, f, is the average of two neighboring regions, one on either side of the test cell. These regions are called the reference cells for f. The region of cells immediately adjacent to the test cell, called the guard region, is not used in the estimation; it prevents a target (that has a finite frequency extent) from inhibiting its own detection by raising the estimated interference power. Various combinations of sizes for the DFT, and the reference and guard regions, were tested.

Doppler CFAR is quite sensitive to dwell time. Figure 9.35 shows the ROCs for various dwell times, that is DFT sizes, each for its best combination of reference and guard-region sizes. The PRF was 1000 Hz, so that input data samples were 1 ms apart. Note that performance improved as dwell time increased up to 0.25 or 0.50 s (DFT size of 256 or 512), and then it degraded because a dwell time of 0.25 or 0.50 s gives the smallest, estimated spectral width for the target (and thus target power is integrated most advantageously). For dwell times shorter than 0.25 s, Doppler resolution limitations increased the apparent spectral width of the target. For well times longer than 0.50 s, the acceleration of the targets over the dwell time (caused by wave motion) increased the estimated spectral width.

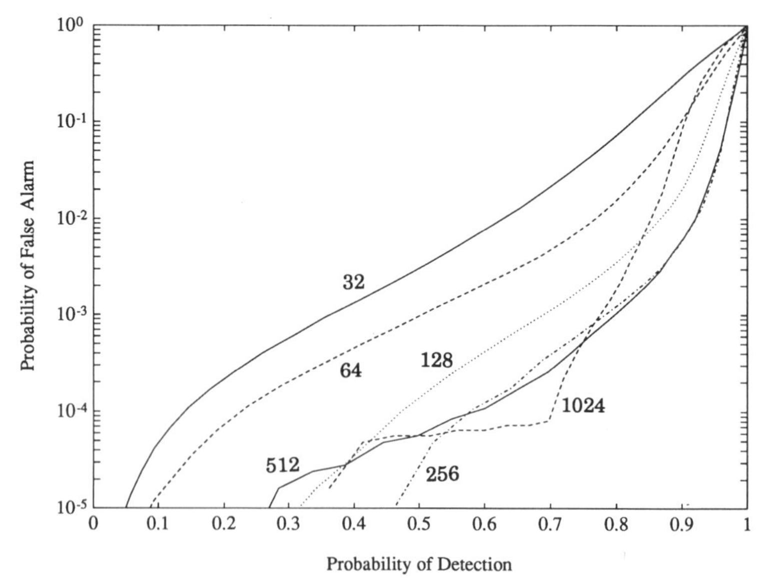

Figure 9.35 Effect of dwell time, hence DFT size, on detection performance. Data samples are collected at a PRF of 1000 Hz. Thus a DFT of size 256 processes data from about a 0.25 s of dwell time. Detection performance improves as the dwell time is increased and the Doppler spectral resolution increases, but degrades beyond 0.5 s as movement of the target begins to smear the target's spectral peak.

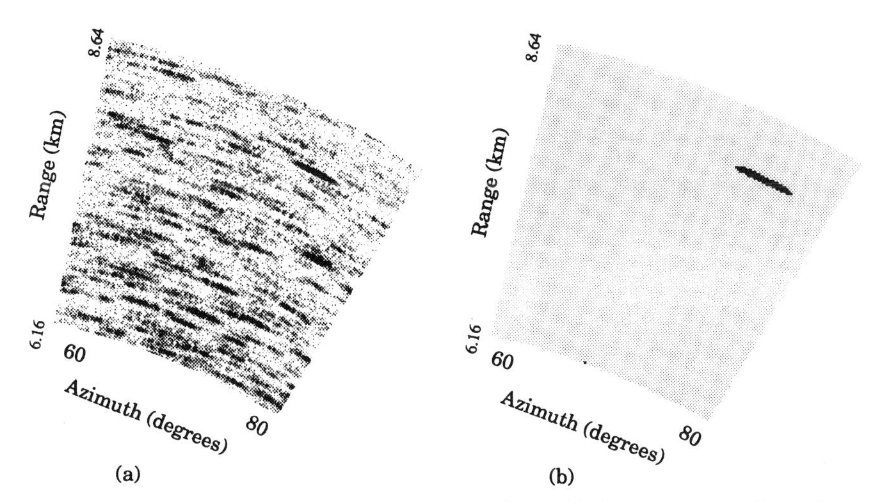

Figure 9.36 Example of Doppler-based target detection, with the antenna scanning at 3 rpm. Image (a) shows the return amplitude over a 20° sector containing a moored radar reflection, with a radar cross section of 2 m^2. The ocean had a significant wave height of 2.1 m and a period of 8 s. Image (b) shows the result of Doppler CFAR processing, in which the reflector is clearly detected, with no false alarms.

Using 10 to 20 total reference samples is considered to be more than enough for the test. The number of reference samples was chosen to ensure good statistical mean estimation, while avoiding straying outside the clutter spectral width. Guard regions were chosen just large enough to avoid self-masking of the targets. Bias is not introduced to the shorter dwell times by using fewer reference samples, as performance is worse with a larger number because of the finite width of the clutter spectrum.

As an example of the Doppler CFAR processor applied to a scanning antenna data set, Fig. 9.36 shows the result for the reflector in sea clutter. The antenna was rotating at 3 rpm, yielding about 111 hits per 3-dB beamwidth (alternating polarization). At a given azimuth, for each range cell, 128 successive pulses were used to calculate the Doppler spectrum, then the CFAR process was used to set the threshold, and to declare a target. The process worked very well.

Spectral-Width Detectors

Figure 9.3(b) showed a Doppler spectrum for a growler in sea clutter; there are two major spectral components: (1) sea clutter, which has a distributed frequency content, thus a fairly large spectral width; and (2) the target, which more coherently integrates into a much narrower spectral peak. In the case of sea clutter alone, the spectrum consisted of one broad sea clutter peak.

Assume that each of these spectral peaks can be fit reasonably with a Gaussian model. For the case of sea clutter alone, there will be one such model, whereas for target and clutter, there will be two models per spectrum. The target can be detected by contrasting the width of its fitted model with that of the clutter. This technique, called the Gaussian spectral width (GSW) detector, operates as follows. A spectrum estimate is computed using 0.5 s of data for each estimate. A single Gaussian model and a double Gaussian model are fitted to the data. Each Gaussian has three free parameters that describe it: the center frequency, the amplitude, and the spectral width. These parameters are optimized with respect to the data in a minimum least-squares error sense. The model that provides the best fit to the data is chosen. If a single Gaussian model is chosen, then its spectral width is used as the sufficient statistic. Otherwise, for a double Gaussian model, the smaller spectral width is used. The sufficient statistic is then compared to a threshold. If the threshold is exceeded, the data set is said to be sea clutter; otherwise, a target is declared.

Another technique, the autoregressive largest pole magnitude (ARLPM) detector, was studied. It also discriminates clutter and targets based on modeled spectral width. For each 0.5 s of data, a sixth-order autoregressive model was evaluated, and the six poles of the model were determined. Each pole was contained within the unit circle. If a target was present, with its attendant narrow frequency content, one of the poles tended to cluster very close to the unit circle, as the target was narrow band. For sea clutter or noise only, all the poles tended to lie further inside the unit circle. The detector compared the magnitude of the largest pole to a threshold, declaring a target if the threshold was exceeded; no target was declared if the threshold was not exceeded.

As a basis of performance comparison, a simple noncoherent detector, the integrated amplitude (IA) detector, was evaluated. This detector consisted of noncoherently integrating all the data within the 0.5-s record, and comparing the sum to a threshold. If the threshold was exceeded, a target was declared.

The coherent detectors were found to be far superior to the noncoherent detector. Of the two coherent detectors, the GSW detector proved to be far more robust to changes in the target-to-clutter ratio, and is, therefore, favored. Because of the relatively small numbers of data sets available, performance at low P_{fa}s could not be evaluated. However, as an indication of relative performance at a P_{fa} of 5%, the noncoherent detector had a P_d of 60%, whereas the GSW detector had a P_d of 96%. Further, if five successive GSW detection decisions were combined using a three-of-five rule, a P_d of 98% at a P_{fa} of 0.1% was possible.

The following figures show some typical results (Nohara [17]). First, Fig. 9.37 shows how the instantaneous ratio of growler power to clutter power can vary over a wide margin within tens of seconds. This variation affects the relative heights, and even the presence, of the growler and clutter spectral peaks. The figure is based on the same data set as Fig. 9.3. Figure 9.38 shows the two Gaussian models fit to data from four types of spectra, which may arise in a cell containing both a growler and sea clutter. The spread of these models is used as the detection criterion. Figure 9.39 shows a sample plot of the autoregressive poles for the growler in sea-clutter data. Note the cluster on the right side, near the unit circle, caused by the growler.

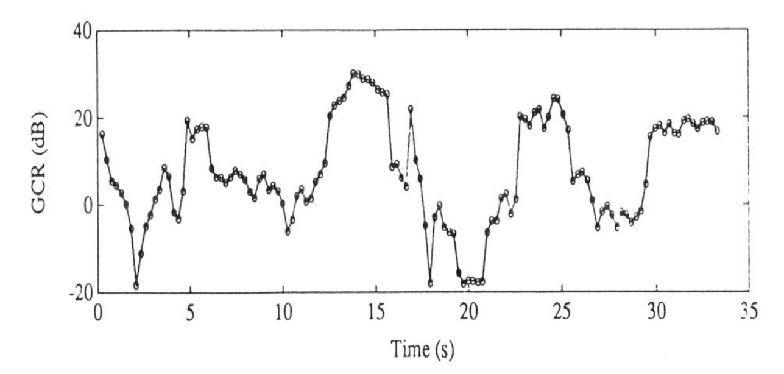

Figure 9.37 Growler-to-clutter power ratio (GCR) estimates as a function of time, calculated from the parameters of a two-peak Gaussian model, applied to growler-in-clutter spectral estimates. Note that wide variations can occur over short periods of time.

Figure 9.40 compares the performance of the three detectors using growler and clutter data from 17 files. There is only enough data to give indicative comparisons at P_{fa}s of 5% or higher. Also shown is the substantial improvement in performance permitted with the combination of the GSW detector and a dual binary integrator (DBI), using a three-of-five decision rule.

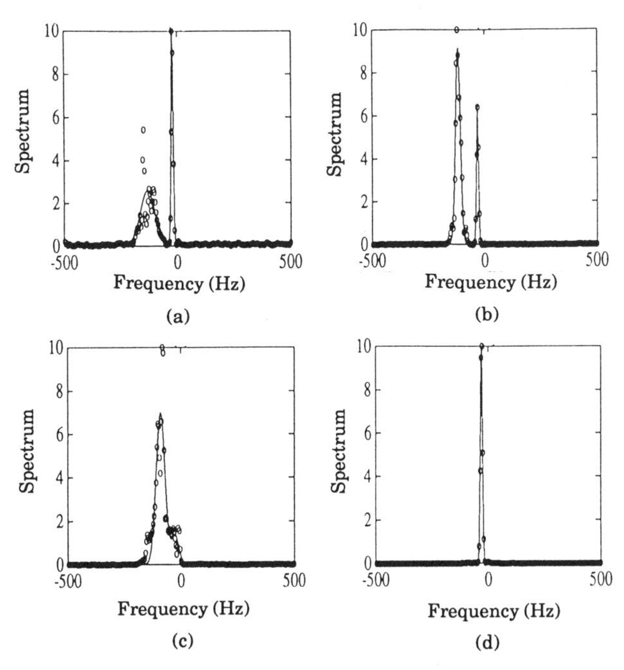

Figure 9.38 Gaussian spectrum models fitted to the Doppler spectral estimates for typical growler in sea clutter data. The Doppler estimates are indicated by "o," and the models by solid lines. Four typical cases are shown: (a) growler and sea clutter present, growler is stronger; (b) growler and sea clutter present, clutter stronger; (c) sea clutter only, growler is essentially absent; and (d) growler only.

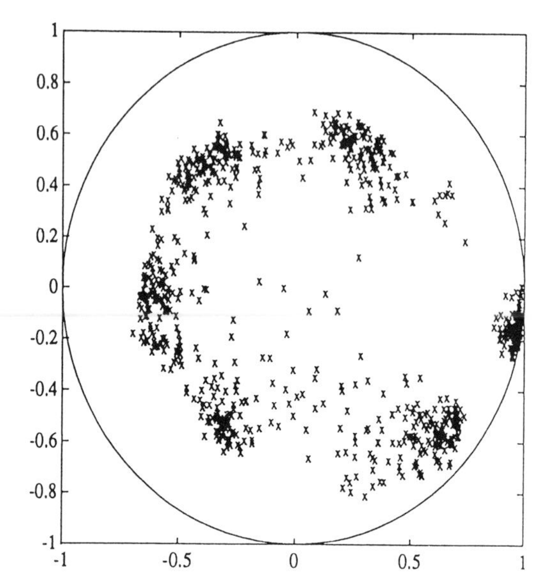

Figure 9.39 Scatter plot of the poles from a sixth-order autoregressive parameterization of the growler in clutter data. The cluster of poles at the right side, near the unit circle, model the narrow-band spectral component of the growler. The lower-right cluster, next closest to the unit circle, models the wide-band spectral component of the sea clutter. The remaining four pole clusters model the noise component.

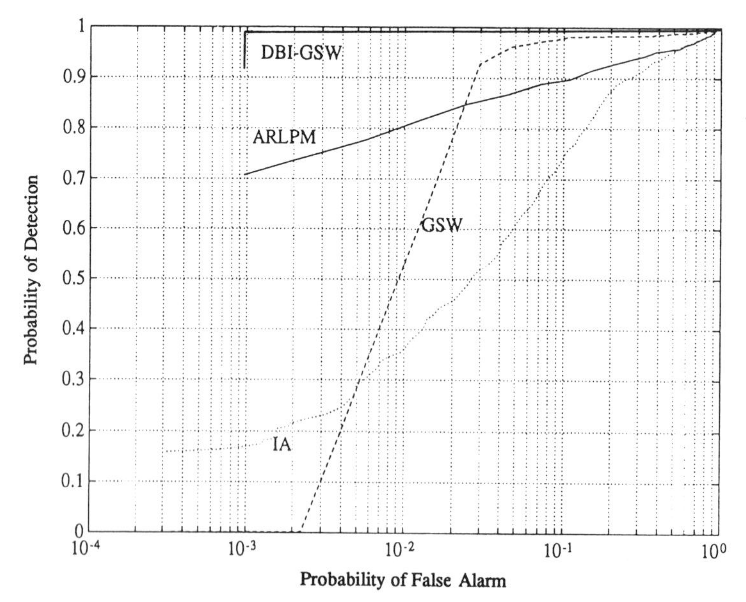

Figure 9.40 Performance comparison of various detectors. Receiver operating curves are given for the integrated amplitude (IA), Gaussian spectral width (GSW), autoregressive largest pole magnitude (ARLPM), and dual binary integrator (DBI-GSW) detectors. The DBI-GSW offers the best performance.

Chirplet Transform

In addition to the above analysis which is based on the use of the Fourier transform, other time–frequency analysis techniques have been investigated. One such investigation led to the development of a new transform, called the Chirplet Transform (Mann and Haykin [14, 15]), which is a generalization of Gabor's logon transform ideas. (The Chirplet transform was invented in 1991, independently by McMaster and Stanford universities.) The Fourier transform assumes a set of frequency basis functions which are continuous in time. Gabor [6] showed that an arbitrary signal can be decomposed onto a set of functions (logons) which are all modulated versions of a single Gaussian envelope. It is possible to adjust the shape of these envelopes to trade off frequency resolution for time resolution. The Chirplet transform uses wavelets as its bases, with each wavelet being a Gabor function. Figure 9.41 depicts a sample basis function graphically for the Fourier transform and for the Chirplet transform, and shows the appearance of the basis function in time–frequency space.

Next, from consideration of how the motion of a floating target is affected by movement of the ocean surface, as found from the modeling and confirmed by the experimental observations, much of the time the target is undergoing acceleration and deceleration. These changes in velocity can be considered linear over short periods, on the scale of the time taken to make a new Doppler frequency spectral estimate. It seems appropriate, therefore, to use a set of basis functions which can correlate more closely to signals whose frequencies (i.e., velocities) may change linearly within the spectral estimation interval. A signal whose instantaneous frequency changes in a prescribed manner (in this case linear), within a time interval of interest, is called a chirp; hence the name "Chirplet transform."

In the same manner in which the continuous-time, wave-basis functions of the Fourier transform were replaced by finite-duration wavelets, the full interval chirp can be replaced by the finite-duration chirplet. Figure 9.41 also compares the chirp and chirplet basis functions, and their appearance in the time-frequency space.

The basis functions of linear chirplets are parameterized by their starting and ending frequencies. From Nyquist limits, the frequencies of interest, normalized by the PRF of the radar, extend from -0.5 to $+0.5$. To produce an image of the transform output, the chirplet snapshot uses the coordinate axes of "starting frequency" and "ending frequency." Each location within this plot corresponds to a single chirplet, and its coefficient's magnitude can be gray-scale coded. For a given snapshot, the mean position of the chirplets within the time interval of interest and the length of the interval are fixed.

Figure 9.42 shows a chirplet snapshot for 1 s of the time series for a growler in sea clutter. Note that the coefficients along the diagonal line, for which the starting and ending frequencies are equal (i.e., no chirp), correspond to the usual Fourier transform. In this example, the much more clearly defined peak, which lies off this diagonal, indicates that the growler did indeed undergo acceleration during this time interval, which smeared the energy in the Fourier

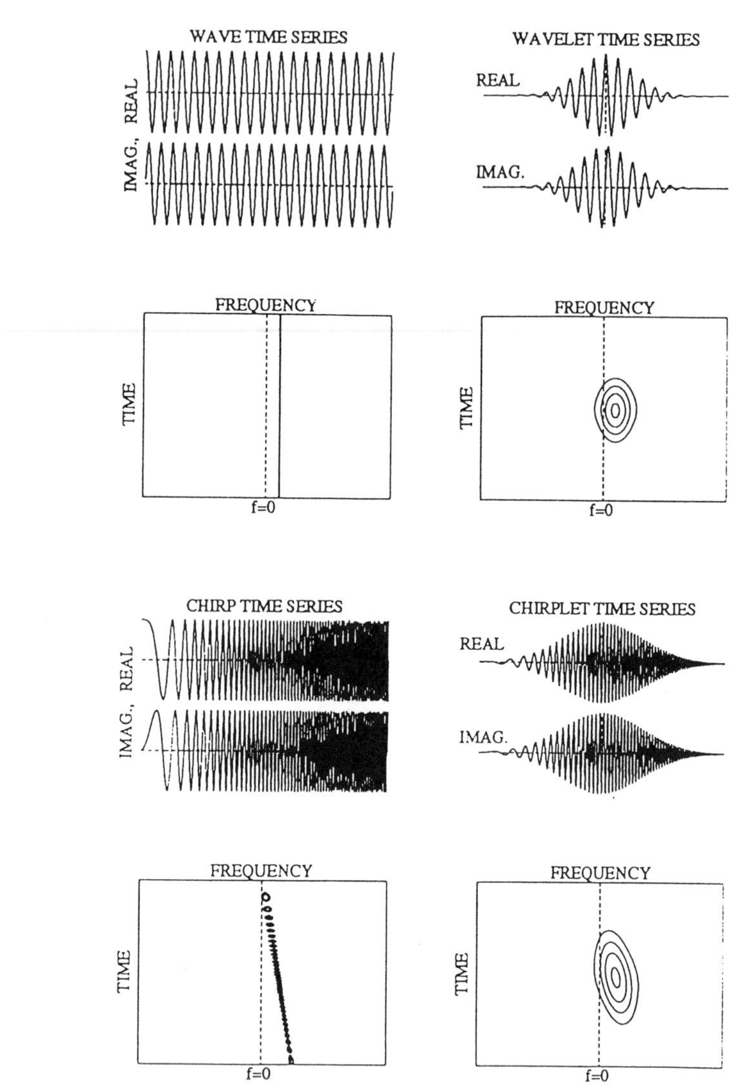

Figure 9.41 Graphical depiction of the relationship between a wave and a wavelet, and a chirp and a chirplet, in terms of time-series and time-frequency magnitudes. Note how the time–frequency behavior of the chirplet can be used to match that of the growler [for example Fig. 9.3(b)] over short periods.

estimate of the spectrum. The narrower and stronger peak in the chirplet snapshot, as shown in Fig. 9.43, offers the possibility of improved target detection.

The use of the chirplet transform (CT) snapshot for target detection was studied using IPIX data. The problem was treated as one of classification, where given a 0.5 s data record from a particular range cell, the decision of whether or not a target was present was required (two-class problem). Based

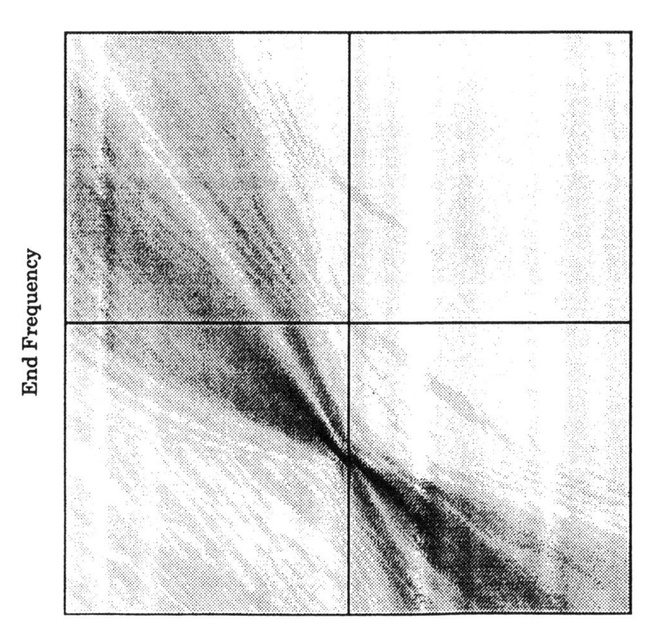

Figure 9.42 Chirplet transform of 1 s of a time series for a growler in sea clutter; the same data file as for Fig. 9.3. The usual Fourier transform would lie on a diagonal from lower left to upper right. The location of the dark peak, off the diagonal, indicates that the growler was undergoing acceleration within this 1 s interval.

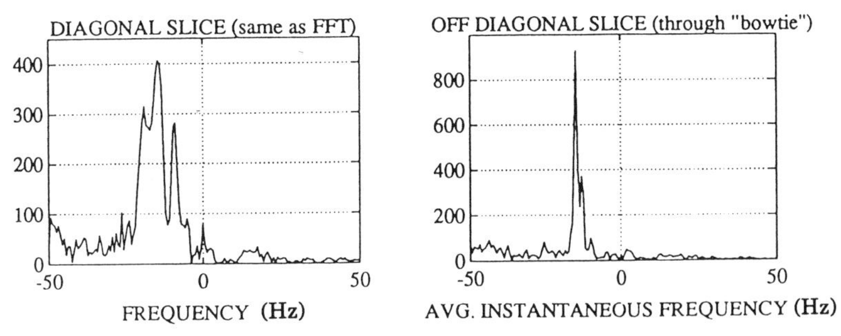

Figure 9.43 Slices through the chirplet transform of Fig. 9.42: (a) along the diagonal corresponding to the Fourier transform, and (b) an off-diagonal slice which passes through the peak of the transform. The latter yields a much sharper and stronger peak, permitting improved detection.

on inspection of the typical CT snapshot, three features were used in the classification:

1. Entropy: the sum of pixel brightness times the logarithm of these brightnesses. This quantity is high for no target present, low with target present.

2. Extent: target CT snapshots are much more compact, clustered about the mean epoch of the energy distribution, whereas clutter is more spread out. A bivariate Gaussian distribution is fit to the data, allowing the quantification of the spread through the determinant of the covariance matrix, S.
3. Slenderness: the ratio of the largest to smallest eigenvalue of S. Targets give rise to slender peaks.

Six different methods were used to perform the classification, resulting in classification accuracies, hence detection performance, of 91% to 96%. Details can be found in Mann and Haykin [14].

Polarimetric Enhancements

Section 9.2.4 presented plots based on the cross-spectra of HH and VV time–frequency data. Plots were shown for the ratio of the power in the two polarizations, and for the phase difference between them. The major observations are summarized here for reference, as a possible basis for enhanced detection algorithms.

Power Ratios. From the measured power ratios, the following differences exist between targets and clutter:

1. The cross-polarization ratios (|HV|/|HH|, |VH|/|VV|) are greater for sea clutter than for ice targets; thus the sea surface depolarizes somewhat more. Observed mean cross-polarization ratios are −10 dB for sea clutter and −16 dB for a growler.
2. The |HH|/|VV| Doppler ratio is near unity for ice targets. This ratio is near +3 dB for sea clutter if the powers are averaged over the whole clutter spectrum. However, at Doppler frequencies nearer 0 Hz, the ratio is very small; the VV spectrum, although less powerful, extends further into these low frequencies. Thus, the HH spectrum appears shifted in frequency from the VV spectrum. For a target, the ratio does not vary much from unity, regardless of frequency.

Phase Differences. Another potential discriminant between targets and clutter are the phase differences between scattering matrix elements. The phase difference, if desired as a function of Doppler, must be computed before the squared-magnitude operation in forming the power spectrum. The following observations have been made:

1. The phase difference between HH and VV spectral samples is more random for sea clutter than for a target, indicating less correlation between the two channels for sea clutter.
2. The phase difference between HH and HV (or VV and VH) appears to be completely random for targets or sea clutter, thus containing no discriminating information.

Polarimetric Detection. For enhanced detection which exploits polarization diversity, the challenge is to combine the outputs from the four polarimetric filter banks so that information contained in their relative powers and phases influences the detection decisions. A simple example of such a combination would be to multiplicatively bias the Doppler CFAR likelihood ratio by an ad-hoc approximation to the likelihood ratio for the phase difference. An assumed phase-difference density could be a Gaussian centered about 0°. Such a scheme would enhance detection when the HH-VV phase difference was near zero, as is the case for ice targets.

Application of Chaos Theory

Radar clutter has a history of being modeled as a stochastic process. One of the reasons for using such a model is that radar clutter appears visually to be random. Given this stochastic assumption, radar detection has been based on statistical decision theory. However, during the past three decades, physicists, mathematicians, biologists, and scientists from many disciplines have developed a new way of looking at complexity. This new way is deterministic in nature; it has been termed chaos theory (Lorentz [13]).

Chaos theory is the theory of dynamical systems that generate paths of evolution that are seemingly random (pseudo-random), as shown by conventional statistical tests. Although chaos puts a limit on long-term prediction, it implies predictability over a short term. Basic to the description of a chaotic phenomenon is the idea of a strange attractor (Leung and Haykin [11]). The term attractor means that all initial states of a dynamical system evolve with time such that they are attracted to the equilibrium state of the system.

In simple terms, a dynamical system is a system whose time evolution, from some unknown initial state, can be described by a set of rules. These rules may be expressed conveniently as mathematical equations. The evolution of such a system is best described by considering trajectories in the phase space (state space) of the system. A chaotic attractor causes nearby trajectories to diverge. Two nearby trajectories do not stay close to each other. Rather, they soon diverge and follow totally different paths on the attractor. In other words, the evolution of a chaotic dynamical system is very sensitive to initial conditions, and the system is said to have generated randomness. One may state that there do exist dynamical systems, called chaotic systems, that can be described by simple deterministic rules, and which can give rise to seemingly random behavior.

A related issue is that of model reconstruction. In model reconstruction, the attempt is to develop a model by seeking an algorithm that processes a time series of data observed from some physical process and, hence, to generate a model to describe the future behavior of the process. If the data were chaotic rather than random, a deterministic model (differential or difference equation) can be used as a model for the physical system. Such an operation is called model reconstruction or reconstruction of dynamics. Although the dynamical system for sea clutter may depend upon many components such as wind speed, wave motion, and temperature, Takens Embedding Theorem (Takens [22]) states that model reconstruction using just one component, such as time series data, should succeed to a certain extent. Further, in principle, one may approximate on the reconstructed phase space using a time series. One interesting point is that the embedded dynamical system can, in principle, extract the underlying dynamic from the data.

An important parameter of interest in a chaotic system is the dimension of the attractor. Although the exact dimension of the attractor is not strictly equal to the number of dynamical variables, the number of dynamical variables needed can be guessed by knowing the dimension of the attractor.

Having obtained time-series data, it is necessary to reconstruct the state space of the system, so as to reinterpret time signals as multidimensional geometric objects. The observed signal is interpreted as the projection of a multidimensional phase-space trajectory. One definition of the attractor dimension in the phase space of the system is the correlation dimension (Grassberger and Procaccia [7]), defined in relation to a sphere of radius, r, in N-dimensional phase space. The details of the calculation are given in Leung and Haykin [11], and actually involve calculation of the cumulative correlation $C(r)$, which is related to the correlation dimension by a power law. Plotting $C(r)$ versus r on log–log paper reveals the correlation dimension as the slope of the linear part of the curve. Figure 9.44 shows such a plot for sea clutter. Also shown is the resulting attractor dimension versus the embedding dimension. Analysis of five data sets revealed attractor dimensions between 6 and 9.

The next problem is in applying the theory of dynamical systems to identify the deterministic component in a set of observations, and in distinguishing this component from the effects of ever-present measurement uncertainty, extrinsic noise, and uncontrolled degrees of freedom. This situation is the inverse problem in nonlinear dynamics: inferring the deterministic equations of motion given observed random behavior in physical systems. The goal is to detect and to model deterministic structure in noisy data, which is where learning techniques can be applied. One such technique involves the use of neural network theory.

Neural Network Implementation. The discovery of the deterministic behavior of sea clutter, and its inherent (short-term) predictability, combined with the ability of neural networks to learn using their inherent nonlinearity, makes for a powerful combination (Haykin and Leung [8]).

A simple three-layer neural network, of the form shown in Fig. 9.45, was studied for modeling sea clutter, using IPIX experimental data (Haykin and Leung [8]). Based on analysis of the correlation dimension of the attractor, which yielded a value around 6.5, the input layer was chosen to have 7 input neurons. The hidden layer was chosen to have 15 neurons. The purpose of the network was to accept, at the input layer, time samples of sea clutter and to predict future samples. Therefore, there was only 1 output neuron, giving the predicted time sample. The network is characterized as having a 7–15–1 structure. The network was trained using the back-propagation algorithm, where the

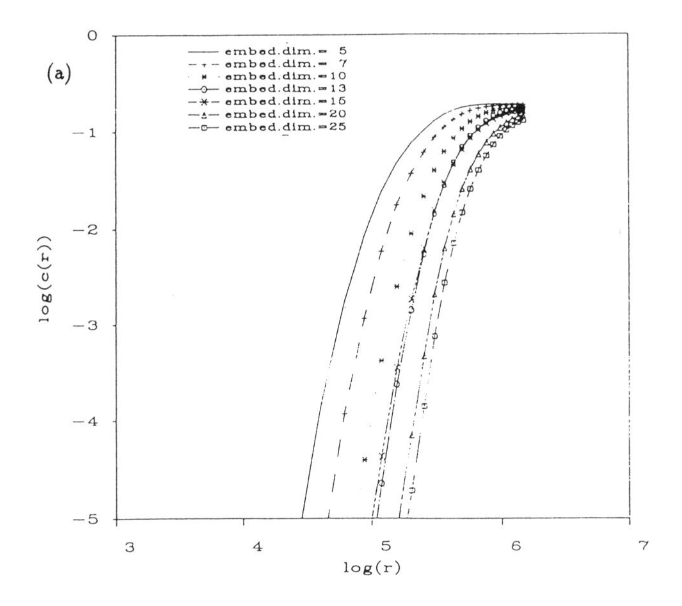

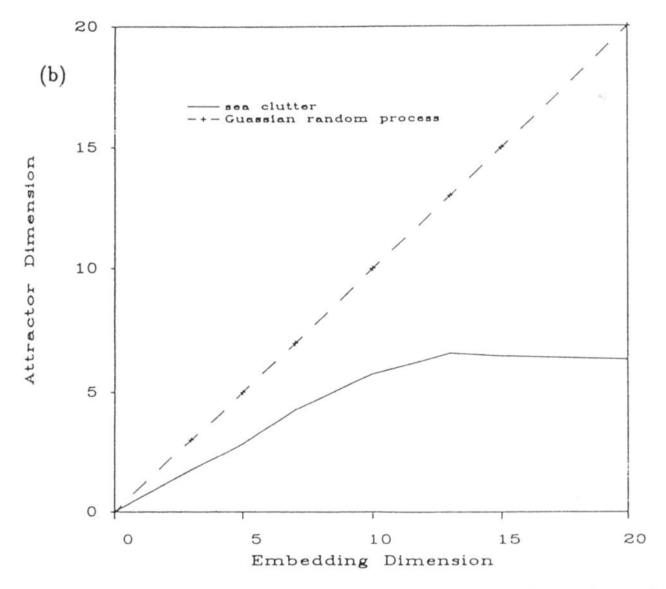

Figure 9.44 Confirmation of the chaotic nature of sea clutter. Graph (a) shows the cumulative correlation, $C(r)$, versus radius, r, of a sphere in N-dimensional space, where the slope of the linear part of the curve reveals the attractor dimension. Curves are shown for various embedding dimensions. Graph (b) shows the attractor dimension versus embedding dimension as determined from (a). Chaotic behavior is confirmed because the correlation dimension reaches a constant value as the embedding dimension increases. The dashed line shows the behavior for a Gaussian, truly random process.

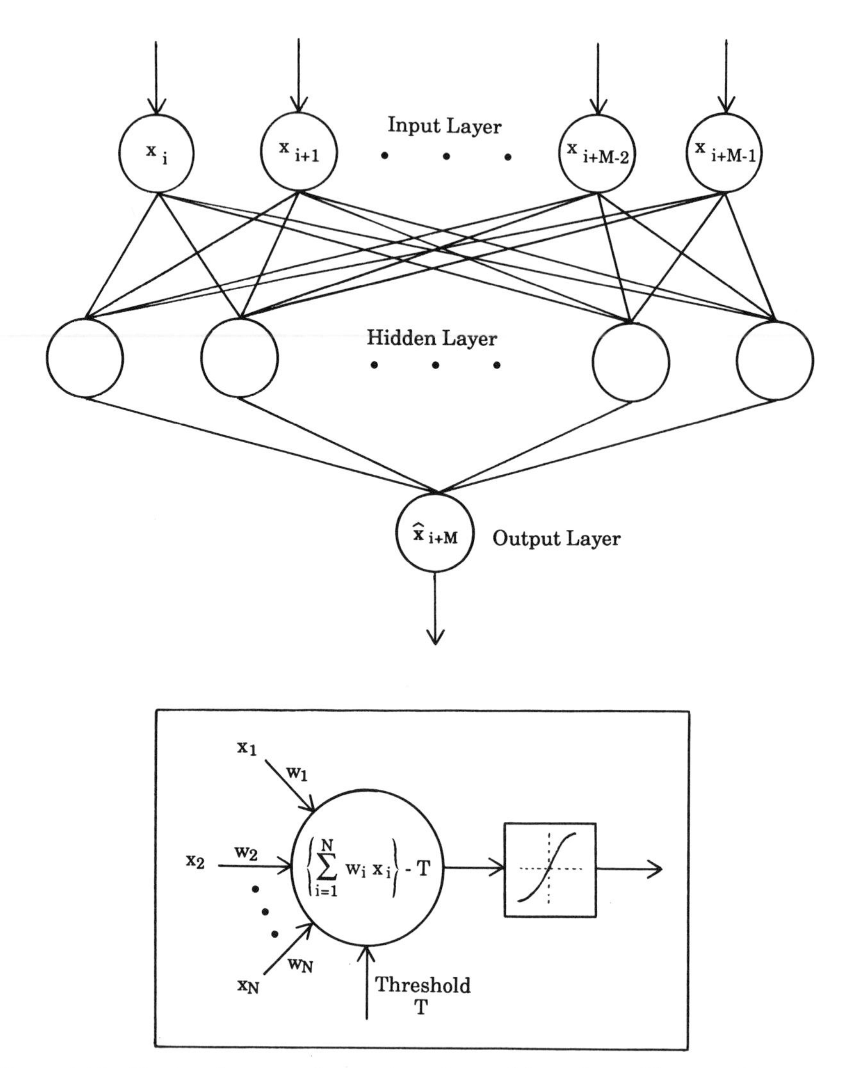

Figure 9.45 The three-layer neural network used to model sea clutter. The network is presented with seven data samples at the input layer, and (once trained) predicts the next sample. The form of the individual neurons in the hidden layer is shown in the blowup. Each neuron compares the weighted summation of its input to a threshold, then produces an output through a nonlinear function. The weights are established during training, while the nonlinearity gives the network its power.

"desired response" was the next sample in the time series. In short, the network was designed to predict the next data sample successively, given the previous seven. Figure 9.46 shows the input time series and the associated neural network output. The latter is almost an exact replica of the original input.

Reducing the degrees of freedom (i.e., the power) of the neural network, by reducing the input layer to three neurons, and the hidden layer to seven

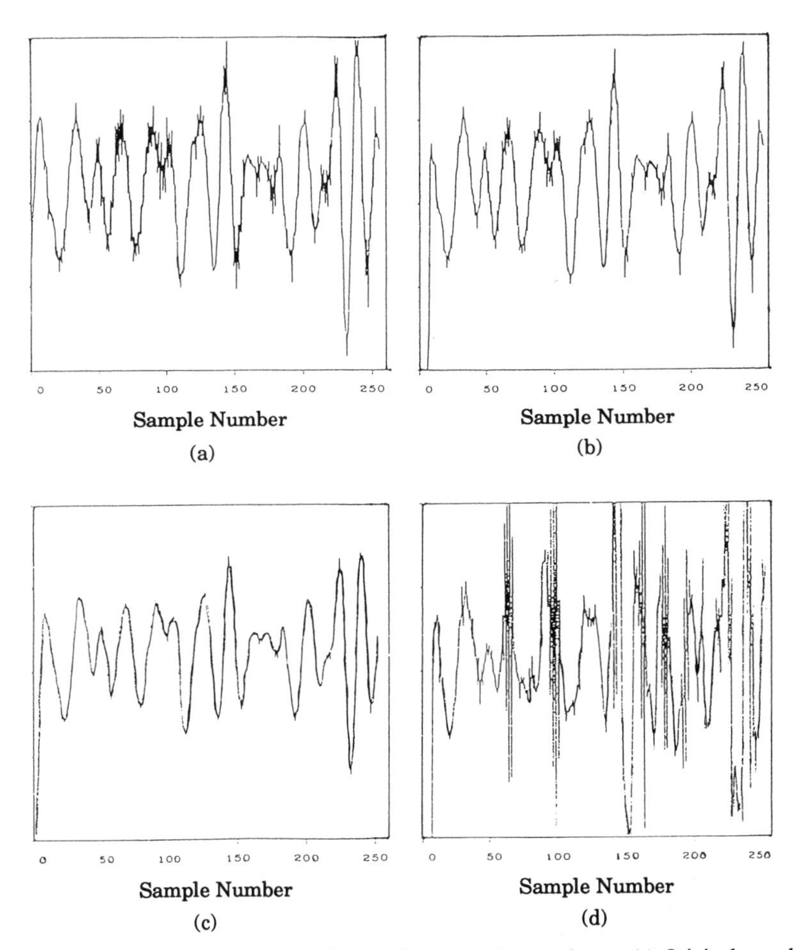

Figure 9.46 Comparison of the modeling performance for sea clutter. (a) Original sea-clutter time series. (b) Nonlinear modeling uses a three-layer neural network with structure 7–15–1, as shown in Fig. 9.45. This network models the clutter very well. (c) Nonlinear modeling using a three-layer neural network with structure 3–7–1. The network is not sufficiently powerful (lacks degrees of freedom) to model the clutter completely. (d) Linear modeling with structure 7–1. The linear network is unable to model the clutter properly.

neurons, shows that the network is no longer capable of modeling the original waveform adequately.

Finally, for comparison, the sea clutter was modeled using a linear modeling network with seven input units and one output. The linear network failed to model the sea clutter wave form.

Detection Possibilities. The correctness of the neural network's model of sea clutter can be assessed by comparing the predicted data samples with the actual data samples, on an ongoing basis. This comparison amounts to monitoring the so-called prediction error. As long as the input data are being taken from the sea clutter only, the model will continue to provide low errors. However, when a target is present in the return signal, which changes the fundamental nature of the signal, the prediction error will suddenly increase. Detecting these sudden increases in prediction error can form the basis for new target detection algorithms, which can take advantage of the power of nonlinear processing. Considered from another point of view, accurate prediction of the sea-clutter samples permits the cancellation of the sea clutter component of the signal through subtraction, leaving only the target signal, if present, for detection.

9.5 CONCLUSIONS

The close agreement between the theoretical and experimental results for a coherent radar, for example, in the nature of the Doppler spectrum for a growler in sea clutter, confirms the usefulness of theoretical models for gaining insight into the detection problem, and in developing new detection algorithms based on predicted behavior. The theoretical model can also be used to extend the studies into target and environmental conditions not yet encountered experimentally.

Using experimental data, the major gain in detection performance afforded by coherent radar was that obtained from Doppler processing, for which the differing properties of ice targets and sea clutter made their separate identification possible. This separation offered a significant detection improvement as compared to noncoherent radar.

A number of possible detection algorithms were examined. The Doppler-CFAR and spectral width detectors exploit the narrowness of the growler spectral peak as compared to the wider sea-clutter spectral component. The Chirplet transform provides a sharper spectral peak for a growler by matching the varying frequency content of the growler's return (caused by the acceleration by the ocean wave). Finally, the use of nonlinear dynamics for modeling the sea-clutter signal was described, implemented with a neural network. This approach showed itself to be very powerful, and its application for detection was discussed.

In addition to coherence, the experimental radar has a number of other features. The one most extensively studied has been dual polarization. Results were presented showing the differing properties of HH and VV for sea clutter, whereas for ice targets and rain clutter the HH and VV returns were highly correlated. This difference could be exploited to provide both the identification of the source of the clutter, and for improved target detection.

Future research includes the examination, through additional experiments, of the properties of the return signal when using the other radar features, such as dual transmit frequency. Having established the usefulness of each of the radar features individually, there remains the challenge of developing algorithms which take full advantage of the properties of all the features, by analyzing them *simultaneously*. Techniques such as nonlinear processing and neural networks hold considerable promise in this area.

ACKNOWLEDGMENTS

The authors would like to acknowledge their past and present colleagues at the Communications Research Laboratory whose research efforts contributed to this chapter: Henry Leung for nonlinear processing, Steve Mann for the Chirplet transform, Tim Nohara for the spectral width detectors, and Peter Weber for Doppler CFAR and polarization analysis. They also wish to thank Larry Bridle, Carl Krasnor, David Hamburger, Oliver Slupecki, and Jerry deBoer for their work on the IPIX radar and participation in the field experiments. Access to the experimental site at Cape Bonavista, Newfoundland, was provided by the Canadian Coast Guard.

The authors also gratefully acknowledge the generous financial support given to this research by the Natural Sciences and Engineering Research Council (NSERC), Litton Systems Canada Limited, the Ontario Ministry of Colleges and Universities, and the Department of Fisheries and Oceans (DFO). In particular, the authors would like to acknowledge the early and valuable guidance provided by the Advisory Committee convened by NSERC under the chairmanship of Jamie Rossiter; the continuing encouragement provided by Ed Lewis, of the DFO, whose research, in fact, sparked the authors' initial interest in the ice detection problem; and the support of Harry McLaughlin and Peter Metherall, both of Litton.

REFERENCES

[1] S. K. Chakrabarti (1987) *Hydrodynamics of Offshore Structures*, Springer-Verlag, Berlin/New York.

[2] B. W. Currie, S. Haykin, and C. Krasnor (1990) "Time-varying spectra for dual-polarized radar returns from targets in an ocean environment," in *Intl. Radar Conf., Radar-90*, 365–369.

[3] B. W. Currie, P. Weber, C. Krasnor, and S. Haykin (1991) "Polarization characteristics of a growler and the ocean," in *14th Canadian Symp. on Remote Sensing*, 301-305.

[4] B. W. Currie, P. Weber, C. Krasnor, S. Haykin, H. McLaughlin, and R. Worsfold (1991) "An improved marine radar," *Intl. Conf. on Ports and Ocean Engineering under Arctic Conditions*, 964–976.

[5] M. A. Donelan and W. J. Pierson, Jr. (1987) "Radar scattering and equilibrium ranges in wind-generated waves with applications to scatterometry," *J. Geophys. Res.*, **92** (5), 4971-5029.

[6] D. Gabor (1946) "Theory of communication," *J. IEEE,* **93** (Part III), 429–457.

[7] P. Grassberger and I. Procaccia (1983) "Measuring the strangeness of strange attractors," *Physica 9D*, 189–208.

[8] S. Haykin and H. Leung (1989) "Chaotic model of sea clutter using a neural network," in *Proc. SPIE Conf.*

[9] S. Haykin and H. Leung (1991) "Neural network modeling of radar backscatter from an ocean surface using chaos theory," *IEEE Conf. on Neural Networks in Ocean Eng.*

[10] S. Haykin, C. Krasnor, T. J. Nohara, B. W. Currie, and D. Hamburger (1991) "A coherent dual-polarized radar for studying the ocean environment," *IEEE Trans. on Geosci. and Remote Sens.*, **29** (1).

[11] H. Leung and S. Haykin (1990) "Is there a radar clutter attractor?," *Appl. Phys. Letters*, **56** (6), 593–595.

[12] J. H. Lever, E. Reimer, and D. Diemand (1984) "A model study of the wave-induced motion of small icebergs and bergy bits," in *3rd Intl. Symp. on Offshore Mechanics and Arctic Engineering*, Vol. III, 282–290.

[13] E. N. Lorentz (1963) "Deterministic nonperiodic flow," *J. Atmos. Sci.*, **20**, 130–141.

[14] S. Mann and S. Haykin (1991) "The Chirplet transform: A generalization of Gabor's logon transform," in *Vision Interface '91.*

[15] S. Mann and S. Haykin (1992) "'Chirplets' and 'warblets': Novel time-frequency methods," *Electron. Letters*, **28** (2), 114–116.

[16] G. Minkler and J. Minkler (1990) *CFAR*, Magellan Book Co., Baltimore, MD.

[17] T. J. Nohara (1991) Detection of growlers in sea clutter using an X-band pulse-Doppler radar, Ph.D. thesis, McMaster University, Hamilton, Ontario.

[18] A. J. Palmer (1990) "Dual-frequency radar: basic theory and surface current mapping performance for land-based systems," NOAA Tech. Memo. ERL WPL-186, Wave Propagation Laboratory, Boulder, CO.

[19] O. M. Phillips (1966) *The Dynamics of the Upper Ocean*, Cambridge University Press, London.

[20] W. J. Pierson and L. Moskowitz (1964) "A proposed spectral form for fully developed wind seas based on the similarity theory of S. A. Kitaigorodskii," *J. Geophys. Res.*, **69** (24), 5181–5203.

[21] D. L. Schuler, W. C. Keller, and W. J. Plant (1991) "A three-frequency scatterometer technique for the measurement of ocean wave spectra," *IEEE J. Oceanic Eng.*, **16** (3).

[22] F. Takens (1981) "Detecting strange attractors in turbulence," in D. A. Rand and L. S. Young (Eds.), *Dynamical Systems and Turbulence, Warwick 1980*, Lecture Notes in Mathematics, Vol. 898, Springer-Verlag, New York, 366–381.

[23] D. B. Trizna (1985) "A model for Doppler peak spectral shift for low-grazing angle sea clutter," *IEEE J. Oceanic Eng.*, **OE-10** (4).

[24] L. B. Wetzel (1977) "A model for sea backscatter intermittency at extreme grazing angles," *Radio Sci.*, **12** (5), 749–756.

[25] J. W. Wright (1968) "A new model for sea clutter," *IEEE Trans. Antennas and Propagation*, **AP-16**, 217–223.

10

OPERATIONAL AIRBORNE RADARS

RAYMOND T. LOWRY

Intera Information Technologies (Canada) Limited
Calgary, Alberta, Canada

10.1 INTRODUCTION

This chapter presents an overview of the operational airborne radar capability that has been developed since World War II and explores the needs that brought about these developments. Operational implies that the system is used in active support of marine operations in ice, on a routine basis. It can also include research radars, depending on the application. The nature of the problems addressed by operational airborne radars is reviewed including the particular needs of marine operators. For clarity, the radar terminology used in commercial remote sensing of sea ice is defined.

In an international context, Canada has emerged in a leadership role in the development of operational ice reconnaissance systems. The reasons for this are many and complex, but a few factors can be identified. First, the northern oceans are the areas most requiring ice reconnaissance. Canada and Russia share the largest part of these waters. Second, technical developments in the former Soviet Union were, until very recently, kept secret from western nations because of the strong military connection of radar. This also reduced the availability of sophisticated technology to the commercial marine sector. Therefore, although shipping volume in the Russian Northern Sea Route is much greater than in the Canadian Northwest Passage, and the Russian operational capability is much larger, Russia has lagged in some key areas of technology.

In Canada, the combination of oil and gas exploration in the offshore, and the government mandate to provide comprehensive ice information to shipping, created a market for ice information. In the scientific, technological, and busi-

Remote Sensing of Sea Ice and Icebergs, Edited by Simon Haykin, Edward O. Lewis, R. Keith Raney, and James R. Rossiter.
ISBN 0-471-55494-4

ness atmosphere of western nations, this demand for ice information has led to a variety of commercial ventures to satisfy the requirement. As a result, operational airborne Synthetic Aperture Radars (SARs) with on-board digital processing of the image to provide a real-time product, have been available in the Canadian private sector since 1983, but in no other part of the world (at the time of writing). Canadian efforts have been, and continue to be, at the forefront of providing operational ice reconnaissance capability to offshore operations.

As described in Chapter 6, a radar produces an image by two-dimensional scanning. In range this is done by measuring the time taken for energy to return from an illuminated point. In the other direction (azimuth), the antenna which directs the energy in one particular direction is moved. There are two principal means of moving the antenna of an airborne radar. The first is to rotate the antenna, as is done with search radars. The second is to point the antenna to the side of a moving platform and use the motion of the platform to scan the scene. Rotating antennas have a number of distinct advantages including (a) the ability to work from platforms with very slow or erratic motion such as ships or helicopters, and (b) the ability to image a given area repeatedly, which allows a time series to be built up. This is a tremendous advantage for applications such as imaging icebergs in sea clutter. The principal disadvantage is that the antenna must be rather short to allow it to be rotated on an aircraft, which results in a broad azimuthal beamwidth. A fixed, side-looking antenna, by contrast, can be as long as the aircraft, which allows a much narrower azimuth beamwidth to be achieved. This class of radar is referred to as side-looking airborne radar (SLAR). A further development of the SLAR is to process the data using coherent techniques to produce a much larger synthetic aperture and hence a much narrower effective azimuth beamwidth. The resulting radar is referred to as a synthetic aperture radar (SAR); see Chapters 6 and 11.

There are three kinds of offshore activities that use ice reconnaissance data: transiting ships, stationary platforms, and operations based on the ice surface. This last group is divided into those using ice as a mobile platform and those using fast ice as a substitute for solid ground. Fast-ice operations have not been an important driver in the development of, or use of, operational radar capability, and the use of ice as a mobile platform can be regarded as a subset of transiting ships. Therefore, this discussion is confined to stationary and transiting maritime operations.

The two main categories of ice of interest for ice reconnaissance are floating sea ice and icebergs. As described in Chapter 2, the properties of these two categories are rather different. From an ice-reconnaissance perspective, the two problems are also somewhat different. Sea-ice reconnaissance is more of a mapping and identification problem. Iceberg reconnaissance requires the detection of point targets, often in very heavy sea states, which gives rise to confusing clutter. From a marine operations perspective, both icebergs in the open sea and floating sea ice can be hazardous unless proper precautions are taken, and therefore accurate and current ice reconnaissance information is of high value. For the most part, unless otherwise specified, this chapter deals with the reconnaissance of floating sea ice.

Ice reconnaissance information is always used in a multidimensional context by the responsible marine operator (such as a ship's captain or a drill master) because a variety of information is available. Operations are typically carried out under many constraints, not all of which are related to the ice. A ship's master, therefore, will use the information gleaned from ice reconnaissance as part of a complex decision-making process. Traditionally, ice information was available only in the form of charts at a coarse scale (see Section 12.3.5). These would include information from airborne reconnaissance, as interpreted by an ice observer who was probably not serving on the ship. In recent years, direct communication downlinks between the ice reconnaissance aircraft and the ship have become commonplace. This communication allows the interpretation task to be performed on the ship so that it can include the contextual information so important to the vessel's operation. It would be difficult to overemphasize the significance of this change to marine operations in ice-infested waters.

One very important driver in the development of ice reconnaissance systems has been the need for timely data (Section 8.1.1). What constitutes timely is somewhat variable and depends on the application, the ice conditions, and in particular, the speed at which the ice is moving. Once ice information has aged, it loses its tactical value. As a result, radars without a real-time image display on the aircraft were not useful in an operational role. Until digital electronics with large data-handling capacity became available in the early 1980s, the use of SAR systems was limited by the lack of practical real-time displays.

10.1.1 Historical Overview

During World War II, airborne search radars were developed to the point where small targets such as icebergs could be detected, and plan position indicator (PPI) displays were developed that gave a crude image of floating ice. During the 1940s and 1950s, a number of long-range surveillance aircraft were deployed, equipped with sensitive search radars. Although these were military systems, the ice information so obtained began to be used to support not only naval operations but also civilian shipping by agencies such as national coast guards.

In the 1950s, radar technology advanced rapidly and gave rise to practical SLAR systems which were tested for ice reconnaissance on a research basis. However, by the 1960s, substantial operational trials of SLARs for ice reconnaissance were being organized by the U.S. (Anderson [3]) and Canadian Coast Guards, by the International Ice Patrol, and in the former Soviet Union, by the Arctic and Antarctic Research Institute (AARI) in Leningrad. The 1970s saw the introduction of operational SLAR ice reconnaissance systems and the trials, in the civilian domain, of fine-resolution SAR systems. By the early 1980s, commercial SAR ice reconnaissance data were being collected routinely, and by the end of the decade, comprehensive SAR and SLAR ice reconnaissance systems were being flown operationally.

A pivotal event in the history of ice navigation in the Canadian Arctic was the 1968 sinking of Panarctic Petroleum's supply barges (Kennedy [19]). This event led to much greater use of aircraft for resupply and demonstrated to the exploration industry the incredible power of ice in the Arctic. This resulted in an acknowledgement of the need for ice reconnaissance and tactical ice information.

In contrast to the experience of Panarctic Petroleum, the 1969 sailing of the SS *Manhattan* through the Northwest Passage was considered something of a success for ice reconnaissance. On an experimental basis, a Philco-Ford AN/UPD-2 *Ku*-band (16.5-GHz) radar was flown on a U.S. Coast Guard (USCG) C-130 ice reconnaissance aircraft. The purpose was to determine if ice concentration, floe size and number, ice-surface topography, age, and deformation could be identified from the imagery. A detailed analysis of these data was made (Johnson and Farmer [18]) and the results were compared with surface data taken by scientists on board the SS *Manhattan* and the Canadian Coast Guard (CCG) icebreakers *Sir John A. MacDonald*, and *Louis St. Laurent*. The results were very encouraging as most of the objectives were realized.

The International Ice Patrol (IIP) was set up in response to the sinking of the RMS *Titanic* and has been operated by the USCG since 1912. Before World War II icebergs were monitored by a combination of ships stationed in the area and iceberg sightings compiled from shipping reports. After a pause during the war, the service was resumed using aircraft. Early work on the use of radar for iceberg detection was very encouraging, which led the IIP to make extensive use of early search radars and, subsequently, SLAR systems mounted on board their aircraft (Super and Osmer [37]). To date, the IIP has not used a SAR system, partly because of the multirole tasking of their aircraft.

Although historically the development of operational ice reconnaissance using airborne radars has been carried on internationally, much of the leading work has been done in Canada. Therefore, in the next section, developments in Canada will be covered in greater depth.

10.2 CANADIAN INDUSTRIAL CAPACITY DEVELOPMENT

The conditions in Canada during the late 1970s and early 1980s were particularly favorable to the development of an operational sea-ice remote sensing capability. There were tax incentives in place that made frontier oil exploration very attractive. The Canada Centre for Remote Sensing (CCRS) of the Department of Energy Mines and Resources was leading research into radar remote sensing. Canada was also fortunate in having close proximity to the United States, where a great deal of civilian sensor development and scientific research, in large part stimulated by the U.S. National Aeronautics and Space Administration (NASA), was taking place. As a result, Canada was well placed to develop a substantial industrial ice reconnaissance capability. Although this discussion concentrates on SAR developments, real-aperture radars (both with side-looking and rotating antennas) have played, and continue to play, an important role in the gathering of operational ice data.

When advanced military reconnaissance technologies became available to the civilian community in the 1970s, the area of remote sensing began to develop rapidly. With the launch of the first earth-resources satellite (*Landsat 1*), remote sensing technology began to outstrip applications. Viable commercial applications, especially in airborne remote sensing, were difficult to develop. Several commercial initiatives were launched, particularly in the United States, but most foundered for lack of solid commercial applications. In this climate, CCRS requisitioned a broad study of the commercial applications of remote sensing which resulted in several observations on the commercial viability of remote sensing, most of them rather bleak (Intera [17]). This study identified ice reconnaissance as one of the few areas where airborne remote sensing had true commercial potential. Remote sensing satellites have to date been launched using funds from government or multigovernment agencies. In commercial airborne remote sensing, a company must pay for both the development and operation of the sensor, as well as making a reasonable profit. Therefore, the data must have relatively high value to the end client, who in turn must find a profitable use for the information.

Ice reconnaissance has two aspects that make it a candidate for commercial remote sensing. First, information on floating ice conditions can have very significant impact on the cost of marine operations. Not only have the spectacular cases such as the sinking of the RMS *Titanic* resulted in loss of lives and property, but the more routine costs of operating in ice can be significant. If good ice information is available, it is often possible to avoid ice completely, thereby reducing both fuel and maintenance costs. Second, ice conditions may change rather rapidly and therefore repeated coverage is required. Ice reconnaissance has the capability to deliver information with high commercial value, and that information needs to be updated regularly. Ice reconnaissance, therefore, represents a true commercial opportunity for remote sensing. Further, since ice is often found in dark and cloud-covered conditions, it is obvious that only radar remote sensing has the capability of providing reliable, current information.

The recognition that ice reconnaissance could be a viable application of radar remote sensing was delineated in a 1976 study (Intera [17]). At that point it was not clear whether satellite or airborne systems or a combination of the two would meet Canadian needs. However, it was obvious that a concerned study was needed to develop the requirements for a system and that this study would need real data to be effective. The Canadian Surveillance Satellite Project was undertaken to fulfill this need (VanKoughnett et al. [38, 39]) and combined both airborne data from a research radar system and spaceborne data from the

L-band satellite, Seasat. This work is described further in Section 10.2.3. However, before discussing the SAR developments, it is worth describing in more detail the development of both rotating antenna radar and side-looking, real-aperture radar for ice reconnaissance. These are dealt with in Sections 10.2.1 and 10.2.2, respectively.

10.2.1 Airborne Rotating Antenna Radars

Airborne ice reconnaissance in the Arctic became practical after World War II. Agencies such as the Royal Canadian Air Force had aircraft capable of operating in the High Arctic which were equipped with navigation radars. These could be used to identify coastlines and other large features. These radars could also be used to map ice features even though they had rather coarse resolution. A good example of the work that could be done with such a radar is the pioneering work done on the ice transport in Nares Strait by scientists at the Defense Research Establishment Ottawa (DREO), using, in part, the standard ARGUS ASV 21 antisubmarine radar. At the south end of Nares Strait is an area known as the North Water. It remains open or develops only a thin ice cover during the entire winter. It was first observed by early explorers and whalers but was not studied until the late 1960s and early 1970s because of its remoteness. Using remote sensing from aircraft, combined with later field work, the mechanisms of the North Water formation were studied and finally were understood (Dunbar [9]).

The Canadian ice reconnaissance service, Ice Branch of the Atmospheric Environment Service (AES), beginning in the early 1970s, used Electra aircraft equipped with navigation radars with considerable success. The data collected, although not suitable for stand-alone use, were of great value. By using a projector system to overlay the search radar image on a map, and combining the radar data with visual observations, quite accurate charts of larger ice features were made. These charts were transmitted by facsimile to the CCG icebreakers, and by the late 1970s were also being relayed to the Ice Centre in Ottawa. A series of technical developments by AES, in both airborne sensors and telecommunications, led to the development of the IRDNET data communications system, which is discussed in Section 10.2.4.

From the beginning of its airborne operations, the IIP used rotating radars to detect icebergs. However, they relied on visual reconnaissance and air photography to provide confirmation of the iceberg counts compiled from the radar data. The IIP began to experiment with SLAR systems in 1957 because the coarse azimuth resolution of scanning radars did not permit target tracking at long range. After several experiments with old and unreliable systems, an APS-94/C SLAR was put into service in 1975. Since that time, the IIP has relied almost exclusively on SLAR.

Recent developments in radar technology have improved the performance of modern search radars compared to that of older models. The use of pulse compression has resulted in simultaneously achieving finer range resolution and longer range capability. (For example, such radars can achieve useful target detection to beyond 100-km maximum range, with range resolution finer than 5 m.) The use of improved navigation technology to provide ground-stabilized displays allows long-term integration of the returns from ice and icebergs. This results in improved target to sea-clutter performance while giving a stable image similar to a SLAR. By using technology developed for antisubmarine warfare, systems such as the Litton V5 (Cantrill and Gordon [7]) have been shown to be very effective for both iceberg detection and ice-edge mapping (Rudkin et al. [32]). However, it remains a difficult task to produce an image with fine gray-scale resolution with such systems, and the short antenna used precludes fine azimuth resolution. Search radar systems are therefore not favored for general ice reconnaissance (such as is needed in the Arctic) or in support of marine activities in ice-covered waterways (such as the Great Lakes and the Gulf of St. Lawrence). However, they are used regularly for iceberg reconnaissance in the North Atlantic (Ryan [33]) and are discussed further in Section 10.2.6.

10.2.2 Side-Looking Airborne Radars (SLAR)

AES SLAR Development

The AES mandate is to provide ice information and forecasts (primarily to the Canadian Coast Guard) in all ice-infested Canadian waters (Section 12.3.2). The CCG in turn provides both ice-breaking support and ice information by radio facsimile and other means to all shipping in Canadian waters. In the spring of 1969, AES organized a test of a variety of remote sensing technologies including a *K*-band SLAR mounted in a DC-6 (AES [4]). This study demonstrated that airborne radar could operate in the areas of interest and provide reliable information regardless of light conditions. However, it also highlighted what remains the most significant problem with SLAR systems, namely the unreliable interpretability of the imagery. This is because of the poor radiometric and spatial resolution typical of SLAR systems. At that time this was a particular problem with film products processed on board the aircraft.

The early SLAR displays consisted of a cathode ray tube (CRT) which was used to expose a strip image on a film. This film was processed in a monobath chemical processor and displayed on a small light table. This produced a film with marginally acceptable radiometric performance, but it involved the use of caustic chemicals aboard aircraft. A heat-developed, dry silver photographic product became available in the mid-1970s, in both a film transparency and paper form. This requires no chemicals to be used aboard the aircraft and can produce an image in near real time. Dry silver paper became widely used in real-time displays because it was safe and easy to use, although the resultant imagery had, at best, only 15 usable gray levels.

A later trial of the Goodyear UPD-10 SAR system, in the Gulf of St. Lawrence, demonstrated the superiority of SAR in producing imagery with well-balanced gray scales and fine, range-independent spatial resolution. Unfortunately, only SLAR offered real-time operational capability at the time. To focus the azimuth image requires considerable computation and until the early 1980s, this was done almost exclusively using coherent optical devices which could not be practically operated in aircraft. Therefore, for ice reconnaissance, no real-time processors were available for SAR.

In 1986, AES commissioned a new reconnaissance system into its fleet, a de Havilland Dash-7 equipped with a new generation SLAR system (Ramplee-Smith and Hall [29]). The airframe was modified to provide for extra payload and endurance, thus allowing it to undertake missions in the Arctic. It has a highly automated navigation system to allow operation with a two-person air crew (no navigator). In addition to the SLAR, the Dash-7 has a 9-in. mapping camera, a PRT-5 radiation thermometer, a data-management computer, downlinks for transmission of SLAR data to ships and, via IRDNET, to the AES Ice Centre, and a 10-in. drop chute for launching bathythermographic buoys.

The SLAR on the Dash-7 is a Canadian Astronautics Ltd. (now CAL Corp.) SLAR-200 system. This system, with a peak power of 250 kW, provides a 100-km swath on either side of the aircraft with a relatively fine range resolution of 36 m (0.24-μsec pulse). The antenna is a 4-m (0.45° beamwidth) fixed antenna, and yaw and motion compensation are provided electronically. Data handling aboard the aircraft is accomplished using digital hardware and, although the primary real-time display is dry silver paper, better radiometric performance is achieved than with older SLAR systems. Once recorded on the data-management system, the data can be transmitted using an *S*-band (2.4-GHz) downlink. All major CCG icebreakers are equipped with STAR-VUE display systems (Section 10.2.4) to receive the data, either in real time or via a delayed playback, if the aircraft is out of line-of-sight of the ship when the data are collected. A second channel on the downlink allows data to be communicated to the AES Ice Centre, via the IRDNET system (Section 10.2.4).

DND SLAR Research

In the early 1970s, the Canadian Department of National Defense (DND) undertook an analysis of radar reconnaissance as it applied to the DND mandate. As part of this evaluation, an APS-94D SLAR was purchased and mounted on an ARGUS long-range patrol aircraft, operated by the Maritime Proving and Evaluation Unit (MP&EU) out of Summerside, Prince Edward Island. The radar was evaluated for a variety of military reconnaissance tasks being undertaken by DND during regular reconnaissance flights in the Arctic. The SLAR was often used to support projects of the Earth Science Group at DREO.

DREO used this system with some success for ice reconnaissance studies, particularly in high northern latitudes (Dunbar [10]). Studies were also undertaken in the Beaufort Sea, Eastern Arctic, and Gulf of St. Lawrence areas. Considerable experience was gained in the use of SLAR for ice reconnaissance. Radar scientists at the Communications Research Centre (CRC), Ottawa, also conducted a technology development program based on this system. A modification based on a form of aperture synthesis called Doppler beam sharpening

(DBS) was implemented on the DND SLAR (Barnes et al. [5]). (Using this technique, the azimuth resolution is only a weak function of range. This contrasts with an SLAR, which is a true real-aperture radar, and the azimuth pixel size increases linearly with range.) Whereas this development achieved the aim of improving azimuth resolution markedly, it did not improve the ability to extract ice information from the data, as had been anticipated (Dunbar [8]). This was because the degree of along-track multilooking had fallen from the traditional SLAR value of several hundred looks to the DBS value of a single look. This demonstrated the value of both resolution and image quality (as represented by incoherent averaging) to the ice community.

In 1978, control of this radar system was transferred to AES and it was installed and flown operationally on an Electra with good results (Hengeveld [13]). The SLAR system helped AES to develop an expanded and enhanced all-weather ice reconnaissance system.

Commercial SLAR Capabilities

In 1979, F. G. Bercha and Associates (now The Bercha Group) of Calgary purchased a Motorola APS 94/D SLAR mounted in a Gulfstream G159 aircraft. The system was first used for studies of ice breakup in the U.S. Beaufort Sea in July 1980 and was subsequently used for routine support of drilling operations in the Beaufort, Chukchi, and Bering seas. As is typical with this type of radar, the system produces an analog image of the ice on either side of the flight track, with swath widths of 25, 50, or 100 km per side. In 1984, a digital processor was incorporated into the SLAR system which included a direct digital downlink. The system was used during the mid-1980s off the Canadian East Coast for iceberg and ice-edge mapping in support of drilling in the Hibernia field (Shaw [35]).

In 1989, the radar was installed in a Fairchild Metro aircraft which has endurance of about 5 h, giving the system a productivity of more than 150,000 km^2 of radar imagery per flight. It has flown periodically in support of drilling activities in the Canadian and American Beaufort Sea. As of 1994, it is not installed in an aircraft.

In the early 1980s, a lightweight, digital SLAR was developed by INTERA Environmental Consultants Ltd. to respond to the need for a small, inexpensive radar system that could distinguish ice from open water and downlink the data to the end user. Based on a TERMA (Denmark) marine radar, with a 4.8-m long antenna, the system was the first commercial radar to use an electronic yaw correction system to allow the data to be downlinked in georeferenced format (Inkster et al., [15]). The imagery was simultaneously recorded on a standard VHS video cassette, a 9-track digital computer-compatible tape (CCT) recorder, and on dry silver paper. It operated for several years off the western coast of Alaska, supporting jack-up rigs which had no ice capability; in the Beaufort Sea, assisting seismic ships which need ice-free water; and doing oil-spill monitoring in various areas. As of 1994, it is not installed in an aircraft.

10.2.3 Synthetic Aperture Radar

The SURSAT Program

A turning point in the development of airborne SAR systems in Canada came in 1977 when CCRS received an Unsolicited Proposal to develop an airborne SAR capability. Canadian SAR interests were organized at the time primarily through the Surveillance Satellite Program, known as the SURSAT Program, which had been started in 1976 in response to an interdepartmental study on Satellites and Sovereignty (Government of Canada [12]). The industry proposal was to import an existing experimental radar, install it on the CCRS Convair 580 aircraft, and conduct applications experiments as part of the SURSAT Program. The radar proposed was the ERIM X/L radar, at the time the best available nonmilitary research radar. It was developed by the Environmental Research Institute of Michigan (ERIM) and was a dual-polarized X- and L-band 4-channel SAR (Rawson and Smith [30]). The combination of the CCRS aircraft and the ERIM X/L SAR became known as the SAR-580 system (Inkster et al. [14]). This system played a crucial role in the SURSAT Program.

A prime objective of the SURSAT Program was the development of user requirements for radar remote sensing in Canada. Over the period of the SURSAT Program, the SAR-580 collected data for a wide variety of users, in most parts of Canada. In the ice community, both government and commercial interests became involved in the analysis of SAR data for ice reconnaissance. Experiments were planned and conducted in various places in the Arctic. Specifically, there were experiments in the Beaufort Sea, the Arctic Islands as far north as Alert, in Baffin Bay, and off the coasts of Labrador and Newfoundland. These experiments demonstrated the advantages of SAR's finer spatial resolution and better radiometric resolution compared to SLAR systems then available (Lowry and Miller [22]). In particular, the experiments in the Beaufort Sea demonstrated that an SLAR, even when equipped with improved data recording and processing equipment, such as the NASA Lewis SLAR (Schertler et al. [34]), could not reliably distinguish old ice in a new first-year ice matrix. Figure 10.1 shows an area of first-year and old ice and the ice island T 3 imaged by the NASA Lewis ADP-94D SLAR in March 1989. Figure 10.2 shows a small portion of ice island T 3 and a portion of the pack ice near the ice island, imaged by the SAR-580 four days later. A number of groups studied the data from this and other experiments and reported on the results in the SURSAT Final Report (VanKoughnett et al. [38, 39]).

These experiments led the ice community to accept that SAR data were more useful than SLAR data for support of offshore activities in pack ice. SLAR imagery could not be as reliably interpreted, even by very skilled and experienced interpreters. For nonspecialists such as ship captains, SLAR data were seen to be potentially confusing. The SURSAT experiments demonstrated the value of shorter wavelength radars, with L-band generally containing less useful information than X-band imagery for operational ice reconnaissance. The enhanced contrast between old and first-year ice on cross-polarized channels was also demonstrated. However, it was noted that this did not improve the inter-

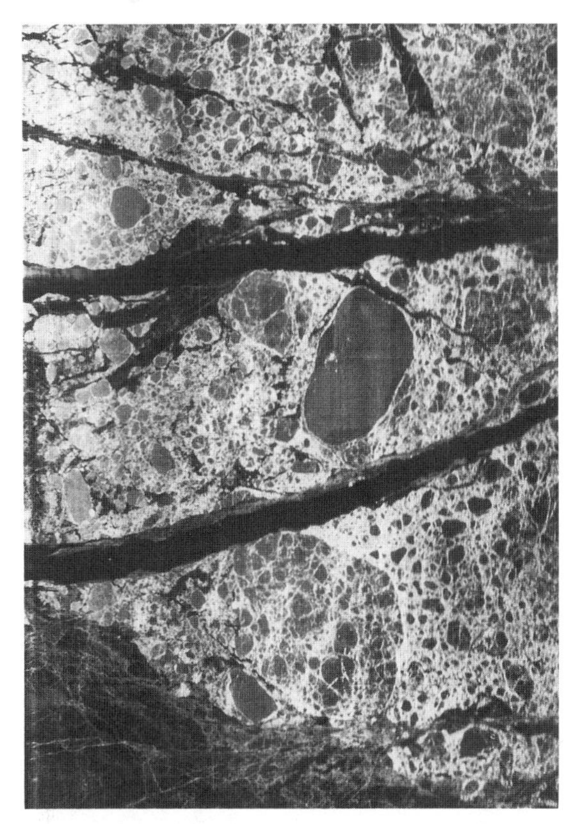

Figure 10.1 Imagery produced by the NASA Lewis APS 94-D SLAR, 12 March 1989. The large kidney near the center of the scene is Ice Island T3. It sits in a mosaic of old ice, between two active leads. The bright spot on T3 is the abandoned camp. T3 was some 22 km in length at the time. (Image courtesy of R. Schertler, NASA Lewis)

pretability of the imagery, as the contrast on the like-polarized *X*-band channel was already quite good at the incidence angles typical of airborne radars.

One of the aspects of the SAR-580 that supported development of sophisticated users was the availability of real-time imagery. By 1978 a real-time digital processor had been developed by MacDonald Dettwiler and Associates Ltd. (MDA) under contract to CCRS, as part of the SURSAT Program. The output image, available only a few seconds after the data had been collected, was printed onto 20-cm wide dry silver paper as a continuous strip image. Although it initially covered only a few kilometers of swatch, it allowed potential users to take the imagery to the field, within a few hours, and to establish by direct observation a correlation between features on the surface and their radar image. Since ice is very dynamic and features can vary rapidly with temperature and wind changes, real-time imagery was a tremendous advantage to the researchers. Because of the availability of real-time imagery, users from several major oil companies involved in offshore development were able to

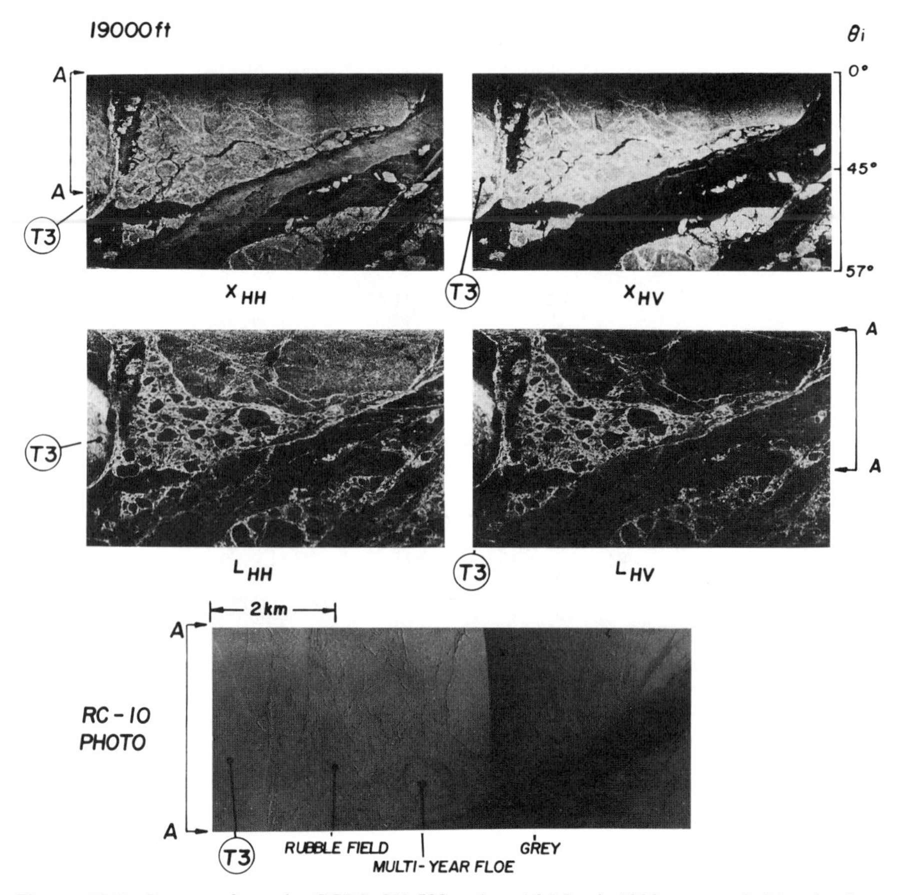

Figure 10.2 Imagery from the CCRS CV-580, taken 16 March 1989 as recorded by the four channels of the radar, *X-HH*, *X-HV*, *L-HH*, and *L-HV*. The edge of ice island T3 can be seen, together with old ice in a mosaic of first year ice. (Imagery courtesy of L. Grey, CCRS)

develop an advanced level of expertise interpreting fine-resolution SAR data. As a direct result of this expertise, several operational demonstration programs were undertaken by the SAR-580 system.

One such program, Operation Ice Map (Mercer et al. [26]) demonstrated that real-time imagery telemetered to a drill ship could be a valuable operational tool. In this test, the SAR-580 flew regular reconnaissance missions over a drilling operation in late-season ice conditions (November 1979). Imagery was produced on the aircraft and was transmitted to the ship via a slow-scan TV system. This demonstrated the utility of SAR data in support of operations in heavy ice and helped extend the operational season. This in turn made the drilling operation more efficient as expensive drill ships could be used for longer periods each year. Further, it showed that timely imagery could be used to find even small leads that could be used to allow ships to transit in almost open-water conditions in spite of nearly continuous ice cover.

The SURSAT Program was one of the most successful user development remote sensing programs ever conducted (Inkster et al. [16]). In particular, the need for quality, reliable, and timely remote sensing services was established with industrial users at a time when offshore exploration was in a very active phase. The enthusiasm generated, combined with the availability of a real-time processor, opened the way for the development of a real-time SAR for ice users. The radar that was developed to fill this need (the first fully commercial SAR ever built) was called STAR-1 (Nichols et al. [28]). The industrial team that put together the system consisted of INTERA Environmental Consultants Ltd. the system integrators and owners, ERIM, MDA, and Millar Communications Ltd.

STAR-1 Development

The first problem confronting the designers of the STAR-1 SAR was to establish an optimum specification for the radar that simultaneously combined fine image quality with low cost, both for operations and capital expenditure. Fortunately, the SURSAT experiments and previously existing knowledge provided the rationale for the choice of frequency and polarization. Since one of the most important tasks of STAR-1 was the mapping of old ice in young and first-year ice, higher frequencies were preferred. *X* band was traditionally the frequency of choice when higher frequency was desired and weather effects were to be minimized. As a result, a great deal of *X*-band equipment is available, and a substantial price premium would have been paid for any other frequency. As the compromises to be made in the design of STAR-1 were the same, that is, higher frequency was considered desirable until weather effects became a problem, the choice of *X* band was straightforward.

On the question of polarization, the choice was less clear. It was easily demonstrated that cross-polarized radars were superior for detecting old ice in a young and first-year ice matrix, but additional transmitter power would have been required to overcome the overall reduction of 10 dB in the signal level with cross-polarized radars. However, whereas the contrast was improved, the contrast between young and first-year ice and old ice was judged acceptable with like-polarized signals, without a 10-dB penalty in overall signal to noise. To keep both the weight and power as low as possible, it was decided to use like-polarized transmission and reception. The choice of horizontal transmit and receive (HH) over vertical transmit and receive (VV) was based as much on finding an off-the-shelf antenna with suitable characteristics as it was on the assumed superiority an HH-polarized radar would have over a VV radar when searching for ice and icebergs in sea clutter.

One of the most difficult design decisions was the trade-off between resolution, numbers of looks, and swath width. The design of the radar was based on a 4096-pixel, real-time SAR processor. This processor was then under development at MDA for the new CCRS SAR-580 *C*-band radar. Therefore, the range resolution and swath width decisions were linked together. Fortunately, at this time the Remote Sensing group at the University of Kansas was studying the effects of resolution and degree of incoherent averaging on the interpretability of imagery (Moore et al. [27]). Their findings demonstrated that the product of range and azimuth resolution, and not one or the other, determined the information content of the image. As a result, it was realized that the range resolution, which determined swath width, given 4096 range cells, and the azimuth resolution had to be seen together. Further, it was found that the relationship between number of looks and resolution meant that, at least for small numbers of looks, it was possible to trade incoherent averaging and resolution without substantial loss of information in the image. This result meant that, within some limits, it was possible to have rather coarse range resolution, and to trade the potential fine azimuthal bandwidth available with a short antenna for incoherent averaging, and still be able to preserve image interpretability (Lowry et al. [23]).

To validate these results and to study their applicability to ice interpretation, a series of tests was conducted with SURSAT data of sea ice. The data were reprocessed in an optical correlator to a variety of resolution/look combinations, and a group of experienced AES ice observers assessed the resulting image samples (Lowry and Hengeveld [21]). Three sets of tests were conducted. The first studied the interpretability of ice imagery with square pixels, processed to one look and with resolutions from 2 m to about 25 m (in both range and azimuth). For a variety of ice interpretation tasks, it was found that no significant loss of information occurred as the resolution was reduced from 2 m to 8 m. At coarser resolutions, there was an apparent loss of information. The results were, of course, somewhat subjective, but the assessments were made by people whose training and experience qualified them to make these judgments. The principle was established, however, that there was a level of detail that was not required to make ice classifications needed for marine operations. Although finer resolution allowed a greater level of detail of ice interpretation, the extra information was not essential. More generally stated, the remote sensing maxim that "finer resolution is always better" does not always apply.

The second test examined the relationship between interpretability and the shape of the pixel. Using imagery with 10×10 m resolution as the reference, a series of images was constructed in which the product of range and azimuth resolution was 100 but the ratio of azimuth to range resolution varied from 50:2 to 2:50. Surprisingly, there did not seem to be a great deal of difference in the amount of information that could be gleaned from the imagery, regardless of the ratio, even if image esthetics were poor with grossly nonsquare pixels. From this was learned that the design of a radar did not need to be based on square pixels. Rather, it was the area of the pixel that determined the usefulness of the image for this application.

The third set of tests examined the question of multilooking, or incoherent averaging versus resolution. In any radar system, the degree of coherent speckle can be reduced by low-pass filtering of the image after the SAR processing has produced a real image from the complex radar video. Such filtering results in coarser resolution but a statistically smoother image. Moore's [27] results indicated that for a variety of remote sensing interpretation tasks, four-look data were more interpretable than data with fewer or more looks. Moore defined a parameter that he called gray-scale resolution that was optimum at four looks. The trials in support of STAR-1 design supported this conclusion. However, it also showed that the loss of information with reduced resolution was nearly offset by improved gray-scale resolution for several more than four looks. The results of these tests were that the overall pixel area should be 64 m^2 or less, that the pixel need not be square, and that the available bandwidth should be used to provide for at least four looks. This set of results was then taken by the STAR-1 design team and considered with a series of hardware and system constraints to produce the design for the STAR-1 radar.

The development of the STAR-1 SAR was a study in compromise between the requirements for low cost, small size, light weight, and high reliability. It was also an excellent example of the productivity of a cooperative relationship between government and industry. The research was conducted and funded at CCRS, combined with the industrial research of the SURSAT program, and coincided with the need for commercial ice reconnaissance services in the Arctic. This created a unique opportunity for STAR-1.

A block diagram of the STAR-1 system is shown in Fig. 10.3. The hardware design of STAR-1 was carried out by ERIM (Nichols et al. [28]), based in large part on the availability of the real-time processor (RTP) developed by MDA for the CCRS radar (Chapter 11). With a weight of 340 kg and a power consumption of less than 5 kW, it was possible to install and operate the radar in a Cessna Conquest. This aircraft has low operating and capital costs typical of small private aircraft, and the reliability and performance of a turboprop that allows the system to fly at 9-km (30,000-ft) altitude at speeds of up to 350 knots for 2000 h or more per year. Table 10.1 lists the primary characteristics of the STAR-1 system. One of the key features of STAR-1 is the low operating cost of the system. By using a Cessna Conquest (Fig. 10.4), marginal operating costs (such as fuel, aircraft maintenance, and so on) were kept very low. As can be seen in Fig. 10.5, the aircraft interior was full and cramped, using the technology available at the time. However, the system was capable of producing high-quality imagery that met the needs of the clients. Figure 10.6 is an example of sea-ice imagery generated by the system in 1983. The system has been upgraded regularly since 1983 and consistently produces imagery of top quality. Modern electronics has been used to further reduce the weight while increasing system capability.

The RTP used for STAR-1 was limited to 4096 range bins, making the trade-off between swath width and resolution rather clear. The other limit on swath width was the system radiometric performance. STAR-1 needed to be able to map first-year sea ice, above the noise threshold, at maximum range.

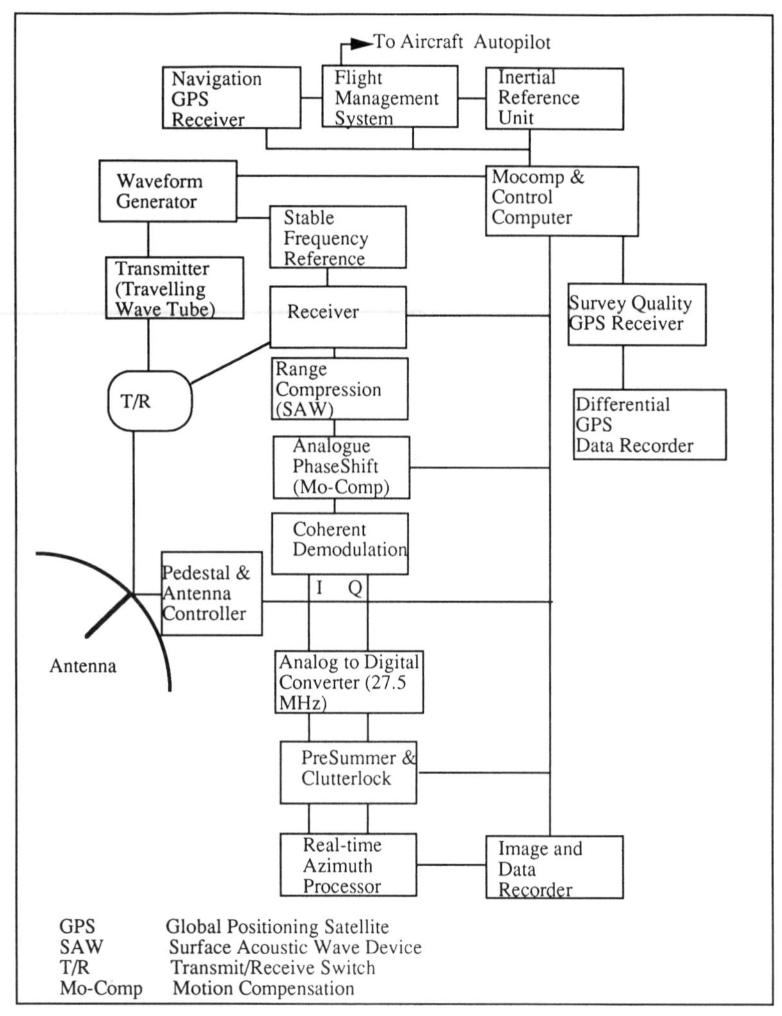

Figure 10.3 A conceptual block diagram of the STAR-1 system showing the key features of the system as it exists at the time of writing. Not shown is a downlink transmitter which is used when the system is actively supporting navigation in ice-infested waters. (Courtesy of Intera Technologies)

Little accurate information was available in 1982 on the reflectivity of first-year ice in the range of incidence angles typical of the far range of an imaging radar (5° to 10°). However, by piecing together data from the NASA Lewis SLAR and the University of Kansas scatterometer system, both of which took part in the SURSAT experiments, it was possible to develop a model of the reflectivity of sea ice in the depression angle range of 5° to 40°, the range of interest for an ice reconnaissance radar. This model was then used to optimize

TABLE 10.1 Typical STAR-1 System Parameters

Parameter	Value
Altitude	29,000 ft (8840 m)
Along-track velocity	280 knots (144 m/sec)
Center frequency	9375 MHz
Range pixel size (4096 pixels)	6 or 12 m
Azimuth pixel size	4.2 m
Swath width	24 or 48 km
Speckle reduction	7 independent "looks"
Maximum far range	70 km
Incidence-angle midswath	73° or 78°
Transmitted pulse length	30 μsec
PRF (scaled to ground speed)	960 Hz (1200 Hz max)
Synthetic aperture length (midswath)	480 or 640 m
Duty factor	4.0%
Peak transmitted power	2 kWatts
Assumed transmitter losses	0.5 dB
Assumed receiver losses	1.0 dB
Assumed radome losses	0.5 dB
Assumed antenna efficiency	1.2 VSWR*
Receiver noise figure	3 dB
Antenna gain	30 dB
Scene dynamic range	$>$40 dB
Minimum detectable signal	<-30 dB σ° noise equivalent at 70-km max. range
Geometric distortion	with GPS, $<$3 pixels plus terrain height effects
Slant-to-ground range conversion	selectable on or off
Peak range sidelobe	−20 dB
Peak azimuth sidelobe	−25 dB
Range to near side (typical)	15 or 20 km
Range to far side (typical)	40 or 70 km
Image formation time	real time(10–20-s delay)
System calibration repeatability	1.2 dB
System weight	450 kg (1000 lb)
Power requirements	6 KVA* (28 V)
Bandwidth of recorded signal	ground speed dependent, max. 250 kbytes/s
Antenna size	1.2 × 0.3 m
Radome size (approx. L, W, H)	3 × 1.5 × 0.6 m
Antenna gimbal requirements	roll, pitch, and yaw synchros from INS*
Antenna polarization	HH (horizontal transmit and receive)

*VSWR = Voltage Standing Wave Ratio
KVA = Kilo Volts Amps
INS = Inertial Navigator Systems

Figure 10.4 The Cessna Conquest aircraft which carries the STAR-1 System. The radome which protects the SAR antenna can be seen directly below the wings, at the center of the aircraft. (Photo courtesy of Intera Technologies)

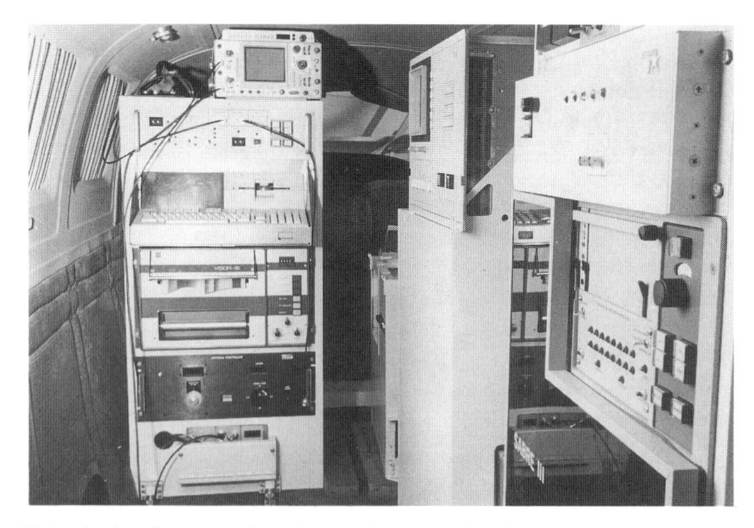

Figure 10.5 An interior view of the Cessna Conquest aircraft showing the STAR-1 operator's console. The system is operated by one technician, who is responsible for the radar and downlink operation, and GPS data acquisition. (Photo courtesy of Intera Technologies)

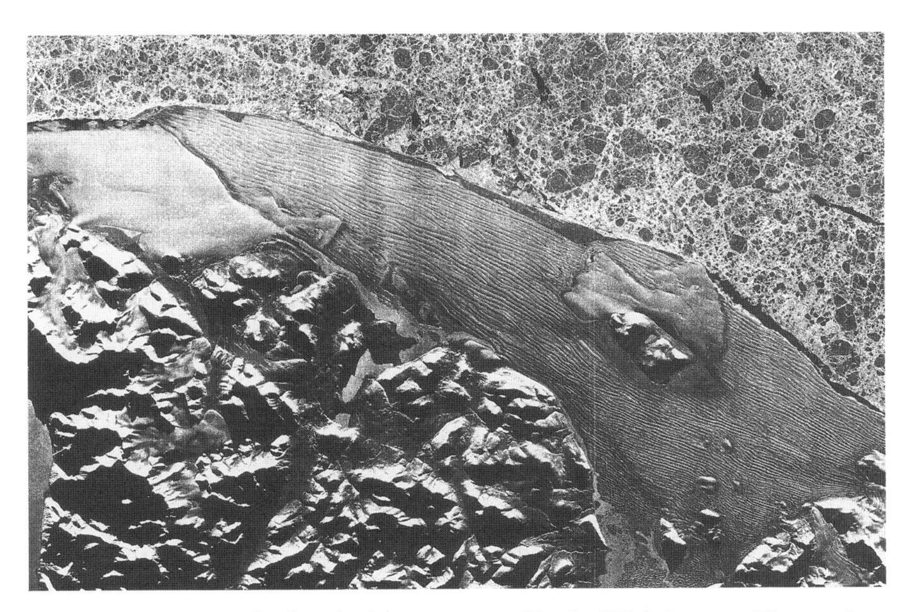

Figure 10.6 An example of sea-ice imagery presented by the STAR-1 system. The area shown is on the north shore of Ellesmere Island including part of the Ward Hunt Ice Shelf. This ice shelf is occasionally the source of ice islands such as ice island T 3 which is shown in Figs. 10.1 and 10.2. (Image courtesy of Intera Technologies)

the design of the STAR-1 system, and to ensure that the radiometric response would provide for a high-quality image over the entire swath. The swath width of the system was established as 50 km, with a minimum range of 65 km. This was based largely on the specification of a −30-dB noise equivalent sensitivity, which was assumed to be the reflectivity of undeformed first-year ice at an incidence angle of 8°.

The RTP was also the primary limit to the azimuthal data rate that could be processed. There was a complex trade-off between the resolution, the number of incoherent looks, and the maximum speed of the aircraft that could be supported. By using a coarser resolution, a greater number of looks could be processed as the size of the azimuth correlation window was reduced. It was not possible to process the data to the full theoretical resolution of 0.5 m (half the length of the antenna) because the line rate would be very high and the correlator size would be too great. The final result was an output line spacing (pixel width) of 4.2 m (42-Hz line rate at maximum velocity), with seven looks per pixel. With 6- or 12-m range pixels, this was well within the criteria set for the system of a pixel area less than 64 m^2 and four looks. (STAR-1 has a choice of 25 or 50 m^2 pixel area with seven looks.)

One of the features of the system that differentiated STAR-1 from all other SARs was its integrated digital downlink. The carrier frequency adopted, 219.5 MHz, had previously been used in the CV-580 program and with SLAR systems. To fit the increased data rate into the bandwidth available, a quadraphase modulation scheme was adopted, resulting in an effective bit rate of 1.4 Mbit/s. In spite of this high rate, an effective bit error rate of less than 10^{-5} was achieved over a range greater than 300 km, at an altitude of 9 km (30,000 ft). This capability made it practical to support transiting vessels and to transmit information to an area ice-management operation, such as the facility located at Tuktoyaktuk, Northwest Territories, where Dome Petroleum and Gulf Canada had their bases of Arctic operations. The effect of downlinks, and later satellite telemetry, on marine operations was very significant and will be described in Section 10.2.4.

10.2.4 Data Delivery and Exploitation Systems

The availability of fine-resolution data on board an aircraft does little to help users on ships navigate through ice-infested waters. To communicate the information on ice conditions rapidly, it is essential to have direct radio link between aircraft and user. From the earliest days of the AES ice reconnaissance program, aircraft were equipped with high-frequency (HF) radio link facsimile systems to allow ice charts to be sent to ships dispersed over wide geographic areas. However, as the radar data quality improved, the demand for the best quality information on the ship grew as well. The development of high-quality digital downlink systems, and the shipborne data-handling systems to exploit these data, was the logical outcome of this demand. The most advanced of this kind is the STAR-VUE system being used by the CCG.

Comprehensive ice charts and forecasts require more information than is available from one observing aircraft. The AES has been combining data available from ice reconnaissance aircraft, from satellite imagery, from ship reports, and from other sources into comprehensive ice information reports and forecasts. To expedite this service, it was necessary to develop a high-speed means of telemetering the data from the aircraft to the Ice Centre in Ottawa. The system is known as IRDNET (for Ice Reconnaissance Data NETwork).

A Canadian Commercial bulk tanker, the MV *Arctic*, is arguably the foremost user of ice navigation information in the world. With a mandate to demonstrate commercial shipping in the far north, this vessel continues to set new records for both early and late voyages into the Arctic. An integral part of the operation is ice information which comes from a variety of sources, including airborne radars. The system developed to exploit this information optimally is known as SINSS (Sea Ice Navigation Support System).

STAR-VUE

During the SURSAT Program and in the early operation of STAR-1 in the Beaufort Sea, considerable experience was gained in the use of direct digital data downlinks. The equipment used had tape recorders and film writers and so was capable of receiving, storing, printing, and reprinting data. The availability of the information in hard copy allowed users to see the general layout of the ice by laying strips of imagery together, and to study an area intensively by magnifying the film in a projector or by creating a photographic enlargement. Several problems with this approach may be circumvented with a soft-copy (TV type) display. First, it is not always convenient or even possible to have access to darkroom facilities, thereby limiting the display to dry silver products. Although these are adequate for some purposes, they result in a considerable loss of information from the high-bandwidth SAR image. Second, the soft-copy technology allows much greater access to the data, to georeference and to differentially reference the data for either a transiting ship or a stationary drilling platform. In addition, with a dedicated work station, a large number of specialized software routines can be developed to meet a specific user's needs.

In 1984, a study was begun to develop an SAR data work station which would display imagery from the STAR-1 system in an optimum and flexible manner. The resulting system, STAR-VUE (Lowry et al. [20]), was the first dedicated radar ice-imagery work station and was developed by Intera Information Technologies (Canada) Ltd. for the Transport Development Centre of Transport Canada and the Canarctic Shipping Company (operators of the MV *Arctic*) in the period 1986 to 1989. It was redesigned in 1989–1991 and tailored to the requirements of the CCG for use on their icebreakers. The system was designed to be user friendly for ship personnel who are experts in navigation and ice interpretation but are not expected to be highly computer literate. It relies on a split-screen concept with an overview and detail window, allowing the user to have both the strategic view normally available from film and the detail available from an enlargement, but with accurate georeferencing applied by the system. Figure 10.7 shows the layout of a typical display on the STAR-VUE screen. Imagery may be viewed in either aircraft flight-path orientation or "north up" modes to facilitate access by the ship's crew. The software also supports a number of navigation functions that allow the crew to track the ship's progress both relative to the ice and in earth coordinates.

The system allows for the reception, storage, and display of data from either the AES SLAR or the STAR-2 SAR (Section 10.2.5). A number of route-planning and course-tracking tools are available to the user. To provide the most flexibility in data interpretation, a hard-copy image printer, using continuous strip, dry silver paper, is available as part of the system and can be used to print either the raw imagery or a variety of other products, including enlargements of the imagery and line drawings from the charting facility.

IRDNET

The problem of transmitting radar data from the aircraft to the AES ICE Centre (ICEC) in Ottawa has been solved effectively with the development of IRDNET. This system is based on dedicated ANIK satellite links leased from TELESAT Canada, and downlinks from the AES Dash-7 and the STAR-2 systems to ground stations at eight locations in the areas of interest for ice reconnaissance (Chapter 12).

(a)

(b)

Figure 10.7 A typical display (a) on the STAR-VUE screen. The screen is divided into two parts: an overview of the entire data file and a detailed display. The overview, shown on the left, displays a portion or all of a flight line at a scale determined by the operator. The operator then selects a portion of the scene to be displayed in greater detail on the right hand portion of the screen, also at a scale determined by the operator. Various specialized software functions (b) are available to support navigating in ice. (Photo courtesy Canarctic Shipping Co. Ltd.)

On board the aircraft, the data are first stored on computer disk and prepared for transmission in "burst" mode to the ground station. The preparation consists of reducing the resolution of the data to 100-m pixels which reduces the total quantity of data to be transmitted and represents a compromise between data quantity and quality. The data are then coded using an absolute moment block truncated code (AMBTC) coding scheme to further reduce the data rate by a factor of 4. Once the aircraft is in direct line-of-sight transmission range of a ground station, the coded and compressed data are transmitted to the ground station over an *S*-band (2.4-GHz) radio link of 700 kbits/s.

There are eight ground stations in the IRDNET system. The location of these stations is shown in Fig. 12.3. It is usually possible, using the "burst" mode communications available with the compressed data, to transmit all the gathered data back to ICEC without the aircraft having to divert from normal reconnaissance or loiter during transmission. Each ground station connects directly with a *C*-band uplink to the ANIK satellite for direct transmission to ICEC. Once there, the data are incorporated into the AES data processing and forecasting system and stored in the ICEC data base. A more complete discussion of the use of these data is given in Chapter 12.

SINSS (Sea-Ice Navigation Support System)

Canarctic Shipping Company Ltd. is the operator of the icebreaking oil-ore-carrying ship, the MV *Arctic*. The company was created in the mid-1970s in response to the development of new mining ventures in the Canadian Arctic, specifically the Nanisivik mine on northern Baffin Island and the Polaris mine on Little Cornwallis Island. There was a recognition by both government and industry at the time that to sustain economic development in the north, experience in the development and operation of a commercial icebreaking cargo ship would be required, together with extension of the navigation season. To achieve this objective, the conceptual design of the MV *Arctic* was formulated and Canarctic Shipping was created to build and operate the vessel. Canarctic is a joint government-industry consortium [with the Federal Government of Canada owning 51% of shares (administered by Transport Canada) and the remaining 49% owned by a number of Canadian shipping companies]. The MV *Arctic* was constructed at the Port Weller Dry Docks in St. Catherine's, Ontario, in 1978 to Arctic Class 2 specifications. The original mission of the vessel was to carry the lead-zinc concentrates from the Nanisivik and Polaris mines to northern Europe. In 1985 and 1986, the vessel was upgraded to an ore, bulk, oil carrier (OBO), thus enabling it to carry petroleum products in addition to bulk cargoes. A new bow and side strengthening were added to the vessel to improve its icebreaking performance. These upgrades were required to lengthen the shipping season to the mines, as well as to enable the MV *Arctic* to transport crude oil from the Panarctic Oils Bent Horn field on Cameron Island, the northernmost producing oil well in the world and the first frontier oil field to be brought into production in Canada. The MV *Arctic* is the only tanker in the world capable of reaching the site because high concentrations of old ice persist throughout the year in the access channels. The operation of the MV *Arctic* has extended the traditional shipping season in the high Arctic from the original 6 to 10 weeks in late summer to almost 24 weeks. The MV *Arctic* holds all the earliest and latest arrival dates to Canadian Arctic ports.

Canarctic has had to develop new approaches to ice navigation. The extension of the Arctic navigation season was not achieved only with improved vessel design. The development and use of remotely sensed data made a critical contribution. The Canarctic approach, known as SINSS, relies heavily on airborne and shipborne radar data (Chapter 8) to provide both strategic and tactical ice-cover information (Canarctic [6]; Sneyd et al. [36]). On board the vessel is a standard HF facsimile system which receives copies of ice charts broadcast by the CCG from data provided by AES. In addition, there is a satellite receiver for visible or infrared data from either the NOAA or *METEOR* satellite systems (Gorman and Sadowsky [11]). Automatic picture transmission (APT) reduced resolution data can be received directly from the satellites, using a VHF stub antenna in place of the larger dish antenna required for the full bandwidth transmissions. The pixel size is, however, degraded to 4 km from 1 km to make this possible. With these two sources of data, the ship's crew can decide well before entering heavy ice areas on their planned voyage what general route to take (for instance, should they choose the north or south side of Baffin Bay to approach Lancaster Sound) and whether airborne SAR data will be necessary. In heavy ice conditions, in the late season, and in the access route to Bent Horn, the ship's operating orders require airborne radar data to support operations.

The MV *Arctic* operation makes extensive use of SAR data for tactical navigation. Before entering an area of heavy ice conditions, the ship will receive, via a downlink, SAR imagery of the ice in the area to be transited. This image forms an "ice map" on which the ship's course is plotted. Using the marine radar to track the forward progress of the ship, the crew track the ship's location very closely on the SAR imagery. The ship's position is plotted relative to the ice map. The absolute position of the SAR image is thus of reduced importance. As the ice pack often maintains a relatively stable pattern as it drifts, this procedure allows the SAR ice map to be of use for a longer period. Depending on the activity in the ice, this period varies from half a day to an entire season.

By using this set of procedures, the MV *Arctic* can avoid much of the increased fuel consumption and wear and tear the ship would normally experience, by routing around heavy ice conditions, sometimes on a floe by floe basis, instead of breaking the ice. It is often possible to sail in effectively open-water conditions, in spite of almost complete ice cover, by finding and following the few leads available.

The SINSS equipment consists of satellite receiver, STAR-VUE downlink receiver, and special marine radar display which facilitates comparison of the marine and SAR radar data. It represents one of the more ambitious and successful applications of radar remote sensing to the problem of shipping in ice-infested waters. Whereas the techniques of the operation were worked out using dedicated STAR-1 flights, the ongoing operation is now supported by the STAR-2 system which is flown under the AES ice reconnaissance operation. Figure 10.8 shows the equipment in use on the MV *Arctic*.

The SINSS equipment is undergoing continuous upgrading and improvement. In addition to the use of airborne radar data for navigation, the conventional marine radar has been extensively modified for ice detection. Following the 1986 demonstration by scientists from the Department of Fisheries and Oceans and McMaster University of the effectiveness of a dual-polarized antenna (H transmit, H and V receive) on board the MV *Arctic*, Canarctic modified the radar on board the vessel. The existing SPERRY RASCAR radar was modified by the addition of a dual-polarized antenna capable of transmitting in HH and receiving in HH and HV. To handle the HV signal channel, a second wave guide and receiver were installed. The performance of the HV signal in detecting glacial and old ice within first-year pack ice is so superior to the HH channel that the officers of the vessel use this channel exclusively when navigating in such conditions. An example of this is shown in Fig. 10.9 which compares the conventional HH-polarized image and the HV- or cross-polarized image. (Chapter 8 deals in much greater depth with the use of marine radars for ice navigation.) Other developments in the SINSS equipment include the upgrades of the SAR video processing and displays to take advantage of the continuing improvements in computer technology and the development of an integrated Arctic navigation system.

Figure 10.8 A photograph showing the SINSS equipment on the bridge of the MV *Arctic*. The data displayed on the monitors are, from right to left, the marine radar display (MARINE-VUE), the SAR display (STAR-VUE), a computer control screen, and weather satellite data (NOAA or METEOR). Hard copy of SAR data can be displayed on the light table directly in front of the ship's officer. (Photo courtesy Canarctic Shipping Co. Ltd.)

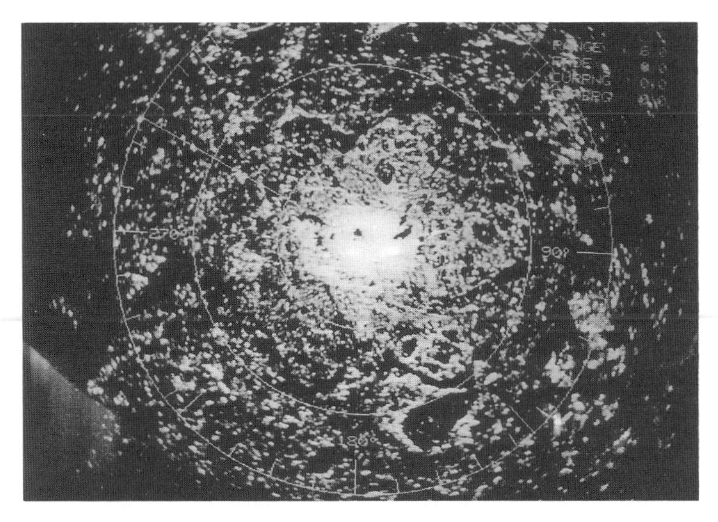

Figure 10.9(a) This image, taken of the MARINE-VUE display screen, shows the scene around a ship in heavy ice, approximately 5/10ths heavy ridged first year ice, and 5/10ths old ice. The radar responds to the ice roughness and it is not possible to distinguish between the more dangerous old ice and the equally bright, rough first-year ice. (Photo courtesy Canarctic Shipping Co. Ltd.)

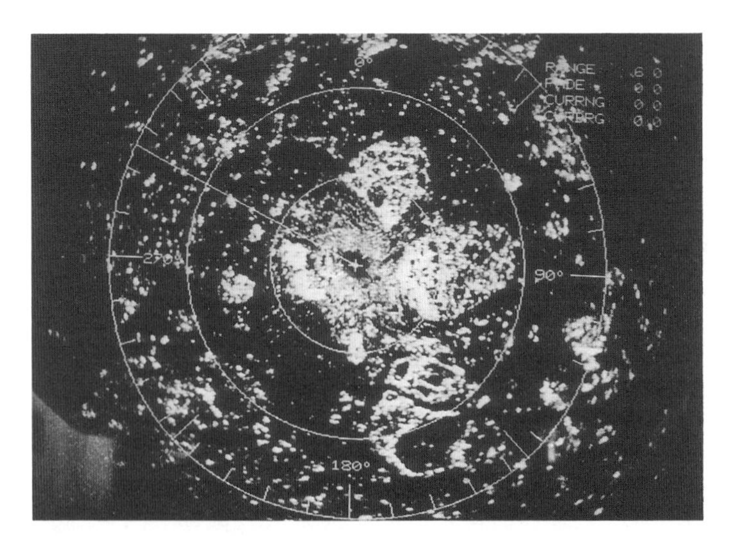

Figure 10.9(b) This image, taken of the MARINE-VUE display screen, shows the same scene but from the HV-polarized channel of the radar. Note that the old-ice forms now are easily detected, in the first-year ice matrix. This makes the HV radar very popular with the ship's officers, as the old ice is very hard, and potentially hazardous, even to the MV *Arctic*. (Photo courtesy Canarctic Shipping Co. Ltd.)

10.2.5 STAR-2: A Fully Operational SAR

In the autumn of 1989, a new level of operational sophistication was achieved when the STAR-2 system went into operation in support of the AES ice reconnaissance activity. It is carried on a Canadair Challenger 600 executive-class jet, which gives both high-altitude and long-range performance. Figure 10.10 is a block diagram of the STAR-2 system. It has two independent SARs looking out on either side of the aircraft, together with highly developed data processing, transmission, and storage capability. It is capable of processing and storing data while downlinking to a ship and of sending data to Ottawa via IRDNET. Unlike previous systems developed by AES, STAR-2 is privately owned and is contracted by AES to provide ice reconnaissance data. Typical parameters for the system are given in Table 10.2.

The radars used in the STAR-2 system are IRIS radars built by MDA (Akem et al. [1]), modified to meet the needs of the AES operation (Mercer, [24]). The SARs have two modes of operation which provide different swath widths and resolutions. The wide swath mode uses 25-m pixel data to achieve a 105-km swath on each side of the aircraft. The narrow swath mode covers 63 km on each side of the aircraft with a 15-m pixel size. Both modes have a large amount of incoherent averaging (22 or more looks), resulting in very good radiometric resolution.

The system is capable of mapping first-year ice (of $\sigma^\circ = -30$ dB) above the system noise out to the far edge of the 105-km swath. This capability is achieved with a 6-kW TWT (traveling wave tube) transmitter, an antenna with a gain of 31 dB (one way), and a low-noise amplifier in the front end of the receiver. It uses mechanical stabilization of the antenna and a Litton LTN 90 Laser Ring Gyro inertial reference system to provide short-term motion data. Navigation equipment includes a dual-flight management system (FMS) with an inertial navigation system (INS), Loran C, and global-positioning system (GPS) input data.

The heart of the image data handling system aboard the aircraft is the data management unit, which stores and formats the data from both radars in real time on disk and later transfers it to 8-mm format tapes. Data are also displayed on fine-resolution graphic monitors for quality assurance. The data are available on disk for downlink transmission to waiting vessels (if needed) at the resolution at which they were collected. The system has a total of 2 Gbytes of disk storage on line. All data are processed to 100-m pixel size for transmission to Ottawa on the IRDNET. The 100-m data for IRDNET transmission is further compressed by absolute moment block truncation coding before transmission to ensure that minimum satellite link time is expended. Both types of transmission are sent framed in SDLC protocol (see Section 12.3.2).

The downlinks used for both transmissions are *S* band (2.4 GHz) at 700 kbps, which use the pulse code modulation with a linear phase modulation (PCM/PM) scheme. Before transmission, the data are Viterbi 1/2 Rate Convolution encoded to improve link margin. Transmissions are circularly polar-

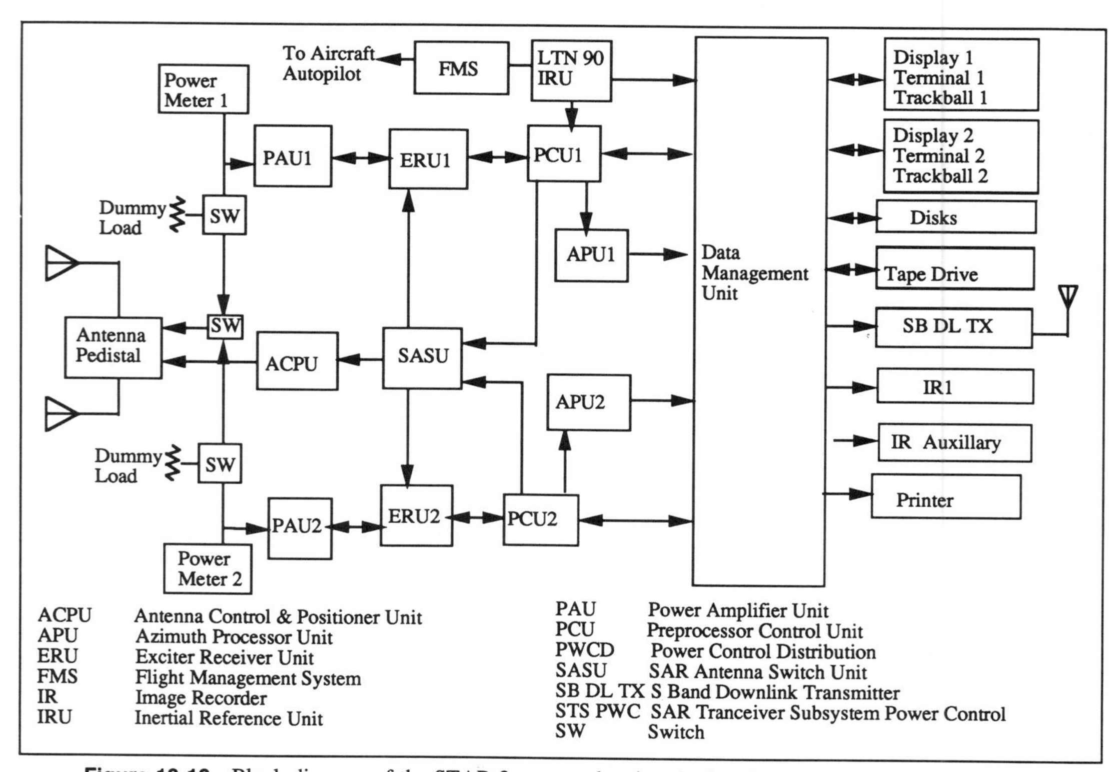

Figure 10.10 Block diagram of the STAR-2 system showing the key features of the system. The two radars are independent except for the shared antenna pedestal, and the Data Management Unit, which records the data from both radars, and produces the various products required by both transiting ships and the Ice Centre in Ottawa. (Courtesy of Intera Technologies)

TABLE 10.2 Typical STAR-2 System Parameters

Parameter	Value
Altitude	35,000 ft (10,600 m)
Along-track velocity	400 knots (205 m/s)
Center frequency	9375 MHz
Range pixel size	15 or 25 m
Azimuth pixel size	15 or 25 m
Speckle reduction	$>$20 looks
Swath width	65 or 100 Km
Incidence-angle midswath	79°/82°
Transmitted pulse length	30 μsec
PRF (scaled to ground speed)	max. 666 Hz @ 450 knots (230 m/s)
Synthetic aperture length (midswath)	900 or 1200 m
Duty factor	2%
Peak transmitted power	6 kWatts
Assumed transmitter losses	0.5 dB
Assumed receiver losses	1.0 dB
Assumed radome losses	0.5 dB
Assumed antenna efficiency	1.2 VSWR
Receiver noise figure	3 dB
Antenna gain	32 dB
Scene dynamic range	$>$40 dB
Minimum detectable signal	<-30 dB σ° noise equivalent at 120-km max. range
Geometric distortion	with GPS, $<$3 pixels plus terrain height effects
Slant-to-ground range conversion	selectable on or off
Peak range sidelobe	−20 dB
Peak azimuth sidelobe	−25 dB
Range to near side (typical)	20 km
Range to far side (typical)	84 or 125 km
Image formation time	real time (10–20-s delay)
System calibration repeatability	1.2 dB
System weight (two radars)	1130 kg (2500 lb)
Power requirements (two radars)	12 KVA (28 V)
Bandwidth of recorded signal	max. 2 × 250 kbytes/s
Antenna size	1.35 × 0.2 m
Radome size (approx. L, W, H)	5.5 × 1.5 × 1.0 m
Antenna gimbal requirements	roll, pitch, and yaw ARINC 429 digital data from LTN 90 INS
Antenna polarization	HH (horizontal transmit and receive)

ized. The overall link margin allows operation to over 400 km in range (line-of-sight), with a bit error rate of better than 10^{-5}.

As part of the STAR-2 system integration into the AES Ice Branch operation, a number of changes were made to the ice information system, most notably the installation of STAR-VUE receiver systems on board 10 CCG icebreakers. These vessels use the most recently available ice information in their role for all support ships in Canadian ice-infested waters. Figure 10.11 shows a sample of the wide swath imagery taken from one side of the STAR-2 system.

10.2.6 Iceberg Reconnaissance

As mentioned in Section 10.1.1, iceberg reconnaissance tends to present a separate class of problems as compared to floating sea-ice reconnaissance. In northern latitudes, icebergs are most common in Baffin Bay and the North Atlantic and off the east coast of Greenland. The source of these icebergs is primarily the Greenland icecap. However, there are numerous other glaciers on the islands in the Arctic which, although not as prolific as Greenland,

Figure 10.11 A sample of the wide-swath SAR imagery from one side of the STAR-2 system. Shown is the Humbolt Glacier on the lower right, part of the coast of Greenland, and first-year ice with small pieces of old ice in the Kane basin. The radar returns from the glacier shows the cloud-like featureless signature typical of the tops of cold glaciers. This is thought to be due to the surface being loose, cold snow, which allows considerable energy to penetrate the glacier, giving returns from a variety of levels within the glacier. (Image courtesy of Intera Technologies)

produce icebergs which pose a problem for shipping and therefore are of interest to operational ice reconnaissance. When included in a sea-ice matrix, icebergs can present a very serious ice hazard. For an airborne ice reconnaissance operation, even relatively small icebergs are detectable when held within sea ice, as they have rather high contrast from young and first-year ice. However, it is when they are floating free in relatively open water and driven by high sea states that they present the most formidable ice reconnaissance problem. Returns from the sea can obscure the returns from the iceberg, making it very difficult to detect. The high sea state can also give the iceberg considerable velocity, making it a more serious threat to a stationary platform or a moving vessel.

For operational reconnaissance, there is an important distinction between the strategic reconnaissance needed to detect the presence or absence of icebergs and the tactical reconnaissance needed to navigate around specific pieces of ice. Small icebergs (5–15 m), bergy bits (1–5 m), and growlers ($\cong 1$ m) are calved from larger icebergs and drift with the currents. However, they erode reasonably rapidly in the sea and thus are seldom found far from larger icebergs (>15 m). This distinction separates the roles of strategic iceberg reconnaissance such as is conducted by the IIP and direct operational support (Section 10-2-1).

In 1984, a series of tests was conducted on a number of operational airborne radar systems. Known as Bergsearch 84, the experiment provided a degree of surface verification of the location and size of some 55 icebergs in the North Atlantic, off the coast of Newfoundland (Rossiter et al. [31]). Two SAR systems (STAR-1 and the CCRS CV-580) and three SLAR systems (the AES, IIP and Bercha APS 94 SLARs) were flown on several different days. In so far as was possible within operational constraints, the systems all imaged the same icebergs under a variety of conditions.

While the results were not totally conclusive, some general observations were possible. The study showed that SLAR was adequate for strategic reconnaissance of the larger icebergs in spite of the poorer spatial and radiometric resolution, SAR systems, however, were able to detect the smaller icebergs in higher sea states. For example, SAR could reliably detect small icebergs in sea state 5 whereas SLAR could not. Neither system performed well enough to provide tactical level information. No radar system tested was able to distinguish between icebergs and ships, a serious problem in iceberg reconnaissance.

One of the limiting factors of the SAR systems tested was the smear associated with the interaction of the sea and iceberg. SAR azimuthal imaging mechanism normally presupposes that the returns from the ground are from stationary scattering centers, and the Doppler associated with a particular points return can be used to place it in the azimuth direction (see Section 6-3). This is clearly not true of sea returns, as the surface is in constant, apparently random, motion and the returns often fill the Doppler spectrum of an SAR. This is emphasized by the interaction between an iceberg and the ocean where the breaking of the waves against the iceberg creates strong returns, with a very wide Doppler spectrum. For large icebergs, this effect is again accentuated by the corner reflector made by the iceberg and the ocean. The vertical motion of the ocean around the iceberg is translated by this into a Doppler modulation, and the magnitude of the return can be quite large. The overall result of these mechanisms is to introduce a smear around the return from an iceberg, which can be as wide as the real aperture of the radar. This makes it difficult to count the number of icebergs present in a confined area and even to detect smaller icebergs in heavy sea conditions.

In the final analysis, SAR was found to have a distinct advantage over SLAR, especially for the smaller icebergs. SLAR systems have a very large resolution cell at far range. This results in a great deal of energy from the sea clutter being integrated into the return. For smaller icebergs, this effectively masks their return, making them very difficult to detect. However, this effect is less pronounced for the larger icebergs, and the relative advantage of SAR over SLAR has not been sufficient for SAR to become the sensor of choice for iceberg reconnaissance, at least up to the time of writing.

Airborne search radars, especially with ground-stabilized displays, have been shown to have much greater ability to detect small icebergs (Ryan [33]). By integrating the returns from each resolution cell on the ground, for times up to several minutes, the sea clutter can be effectively suppressed and small icebergs detected. Coherent radars, with the smaller resolution cell, perform much better than noncoherent radars in this respect, and one such system is in current operation off the Atlantic Coast of Canada (Cantrill and Gordon [7]).

10.3 INTERNATIONAL ICE RECONNAISSANCE CAPABILITY

In the former Soviet Union, ice reconnaissance was deemed to be very important because the Northern Sea Route is a major supply route for the northern ports. It is easier to navigate than the Northwest Passage and its east and west terminals are within the former USSR. As a result, both conventional airborne SLARs and spaceborne reconnaissance systems, based on nuclear-powered satellites with real-aperture sensors, were developed and used extensively. Airborne ice reconnaissance has played a vital role in supporting military and commercial shipping. Visual reconnaissance and the production of ice maps have been undertaken on an increasingly regular basis since World War II. The use of radar in ice reconnaissance began with the development of imaging radars in the 1950s. The first operational imaging SLAR system, TOROS, was put into service in October 1967. This real-aperture, *X*-band system produced an image with 90 × 110 m resolution, mapping 37.5 km to either side of the flight track. Data were recorded on optical film. Four systems were built and flown on regular reconnaissance missions. The imagery was available to the Arctic and Antarctic Institute, Leningrad, for use in the production and distribution of ice charts.

The current operational SLAR system, the "Nit" came into production in

1978, with the primary product being hard-copy strip maps at a variety of scales. The radar operates at *X* band but has better resolution (30 × 50 m) than TOROS. Ten systems were developed: two were installed on AN-24H aircraft and eight were installed on Tupolov TU-134CH aircraft. Since 1991, the Nit system has been equipped with an IBM AT computer, and the variety of products has been expanded to include downlinking of the data to ships and ground stations and electronic ice charts.

In the United States there is little routine airborne ice reconnaissance of floating ice. However, the U.S. Coast Guard operates radar-equipped aircraft for the IIP. The current operation (Alfultis and Osmer [2]) is based on the Motorola AN/APS 135 SLAR known as "SLAMMR." This radar is similar to the older APS/94D but has a peak transmitter power of 200 kW. Two C-130-H aircraft are equipped with the APS 135 radars, with one 16-ft antenna on either side of the fuselage. In addition, the IIP has access to two Falcon HU25B jets, with a single 8-ft antenna and SLAMMR SLARs. This configuration is designated an APS/131. The Falcon systems are equipped with a digital data-handling system but, like the nondigital C-130 operation, the primary analysis continues to rely on the dry silver film. The performance of this radar in detecting icebergs is judged to be very good by the IIP. The main problem with the radar is its inability to distinguish the returns of fishing vessels and those of icebergs.

In the future, the IIP hopes to procure upgraded radar systems. Trials are ongoing on a Texas Instruments radar used by the U.S. Navy in antisubmarine operations in P-3 aircraft. Designated an APS-137, this radar has an SLAR mode and a Spotlight SAR mode. It is hoped that this will help the IIP to discriminate vessels and icebergs.

In the Baltic, the Swedish Meteorological and Hydrological Institute produces ice charts based largely on satellite data. From time to time, they receive data from the Swedish Coast Guard aircraft equipped with an Ericcson SLAR. For the most part, however, this SLAR is used for oil pollution monitoring and not for ice reconnaissance.

The Danish Coast Guard conducts regular ice reconnaissance missions in the waters around Greenland, using a Twin Otter equipped with a small search radar. In addition, the Royal Danish Air Force has a Gulfstream G-3 equipped with a TERMA SLAR for use in ice and iceberg mapping around Greenland. Some consideration has been given to using the SAR under development at the Technical University of Denmark for this task, but at the time of writing, the SAR was not yet available for testing over ice or icebergs.

10.4 SUMMARY

This chapter provides an overview of the development of operational airborne remote sensing of floating ice, with particular attention to those developments which have taken place in Canada. The history is provided to help the reader understand the driving forces that led to the development of the world's only dedicated dual-SAR ice reconnaissance system and to the planned launch of the world's first fully operational SAR satellite, RADARSAT.

ACKNOWLEDGMENTS

The author gratefully acknowledges the input to this Chapter from the following:

R. W. Gorman of Canarctic Shipping Company Ltd., Ottawa
B. Bullock, E. Krakowski, E. Leavitt, and J. Sutton of Intera Technologies, Calgary
C. Ramplee-Smith of AES, Environment Canada, Downsview
F. G. Bercha of The Bercha Group, Calgary

REFERENCES

[1] W. Akem, R. O. Deane, M. Sartori, R. T. Lowry, and J. B. Mercer (1988) "An integrated radar imaging system for the STAR-2 aircraft," in *Proc. IEEE Radar-88 Symp.*

[2] M. A. Alfultis and S. R. Osmer (1988) "International Ice Patrol's Side Looking Radar Experiment (SLAREX) 1988," Report of the International Ice Patrol in the North Atlantic, 1988 Season, Bull. No. 74, Int. Ice Patrol, Groton, CT.

[3] V. H. Anderson (1966) "High-altitude side-looking images of sea ice in the Arctic," in *Proc. 4th Int. Symp. on Remote Sensing of the Environment*, 847–857.

[4] Atmospheric Environment Service (AES) (1969) *ICEMAP 1 Remote Sensing Trials, Spring 1969*, Limited Edition Publication.

[5] D. C. Barnes, G. E. Haslam, and D. J. Newman (1977) "The CRC/DND experimental side-looking radar facility," in *Proc. Remote Sensing Sci. Tech. Symp.*

[6] Canarctic Shipping Company Ltd. (1987) "SINSS winter commercial probe 1986," Transport Canada Rept. No. AMN8103-88.

[7] D. C. Cantrill and P. A. Gordon (1989) "Lightweight airborne radar for maritime surveillance," in *Proc. IGARSS 1989*, 2197–2200.

[8] M. Dunbar (1978) "Interpretation of ice imagery from original and modified versions of a real aperture SLAR," DREO Tech. Rept. No. 770.

[9] M. Dunbar (1973) "Ice regime and ice transport in Nare's Strait," Arctic, **26** (4).

[10] M. Dunbar (1975) "Interpretation of SLAR imagery of sea ice in Nares Strait and the Arctic Ocean," *J. Glaciology* **15** (73), 193–213.

[11] R. W. Gorman and A. Sadowsky (1986) "Operational use of direct receive NOAA and Meteor Satellite APT images for strategic route planning of an Arctic commercial vessel," in *Proc. 10th Canadian Symp. on Remote Sensing*, 775–784.

[12] Government of Canada (1977) ''Satellites and Sovereignty, Report of the Interdepartmental Task Force on surveillance satellites.'' (This report can be obtained through the CCRS Library, RESORS No. 1013374).

[13] H. G. Hengeveld (1980) ''Utilization and benefits of SLAR in operational ice data acquisition,'' *Proc. 6th Canadian Symp. on Remote Sensing*.

[14] D. R. Inkster, R. K. Raney, and R. F. Rawson (1979) ''State-of-the-art in airborne remote sensing radar,'' in *Proc. 13th Int. Symp. on Remote Sensing of the Environment*, 361-381.

[15] D. R. Inkster, B. M. Sorenson, and D. Grant (1983) ''A new digital *X*-band SLAR,'' in *Proc. 17th Int. Symp. on Remote Sensing of the Environment*.

[16] D. R. Inkster, R. T. Lowry, N. A. Prout, R. K. Raney, and K. P. B. Thomson (1980) ''The airborne SAR project—Conclusions and applications,'' in *Proc. 6th Canadian Symp. on Remote Sensing*, 119-126.

[17] INTERA Environmental Consultants Ltd. (1976) ''Transfer plan report: A detailed plan for completion of the transfer of the CCRS airborne production system from government to industry.'' (This report can be obtained through the CCRS Library, RESORS No. 1050755).

[18] J. D. Johnson and L. D. Farmer (1971) ''Use of side-looking airborne radar for sea ice identification,'' *J. Geophys. Res.*, **76** (10) 2138–2155.

[19] T. Kennedy (1988) *QUEST Canada's Search for Arctic Oil*, Reidmore Books, Edmonton, Alberta.

[20] R. T. Lowry, S. D. Thornton, and J. G. McAvoy (1986) ''STAR-VUE tactical ice navigation workstation,'' in *Proc. IGARSS'86*, 1443–1448.

[21] R. T. Lowry and H. G. Hengeveld (1980) ''Optimizing imaging radar parameters for ice reconnaissance,'' in *Proc. 6th Canadian Symp. on Remote Sensing*.

[22] R. T. Lowry and J. Miller (1983) ''Iceberg mapping in Lancaster Sound with synthetic aperture radar,'' in *Proc. 8th Canadian Symp. on Remote Sensing*, 239–246.

[23] R. T. Lowry, G. J. Wessels, and D. R. Inkster (1984) ''Optimizing image quality for SAR,'' in *Proc. IGARSS'84*, 579–594.

[24] J. B. Mercer (1989) ''A new airborne SAR for ice reconnaissance operations,'' in *Proc. IGARSS'89 Symp*.

[25] J. B. Mercer, R. T. Lowry, and E. D. Leavitt (1985) ''An integrated ice monitoring system for offshore drilling,'' *Proc. Arctic '85 Assoc. Civil Engineers*.

[26] J. B. Mercer, R. T. Lowry, and S. K. Leung (1980) ''Experimental use of real time SAR imagery in support of oil exploration in the Beaufort Sea,'' in *Proc. 6th Canadian Symp. on Remote Sensing*, 143–152.

[27] R. K. Moore (1979) ''Tradeoff between picture element dimensions and noncoherent averaging in side-looking airborne radar,'' *IEEE Trans. Aerospace and Electronic Systems*, **AES-15** (5), 697–708.

[28] A. Nichols, J. W. Willhelm, T. W. Gaffield, D. R. Inkster, and S. K. Leung (1986) ''A SAR for real time ice reconnaissance,'' *IEEE Trans. Geosci. and Remote Sens.*, **GE 24** (3).

[29] C. Ramplee-Smith and R. B. Hall (1991) ''The AES Dash-7 ice reconnaissance aircraft,'' in *Proc. WMO Operational Ice Remote Sensing Workshop*.

[30] R. Rawson and F. Smith (1974) ''Four channel simultaneous *X-L* imaging SLAR,'' in *Proc. 9th Int. Symp. on Remote Sensing of the Environment*.

[31] J. R. Rossiter, L. D. Arsenault, E. V. Guy, D. J. Lapp, E. Wedler, B. Mercer, E. McLaren, and J. Dempsey (1985) "Airborne imaging radars for the detection of icebergs," Environmental Studies Revolving Funds, Rept. No. 016. (This report can be obtained through the CCRS Library, RESORS No. 1061044).

[32] P. Rudkin, H. Ripley, and K. Ludlow (1989) "Ice management for the offshore fishery," in *Proc. of IGARSS 1989*, 2201–2202.

[33] J. Ryan (1991) "Iceberg detection using microwave radar" in *Proc. Offshore '91*.

[34] R. J. Schertler, R. A. Mueller, R. J. Jirberg, D. W. Cooper, T. Chase, J. E. Heighway, A. D. Holmes, R. T. Gedney, and H. Mark (1975) "Great Lakes all-weather ice information system," in *Proc. 10th Int. Symp. on Remote Sensing of the Environment*.

[35] V. L. Shaw (1986) "Iceberg detection using SLAR," in *Proc. 10th Canadian Symp. on Remote Sensing*, 767–773.

[36] A. R. Sneyd, M. P. Luce, R. W. Gorman, D. Deer, and D. Miller (1986) "Operational use of remote sensing for commercial Arctic class vessel navigation," in *Proc. 10th Canadian Symp. on Remote Sensing*, 273–280.

[37] A. D. Super and S. R. Osmer (1975) "Remote sensing as it applies to the International Ice Patrol," in *Proc. 10th Int. Symp. on Remote Sensing of the Environment*.

[38] A. L. VanKoughnett, E. J. Langham, and R. K. Raney (1980) "SURSAT Program—Part I—Executive Summary: Final Report." (This document and the full SURSAT Final Report can be obtained through the CCRS Library, RESORS No. 1027378.)

[39] A. L. VanKoughnett, R. K. Raney, and E. J. Langham (1980) "The surveillance satellite program and the future of microwave remote sensing," in *Proc. 6th Canadian Symp. on Remote Sensing*, 9–16.

11

SYNTHETIC APERTURE RADAR IMAGES OF SEA ICE

CHARLES E. LIVINGSTONE
Canada Centre for Remote Sensing
Natural Resources
Ottawa, Ontario, Canada

Synthetic aperture radars (SARs) have been used increasingly as imaging tools to monitor and measure sea ice in the Arctic Ocean. Current and future satellite SAR systems will make this sensor class one of the primary tools for polar ocean surveillance and research in the 1990s. This chapter summarizes research results that show how the radar images present information concerning the ice surface and its physical characteristics, and reviews issues that must be considered by analysts working with SAR images to separate instrument properties from measurements of the target medium. Complementary discussions may be found in Carsey [12] and Shuchman et al. [76].

This chapter is divided into two main parts. The first part examines synthetic aperture radar as an imaging and measurement tool. Sensor properties which can influence the interpretation and analysis of SAR data are identified and concepts introduced in Chapter 6 are expanded to provide further detail. Specific topics discussed are: image geometry; impulse response; image formation; motion compensation; pixels as SAR image samples; speckle and speckle reduction; the radar equation and SAR image radiometry; and radiometric calibration of SAR systems.

The second part of the chapter discusses various examples of SAR imagery of sea ice and relates features observed in the imagery to sensor properties, surface properties, and observation geometry. Like any radar, SAR observes electromagnetic scattering from a reflecting surface. Geophysical properties of

Remote Sensing of Sea Ice and Icebergs, Edited by Simon Haykin, Edward O. Lewis, R. Keith Raney, and James R. Rossiter.
ISBN 0-471-55494-4

the surface must be deduced from these observations, taking into account the properties of the radar system and its imaging geometry. Relationships between sea-ice properties and observations of microwave scattering (made by surface and airborne profiling scatterometers) are discussed in Chapter 2. These are augmented by quantitative measurements made by SAR systems to form the physical basis for the image interpretations presented in this section. Specific topics addressed are: the inhomogeneity scales observed in sea-ice images; SAR observations of sea-ice seasons and marginal ice zones; and the influence of incidence angle, radar frequency, and radar polarization on SAR observations of sea ice.

11.1 SYNTHETIC APERTURE RADAR IMAGES

Chapter 6 provides a review of the principles underlying the radar imaging process for both real aperture and synthetic aperture radars. This section focuses on aspects of SAR that influence the properties and content of SAR imagery. Specific attention has been paid to topics that often are difficult to appreciate by novice users of SAR data.

11.1.1 Image Geometry

All radars measure the strength and time delay (distance) of reflected electromagnetic fields. The natural coordinate system of a radar image is, thus, the "slant range plane." In a slant range image, each sample in the range direction provides a relative measurement (with respect to the first range sample) of the distance between the radar and the imaged terrain element. The range gate delay (the time between pulse transmission and the start of radar reception) of the radar defines the distance between the radar and the first image sample and allows the absolute range of each sample to be computed. The mapping between the illuminated surface and the radar image occurs along constant range arcs centered at the radar. Conceptually, for the simplest airborne SAR case (when the range vector is normal to the SAR velocity), it helps to visualize the surface-to-image transformation as a series of nested, generalized, circular cylinders whose axis lies along the spatial trajectory of the radar. Each surface element can be visualized as being transported along the appropriate cylinder surface to the image "plane." A similar geometric visualization for the more general case, where the range vector is squinted with respect to the SAR trajectory (this includes satellite SARs), can be based on a family of nested, concentric spheres centered at the radar.

It follows that high-relief terrain, when imaged at small incidence angles, may not have a unique image point for each surface element. A single image point can correspond to many points in the scene, Fig. 11.1. Conversely, the effect of terrain relief at large incidence angles is to produce radar shadows (range samples which have no corresponding surface element).

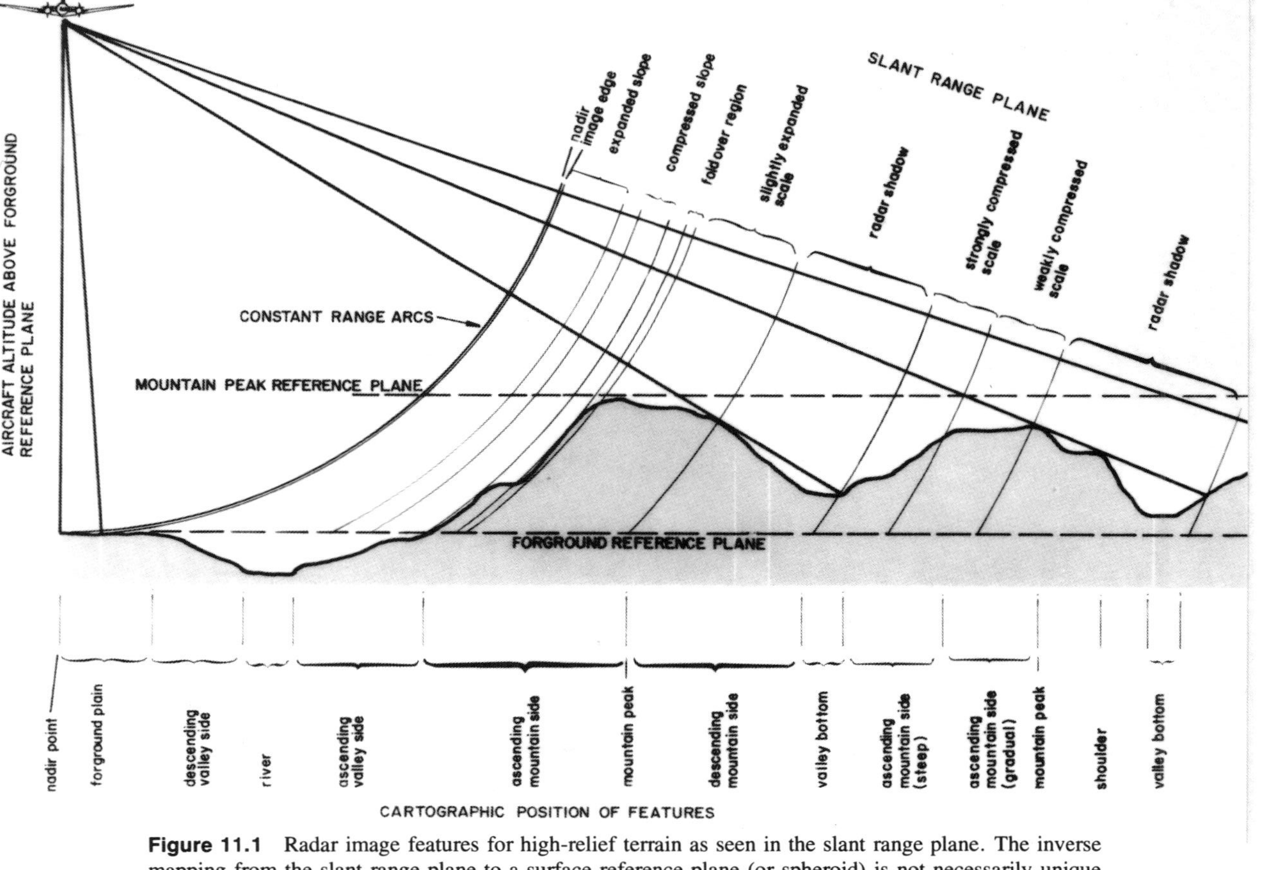

Figure 11.1 Radar image features for high-relief terrain as seen in the slant range plane. The inverse mapping from the slant range plane to a surface reference plane (or spheroid) is not necessarily unique as can be seen for hill slopes facing the radar. An incidence angle dependent projection distortion is always seen in slant range radar imagery.

Fortunately for sea-ice imaging, the surface being imaged lies very close to the local geoid surface and local relief is small. With the exceptions of icebergs and some ice ridges, "many to one" mappings are not found and the spatial range transformation between the image plane and the surface, although not linear, is relatively simple. Image distortions produced by local relief on the ice surface (hummock fields and ridges) are, at worst, both local and small. Airborne SAR systems which image the ice surface at large incidence angles often generate images that contain radar shadows of icebergs and of large deformation features (large ice ridges). For some applications, such as iceberg mapping, these radar shadows are helpful clues for feature identification.

As the surface being mapped does not lie in the natural coordinate system of the radar image, projection from the image to the surface requires a model in which the relative altitudes of the radar and the terrain must be known accurately. For airborne SAR systems, the most accessible measurements of aircraft altitude can contain large errors, and imaging procedures that provide auxiliary measurements, or additional altitude measurement equipment, are required when spatial accuracy is important. For satellite SAR systems, the satellite altitude is generally known to sufficient accuracy; however, the steeper incidence angles imply that the transformation is an essential part of image product generation.

11.1.2 Impulse Response

In Chapter 6, (6-20), (6-26), and (6-27), the impulse response function of the radar is shown to define the smallest resolvable element in a radar image (a point). The half-power contour of the impulse response function defines the Rayleigh resolution cell (the minimum separation of two point features that appear to be separated in the image) of the radar. By extension, all surface elements whose slant range projections are separated by less than a resolution cell width (Fig. 11.2) are fundamentally indistinguishable by the radar. It was further noted in Chapter 6 that the scattered electromagnetic fields which contribute energy to any specific radar resolution cell add in a coherent, vector sense at the radar receiving antenna. The physical geometry of scattering elements at a subresolution cell spacing can alter only the resolved vector components and the phase structure of this sum.

The schematic representation of a resolution cell shown in Fig. 11.2 provides an illustration of these ideas. The "cell" is defined in terms of a range impulse response width, r_{sr}, and an azimuth impulse response width, r_a, at slant range R and along-track position x. Implicit in this definition is the concept that all points along all constant range arcs, which pass through the slant range plane representation of the cell, belong to the resolution cell. The cell is thus unbounded above the nominal surface being imaged. The intersection of the cell with the terrain "surface" is represented by the x,y plane. For most natural materials this surface is more of a mathematical convention than a physical reality, as dielectric discontinuities below the surface (to distances on the order

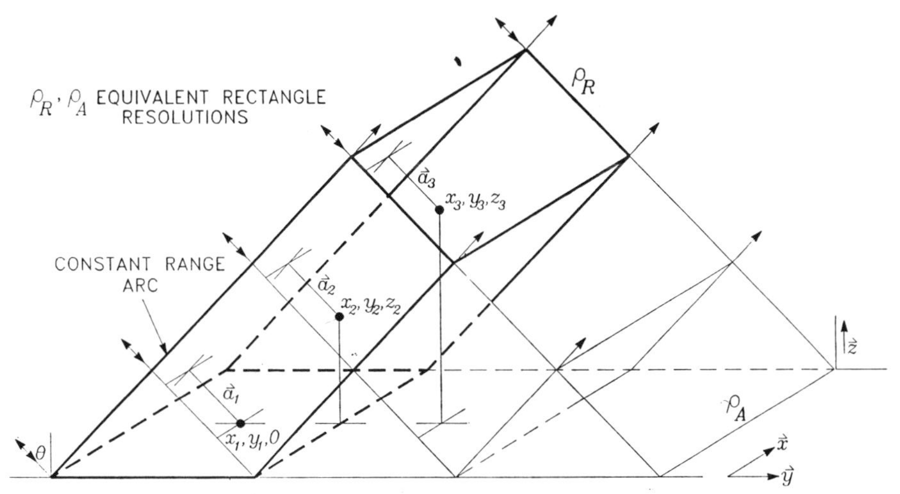

Figure 11.2 Geometry of a SAR resolution cell. The range and azimuth increments are represented by the energy equivalent rectangle resolutions ρ_R and ρ_A. A reference "surface," the x,y plane, is illuminated at incidence angle θ, by a SAR at slant range, R. Three scatterers, a_1, a_2, a_3, and their coordinate plane projections, are shown within the resolution cell volume.

of its skin depth, see Chapter 2) can contribute to the scattered fields. Three scatterers, a_1, a_2, a_3, and their coordinate plane projections are shown within the resolution cell volume. The scattered electric fields from these objects can be considered (from the viewpoint of the radar) to add vectorially on the surface defined by the family of constant range arcs facing the radar. The phase of the scattered field from each object at the summing surface is the phase shift of the backscattered component of the scattered field at the surface of the object plus a propagation path contribution of $4\pi d_j/\lambda$ (for radar wavelength λ). All combinations of scatterers and scatterer positions which produce the same vector sum of scattered fields are equivalent at the output of the SAR. *The signal seen by any radar receiver is the coherent sum of the electromagnetic fields scattered from each illuminated element of the surface.*

For remote sensing radars, the phase-frequency elements of the electromagnetic field summation are modified by each nonlinear element in the radar receiver/processor signal path. In modern instruments, the first nonlinear element is the detector. It is here that the potential output resolution of a radar image is frozen and subsequent signal processing cannot improve upon it. In Chapter 10, it is noted that the area of the radar resolution cell, combined with the variance of the signal, is a significant measure of image interpretability.

As was pointed out in Chapter 6, the minimum widths of the impulse response function of any SAR system (including all signal processing stages) is determined by the range and azimuth bandwidths of the received and processed signals. To first order, this two-dimensional function can be written (Subsection 6.3.6) as the product of a range impulse response function and an azimuth impulse response function.

The range impulse function is determined by the form and structure of the transmitted radar pulse and by the processing applied to the received signal to produce a range-focused image. Most SARs transmit an internally encoded long pulse, or chirp, to minimize peak radiated power requirements. Such pulses require range pulse compression in the image formation process (Subsection 6.2.7). The Russian *ALMAZ* SAR actually transmitted a short, uncoded, range pulse (W. McCandless, personal communication) and applied no postreception range processing. In the SAR realizations used to build most existing equipment, digital signal processing technology is employed extensively in data acquisition and in image processing. Some Russian and Chinese airborne SARs still use analogue technology for optical recording and image formation. In the digital radars, a quantized form of the range impulse function results either from the application of a pulse compression (range focusing) algorithm to the digitized received signals, or from digitized, compressed (focused) signals from an analogue pulse compression network. The Nyquist theorem constrains the digitizer sampling rate to be no less than the bandwidth of the transmitted pulse. Often in the pulse compression process some of the signal bandwidth is sacrificed to minimize pulse compression sidelobes, and thus to minimize the integrated sidelobe ratio of the radar. The resulting impulse response width in the range direction is somewhat larger than would be predicted based on a knowledge of its bandwidth alone.

The azimuth components of the SAR signal are sampled by the transmitted pulse train, as is indicated in (6-31), and thus are subject to the constraints of the Nyquist sampling theorem. In existing radars, the azimuth beam shape, defined by the SAR antenna, forms a spatial filter which weights the azimuth (Doppler) spectrum of the radar returns from the earth's surface. The ratio of the pulse repetition frequency to the SAR velocity (with respect to the terrain) determines the spectral sampling rate, which is chosen usually to place the Nyquist folding frequency outside of the -6 dB point of the spectral envelope (defined by the -3 dB point of the one-way azimuth antenna pattern). The relationship between the shape of the spectral envelope and the Nyquist frequency determines the fraction of the received energy which is aliased into the central portion of the spectrum by the sampling process. The azimuth ambiguity content of the radar image is determined by the fraction of the spectral bandwidth (and thus the fraction of the aliased signal) which is processed to form the SAR image. Figure 11.3 provides a schematic view of the relationships between these quantities.

For both the range and azimuth signals, the Nyquist sampling theorem defines the lower bounds of the sampling rates used in any radar realization. Provided that these bounds are respected in the design of the equipment, the range and azimuth sampling rates implemented in a specific radar are engineering choices governed by other system design constraints. They are not simply related to the widths of the range and azimuth impulse response functions.

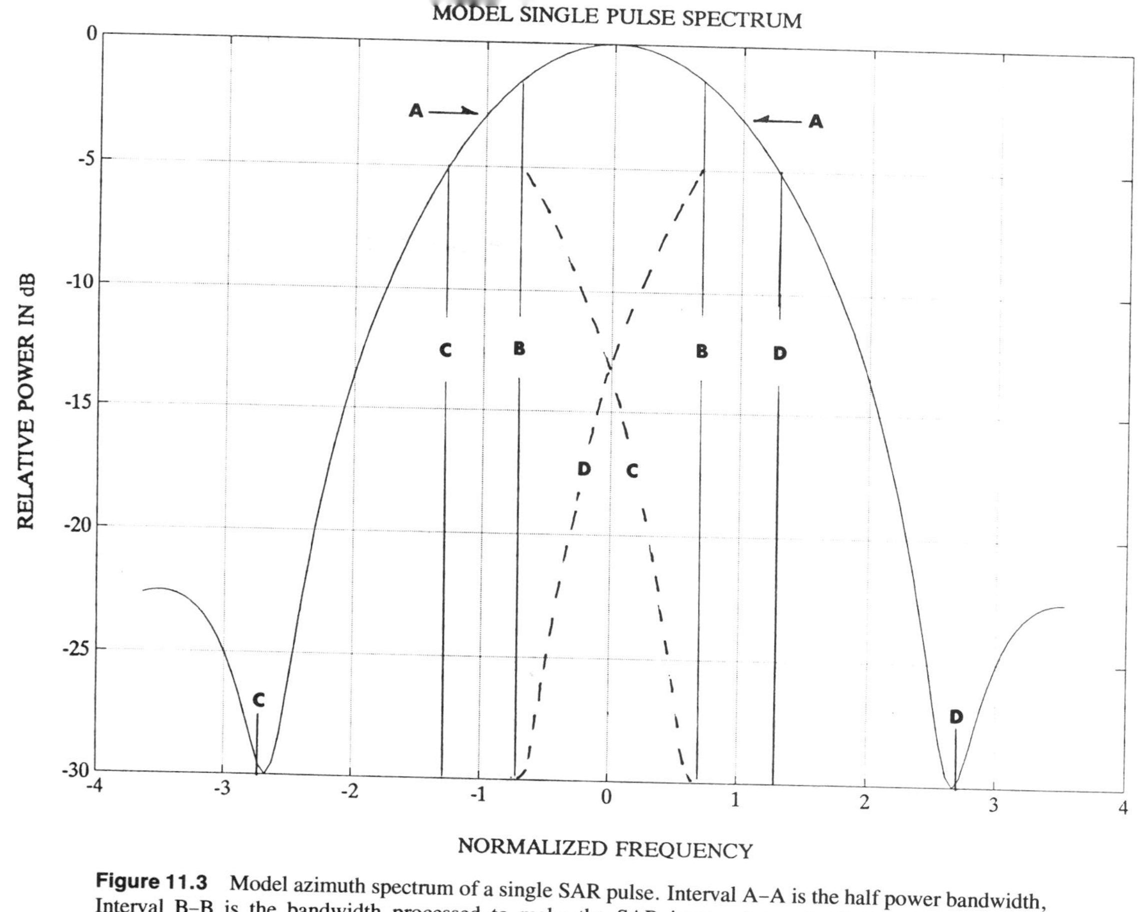

Figure 11.3 Model azimuth spectrum of a single SAR pulse. Interval A–A is the half power bandwidth, Interval B–B is the bandwidth processed to make the SAR image. Intervals C–C and D–D contain ambiguous image data when the normalized sampling rate is 2. For the model case the azimuth ambiguity ratio will be approximately -12 dB.

11.1.3 SAR Image Formation

Conceptual "Signal" Geometries

The SAR azimuth focusing process (6-26) follows principles similar to those of the range focusing process discussed in Subsection 6.2.7. In the azimuth case, however, the phase coding of the received signal is not applied by the radar transmitter. It is a natural result of the relative motion of the radar and the imaged terrain. The constant range surfaces of a radar are members of a family of spheres centered at the radar. From the viewpoint of the sampled, range-compressed receiver output, these surfaces are separated in range by one-half of the range sample spacing (times the speed of light). As seen on the surface of a smooth earth, the contours of constant range form a family of concentric circles (constant range arcs) centered at the radar's nadir point. When earth rotation effects are unimportant (airborne SAR), or have been compensated for (*ERS*-1 SAR), the contours of constant Doppler frequency (rate of change of phase) for any single radar pulse are the intersections of the earth's surface with a family of cones whose axis is aligned along the radar antenna's trajectory.

For the simple case of an airborne SAR, the earth's surface can be modeled locally as a plane and the constant Doppler shift contours can be approximated as a family of hyperbolas, which degenerate to straight lines (zero Doppler shift) along the normal to the trajectory. The isorange/isoDoppler contours for each transmitted pulse, Fig. 11.4, define a coordinate system in observation space which cannot, in general, be mapped into mathematically separable (uncoupled) coordinates in a rectangular (range, azimuth) signal space. This leads to range/azimuth coupling terms (range cell migration) in the SAR signal data. Modern SAR processors fully compensate for these effects during the process of image formation.

Phase History

When data are accumulated to form a SAR image, a point on the earth's surface that is imaged by the radar can be considered to pass through the radar antenna's illumination pattern, along a trajectory that is parallel to the ground projection of the radar's path. In the ideal case of a straight flight line, the range to the point from the radar varies hyperbolically in time, according to the relationship

$$R(t) = (R_0^2 + V^2(t - t_0)^2)^{1/2}$$

where

R_0 = the range of closest approach of the radar to the point
V = the radar velocity
t_0 = the time of closest approach
$V((t - t_0)$ = the along-track distance of the radar from the point of closest approach

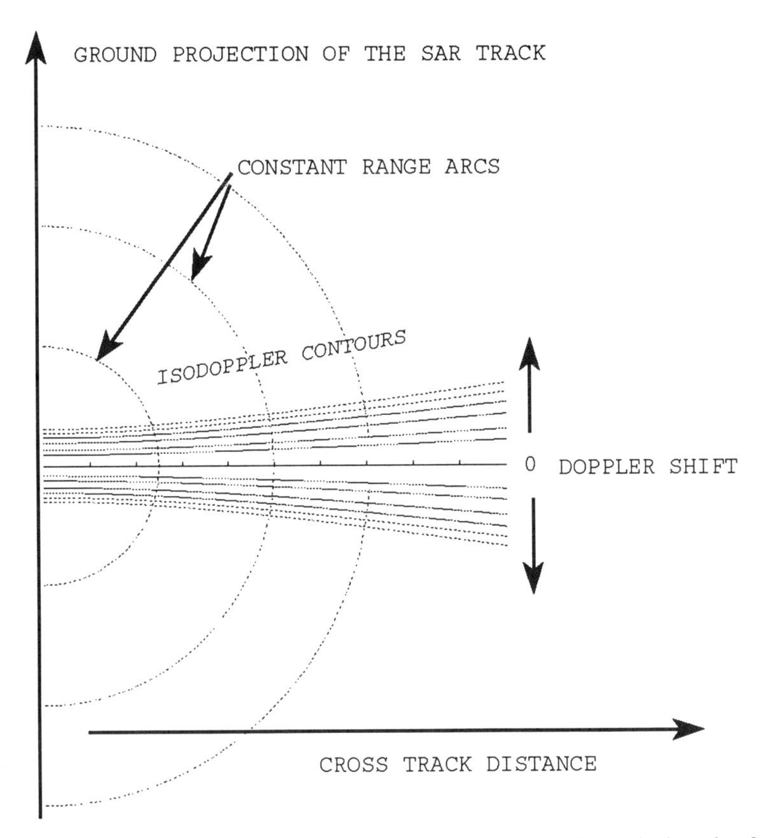

Figure 11.4 Isorange/isoDoppler shift contours on a smooth earth. A single pulse from an airborne SAR can be described on the surface of a smooth, plane earth in terms of a family of constant range arcs which intersects a family of isoDoppler hyperbolas. The figure provides a schematic view of the resulting range-Doppler coordinate system.

The phase, φ, of the signal returned to the radar is related to the range, R, by

$$\varphi(t) = 4\pi R(t)/\lambda$$

where λ is the radar wavelength. The azimuth illumination pattern of the antenna, $H(\phi - \phi_0)$, selects the SAR data segment that contains sufficient radar return energy to form an image and, thus, the portion of the target range hyperbola that can be processed usefully. A schematic view of the illumination and processing geometry is shown in Fig. 11.5. For antennas used in radar remote sensing, the illuminated segment of the trajectory of the point, $D = V(t_{max} - t_{min})$, is much smaller than R_0. Hence, the phase of the signal can be expanded about the center of illumination using a classic power series. The standard parabolic expression for the target phase history emerges as the qua-

Airborne SAR With Small Squint

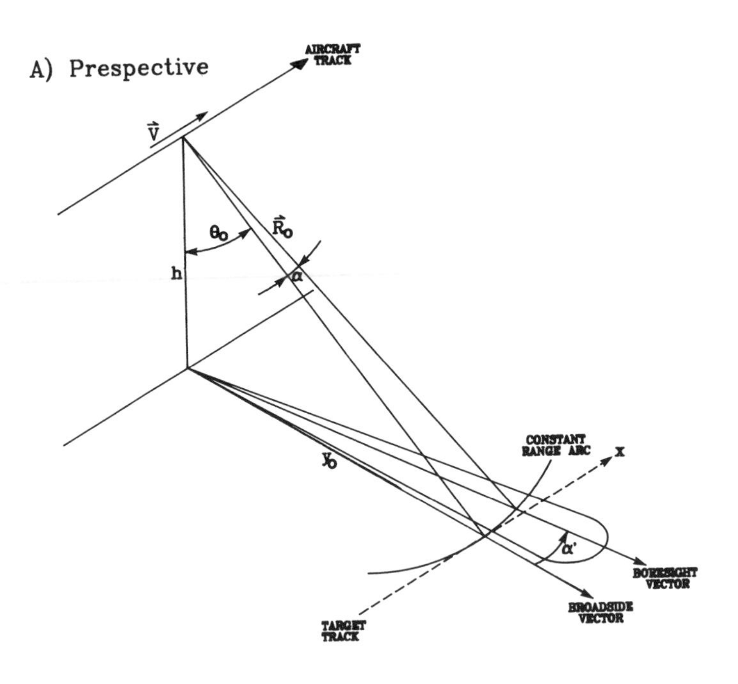

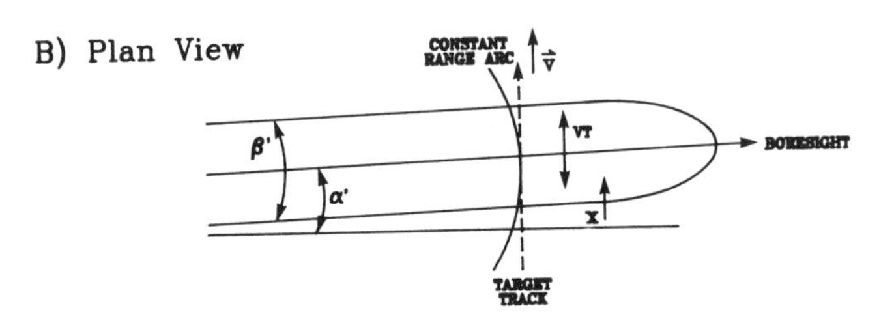

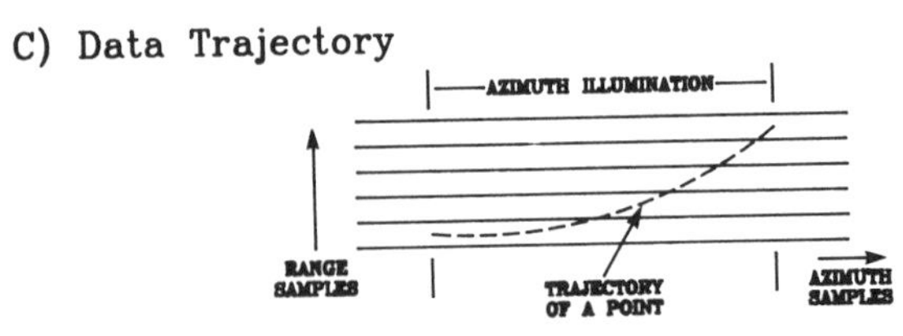

Figure 11.5 Illumination and processing geometry for an airborne SAR with a small squint, (A) in perspective and (B) in plan view. (C) is a schematic view of the range-Doppler space as seen by a SAR processor.

dratic term in time:

$$\varphi(t) = \frac{4\pi}{\lambda} R_0(1 + \frac{V^2(t - t_0)^2}{2R_0^2} - \frac{V^4(t - t_0)^4}{8R_0^4} + \frac{V^6(t - t_0)^6}{48R_0^6} - \ldots \quad (11\text{-}1)$$

Equation (11-1) also can be used to define approximations to the point trajectory in signal space. The primary task of an SAR processor is to match the phase history of the received radar signals and by so doing, to focus the signals in azimuth. The phase matching is a function of target range, R_0, and the velocity of the radar platform over the imaged terrain.

SAR image formation also must compensate for the signal range change for each surface element over the processed portion of the real aperture of the radar. These are known as range walk and range curvature corrections. Although there are many alternatives, two different formalisms, range-Doppler and wave equation, are the most popular for SAR processor design.

In the range-Doppler class of processors, the SAR data range and azimuth coordinates are treated as being "almost separable." High-resolution data are processed by using a model of the data acquisition process to track the azimuth returns of range-processed target elements in signal space (memory) and to map these into data sequences for processing in the azimuth focusing filter (Bennett and Cumming [6]; Brusmark et al. [7], Curlander and McDonough [17]; Smith [77]).

In variations of the "downward continuation" geophysical algorithms (wave equation algorithms) (Rocca [72]; Cafforio et al. [9]; Raney and Vachon [69]), the processor algorithm is tailored to the two-dimensional, frequency domain, signal-space structure of the data to provide a better physical link between the data acquisition and the data processing. The fundamental equivalence between, and the strengths and weaknesses of, the range-Doppler and the wave equation formulations of the SAR focusing algorithms are discussed by Scheuer and Wong [73] and by Bamler [3].

In any case, the impulse response of the radar system, as it is imbedded in any SAR image, is influenced by the degree of accuracy with which the data acquisition model matches the complex structure of the data. Mismatches between the model and reality result in misfocused data and, thus, a broadening of the impulse response (usually in the azimuth direction). At worst, this class of model mismatch causes the radar image resolution to vary slowly over an image. Even when there is misfocus, however, SAR provides a robust estimate of the mean reflectivity of the scene for distributed scatterers such as sea ice.

11.1.4 Motion Effects and SAR Trajectory Errors

Airborne SARs with good motion compensation routinely produce images in which the theoretical range and azimuth resolutions are attained under most flight conditions. SAR signal models usually assume an idealized radar trajectory to define the processing functions. Radars do not fly along idealized trajectories but rather along paths dictated by their operating environments. Spaceborne SARs are constrained to the detailed geometry of their orbits; furthermore, SAR image formation is influenced by earth rotation. The illuminated portion of the earth's surface is determined by the spacecraft attitude control and by antenna beam-pointing controls. Airborne SARs operate at the mercy of atmospheric turbulence and the limitations of physical navigation systems.

In both cases, the actual radar trajectories must be estimated to remove vehicle motion artifacts from the data, either before or during image formation. Vehicle trajectory and attitude compensations that need to be executed in "real time" during data acquisition include: radar sample rate (pulse repetition frequency) corrections for along-track velocity changes, radar beam steering, and receiver range gate delay adjustments.

Airborne SAR Motion Compensation

The elimination of motion effects for airborne SARs is difficult to achieve because variations in the platform motions are dynamic and unpredictable. The data phase compensation used to correct for nonideal motion is called motion compensation (MC). It is implemented by computing the deviations of the radar from its ideal trajectory, using measured motion data. A time-history of the error vectors in radar position forms a trajectory error model, which is combined with a range/illumination angle model to estimate radar range errors to each imaged point. Range error compensations, $\Delta R(t)$, are applied to the radar azimuth data as range-dependent phase rotations, $4\pi\Delta R(t)$, prior to image formation.

These corrections can be made either-in-flight, based on real-time motion measurements, or with a ground-based signal processor using recorded motion and position data. For example, the airborne SARs operated by Canada Centre for Remote Sensing (CCRS) allow either real-time or postflight MC to be selected (Livingstone et al. [51]). The airborne SARs operated by the Jet Propulsion Laboratory (JPL) use only postflight MC (Freeman et al. [26]; Held et al. [31]).

Mismatches, between the MC models used to correct the data and the phase modulation of the data by the motion errors, result in a correspondence processor mismatch to the received signal. The signal/processor mismatches cause phase errors that broaden the radar impulse response locally in the azimuth direction, as shown in Fig. 11.6. The most obvious signature of serious MC errors in a detected image are defocused zones in the image that increase in width as the synthetic aperture length of the radar grows with increasing range. An example of this artifact is shown in Fig. 11.7. Both radiometric and geometric artifacts can be created by MC errors.

Real Time Versus Postflight Motion Compensation

Real-time MC relies on motion data measured in flight by some combination of inertial navigation systems, Global Positioning Satellite (GPS) receivers,

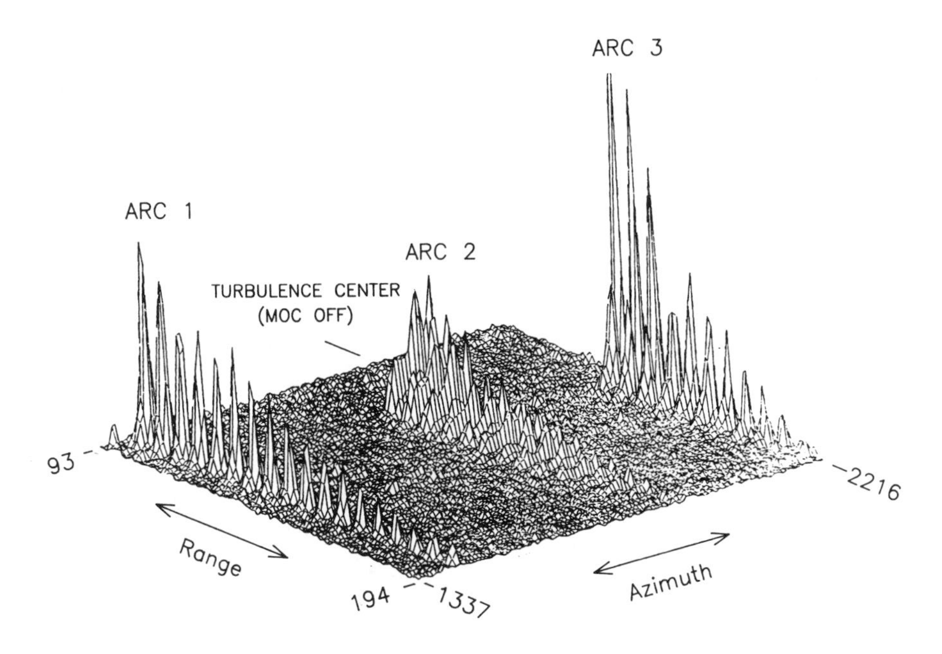

ARC CROSS SECTION - $\sigma_1 = \sigma_2 = 33.3\text{dB}$; $\sigma_3 = 36.3\text{dB}$
PEAK RESPONSE ARCS-1&2 20 Log $(C_{p1}/C_{p2}) = 6.4\text{dB}$
INTEGRAL RESPONSE ARCS-1&2 20 Log $(C_{i1}/C_{i2}) = -0.4\text{dB}$
PEAK RESPONSE ARCS 2&3 - 20 Log $(C_{p3}/C_{p2}) = 9.8\text{dB}$

Figure 11.6 The impact of motion compensation error on Active Radar Calibrator (ARC) responses in a SAR image. The echos from ARC 2 in a group of three recirculating ARCs have been defocused by uncompensated SAR motion. The impulse response of ARC 2 has been broadened and the intensity of the return has been reduced. Arcs 1 and 2 have identical characteristics.

radio navigation beacons, and autofocus algorithms. Its accuracy is constrained by the chaotic nature of aircraft motion, by the accuracy/precision limitations of the motion sensors, by the time required to make motion measurements, and by the time required for motion data processing. At the radar resolutions used in remote sensing, real-time motion compensation is adequate for image formation under most conditions and allows the radar to produce large volumes of good-quality processed imagery.

In principle, postflight MC is more precise than real-time MC because data from multiple sensors (inertial navigation systems, radio location devices, and GPS receivers) can be integrated as a priori known random variables. Postflight MC is valuable for processing images which have very high azimuth resolution requirements and for processing interferometric SAR images when the radar data were acquired under turbulent flight conditions. Postflight motion compensation imposes an additional computational burden on the image-processing

Figure 11.7 Single-look, airborne SAR image showing artifacts produced by uncompensated SAR motion.

facility and, at present, is not suited for the routine production of large volumes of SAR image data.

11.1.5 Digital Images: The Pixel

In digitally generated imagery from any source, the smallest observable element in the image (1 pixel) is displayed as a rectangle which has a single intensity value or brightness. The pixel area for some image types (optical imagery from many sensors) can be mapped onto the imaged terrain as a meaningful "information integration" area. This relationship is not valid for radar data.

Behind the visual representation, a digital radar image is a large matrix, an array of numbers, which represent samples of a space. The separation between adjacent numbers has the image world connotation of a pixel spacing and is frequently extended to the physical world by means of the image scale factor. However, the pixel spacing is determined by the sampling process used to create the data array and has no other implicit meaning.

The impulse response function of the radar defines the minimum scale size (radar resolution cell) that can be associated with the spatial distribution of information in a focused SAR image. Invoking the Nyquist theorem, it is desirable that there are at least two samples (pixels) per minimum impulse-response width in each of the range and azimuth dimensions to prevent aliasing of spatial information in the image data. A larger number of samples continues to preserve the spatial content but does not add information. The correlation between the numeric values of adjacent samples is determined by the impulse-response function, not by the pixel spacing. Thus, spatial correlations between adjacent image elements provide more information about the radar than about the scene (Raney and Wessels [71]). If a SAR image is subsampled to below the Nyquist limit, without first applying an appropriate low-pass spatial filter, aliasing will occur.

11.1.6 Speckle and Speckle Reduction

Origins of Speckle

In Chapter 6, the phenomenon known as speckle was introduced as an essential by-product of image formation with a coherent radar system. The origins of speckle are understood easily by considering a uniform, ideal, distributed target. This class of surface has a constant, mean-scattering cross section. It is composed of a random distribution of a large number of scattering elements, whose radar reflectivities are distributed statistically in a Gaussian manner and whose phase shifts on reflection are distributed uniformly. For each point, the vector addition of scattered fields seen at the radar, has a maximum power if all scattered contributions add in-phase and has a minimum power equal to zero if the complex sum of all fields is 0. The image power, seen as the resolution cell level, is distributed exponentially for detected imagery when no averaging has been used. For such data, the most probable value for the power (per pixel) is zero, although the average value expected is proportional to the average reflectivity of the scene.

The foregoing conceptual definition of speckle can be extended to include scenes which are not uniform but whose elements, seen at the scale of the radar resolution cell, contain a large number of scatterers (Oliver [65]). In this case, the radar cross section of the imaged area is not constant but varies spatially. The focused SAR scene, thus, has a fundamental radar texture associated with the radar cross-section variations of the surface and may contain recognizable geometric features (clear cuts, field boundaries, roads, etc.). Scatterers contributing to each resolution cell are, however, both numerous and independent so that the phases of their scattered fields are distributed uniformly. The SAR image contains fully developed speckle as a fundamental component of each target area of this type.

In a single-look, detected SAR image, the smallest scale of image intensity variation (finest measurable image texture) is dominated statistically by the radar speckle. This effect occurs at the spatial scale corresponding to the image dimensions of the radar resolution cell (impulse response).

Coherent Target Effects

Coherent targets are natural scene elements (or constructed devices) which return signals to the radar from a single region without randomizing the signal phase. As the phase structure of the scattered signals is deterministic and does not (to first order) contain electromagnetic interference terms, this target class does not produce radar speckle when imaged by a SAR. Coherent targets whose scattering matrices are well known, such as corner reflectors, active radar calibrators, Luneberg lenses, metallic plates, or metallic spheres, often are employed as calibration references for radar systems.

When a coherent target is imaged by SAR, the radar returns from the target are embedded in the radar returns from the background terrain, which is imaged into the same resolution cells. Although the coherent target may dominate the received signal, the vector sum of the scattered fields from the coherent target and background terrain contains random phase elements which contribute a fading noise to the target image. The ratio between the peak of a radar impulse response representing the coherent target alone and the corresponding measure of the background radar return is called the signal-to-clutter ratio, SCR. When the coherent target is employed as a SAR calibration reference and all signals are above the noise floor of the radar, the SCR determines the maximum calibration accuracy obtainable from analysis of the target image (Gray et al. [27]; Ulander [79]). The SCR for a specific target and radar combination can be maximized by processing the radar data to create the highest resolution, single-look image that can be supported by that radar.

Simple Speckle-Reduction Strategies

In cases where the features of interest in a SAR image have dimensions that are significantly greater than the finest available SAR resolution, the speckle content of the image can be treated as spatial noise. In doing this we assume implicitly that the image spatial information content, at the scale of an individual resolution cell, is small and that very little information is lost by trading image resolution for speckle reduction by *incoherently averaging* the image data in an appropriate manner. Two simple strategies for signal energy averaging are commonly used: image domain averaging and frequency domain partitioning and averaging.

The simplest approach to incoherent averaging is to process the SAR image data to full resolution, detect the image to produce power estimates, and then sum blocks of adjacent pixels with appropriate spatial weighting (Gaussian weighting is often used). The reduction in the speckle contribution to the averaged (reduced resolution) image can be expressed in terms of the ratio between the squared mean and the variance of a segment of the image which contains little, or no, structure that can be attributed to radar cross-section variation over the imaged surface. The value of the ratio is known as the effective number of "looks," N_{eff}, of the averaged image. From the discussion in Subsection 11.1.5, it must be noted that N_{eff} is not necessarily the number of pixels, N_{pix}, that are summed. Usually $N_{\text{eff}} < N_{\text{pix}}$. N_{eff} is the number of statistically independent subimages used to form the final radar scene.

A more complex approach to incoherent averaging requires that the signal spectrum be filtered into frequency bands in azimuth and/or range (Fig. 11.8). Each band is processed to a single-look subimage, and each of the single-look

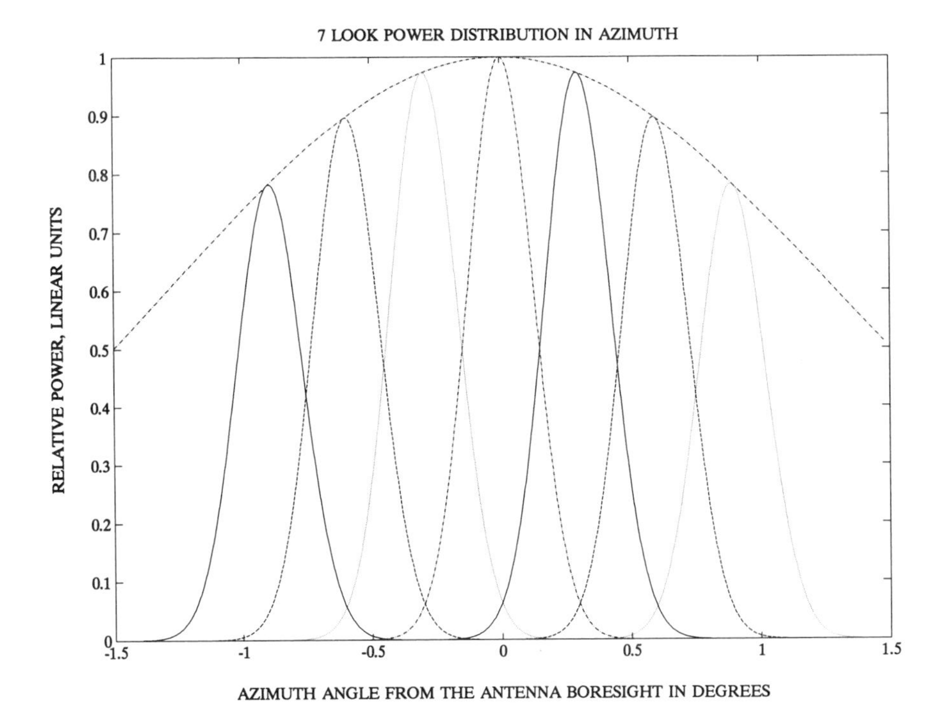

Figure 11.8 Model representation of SAR subbeams produced by a look extraction filter during the formation of a seven look image. The impact of azimuth ambiguities on the resulting multilook image can be judged by comparing this figure with Fig. 11.3.

detected (power) subimages is registered to each other to superimpose returns from the same surface area. The subimages are summed to form a multilook image. The term "look" implies that the information content of each filtered band is statistically independent. This further implies that the filter transfer functions for adjacent looks cannot overlap significantly. If each single-look image is statistically independent of all others, the resolution cells of the summed image will have an $X^2(2N)$ probability density function for an N-look sum. Uniform, textureless target areas in the summed image will have a (mean power)^2/(power variance) ratio of N.

The two approaches to incoherent averaging are fully equivalent. Provided that the two-dimensional bandwidth processed by the two techniques is the same, the number of looks in the output images are the same, and no image energy is discarded in the averaging process, the image information content (in the Nyquist sense) of the averaged images will be the same for both speckle-reduction techniques. Energy integrals used in the averaging process preserve mean parameters (such as the normalized radar cross section, σ^0) in radar images and provide robust tools for defining relationships between these means (Gray et al. [27]).

The Impact of Simple Speckle-Reduction Strategies on SAR Design

Although multilook processing can be done in both range and azimuth, the bandwidth cost of range multilooking is excessive and most multilook radars are designed to perform look summation (or averaging) in azimuth only. This class of radar often has a full bandwidth, single-look, azimuth-resolution cell width which is much smaller than the range-resolution cell width so that the range- and azimuth-resolution cell dimensions of the multilooked image are approximately equal in the slant range plane. Airborne multilook radars designed for remote sensing within a 100-km swath are practically constrained to range resolutions greater than about 2 m, by design trade-offs between swath, range resolution, azimuth sampling rate, azimuth ambiguities, and range ambiguities.

SARs which have been designed for point (cultural) target detection often have nearly square, single-look, resolution cells. Practical constraints on SAR range resolution for remote sensing are defined by the data-recording problems encountered for large image swaths and by radio bandwidth allocation regulations.

More Complex Speckle-Reduction Strategies

For some applications of SAR data, neither single-look images nor uniformly averaged images are optimal for information extraction. In a sea-ice context, such applications could include the identification and measurement of leads, ridges, and regions of homogeneous ice from SAR images. Image preprocessing requirements in these cases demand an adaptive speckle-filtering strategy that reduces speckle in image areas with little texture, minimizes the reduction of scattering texture in regions with larger local variability, and preserves spatial information in very inhomogeneous regions. Filters of this type are collectively known as adaptive filters and, unlike the simple speckle-reduction techniques, cannot be implemented during data acquisition.

Several classes of adaptive filters have been investigated for SAR image analysis and a sufficiently large body of literature has accumulated that a detailed discussion of the properties of all filters of this general type is beyond the scope of this section. A good overview of adaptive filters which use local scene statistics, minimum mean square error (MMSE) filters, is presented by Lopes et al. [56]. These filters have several common properties which are of interest here:

1. The filters use a window, which is much smaller than the SAR image, to examine the local statistical properties of the image. The window is scanned over the image to examine the neighborhood of each pixel.
2. The statistics of the image pixels within the window are analyzed to estimate the local heterogeneity of the data. Heterogeneity measures used include the coefficient of variation (variance divided by the local mean) and directional derivatives.

3. The central pixel within the window is replaced by a filtered representation of the data within the window, where the filter applied is selected according to relationships between the magnitude of the local heterogeneity measure and a set of decision thresholds.
4. If the data are judged to be homogeneous, the value of the central pixel is replaced by a statistical central measure (the median value or the mean value) of the data in the window. If the data are judged to be somewhat heterogeneous, the value of the central pixel is replaced by a weighted mean of the window contents, where the weighting of neighboring points depends on the "heterogeneity index" and the distance from the center. If the data are judged to be very heterogeneous, the value of the central pixel is unchanged.
5. The prefiltering transformations applied to the data, the heterogeneity measurements used, the decision threshold rules, and the weighting functions vary with the filter type.
6. All adaptive filters discussed by Lopes et al. are computationally intensive. When the time required to execute a boxcar averaging filter is used to define a baseline, the filter computation times vary from the baseline to a factor of 10 greater. In general, the most effective filters are the slowest.

Not all adaptive filters for speckle reduction belong to the MMSE class. Crimmins [14] discusses an arguably more effective speckle-suppression filter based on the geometric concept of convex hulls, and Lopes et al. [55] have shown that suitably designed, maximum a posteriori filters can reduce speckle significantly in detected SAR images when they are applied to complex SAR data prior to detection and look summation. The development of adaptive filters and their application to SAR image analysis is still an active research area.

11.1.7 The Radar Equation and Image Radiometry

Equation (6-32) presents a form of the radar equation which states clearly that the signal-to-noise ratio in any processed SAR image is dependent on the ratio of two products containing both systematic and stochastic variables. Those systematic variables that are determined by the imaging geometry are of particular importance in the analysis of SAR imagery. The explicit dependence of the radar antenna gain term, G^2, on the radar polarization, the illumination angle, θ, measured from nadir, and the antenna pointing angle, θ_0, is:

$$G^2 = G_T(\theta - \theta_0)G_R(\theta - \theta_0)$$

where T is the transmitted polarization; R is the received polarization; and G represents the one-way power gain of the antennas. θ is coupled to the range, R, by the imaging geometry including the terrain relief. For sea-ice images, the terrain is effectively at sea level and has no significant relief. The illumination angle can, therefore, be approximated as a deterministic function of slant range for calculations of the large-scale image radiometry. The deterministic illumination ratio, $G_T G_R / R^3$, and the stochastic scattering ratio, $\gamma_L = \sigma_0(\theta_L)/\sin(\theta_L)$, (for local incidence angle, θ_L), together determine the average variation in radar return strength and, thus, the variation in signal-to-noise ratio across the image. Two important quantities are determined by the product of these ratios in (6-32):

1. The dynamic range of the received signal at the radar digitizer as a function of slant range.
2. The variation of image signal-to-noise ratio, SNR, across the scene.

Both of these quantities define radar design requirements which must be met if the radar data are to be useful for measurements of radar reflectivity.

Dynamic Range Limits: Saturation and Underflow

The relationship between the signal strength and the digitizer's operating range has a major impact on the linearity of the radar. If the dynamic range of the signal at the radar digitizer exceeds the dynamic range of the digitizer, information will be lost irretrievably by saturation of the digitizer electronics, by digitizer underflow, or by both processes.

When saturation occurs before the SAR image formation is complete, all unfocused elements within the *real aperture resolution cell*, $r_R r_A$, centered on the saturation event will be affected (Livingstone et al. [50].

$$r_R r_A = R\beta c\tau/2$$

where

R = slant range
β = the equivalent rectangle azimuth beam width (one way, power)
c = the speed of light
τ = the chirp length of the transmitted pulse if saturation (digitization) occurs before range focusing; otherwise, it is the compressed range pulse length, τ_0.

One image manifestation of saturation is the presence of small signal suppression, a dark region around the offending response. If saturation occurs before range focusing, the artifact has significant size in both the range and azimuth directions. Saturation artifacts of range-compressed point targets appear as dark lines or "eyebrows" extending in the azimuth direction on either side of the target, as shown in Fig. 11.9. Image points within the saturation artifact have their brightness reduced nonlinearly by the saturation process.

When the radar signal is smaller in amplitude than the smallest quantization level of the digitizer, the signal data are lost. Statistical measurements on

Figure 11.9 Digital SAR image enlargement showing saturation "eyebrows." Overflow of the analogue-to-digital converter introduces azimuth small-signal suppression artifacts (dark lines) bracketing the targets which caused the saturation. The data were range focused before digitization.

signals close to the quantization floor must take the underflow into account. Radars are normally operated in a dynamic range that minimizes saturation and underflow events. To do this, the receiver gain must be set to accommodate the expected mean brightness and dynamic range of the scene.

Signal-to-Noise Ratio

The variation of the signal-to-noise ratio (SNR) across the radar swath can be of significance in the quantitative analysis of the image data if the SNR becomes sufficiently small. For example, measurements of scattering cross sections of the surface, accurate to ± 0.5 dB, require minimum SNR on the order of 15 dB. Equation (6-32) must exceed the minimum SNR for the desired measurement accuracy at all points of interest across the radar image. To make radiometric measurements from SAR imagery, the mean noise power should be subtracted from the image power estimates.

Impacts on Radar Design: Signal Conditioning

For an SAR to provide quantitatively useful imagery, all points in the radar swath must satisfy all of the following conditions:

1. The SNR exceeds the minimum requirement for the desired measurement accuracy.
2. The maximum signal strength at the radar digitizer is smaller than the saturation threshold.
3. The signals of interest exceed the minimum quantization level at the digitizer.

Depending on the radar imaging geometry, the dynamic range of the product of the systematic and stochastic ratios mentioned above can be so large that the joint condition set cannot be met over the imaged swath unless some signal conditioning strategy is used. Several options are possible, and may be found in currently operating radar systems:

1. Constrain the imaging geometry and imaged swath width sufficiently that the dynamic range of the ratio product is of minor importance in most cases. This approach is used commonly in satellite SARs and in some high-resolution, narrow-swath, airborne SARs (Kasischke and Fowler [40]).
2. Shape the elevation (across-track) radiation pattern of the SAR antenna to minimize the signal dynamic range over an illumination angle interval. This approach is used in the Danish KRAS (Madsen et al. [58]) and the Intera Technologies Inc. STAR-1 (Nichols et al. [64]) airborne SARs.
3. Combine 1 and 2 as has been done in the design for the RADARSAT multiswath SAR (Ahmed et al. [1]).
4. Vary the gain of the radar receiver during the reception time window to compensate the output signal for the systematic and mean stochastic ratios. This approach has been used in the CCRS SARs (Livingstone et al. [51]; Ulander et al. [80]) and is known as sensitivity time control (STC). The STC approach has the advantage that the expected incidence angle dependence of the scene scattering law may be included in the dynamic range compensation of the radar.
5. Combine option 4 with option 2 as has been done in the STAR-2 radar (S. Leung, personal communication; Akam et al. [2]).

The signal dynamic range problem is most severe in airborne SARs, which create images that span a large incidence angle range and have large R_{MAX}/R_{MIN} ratios.

Whichever signal-conditioning strategy is used in the radar system, the image will contain radiometric artifacts that are specific to that radar and the imaging geometry used (see, for example, Shuchman et al. [74]). These artifacts and their models must be known to the analyst if the image radiometry is to be used quantitatively.

11.1.8 Radiometric Calibration of SAR Systems

Radiometric calibration of SAR images is the process of removing radar-dependent parameters from the radar images so that image intensity (represented as digital numbers) can be quantitatively related to the radar cross section of the imaged terrain, to within a known error. Calibration provides a connection between the radar image and the physical properties of the measured surface.

Calibration by Radar Equation Inversion

Simply stated, calibration is the process of inverting (6-31) to extract an estimate of σ^0 for each resolution cell in the scene. To accomplish the inversion successfully, accurate models are required for imaging geometry, antenna gains, radar signal conditioning functions, and processor gain (Ulander et al. [80]; Ulander [79] Livingstone and Drinkwater [49]). The radar equation inversion approach to radiometric calibration requires that the radar system has been designed and implemented with this end in mind. These instruments must have: well-characterized, temporally stable parameters; a large, linear dynamic range; operating states that are accurately repeatable; and stable, internal reference generators which allow routine performance measurements to be made over the largest possible portion of the radar signal path. In addition to the use of a correctly designed radar, radiometric calibration of radar data requires that the sensor is operated appropriately. Engineering data bases describing the calibration model parameters must be maintained and updated regularly. A complete description of the radar operating state must accompany the data and the internal reference generator measurements must accompany the radar image data routinely.

Calibration by External Reference Targets

An alternative approach to radiometric calibration, which can be used with any SAR, makes use of reference targets (whose radiometric properties are known) within a scene to measure the radar cross section of other scene elements using comparative techniques (Larsen and Maffet [45]; Larsen et al. [44]; Gray et al. [27]; Freeman [25]). In this case, most of the radar parameters that influence the instrument calibration are implicitly contained in the radar returns from the reference targets. The calibration obtained from these measurements is valid only for the target measurement geometry used and for the operating state of the radar at the time that the target data were acquired. Extension of the calibration points to the neighboring terrain requires models of the antenna elevation pattern and of the imaging geometry over the extended scene. If the radar system is sufficiently stable, reference target calibration can be extended to other scenes within the same data acquisition pass, without performing a full radar equation inversion.

Calibration of Early Remote Sensing SARs

The airborne SARs used for remote sensing prior to 1984 all contained a number of analogue control systems which severely limited their long-term stability.

In addition, most of the data acquired were recorded and processed optically, which imposed additional repeatability and linearity constraints. Two older systems, the JPL SAR and the CCRS/ERIM (Environmental Research Institute of Michigan) SAR, contained built-in reference signal generators, which could be used to monitor the radar gains, and both had a digital signal recording capability. Both radars could be calibrated locally, using reference target comparison techniques, but neither system was stable enough to transfer the calibration far beyond the reference scenes.

Calibration of Recent Remote Sensing SARs

The first of a modern series of remote sensing SARs, the Intera STAR-1 radar discussed in Chapter 10 (Nichols et al. [64]), began service as a sea-ice imaging tool in early 1984 to support arctic oil exploration and shipping activities. Although this instrument is calibratable (in the radar equation inversion sense), the data acquisition procedures used preclude absolute (and often relative) calibration of its data. Since 1984, a number of calibratable airborne, remote sensing SARs have been commissioned. The CCRS *C*-band SAR, commissioned in early 1987, and the CCRS *X*-band SAR, commissioned in 1988, are two of these instruments (Livingstone et al. [51]). Other calibratable, modern, airborne SARs are: the JPL AIRSAR commissioned in 1987 (Held et al. [31]; Way and Smith [83]), the E-SAR commissioned in 1988 (Horn [37]; Way and Smith [83]), and the KRAS under development in 1993, with its basic mode commissioned 1990 (Madsen et al. [58]).

An extensive series of radar characterization and calibration studies performed on the CCRS *C*-band radar resulted in a set of models (and data acquisition procedures) which can be used routinely to calibrate its radar images (Ulander et al. [80]). The study results have shown that the models developed are sufficiently accurate to yield calibration errors smaller than 1 dB for most imaging geometries. Not surprisingly, the data acquisition procedures required for the creation of calibrated images are as important as the model accuracies in generating precision radar data products. At the time of writing, work is still in progress to improve the calibration accuracy of the CCRS radars and to move the data calibration procedures from the development laboratory to a data-production environment.

To demonstrate the utility of radiometrically calibrated SAR, the *C*-band SAR has been used as an "imaging scatterometer" to determine the quantitative scattering properties of sea ice (Livingstone and Drinkwater [49]; Lukowski et al. [57]). The relationships between a calibrated airborne SAR and a calibrated spaceborne SAR were illustrated in a cross-calibration experiment. The CCRS *C*-band SAR system was used to measure calibration functions for the European ERS-1 *C*-band SAR system (Livingstone et al. [52]). In this study it was shown that the relative calibration errors of the airborne and satellite sensors were within the tolerances established for the satellite radar by other calibration techniques.

SAR Calibration History and SAR Measurement of Sea-Ice Properties
Many of the SAR images of Arctic sea ice that exist in the research archives were acquired during a period of intensive investigation (1978–1984) of the use of remote sensing technology to support arctic marine activities. Very little of this historical SAR imagery has been (or can be) radiometrically calibrated. Phenomena observed in this imagery are often supported by quantitative airborne profiling or surface sensor measurements (discussed in Chapter 2) of subsets of the imaged surfaces. However, discussions of image observations using these data are necessarily qualitative. In the discussions of SAR imagery of sea ice, in Section 11.2, only SAR images acquired after March 1987 can be analyzed quantitatively in radiometric terms due to the lack of suitable calibrated sensors prior to that time.

There are now several calibratable SAR systems that can be used for sea-ice research, as well as other quantitative applications. Modern airborne systems offer the possibility of investigating the significance of a wide range of electromagnetic parameters. The major limitation of airborne sensors is that to conduct several missions in different seasons to observe temporal variations of sea-ice scattering signatures is both difficult and expensive.

Modern spaceborne SAR systems (the ERS-1 SAR is the first of these) offer the capability of monitoring the temporal evolution of sea-ice signatures over a narrower range of electromagnetic parameters than are accessible to airborne radars. Recently reported results of calibration studies for the ERS-1 SAR (Laur and Doherty [46]) show that this instrument is stable and can be calibrated accurately. Its imagery can be used quantitatively to obtain lengthy time sequences of observations at the same site. Sea-ice research programs based on the ERS-1 SAR are in progress at the time of writing and some early results have been published (Cunningham et al. [15]; Kwok and Pang [41]).

11.2 SEA ICE AS A RADAR TARGET

Historically, sea-ice reconnaissance, in support of Arctic resource exploration and shipping operations, was one of the primary factors that drove the development of SAR technology in Canada. Discussions in Chapters 1, 10, and 12 trace the evolution of the role of SAR in addressing sea-ice reconnaissance. The operational ice reconnaissance requirement continues to this day and is one of the major commercial uses of SAR technology.

By the mid-1970s, SAR was recognized as a potentially suitable sensor class for all-weather sea-ice mapping but the knowledge of sea-ice properties that could be observed by radar systems was insufficient, at that time, to design dedicated ice reconnaissance radars. The combination of need and ignorance triggered a long series of research projects to investigate the physical structure of sea ice, its evolution, and its electromagnetic signatures. Much of the early work was focused on transportation issues related to ship navigation in ice-infested waters and the design of stationary structures for this environment. The radar remote sensing component of these studies was driven by a desire to know the optimum set of radar parameters and observation scales for ice reconnaissance. As the SARs of that period were not calibratable instruments, information extraction relied on visual interpretation of image features within a small set of images. Quantitative measurements, made on the ice surface or acquired by calibrated, nonimaging sensors, were treated by SAR image interpreters as interesting, auxiliary data. More recently, SARs have become calibratable instruments and quantitative measurements made by these systems are becoming increasingly important in sea-ice research.

As research progressed, the problems defined by marine transportation issues were recognized to form an evolving subset of the more general problems of understanding sea ice as a material; understanding the interactions between ice, ocean, and atmosphere; understanding the biology of ice-covered oceans; and understanding the relationships between the electromagnetic signatures of the ice/ocean system and its structure. The scientific investigations in these problem areas have provided design data that have been used to develop reconnaissance systems and, conversely, the need for fundamental information inputs to reconnaissance and data interpretation has stimulated scientific research. As is desirable, the relationship between the operational and the research worlds has been symbiotic.

The discussions in this part of Chapter 11 emphasize the role of SAR in the measurement of sea ice. Subsections 11.2.1 to 11.2.3 examine measurement scale, the effects of seasonal evolution, and marginal ice zones. Subsections 11.2.4 to 11.2.6 discuss the influences of incidence angle, radar frequency, and radar polarization on sea-ice observation and measurement.

In many cases, existing knowledge of the relationships between sea-ice properties and their electromagnetic signatures is insufficient to support valid generalizations. Research into electromagnetic scattering models, both in progress and planned, is expected to provide the tools needed to integrate and extend knowledge derived from empirical observations, but much work remains to be done in this area. In the following sections, when current knowledge is insufficient to support generalization, results and observations from specific experiments are juxtaposed to indicate the scope of what is known.

11.2.1 Sea Ice as an Inhomogeneous Radar Target

In early attempts to optimize SARs for ice reconnaissance, the significance of trade-offs between imaged swath and radar resolution was unknown. Implicit in the questions asked was the assumption that all information of practical importance existed at scale sizes greater than some lower bound. For navigation support, this is often true, and the lower bound of significance is related directly to ship capabilities. The SAR measurement of sea-ice properties, however, requires knowledge of the scales associated with parameter inhomogeneity to relate radar measurements to the properties of the material and to define the minimum observation area needed to measure ice processes.

In Chapter 2 the descriptions of ice and snow structures show that both of these media are spatially inhomogeneous (in three dimensions), on scales ranging from a fraction of a millimeter to several meters. Mechanical deformation processes produce structural inhomogeneities with local dimensions ranging from centimeter scales to tens of meters and with linear dimensions extending to several kilometers. Thermally induced metamorphoses produce local, structural changes whose spatial extent ranges from meters to several hundred meters. Depending on the stage of ice development and the mechanical environment, floe sizes range from meter to kilometer dimensions and composite floes whose dimensions are several kilometers have been observed. When ice pack morphology is considered, there is a physical basis for considering natural sea ice to be an inhomogeneous target at all scales of interest.

As expected, electromagnetic measurements of sea-ice properties reveal a continuity of significant scale sizes whose smallest observable scales are determined by the limit of resolution of the sensors. Thus, direct, surface measurements of the complex dielectric constant of snow and sea ice using microwave probes show spatial variability at the scale of the probe measurement volumes (Mätzler [60]; Holt [35]). Surface radiometer transects of natural sea ice show spatial variability at the scale of the antenna beam width (<1 m^2) (Lohanick and Grenfell [54]). Near-surface, helicopterborne, scatterometer transects of natural sea ice show spatial variability at the scale of the scatterometer resolution cell (<4 m^2) (Onstott et al. [67]; Onstott [66]). Large scale SAR mosaics of Arctic and marginal sea ice (Johannessen et al. [39]; Carsey et al. [11]; de Bastiani [18]; Shuchman et al. [74]; Haugan and Preller [30]) show spatial variability in sea-ice returns from radar resolution cell scales (40 m^2) to the scale of the mosaics ($>10{,}000$ km^2). When the entire Arctic Ocean is viewed by spaceborne radiometers (Campbell et al. [10]; Comiso [13]) the spatial variability of the ice cover extends over scale sizes ranging from subpixel (40 km^2) dimensions to the entire Arctic basin.

Within the wide range of scale sizes measured and reported in the literature, there is no indication of an optimum resolution cell size for sea-ice signature measurements so that optimization criteria sought by operational agencies are based on other considerations. For SARs designed to support sea-ice research, desirable resolution cell dimensions are determined by the particular processes or signatures to be investigated.

Electromagnetic Scattering Signatures: Mean Parameters

In spite of the inhomogeneities observed within sea-ice data sets, it has been found that there are electromagnetic scattering signatures observed in the microwave frequency regime which are characteristic of the various stages of ice development. Frequency, polarization, and incidence angle dependence of radar cross sections have been empirically coupled to ice type under many conditions. As discussed in Chapter 2, much of our knowledge of these ensemble mean signatures comes from surface measurements and measurements with profiling sensors. For this type of data, inhomogeneities within an ice type appear in the ensemble variance of the measured parameters, and relatively little spatial information is available.

Sea-ice signatures based on radar scattering cross sections are subject to seasonal variations and are not robust in melting conditions. As the free water content of the snow cover and ice surfaces changes, the ice properties that are measured by a radar also change. Under all conditions, however, a SAR image provides a spatial map of electromagnetic scattering and the spatial distribution of electromagnetic scattering properties observed in the image contains information about the surface processes at all observable scales.

Texture in Radar Images of Sea Ice

The statistical signatures of sea-ice inhomogeneities within ice floes have been investigated by a number of researchers who have applied texture analysis techniques to SAR images. Some success has been reported in classifying a limited range of sea ice types from various texture measurements (Holmes et al. [34]; Barber [4]; Shuchman et al. [75]). Ice-type classification using texture measurements as feature vector elements is still a subject of active investigation for all season and regional conditions. A short review of this work can be found in Barber and LeDrew [5].

Radar Observations of Ice Pack Processes

At scales larger than single floe dimensions and extending to ocean basin dimensions, inhomogeneities within the ice pack are determined by ice-type distributions, mechanical deformations, snow cover, thermal history, and ice concentration variations. These parameters vary regionally within the ice pack and both result from and provide information on the pack dynamics.

Algorithms have been developed to use multitemporal image sequences of the same geographic area to extract ice pack motions and produce maps of the form shown in Fig. 11.10 (Hall and Rothrock [29]; Curlander et al. [16]; Fily and Rothrock [23]; Hirose et al. [33]). The techniques rely on the identification of relatively long-lived inhomogeneities within the ice pack to provide recognizable features in the image sequence. The absolute and relative positions of these, as determined from descending-scale spatial correlations, define image feature translations between the images in the set and, thus, the time-varying, mean velocity field of the ice pack. With the advent of the recent generation of SAR satellites (ERS-1, 1991; *RADARSAT*, 1995; ERS-2, 1995) sea-ice motion measured from SAR images is expected to become a major input (Lepparanta and Yan [47]) to operational ice kinematic models used for ice forecasting (Preller and Posey [68]), that are based on the physics of the ice pack (Maykut and Untersteiner [61]; Hibler [32]).

At intermediate size scales, ice-cover inhomogeneities include ridges, rubble fields, cracks, and leads. These features are of significance in determining wind

Figure 11.10 Sea ice motion map of the near-shore ice north of Prudhoe Bay in the Beaufort Sea. The arrows show the ice displacement caused by wind and current forces over a 24-hour interval in May 1990.

drag effects, heat transfer, ice pack mobility, and navigability and they are of interest in their own right as ice information. Their extraction from SAR imagery using an automated texture analysis algorithm is discussed by Vesecky et al. [82].

Ice ridges have been identified as being particularly important in determining the coupling between winds and ice motion. Although the spatial distribution of these features is observed readily in SAR imagery, there appears to be no consensus in the literature on the physical properties that govern the SAR signatures of ridges. The relationships between image contrast, ridge geometry, ridge orientation, and SAR imaging geometry have not been established.

11.2.2 SAR Observations of Sea-Ice Seasons

Sea-Ice Seasons

The discussions in Chapter 2 outline the physical processes that take place within the ice and its snow cover during the growth, aging, and decay periods.

Recognition that the electromagnetic signatures of natural sea ice are tightly coupled to these processes led to the formulation of the sea-ice "season" concept introduced in Chapter 2. Here, the processes which are active within the ice cover, and their influence on the microwave scattering and emissivity signatures of the ice, define the ice seasons. The following discussions illustrate the ice season concept with SAR images.

Freeze-Up

The seasonal cycle beings with freeze-up. A feature space representation of Arctic first-year ice signatures was shown in Fig. 2.33 (freeze-up) and companion SAR imagery from this season in the Beaufort Sea can be found in Figs. 11.11 (for *X* and *C* band) and 11.12 (for *X* and *L* band).

The beginning of the freeze-up period is shown in the 1991 *ERS*-1 *C*-band SAR image in Fig. 11.13 at the mouth of Admiralty Inlet. The open-water returns from Lancaster Sound, along the coast of Brodeur Peninsula, are modulated by dark (low return) patches of new ice (probably nilas) and streamers of frazil ice. An ice edge of consolidated young ice crosses the mouth of Admiralty Inlet. Within the Admiralty Inlet, bands of young-ice fragments and small floes appear as brighter spots. A patch of open water shows signs of grease ice formation in patches and streamers.

As freeze-up progresses and the freezing horizon in old ice descends into

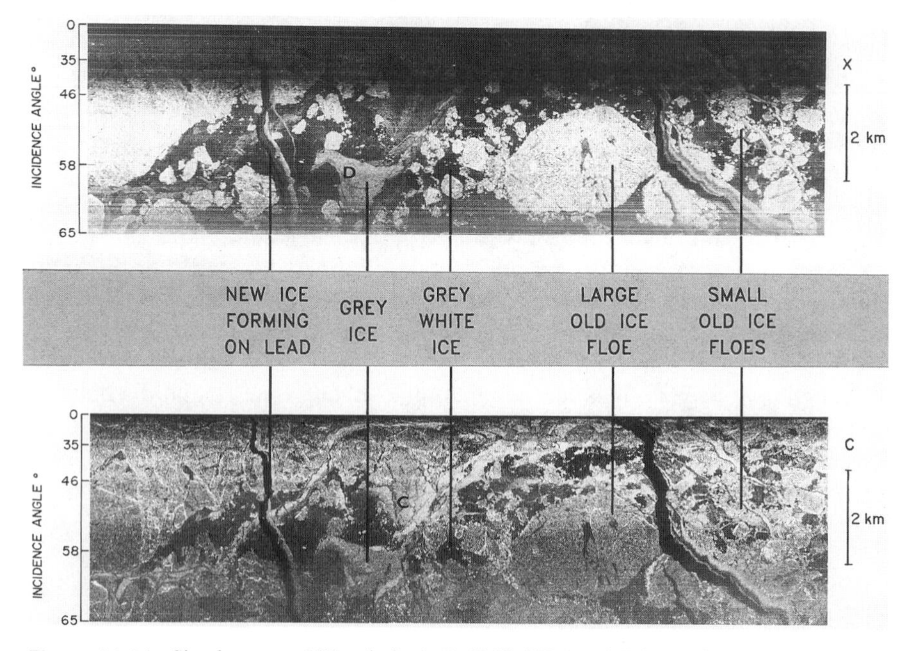

Figure 11.11 Simultaneous, HH polarized *C*- (5.35 GHz) and *X*-band (9.35 GHz) images of sea ice in the Beaufort Sea, October 28, 1981. These data were recorded and processed optically.

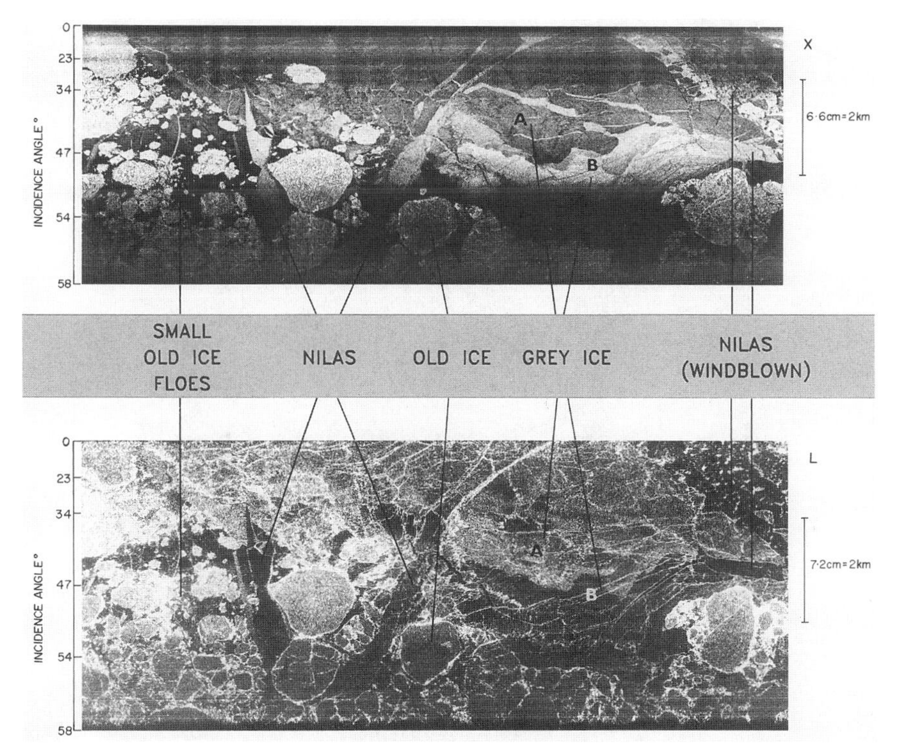

Figure 11.12 Simultaneous, HH-polarized *L*- (1.25 GHz) and *X*-band (9.35 GHz) images of Beaufort Sea ice acquired on November 1, 1981. These data were recorded and processed optically.

the ice volume, the characteristic volume scattering signature of this ice type moves from high radar frequencies (above *Ku* band) to lower radar frequencies (below *C* band). As the season progresses and the temperature decreases, ice-formation rates increase, as does the ice-transformation rate through its early growth stages. Feature-spaces for ice signatures migrate towards their winter forms as freeze-up progresses.

Winter

The onset of the winter or cold ice season is marked by the development of first-year ice. Available data suggest that microwave sea-ice signatures are quite stable throughout this season in polar regions. Figure 2.34 (winter) shows a feature space representation of arctic winter ice signatures. Companion *X*- and *L*-band SAR images from the edge of the Beaufort Sea polar pack are shown in Fig. 11.14 (and later in Fig. 11.33). *C*- and *X*-band SAR images in Fig. 11.15 show first-year ice in late winter at the active shear zone near Prudhoe Bay. The air temperature at the time of acquisition was $-3°C$ and no preceding melt events had been observed. The SAR used to generate these images is a

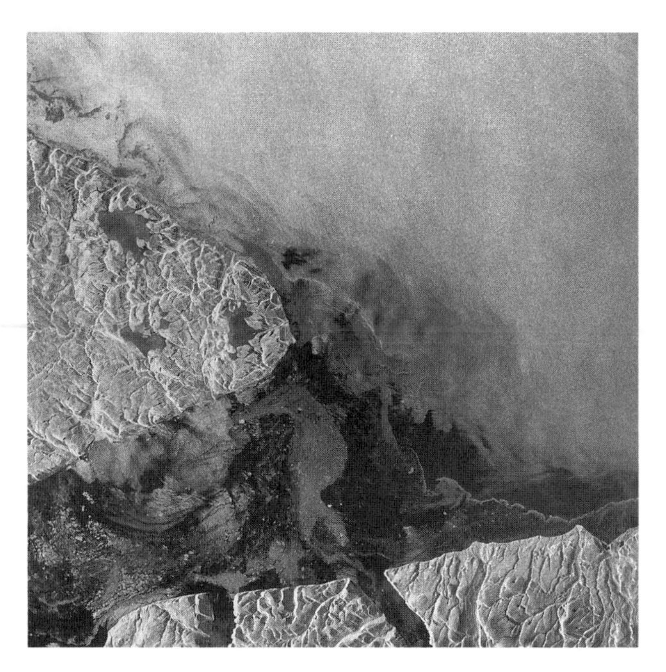

Figure 11.13 *ERS*-1 C-band SAR image of the mouth of Admiralty Inlet, September 1991. Note the new ice formation in Admiralty Inlet and the open water (bright, diffuse returns) in Lancaster sound.

calibrated instrument and the images shown here are quantitatively relatable. As expected for rough-surface scattering, the HH- and VV-polarized radar cross sections at both frequencies show no polarization dependence (to within half of the subscene standard deviations measured). Although all ice in this scene is first-year ice at least 1.5-m thick, a range of returns can be observed in the image data. C band σ^0_{HH}, and σ^0_{VV}, measured from the SAR returns near 40° incidence angle vary from -17.6 ± 1.8 dB to 22.1 ± 1.8 dB. The differences in cross section for undeformed ice surface were related to ice surface roughness (measured at test sites within the image). Continuous pack deformation was observed on the outer (upper) side of the outer (upper) shear zone ridge. Ridge heights in the shear zone exceeded 10 m in many places.

Early Melt Season

At winter's end, when the diurnal temperature cycle peaks exceed the freezing point, moisture begins to accumulate in the snow pack and a period of extensive snow pack recrystallization begins. In this early melt season, the snow cover contributions to the sea-ice scattering signatures become stronger as scatterers within the snow pack increase in size and as accumulating moisture within the upper snow layers increases the loss tangent of the snow. Feature space clusters

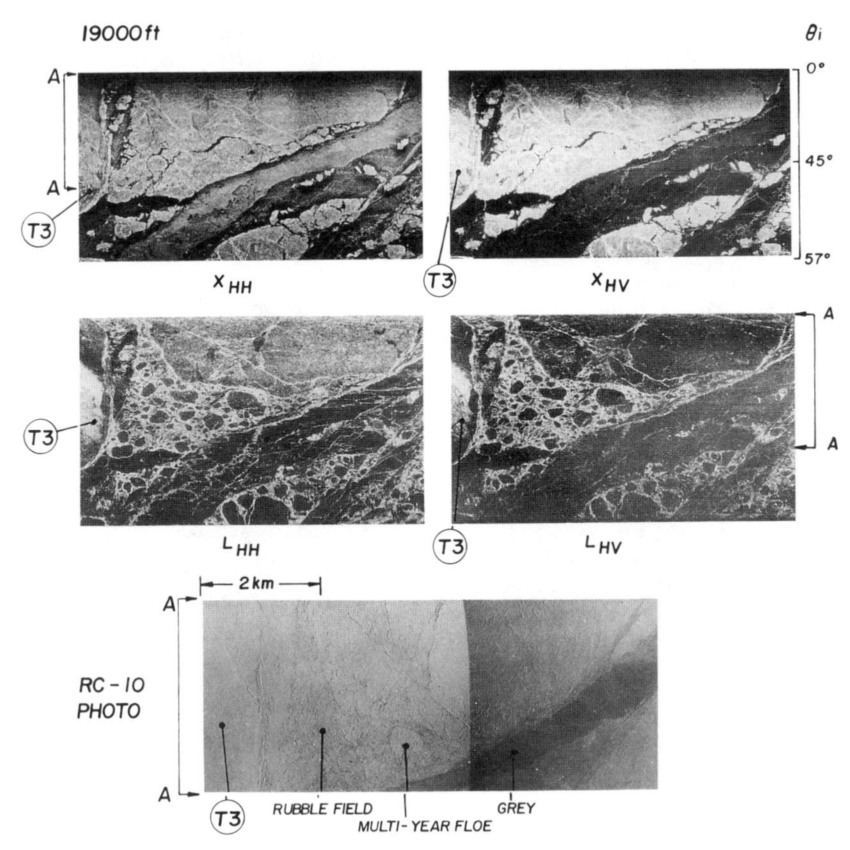

Figure 11.14 Four-channel, steep depression angle SAR imagery of Beaufort Sea ice. These data were acquired on March 16, 1979, by an uncalibrated SAR. Optical data recording and optical SAR processing technologies were used to create the images.

which had remained stable throughout the winter begin to migrate within the space, as was shown in Figure 2.35 (early melt season). These data were acquired by profiling sensors and no companion SAR imagery that unambiguously belongs to this season is known to exist.

Melt Onset

During melt onset, wet snow and snow-base remnants dominate radar returns from the ice surface. The feature space representation of this season, shown in Figure 2.36 (melt onset) reveals that almost all ice-type dependencies have vanished. Figure 11.16 shows an *X*-band SAR image acquired in Crozier Channel during the melt onset season. Of particular interest is the emergence of snow as a microwave absorber (especially along the old-floe boundary ridges). The radar signatures of bare first-year ice are dominated by superimposed ice

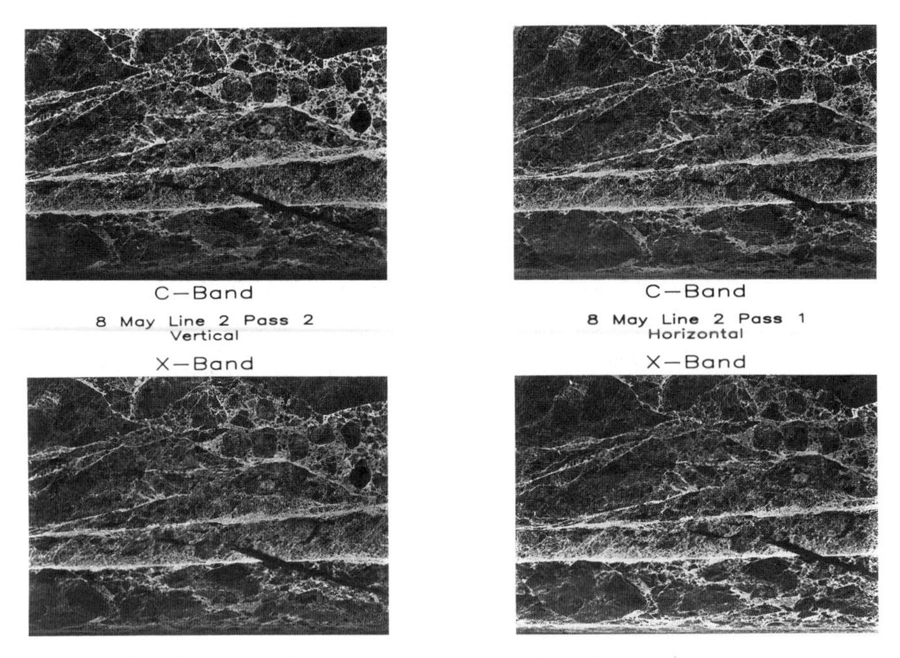

Figure 11.15 First-year sea ice in the Beaufort Sea in May 1990; data acquired by a calibrated SAR and processed digitally. The conditions shown are typical of the area north of Prudhoe Bay at the end of winter.

patches which mark the locations of small ablating snow drifts. Although the SAR used to make the image in Fig. 11.16 was uncalibrated, a calibrated scatterometer profile was flown along the SAR swath within 30 min of SAR image acquisition and nearly simultaneous helicopterborne scatterometer data were acquired in the vicinity of the floe marked "E." Livingstone et al. [53] discuss the integration of the scattering data sets and demonstrate the use of the scatterometer data to provide a SAR calibration so that image features can be quantified.

Advanced Melt

The complexity of SAR signatures of sea ice during the advanced melt season arises from the rapid response of the ice surface to thermal conditions and from the instability of summer weather in arctic regions. The early part of advanced melt is marked by flooding and subsequent drainage of the ice surface. As the season advances, other flood/drain cycles may occur, interspersed by periodic freezing. In the feature space representation of this season, Figure 2.37 (advanced melt) the standard deviations of the ice-type signatures are seen to be large.

The flood part of the melt cycle is illustrated by *X*- and *L*-band SAR images

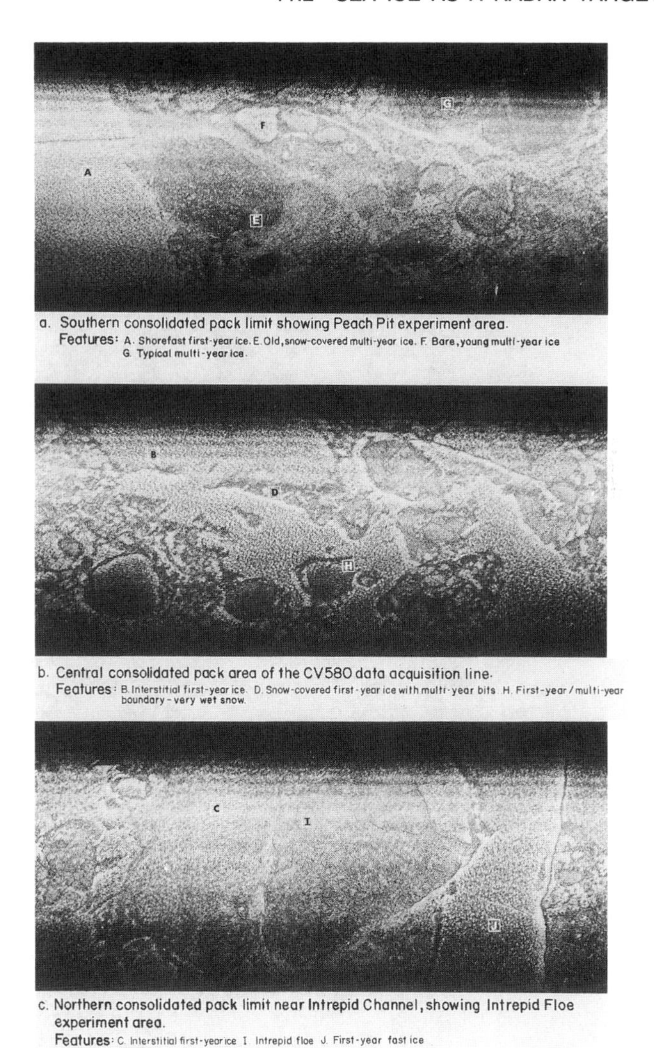

Figure 11.16 Melt onset in Crozier Channel, 1982. These optically recorded, optically processed data were acquired by an uncalibrated SAR.

acquired in Crozier Channel in 1982, Fig. 11.17. The flat and low-lying ice surfaces are covered by a layer of water through which the high points on the ice surface project. High radar returns are obtained from the rough-ice projections and from drained ice-bordering cracks and seal holes. Melt ponds on old ice are large and are visible in the SAR imagery. The *L*-band radar cross section decreases rapidly with incidence angle and slope modulation effects are visible on ridge remnants. A good case study for these conditions can be found in Holt and Digby-Argus [36].

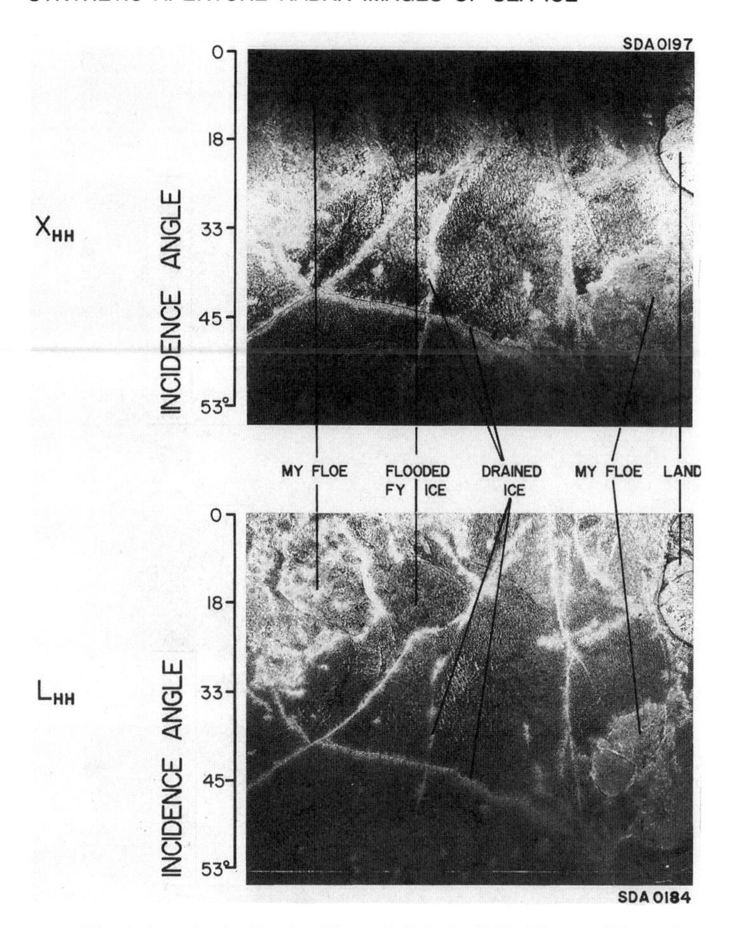

Figure 11.17 Flooded sea ice in Crozier Channel, July 7, 1982. The conditions shown followed a rapid melting of the snow cover on the ice at the start of the melt season. Multiyear (old) sea ice is marked MY and first-year ice is marked FY. The SAR was uncalibrated.

In Fig. 11.18, the ice in the Beaufort Sea polar pack has passed its flooding peak but large melt ponds remain on all ice types. At steep incidence angles, wind-roughened water on melt ponds yields high radar returns at *L* band. Smoothing of flow surfaces during the melt process increases the sensitivity of the *L*-band radar cross section to incidence angle variation.

In Fig. 11.19, the snow cover on the Beaufort Sea pack ice has melted and much of the first-year ice has thinned. A recent freezing event has left patches of smooth nilas on the open lead on the left-hand side of the photo mosaic. The *L*-band SAR returns are less sensitive to incidence angle variation than those in Fig. 11.18. Water surfaces roughened by a light breeze have a radar cross section close to that of the ice at this frequency. The high *X*-band radar returns are caused by frost flowers (just visible in the photography) on the recently refrozen surface of some thinned first-year ice. The cross-polarized

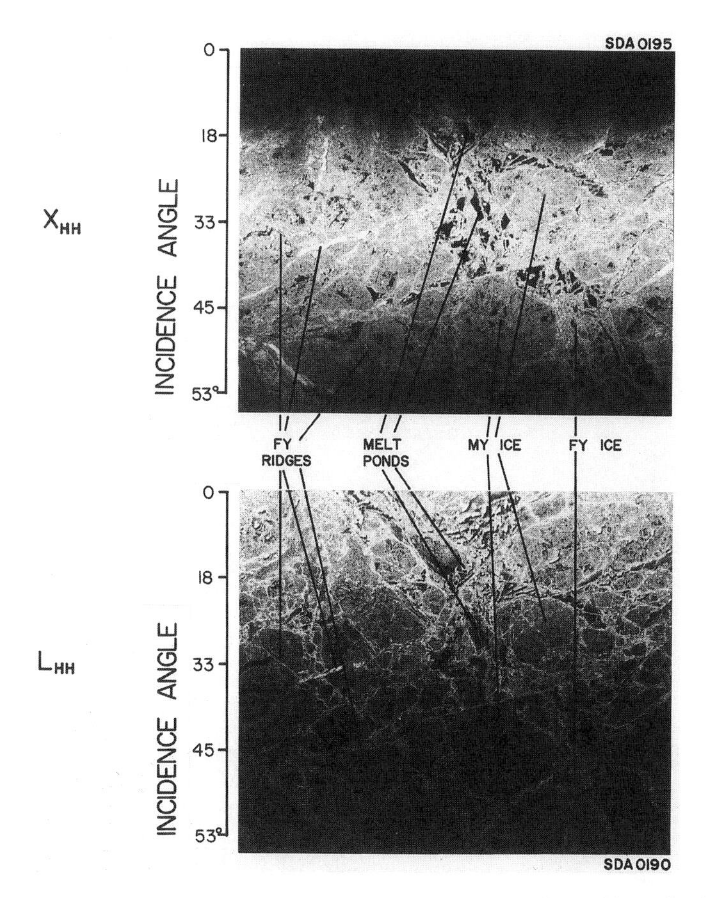

Figure 11.18 Wet, drained sea ice in the Beaufort Sea, July 1982. Multiyear ice is marked MY and first-year ice is marked FY.

radar returns are very weak, as can be seen by the poor SNR, but they do emphasize some structures not seen clearly in the like-polarized imagery.

A final example of the variability of sea-ice scattering in the melt season is seen in two *Seasat L*-band SAR images of the mouth of the Prince of Wales Strait shown in Figs. 11.20 and 11.21. In the first of these, acquired on July 11, 1978, the radar cross section of old ice and low-relief first-year ice in M'Clure Straight is low, possibly indicating water-saturated surfaces. The returns from both ice types in Prince of Wales Strait are high and the radar signatures are reminiscent of the refrozen Beaufort Sea ice in Figure 11.19. In the second image, acquired 17 days later on July 28, the tonal contrasts are reversed. Careful examination of both images reveals that there has been very little ice motion between the two satellite images and that the signature changes represent changes in local conditions. Unfortunately no surface information exists to validate any hypothesis that can be made.

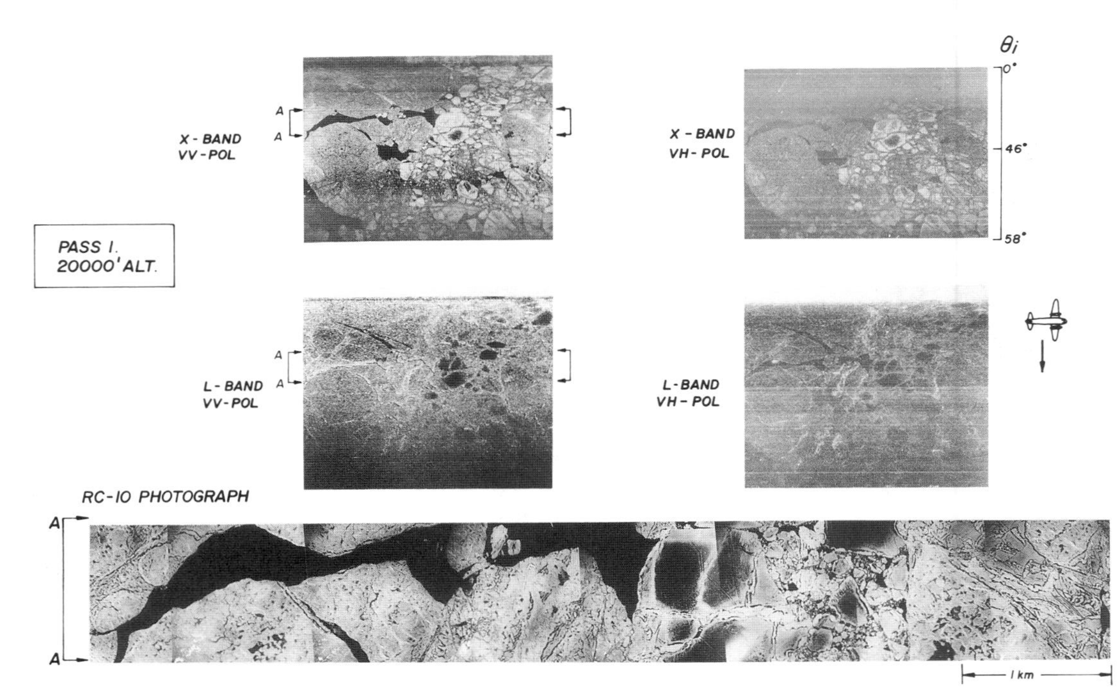

Figure 11.19 Melt-season sea ice in the Beaufort Sea, June 27, 1980. This imagery was acquired following a freezing event. The data were optically recorded and optically processed.

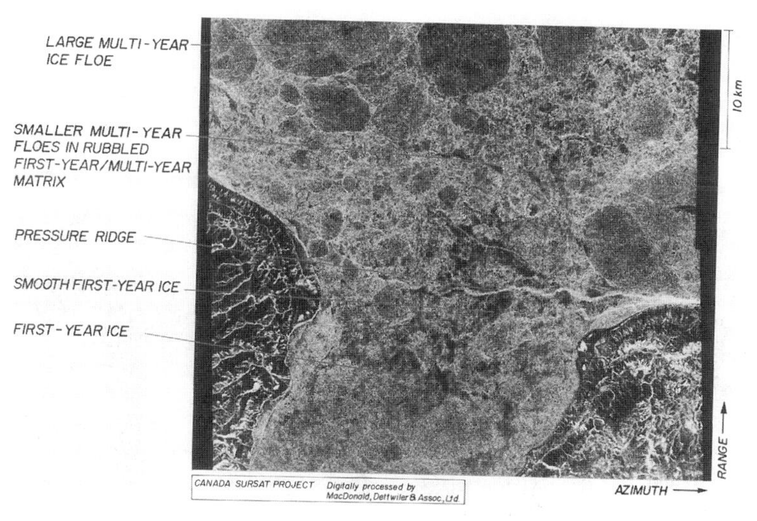

Figure 11.20 SEASAT *L*-band SAR image of the mouth of Prince of Wales Strait at McClure Strait, July 11, 1978.

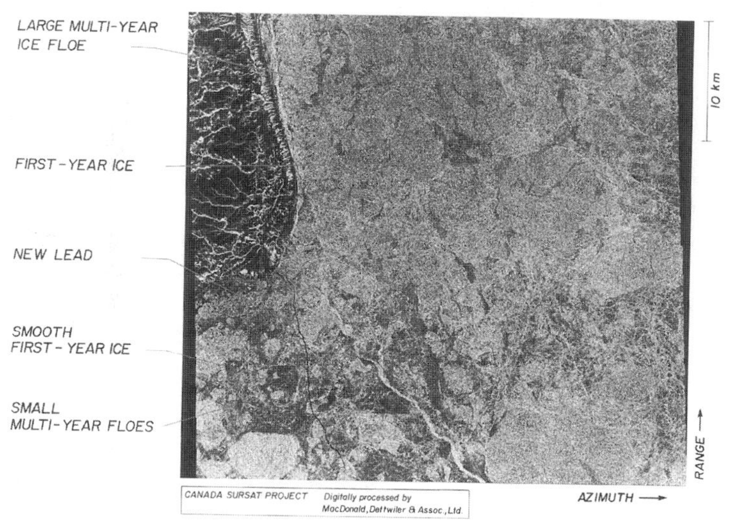

Figure 11.21 SEASAT *L*-band SAR image of the mouth of Prince of Wales Strait at M'Clure Strait, July 28, 1978.

Summary

Knowledge of the seasonal evolution of sea-ice scattering signatures is a composite built on observations accumulated over more than a decade. The airborne SAR imagery represents snapshots in time and space, whereas the *ERS*-1 SAR satellite has produced its first full year of imagery. Future analysis of data from this sensor, and from future SAR satellites, will yield a much more complete picture of sea-ice seasons and the factors that control them.

11.2.3 Marginal Ice Zone Effects

Background

An oceanic ice margin is the dynamic interface between an ice pack and the open ocean. These are regions of complex interactions between the ice pack, the ocean, and the atmosphere. Since 1983 an extensive body of literature has accumulated on marginal ice zone observations and processes. A thorough review of this work would be a major study and is beyond the scope of this book. More detail is found in three special editions of the *Journal of Geophysical Research*: "Marginal Ice Zones" [62], "Marginal Ice Zone Research" [63], and "Marginal Ice Zone Experiment III" [59]; in a special edition of the *IEEE Transactions on Geoscience and Remote Sensing*; "Labrador Ice Margin Experiment (LIMEX), and the Labrador Extreme Waves Experiment (LEWEX)" [42]; and in a special edition of *Atmosphere–Ocean*: "LIMEX" [43]. These issues consolidate much of the marginal ice zone research conducted in Fram Strait, the Antarctic, the Bering Sea, and the Labrador Sea. Many other significant papers can be found in other journals.

The size scales of many ice and ocean features found in marginal ice zones vary from a few meters to tens of kilometers. Sea-ice distributions near the edges of marginal zones are often complex and are suited to observation and measurement by SAR.

Ice Growth in Marginal Ice Zones

During ice-formation periods at an ice margin, cold, dense, highly saline water (produced by the rejection of salts from the consolidating ice sheet) sinks below the growing ice and drives oceanic convection cells which import less saline water into the ice-formation region. A thick, mixed layer, whose temperature is near the freezing point, forms beneath the ice growth region and, as is discussed in Chapter 2, becomes an integral part of the ice growth process.

When the growing ice margin exists in a relatively calm and shallow sea, such as the Beaufort Sea, grease ice growth suppresses the wind–sea interaction (Guest and Davidson [28]) and the local wave field does not, on the whole, control the structure of the growing ice. In this region, new ice forms in the annual shore leads and quickly forms a continuous ice sheet. Subsequent ice formation occurs in leads within the ice pack.

When the growing ice margin exists in a mechanically dynamic environment, such as the Labrador Sea, the oceanic wave field plays a dominant role in defining the physical structure of the forming ice. In this environment, ice forms into small pancakes with raised edges and multiple-rafting of pancake floes is common at all stages of ice growth. A very rough ice surface results, with surface relief elements ranging from submillimeter to meter scales. Fig. 11.22. Even when there is 100% ice cover, long-period ocean waves propagate

Figure 11.22 Two sequential frames from a fixed camera showing wave motion in sea ice in the Labrador Sea marginal ice zone. The sensor package is about 1-m high and the upended ice block in the background is about 2-m high.

through the ice pack. Swell penetration distances exceeding 100 km have been recorded (Liu et al. [48]). Throughout the ice season there is a continuous transport of ice southwards along the Labrador coast, driven by winds and currents. Transport rates of up to 60 km/day have been recorded (Tang [78]).

Unlike the Labrador Sea marginal ice zone, in which most of the ice which is transported through the zone has formed within it, the marginal ice zone in Fram Strait and along the East Greenland coast is fed continuously by an effluence of ice from the Arctic Ocean. During the ice-formation period, the Fram Strait marginal ice zone contains a mixture of imported ice of all ages and young ice grown in situ. Young ice growing near the ice edge is subject to a mechanical environment similar to that found in the Labrador marginal ice zone. Measured ice transport rates near the center of the Strait are about 17 km/day (Emery et al. [22]).

Ice Decay in Marginal Ice Zones

During the ice decay period, ice in the marginal ice zones becomes a source of relatively fresh water for the surrounding ocean. The details of the air–ice–ocean interactions are complex and are not discussed here; however, a few general ideas are introduced. The surface roughness of the ice can be described in terms of an air drag coefficient, which couples the ice surface to a wind field and generates stresses within the ice pack. The resulting differential ice motion fractures the ice and speeds breakup. Periodic stresses, applied by waves propagating beneath the ice, create strains within the ice sheet and cause it to fracture (Fox and Squire [24]). Floe-to-floe abrasion has been observed to create a slush layer between the floes and speeds the melting process.

In relatively calm, shallow oceans, such as the Beaufort Sea, the main mechanical contributor to breakup is the large-scale stress field resulting from the interaction of the motion of the Beaufort gyre and the fast ice. As the shore lead develops, some wave action is generated but the high-energy, long-period swell waves (common in the open ocean) cannot be generated. Direct solar input and wind-aided heat transport from the atmosphere to the ice appear to be the main factors which generate the summer marginal ice zone at the edge of the polar pack.

In more dynamic environments, such as in the Labrador Sea and Fram Strait marginal ice zones, wave action plays a large role in ice decomposition by fracturing large floes into small, mobile fragments. The increased surface area exposed to heat inputs from the atmosphere and ocean speeds melting. Model calculations of the thermodynamics associated with the advance and retreat of the Labrador Sea seasonal ice (Tang [78]) suggest that the ocean is the primary heat source controlling ice melt and retreat of the ice margin. The annual ice-edge retreat clears the Labrador coast of sea ice and the retreat progresses to the northward to Davis Strait. In Fram Strait and along the East Greenland coast, ice is continually replenished by an outflow from the Arctic Ocean and the marginal ice zone is a permanent feature whose characteristics change with the seasons.

SAR Images of Marginal Ice Zones

SARs have been used extensively to map and measure marginal ice zone phenomena. As many of the processes which occur at and near the ice margin have scales of many kilometers, SAR image mosaics are needed to provide adequate spatial coverage, while still preserving the fine observation scales needed to investigate local phenomena.

In the earlier studies of the Fram Strait marginal ice zone, uncalibrated SARs were used to map the distribution and structure of the ice pack across the ice edge. Previously unknown structures within the ice pack and along its edges were identified and investigated in detail. SAR imagery provided major insights into the mesoscale oceanography associated with these regions.

Fram Strait Mosaic. Figure 11.23 shows a well-known example of a marginal ice zone SAR mosaic. Optically processed L-band SAR data at 3-m resolution (single look) were collected under summer ice conditions. At the time of data acquisition, the air temperature at the surface was above freezing. The large eddy at the center of the image has been associated with the ocean current structure over the Molloy Deep (Johannessen et al. [39]) and is a commonly found feature at this location. The large old-ice and first-year ice floes seen

Figure 11.23 Fram Strait marginal ice zone mosaic. This mosaic was constructed from *L*-band SAR data acquired on July 5, 1984. The large eddy in the upper half of the image is a common feature associated with the Molloy Deep. The SAR was uncalibrated and the data were acquired and processed using optical technology.

within the ice pack are Arctic ice which is being advected southward by the surface current. Also visible in the image are other ice-edge eddies and a current jet.

Provided that ice floes form a single layer (no significant rafting) on the ocean and provided that the scattering cross section of the ice does not vary with ice type (wet ice surfaces), the magnitude of the radar returns from a cluster of small floes can be interpreted in terms of ice concentration. Burns et al. [8] analyzed SAR data and aerial photography of summer ice in the Fram Strait to show that ice concentration accuracies on the order of 14% can be obtained from algorithms based on SAR image intensities.

Labrador Sea Mosaic: Open Pack. Figure 11.24 is a mosaic of *C*-band SAR imagery of the southern edge of the Labrador marginal ice zone during LIMEX'89. The radar used for these measurements was a calibrated, digital imaging system operated in wide-swath mode (20-m resolution at seven looks) and the images which form the mosaic were processed in real time. At the time

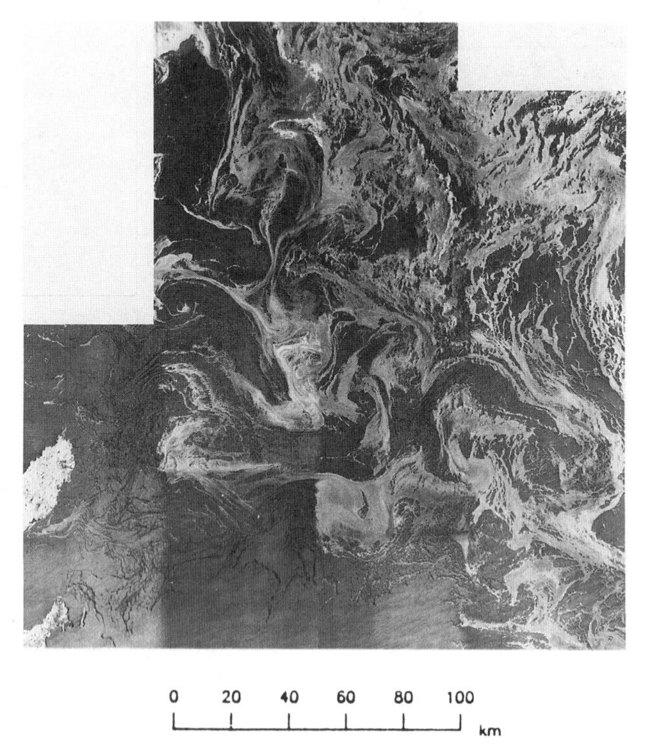

Figure 11.24 Southern edge of the Labrador Sea ice pack, March 19, 1989. This LIMEX mosaic shows ice streamers in open water near Cape de Verde (on the left-hand side of the image Cape Bonavista is the upper point of land and Cape de Verde is the lower), Newfoundland. Ice growth is evident in the dark streamers near the bottom of the image. This C_{VV} mosaic was made using digital data from a calibrated SAR.

of data acquisition, the air temperature was below freezing (as can be seen by the suppression of radar returns from the ocean in the near-shore slush streamers) and the wind was blowing towards the shore (note the wind shadow of Baccalieu Island off Cape de Verde). The sea-ice cover was small floes (<20-m diameter) of first-year and younger ice, being transported southwards by winds and currents. The complex flow patterns seen in this image were produced by wind-driven surface currents (Ikeda [38]). Since the ice floes are all smaller than the resolution cell of the radar, each image pixel contains contributions from both the ice and the surrounding ocean. Similar conditions existed near the ice edge shown in Fig. 11.23.

Labrador Sea Mosaic: Compressed Pack. Figure 11.25 is a mosaic of *C*-band SAR imagery of a compact and highly compressed ice margin, recorded along the Newfoundland coast in March 1987. The SAR data were processed in real time to 6-m resolution, seven-look images. The ice pack had been driven onshore by winds and was rafted up to three layers in thickness. Compression forces were sufficient to trap vessels not designed for icebreaking operations. Daytime air temperatures were just above freezing and the ice surface was wet. Bright, linear features in the mosaic depict ice rubble from the action of shear forces, marking boundaries between regions that are moving differentially within the ice pack (Drinkwater and Squire [19]). No ridges existed within the ice cover. The large, rounded "floes" are patches of weakly consolidated small rough floes which were composed of consolidated pancakes of first-year ice. A mesoscale eddy feature can be seen at the ice edge and careful examination of the image shows wave modulation of the sea-ice returns within the ice rubble at the ice edge.

Waves in Ice. The previous discussion has concentrated on the use of SAR imagery to study large-scale features in the marginal ice zone. Figure 11.26 uses the full resolution of the CCRS SAR to examine sea-ice features created by ice motion during the imaging process. In this scene, the azimuth velocity component of the ocean wave field alters the SAR image focusing to create a condition known as velocity bunching, in which the spatial position of the focused radar reflection in the scene shifts periodically along the SAR's direction of motion. This is the principle mechanism which renders ocean waves travelling in the azimuth direction visible to SARs.

The sea ice in Fig. 11.26 is the high-return (bright) area in the image and the open-water surface is the low-return area. The ice is part of a streamer of small (<10-m diameter) mobile floes with spacing less than a radar resolution cell. Although the ocean wave field is the same for the open water and the ice-covered surface and the SAR image displays recognizable wave modulation for both surfaces, the appearance of the wave field for the two regions is very different. Wave modulation in the ice-covered area produces a complex image structure, characterized by sharply defined spatial intensity variations which are products of nonlinearities in the velocity bunching mechanism (Vachon et

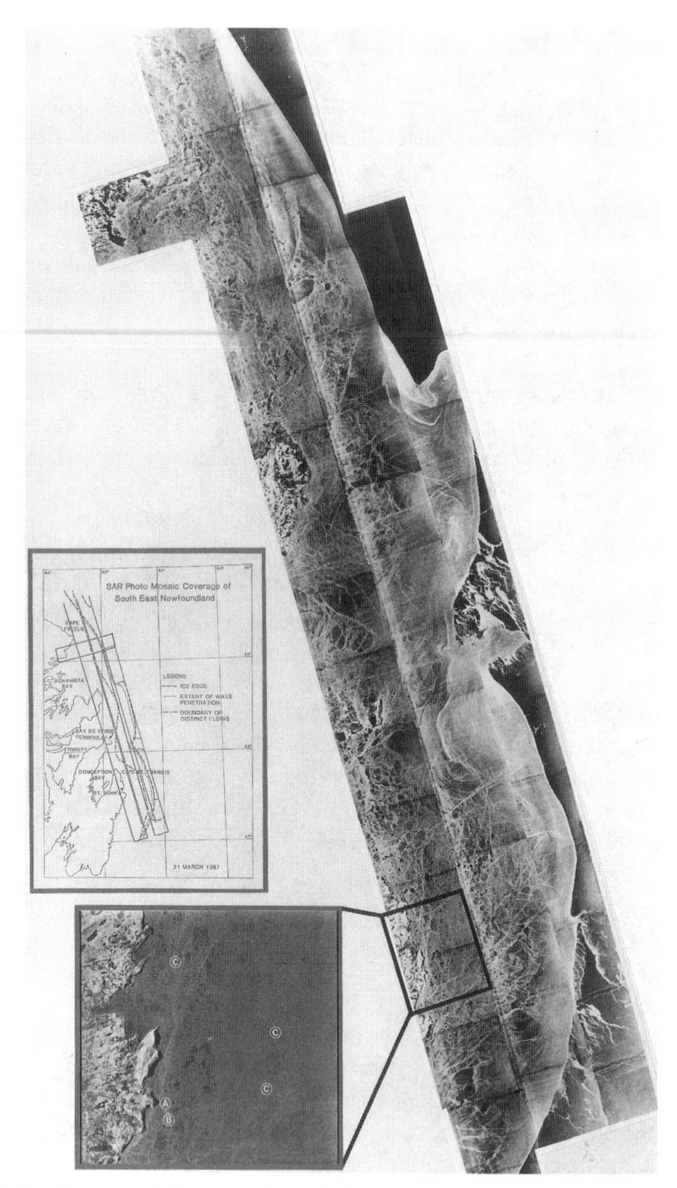

Figure 11.25 Compressed ice pack along the Newfoundland coast during LIMEX'87. Shear motion of the ice causes the bright, linear rubble features in the digital C_{HH} images used to create the mosaic.

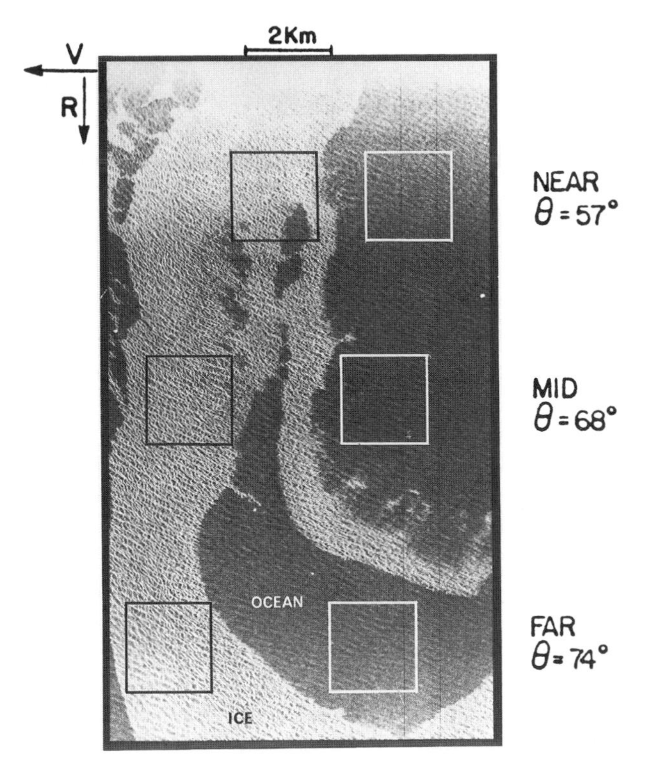

Figure 11.26 Wave signatures in sea ice and oceans as seen in C_{HH} SAR imagery during LIMEX'87. The rectangles in the image highlight changes in the radar signature with incidence angle.

al. [11]; Raney et al. [70]). Wave modulation of the open-water area lacks the fine structure visible for the ice surface and the spatial variation of image intensity is more gradual. Vachon et al. explain the difference in behavior of the ice and open-water surfaces in terms of the scene coherence time since the velocity bunching nonlinearities become more evident when the lifetime of scattering centers (scene coherence time) is long with respect to the SAR aperture time. For sea ice, the scattering centers are defined by the surface structure of the ice and are invariant over times much longer than either the SAR aperture time or the ocean wave period. The ice also acts as a natural low-pass filter and only responds to the lower frequency components of the wave height spectrum. For open water, the scattering centers are patches of short wavelength wind waves (Bragg scatterers) whose lifetimes are much shorter than the SAR aperture time and the high-frequency components of the waveheight spectrum are constrained only by hydrodynamics. The motion spectrum of the open-water surface is much broader than the motion spectrum of the ice surface. Other scenes of waves propagating into the ice cover of a marginal ice zone have been used to compute the wave attenuation coefficient caused by the ice (Liu et al. [48]).

11.2.4 Incidence Angle Dependence of Sea-Ice Scattering Cross Sections

Background

The normalized scattering cross section of a surface, σ^0, expresses the power returned to a radar, per unit surface area, as a fraction of the power that would have been returned by an ideal, isotropic scatterer. Therefore σ^0 describes distributed targets in a manner that is characteristic of the surface material and its wavelength normalized geometry. It is a dimensionless quantity that varies with radar frequency and incidence angle. The relationship between σ^0 and the complex radar reflectivity, Γ, is defined in Chapter 6 and will not be pursued here.

In making radar measurements of the earth's surface, the backscattered energy from the spatial volume describing each radar resolution cell, Fig. 11.2, is interpreted in terms of an effective σ^0 on a locally plane surface at incidence angle θ. With this approximation, it is possible to invert (6-31) to extract an effective scattering cross section, $\sigma^0(\theta)$ (Livingstone and Drinkwater [49]; Ulander et al. [80]), which is characteristic of the scattering physics of the surface but contains an implicit dependence on the local terrain slope. For sea-ice scenes, the surface relief is relatively small and the slope effects are localized to the immediate vicinity of surface relief features. For example, when ocean waves propagate through a marginal ice zone, waves travelling in the range direction are made visible by their slope modulations.

Most of our knowledge of the relationships between radar frequency, radar polarization, ice type, and incidence angle dependence comes from measurements made by surface and profiling scatterometers. An overview of this information may be found in Chapter 2.

Scattering Regimes

In Chapter 2, the principles underlying the incidence angle dependence of radar waves scattered from sea ice are discussed. The concepts of surface and volume scattering are introduced and rough and smooth surfaces are defined in terms of radar wavelength and ice structure.

Looking at a surface which has an arbitrary roughness distribution, the incidence angle dependence of the scattering cross section can be subdivided into three regimes, as shown in Fig. 11.27:

1. The normal, or vertical, incidence angle regime spans the incidence angle range from 0° to about 20° for a rough ice surface and 0° to about 10° for smooth ice surfaces. In this regime, σ^0 is dominated by returns from those surface elements whose local (surface normal) incidence angles lie closest to 0°. For very smooth or very regular surfaces, the phase struc-

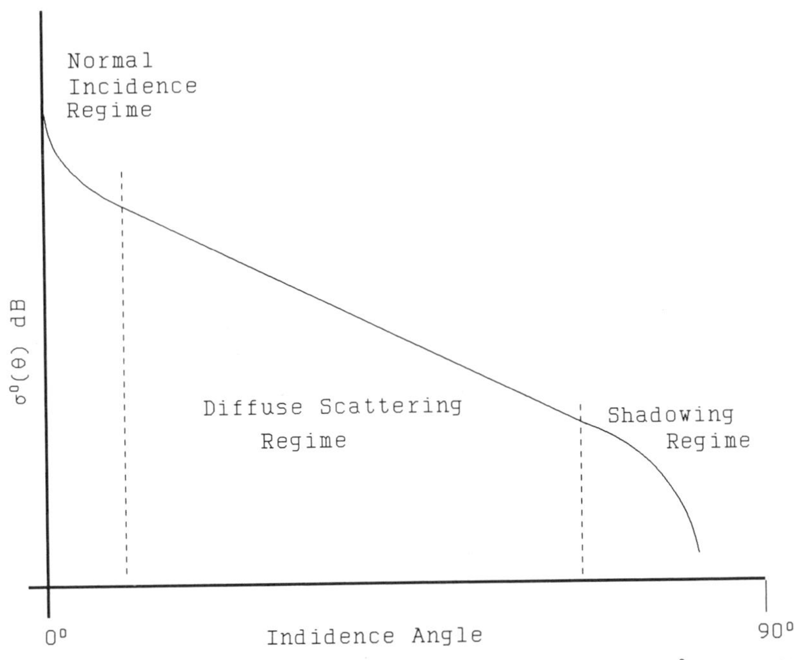

Figure 11.27 The variation of normalized radar scattering cross section, σ^0, with incidence angle and scattering regimes.

ture of the signal is preserved on reflection and the dominant surface reflections are specular.

2. Over the incidence angle range from about 20° to about 80°, σ^0 is dominated by diffuse scattering in which all surface elements contribute to the background field, according to the surface roughness statistics. Here, the radar cross section varies relatively slowly with incidence angle for rough surfaces. It is this regime in which Bragg scattering effects are observed.
3. For incidence angles greater than about 70° for rough ice or about 80° for relatively smooth ice, some parts of the surface are shadowed by adjacent ice and do not contribute to the scattered energy. In this region, σ^0 decreases more rapidly with incidence angle than in the diffuse scattering region and large-scale surface deformation features (ridges, blocks, rubble, etc.) become increasingly significant. At incidence angles greater than 85°, the deformation features may dominate the radar returns from the ice.

Observational Evidence From Historical SAR Data

A qualitative feel for the dependence of the scattering cross section of sea ice with incidence angle can be gained from examination of the SAR image sequence shown in Fig. 11.28. These images were acquired during the winter of

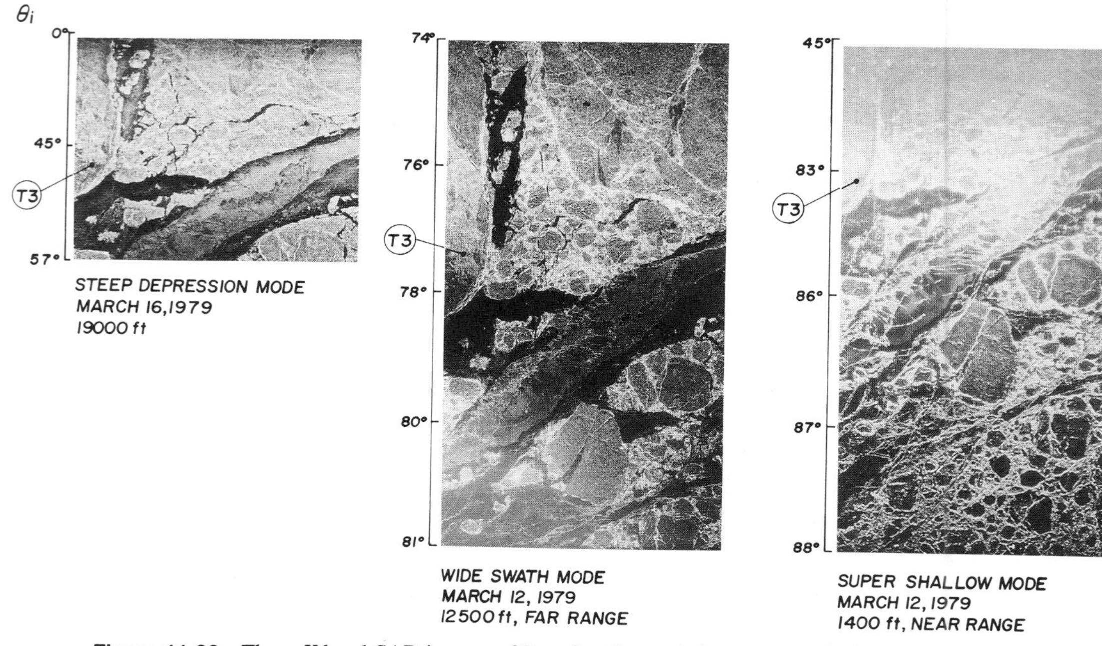

Figure 11.28 These *X*-band SAR images of Beaufort Sea pack ice were acquired over a 4 day period to investigate the effect of incidence angle on X_{HH} SAR imagery of winter sea ice in the Beaufort Sea. These 1979 SAR data are uncalibrated.

1979 in south-central Beaufort Sea by an uncalibrated X-band SAR, for which data were recorded optically on photographic film and later were processed digitally. In addition to the inherent nonlinearities of optical recording, several radar artifacts are present in these images:

1. The image intensities are weighted in range by the elevation pattern of the radar antenna.
2. The steep depression mode image contains radar saturation artifacts which are seen as intensity graded, suppressed returns near some first-year/old ice boundaries.
3. The super-shallow mode image is noise limited for incidence angles greater than 84°.
4. All images are recorded in slant range and contain the characteristic cylinder distortions mentioned in Subsection 11.1.1. These are obvious in the image set as geometric compressions in the range direction for image elements between 0° and 45° incidence angle.

Because of the artifacts and as the radar was not calibratable, the characteristic variation of σ^0 with incidence angle cannot be extracted from the images. However, at any given incidence angle, the relative variation of σ^0 between ice types, and their relation to deformation features, can be discerned and evaluated qualitatively. All three images contain a large number of common surface features consisting of cold ($< -25°C$) sea ice which was both mechanically and thermally stable over the observation period.

The ice island, T 3, at the upper left-hand side of all images, had an embayment (darker image tone) at the top of the image which consisted of very old (>15 years) multiyear ice. To the right of T 3 is first a narrow band of relatively undeformed first-year ice and then a densely packed group of large to vast old ice floes set in a matrix of rubbled first-year ice. Below T 3, and extending diagonally across the imagery, is a refrozen lead covered by first-year ice (dark returns) and gray ice at two different development stages (medium gray tone with numerous small fractures and finger rafts). The lower part of the imagery contains old ice floes of various sizes in a first-year ice matrix.

Starting with the steep depression mode image (Fig. 11.28), the unsaturated first-year ice returns vary more rapidly with incidence angle than the old-ice returns, as is expected when a surface scattering medium is compared to a volume scatterer. Some detail is visible within the old ice floes, as a result of their hummock and melt pond distribution, but these textural contributions are small over this incidence angle range. The tonal differences of the gray ice are related to differences in the ice material: the youngest ice had the highest frost flower concentration and produced the brightest radar returns.

For incidence angles smaller than 50°, the radar returns from deformed first-year ice do not vary significantly from the surrounding old-ice floes. At incidence angles exceeding 50°, the deformation features are beginning to become more prominent, as expected from local slope effects, but all ice types are still in their diffuse scattering regime at 57° incidence angle. In this image, the scattering enhancements associated with the normal incidence regime are masked by the elevation gain pattern of the radar antenna.

In the wide swath mode image (Fig. 11.28), the incidence angle ranges from 74° to 81°. Beyond 78° incidence angle, the first-year ice surface returns are dominated by deformation features (grazing angle regime) as is expected for a smooth, surface scattering medium. The texture visible on the surface of the old-ice floes becomes more prominent with increasing incidence angle as the hummock slopes begin to dominate the radar returns. Ridges and rubble areas become increasingly prominent as the incidence angle increases. At first glance, the incidence angle sensitivity of the gray ice region is somewhat surprising, in that the smooth, flat, low-relief (rafting heights are less than 20 cm) material remains in the diffuse scattering regime at incidence angles exceeding 80°. However, the gray ice was densely covered by frost flowers (see Subsection 2.1.1), which totally dominate the radar scattering at *X* band (9.35 GHz). These features project a few centimeters above the ice surface and are sufficiently sparse that they do not shadow each other. However, elements of their dendritic structures are resonant scatterers at this radar frequency and they are very efficient radar targets.

In the super shallow mode band image (Fig. 11.28), details at incidence angles less than 84° must be ignored due to the poor SNR in these data. The grazing angle scattering regime for gray ice and old ice starts at about 85°. At larger incidence angles, only hummock tops and ridges are detected and surface deformation features dominate the imagery.

At radar frequencies below *X* band, the grazing angle "knee" of the scattering cross section for each ice type shifts to smaller incidence angles as frequency decreases. In Fig. 11.14, the diffuse scattering regime for old ice at *L* band only extends to approximately 50°, a shift of 36° from the diffuse scattering regime boundary at *X* band. The incidence angle dependence of the scattering cross section for gray ice shifts towards that of first-year ice as the radar wavelength increases and the resonant scattering contribution of frost flower dendrites becomes increasingly rare.

Direct Measurement of Incidence-Angle Effects by SAR

A quantitative view of the incidence-angle dependence of gray ice scattering at *C* band is shown in Fig. 11.29. The data were acquired by a radiometrically calibrated SAR, which imaged gray ice in the Labrador Sea in 1989. The normal incidence regime is seen clearly as an increase in scattering cross-section slope as the incidence angle decreases below 10° (the reduction in σ^0 between 0° and 3° is a measurement artifact). The diffuse scattering regime occupies the incidence angle range 15° to 60° and the knee of the grazing angle regime is found at about 65°. As is expected from the radar frequency dependence discussed in Chapter 2 (Section 2.4), the incidence angle boundary between the diffuse scattering and the grazing angle regimes for gray ice at *C* band (5.3 GHz) occurs between that for *X* band (9.25 GHz) and that for *L* band (1.245

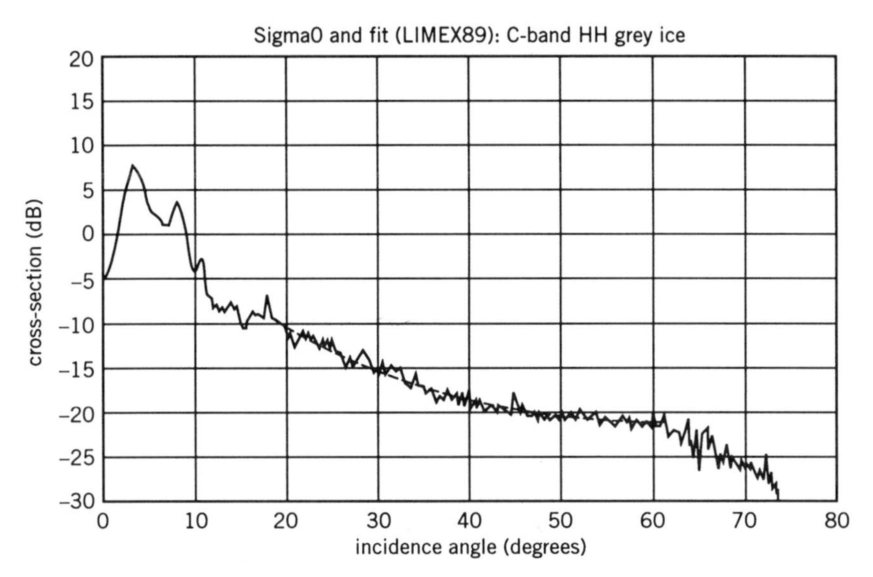

Figure 11.29 C_{HH}-band radar cross section of cold, rough, gray ice in the Labrador sea, 1989. These data were measured by a calibrated SAR.

GHz). As the *X*- and *L*-band radars that produced the data shown in Fig. 11.14 were uncalibrated, no quantitative relationship can be established between the radar frequency and the incidence angles at the scattering regime boundaries from these data.

Figure 11.30 shows an extension of the incidence angle dependence of radar scattering to rough first-year ice, as detected by a calibrated SAR. In this case, the ice was very rough with extensive rafting (Fig. 11.22), typical of the Labrador Sea marginal ice zone. The normal incidence scattering regime is found at incidence angles less than 10° and the grazing incidence regime starts at about 75° incidence angle. From the discussions in Chapter 2, both of these results are expected for this ice condition.

11.2.5 Frequency Dependence of Sea-Ice Scattering Cross Sections

Background

From the discussions of the small-scale structure of ice and snow in Chapter 2, it is evident that there is no "characteristic" scattering element dimension, a_0, within the volumes, or on the surfaces, of these materials. Rather, the scale sizes corresponding to scattering centers exist as a continuum spanning sub-millimeter to meter dimensions. The scattering element populations depend on the type of material, its evolutionary stage, and its degree of deformation. As different radar wavelengths will measure different scale size intervals in the scatterer populations, different features will be observed at different frequen-

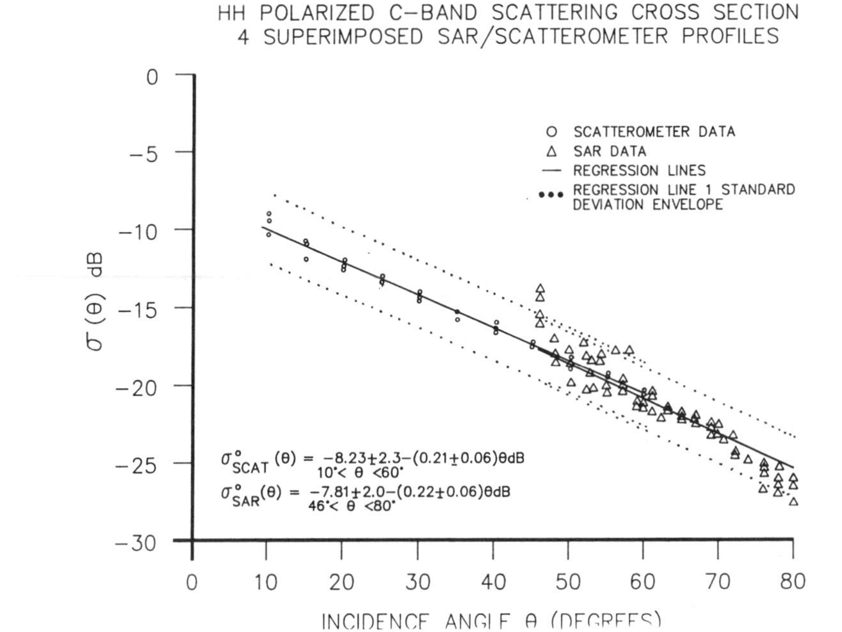

Figure 11.30 C_{HH}-band radar cross section of wet, rough first-year ice in the Labrador Sea, 1987, as measured by a scatterometer and a calibrated SAR.

cies. Information extracted from radar images which were acquired at different frequencies is expected to be complementary. At any frequency, spatial structures associated with local variations in scattering properties are evident in the radar images only if the radar has a sufficiently high resolution. Usually SAR is required to observe these features.

Accumulated empirical observations have shown that imaging radar systems operating in the frequency range from 430 MHz (*P* band) to 37 GHz (*Ka* band) are useful for sea-ice measurements. Under cold conditions, radar frequencies above 10 GHz (*X* band) primarily provide information on ice surface roughness and on volume scattering in the ice and snow cover, whereas *L* band and lower radar frequencies primarily provide information on the mechanical deformation features of the ice. Volume scattering contributions to the microwave backscattering cross section of sea ice are more significant at higher radar frequencies than at low frequencies, due to the relative scarcity of physically large, volume-distributed scatterers within the upper layers of the ice. During the series of material transformations leading to and coincident with ice melt, lower radar frequencies are believed to have some advantages. From the experimental and theoretical data available to date, there is no "best" radar frequency from the viewpoint of sea-ice measurement.

Sea-Ice Signatures in Historical SAR Data

The pairs of simultaneously recorded images in Figs. 11.11 and 11.12 provide a qualitative view of the frequency dependence of sea-ice scattering during freeze-up conditions in the Beaufort Sea. The data were acquired in late October and early November 1981 by an uncalibrated SAR. The scattering cross sections cannot be computed from the radar data and quantitative relationships between the four images cannot be established. However, within each image, the relative returns for each of the ice types present can be estimated at a given incidence angle. In these images, antenna elevation pattern artifacts are evident in the *X*-band data as both a range-dependent intensity modulation and an SNR degradation at the edges. The *C*-band image suffers from SNR degradation at the far edge of the swath.

For the *X*- and *L*-band images in Fig. 11.12, it is evident that the *X*-band radar provides more detailed information about the ice surface and the *L*-band radar preferentially maps the fractures and deformation zones, as expected from the preceding discussion. At *X* band the radar returns from very young ice covered with frost flowers (surface scatterers) and old ice (volume scattering) are within 2 dB of each other, indicating that the frost flower crystals are efficient scatterers at *X* band. The low *L*-band returns from the areas with frost flowers shows that the crystal dimensions are much less than the wavelength (24 cm) at this frequency. The young ice marked "gray-white ice" on the right-hand side of each image has two zones which are resolved readily when the *X*- and *L*-band data are compared. The upper portion of this area (at A) appears darker at *X* band and lighter at *L* band. It is the thicker (older) ice and may be at the gray-white stage of development. The *X*-band image shows surface structure that suggests that this ice was fractured and wind-blown while still very thin. The *L*-band returns from this area suggest that the surface was rough at centimeter scales. The younger ice returns (at B) give evidence of several growth stages and frost flower coverage (high cross section at *X* band and low cross section at *L* band). The *L*-band radar returns from this area detect only minor deformation during the growth stages.

The *X*- and *C*-band image pair in Fig. 11.11 were collected a few days earlier than the imagery in Fig. 11.12 but under the same thermal and environmental conditions. Prominent features in the scene are two vast old-ice floes (one on the left and one to the right of center), two actively opening leads, and a prominent region of gray-white ice to the left of center. The remainder of the scene shows young-ice growth stages up to and including gray-white ice. Old-ice floes are present in a wide range of sizes. In the *X*-band image the most prominent features are the old-ice floes and the young-ice growth pattern bordering the left-hand lead. The relatively high returns from the old-ice floes and suppression of surface deformation features at *X* band (9.35 GHz) are clear evidence that volume scattering processes are dominating the old-ice signatures at this frequency.

In the *C*-band image, the old ice floes have scattering cross sections similar to those of gray ice, whereas in the *X*-band image the scattering cross section for the old-ice floes is significantly larger than that of gray ice. Deformation features on and between old-ice floes are more visible at *C* band than at *X* band. Volume scattering is, thus, a smaller component of the old-ice signatures at *C* band than at *X* band in this data set.

From the low-temperature, empirical, frequency dependence relationships discussed in Chapter 2, we expect the old-ice returns at *C* band to be about 5 dB smaller than at *X* band and we expect the *C*-band gray ice returns to be about 3 dB smaller than at *X*-band. The image contrast between these ice types should thus change by about 2 dB. As estimated from the optically processed images (this is an analogue data set), the multiyear ice contrast shift from *X* band to *C* band is about 6 dB. A probable explanation for the discrepancy is that the freezing horizon in the old ice has not yet reached its winter depth and the *C*-band volume scattering component is not developed fully. If this hypothesis is true, the seasonal variation of microwave scattering signatures for sea ice has a frequency-dependent aspect that has not been investigated quantitatively to date.

Examination of the gray ice region in Fig. 11.12 reveals that the *X*- and *C*-band SARs provide complementary information about the growth and structure of young ice. The differences in appearance of gray ice in the two bands are a result of *X*-band sensitivity to ice surface features and *C*-band penetration into the ice. The patterns observed at *C* band at "C" are reminiscent of convection patterns where frazil ice is concentrating to form a grease ice sheet under calm conditions. This effect is discussed briefly in Chapter 2. In the area of gray ice at "D," both the *X*- and *C*-band SARs are responding to frost flowers on the ice surface. The sequence of SAR gray tones in the lead on the right-hand side of the *X*-band image is a typical congelation pattern for ice forming from frazil slush in the presence of a light wind. In the *C*-band response to this pattern, the earliest growth stages are imaged as a smooth surface.

Generalizations

Some generalizations follow about the complementariness of radar frequencies for freeze-up and winter conditions:

1. A two-frequency radar system, in which the lower frequency is above *L*-band, has better potential for ice type classification than a single-frequency radar. Significant differences in sea-ice signatures are often observed even if the two frequencies are as close as *X* band (9.25 GHz) and *C* band (5.3 GHz).
2. A two-frequency radar system, in which the lower frequency is at, or below, *L* band, is a powerful tool for the study and measurement of ice deformation features.
3. A larger frequency spread is required at steep incidence angles (16° to 30°) than at shallower incidence angles ($> 45°$) to separate ice type and deformation effects.

4. From other studies presented in Carsey [12], radars at frequencies above *Ka* band (37 GHz) may be influenced too strongly by snow cover to be useful for ice property measurements. Very little sea-ice scattering data exist at these frequencies.

11.2.6 Polarization Effects

Background: Concepts, Polarimetry, and Polarization Diversity
Synthetic aperture radars transmit and receive polarized radiation (usually linearly polarized). The response of the imaged terrain to the polarization of the incident (illuminating) wave provides information about the dominant scattering processes on, and within, the imaged surface. From the discussions in Subsection 11.2.5, we expect the polarization-state matrix, which describes the scattered signals from a sea-ice surface (6-35), to be frequency dependent. For natural targets, the ensemble means of the offdiagonal terms, Γ_{HV} and Γ_{VH}, in (6-35) have equal magnitudes.

Two classes of SAR are used to provide information about the polarization sensitivity of sea-ice scattering processes:

1. Polarization diversity radars transmit a selected polarization and receive both like- and cross-polarized returns as mutually coherent or incoherent data.
2. Polarimetric radars transmit time-multiplexed, orthogonal polarizations and receive like- and cross-polarized returns in a mutually coherent form.

The discussions in this subsection will use the linear polarization convention followed for most remote sensing SARs and will use the subscript notation defined in Subsections 6.4.1 and 6.4.2 to describe the polarization state of the scattered signals.

The polarization sensitivity of sea-ice scattering provides information on the scattering processes (surface scattering and volume scattering) that are active at the ice surface and identifies cases where preferentially oriented, asymmetric scatterers (frost flower dendrites, hoar crystals, drainage channels, etc.) make significant contributions to the radar returns. In quantitative discussions of the polarization dependence of radar waves from sea ice, two derived parameters are of particular interest:

- the copolarization ratio, $r_p = \sigma^0_{HH}/\sigma^0_{VV}$
- the cross-polarization ratio, $r_c = \sigma^0_{HV}/\sigma^0_{HH}$

The discussions about electromagnetic wave scattering in Chapter 2 provide indications of the expected behavior of r_p and r_c for sea ice. Surface scattering from ice surfaces that are randomly rough at the radar wavelength yields ensemble mean estimates of $r_p \approx 1.0$ and $r_c \ll 0.1$. The presence of volume scattering within the ice or the presence of multiple surface scattering at large deformation features increases r_c to values greater than 0.1. Multiple-bounce surface scattering should yield mean values of $r_p > 1.0$ but the statistical distribution function describing r_p measurements may be broad. Vertically oriented scattering elements and preferential coupling of V-polarized radiation into smooth ice surfaces yield $r_p < 1.0$. Similar results are expected at low radar frequencies, when the penetration depth of the ice (Chapter 2) is sufficient that the dielectric tensor of the columnar ice layer influences the observed scattering.

SAR Measurements of Sea-Ice Polarization Sensitivity

At the time of writing, the only polarimetric SAR data sets for Arctic sea ice which are described in the literature, were collected in the Beaufort, Bering, and Chukchi seas by the JPL *C-L-P*-band radar in March 1988 (Drinkwater et al. [20, 21]). As these data are limited to a few areas and a narrow range of conditions, most of our knowledge of the polarization sensitivity of sea-ice scattering signatures comes from polarization diversity instruments (surface and airborne scatterometer and airborne SAR measurements).

The study of the polarimetric signatures of sea ice remains an active research area and no firm conclusions about the relative merits of polarization diversity and polarimetric SAR instruments for sea-ice surveillance can be made at this time. The remainder of this section presents examples of the polarization sensitivity of sea-ice signatures as seen in SAR image data and relates these to the electromagentic scattering section of Chapter 2.

Polarimetric SAR Measurements of Sea Ice

Drinkwater et al. [20, 21] present results of their analysis of two Beaufort Sea scenes and one Bering Sea scene from the three-frequency polarimetric SAR data set referred to previously. The data were acquired under winter ice growth conditions and contain no surface verification sites. Their results are in general agreement with the expectations outlined above and provide some quantification of these relationships.

At C band, measurements of the copolarization ratios, r_p, for thick first-year ice and for old ice show the expected behavior for surface scattering from a rough material and for scattering from a medium which has both surface and volume scattering contributions ($\lambda = 5.67$ cm, $r_p = 1.06 \pm 0.67$ for first-year ice; $r_p = 0.94 \pm 0.43$ for old ice). The mean cross-polarization ratios, r_c, for these two cases were measured at 0.045 ± 0.011 (-13.5 dB) and 0.110 ± 0.005 (-9.6 dB), respectively (for the three examples of each class analyzed). Two thin-ice scattering categories were reported: one with $r_p = 0.96$ (-0.18 dB) and $r_c = 0.034$ (-14.7 dB); and the other with $r_p = 0.33$ (-4.8 dB) and $r_c = 0.082$ (-10.9 dB).

The reductions observed in r_p, when L band or P band are used instead of C band, are consistent with the expected behavior of scattering signatures when the penetration depths are large. For L band: $\lambda = 23.3$ cm, $r_p = 0.33 \pm 0.43$ for first-year ice; $r_p = 0.43 \pm 0.03$ for old ice. For P band: $\lambda = 69.8$ cm, $r_p = 0.39 \pm 0.17$ for first-year ice; $r_p = 0.54 \pm 0.06$ for old ice. At the longer wavelengths, the scatterers on the ice surface and throughout the ice volume are small with respect to the radar wavelength. The polarization sensitivity of the Fresnel reflection coefficients combines with the dielectric constant anisotropy of the deep (columnar) ice to dominate the scattering signatures.

Extremely rough, deformed ice in the Bering Sea scene exhibits scattering properties consistent with strong contributions from multiple scattering. For this ice, Drinkwater et al. [21] report a mean L-band r_p of 0.80 from a statistical distribution whose tail extends up to $r_p = 2.00$. The corresponding r_c is reported to be 0.120 and when combined with the copolarization ratio suggests the presence of multiple scattering on the ice surface. More typical rough first-year ice signatures from this area have r_p about 0.54 and r_c about 0.091. No C-band results are quoted for this scene.

Although Drinkwater et al. have analyzed only a small portion of the available data set, their results indicate that further work to determine the polarimetric statistics of sea ice will be of value. To evaluate SAR polarimetry fully as a tool for sea-ice reconnaissance, many more data sets must be acquired and analyzed.

Polarization Diversity in SAR Measurements of Sea Ice

During the LIMEX'89 experiment, conducted in the Labrador Sea in March 1989, the CCRS CV 580 SAR system was operated as a calibrated polarization diversity radar. On 14 March 1989 a pair of flight lines was used to generate a set of SAR scenes in which a common strip of Labrador pack ice was imaged with both H- and V-transmit polarizations. As the ice was stationary and the environmental conditions were stable over the imaging interval, the data can be combined to estimate r_p (the measurement reported was r_p^{-1}) and r_c. the registered pair of r_p^{-1} and r_c polarization ratio images are shown in Fig. 11.31 to illustrate the range of detail found in these parameters. The scenes show gray ice (right-hand two-thirds of the images) and nilas (left-hand one-third of the images plus interstitial patches throughout the scene). The ice had formed under relatively quiet conditions in a slowly diverging ice pack. Its characteristics were measured a few hours before the SAR flight by the ice research vessel, MV *Terra Nordica*, whose track is visible as a dark horizontal line along the bottom of both images. All the bright areas in the r_p^{-1} image are nilas covered with frost flowers, for which the VV-polarized radar returns are correlated with the spatial distribution of frost flowers. As the incidence angle range for the VV-polarized data is 41° to 62°, VV return enhancement due to Brewster angle effects may contribute to r_c. The ratio r_p^{-1} is approximately 1.00 in the dark areas, corresponding to the ship track and the rafted ice edges. It has a mean value of approximately 4.00 in the bright regions in the lower third of this image. The mean value of r_c is 0.050 in the gray ice region. The numeric values of the polarization ratios estimated from these data must be treated with caution because they contain a correction for the 10° incidence angle offset between the H-transmit and V-transmit scenes used to generate the images.

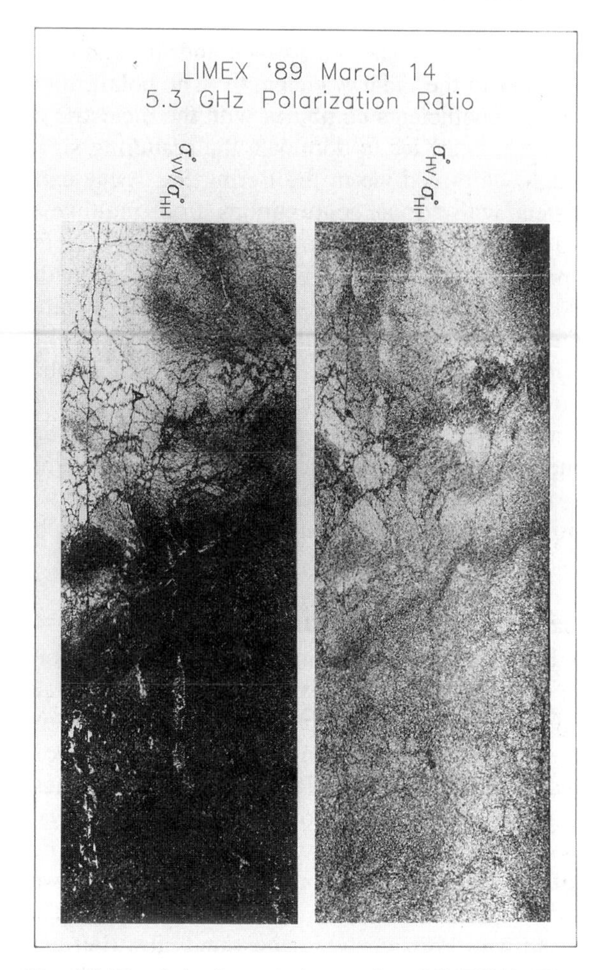

Figure 11.31 *C*-band SAR polarization ratio images from LIMEX'89. The symbol, A, in the copolarization image identifies a point in the overlap between this figure and Fig. 11.32. Bright regions in the imagery correspond to large polarization ratio values.

Figure 11.32 shows C_{HH}, C_{VV}, and X_{VV} imagery from the same data set as Fig. 11.31 to illustrate the combined effects of polarization and radar frequency on the imaging process for young ice. These images overlap the data presented in Fig. 11.31 near the ends of both scenes. The vertically oriented rafting pattern seen "A" in the C_{HH} image of Fig. 11.32 is found at "A" in the r_p image of Fig. 11.31. Of particular interest in Fig. 11.32 is the relative contrast between the material types and their radar signatures:

1. Nilas are recognized readily in the C_{HH} image by their rafting patterns. σ^0_{HH} varies with incidence angle and the age of the nilas from -18 to -25 dB across the scene. In the C_{VV} image, the raft edges are not discernible in the radar returns from the ice surface and σ^0_{VV} varies from -15 to -19 dB across the scene. In the X_{VV} image, the radar returns

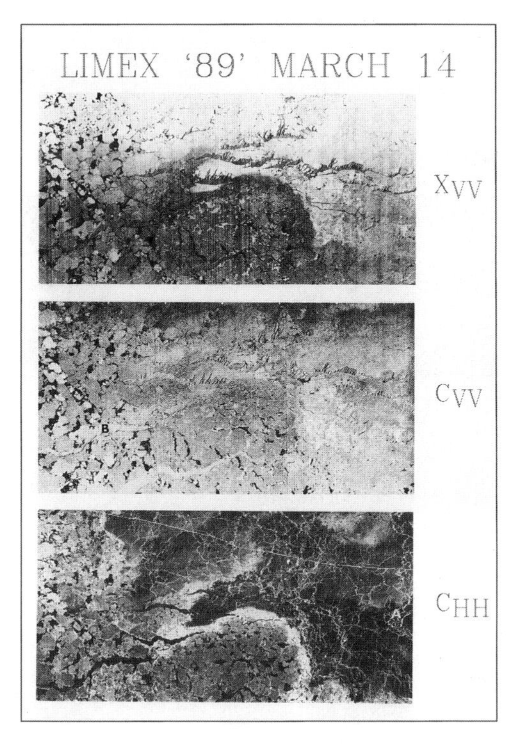

Figure 11.32 Polarization and frequency effects on the SAR signatures of young sea ice under cold (−9°C) conditions.

from the nilas region are the highest in the scene and σ^0_{VV} varies from −13 to −16 dB across the scene.

2. The highest radar returns in the C_{HH} image are produced by nilas fragments which have reconsolidated as ice sheets along the edge of thicker ice. In the C_{VV} image, this ice is indistinguishable from undeformed nilas and in the X_{VV} image its cross section is very similar to that of the adjacent gray ice.
3. The highest radar returns in the C_{VV} image are produced by open water in leads, such as at "B." In contrast, the open water has the lowest returns of any target in the C_{HH} image and has signatures very similar to those for ice on the X_{VV} image.
4. Where the nilas surface has buckled under pressure and has recently flooded, the radar returns are suppressed in the X_{VV} image. These areas are still visible as features in the C_{VV} image and are very difficult to detect in the C_{HH} image.

Historical SAR Observations of Polarization-Sensitive Scattering

The SAR image data shown in Fig. 11.33 were acquired by an uncalibrated SAR system in the Beaufort Sea during April 1982. The *X*- and *L*-band scene

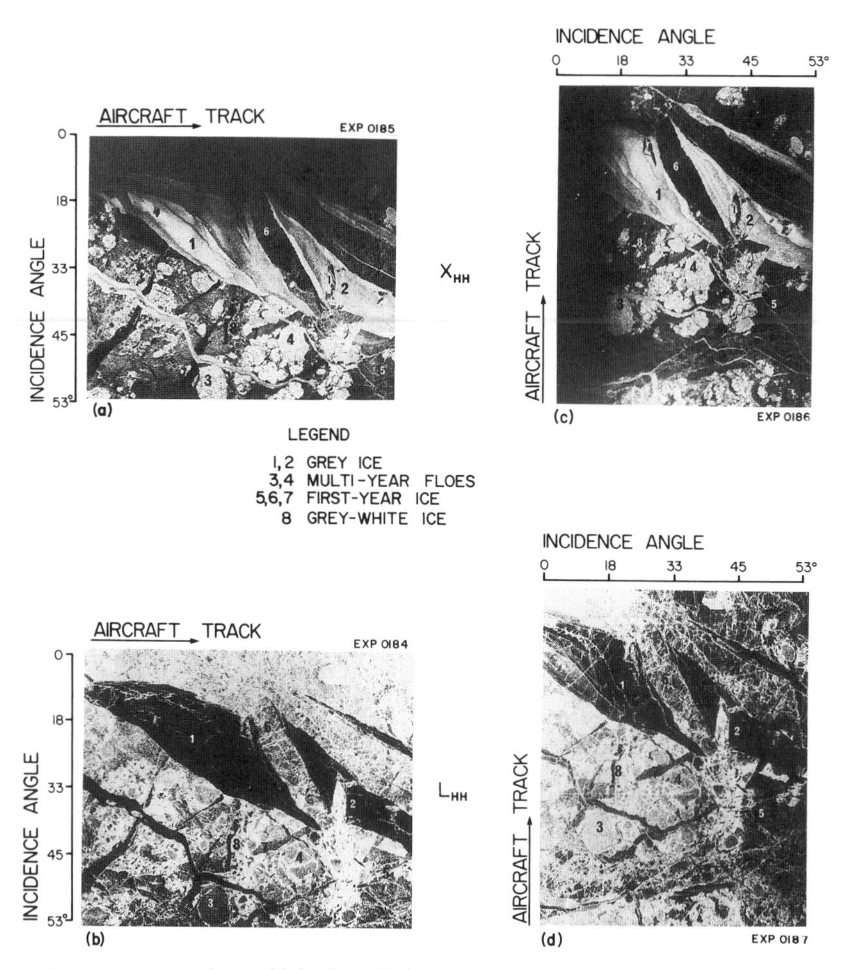

Figure 11.33 Aspect angle sensitivity in SAR imagery of Beaufort Sea ice under winter conditions. These uncalibrated SAR images were acquired on April 15, 1982.

pairs show the same ice from orthogonal viewing directions (90° aspect angle variation) over the same range of incidence angles. Aerial photography of the area showed that nilas and gray ice (forming as the leads widened) were covered by frost flowers. It is interesting to note that the young-ice returns at *X* band are similar in magnitude to the old-ice returns. The young-ice signatures at *X* band are aspect angle dependent but the returns from the surrounding old ice and first-year ice are not. As expected, the young-ice returns at *L* band are much lower than either the first-year or the old-ice returns. No aspect angle sensitivity is seen at *L* band. The aspect angle sensitivity for young ice at *X* band has been interpreted as a prevailing wind effect on the frost flower growth pattern.

Figure 11.14 shows simultaneous *X*- and *L*-band SAR imagery of Beaufort Sea pack ice acquired in March 1979. The absolute scene brightness was shifted to match the response of photographic paper, so only the relative radar returns within each image are significant. In this case the radar was uncalibrated and quantitative relationships between the images cannot be established. When the X-band HH and HV images are compared, it is seen that the contrast between the volume scattering medium, old ice, and the surface scattering materials, first-year ice and gray ice covered with frost flowers, is enhanced in the cross-polarized image. In the L-band image pair, only the ice island shows volume scattering characteristics. The L-band cross-polarized image, however, contains the first-year ice ridge features, probably due to a combination of volume scattering from large scatterers in the fractured ice and multiple-reflection surface scattering. Details of the gray-ice rafting pattern are compatible with single-layer rafts (the edges form distorted dihedral reflectors); they are visible in the L_{HH} image but are suppressed in the L_{HV} image.

Generalizations

Qualitative relationships between sea-ice scattering signatures and the radar polarizations used to image them are well established from ice imagery acquired by uncalibrated polarization diversity SARs. In particular, it is well known that volume scattering effects in old ice and combined volume scattering and multiple-reflection surface scattering in ice ridges enhance cross-polarized radar returns for radar frequencies above C band. It has also been shown that frost flowers in the surface of smooth, young ice enhance VV-polarized C- and X-band SAR returns. Old-ice volume scattering effects and enhanced scattering from frost flowers are not observed at L band. Quantitative relationships corresponding to these historical SAR data sets have been obtained by calibrated profiling sensors for a number of radar frequencies and ice conditions but their generality has been limited by the relatively small sample sizes used. Some of these results are discussed in Chapter 2.

Further study of the polarization sensitivity of sea-ice scattering signatures has been made possible since 1987 by the availability of calibrated polarization diversity SARs and polarimetric SARs. Two useful quantities, r_p, and r_c, can be computed from calibrated data outputs of either instrument. The values of r_p and r_c and their variation over a sea-ice scene provide information on the scattering mechanisms which make significant contributions to measured SAR signatures of sea ice. The ability to measure the r_p is particularly valuable since this parameter is not extracted easily from earlier profiling sensor data, due to spatial registration difficulties. Results obtained from the limited set of quantitative SAR data currently available provide qualitative confirmation of deductions based on present knowledge of the dielectric properties of sea ice:

1. When $r_p \approx 1.0$ and $r_c \leq 0.1$, scattering from a rough surface dominates the SAR signatures.
2. When $r_p \approx 1.0$ and $r_c > 0.1$, there is a significant volume scattering component in the SAR signatures.
3. When $r_p < 1.0$ and $r_c \ll 0.1$, surface scattering dominates and the

surface either has a large population of vertically oriented scatterers or is smooth at the radar wavelength employed.

4. When the statistical distribution of r_p is broad, a significant number of r_p samples is greater than 1.0. When $r_c > 0.1$, multiple-scattering events on a deformed ice surface are probable.

The data available are insufficient to determine the statistical significance of these observations or their sensitivity to ice conditions. In addition, existing electromagnetic scattering models are still too primitive to be useful for quantitative evaluations of the empirical results.

Although polarimetric SAR has the potential to provide more information on the electromagnetic scattering processes at the ice surface than can be obtained from polarization diversity SAR, the data available are insufficient to evaluate the relative merits of these two classes of radar for sea-ice remote sensing. Further research is required in this area.

11.3 SUMMARY

In 1993 the scientific community has a reasonably good qualitative understanding of sea ice: its development and decay cycle, and its physical and electromagnetic properties. Historically, SAR systems have played important roles in the research that has yielded this knowledge. However, large gaps exist in the quantitative understanding of much of the preceding. In particular, our ability to model radar signatures of sea ice accurately from measurements of the ice structure is still very limited. Many believe that, in spite of the enormous effort that has been expended in sea-ice research in the past decade, as much work remains as has already been done. Calibratable airborne and spaceborne SAR sensors will be primary measurement tools in future sea-ice research. It is expected that quantitative SAR measurements of the ice cover will play increasingly significant roles in operational sea-ice reconnaissance.

The references used in this chapter are indicative of the scope of past research and identify many of the important papers in this field. The list is not exhaustive and many equally important papers have been omitted for lack of space to do them justice.

ACKNOWLEDGMENTS

The author is indebted to R. K. Raney, J. R. Rossiter, and E. O. Lewis for their patience and perseverance in criticizing the drafts of this chapter and guiding it to its final form. The author's understanding of SAR and its use for sea-ice remote sensing has evolved over more than a decade as a result of shared insights, criticism, and argument from many research colleagues; W. J. Campbell (deceased), USGS; F. D. Carsey, M. R. Drinkwater, B. Holt, S. A. Digby, JPL; A. Stogryn, Aerojet Electrosystems; T. C. Grenfell, U. Washington; S. P. Gogineni, U. Kansas; R. A. Shuchman, R. G. Onstott, R. F. Rawson, R. W. Larson, ERIM; J. Askne, L. Ulander, U. Goteberg; S. N. Madsen, C. Lintz, N. Skou, J. Dahl, TUD; J. Comiso, NASA Goddard; M. Shokr, J. Falkingham, AES; R. Olsen, Satlantic Inc.; W. F. Weeks, U. Alaska; L. D. Farmer; A. L. Gray, R. K. Hawkins, P. W. Vachon, CCRS; M. Wong, Varian Associates; and many, many others. The majority of the Canadian research discussed in this chapter either has been carried out at the Canada Centre for Remote Sensing or has received support from that agency.

REFERENCES

[1] S. Ahmed, H. R. Warren, M. D. Symonds, R. P. Cox (1989) "The RADARSAT system," in *Proc. IGARSS'89*, 213–217.

[2] W. Akam, R. Deane, M. Sartori, R. Lowry, B. Mercer (1988) "An integrated radar imaging system for the STAR-2 aircraft," *Proc. 1988 IEEE Nat. Radar Conf.*, 28–32.

[3] R. Bamler (1991) "A systematic comparison of SAR focusing algorithms," in *Proc. IGARSS'91*, 1005–1009.

[4] D. G. Barber (1989) "Texture measures for SAR sea ice discrimination: an evaluation of univariate statistical distributions," Earth Observation Lab. Tech. Rept. Ser. ISTS-EOL, TR89-005 (Deprt. Geography, Univ. Waterloo, Waterloo, Ontario).

[5] D. G. Barber and E. F. LeDrew (1991) "SAR sea ice discrimination using texture statistics: a multivariate approach," *Photogrammetric Eng. and Remote Sens.*, **57** (4), 385–395.

[6] J. R. Bennett and I. G. Cumming (1979) "Digital techniques for the multi-look processing of SAR data with application to SEASAT A," in *Proc. 5th Canadian Symp. on Remote Sensing*, 506–516.

[7] B. A. Brusmark, A. Gustavsson, and A. Nealander (1985) "Digital signal processing for spaceborne synthetic aperture radar," FOA (Sweden) Rept. C 30381-E1.

[8] B. A. Burns, D. J. Cavalieri, M. R. Keller, W. J. Campbell, T. C. Grenfell, G. A. Maykut, and P. Gloersen (1987) "Multisensor comparison of ice concentration estimates in the marginal ice zone," *J. Geophys. Res.*, **92** (C7), 6843–6856.

[9] C. Cafforio, C. Prati, and F. Rocca (1988) "Full resolution focusing of SEASAT SAR images in the frequency-wave number domain," in *Proc. EARSel Workshop*.

[10] W. J. Campbell, P. Gloersen, and J. Zwally (1984) "Aspects of Arctic sea ice observable by sequential passive microwave observations from the Nimbus-5 satellite," in I. Dyer and C. Chryssostomidis (Eds.) *Arctic Technology and Policy*, Hemisphere Publishing Co., New York, 197–222.

[11] F. D. Carsey, S. A. Digby-Argus, M. J. Collins, B. Holt, C. E. Livingstone, and C. L. Tang (1989) "Overview of LIMEX'89 ice observations," *IEEE Trans. Geosci. and Remote Sens.*, **27** (5), 468–482.

[12] F. D. Carsey (Ed.) (1992) *Microwave Remote Sensing of Sea Ice*. AGU Geophysical Monog. 68.

[13] J. C. Comiso (1990) "Arctic multiyear ice classification and summer ice cover

using passive microwave satellite data,'' *J. Geophys. Res.*, **95** (C8), 13411-13422.

[14] T. R. Crimmins (1985) ''Geometric filter for speckle reduction,'' *Appl. Optics*, **24** (10), 1438–1443.

[15] G. Cunningham, R. Kwok, and B. Holt (1992) ''Preliminary results from the ASF/GPS ice classification algorithm,'' in *Proc. IGARSS'92*, 573–575.

[16] J. C. Curlander, B. Holt, and K. J. Hussey (1985) ''Determination of sea ice motion using digital SAR imagery,'' *IEEE J. Oceanic Eng.*, **OE-10** (4), 358–367.

[17] J. C. Curlander and R. N. McDonough (1991) *Synthetic Aperture Radar Systems and Signal Processing*, Wiley, New York.

[18] P. deBastiani (1990) *Canadian Marine Ice Atlas—Winter 1987/88*, Canadian Coast Guard, Ottawa, Ontario.

[19] M. R. Drinkwater and V. A. Squire (1989) ''*C*-band SAR observations of marginal ice zone rheology in the Labrador sea,'' *IEEE Trans. Geosci. and Remote Sens.*, **27** (5), 522–534.

[20] M. R. Drinkwater, R. Kwok, and E. Rignot (1990) ''Synthetic aperture polarimetry of sea ice,'' in *Proc. IGARSS'90*, 1525–1528.

[21] M. R. Drinkwater, R. Kwok, D. P. Winebrenner, and E. Rignot (1991) ''Multifrequency polarimetric synthetic aperture observations of sea ice,'' *J. Geophys. Res.*, **93** (C11), 20679–20698.

[22] W. J. Emery, C. W. Fowler, J. Hawkins and R. H. Preller (1991) ''Fram Strait satellite image derived ice motions,'' *J. Geophys. Res.*, **96** (C3), 4751–4768.

[23] M. Fily and D. A. Rothrock (1986) ''Extracting sea ice data from satellite SAR imagery,'' *IEEE Trans. Geosci. and Remote Sens.*, **GE-24** (6), 849–854.

[24] C. Fox and V. A. Squire (1991) ''Strain in shore-fast ice due to incoming ocean waves and swell,'' *J. Geophys. Res.*, **96** (C3), 4531–4548.

[25] A. Freeman (1992) ''SAR calibration: an overview,'' *IEEE Trans. Geosci. and Remote Sens.*, **30**, 1107–1121.

[26] A. Freeman, C. Werner, and Y. Shen (1988) ''Calibration of multipolarization imaging radar,'' in *Proc. IGARSS'88*, 335–339.

[27] A. L. Gray, P. W. Vachon, C. E. Livingstone, and T. I. Lukowski (1990) ''Synthetic aperture radar calibration using reference reflectors,'' *IEEE Trans. Geosci. and Remote Sens.*, **28** (3), 374–382.

[28] P. S. Guest and K. L. Davidson (1991) ''The aerodynamic roughness of different types of sea ice,'' *J. Geophys. Res.*, **96** (C3), 4709–4722.

[29] R. T. Hall and D. A. Rothrock (1981) ''Sea ice displacement from SEASAT synthetic aperture radar,'' *J. Geophys. Res.*, **86** (C11), 11078–11082.

[30] P. M. Haugen and R. Preller (1990) ''Sea ice modelling in the Barents Sea during SIZEX'89,'' in *Proc. IGARSS'90*, 1509–1512.

[31] D. N. Held, W. E. Brown, A. Freeman, J. D. Klein, H. Zebker, T. Sato, T. Moller, Q. Nguyen, and Y. Lou (1988) ''The NASA/JPL multifrequency, multipolarization SAR system,'' in *Proc. IGARSS'88*, 345–349.

[32] W. D. Hibler, III (1985) ''Modelling sea ice dynamics,'' *Adv. in Geophys.*, **28**, 549–578.

[33] T. Hirose, L. McNutt, and M. Manore (1992) "Automated sea ice tracking results for LIMEX'87 and 89 data," *J. Atmosphere—Ocean*, **30** (2).

[34] Q. A. Holmes, D. R. Neusch, and R. A. Shuchman (1984) "Textural analysis and real time classification of sea-ice types using digital SAR data," *IEEE Trans. Geosci. and Remote Sens.*, **GE-22** (2), 113–120.

[35] B. Holt (1989) "Dielectric measurements of snow and ice LIMEX'89," LIMEX'89 Data Report, AERDE, 1-27.

[36] B. Holt and S. A. Digby-Argus (1985) "Processes and imagery of fast first-year sea ice during the melt season," *J. Geophys. Res.*, **90** (C3), 5045–5062.

[37] R. Horn (1989) "*C*-band SAR results obtained by an experimental airborne SAR sensor," in *Proc. IGARSS'89*, 2213–2216.

[38] M. Ikeda (1991) "Wind-induced mesoscale features in a coupled ice-ocean system," *J. Geophys. Res.*, **96** (C3), 4623–4630.

[39] J. A. Johannessen, O. M. Johannessen, E. Svendsen, R. Shuchman, T. Manley, W. J. Campbell, E. G. Josberger, S. Sandven, J. C. Gascard, T. Olaussen, K. Davidson, and J. VanLeer (1987) "Mesoscale eddies in the Fram Strait marginal ice zone during the 1983 and 1984 marginal ice zone experiments," *J. Geophys. Res.*, **92** (C7), 6754–6772.

[40] E. S. Kasischke and G. W. Fowler (1989) "A statistical approach for determining radiometric precisions and accuracies in the calibration of synthetic aperture radar imagery," *IEEE Trans. Geosci. and Remote Sens.*, **27** (4), 410–426.

[41] R. Kwok and A. Pang (1992) "Performance of the ice motion tracker at the Alaska SAR Facility," in *Proc. IGARSS'92*, 588–590.

[42] "The Labrador ice margin experiment (LIMEX) and the Labrador extreme waves experiment (LEWEX)," Special issue: *IEEE Trans. Geosci. and Remote Sens.*, **27** (5), 1989.

[43] "LIMEX," Special Issue: *J. Atmosphere-Oceans* **30** (2), 1992.

[44] R. W. Larsen, P. L. Jackson, and E. S. Kasischke (1988) "A digital calibration method for synthetic aperture radar systems," *IEEE Trans. Geosci. and Remote Sens.*, **26** (6), 753–763.

[45] R. W. Larsen and A. L. Maffett (1984) "Calibration model for SAR systems," *Electromagnetics* **4**, 277–293.

[46] H. Laur and G. M. Doherty (1992) "ERS-1 SAR calibration history and status," in *Proc. CEOS Workshop on SAR Calibration*.

[47] M. Lepparanta and S. Yan (1991) "Use of ice velocities from SAR imagery in numerical sea ice modelling," in *Proc. IGARSS'91*, 1233–1237.

[48] A. K. Liu, B. Holt, and P. W. Vachon (1991) "Wave propagation in the marginal ice zone: model predictions and comparisons with buoy and synthetic aperture radar data," *J. Geophys. Res.*, **96** (C3), 4605–4622.

[49] C. E. Livingstone and M. R. Drinkwater (1991) "Springtime *C*-band backscatter signatures of Labrador Sea marginal ice: Measurements versus modelling predictions," *IEEE Trans. Geosci. and Remote Sens.*, **29** (1), 29–41.

[50] C. E. Livingstone, D. Hudson, J. D. Lyden, C. Liskow, R. Shuchman, and R. Lowry (1983) "Gain compression effects in SAR imagery," in *Proc. 17th ERIM Symp*.

[51] C. E. Livingstone, A. L. Gray, R. K. Hawkins, and R. B. Olsen (1988) "CCRS

C/X-airborne synthetic aperture radar: An R&D tool for the ERS-1 time frame,'' *Aerospace and Electr. Systems Mag.*, **3** (10), 11–20.

[52] C. E. Livingstone, D. Maxwell, and J. R. C. Lafontaine (1993) ''The ERS-1/CV 580 SAR Cross calibration experiment: Sault Ste. Marie, October 1991,'' in *Proc. First ERS-1 Symp.*, 167–171.

[53] C. E. Livingstone, R. G. Onstott, L. D. Arsenault, A. L. Gray, and K. P. Singh (1987) ''Microwave sea ice signatures near the onset of melt,'' *IEEE Trans. Geosci. and Remote Sens.*, **GE-25** (2), 174–186.

[54] A. W. Lohanick and T. C. Grenfell (1986) ''Variations in brightness temperature over cold first-year sea ice near Tuktoyaktuk, Northwest Territories,'' *J. Geophys. Res.* **91** (C4), 5133–5144.

[55] A. Lopes, E. Nezry, S. Goze, R. Touzi, and G. Aarbakke Solaas (1992) ''Adaptive processing of multilook complex SAR images,'' in *Proc. IGARSS'92*, 890–892.

[56] A. Lopes, R. Touzi, and E. Nezry (1990) ''Adaptive speckle filters and scene heterogeneity,'' *IEEE Trans. Geosci. and Remote Sens.*, **28** (6), 992–1000.

[57] T. I. Lukowski, C. E. Livingstone, N. M. August, P. A. Nordstrom, and L. D. Farmer (1992) ''Beaufort Sea ice—1: Selected SAR results,'' in *Proc. IGARSS'92*, 131–133.

[58] S. N. Madsen, E. L. Christiansen, N. Skou, and J. Dahl (1991) ''The Danish SAR system: design and initial tests,'' *IEEE Trans. Geosci. and Remote Sens.*, **29** (3), 417–426.

[59] ''Marginal Ice Zone Experiment III,'' Special Issue: *J. Geophys. Res.*, **96** (C3), 1991.

[60] C. Mätzler (1987) ''Time variations of microwave signatures and surface properties of sea ice observed from Polarstern during MIZEX 1983,'' in *Proc. Inst. Appl. Phys. Workshop*, 54–65.

[61] G. A. Maykut and N. Untersteiner (1971) ''Some results from a time-dependent thermodynamic model of sea ice,'' *J. Geophys. Res.*, **76** (6), 1550–1575.

[62] R. D. Muench (Ed.) (1983) ''Marginal ice zones,'' Special issue: *J. Geophys. Res.*, **88** (C5).

[63] R. D. Meunch, S. Martin, and J. E. Overland (Eds.) (1987) ''Marginal ice zone research,'' Special Issue: *J. Geophys. Res.*, **92** (C7).

[64] A. Nichols, T. Gaffield, J. Wilhelm, R. Inkster, and S. Leung (1984) ''A SAR for real time ice reconnaissance,'' in *Proc. IGARSS'84*, 71–76.

[65] C. J. Oliver (1982) ''Fundamental properties of high-resolution sideways-looking radar,'' *Proc. IEE*, Part F, **129**, 385–402.

[66] R. G. Onstott (1990) ''Near surface measurements of Arctic sea ice during the fall freeze-up,'' in *Proc. IGARSS'90*, 1529.

[67] R. G. Onstott, R. K. Moore, S. Gogineni, and C. Delker (1982) ''Four years of low-altitude sea-ice broadband backscatter measurements,'' *IEEE J. Ocean Eng.*, **7**, 44–50.

[68] R. H. Preller and P. G. Posey (1989) ''A numerical model simulation of a summer reversal in the Beaufort Gyre,'' *Geophys. Res. Let.*, **16** (1), 69–72.

[69] R. K. Raney and P. W. Vachon (1989) ''A phase preserving SAR processor,'' in *Proc. IGARSS'89*, 2588–2591.

[70] R. K. Raney, P. W. Vachon, R. A. DeAbreu, and A. S. Bhogal (1989) "Airborne SAR observations of ocean surface waves penetrating floating ice," *IEEE Trans. Geosci. and Remote Sens.*, **27**, 492–500.

[71] R. K. Raney and G. J. Wessels (1988) "Spatial considerations in SAR speckle simulation," *IEEE Trans. Geosci. and Remote Sens.*, **26**, 666–672.

[72] F. Rocca (1987) "Synthetic aperture radar: A new application for wave equation techniques," Stanford Exploration Project Rept. SEP-56, 167–189.

[73] T. E. Scheuer and F. Wong (1991) "Comparison of SAR processors based on a wave equation formulation," in *Proc. IGARSS'91*, 635–639.

[74] R. A. Shuchman, L. L. Sutherland, C. C. Wackerman, O. M. Johannessen, J. A. Johannessen, and L. H. Pettersen (1990) "Geophysical information on the winter marginal ice zone obtained from Cearex SAR data," in *Proc. IGARSS'90*, 1505–1508.

[75] R. A. Shuchman, C. C. Wackerman, A. L. Maffit, R. G. Onstott, and L. L. Sutherland (1989) "The discrimination of sea ice types using SAR backscatter statistics," in *Proc. IGARSS'89*, 381–385.

[76] R. A. Shuchman, C. C. Wackerman, and L. L. Sutherland (1991) "The use of synthetic aperture radar to map the polar oceans: A remote sensing manual," Environmental Research Institute of Michigan Report.

[77] A. M. Smith (1991) "A new approach to range-Doppler SAR processing," *Intl. J. Remote Sens.*, **12**, 235–251.

[78] C. L. Tang (1991) "A two-dimensional thermodynamic model for sea ice advance and retreat in the Newfoundland marginal ice zone," *J. Geophys. Res.*, **96** (C3), 4723–4738.

[79] L. M. H. Ulander (1991) "Accuracy of using point targets for SAR calibration," *IEEE Trans. Aerospace and Electr. Syst.*, **27** (1), 139–148.

[80] L. M. H. Ulander, R. K. Hawkins, C. E. Livingstone, and T. I. Lukowski (1991) "Absolute radiometric calibration of the CCRS SAR," *IEEE Trans. Geosci. and Remote Sens.*, **29** (6), 922–933.

[81] P. W. Vachon, R. B. Olsen, C. E. Livingstone, and N. G. Freeman (1988) "Airborne SAR Imagery of ocean surface waves obtained during LEWEX: some initial results," *IEEE Trans. Geosci. and Remote Sens.*, **26**, 548–561.

[82] J. F. Vesecky, M. P. Smith, and R. Samadani (1990) "Extraction of lead and ridge characteristics from SAR images of sea ice," *IEEE Trans. Geosci. and Remote Sens.*, **28** (4), 740–744.

[83] J. Way and E. A. Smith (1991) "The evolution of synthetic aperture radar systems and their progression to the EOS SAR," *IEEE Trans. Geosci. and Remote Sens.*, **29** (6), 962–985.

12

RADARSAT AND OPERATIONAL ICE INFORMATION

R. KEITH RANEY
Canada Centre for Remote Sensing
Energy, Mines and Resources
Ottawa, Ontario, Canada

JOHN C. FALKINGHAM
Ice Branch, Atmospheric Environment Service
Environment Canada
Ottawa, Ontario, Canada

12.1 INTRODUCTION

RADARSAT will be the first high-resolution imaging system on a spacecraft to provide complete global coverage, including, in particular, the North and the South polar regions. Observations of the Arctic are of operational importance, particularly to Canada. Observations of Antarctica are of scientific importance and are a keystone of the cooperative agreement between Canada and the United States supporting the spacecraft launch. This chapter describes the *RADARSAT* mission and instrument, in the context of polar ice observations. Emphasis is placed on the capabilities of the imaging system and on the use of *RADARSAT* data in the context of operational ice monitoring in Canada. The two major sections of the chapter develop these themes.

Satellite synthetic aperture radar (SAR) has several advantages and disadvantages when compared to a nominally similar airborne system. Satellite SARs use a much smaller span of incidence angles than do aircraft SARs to image a given swath width. This implies that ice signatures in the imaged swath have less variation in tone and contrast than for comparable coverage from the much

Remote Sensing of Sea Ice and Icebergs, Edited by Simon Haykin, Edward O. Lewis, R. Keith Raney, and James R. Rossiter.
ISBN 0-471-55494-4

lower altitude of aircraft. Satellite radars, by their very nature, provide systematic and periodically repeatable coverage, relatively independent of local weather conditions. Both of these characteristics are helpful when designing an operational ice surveillance information system. In general, aircraft SARs have higher resolution than is available from satellite radars, although this characteristic has more experimental than operational significance. Indeed, as is discussed later in this chapter, operational applications often dictate an intentional compromise in resolution in order to gain large area coverage. Although there is a very large initial cost, satellite systems provide lower overall costs for large-scale repeat coverage of critical ice regions than is available from aircraft alone. The primary disadvantage from an operational point of view is that once launched, a satellite may not be steered to obtain data from a specific area at an arbitrary time: its orbit, together with imaging mode selections if such are available, determines its site coverage potential.

The history of civilian SARs in space is short, but is being considerably enhanced in the decade of the 1990s (Li and Raney [21]). An overview of all civilian SAR satellite systems is provided in Table 12.1. Only two satellite SARs saw operation before 1990, and those were primarily experimental demonstration missions. The *L* band *Seasat* SAR (the United States) operated for 3 months in 1978, and the *S* band *Kosmos 1870* (the Soviet Union) operated for 2 years (1987–1989), although data were classified until after the mission was completed. There have been three *L* band shuttle imaging radar missions sponsored by NASA (*SIR-A*, 1981, *SIR-B*, 1984 and *SIR-C*, 1994), although their orbits and 1-week duration precluded significant observation of ice phenomena. The current decade is witnessing more satellite SAR activity. *Almaz* (*S* band, Russia) was launched by the Soviet Union in March 1991, and *ERS-1* (*C* band) was launched by the European Space Agency in July 1991. (The Almaz mission terminated in October 1992 when the spacecraft descended into the Pacific Ocean.) Ice imagery from both satellites is excellent; an example of *ERS-1* data appears later in this chapter. *J-ERS-1* (Japan, *L* band) SAR began operation in April 1992. Before the end of the decade, the launches of *Almaz II* and *ERS-2* are planned. Other space-based radars are under con-*RADARSAT*, expected to be operational for the last half of the decade, fits nicely in this important class of systems. It is reasonable to anticipate that space-based SARs will become important operational systems for observation of the ice environment.

12.2 RADARSAT

On 13 September 1989, the Government of Canada announced full commitment to build and operate *RADARSAT*, Canada's first earth resources remote sensing satellite (Fig. 12.1). It is being prepared for a launch in 1995 and is designed for 5 years of service in orbit. The only payload instrument is an SAR (e.g., Brown and Porcello [4]; Livingstone, Chapter 11 of this book) using horizon-

TABLE 12.1 Selected Parameters of All Civilian SAR Satellites

	Seasat USA	*ALMAZ*[a] USSR	*ERS-1* ESA	*J-ERS-1* Japan	*SIR-C*[b] USA	*ERS-2*[c] ESA	*RADARSAT* Canada	*EOS*[d] USA
Operation	1978	91–92	91–94	92–94	94, 94, 96	95–96	95–99	99(?)–
Radar bands	*L*	*S*	*C*	*L*	*C, L, X*	*C*	*C*	*C, L (X)*
Polarization	*HH*	*HH*	*VV*	*HH*	QUAD[e]	*VV*	*HH*	QUAD[e]
Swath width (km)	100	20–45	100	75	15–60	100	50–500	50–500
Resolution (m)	25	15–30	~25	18	~30	~25	10–100	10–100
Incidence angle (°)	~22	30–60	~23	~39	15–55	~23	<20–>59	15–55
Max. N latitude (°)	72	73	82	83	57	82	90	· · ·
On-board recorder	· · ·	yes	· · ·	yes	yes	· · ·	yes	(yes)

[a]Recorded data were selected from an accessible swath width of 350 km, but only on one side rather than both sides as was originally planned. *Almaz*' mission terminated in October 1992 with the descent of the spacecraft from orbit into the Pacific Ocean.
[b]Shuttle radar flights (SIR) each last about 1 week, as did *SIR-A* (1981) and *SIR-B* (1984), and are experimental rather than operational.
[c]*ERS-2* parameters are essentially the same as those of *ERS-1*.
[d]Current NASA planning has deleted the SAR as a facility instrument on *EOS* and is seeking approval for a separate but related SAR satellite. System parameters of the proposed SAR are not established, but are likely to be rather close to those listed above.
[e]Quadrature polarimetry (see Chapter 6).

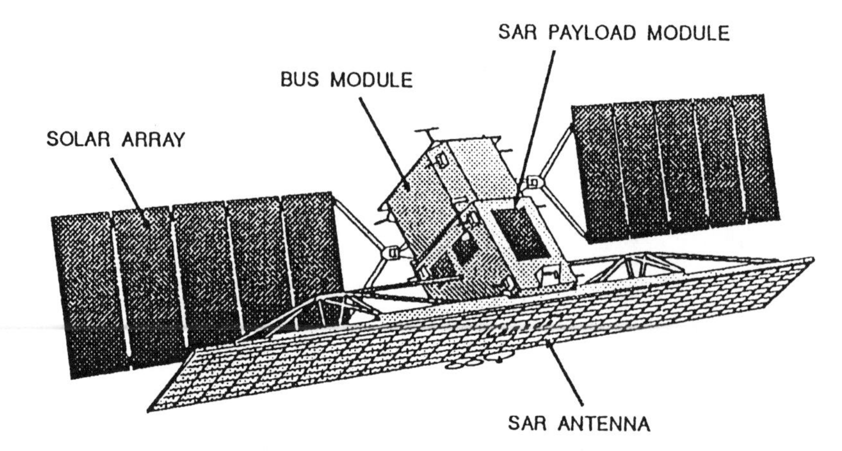

Figure 12.1 *RADARSAT* configuration, showing the antenna aligned with the solar panels, consistent with the dawn–dusk orbit.

tally polarized, *C*-band microwave radiation (wavelength of 5.6 cm). A variety of resolution, image swath width, and incidence angle parameters are planned that may be selected through ground command. The design described in this chapter is based on more than 15 years of mission study and preparation in Canada, on extensive domestic experience with airborne radar systems and radar data processing, and on Canada's participation in, and heritage from, the *Seasat* and *ERS-1* satellite missions.

12.2.1 Objectives

RADARSAT mission objectives are based, primarily, on Canadian national requirements for information to support domestic resource management and environmental monitoring (*RADARSAT* Project Office [32]). These objectives provide justification for the large national investment now committed. Applications of importance to Canada include mapping of ice and northern regions; monitoring of agricultural, forest, and geological resources; observing oceans, floating ice, and coastal zones; and maintaining an all-weather capability for Arctic observations.

Although the *RADARSAT* rationale and design are based almost exclusively on identifiable domestic benefits, Canada recognizes the potential importance of this system at the global level. Coverage of extensive, environmentally sensitive areas (such as South Polar as well as North Polar ice; the earth's forests, particularly in the tropics; large-scale ocean features; and desert expansion) represents an important, potentially significant, use of radar satellite data. Whereas it is known that a *C*-band SAR is not necessarily the optimum remote sensing instrument for all such applications, the resulting data should be useful for any information system designed to monitor these environmental regimes (Brundtland [5]) or for studying global change (NASA [28]). An Announcement of Opportunity for investigators and agencies worldwide who wish to participate in these aspects of the *RADARSAT* mission is being developed and is expected to be published by the Canadian Space Agency before the launch.

Whereas the current objectives for *RADARSAT* cover a broad spectrum, the genesis of *RADARSAT* was based on ice applications, and they continue to be of prime value to the science community. These aspects are developed in the following sections.

Ice as RADARSAT Rationale

Early studies (CCRS [6]) established for Canadian Arctic surveillance that an imaging radar satellite system would be the only acceptable sensor technology for reliable and timely information, and would serve as an additional data source to complement optical data, where such were available. This conclusion was driven by the harsh realities of Arctic observation, coupled with the economic and sovereign importance of Canada's remote northern regions. The original interest by the Canadian Government to support *RADARSAT* was fueled by the expectation, in the early 1980s, that large-scale development of Arctic petroleum reserves would require reliable data on ice conditions, ice movement, navigational hazards, and human activities.

Given that ice observation was the most important economic justification for *RADARSAT*, the original system design was chosen in response. With orbit inclination fixed by the requirement to be sun-synchronous, regular observation of shipping routes and resource areas of the Arctic (Southern Beaufort Sea and the Northwest Passage) suggested that the radar should look to the *south* of the orbit plane. Four subswaths, of nominally 100-km width each, were needed to cover the regions of interest on a frequent basis. The frequency, *C*-band, was selected as a compromise between the constraints of hardware implementation (which are more rigorous at shorter wavelengths) and ice discrimination sensitivity (which is improved at shorter wavelengths). Other potential applications of space radar data seemed to offer less persuasive arguments for choice of radar wavelength.

By about 1984, however, it became clear that justification for *RADARSAT* could no longer be based on economic development in the Arctic. In response, the design of the radar was refined so that additional modes would be available, thus offering rather better data for a wider variety of applications. Full coverage of all Arctic areas became more important than observation of the Northwest Passage, so that the orientation of the radar was changed to look northward. The decision was taken to have software control of the antenna beams, thus making many imaging modes possible. Finally, with the removal of other instruments from the payload (because of withdrawal from the program by the British), the orbit was changed to maximize solar illumination of the spacecraft, thus lengthening the operation time budget for the radar.

The final *RADARSAT* design is a more powerful ice observation platform than the earlier versions, largely due to the extensive coverage and resolution options. Using the 500-km ScanSAR swath, *RADARSAT* will provide imagery on a daily basis for the entire polar region north of 79°N (see Fig. 12.2). With beam steering, any given point at Canadian latitudes can be observed within 3 days. Fine-resolution performance (10-m resolution) is available on command for areas of detailed interest. Although the orbit repeat period is 24 days, with the 500-km wide swath, substantially complete surface coverage is available after only 4 days, even at equatorial latitudes. A more complete description of the *RADARSAT* imaging modes is given later in the chapter.

Antarctic Coverage

As noted above, the nominal configuration of the spacecraft has the SAR pointing to the north side of the orbital plane. This orientation makes possible regular coverage of the Arctic as far as the pole, but coverage of the central Antarctic region is not possible with this geometry. However, for two periods (of 2 weeks each) during the first 2 years of the mission, the satellite will be rotated 180° about its yaw axis, thus directing the radar antenna beam to the south side of the orbital plane. This maneuver will allow a complete SAR map of Antarctica to be obtained at the times of maximum and minimum ice cover. (To the extent that spacecraft thermal budgets allow, this will also allow opposite-side viewing for selected regions north of the Equator during these periods.) The yaw maneuver has been accommodated in response to science requirements on *RADARSAT* as expressed by NASA (National Research Council [29]; McMurdo Ad Hoc Science Team [25]).

Coverage of Antarctica available from *RADARSAT* in normal configuration and in Antarctic configuration is illustrated in Fig. 12.3. Normal configuration is satisfactory for regular coverage of most of the Antarctic coastal and marginal ice zones. The Antarctic configuration is essential for coverage of the interior regions of the continent. Science plans require mapping and high-resolution mode coverage of selected features in this region, as well as full coverage from the ScanSAR mode (McMurdo Ad Hoc Science Team [25]).

Antarctic coverage from *RADARSAT* offers the science community the first high-resolution seasonal observations of the continental interior, an unprecedented opportunity [25]. Optical sensors, such as those on board *Landsat*, have no access to regions closer to the poles than 81°, and are further compromised by clouds and darkness. Other space radars, such as the SAR on the European *ERS-1* satellite, have no ability to look to the south. Although the *Almaz* radar was intended to look to both sides of the orbital plane, only one antenna was operational during the mission, and *Almaz* had an orbital inclination such that Antarctic coverage was incomplete (Li and Raney [21]).

Objectives for Antarctic coverage include ice sheet mapping (including mass balance and dynamic studies), marginal ice zone monitoring, meteorology, and geology. The interior of Antarctica remains the least known land mass on earth: *RADARSAT* data will allow significant new information to be derived for this region. Furthermore, changes in the ice structure of Antarctica are expected to be one of the most sensitive indicators of global environmental change [25]. From the point of view of the Antarctic ice community, *RADARSAT* coverage

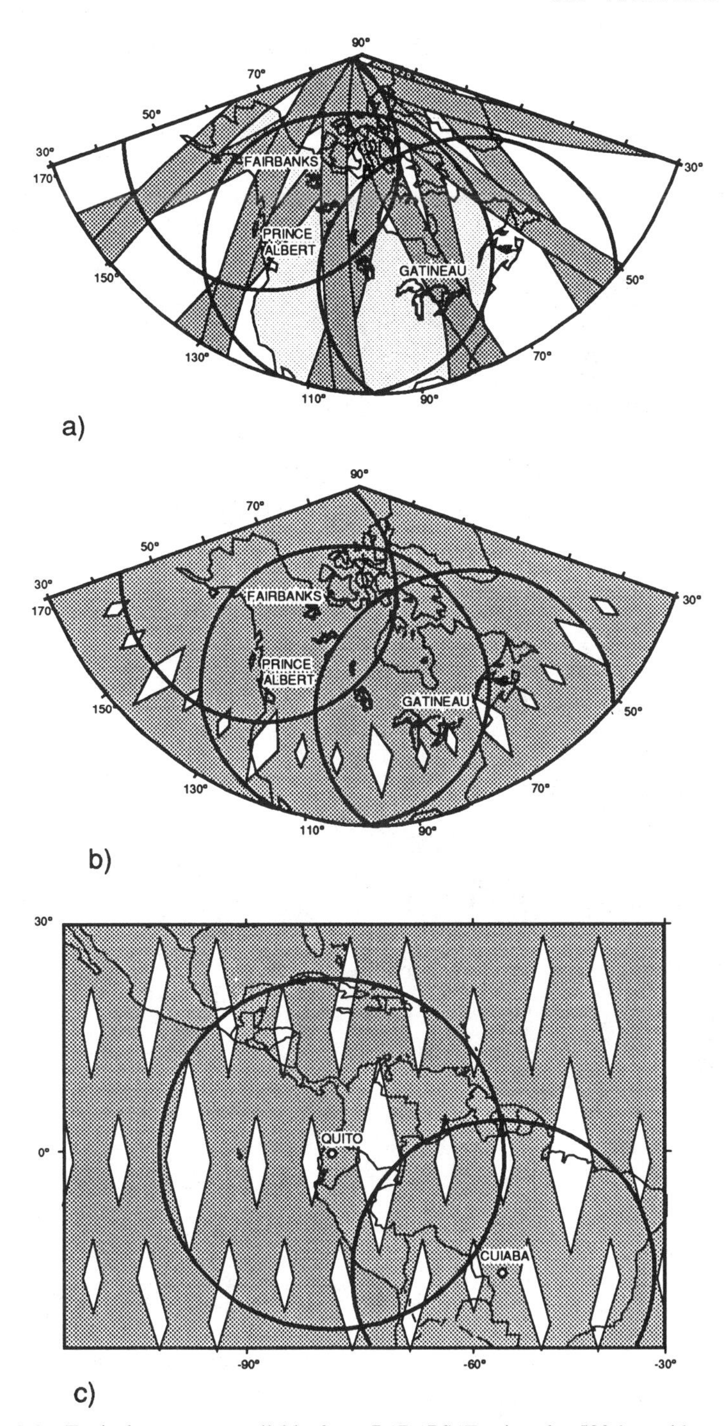

Figure 12.2 Typical coverage available from *RADARSAT* using the 500-km wide swath. (a) North America, 1-day coverage; (b) North America, 3-day coverage; and (c) Amazonia, an example of equatorial coverage for 3 days. In all cases, available image data are shown as shaded.

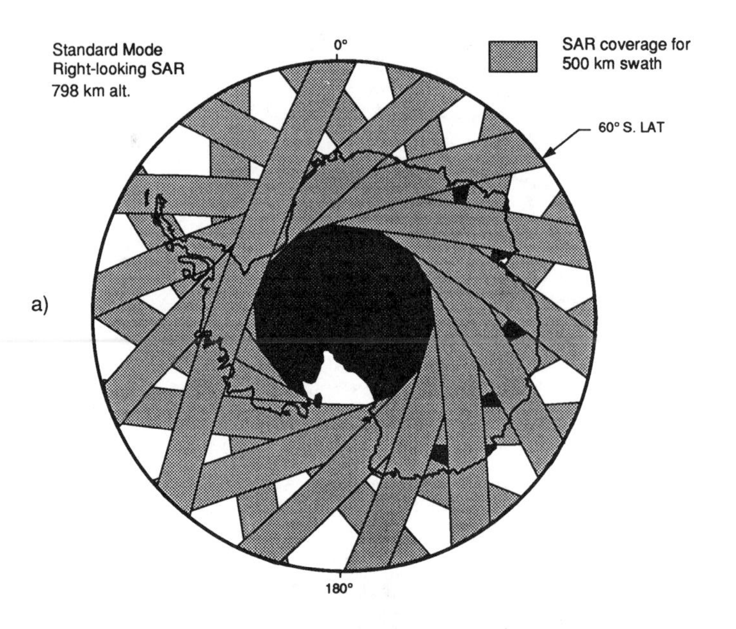

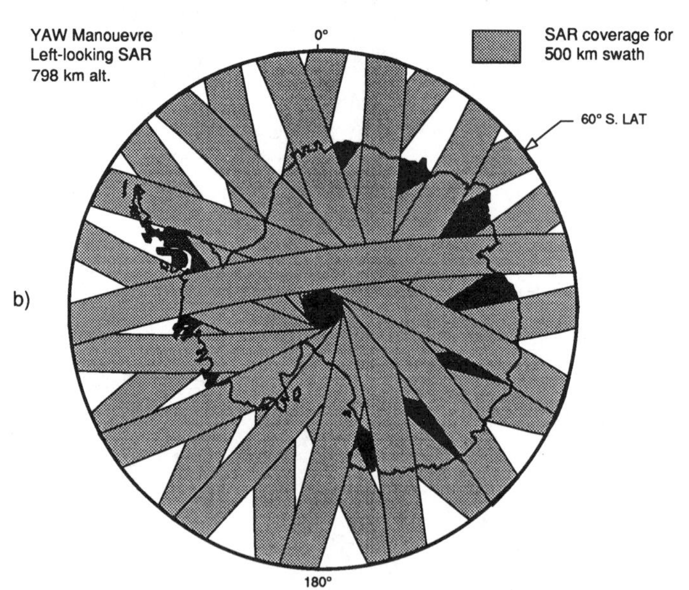

Figure 12.3 Typical coverage of Antarctica for 1-day using the ScanSAR mode for (a) normal spacecraft orientation, and (b) Antarctic orientation. In both cases, land areas shown in dark shading are not imaged in the time interval selected. Full polar coverage may be achieved using the extended beam mode.

of that polar region promises to offer the most exciting science opportunity of the decade.

12.2.2 Mission

RADARSAT, as an operational remote sensing system, implies many considerations beyond the radar instrument itself. Pertinent aspects of the *RADARSAT* mission are outlined in this section (Raney et al. [34]). *RADARSAT* data as they will be used in an operational scenario is described in Section 12.3.

Mission Profile

NASA is a major partner in the *RADARSAT* mission, contributing services for the planned 1995 launch from Vandenburg AFB, using a medium-class expendable vehicle (McDonnell Douglas Delta II 7920-10). The satellite payload will consist of the SAR described below, and its associated downlink transmitters, tape-recorders, and command and control computer. The spacecraft is being procured from Ball Aerospace (United States), whereas SPAR Aerospace (Canada) has prime system responsibility under contract with the Canadian Space Agency.

For a satellite using a radar sensor, good solar illumination of the spacecraft is more important than sunlight on the earth's surface. For this reason, *RADARSAT* will use a sun-synchronous, *dawn–dusk* orbit (Table 12.2). In this orbit, the spacecraft is illuminated by the sun throughout each orbit, as indicated in Fig. 12.4. (There are short periods at southern latitudes during the Austral winter during which the spacecraft is in eclipse.) Perhaps the greatest operational advantage of this orbit is that the SAR can be fully dependent on solar rather than battery-stored power. This means that there is no limiting distinction between ascending and descending passes from an applications point of view. Nearly twice as many viewing opportunities are available to the mission than would otherwise be available. Another operational advantage is that the data downlink periods at *RADARSAT* ground stations will not conflict with other remote sensing satellites, most of which use near-midday orbit timing.

RADARSAT will carry tape recorders with sufficient capacity for more than

TABLE 12.2 Orbit Parameters for *RADARSAT*

Parameter	Value
Altitude (local)	793–821 km
Inclination	98.6°
Ascending role	1800 h
Period	101 min
Repeat cycle	24 d
Subcycles	7 and 17 d
Reobservation	3+ d

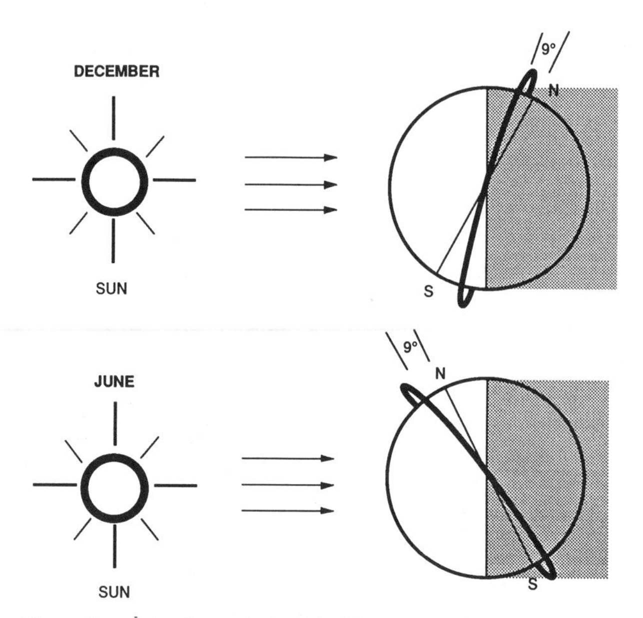

Figure 12.4 Illustration of the dawn–dusk orbit. The spacecraft stays near the solar terminator, thus is it illuminated throughout each orbit (with the exception of short periods during the southern hemisphere winter).

10 min of full-quality SAR data. As a consequence of the orbit yaw maneuver and the on-board recorders, *RADARSAT* will be the first fine-resolution satellite system capable of complete global coverage. This feature of *RADARSAT* is relevant to global environmental monitoring, as well as being of scientific value.

Operations

Mission operations will be coordinated by a Mission Management Office which will serve as the interface between the user community, the Mission Control Facility for commanding the satellite, and the ground reception and processing facilities (Sack et al. [37]; Lim et al. [22]). In Canada, the ground facility now in place for *ERS-1* will be upgraded to handle the increased operational throughput expected from *RADARSAT*. In addition to coordinating and scheduling requests for SAR data acquisition, the Mission Management Office will monitor the entire data distribution system and will serve as the executive office for the mission.

Partnerships

Unique among remote sensing initiatives worldwide, the *RADARSAT* Program is supported through funding partnerships with agencies in the United States, Canadian provincial governments, and the private sector. The agreements are in the form of Memoranda of Understanding. Canada is responsible for the design and integration of the overall system, for construction of the radar, for the provision of the satellite platform, for control and operation of the satellite in orbit, and for operation of the data reception stations in Prince Albert, Saskatchewan, and Gatineau, Quebec.

NASA will provide the launch services and operate a reception station in Fairbanks, Alaska, in exchange for radar data for research programs. The National Oceanographic and Atmospheric Administration of the United States (NOAA) will facilitate the participation of the American private sector in the distribution of data and will conduct pilot projects for applications in support of its mandate.

All Canadian provinces have participated in planning the *RADARSAT* Program. Quebec, Ontario, British Columbia, and Saskatchewan share in the capital costs in proportion to technology development within their industries. An agreement has also been developed with other provinces for their participation in the operational phase of the mission.

RADARSAT International (RSI) has been formed, a private sector corporation in Canada, with an international mandate. RSI has the right to distribute the data and to request data acquisition when the available SAR operating time in orbit is in excess of the international partners' governmental requirements. In return, they will invest in developing the international market and, from the resulting sales, provide royalty payments to the Canadian Government to help defray the operating costs of the system. Value-added processing will be done in the private sector, as is the practice for most remote sensing satellite missions.

12.2.3 SAR Payload

RADARSAT has been designed in response to user requirements that demand a variety of incidence angles (from about 20° to 50°) in the standard imaging modes. An antenna with electronic beam steering (see below) is part of the baseline *RADARSAT* design. Although this enables user requirements to be met, it does add further complexity to the entire system. To provide a (nominally) constant ground-range resolution over the range of incidence angles, three different pulse bandwidths are needed. It also follows that very fine control of the transmitter pulse repetition frequency (PRF) is required. Selected characteristics of the radar are listed in Table 12.3, and a simplified block diagram of the radar system is shown in Fig. 12.5.

Having an antenna and control system with the flexibility (and complexity) necessary to support standard imaging modes at a variety of incidence angles enables several additional imaging modes to become available at rather small marginal cost. The design philosophy for these extra modes has been to base the system specifications on the standard imaging modes, and to optimize the performance of the additional modes within the design envelope determined by the standard modes. Image quality in the additional modes is predicted to be comparable to that of the standard modes.

TABLE 12.3 Selected Parameters of the *RADARSAT* SAR System

Parameter	Value
Frequency	5.3 GHz
Wavelength	5.6 cm
Polarization	Horizontal
Pulse bandwidths	11.6, 17.3, or 30.0 MHz
Pulse length	42.0 μsec
PRF	1270–1390 Hz (2-Hz steps)
Peak power	5 kW
TWT output power (avg)	300 W (nominal)
Maximum on time	28 min per orbit
Antenna size	15.0 m × 1.5 m
Beam-pointing stability	$\leq$0.2° (boresight, both axes)
Preset elevation beams	20 (programmable)

Radar operation time is constrained primarily by the spacecraft power system and by the thermal response of the high-power transmitter. Typical continuous on-time of the radar is 10–15 min per orbit, out of 28 min per orbit maximum on-time when in normal orientation. In the Antarctic orientation, the maximum imaging time is 15 min per orbit. The minimum radar on-time is 3 min. Up to seven on/off cycles are available per orbit. On-time is not dependent on the active mode, a variety of which may be preselected through ground command for use during each orbit.

Imaging Modes

Imaging modes for *RADARSAT* include Standard, Wide Swath, Fine Resolution, Extended, and ScanSAR (see subsection below). First-order image characteristics of those modes are given in Table 12.4. In each mode, data are

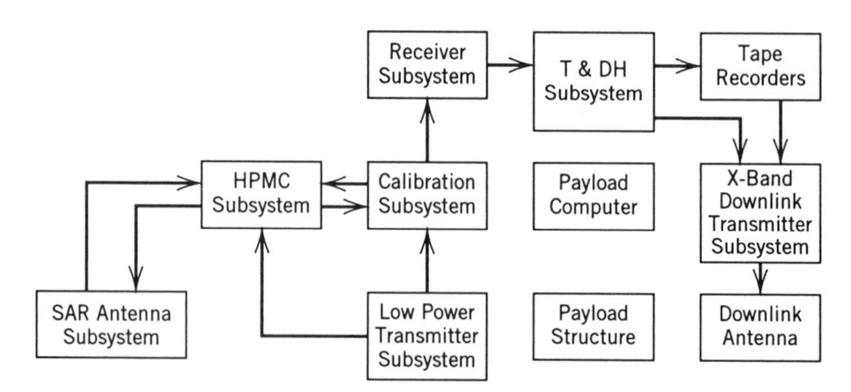

Figure 12.5 Functional block diagram of the radar. Only one string of the two on-board systems is shown. T&DH: Timing and Data Handling. HPMC: High Power Microwave Circuit.

TABLE 12.4 Characteristics of *RADARSAT* Imaging Modes

Mode	Resolution ($R^a \times A^b$, m)	Looks[c]	Width (km)	Incidence Angle (°)
Standard	25 × 28	4	100	20–49
Wide swath (1)	48–30 × 28	4	165	20–31
Wide swath (2)	32–25 × 28	4	150	31–39
Fine resolution	11–9 × 9	1	45	37–48
ScanSAR (narrow)	50 × 50	2–4	305	20–40
ScanSAR (wide)	100 × 100	4–8	510	20–49
Extended (high)	22–19 × 28	4	75	50–60
Extended (low)	63–28 × 28	4	170	10–23

[a]In range direction, resolution on the surface varies with range.
[b]In azimuth direction.
[c]Nominal; range and processor dependent.

collected continuously along a swath parallel to the subsatellite path, as shown in Fig. 12.6. Swath length is limited only by the duration of continuous radar operation, and may be thousands of kilometers long. Swath widths and positions are determined by the antenna elevation beam patterns (and the radar range gate control), and have been chosen for the standard modes so that there is at least 10% overlap between adjacent swaths. Range resolution, when projected onto the earth's surface, varies with incidence angle and, hence, with ground range. Three range bandwidths are available to allow choice in ground-range resolution to be achieved at each incidence angle. Nominal range reso-

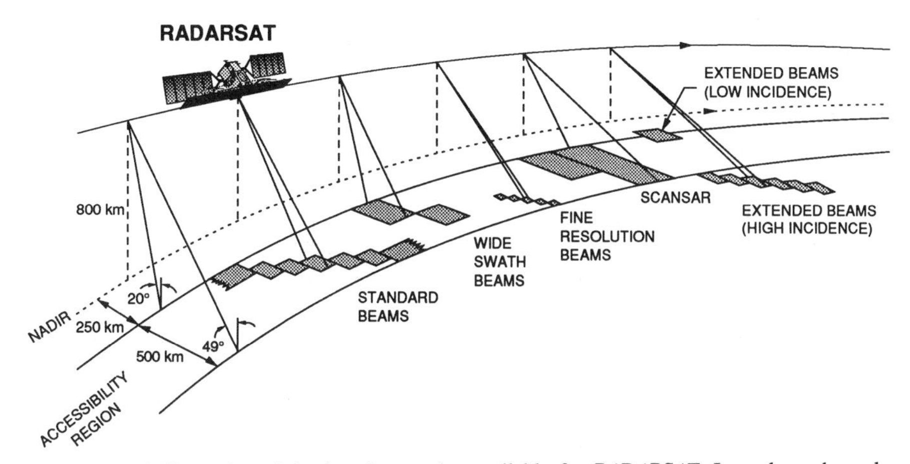

Figure 12.6 Illustration of the imaging modes available for *RADARSAT*. In each mode, only one of which is available at one time, the swath for collected data is continuous and parallel to the subsatellite track, although data products generally are to be available on a frame-by-frame basis.

lution for the standard beams has been specified at ground ranges of 400 km and 675 km for the subsatellite locus.

The additional modes are generated by appropriate choices of antenna beam and range pulse bandwidth. Time lapse between modes is limited only by the necessity to form full synthetic apertures for the adjacent configurations, so that a gap of several kilometers will occur at the time of changeover.

The fine resolution mode is achieved by selecting the widest available bandwidth (30.0 MHz), and using a narrow beam in elevation, at incidence angles larger than (nominally) 45°. A narrow swath results from the requirement to minimize beamwidth (to maintain good signal-to-noise ratio), and from the necessity to maintain data rates consistent with downlink channel capacity. Wide swath modes are supported by wider antenna beamwidths than normal and at steeper incidence angles. They use the smallest available range pulse bandwidth (11.6 MHz), leading to coarser ground-range resolution. Signal-to-noise ratio and data bandwidth arguments apply in these modes that are counterparts to those for the fine-resolution modes, but with the result of broadening the usable swath width. Extended modes result from selection of beams outside of the nominal 500-km accessibility region, either closer to nadir (steeper incidence angle) or further away (more shallow, or grazing, incidence angle). ScanSAR is of sufficient importance for Canadian ice monitoring requirements that it is discussed in a separate subsection.

Data will be processed on a frame-by-frame basis, leading to individual image products for each mode. (Continuous image generation to produce strip maps is under consideration for selected modes, such as ScanSAR.) Pixel spacing is to be consistent with standard map scales. Image quality typical of a standard imaging mode is summarized in Table 12.5. Note that relative calibration is required between modes as well as within each mode (see subsection below). The baseline image data format is that specified by the International Committee on Earth Observation Satellites.

System Implementation

The radar system is being developed in Canada by SPAR Aerospace Ltd., with major subcontracts to CAL Corporation and COMDEV Ltd. All subsystems (except the antenna and its feed subsystem) are designed with redundant components and parallel networks to maintain the required reliability over the 5-year design life of the system. The *C*-band transmitter uses a travelling wave tube (TWT) and is based on the high-power amplifier of Dornier Systems/AEG for *ERS-1* (Attema [1]). The transmitter required only minor modifications for the *RADARSAT* application. The grid-modulated TWT is a three-stage collector design, using a molybdenum ring and bar delay line and 15-kV cathode voltage. Life testing of this family of tubes has shown excellent results.

Signal generation is digital, using eight bits (in-phase, *I*, and quadrature, *Q*), followed by digital to analog conversion. A programmable pulse-generation system was determined to be more reliable and cost effective than analog tech-

TABLE 12.5 Image Quality Parameters for *RADARSAT*

Parameter	Value
Noise equivalent σ^0	< -23 dB
(Design value, beam edge, with margins)	< -18.5 dB
Total signal-dependent noise ratio[a]	< -10 dB
Azimuth ambiguity ratio[b]	< -22 dB
Range ambiguity ratio	< -22 dB
Peak sidelobe ratio	< -20 dB
Global dynamic range	> 30 dB
Relative radiometric accuracy	
100 km × 100 km scene	< 1 dB
During one orbit	< 1.5 dB
Over 3 days	< 2 dB
Satellite lifetime	< 3 dB
Absolute scene location[c]	< 1500 m
Geometric distortion[d]	
100 km × 100 km scene	< 40 m

[a]Includes integrated sidelobe ratio, aliasing effects from sampling, quantization noise, and so on.
[b]Defined using integral over ambiguous regions.
[c]As specified; better precision expected.
[d]Excluding terrain effects.

niques. Three linear frequency modulated and five continuous wave tone burst signals are resident in a read-only memory (ROM) in the pulse generator, and other pulse waveforms may be programmed in flight, as desired. The power level at the output of the low-power transmitter is programmable.

The receive chain includes limiters to protect the subsystem from both normal leakage during pulse transmission and anomalous power levels that could occur under fault conditions. The receiver bandwidth is selectable, so that the three transmit pulse bandwidths may be matched to reduce noise aliasing during receiver analog-to-digital sampling.

System control is maintained through the SAR control computer, which is responsible for setting all engineering and operating parameters appropriately for each imaging period. These parameters depend upon the desired incidence angle and imaging mode, and on the local spacecraft altitude. The SAR computer is also tasked with setting the phase shifters used in the antenna elevation beam-forming network for both transmit and receive. As these phase shifters are nonreciprocal, and to allow different beams to be synthesized on transmit and receive for optimum sidelobe control, the phase shift parameters in general must be different on transmit and receive. In addition, each is temperature dependent, so that the SAR control computer generates the required compensation to maintain correct phase settings under a variety of thermal environments.

On-Board Data Management

Within the receiver subsystem (Fig. 12.5), the signal is demodulated coherently to baseband, and the resulting I and Q channels are digitized. A block-floating point algorithm (Curlander and McDonough [9]) is used. Four digital bits are retained in each channel, and the corresponding gain control setting is included in the telemetry header, so that the signal level is recoverable in the ground image processor. Nominal data rate capacity of these channels is 105 Mbps; thus it is compatible with the *ERS-1* system (Attema [1]), although the rate required to support the specified swath width in most modes is less than 85 Mbps. The extra channel capacity will be used to downlink data from a larger swath, thus allowing antenna elevation beam edges to be located accurately in the images, which aids radiometric calibration.

Two data paths are available for signal recovery through telemetry. The direct signal path routes the data directly to the downlink telemetry system, thence to ground stations whose receiving antennas are able to track the spacecraft in real time. The indirect signal path includes one of two Odetics tape recorders, capable of full retention of all data for most *RADARSAT* modes, for up to 15 min of recording time. For each full set of recorded data, no more than three radar operating intervals may be used. Data rate for tape recorded signals must be less than 85 Mbps. The recorded data are downlinked to a receiving station later during the orbit.

Signal telemetry includes three systems (for redundancy), any two of which may be used in downlink operations, one for real time data, and one for recorded data. Telemetry signal and system parameters are listed in Table 12.6. The telemetry subsystem is very similar to that provided to the *ERS-1* mission in which Canada is a contributing partner.

Calibration

In response to user requirements, the design includes provision for calibration data (Luscombe [24]) in addition to the engineering telemetry record normally available for selected system performance parameters. *RADARSAT* calibration

TABLE 12.6 Selected Parameters for *RADARSAT* Signal Telemetry

Parameter	Value
Carrier frequencies	8.245 and 8.370 GHz
Number of channels	2
Modulation	Quadrature phase shift keying (QPSK)
Signal quantization	4 bits (I & Q; floating point)
Signal record capacity	$<$ 15 min/recorded pass
Data rates	105 Mbps (direct telemetry)
	85 Mbps (recorded)
RF power	22 Watts/channel
Bit error rate	10^{-5}
Spacecraft antenna beamwidth	124°

depends on both internal references and on analysis based on radar data from an external calibration site. Most applications will be satisfied with relative calibration, for which the internal system is sufficient. Quantitative comparison of *RADARSAT* imagery with that from other (space) radars will require cross-calibration exercises using an external site. Verification of the antenna gain, beam shapes, and steering for the various modes also requires external calibration (Dobson et al. [11]).

A schematic diagram of the *RADARSAT* internal calibration subsystem is shown in Fig. 12.7. The main radar signal path passes from the low- and high-power transmitter units (LP Tx and HP Tx, respectively) to the antenna, and after reflection from the scene, through the receiver chain. The components within (and connections to) the dotted line portions in the figure constitute the calibration subsystem. Design features from *ERS-1* and other spacecraft radar systems are used in components required for the calibration implementation (Attema [1]; Naderi et al. [27]).

There is a choice between three different points for signal input to the subsystem: (1) a signal from a calibrated, variable-level, reference noise source; (2) single frequency or modulated signals produced by a digital pulse generator;

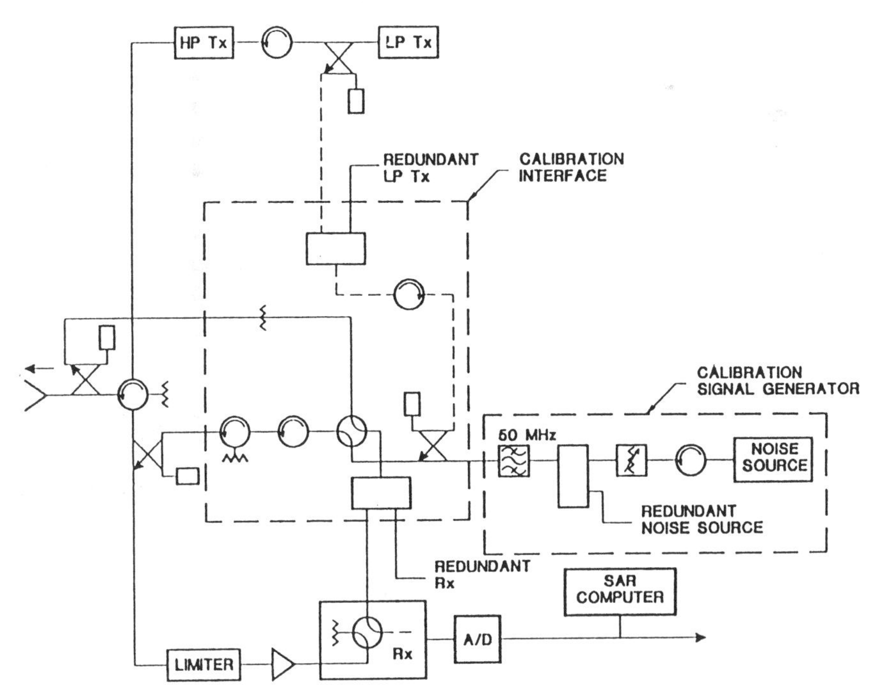

Figure 12.7 Block diagram of the on-board calibration subsystem for the *RADARSAT* SAR. Data from this system will be embedded in the radar signal telemetry, and may be used in the processor for routine system monitoring. (Unmarked boxes represent signal splitters for the redundant channels.)

and (3) low-level examples of the pulses generated by the radar transmitter tapped after the high-power amplifier. These signals are reentered into the main signal path at one of two points in the receiver. Thus, a variety of calibration and verification operations are available on command.

In all cases, the calibration data sent on the downlink consist of a digitized, sampled pulse. Variations observed in signal level, and integrated signal energy from these reference pulses, are used to monitor variations in the gain of elements included in the selected calibration signal path. Baseline reference gain parameters will be determined at times coincident with operation of the radar over an external reference calibration site. Subsequent calibration information may be extrapolated from the baseline data for use with image data from application sites anywhere in the world.

In normal operation, data from calibration signals will be embedded in the signal sequence for all operating modes of the radar. These data should provide a routine reference for signal level monitoring, and should serve also as a basis for relative cross-calibration between the several operating modes of *RADARSAT*.

In addition to routine calibration of *RADARSAT* imagery, the calibration subsystem can be used to generate data for more specialized system performance analysis, such as are required for initial in-orbit acceptance tests, periodic system characterization, and investigation of possible performance anomalies.

12.2.4 Antenna

The standard form of image that the *RADARSAT* SAR system is required to provide overs a 100-km swath (nominal, minimum) positioned within an ''accessibility region'' of over 500 km, which spans a range of incidence angles at the earth's surface from 20° to nearly 50° (Fig. 12.6). To achieve these swath widths at the larger incidence angles without incurring unwanted range ambiguities, it is necessary to operate with a lower PRF than previous satellite systems. A relatively low PRF, in turn, dictates the use of a longer antenna, so as to provide the azimuth beamwidth of 0.2° (nominal) which is needed to avoid azimuth ambiguity problems (Raney and Princz [35]). The 15-m antenna length that has been selected enables this beamwidth to be obtained and still allows a small amount of sidelobe suppression through aperture weighting.

The elevation beams, required to illuminate the standard 100-km swaths, decrease in angular width as incidence angle increases. The antenna, therefore, has to be capable of providing a variety of different elevation beamwidths, as well as variable beam directions. In addition to providing the required mainlobe illumination with each of these beams, the antenna must achieve sufficient control of the elevation sidelobe levels to ensure acceptable range ambiguity performance. Unwanted nadir returns are suppressed by sidelobe control in the nadir direction for each beam, and by PRF and range gate selection, so as to avoid receiving imaging data at times during which returns from nadir are expected.

The most stringent set of requirements on mainlobe and sidelobe patterns arise from the beams at higher incidence angles. An antenna height of 1.5 m was chosen to provide the rapid fall-off in gain that is required for these beams between the far edge of the swath and the near edge of the first range ambiguity. The resulting 15 m × 1.5 m array is about 50% larger in each dimension than the antenna on *ERS-1* (Attema [1]).

The simultaneous requirements for a fixed beam shape in azimuth and rapidly selectable beam shape in elevation led to the adoption of a slotted waveguide, planar-array design for the antenna, with an adjustable phase shifter at the input to each of the waveguide rows (Zimcik et al. [47]). This approach makes use of proven technology and leads to a flexible beam-forming capability in the vertical dimension.

The power-dividing network along each horizontal waveguide row provides the small taper required to give azimuth sidelobe suppression. In the vertical dimension, the elevation beam-forming network produces a fixed power division to a set of 32 waveguide rows, but with adjustable phases provided by 8-bit, digitally controlled, ferrite phase shifters. Tests have shown that the phase accuracy for these devices is better than 1.5° rms. Phase distribution over the set of phase shifters is optimized for each beam to give the required shape and direction. The effect of the phase shifters on system power budgets is minimal, as insertion loss in the signal path is less than 0.6 dB, and no holding current is required after the phase has been set. Beam switching time is less than 15 μs.

Elevation beam control is maintained by the SAR control computer, referenced to one of 20 preset beams available on-board in memory, or to beam shape files up-loaded during mission operations. Thus, full beam-selection flexibility is available, even after launch. A set of example patterns, designed to illuminate the seven swaths which cover the baseline 500-km accessibility region, is shown in Fig. 12.8.

Because of the size of the antenna, the mechanical design presents challenging demands. The 15-m length is split into four panels, with each wing consisting of two panels that are folded during launch to one side of the spacecraft. Once in orbit, each of these wings must be deployed and then held rigidly and accurately in its planar configuration. Deployment and positioning of the antenna panels is accomplished with a graphite fiber-epoxy structure similar to that used on *Seasat* (Jordan [20]). Baseline design material for the radiating slotted waveguide elements is thin-walled aluminum.

Once deployed, the antenna must have minimal thermal or mechanical distortions through the range of environmental conditions that are to be encountered during the mission. These requirements apply not just when the spacecraft is right-looking but also to the periods of yaw maneuver when the opposite side of the antenna is facing the sun. A combination of multilayer thermal blankets and low-absorptivity paint is used to minimize temperature gradients throughout the radiating panels.

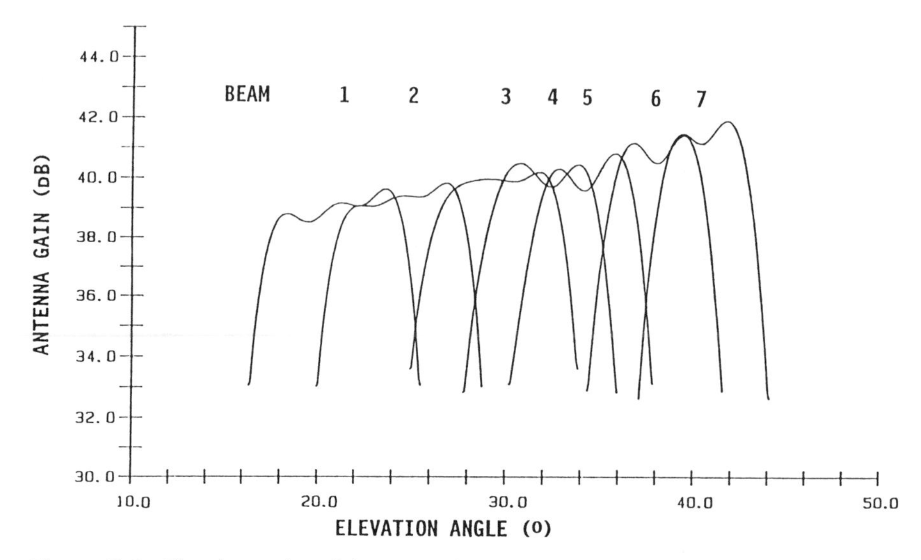

Figure 12.8 Elevation section of the antenna beam patterns to be used in the standard *RADARSAT* imaging modes. (Elevation angle is in spacecraft coordinates, not to be confused with incidence angle at the earth's surface.)

12.2.5 ScanSAR

To allow imaging of a swath much wider than ambiguity limits would normally allow, the *RADARSAT* system has been designed to incorporate an alternative and less conventional mode (Moore et al. [26]; Tomiyasu [42]) known as ScanSAR. In this mode, for which rapid steering of the elevation beam pattern of the antenna is essential, extended range coverage can be obtained using a set of contiguous beams, enabling production of images having total swath width up to about 500 km. This rather wide swath is accomplished at no increase in mean data rate from the radar, but at the cost of degraded resolution of the resulting image. *RADARSAT* will be the first operational space radar system to implement the ScanSAR technique.

The principle of ScanSAR is to share radar operational time between two or more separate subswaths, in such a way as to obtain full image coverage of each. The set of returns used to image a section of one subswath must be from consecutive pulses to provide adequate sampling, and must be of sufficient length to allow formation of the synthetic aperture needed for the subswath at the required resolution. The imaging operations are, therefore, split up into a series of blocks of pulses, each block providing returns from one of the subswaths. Each block is processed to provide an image of a section of the corresponding subswath. The imaging operations cycle around the full set of subswaths sufficiently rapidly that the imaged sections in any one subswath are adjoining or overlapping. The sequence for the simplest form of ScanSAR imaging, with two subswaths, is illustrated in Fig. 12.9.

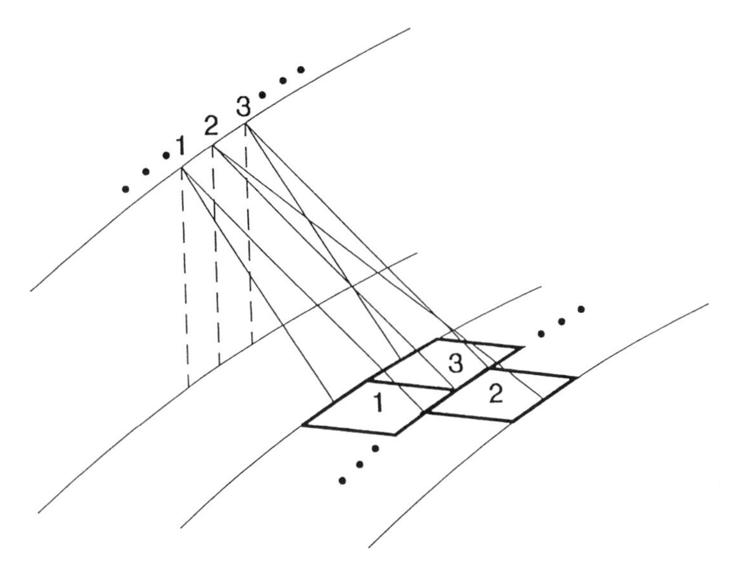

Figure 12.9 Sequence of beam positions used in two-beam ScanSAR. Beam switching is continuously maintained throughout the ScanSAR data take.

With any particular PRF there are certain range positions which cannot be imaged because the signal returns coincide with a later pulse transmission. To obtain the very wide, continuous coverage that is the principal purpose of ScanSAR, it is necessary to use more than one PRF across the set of subswaths. PRF control is maintained by the SAR computer. One consequence of the change of PRF is that a break of a few pulses must be introduced at transitions between blocks to avoid the loss of returns during the time when transmission and reception periods may be out of synchronicity.

As implemented on *RADARSAT*, there are two preprogrammed ScanSAR modes (Luscombe [23]), although the design could realize other ScanSAR configurations using in-flight command sequences to reset phase-shifter references in the SAR computer. For the first ScanSAR mode, imaging alternates between the two wide-swath beams to give total ground range coverage of over 300 km. In the second ScanSAR mode, the coverage of four beams is combined to allow imaging of the full 500-km accessibility region in range. The resolutions associated with these modes is given in Table 12.4.

The main requirements imposed on the system by the introduction of ScanSAR imaging affect antenna beam switching, command and data handling, and processing. To allow effectively instantaneous switching, additional registers are being provided in the antenna for simultaneous storage of the coefficients for four beam patterns. The SAR control computer is being programmed to perform the correct beam sequencing and to coordinate the more complex signal formats delivered in the ScanSAR mode. Provision is being made in the uplink command list for the extra parameters needed to define the ScanSAR operation, and in the downlink data stream to identify the subswath corresponding to each return.

As very rapid phase shift settings are achieved through the *RADARSAT* antenna control, negligible time is lost to ScanSAR operational control functions during data collection. The major limitation derives from the necessity to change radar PRF, which may cause a loss of as many as nine pulses between blocks of data when beam switching occurs. As a block of data in this mode is about 1000 pulses in length, no serious impact on data collection efficiency is expected.

Rather than a continuous stream of data as is normally done, the data from each subswath are received in the form of discrete blocks of echoes and are analogous to the imaging strategy used by the SAR on board *Magellan* when mapping Venus (Johnson [19]). As a result, many of the algorithms that have been developed for processing continuous sequence SAR data are not appropriate. With data used as in conventional imaging, the sets of pulses which are combined to form the synthetic aperture change with the along-track position on the ground: the pulses selected cover the period that the point was receiving maximum illumination by the antenna beam. With ScanSAR data, however, the same block of pulses must be used to image a series of points at different along-track positions. Additional processing requirements follow from the segmented form of the ScanSAR image. Radiometric and geometric matching is required in both along- and across-track directions to knit output data blocks into a continuous image file. Design optimization for the ScanSAR processing algorithm is being developed in the current phase of *RADARSAT* system implementation.

12.2.6 Data Products

Image products planned for *RADARSAT* may be grouped into three classes, generally available both as hard (photographic) copy and as digital files. *Georeferenced* products, the standard image, include systematic geometric corrections with respect to the subsatellite path and are based on satellite ephemeris data. *Geocoded* products are resampled and rotated to conform to a standard map projection. Ground correction points and correction for terrain elevation will be included, if available for the scene in question. *Special* products include unprocessed signal data, single-look (full-resolution) real or complex image files, and analysis of external calibration references.

Extensive data processing is required to form radar images from the data delivered from the spacecraft. For any SAR system this is may be a lengthy and complex operation (Cumming and Bennett [8]; Raney [33]; Curlander and McDonough [9]). Processing for the *RADARSAT* SAR presents its own set of challenges because of its operational requirements and choice of imaging modes. In an earlier phase of this program, a demonstration processor was developed to verify the feasibility of maintaining processing speed and quantity consistent with *RADARSAT* requirements. This processor is being used in Canada's ground station at Gatineau, Québec, for *ERS-1* data. In light of significant advances made over the past few years in radar signal processing, however, a new approach would seem to be justified. Studies are under way to assess both algorithms and specialized hardware to be used in the *RADARSAT* production processor. The main requirement is to produce high quality imagery in near-real time. Data will be sent via satellite links to a central processing facility at 1/4 real-time rate. The initial data processing capability is specified to be 1/10 real time, to be increased to 1/4 real time as demand requires. The requirements stipulate that all data collected will be capable of being processed as received, a "zero data backlog" approach. Archives will be maintained for the received signals and, for selected processed image files, for full resolution single-look complex (SLC) data.

For highly perishable data, such as required by Environment Canada's Ice Centre, an on-line data port is being implemented. The Ice Centre will be able to receive processed SAR data, integrate it with other pertinent data records in their information system, and relay derived user products rapidly to operators in the field. Baseline design for this operational data product delivery system is to deliver data within 4 hours of scene observation by the radar.

12.3 OPERATIONAL ICE INFORMATION

A satellite such as *RADARSAT*, which will be capable of systematic imaging of Canada's ice-covered water, is of little practical use unless the infrastructure is in place to acquire the data, to combine them with all other pertinent sources of data, and to produce information products tuned to the requirements of operational users. Canada has one of the most advanced and comprehensive ice information systems in the world (Falkingham [12]). The following sections describe that system and the role that *RADARSAT* data will play in operational ice information in support of marine requirements.

For operational ice applications, it is most important to obtain an analysis within a few hours of the data take: even a perfect analysis is useless if it arrives in the field too late to be used in decision making. Once an image has been used in an analysis, it is rarely referred to again. The next day will bring another suite of images for analysis; there is no spare time to refine the previous analysis, and little additional surface data with which to verify a refinement.

12.3.1 Ice Information in Canada

Within the Canadian federal government structure, the Atmospheric Environment Service (AES) of the Department of Environment is responsible for the provision of information on atmospheric, sea state, and ice conditions for the safe and efficient conduct of social, cultural, and economic activities. The Ice Branch is the component of the AES that provides information on lake ice, sea ice, and icebergs in the navigable waters of Canada's economic zone. The

Ice Centre, located in Ottawa, is the operational headquarters for the Ice Branch.

Figure 12.10 shows the major areas for which the AES provides ice information at various times of the year. Generally, ice information is available whenever there is significant marine activity in the vicinity of ice. Information is provided for the Beaufort Sea from late June, when breakup begins to permit marine operations, to October when freeze-up prevents further operation. The central Arctic is active from late July to October, the period during which resupply of the local communities occurs and transits of the Canadian arctic take place. Parry Channel (including Lancaster Sound and the approaches to Resolute Bay) has become active as early as May during recent years and shipping continues well into November. For instance, the MV *Arctic*, an icebreaking bulk carrier operated by Canarctic Shipping Ltd., is extending the initiation and the conclusion of the navigable season in this area. In the Queen Elizabeth Islands, there is usually only one voyage, during September, to carry supplies to Eureka.

Further south, ice information is provided for Hudson Bay and its approaches from May through to the end of the shipping season in October. Shipping here includes grain from Churchill and the resupply of the many communities around Hudson Bay and Foxe Basin. Information on the ice-edge location for the Labrador Sea, Davis Strait, and Baffin Bay is provided year-round for the

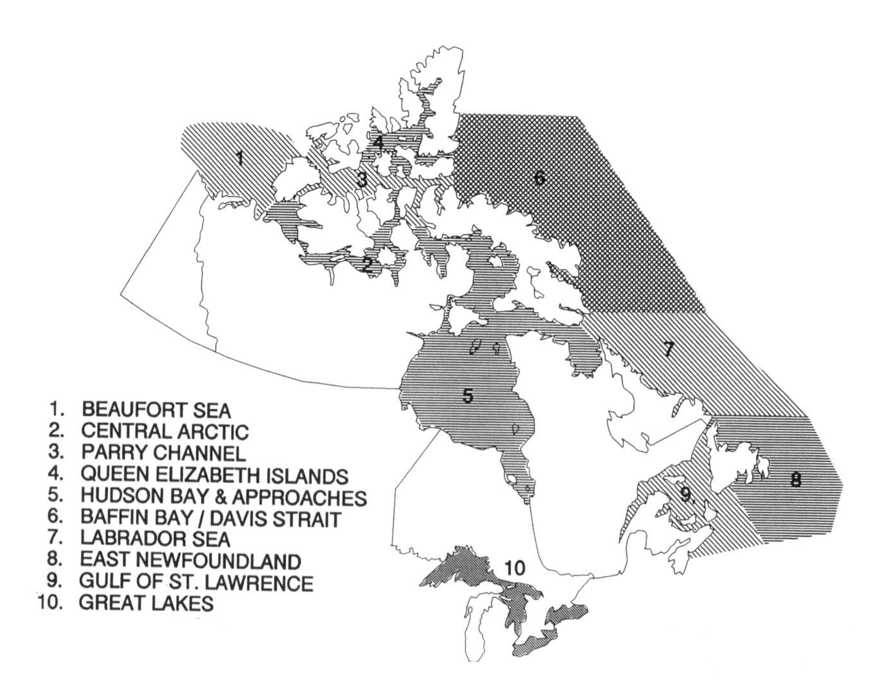

Figure 12.10 Primary Canadian ice areas regularly observed by the Ice Centre. (Figure courtesy of the Arctic Institute of North America)

fishing industry that operates along the edge of the pack ice. Information about the interior of the pack ice is provided from May to November, both for resupply of coastal communities and for access routes to Hudson Strait and Lancaster Sound.

By far the largest number of marine operators receive ice information for the waters around Newfoundland, the Gulf of St. Lawrence, and the St. Lawrence River. The Ice Centre issues a Seasonal Outlook at the beginning of December before the ice begins to form and continues to issue charts and bulletins on the ice situation until the ice disappears, usually in May. Marine activity in these areas includes commercial shipping to and from Montreal and the other major ports in the area, the offshore and inshore fisheries, and oil exploration. Regular ice information is also issued to the small number of marine operators on the Great Lakes between December and April.

The major clients for ice information are the Canadian Coast Guard which operates a fleet of federal government icebreakers, the commercial shipping industry, the fishing industry, and the oil and gas industry. Other clients include the military, the insurance industry, scientists working on, under, or around the ice, and other individuals with a need for up-to-date knowledge of sea ice distribution.

The ice information is distributed in the form of text and radio bulletins and warnings, and ice charts which summarize present and future expected ice conditions. The Ice Centre operates on a primary cycle of one day: each day, new ice data are collected, analyzed, and integrated into previous data to create new products which are distributed to clients in the field. The frequency of this cycle has been chosen as being the most appropriate given the rate of ice movement, the frequency of decisions that must be made by clients, and the availability of new ice data. Other products, intended primarily for longer range planning, are distributed biweekly and seasonally.

12.3.2 Sources of Operational Ice Data

The sources of operational ice data described in this section are summarized in Tables 12.7 and 12.8. Although several different remote sensors have been used in the past and are planned for the future, only those currently in operation at time of writing, and those for which future plans have been committed, are described.

NOAA AVHRR

One of the most useful sources of ice data is the advanced very high resolution radiometer (AVHRR) carried on the polar orbiting U.S. NOAA satellites (Hussey [17]). This satellite program has been in continuous operation for many years and is expected to remain so. Normally two satellites are kept in orbit at any one time, to increase the imaging frequency and to provide failure redundancy. The satellites are in a near-polar, sun-synchronous orbit, at an altitude of about 850 km which results in an orbital period of 102 min. Given two

TABLE 12.7 Remote Sensing Data Sets Used by the Ice Centre

Source/sensor	Typical pixel (m)	Swath (km)	Sensor characteristics
NOAA/AVHRR	1,000	2,000	Passive optical sensor using 5 channels from visible to thermal infrared
DMSP SSM/I	25,000	1,400	Passive microwave sensor using 5 channels from 19 to 85 GHz
Aircraft SAR	25	200	Active microwave sensor using X band, HH polarized
Aircraft SLAR	25–400	200	Active microwave sensor using X band, HH polarized
ERS-1 SAR	50	100	Active microwave sensor using C band, VV polarized (pixel size degraded from original image product)
RADARSAT SAR	50	500	Active microwave sensor using C band, HH polarized (pixel size corresponds to the ScanSAR mode data product)

TABLE 12.8 Remote Sensing Data Sets Used in the Past (But Not Operationally in 1992)

Source/sensor	Typical Pixel (m)	Swath (km)	Sensor characteristics
Airborne laser profilometer	1–3	0.001–0.003	Active optical range-finding device using a single visible channel
Airborne imaging microwave radiometer	25	10	Passive microwave sensor using 5 channels from 19 to 90 GHz
Airborne infrared line scanner	25	10	Passive optical sensor using a single thermal infrared channel

satellites and westward precession of their orbits, it is quite possible to obtain imagery up to 20 times per day at 50°N, although half of these passes will take place at night when only the infrared channels are useful. Further north, because of the convergence of orbits toward the poles, even more frequent coverage is possible. The probability of imaging a particular ice area within 1–3 days is better than 0.5, despite the inability of the sensor to observe the surface through clouds.

The AES operates two high resolution picture transmission (HRPT) ground stations, at Toronto and at Edmonton, to receive NOAA AVHRR data, where the downlinked data are corrected radiometrically for atmospheric distortions and preordered images are converted from 10 bits to 8 bits per pixel. These images are transmitted to the Ice Centre via the Meteorological Satellite Information System, a private satellite communications network operated by the AES. Data reaches the Ice Centre within 2 hours of the satellite overflight, and enters the processing chain as described in the Subsection 12.3.3.

AVHRR data has been available since 1979, and its use for sensing sea ice has been well documented (Condal [7]; Dey and Richards [10]; Flett et al. [13]). The swath is approximately 2000-km wide, and the pixel size is about 1 km^2 which gives an excellent synoptic-scale overview of the ice regime. From this imagery, ice analysts can determine the extent and concentration of sea ice over large areas and can infer relative ice thickness from temperature information contained in the infrared channels. Large-scale ice motion can be determined from lead patterns and eddied along the ice edge observed through sequential images. The major disadvantage of the AVHRR sensor is that it cannot penetrate clouds and so, at any given time, the ice surface may be obscured from view. This disadvantage can be reduced in some circumstances, as described in Subsection 12.3.4.

DMSP SSM/I

The U.S. Defense Meteorological Satellite Program (DMSP) is an ongoing operational program with several satellites, both in orbit and in various stages of launch readiness (Wittman [44]). One of the instruments carried on recent satellites is a passive microwave radiometer called the special sensor microwave/imager (SSM/I) (Hollinger and Lo [16]). (This sensor is described in the context of passive microwave remote sensing of sea ice in Chapter 5.)

The swath width of the SSM/I is about 1400 km and four separate frequencies are sensed. At 19, 22, and 37 GHz, the pixel size is 25×25 km, whereas at 85 GHz it is 12.5×12.5 km. At these frequencies, the sensor can observe the surface radiation through atmospheric cloudiness and almost complete global coverage of the earth's surface is obtained daily. The data are downlinked to several U.S. ground stations and forwarded by communications satellite to the U.S. Navy's Fleet Numerical Oceanography Center (FNOC) in Monterey, California. FNOC converts the data into earth-located microwave brightness temperatures and makes these data available to the operational and research communities.

The Canadian Meteorological Centre in Montreal obtains the multifrequency brightness temperatures daily and computes percentage concentrations of old, first-year, and young sea ice using an algorithm that considers the relative strengths of the various spectral returns (Rubenstein and Ramseier [36]; Swift et al. [39]). The Ice Centre retrieves these ice concentrations via a data telephone line for integration with the other data sets.

Aircraft SAR and SLAR

The AES Ice Branch operates two fixed-wing aircraft which carry SAR or SLAR imaging radars. Detailed description of the SAR systems may be found in Chapter 10. An overview of the role of these systems in the operational tasks of the Ice Centre is included here.

A Challenger jet aircraft, owned and operated by Intera Information Technologies (Canada) Ltd. under a contract with the Ice Branch [18], is the newest ice reconnaissance aircraft. It carries two MacDonald Dettwiler and Associates Ltd. SARs that image a 100-km swath at 25-m resolution on each side of the

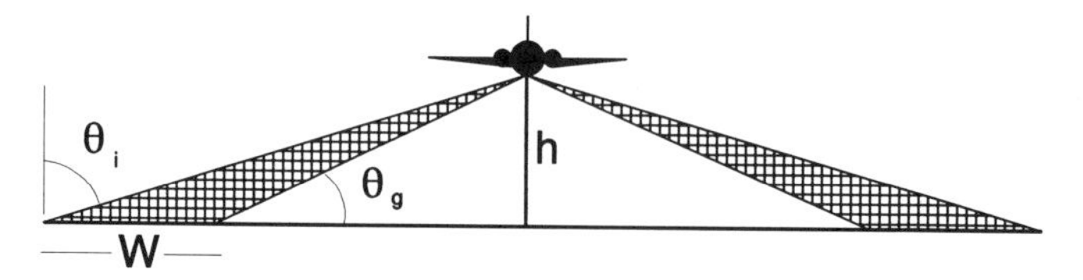

Figure 12.11 Airborne radar geometry, where **h** is aircraft altitude, θ_i is incidence angle, and *W* is swath width. (Courtesy of the Arctic Institute of North America)

aircraft (Fig. 12.11). An optional fine-scale mode on each radar provides imagery at 15-m resolution over a 60-km swath. The image data are displayed on the aircraft in real time, both on hard copy and a display terminal, and recorded by the on-board computer. A telemetry downlink system allows the SAR data to be transmitted to ships and to the Ice Centre while the aircraft is in flight. The normal operating altitude is 11,000 m (35,000 ft) at a speed of 700 km/h. Missions, which usually cover 3500 km and last 5 hr, are conducted every 2 or 3 days.

A Dash-7 turboprop aircraft, owned by the government and operated under contract, complements the Challenger. It carries a CAL Corporation SLAR that images a 200-km wide swath 100 km on each side of the aircraft. The azimuth resolution of the SLAR is variable, spreading from about 40 m at close range to several hundred meters at far range, although the imaged signal is mapped into constant pixels of 37.5 × 37.5 m. The radar data are displayed on the aircraft in real time, and recorded by the on-board computer. A telemetry downlink system, similar to that on the Challenger, allows the SLAR data to be transmitted to ships and to the Ice Centre while in flight. Additionally, this aircraft can carry a laser profilometer to measure the height of ice surface features accurately, and a Zeiss mapping camera for photographic verification of the radar data. The aircraft is outfitted with top and side bubbles to allow visual reconnaissance. The normal operating altitude is 3000 m (10,000 ft) at a speed of 300 km/h. Missions usually extend over 2500 km, last 6 to 7 h, and are conducted every 2 or 3 days.

Both ice reconnaissance aircraft are equipped with data downlink systems to transmit radar image data from the aircraft to the surface. Ships that need a detailed knowledge of the ice, such as Coast Guard icebreakers, are equipped with systems, known as STAR-VUE, that can receive the downlinked data and display the radar images in real time as the aircraft flies overhead. These systems provide additional functions that allow the shipboard operator to enhance the images, to determine the correct geographical position of the images, and to plan the vessel's track through the ice field. The STAR-VUE system is discussed further in Chapter 10.

The Ice Reconnaissance Data Network (IRDNET) employs several ground stations, located at strategic locations along the Canadian coast (Fig. 12.12), to receive the downlinked data from the aircraft (Telesat [40, 41]). These ground stations relay the data through a communications satellite to the Ice

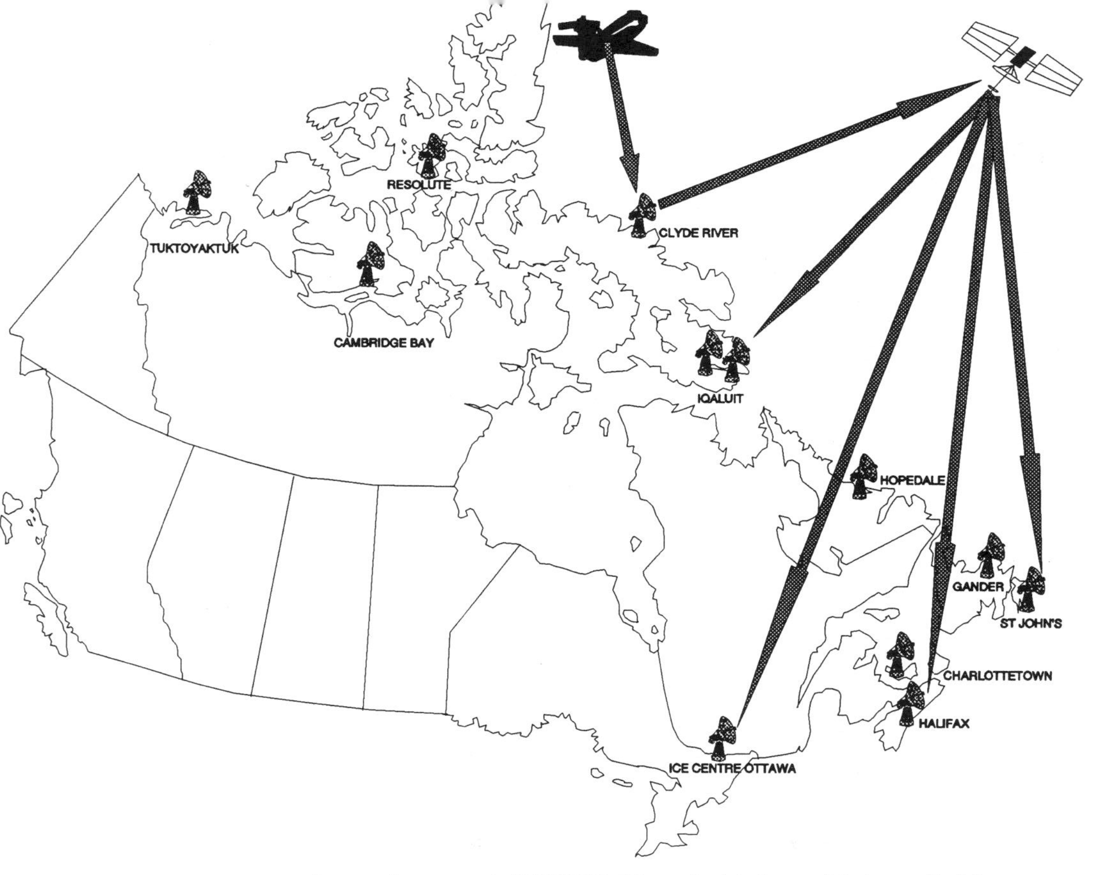

Figure 12.12 Ice reconnaissance data network (IRDNET). Figure depicts image data transmitted from aircraft to the Clyde River ground station, which relays the data via a communication satellite to the four receiving stations. At Iqaluit there are both relay and receive stations. Other relay stations are shown.

Centre. The aircraft downlink operates on a line-of-sight radio frequency which provides a range of up to 400 km. During a mission, the aircraft will collect radar data and record them in the on-board computer. When the aircraft comes within line of sight of one of the ground stations, the recorded data are transmitted at about 30 times the acquisition speed. The ground station receives the data sequence and relays them to the Ice Centre via a communications satellite. In this manner, the Ice Centre receives the radar data within 3 hours of initial recording, and can begin their integration into ice information products.

In Situ Data

The Ice Branch posts an Ice Services Specialist (ISS) on board each of the major icebreakers operated by the Canadian Coast Guard, the major user of operational ice information. In addition to advising the captain about ice conditions that the vessel may encounter, the ISS provides essential observational data to the Ice Centre which assists in their interpretation of remotely sensed data. Much of this is visual reconnaissance data that is obtained by the ISS from the ship's helicopter, which flies local sorties at a very low altitude, generally within about 20 km of the ship.

The ISS records the ice type and concentration, stage of development, floe size, and notable surface features, on a hand-drawn ice chart using the standard international symbology (see Fig. 12.16, inset). The resulting chart is used to brief the captain, and a copy is transmitted to the Ice Centre (via facsimile) for integration with other data. At regular intervals, when time and ice conditions permit, the ISS disembarks to the ice to measure its thickness, using a sample hole bored for the purpose. This information is also relayed to the Ice Centre for comparison with the accumulated data base.

Satellite SAR

Currently, the Ice Centre receives SAR data from the European Space Agency's *ERS-1* satellite. The sun-synchronous, near-polar orbit of the satellite provides a generous sampling of data over Canadian ice areas. For the Ice Centre's real-time requirements, data are acquired by the ground station at Gatineau, Quebec, and are processed into images with 50 $\times$ 50 m pixels. *ERS-1* images are available at the Ice Centre via data communication lines within a few hours of the satellite pass. An example of *ERS-1* imagery from the Canadian Arctic, received in near-real time at the Ice Centre, is shown in Fig. 12.13. In the near future, a communication system should be in place to allow the reception of near real-time imagery from the ground station at Fairbanks, Alaska. Those images will have degraded resolution, having 240 $\times$ 240-m pixels. Data are also received at Prince Albert, Saskatchewan, but are not available at the Ice Centre in near-real time. Although useful to Ice Centre operations, the main reason for using *ERS-1* data at this time is to test the ground processing and communication systems in preparation for *RADARSAT*.

When *RADARSAT* begins operational provision of ice imagery in 1995, the Ice Centre will use the ScanSAR image products almost exclusively. For ice

Figure 12.13 Example of ERS-1 image of Arctic ice as seen on IDIAS. This image of the south end of Peel Sound was acquired on January 15, 1992. Prince of Wales Island is on the left. Multi-year ice floes embedded in a matrix of consolidated first-year ice are clearly visible in the Sound. The image was produced on a quick-look paper recorder from an IDIAS display.

information needs of AES, ScanSAR will provide an ideal mix of large area coverage (500-km swath width), resolution (50-m pixels), and timeliness (4 h).

Ground stations at Gatineau and Prince Albert will receive data for all of the Canadian ice areas. Raw data from Prince Albert will be transmitted to the SAR Data Processing Facility (SARDPF) at Gatineau via communications satellite. The SARDPF will produce the ScanSAR 8-bit image product within 4 hours of satellite overflight. This product will be transmitted from the SARDPF to the Ice Centre via a high-speed communications line, which will deliver the image data directly into the Ice Centre computer network for integration with other data sets.

RADARSAT data are expected to become the cornerstone of ice information used at the Ice Centre on an operational basis. The swath will be wide enough to provide a large-scale view of ice areas, without being subject to cloud cover obscuration. At the same time, spatial resolution of the data will be sufficiently detailed to support all but the most demanding tactical requirements. Unlike aircraft reconnaissance, the satellite platform will be independent of earth-bound constraints such as favorable weather and the availability of personnel. Finally, it is expected that *RADARSAT* data will be considerably more cost-effective for the Ice Centre's purposes than corresponding aircraft data, although it is anticipated that some airborne data will continue to be required for close tactical support and for backup purposes.

12.3.3 Data Integration

A major task of the Ice Centre is to infer actual ice conditions from the representations made by various remote sensors. The data sets used to make this determination are summarized in Table 12.7 which shows that widely varying spatial scales are involved, depending on the sensor geometry and the flight characteristics of the platform (Fig. 12.14). Whereas all of the satellites involved are in near-polar, low-earth orbits, the minor differences between these orbits must be taken into account when locating the imagery on the globe. Aircraft data are oriented parallel to the flight track of the aircraft, which is mostly dependent upon the orientation of the waterways being imaged.

The Ice Data Integration and Analysis System (IDIAS) (Fig. 12.15) at the Ice Centre is the central computer system used to integrate the various remote sensing data sets. A Lambert conformal conic map projection, with standard parallels at 49°N and 77°N, has been chosen as the best compromise for a standard base map for this purpose. It preserves the shape of geographical areas and provides scale errors of less than 0.5% between the standard parallels (Philip A. Lapp [31]).

Each of the data sets contains sufficient information to allow IDIAS to compute an approximate earth location for each image pixel. Because of the large volume of data received in near-real time and the need to analyze these data within a few hours, it is imperative that this georeferencing be accomplished rapidly, and without the need for human assistance. In contrast to the more common geocoding process for which an operator must select known geographic reference points in the remote sensing image, IDIAS uses the geometry of the sensor and information about the platform's flight path to locate the imagery. This "systematic" method of georeferencing the images is considerably faster than human-assisted precision geocoding, and although it is less accurate, the errors incurred are within the nominal 5-pixel tolerance needed for a real-time operational analysis.

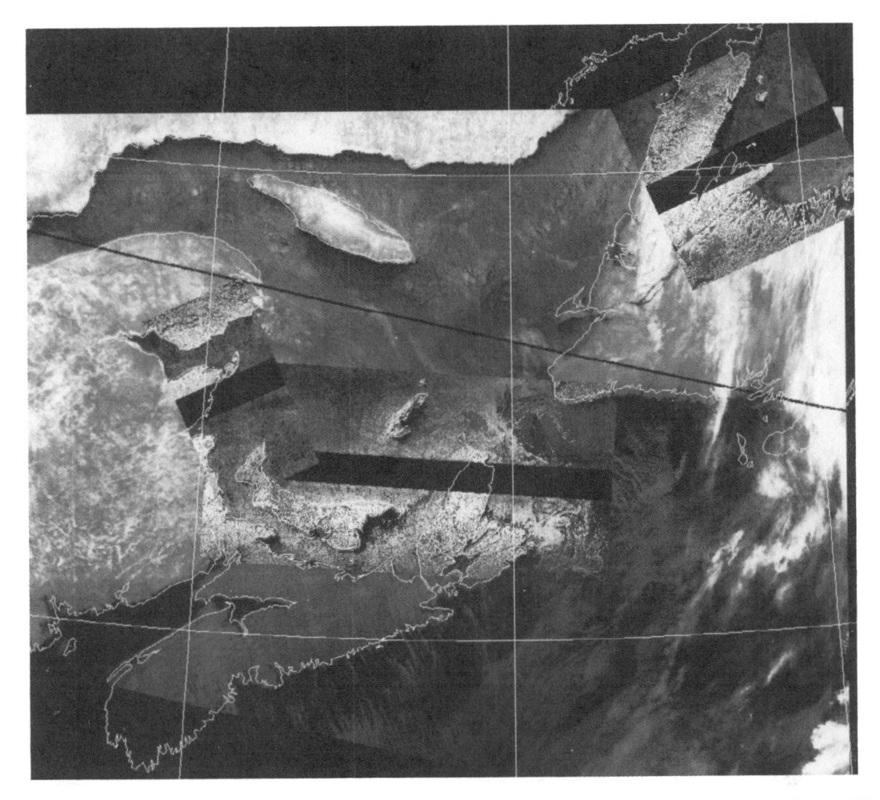

Figure 12.14 Example of IDIAS composite image of the Gulf of St. Lawrence, February 23, 1992. Portions of an aircraft SAR flight are overlayed on an AVHRR image showing the large difference in scale and resolution between these two systems, and the cloud penetration ability of the SAR.

NOAA AVHRR Data

The NOAA AVHRR imagery is located with respect to time, but it is not provided with latitude/longitude information. Through the Global Telecommunications Network, NOAA provides daily information on the estimated position of the spacecraft. Using this ephemeris information, IDIAS must compute the orbital trajectory of the satellite to determine its position in space for every scan line of the image. This position can be fixed relative to a point on the earth's surface, including allowance for the rotation of the earth. Because of the wide swath of AVHRR imagery, earth curvature and the projection geometry of the sensor must be considered both along and across the scan lines. Once the geometry has been computed for every pixel, IDIAS warps the image to fit the standard base map.

SSM/I Data

The DMSP SSM/I data are retrieved from the Canadian Meteorological Centre, with latitude/longitude tags already provided for every pixel, so no georefer-

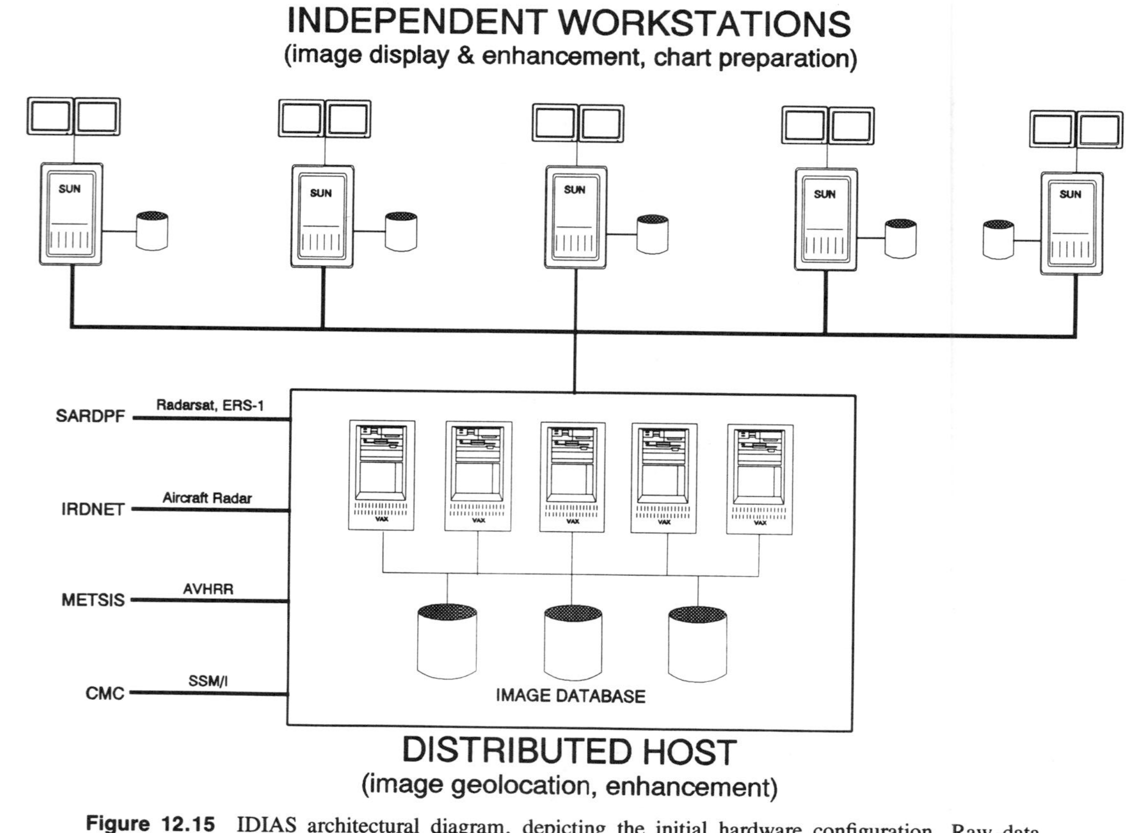

Figure 12.15 IDIAS architectural diagram, depicting the initial hardware configuration. Raw data enter the distributed host (left side of the diagram), are processed, and subsequently stored in the image data base. In addition to the network for interprocessor communication that is indicated, there is a direct link between each work station computer and a host computer to permit high-speed retrieval of image data to the work station, avoiding network congestion. Graphics data are stored on the disks attached to each work station, and shared among the work stations via the network.

encing is required. IDIAS uses these tags to display the numerical pixel values in the correct geographic positions on chart displays.

Aircraft SAR and SLAR Data

Auxiliary data, downlinked with the aircraft SAR and SLAR data, provide information about the imagery, including flight number, date, time, latitude, longitude, range and swath width, and pixel spacings in azimuth and range. Aircraft operating parameters include heading, ground speed, altitude, and track angles, both short- and long-term. These data are updated for every scan line of the image which allows IDIAS to recreate a model of the aircraft flight track. This model, together with the known geometry of the radar (Fig. 12.11), is sufficient to determine the latitude and longitude of every pixel in the image, within the limits of error of the aircraft navigation systems. Using Global Positioning System (GPS) information, image positioning errors are considerably less than 1 km and are well within the tolerances required for real-time ice information. The georeferencing process is simplified somewhat for sea ice because changing surface elevation, usually problematics in airborne radar imagery, can be ignored.

Satellite SAR Data

ERS-1 and *RADARSAT* SAR data products include latitude/longitude positions for the near, center, and far pixels of each scan line which allows the imagery to be georeferenced by a straightforward geometric warping process, thus eliminating the need for orbital modeling.

Integration

Once the data sets have been georeferenced, IDIAS can display them in image (or, in the case of SSM/I, numerical) form on any of its dual-screen workstations. Images are contained in windows that can be placed side by side or overlapped in limitless configurations. Within each window, images can be zoomed to any desired magnification (with pixels being remapped or replicated as necessary) and many graphic displays can be overlaid, including geographical information (coastlines and latitude/longitude lines). Data can be viewed either in *image orientation*, which keeps image boundaries parallel to the screen dimensions and rotates the graphics to allow as much of the image as possible to be viewed at once, or in *chart orientation*, in which images are rotated to align with an approximately north-up chart. In chart orientation, many images can be tiled together onto one display to create a composite image.

At this time, integration of the data sets is strictly a visual process performed by ice image analysts. Information from various remote sensing data sets is integrated onto a chart product, as described below. In the future, and particularly with the coming of *RADARSAT*, it is expected that more and more of the data sets will be integrated automatically by the computer to provide a seamless mosaic of imagery for the ice analyst.

12.3.4 Data Analysis

Following the integration process that allows the various data sets to be intercompared, the task of analyzing them begins, to determine the prevailing ice conditions. Parameters of first-order importance include total and partial concentrations of ice, ice type classification, stage of development (thickness, strength) of the predominant ice types, the range of floe sizes present, and the presence of any significant hazards or aids to navigation, such as rubble fields and major leads. At the present time, the analysis task is accomplished manually by experienced ice analysts, who view the images on the IDIAS workstation screens and make subjective assessments of the ice conditions. To help the ice analysts, several automatic and interactive image enhancement tools are available to highlight the various ice features.

It should be noted that the analysis process used at the Ice Centre differs considerably from that employed by cartographers, geologists, or other specialists. The primary difference arises from the transient nature of the geophysical feature being analyzed. In other fields, it is common practice to work for extended periods on a single image or a small set of images, often obtaining surface information referenced to individual pixels in the image.

To make image enhancements available for display and redisplay as quickly as possible, IDIAS uses automatic preprocessing to produce the most common enhancements required. When a data set arrives at IDIAS, it is recognized by type and submitted to an appropriate production schedule for processing. The production schedule consists of a collection of processing stages that perform georeferencing of the image and create one or more image enhancements, depending on type of image, geographic regions, time of year, and time of day.

Standard Monochromatic Enhancements

There is a limited amount of enhancement that can be (or needs to be) performed on images that have only a single channel (frequency) of information, such as airborne and satellite radar data. A conventional image analysis technique is to apply a contrast stretch algorithm to the image which spreads the distribution of pixel values across the entire gray-scale available on the display. In an operational environment, however, it has been found important to employ a standard contrast stretch for all images. Although this procedure may result in less than optimum enhancements for individual images, it is necessary to maintain as much continuity as possible in both space and time. Many comparisons must be made between image frames: in different locations, at different times, and with data from different sensors; consistency is paramount.

It is common also to apply a speckle-reduction filter which reduces the amount of noise in a SAR image. Such enhancements must be used with care, as small but significant targets (such as icebergs) could be removed from an image through use of an inappropriate filter. For this reason, the original image is retained, and referred to again and again during the analysis process.

An edge enhancement filter is applied to radar images, which helps to delineate leads in the ice pack and the boundaries of floes. To support particular operations, the edge filter, which has a preferred orientation for enhancement, can be applied to highlight features with directional significance, such as leads running parallel to the vessel's desired track that could be used to aid progress, or ridge fields perpendicular to the track that could present a significant obstacle.

Various attempts have been made to use color to enhance brightness levels in monochromatic images but little success has been achieved using this technique in ice applications. Discriminating different types of ice with color keyed to the subtle differences in gray-scale tone has proved to be problematic in practice, as there are far too many ambiguities between tone and ice type, and the resulting patchwork of color serves to mask other gray-scale indicators, such as shape and texture.

Multi-Spectral Enhancements

NOAA AVHRR imagery is useful for ice analysis because the large area covered and the frequent reobservation opportunities allow the analyst to understand the dynamics of the ice fields. Although the surface is frequently obscured by clouds in these optical images, this liability is mitigated by a high frequency of observation and the large difference in the speed of motion between clouds and ice. In addition, the five spectral channels of data allow a wide range of image enhancements to be applied.

A standard contrast stretch or histogram equalization is applied, to one of the visible channels and to one of the thermal infrared channels, to produce the basic image product used in analysis. Standardization of the enhancement rules between images is less of a problem with NOAA AVHRR data than for high-resolution data, because NOAA AVHRR frames cover a much larger area that inherently includes a wide variety of target types.

A standard NOAA AVHRR color enhancement is produced by applying a combination of image channels to the three colors of the work station display as follows:

Red Gun: applies linear look-up table to Channel 1 (visible)

Green Gun: contrast stretch of Channel 2 (near-infrared)

Blue Gun: contrast stretch in Channel 4 (thermal infrared)

The resulting presentation is a very natural-looking image (Fig. 12.16) in which open water is dark blue and clouds and ice are varying shades of white, from creamy-white to blue-white (depending primarily on the temperature as indicated by the strength of the infrared return, see color insert). The advantage of using a color image of this type is that it allows the ice analyst to obtain information contained in several spectral channels from one image.

Another very useful color enhancement that is applied automatically is the ice-cloud discriminator. This enhancement has the form:

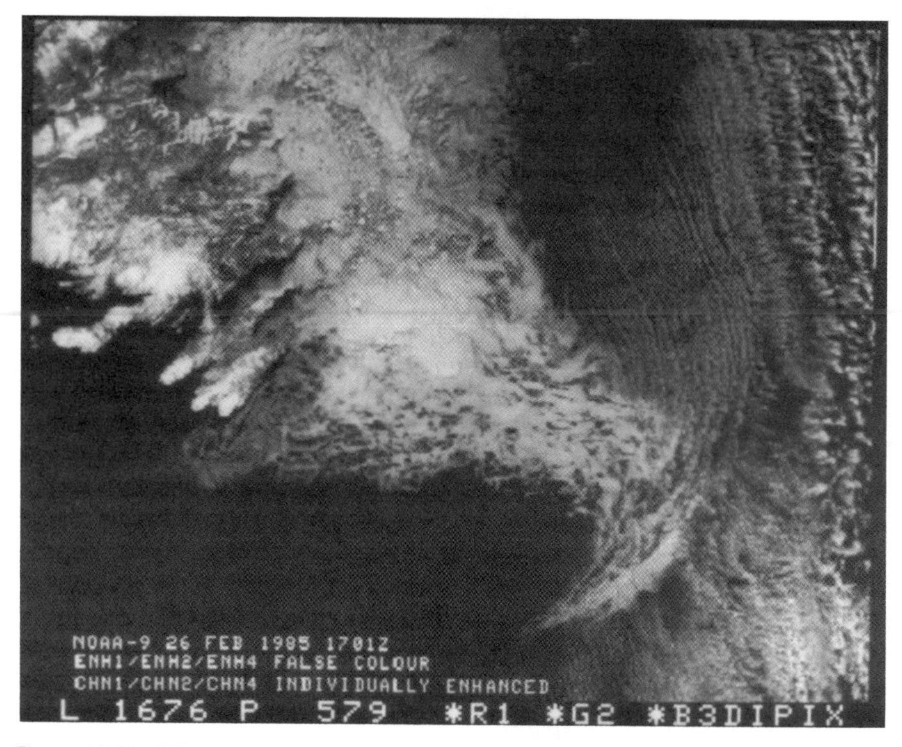

Figure 12.16 NOAA AVHRR image of the Grand Banks. Newfoundland is at the upper left. Because of the enhancement, the Labrador pack ice in the center of the image is clearly distinguishable from the clouds at the right, even where the clouds partially obscure the ice (see color insert).

Red Gun: Channel 1
Green Gun: difference of Channel 1 − Channel 2
Blue Gun: ratio of Channel 3 (water vapor)/Channel 4

Information from both the thermal channel (temperature differences between cloud and ice) and the water vapor channel (water content in clouds) is used to color the basic visible image. The resulting enhancement is a rather garishly colored image, in which thick clouds become green and ice becomes reddish, but this enhancement has the significant attribute of showing ice clearly through thin clouds. It is a frequent occurrence that, where an ice field disappears under a cloud layer, the occlusion is not continuous, particularly near the edge of the cloudy area. As both clouds and ice are white in normal single-channel images, it is impossible to distinguish the difference between them in these situations. The ice–cloud discriminator is valuable because the ice is given a different color from the clouds, and can be seen clearly through gaps in the cloud that are not evident in a monochromatic image. Once the cloud layer becomes thick enough to obscure the surface completely, no enhancement is possible, and microwave imagery must be used.

Interactive Image Enhancements

The enhancements described above are created automatically when an image set is first received by IDIAS. The ice analyst at a work station is notified when the various enhancements are available, and, for speed of display, each enhancement is treated as a separate image. All of the different enhancements of a single image can be displayed simultaneously for intercomparison if it is so desired.

To provide even more assistance to the ice analyst with particularly difficult or critical situations, a standard suite of interactive image enhancements is available on IDIAS. The ice analyst can alter the gray-scale contrast and tone of an image on the screen dynamically, and can render an image into false color coding manually by assigning color ranges to the various pixel intensities present. Although this is easily accomplished with the IDIAS work station interface, it is a feature that is used sparingly because of the tight time constraints facing the ice analysts and the marginal improvement that is realized in practice.

RADARSAT

RADARSAT will bring operational remote sensing of ice to a new plateau because the SAR can penetrate clouds and darkness, and its high resolution is combined with a wide swath and orbital repeat cycle that, taken together, lead to coverage which approaches that of the NOAA AVHRR. To take advantage of these attributes, several changes will be made to IDIAS which are quite fundamental in nature.

At the present time, ice data (with the exception of SSM/I which has very coarse resolution) are treated on an image by image basis. Each frame is displayed, enhanced, and analyzed separately, ranging from a 1000 × 1000 km AVHRR image to a 100 × 100 km SAR or SLAR image. As a result, the ice analyst may have several tens of images to manipulate at any one time. The integration of these images occurs only in the final chart analysis product. The reason for this is primarily technological: when IDIAS was built, the technology to do otherwise was not available. For the number of images that are involved, and because the imagery available on any given day represents only a fraction of the total surface area for which information is provided to clients, the situation is acceptable and workable.

When the *RADARSAT* SAR becomes operational, it will provide an unprecedented volume of imagery. Using ScanSAR, most of the Canadian arctic will be imaged every day, and more southern areas of interest to the Ice Centre will be imaged at least every 3 days. No longer will ice analysts have to complete the data gaps by using numerical forecasts and rules of thumb.

This wealth of SAR data could cause a serious problem if the method of handling images in IDIAS is not changed. Instead of the current workload of tens of images per day, ice analysts would have to manipulate hundreds of images, clearly an unmanageable task. The current thinking is to change the philosophy of image handling in IDIAS to more closely match the manner in which the imagery is collected and the manner in which it is used. With advances in computer hardware and software, it is possible for the computer to mosaic image frames automatically, within a satellite pass and between passes. With this capability, a complete image of the Canadian ice areas could be built up, and continually updated, as new imagery is received. Ice analysts could extract any arbitrary portion from the data set (for example, the area corresponding to a particular ice-analysis chart product) and display, enhance, and analyze it as a single entity, thereby eliminating the need to sort through dozens of images individually.

Automated Ice Analysis of SAR Imagery

In addition to the great gains anticipated by using *RADARSAT* SAR data, automated methods of extracting information from satellite SAR data are under development.

The most well developed of these techniques is the Ice Motion Algorithm (IMA), a completely automated technique in which time-sequential pairs of images are compared. The differential ice motion between them is calculated and displayed as a set of vectors which depict the displacement and rotation of the ice. This result is of considerable use to marine operators interested in short-term ice motion, for climatological studies over a longer term, and to initialize and verify numerical models for predicting ice drift. IMAs have been tested successfully with *Seasat* SAR, airborne SAR, and AVHRR images (Hirose et al. [15]; Heacock et al. [14]). At time of writing, the Alaskan SAR Facility (ASF), developed by NASA at Fairbanks, has an operational IMA being used with *ERS-1* SAR data. This capability is in proof of concept evaluation in advance of operational service using *RADARSAT* data.

Development of automated classification algorithms for SAR data of ice is under way at several centers, and many different approaches are being taken (Barber et al. [3]; Wackerman et al. [43]). A single measure, such as the tone of an SAR image or its texture, is not sufficient to classify the type of ice uniquely. The most promising efforts seem to be in the application of several such measures simultaneously. By the time of the *RADARSAT* launch, a realistic research objective is to be able, with confidence, to classify four ice types: open water, young ice, old ice, and rubbled ice.

A different approach to the automated classification of SAR imagery for ice would use expert systems, or Artificial Intelligence (AI). Ice analysts routinely use a variety of ancillary information that serves to both limit and focus analysis during interpretation, such as season, temperature regime, and geographic location. Such information could be applied also by an AI-based ice classification system. Even further, an expert system could use information from previous days' analyses (as do experienced human analysts) and employ the results of the IMA to detect only changes in the ice conditions.

All of these techniques, and most likely some combination of them, hold the promise of providing rapid, automatic analysis of *RADARSAT* data to create information about ice for marine clients. Ice analysts will be able to spend time more profitably in analyzing those areas where the ice situation is critical, and where the automated results are ambiguous. An effective human-machine mix should provide much more voluminous and detailed knowledge of ice conditions than ever before possible.

12.3.5 Products

The role of the Ice Centre is to provide information about ice conditions in Canadian waters to support decision-making. In most cases, clients do not want large amounts of raw data because they have neither the time nor the expertise to conduct analyses of the ice situation. A ship's captain must consider a multitude of other factors as well as ice conditions when selecting a route through ice-encumbered waters. Bathymetry, sea state, weather forecasts, fuel availability, loading schedules, crew time, navigation aids, and equipment status are all important, and the captain can ill-afford to expend an inordinate amount of time on any one factor to the detriment of another. Ice information products prepared by the Ice Centre are designed to provide a summary of the most important information for the largest group of clients, such that it can be absorbed in a short period of time.

As stated earlier, the total concentration of the ice, including the ice edge, is the single most important element of ice information which allows vessels with no ice capability to avoid ice areas when possible. Second in importance is the stage of development of the predominant ice types present and the partial concentrations of those ice types. This information allows vessels with some degree of ice capability to assess the difficulty or risk involved in entering an ice area. Knowledge of the range of floe sizes present, and the proximity of any significant hazards or aids to navigation, provides a third level of information to further refine the risks.

All of this information is presented in simple form on the daily ice analysis charts, the primary data products of the Ice Centre. These charts employ the World Meteorological Organization's (WMO) international standard symbology (Fig. 12.17, inset) to portray the three elements described above (AES [2]). Standard charts cover areas that are about 1000 × 1000 km (e.g., Gulf of St. Lawrence, Baffin Bay) and, in normal practice, one chart is produced daily for each area in which there is significant marine activity in the vicinity of sea ice or icebergs. The ice charts together with weather charts are broadcast by marine radio stations on a daily schedule for reception by radio-facsimile machines on board ships at sea.

The ice charts are produced directly on IDIAS using graphical editing tools built into the software. The ice analyst displays the set of remote sensing images that are available for the particular chart being produced, together with charts from previous days or other graphical information, on the dual screens of the IDIAS work station. Boundaries are drawn around the various ice areas, manually grouping the appropriate elements of the ice matrix. On ice analysis charts, boundaries delimit open water (< 10% ice concentration), very open ice (10–30% concentration), open ice (30–60% concentration), close and very

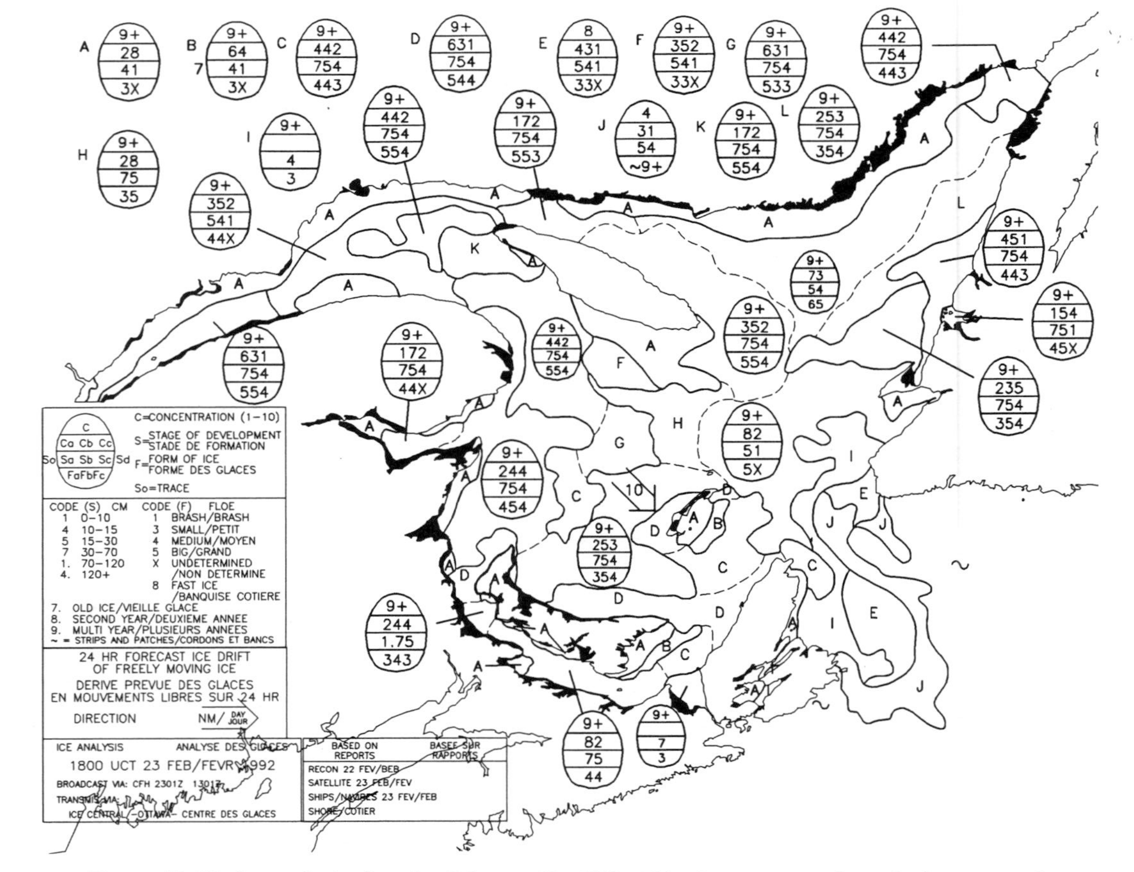

Figure 12.17 Ice analysis chart for February 23, 1992. This chart corresponds to the image set of Fig. 12.14. The inset legend describes the WMO symbology used on the chart.

close ice (60–95% concentration), and consolidated ice (95–100% concentration). Boundaries also delimit significant changes in predominant ice type and may be used to outline significant ice features. Ice codes are placed on the charts, using IDIAS, by entering the appropriate code numbers into a form on the screen and pointing to the desired location for the code symbol. The process is tedious and time-consuming, even with the advantages of convenient interactive computer graphics tools, and a great deal of effort is being expended to make it more efficient without sacrificing accuracy.

This type of ice interpretation chart (Fig. 12.17) has been in use for many years and is familiar to mariners who frequent ice-encumbered waters. A mariner can determine quickly, from the chart, where the difficult ice is and what route must be taken to avoid it or, if necessary, what difficulties are likely to be encountered along a previously chosen route. During 1989, the Canadian Coast Guard began field tests of the new Canadian Arctic Shipping Pollution Prevention Regulations (CASPPR) which are based upon up-to-date ice analysis charts. Unlike the previous regulations, which stipulate fixed entry and exit dates for the various geographic zones of the Canadian Arctic based upon the ice climatology, the new CASPPR allow for the dynamic determination of entry decision criteria, based in large part upon the latest ice analysis of the area (Subcommittee . . . [38]; Norland . . . [30]). It is expected that the new CASPPR will be given legislative force in the 1990s, after a period of fine-tuning based on accumulated experience. The Act will further increase the importance of ice analyses, from being merely information documents to actual law enforcement tools.

In addition to the daily ice analysis charts, the Ice Centre produces several other products conveying much the same information in different formats. Daily text bulletins are produced to accompany each ice analysis chart. These bulletins are relayed to marine radio stations and are read out over the airwaves for the benefit of many smaller vessels (particularly coastal fishing vessels) that do not have radio-facsimile equipment. Composite ice charts cover a much larger area than daily ice analysis charts, but with considerably less detail, and are produced weekly. These are mailed to ship owners, operators, insurance agents, routing companies, research organizations, and a host of other clients who are interested in the general ice situation for planning purposes but have no need of the details. These charts are also the most often used for climatological purposes. When warranted, special analysis charts are produced at larger scales than usual, with considerably more detail, to support particular vessel operations. Ice forecast charts are produced daily, to depict the ice conditions expected within the next few days. Longer range test forecasts provide monthly and seasonal predictions.

The predominant chart format available is the WMO standard radio-facsimile that provides for relatively reliable communication of information over HF radio waves (Yamamoto [46]; WMO [45]). Unfortunately, the quantity of information that can be conveyed by this medium is quite low. Modern satellite communication methods are available and are being improved, and allow for much more information to be packaged into a single product.

12.4 CONCLUSION

Canada has the longest coastline of any country in the world and most of that coastline is ice-bound for at least part of the year. Ice impedes marine transportation to the most populated areas of the country and continues to pose a major obstacle to development of the north. To achieve safe and environmentally responsible economic activity requires an intimate knowledge of the ever-changing ice conditions, a knowledge that can be obtained only with remote sensing technology and supporting information delivery systems.

Over the past 30 years, the national ice program has employed many different remote sensing techniques. This chapter has described the current application of real-time remote sensing of ice for operational decision-making, and the future developments *RADARSAT* will bring.

RADARSAT will provide the single biggest leap in our capability to map the sea-ice areas of Canada. For the first time, it will be possible to determine the ice conditions accurately and reliably throughout Canadian water, on a daily basis and at high resolution. This satellite data will increase our knowledge of the ice environment and ice dynamics in as profound a way as the advent of time-lapse satellite imagery enhanced meteorological studies and forecasting in the 1970s.

The sensor is not sufficient unto itself. For operational use, it is essential to be able to ingest, analyze, and deliver reliable and timely ice information to users. IDIAS has the capability to produce remote sensing images of ice areas complete with a type of WMO analysis superimposed and, if the cost of communicating this type of product to clients at sea can be made sufficiently economical, a single product could not only show the traditional boundaries between ice areas, but also could use the picture quality of the image to indicate that gradations between these boundaries and to highlight features in the ice that are too small, or too numerous, to be transcribed onto an analysis chart. Digital communication of this image plus the analysis product to clients would be error-free, and they could integrate the product into their own geographic information system data bases, containing electronic hydrographic charts or other geophysical parameters of interest to their particular operation. These products are available, or are under development at the present time, and only await the availability of cost-effective communication capability for their introduction into routine usage.

ACKNOWLEDGMENTS

RADARSAT is the result of many years of work of technical teams as well as administrative leadership. The concept grew from the early inspirations of Larry Morley and Lee Godby at the Canada Centre for Remote Sensing, and was artfully steered through difficult political approvals by Ed Shaw. After the transfer of *RADARSAT* from the Department of Energy, Mines and Resources to the newly formed Canadian Space Agency, the Project has been led by Joe McNally. Technical aspects have been the responsibility of Shabeer Ahmed, with irreplaceable support from many Canadian industrial partners. Mission development has been led by Ed Langham. CCRS continues to lead the radar data development activities. Ice applications have benefited from substantial contributions from Lyn McNutt, Sue Digby, and Mike Manore. *RADARSAT* is becoming a reality thanks to these people, and many others over the years, to whom Canada owes a large debt of gratitude.

The satellite system would have little value without a dedicated ground-based operational data-utilization system, particularly for ice applications. All of the staff, past and present, at the Ice Centre Environment Canada deserve recognition and a sincere thank you. It is their dedication that makes Canada proud of the world's finest Ice Service. Bruce Ramsay and his host of coop students did the scientific work upon which work described in this chapter is based. Bruce is to be thanked for his careful reviews and many helpful suggestions. Finally, Don Champ, Director of Ice Services, deserves recognition and appreciation for his foresight and tenacity to make the modern Ice Centre a reality.

REFERENCES

[1] E. Attema (1991) ''The active microwave instrument on board the *ERS-1* satellite,'' *Proc. IEEE,* **79**(6), 791–799.

[2] AES (1989) *MANICE (Manual of Standard Procedures for Observing and Reporting Ice Conditions)*, 7th ed., Atmosphere Environment Service, Environment Canada, Ottawa.

[3] D. G. Barber and E. F. LeDrew (1991) ''SAR sea ice discrimination using texture statistics: A multivariate approach,'' *Photogrammetric Eng. and Remote Sens.* **57** (4), 385–395.

[4] W. M. Brown and L. J. Porcello, (1969) ''An introduction to synthetic aperture radar,'' *IEEE Spectrum,* **6**, 52–62.

[5] G. Brundtland, (Chairman) (1987) *Our Common Future*, World Commission on Environment and Development, Oxford.

[6] CCRS (1977) ''*Report of the Interdepartment Task Force on Surveillance Satellites*,'' Canada Centre for Remote Sensing, Ottawa.

[7] A. R. Condal (1984) ''Automated computer monitoring of sea-ice temperature by use of NOAA satellite data,'' in *Proc. 8th Canadian Symp. on Remote Sensing*, 145–150.

[8] I. G. Cumming and J. R. Bennett (1979) ''Digital processing of Seasat SAR data,'' in *Proc. Cong. on Acoustics, Speech and Signal Processing.*

[9] J. C. Curlander and R. N. McDonough (1991) *Synthetic Aperture Radar, Systems and Signal Processing*, Wiley, New York.

[10] B. Dey and J. H. Richards (1981) ''The Canadian North: Utility of remote sensing for environmental monitoring,'' *Remote Sens. of Environment,* **11**, 57–72.

[11] M. C. Dobson, F. T. Ulaby, D. R. Brunfeldt, and D. N. Held (1986) "External calibration of SIR-C imagery with area extended and point targets," *IEEE Trans. on Geosci. and Remote Sens.*, **24** (4), 453–461.

[12] J. C. Falkingham (1991) "Operational remote sensing of sea ice," *Arctic,* **44,** Supp. 1, 29–37.

[13] D. G. Flett, D. A. Henderson, and B. R. Ramsay (1987) "Evaluation of unsupervised classification for identifying sea ice types in an operational environment," in *Proc. 11th Canadian Symp. on Remote Sensing*, 795–806.

[14] T. Heacock *et al.* (1992) "Sea ice tracking on the East Coast of Canada using NOAA AVHRR imagery," *Ann. of Glaciology,* **12**.

[15] T. Hirose, L. McNutt, and M. Manore, Automated sea ice tracking: results for LIMEX'87 and LIMEX'89. *Atmospheres-Ocean*, (in press), 1992.

[16] J. P. Hollinger and R. C. Lo (1989) *DMSP Special Sensor Microwave/Imager Calibration/Validation Final Report 20 July 1989*, U.S. Naval Research Laboratory, Washington D.C.

[17] W. J. Hussey (1979) *The TIROS-N/NOAA Operational Satellite System*, U.S. National Environmental Satellite Service Pub. No. 297009.

[18] Intera Technologies Ltd. (1987) *Provision of a Comprehensive Ice Reconnaissance Service for the Atmospheric Environment Service*, Doc. No. CP87-001. (Available from Ice Centre, Environment Canada, Ottawa).

[19] W. T. K. Johnson (1991) "Magellan imaging radar mission to Venus," *Proc. IEEE,* **79**(6), 777–790.

[20] R. L. Jordan (1980) "The Seasat-A synthetic aperture radar system," *IEEE J. Oceanic Eng.* **5**(2), 154–163.

[21] F. Li and R. K. Raney, (Eds.) (1991) "Special section on spaceborne radars for earth and planetary observation," *Proc. IEEE,* **79**(6), 773–880.

[22] P. Lim, P. George, P. McConnell, and J.-P. Guignard (1989) "*ERS-1* synthetic aperture radar verification mode processor," in *Proc. Int. Geosci. and Remote Sensing Symp.*, 1694–1696.

[23] A. P. Luscombe, (1988) "Taking a broader view: *RADARSAT* adds ScanSAR to its operation," in *Proc. Int. Geosci. and Remote Sensing Symp.*, 1027–1032.

[24] A. P. Luscombe (199) "Internal calibration of the *RADARSAT* synthetic aperture radar," in *Proc. Int. Geosci. and Remote Sensing Symp.*, 2325–2328.

[25] McMurdo (1990) "*McMurdo SAR Facility Science Plan*," Rep. of the Ad Hoc Science Working Team (draft) (courtesy of K. Jezek, Ohio State Univ.)

[26] R. K. Moore, J. P. Claasen, and Y. H. Lin (1981) "Scanning spaceborne synthetic aperture radar with integrated radiometer," *IEEE Trans. Aerospace and Electron. Syst.* **17**(3), 410–421.

[27] F. M. Naderi, M. H. Freilich, and D. G. Long (1991) "Spaceborne radar measurements of wind velocity over the ocean—An overview of the NSCAT scatterometer system," *Proc. IEEE,* **79**(6), 850–866.

[28] NASA (1986) *Earth System Science Overview: A Program for Global Change*, NASA Advisory Council, Washington D.C.

[29] National Research Council (1989) *Prospects and Concerns for Satellite Remote Sensing of Snow and Ice*, Ad Hoc Panel on Remote Sensing of Snow and Ice, Polar Research Board, National Academy Press, Washington, D.C.

[30] Norland Science and Engineering Ltd. (1992) ''*Operational Implementation of the Proposed Ice Regime System for Shipping Safety Control*,'' Report prepared for the Canadian Coast Guard, North.

[31] Philip A. Lapp Ltd. (1983) ''*Base map specifications for AES ice charts*,'' CAES-540-83-003, unpublished (available from the Ice Centre, Environment Canada, 373 Sussex Dr., Ottawa).

[32] *RADARSAT* Project Office (1982) ''*Mission requirements document*, 82-7,'' Ottawa.

[33] R. K. Raney (1982) ''Processing synthetic aperture radar data,'' *Int. J. Remote Sens.* **3**(3), 243–257.

[34] R. K. Raney, A. P. Luscombe, E. J. Langham, and S. Ahmed (1991) ''*RADARSAT*,'' *Proc. IEEE,* **79**(6), 839–850.

[35] R. K. Raney and J. Princz (1987) ''Reconsideration of azimuth ambiguities in SAR,'' *IEEE Trans. Geosci. and Remote Sens.* **25**(6), 783–787.

[36] I. G. Rubenstein and R. O. Ramseier (1985) ''Scientific application of passive microwave satellite data for ice monitoring and research,'' in *Proc. Conf. on the Use of Satellite Data in Climate Models,* SP-244, European Space Agency, 117–123.

[37] M. Sack, J. Ward, and J. G. Princz (1989) ''Overview of Canada's *ERS-1* SAR data processing facility,'' in *Proc. Int. Geosci. and Remote Sensing Symposium,* 1791–1794.

[38] Subcommittee for the Revision of the Arctic Shipping Pollution Prevention Regulations (1989) ''*Proposal for the revision of the Arctic shipping pollution regulations*,'' Report prepared for the Canadian Coast Guard, Northern, TP-9981, 2 vol.

[39] C. T. Swift, L. S. Fedor, and R. O. Ramseier (1985) ''An algorithm to measure sea ice concentration with microwave radiometers,'' *J. Geophys. Res.*, **90**(C1), 1087–1099.

[40] TELESAT Canada (1989) ''*AES Ice Branch IRDNET Project Design Review, February 1989*,'' Project 2607–28, Telesat Canada, Gloucester, Ontario.

[41] TELESAT Canada (1990) ''*IRDNET Project Phase II Design Review, February 1990*,'' Telesat Canada, Gloucester, Ontario.

[42] K. Tomiyasu (1981) ''Conceptual performance of a satellite borne, wide swath synthetic aperture radar,'' *IEEE Trans. Geosci. and Remote Sens.* **19**, 108–116.

[43] C. C. Wackerman, R. R. Gentz, and R. A. Shuchman (1988) ''Sea ice type classification of SAR imagery,'' in *Proc. Int. Geosci. and Remote Sensing Symp.* 425–428.

[44] G. D. Wittman (1984) *Defence Meteorological Satellite Program (DMSP)*, U.S. Air Force Air Weather Service Pub., Scott AFB IL.

[45] WMO (1970) *Radio Facsimile Transmission of Weather Charts for Ships*, World Meteorological Organization, Geneva, Switzerland.

[46] R. Yamamoto (1977) *Facsimile Transmission of Ice Charts via HF Radio Station CHF Halifax*, Transport Canada, Canadian Coast Guard, Ottawa.

[47] D. G. Zimcik, L. Martins-Camelo, and P. R. Cowles (1988) ''Development of the *RADARSAT* antenna,'' *Canadian Aeronautics and Space Inst. J.* **34**(2).

13

SUPPLEMENTARY TOPICS AND FUTURE DIRECTIONS

SIMON HAYKIN
Communications Research Laboratory
McMaster University
Hamilton, Ontario, Canada

R. KEITH RANEY
Canada Centre for Remote Sensing
Energy, Mines and Resources
Ottawa, Ontario, Canada

13.1 INTRODUCTION

An important issue is the remote sensing of sea ice is how to exploit fully the information content available in the received signal of a sensor. The signal processing procedures described in the previous chapters have followed mainly traditional lines. Notwithstanding the practical values of these well-tested methods, there is something to be said for the use of a new generation of signal-processing techniques that offer the potential of improved system performance. In Section 13.2, a radar vision system is described for the detection of growlers in an ocean environment, which offers performance far in excess of that attainable with conventional coherent radar signal processing.

In previous chapters, detailed expositions have been presented of various active and passive devices for the remote sensing of sea ice and icebergs. Each of these sensors is designed to exploit selected attributes of the observed scene. The specific details of the output of a particular sensor naturally depend on the characteristics of the sensors being considered and the application of interest. Recognizing that the capital investment made in the development of sensors is

Remote Sensing of Sea Ice and Icebergs, Edited by Simon Haykin, Edward O. Lewis, R. Keith Raney, and James R. Rossiter.
ISBN 0-471-55494-4

usually immense, it would be highly desirable to combine the outputs of several sensors in order to realize a level of performance that is beyond the reach of any individual sensor. An elegant response to this basic objective lies in the use of multisensor fusion, considered in Section 13.3.

In Section 13.4, the discussion is continued of improving remote sensing performance by adding new degrees of freedom to the system so as to expand the pool of information available on an observed scene. The discussion is presented in the context of synthetic aperture radar. Three approaches are outlined: polarimetry, interferometry, and coherence assessment. The common feature of these approaches is that they involve the use of two or more data channels that are mutually coherent, and they generate new degrees of freedom by taking advantage of interchannel phase differences.

The chapter concludes with a discussion of certain spin-off benefits that have emerged from the various Canadian research initiatives that were launched in the 1970s and 1980s for the remote sensing of ice and icebergs.

13.2 RADAR VISION

The conventional signal processing system incorporated into an ice surveillance radar system has at least two performance-limiting constraints:

1. Little use is made of the environmental operating conditions that characterize the radar.
2. Potentially valuable information about the surveillance application is disregarded during system design, chiefly because the mathematical models cannot accommodate the needed parameters.

The purpose of radar vision is to overcome these limitations and to strive towards generality until the very last stages of either feature detection or parameter estimation. Moreover, the system is enabled to develop cognition of the surrounding environment by exploiting recent developments in time–frequency analysis, nonlinearly separable pattern classification, and sequential adaptive decision rules.

Two of the distinguishing features of radar vision are preservation of the full information content of the radar returns and presentation of ice information to the classifier part of the system in a way that eases signal separation or discrimination. Classically, signals are treated either in the time domain or in the frequency domain. Yet, there is an important class of signals that cannot be separated in only one of these domains. Time–frequency representation is one possible way to separate such signals. Basically, this implies transforming the time series representing the radar return (obtained by dwelling along a particular azimuthal direction of interest) into a two-dimensional, time–frequency image. For a detailed treatment of this subject, see Boshash [2]. However, as alluded to earlier, the input being presented to the classifier requires careful selection. Hence, it is worthwhile to dig a little more deeply into time–frequency signal representations.

Given any realistic signal $s(t)$, its Fourier transform $S(f)$ is defined by

$$S(f) = \int_{-\infty}^{\infty} s(t)e^{-j2\pi ft}\, dt$$

The two complementary representations, $s(t)$ and $S(f)$, are sufficient to represent uniquely any nonstationary behavior of the signal. The main problem is the complex nature of $S(f)$. The nonstationary behavior of $s(t)$ is expressed in the phase of $S(f)$, a quantity that often is difficult to interpret.

An alternate representation is to use a time–frequency energy distribution $T_s(t, f)$ formally defined by two marginal properties: energy norm and power norm.

Energy Norm. As a formal expression of the conservation of energy required for any signal transformation, the distribution should satisfy the condition

$$\int_{-\infty}^{\infty} \int_{-\infty}^{\infty} T_s(t, f)\, dt\, df = E_s$$

where E_s is the energy in the signal $s(t)$.

Power Norm. Likewise, signal power, expressed either in the time domain or the frequency domain, should be preserved through the transformation. Thus, the distribution should satisfy the conditions

$$\int_{-\infty}^{\infty} T_s(t, f)\, df = |s(t)|^2$$

and

$$\int_{\infty}^{\infty} T_s(t, f)\, dt = |S(f)|^2$$

The conventional method of doing this is through the use of the spectrogram defined as

$$|F_s(t, f)|^2 = \left| \int s(\tau)w^*(t - \tau)e^{-j2\pi f\tau}\, d\tau \right|^2$$

where $F_s(t, f)$ is the short-time Fourier transform of $s(t)$, and $w(t)$ is a window function localized in time and centered at time t. It is trivial to see, although the spectrogram does satisfy the energy norm property for any window function of unit energy, that the power norm properties impose conflicting requirements on $w(t)$.

A more attractive transformation, from a signal theoretic point of view, is the Wigner-Ville distribution (WVD), defined as

$$W_s(t, f) = \int s(t + \tau/2)s^*(t - \tau/2)e^{-j2\pi f\tau}\, d\tau$$

which is a time–frequency energy distribution.

The WVD has two properties of importance to applications: signal conservation and ambiguity conservation. These each deserve discussion.

Signal Conservation. Given the WVD of a signal $s(t)$, the signal may be recovered uniquely, to within a phase constant, by a linear transformation

$$s(t) = \frac{1}{s^*(0)} \int_{-\infty}^{inf\, W_s} (t/2, f)e^{j2\pi ft}\, df$$

where $s^*(0)$ is the complex conjugate of $s(t)$ evaluated at $t = 0$. It follows from this statement that the WVD representation preserves the structure, and hence the information content, of the underlying signal.

Ambiguity Conservation. The two-dimensional Fourier transform of the WVD of a signal $s(t)$ is the *ambiguity function* of the signal, a common definition for which is

$$A_s(\tau, \nu) = \int_{-\infty}^{\infty} s(t + \tau/2)s^*(t - \tau/2)e^{-j2\pi\nu t}\, dt$$

where τ is delay and ν is Doppler shift, a formulation introduced by Woodward [22], based upon the pioneering work of Wigner. The ambiguity function is closely allied to the uncertainty principle of classical physics; it expresses the fact that simultaneous sharpness of a distribution in both delay and Doppler is not possible. Together, these two quantities determine the "information," defined in the formal Nyquist sense, that may be carried by the signal. Note that the Fourier transform relationship $W_s(t, f) \leftrightarrow A_s(\tau, \nu)$ is completely reversible. This latter property has been exploited in Haykin and Bhattacharya [10, 11], where it is shown for certain applications that feature extraction from the ambiguity space is easier than in the original Wigner-Ville space.

Transformations that derive from signal products, such as the Wigner-Ville distribution, often are criticized for the appearance of cross terms that arise with multicomponent signals. Specifically, if

$$s(t) = s_1(t) + s_2(t)$$

then

$$W_s(t,f) = W_{s1}(t,f) + W_{s2}(t,f) + W_{s1s2}(t,f) + W_{s2s1}(t,f) \quad (13\text{-}1)$$

The last two terms in (13-1) are the so-called "cross terms." Without going into details, it is easy to see that these terms indicate the time–frequency organization of the two components and their correlation/interaction. For the specific application of detection of targets in clutter, the target echo and the clutter component comprise two classes of interest. In this case, the cross terms inherent in the WVD provide valuable information about the radar operating environment. The cross terms may therefore be used to further enhance the detection of targets. Seen in this way, the cross terms may be viewed as an integral part of the information-preserving property of the transformation.

Thus, the first step in radar vision is to use a suitable information-preserving transformation, such as the WVD, to map the one-dimensional time series of the incoming signal into a two-dimensional time–frequency image. The second step in radar vision involves the use of a neural network to perform adaptive pattern classification (i.e., target detection), with the time–frequency image providing the input applied to the input layer. We are not suggesting here that a neural network is the only way of performing adaptive pattern classification; rather, it is a natural way for this kind of an application by virtue of the following properties of a neural network (Haykin [9]):

1. Nonlinearity built into the composition of each computational (processing) unit.
2. Hidden computational units, hidden in the sense that they are not directly connected to the input of the network nor to the output.
3. Ability to learn from, and therefore adapt to, the environment by undergoing training through the use of examples representative of the environment.

Owing to the large amount of data contained in a two-dimensional image, the design of the neural network in this application requires careful attention for the network to be of a manageable size. A specific architecture that is well suited for radar vision involves the use of three techniques (Le Cun et al. [14]):

1. Employing two (or more) hidden layers, with the computational units in each of these layers being assigned "receptive fields" of their own, thereby making the network partially connected.
2. Making the receptive fields "overlap" each other.
3. Fully connecting neurons in the output layer.

Figure 13.1 (Haykin and Bhattacharya [10, 11]) shows that the two-stage processing described herein can outperform a traditional constant rate false alarm (CFAR) Doppler processor in a significant way.

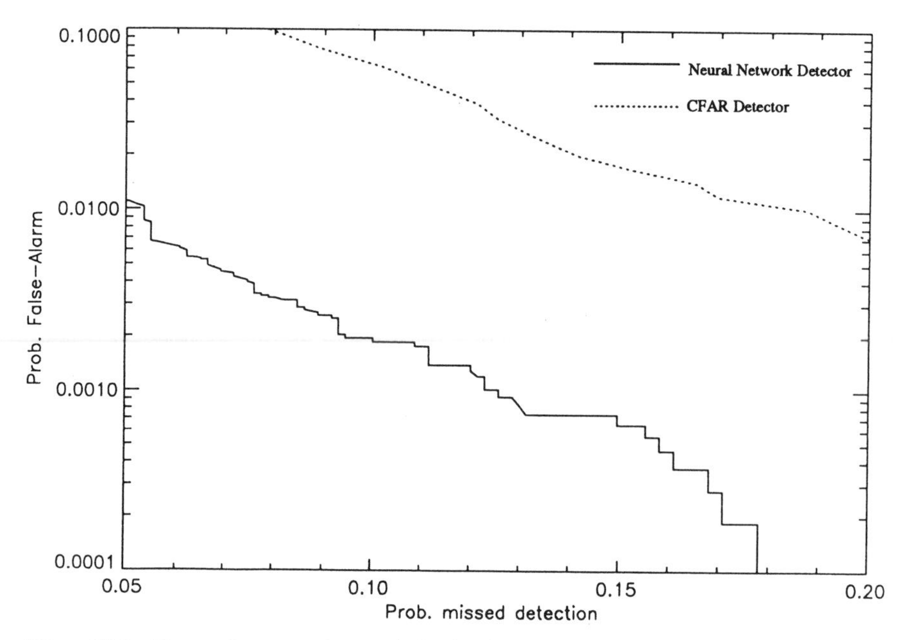

Figure 13.1 The receiver operating curve for the classification of a growler using a two-hidden-layer, locally connected, neural network and a Doppler CFAR detector.

13.3 MULTISENSOR FUSION

To appreciate the practical value of multisensor fusion, one need only observe how a human uses several sensory organs (eyes, ears, nose, and fingers) to characterize the immediate environment. Even more striking is the relative ease with which the outputs of our visual, auditory, olfactory, and tactile systems are appropriately combined by the nervous system to provide an overall assessment of the environment. Needless to say, any one of these systems working alone is usually inferior if compared to their combined operation.

In the context of the remote sensing of sea ice and icebergs, each sensor tends to accentuate those features of the environment that match its design capability. Typically, we find that different sensors are complementary in their coverage of the environment. For example, a satellite synthetic aperture radar provides *global* coverage *periodically*, whereas a surface-based radar provides *local* coverage of the environment *continuously*. Clearly, much can be gained by combining (fusing) the outputs of these two sensors, yielding a performance that would be beyond the reach of either sensor alone. Moreover, human observations of the environment and prior knowledge can be incorporated into the integrated system, thereby making the output that much more informative and error-free. The key question is: how can the fusion of multisensor data be organized to exploit the inherent synergism across the sensor suite in a robust manner? The answer to this question is complicated by several factors:

1. Sensors usually operate in a nonsynchronous fashion with respect to each other—temporal differences.
2. Sensors exhibit different resolution, scale, and geometric capabilities—spatial differences.
3. Observations produced by the sensors are subject to differing measurement errors that usually vary with environment changes—precision, tolerance, and calibration differences.

The discussion of multisensor fusion up to this point has been in general terms. More specifically, there are two different ways in which such fusion may be achieved. We may integrate the raw data extracted by the various sensors into a common channel and then perform the necessary signal processing operations on the resultant output. Alternatively, each sensor could include an appropriate preprocessor, designed to match the sensor output to the requirements of the multisensor system. There are two important reasons for preferring the latter method:

1. Ordinarily, the different sensors are built with their respective signal processors optimized according to some criterion.
2. Based on experience, it is preferable that preprocessing be used on a sensor-by-sensor basis and that integration take place towards the end of the process when the outputs of the individual sensors are available.

In the context of the approach advocated here, the system integration may be performed in one of two basic ways, as illustrated in Figs. 13.2 and 13.3 for the case of target (e.g., growler) detection. In Fig. 13.2, each sensor system is complete until the decision-making stage. In such a case, system integration may be achieved simply by taking a majority vote among the individual decisions made by the set of sensors. In a more sophisticated form of this approach, some form of weighting may be applied to the individual decisions prior to their integration, with the weighting being determined in accordance with the levels of confidence appropriate for each of the various decisions. The disadvantage of this approach, irrespective of how the overall decision is made, is that it violates an important lesson learned from information theory. According to this lesson, no useful information should be thrown away until the final decision-making process (Viterbi [21]). Although the individual components of the multisensor fusion scheme shown in Fig. 13.2 may satisfy this requirement, it is clearly violated in an overall sense. This limitation may be circumvented by using the scheme shown in Fig. 13.3, in which the signal processing associated with each sensor stops at the computation of a sufficient statistic. Multisensor fusion takes place by means of a fusion network that combines the sufficient statistics computed by the preprocessors of the individual sensors.

As an alternative to the method of sufficient statistics shown in the scheme of Fig. 13.3, the preprocessor associated with each sensor may be designed to

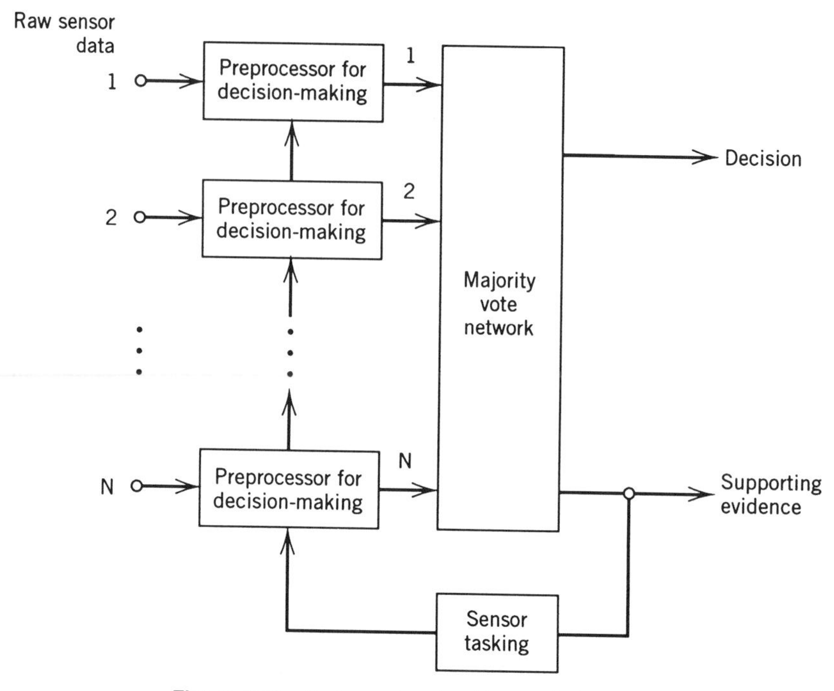

Figure 13.2 Fusion network using a majority vote.

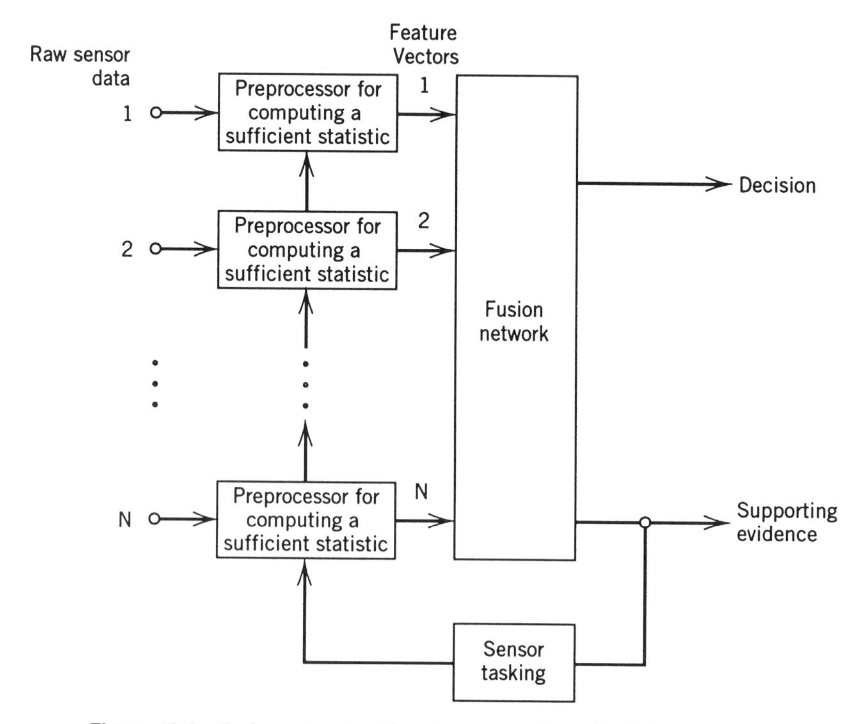

Figure 13.3 Fusion network with prior computation of sufficient statistics.

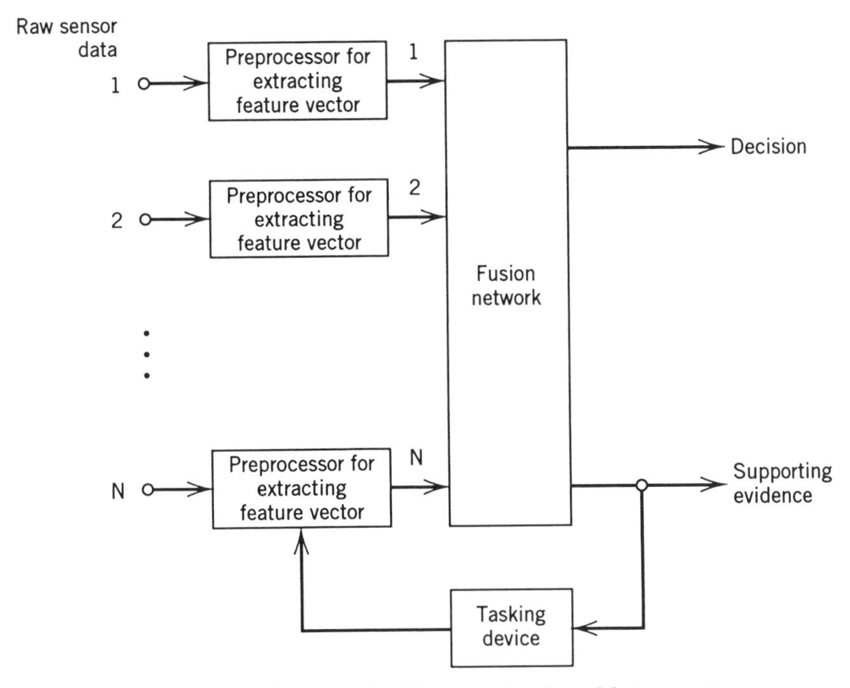

Figure 13.4 Fusion network with prior extraction of feature vectors.

construct a feature vector of the raw sensory data, as depicted in Fig. 13.4. Types of preprocessing that may be used in ice applications include multispectral analysis and principal components analysis, the purpose of which is to compress the information content of the source data to its most efficient form. The set of sufficient statistics in Fig. 13.3, or the set of feature vectors in Fig. 13.4 from the individual sensors, are combined to generate an overall decision. Moreover, together with the decision output, an analog measure of confidence in the decision is derived. This confidence measure in turn is utilized in a feedback loop to redirect sensor taking, which improves the system performance for the task required (Eggers and Kuhon [4]). The structures described in Figs. 13.3 and 13.4 lend themselves nicely to the use of artificial neural networks.

It may be rewarding to explore novel algorithms for multisensor fusion derived from biology. In other words, there is potential benefit from a deeper understanding of the way in which the human nervous system performs its own multisensory fusion.

13.4 MULTICHANNEL COHERENT RADAR TECHNIQUES

Three radar approaches are available that use combinations of two or more data channels which are mutually coherent. These techniques take advantage of interchannel phase differences to derive new degrees of freedom for extracting information about an observed scene. The first method, quadrature polarimetry, a specialized development applicable to polarimetric radars, has seen rather little exploration in ice applications. The state of the art and future outlook of polarimetry are described in this section. The second method, interferometry, is a topic of active research, from which have come several specialized versions that take advantage of both space and time baselines. Together, quadrature polarimetry and interferometry are the two most important techniques that have emerged during the past decade in the field of imaging radar. The third method, scene coherence assessment, is a specialized aspect of interferometry and has received rather little attention in radar research to date. This technique shows promise for specific problems of sea-ice characterization. All three of these multichannel radar methods are likely to attract increasing attention from the ice remote sensing community in future years.

13.4.1 Polarimetry

Use of polarimetric discrimination for sea ice has been widely exploited (Chapters 6, 8 and 11). Conventional radars that rely on polarimetric diversity are based on systems that have available only one or two polarization states. In contrast, radars that use quadrature polarimetric synthesis techniques are able to emulate any arbitrary combination of transmit and receive polarization states (Chapter 6). Quadrature polarimetry, although proven in principle and successfully demonstrated for certain land applications, has seen rather little application to problems in sea-ice observation. The ability to select a preferred transmit and receive polarization combination adaptively at the time of data analysis should yield immediate benefits for situations in which the geometry of structural detail within the scene favors certain orientations of the electromagnetic field. Sea ice, by definition, is characterized by physical properties that are different in the three spatial dimensions, so that polarization discrimination in many ice applications is a reasonable expectation. The preferred polarization state may change, depending upon the ice types to be encountered and the prevailing environmental conditions. Polarimetric synthesis deserves to see increased experimental and operational utilization for both the surface-based (coherent) radars as well as airborne and spaceborne imaging radars.

Early attempts at using quadrature-polarimetric radar techniques have been limited to airborne SAR systems. Results have been published by Drinkwater et al. [3] and Israelsson and Askne [12] that show only partial success with polarimetric synthesis for ice discrimination: in the experiments reported therein, wavelength diversity was found to be a more powerful approach. With a single wavelength radar, however, optimization of signal selectivity available through polarimetric synthesis has been proven to be of value. Results should be expected that are better than those available through the more limited parameters derived through polarization ratios, for example, as discussed in Chapter 11.

With coherent multipolarization of surface-based radars (Chapter 9), polarization synthesis using quadrature-polarimetric signal combination has yet to be investigated. It is reasonable to expect, particularly for the grazing incidence angle so characteristic of such radar geometries, that new and useful results should be obtained. Quadrature polarimetry for nonimaging surface-based radars will be investigated in the immediate future.

The set of output products from a polarimetric synthesis radar may be viewed as deriving from different channels or, conceptually, even from different sensors. Such signals could form a basis for processing techniques previously developed and applied to multisensor data sets. For example, one might choose one particular polarization combination, A, that minimizes the reflectivity from the ocean surface, leading to an image $\sigma_A^0(x, y)$. Likewise, a second polarization combination, B, could be chosen to maximize the returns from a small ice feature such as a growler, leading to an image $\sigma_B^0(x, y)$ in which such features are emphasized. These two data sets could be treated as logically orthogonal; in this case, the polarimetric analog to the corresponding eigenvectors over the set of all possible polarization states is obtained. Using this construct, application of multichannel algorithms to the polarimetric set $\{\sigma_A^0(x, y), \sigma_B^0(x, y)\}$ should yield more robust detection of growlers than conventional algorithms applied to single-channel radar data, even if the single channel were a well-chosen polarization state. There are many related avenues of investigation in this special field that remain for future exploration.

13.4.2 Interferometry

Principle of Interferometry

The basic interferometric environment is one in which there is a pair of mutually coherent signals having the same envelope, carrier frequency, and phase modulation, but which have different reference phase. Signal processing is designed to estimate the phase difference, and to deduce geophysical parameters of interest from the resulting measurement. In its finest forms, interferometry is able to observe physical movements or other changes in the observed scene that are at the scale of the wavelength of the radar being used, typically on the order of centimeters.

The signal model is simple in concept. For the complex return from one scattering element, let the input signal pair be described by

$$s_1 = \Gamma_{t_1} a \exp\{-j\phi\}$$

and

$$s_2 = \Gamma_{t_2} a \exp\{-j\phi + j\varphi(r, t_2 - t_1)\}$$

where the subscript t suggests that the two signals, in general, are obtained at two different points in time. For completeness, an arbitrary phase ϕ is included in the formulation. The objective is to estimate the phase $\varphi(.,.)$, which describes a systematic phase difference between the two observations. There are two methods of phase estimation employed: correlation and sum-and-difference linear combination. Correlation, being the more generally applicable method, is based on the statistical norm

$$\mathbf{E}[s_1(t_1)s_2^*(t_2)] = R_{12}(t_1, t_2)a \exp\{-j\varphi(r, t_2 - t_1)\}$$

where $\mathbf{E}[.]$ is the expectation (averaging) operator, and the exponential asterisk (*) denotes complex conjugate. This expression is presented here as it would be calculated using complex image data. The method always includes the cross correlation $R_{12}(.,.)$ of the two scene functions as well as the systematic phase difference $\varphi(.,.)$ between them.

The observed phase difference may be due either to geometric or to temporal differences between the two observations. These possibilities are discussed further below. Any approach to phase-difference measurement is subject to the fundamental 2π ambiguity characteristic of phase-estimation algorithms. In many radar situations, knowledge of the physical constraints of the situation, coupled with phase unwrapping algorithms, is sufficient for the purpose. More on the phase unwrapping problem as encountered in radar interferometry may be found in Prati et al. [17].

Successful interferometric phase estimation depends on the cross correlation $R_{12}(t_1, t_2)$ of the scattering function Γ, which becomes by default an observable of the measurement. The normalized cross-correlation function, a useful form of the same statistic, is the mutual coherence function

$$\frac{R_{12}(t_1, t_2)}{\sqrt{\mathbf{E}[|s_1|^2]}\sqrt{\mathbf{E}[|s_2|^2]}}$$

which frequently is encountered in physical optics (Born and Wolf [1]). Either the cross correlation or the mutual coherence function allows quantitative estimates of both the spatial and the temporal loss of correlation between the two observations. Its magnitude approaches unity for highly correlated signals, and zero for weakly correlated signals.

The three classes of interferometry, and variations on these main themes, are distinguished by conditions on the spatial geometry and temporal properties of the signals used in the measurement, and by conditions on scene coherence.

Differential Range Vector Interferometry

A single-channel radar can measure only the scalar range, or distance, to each scattering object. One method of measuring the range vector, which includes information on both the distance and the relative elevation angle of the scatterer, is for the radar to use two antennas rather than one, leading to the so-called *delta-R* interferometric configuration. The two antennas, in general, are both in a plane orthogonal to the motion vector of the radar, but they must be separated spatially in range and elevation, as in Fig. 13.5. The range-varying

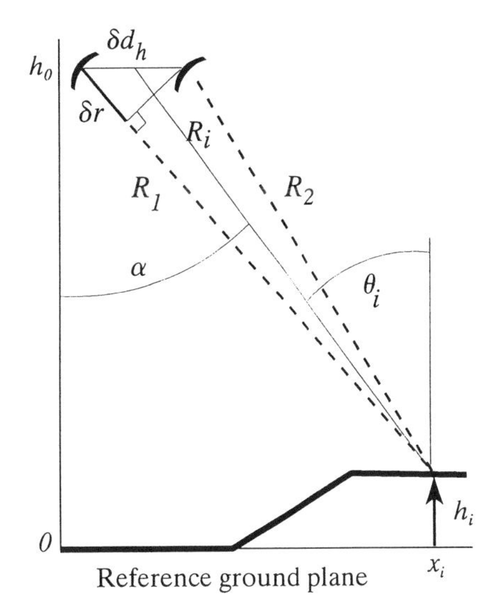

Figure 13.5 Geometry of interchannel phase difference arising in a delta-R interferometer configuration. In general, there is a differential position both in the vertical direction and in the horizontal direction. For the simpler case shown, and assuming that signal transmission is from the left antenna, the phase difference between the two received signals is $k\delta r$, where $\delta r = \delta d_h \sin \alpha(R_i, h_i; h_0)$, in which the altitude of the SAR above the reference ground plane and the range and relative scatterer altitude are included explicitly.

phase difference between the two channels is found by forming the cross-correlation between the two received signals. Note that the two channels are operated simultaneously, so that in this case the mutual coherence function is always unity. From any given transmission, the two signals received at the two antennas will have a phase difference determined uniquely by the geometry and, of most interest, by the combined range and elevation angle to each scattering element. Through interferometric signal analysis, a vector range record of reflectivity is obtained. This record may be used to plot an orthographic terrain map, free of any terrain-elevation distortion effects. Of more importance in sea-ice mapping, the signal pair may be processed to obtain an estimate of the height of each element of the scattering surface with respect to a datum plane. With this method, i.e., using only one pair of signal sequences, obtainable precision is on the order of the range resolution of the radar (Prati et al. [18]). The technique applied to SAR was investigated first by Graham [7].

From geometric constraints, the spacing between the antennas should be on the order of 1 m for aircraft radars. There are several such systems in operation, including the SAR operated on the CCRS Convair 580, described in Chapter 11. However, for orbital altitudes, the same configuration, in principle, would lead to antenna separation on the order of 1 km, which is not practicable for a single spacecraft. There is a way around the problem, however, as discussed next.

Double-Pass Delta-R Interferometry

In the mid-1980s, a brilliant innovation was proposed by R. Goldstein (Zebker and Goldstein [23]) to circumvent the antenna spacing problem for satellite interferometric radars. Assuming that data in the two signals were mutually coherent, spatial interferometry could be obtained from a pair of signals collected (and recorded) from one spacecraft observing the scene from two different but neighboring positions, so that the needed data would be collected on different orbital passes. The technique has been proven with spacecraft SAR data (Li and Goldstein [15]). The two-pass (delta-R) interferometry technique also has been demonstrated with airborne SAR data (Gray and Farris-Manning [8]).

There are two caveats in the approach. The first is imbedded in the idea of "neighboring" positions. The two radar passes should emulate the simultaneous geometry of Fig. 13.5. One may show that for statistically stable scene reflectivity, spatial correlation of the pair of return signals is maintained for separations as large as the radar pulse length. For most systems, this is on the order of several kilometers, hence the desired spatial separation requirement may be satisfied. For absolute elevation maps to be derived, however, precise knowledge of the difference between observation tracks is required.

The second has to do with scene coherence: in order for the two signals to act as an interferometric pair, their respective phase structures must be robust over the time interval between satellite observations. In short, there must exist mutual coherence between the signals, even though they are observed at different times. This requirement is satisfied readily for short interopportunity intervals, such as from the 3-day repeat orbit first used with *Seasat-A* data to prove the concept, and for stable terrain features, such as unvegetated rocky mountain slopes. It is not necessarily satisfied for scenes, such as ice cover, that may undergo changes in the details of reflection and scattering between observations. (For example, in one unpublished experiment, elevation contours were derived from two-pass *ERS-1* SAR data for a mountainous region of western Canada, except for a glaciated slope for which the value of the mutual coherence function approached zero.) Implications of the coherence sensitivity of interferometric measurements is developed further below.

Differential Delta-R Interferometry

Consider two pairs of delta-R interferometric measurements which share a common reference phase. Such a data set could be formed by using three single passes, assuming suitably stable scene correlation, or by two passes of a two-channel delta-R interferometer. In either case, the phase measurement

$$\Delta\varphi = (\varphi_1 - \varphi_0) - (\varphi_2 - \varphi_0) = \varphi_1 - \varphi_2$$

in effect uses the common phase reference φ_0 as a transfer standard of range/elevation, with a precision that is on the order of the illuminating wavelength (Prati et al. [18]). The technique has been demonstrated and was used to indicate centimeter scale displacement of objects (Gray and Farris-Manning [8]) and of distributed terrain features (Gabriel et al. [5]).

It is of interest to consider what could be accomplished by application of this technique to sea ice. If the reflectivity of ice changes with time, then three-pass interferometry would not work, as the respective mutual coherence functions would not support the needed phase estimation. However, even if the coherence of the reflected field is not maintained in time, two passes of a two-beam interferometer could be used to form a differential delta-h estimate of surface deformations with a precision on the order of the radar wavelength. The technique deserves development for sea-ice observation, and it would be useful for studying detailed changes in ice surface topography under a variety of conditions.

Time-Difference Interferometry

A systematic phase difference between two otherwise identical signals will occur if they are gathered at different times and there is movement in the observed scene. A delta-t, or time difference, interferometer is based on the use of two antennas separated along the motion vector of the radar, leading to a difference δt in the observation time of the scene. If there are elements of the scene with velocity v_{rad} towards the radar, then a phase difference $\varphi_{12} = 2k\delta t v_{rad}$ will occur. The phase shift may be used to detect small motions of scatterers from a high-speed moving platform (Urkowitz [20]), even in a high-resolution SAR context (Raney [19]). Delta-t interferometry has been used successfully to measure ocean surface currents (Goldstein and Zebker [6]), and it has been proposed for improvement of directional wave estimation (Lyzenga and Bennet [16]). It has many as yet untapped advantages for radar observation of ice and water environments.

The method is very similar in concept to range-Doppler motion detection algorithms used in surface-based coherent radars (Chapter 9). The interesting difference encountered in the SAR case is that the technique eliminates disturbing effects due to the motion of the sensor platform. Doppler modulation, so essential for SAR image formation, couples into the scene such that small object motions are not discernable unless the platform motion is removed from the signal. Furthermore, in either the land-based case or the moving platform case, by using delta-t interferometry, the phase difference may be used to estimate the radial velocity of the scatterers.

Just as for other forms of interferometry, there must be mutual coherence of the respective scattering functions in order to estimate their phase difference successfully. If there is loss of coherence, then there will be a loss in interferometric discrimination.

13.4.3 Scene Coherence Assessment

One of the problems prevalent in sea-ice observation is differentiation of ice from water, often a difficult task under light wind and new ice conditions. Although both of these classes of scatterers may have the same mean reflectivity σ^0, they may be distinguished using the cross-correlation function (or the mutual coherence function) derived from two different times of observation. For a time base on the order of a second, the mutual coherence function for the ice should be close to unity, whereas for the water it should approach zero. The mutual coherence function approach has been used to estimate the "coherence time'' of the ocean's surface, although the complementary experiment suggested here would seem to be new.

This example suggests a new class of instrumentation and data processing that should be of value in sea-ice observation. Subtle changes in scene reflectivity may be explored by mapping the mutual coherence function onto the image plane. In general, the mapping should be derived using two (complex) data sets, constrained such that unity mutual coherence would be expected unless a time- or space-dependent change in reflectivity occurs.

13.4.4 Summary

For radar systems in general, and SAR in particular, multichannel coherent techniques are emerging that have proven to set new quantitative standards for remotely sensed scene analysis. Specific and focused application of these and related techniques to the observation of sea ice and icebergs offers exciting possibilities. The most promising techniques include:

1. Extension of quadrature polarimetric synthesis, now proven in land applications of suitable SAR data, to ice applications using coherent surface-based systems.
2. Application of delta-*t* interferometry SAR techniques to ice movement mapping.
3. Use of differential SAR measurements to study area deformation of ice distributions.
4. Application of the mutual coherence function to differentiate ice and open water, and to study time and space variations of ice morphology.

13.5 SPIN-OFF BENEFITS

This final section of the book examines where the critical mass of intellectual capability that emerged during the 1980s for the remote sensing of ice is now headed. By the early 1990s, the anticipated development of Canada's Arctic and east coast regions had come to a virtual standstill. No oil or gas exploration or development drilling is taking place in the Canadian Arctic. Off the east coast, the substantial Hibernia find is under production development but is proceeding far more slowly than previously anticipated, because of lack of financial incentives, even though jurisdictional disputes have been settled. In addition, incentives to develop Arctic surveillance technologies that might have been driven by military requirements have also lost momentum due to the end of the Cold War. In this environment, one might conclude that the technological developments had been in vain. It is our belief that nothing could be further from the truth.

One of the advantages of a multiagency, diversely funded, activity is its robustness. As there is no central authority which can ''0-line'' a budget, a variety of activities continue based on their own merits rather than on financial whim. The results are twofold. First, Canada now has the ability, sought since the early 1970s, to monitor its northern and offshore regions. Although not all of the operational aspects are in place as yet, there would seem to be no interest in program cancellations, in spite of governmental constraints. For example, the development and deployment of *RADARSAT* is continuing and is expected to be launched on schedule. Second, the technologies which have been described in this book are being developed further and adapted for a wide range of other applications worldwide. It is our belief that some of the ongoing activities, outlined in this section, will have substantial long-term benefit to Canada and beyond.

13.5.1 Ice Thickness

The Canadian ice-thickness sensing work has led directly to two quite different kinds of development. First, the EM technology has been developed for bathymetric sounding through ice. The ability to measure water depth in ice-covered oceans has long posed a serious problem for the Canadian Hydrographic Service, and this technology appears to provide a solution. Because Canada has large regions of ocean which are ice-free for only a short part of the year, this technology will lead to safer Arctic shipping and to greater knowledge of the sea bed generally.

The second activity is further development of ground-penetrating radar. Although the pertinent technology, as described in Chapter 4, is not the appropriate one for some types of sea-ice–thickness mapping, its development has proceeded for a variety of environmental problems. Personnel involved in ice-thickness sensor research have taken their expertise and adapted it to applications as varied as contaminant detection in soils, geoengineering evaluations, and the detection of deterioration of roads and bridges. These capabilities are now being demonstrated worldwide.

13.5.2 Over-the-Horizon Radar

This technology, although one of the later ones to be developed for Arctic applications, is moving rapidly towards commercial exploitation. The technology development program, initially carried out with a Canadian university research center, has been transferred successfully to a start-up company, and is being developed commercially through a multinational joint venture. The expectation is that this technology will be used for a variety of coastal surveillance applications, such as monitoring of fishing fleets and contraband activities. Increasing interest in this kind of capability is expected from around the world.

13.5.3 Passive Microwave

This technology has long been recognized as well suited to provide information on global synoptic conditions. Currently, it is being investigated further to study winds and clouds. These avenues are expected to improve our understanding of weather patterns and therefore improve weather prediction.

13.5.4 Marine Radar

Although marine radar has been used extensively for years as a shipboard aid to navigation, it is only with the advent of coherent radars and sophisticated signal processing that fundamental aspects of understanding ocean conditions have emerged. Here, particular mention should be made of the mobile radar research laboratory, called the IPIX radar, which has been designed and developed at McMaster University. Results already demonstrated with the IPIX radar will lead to a new generation of marine radars which, although no more costly than current radars, will be better able to detect features, including ice, than current systems. As a case in point, the noncoherent dual-polarization radar installed on the M.V. *Arctic* (see Chapter 10) has proven to be very effective. However, it is not yet known how some of these new and exciting discoveries might be commercialized.

The use of the IPIX radar is presently being redirected toward the study of atmospheric phenomena, an application for which it is rather well suited by virtue of the Doppler discrimination and the polarimetric capability built into its design (Kezys and Haykin [13]).

13.5.5 Synthetic Aperture Radar

Synthetic aperture radar was one of the first technologies to be developed commercially for ice reconnaissance. It is now an operational tool used for the Canadian Atmospheric Environment Service and has for years proved its utility. This technology is now finding use all over the world for a variety of resource-mapping applications, including discovery of minerals, forestry mapping, agriculture, and so on. It is being investigated for topographic mapping and should provide a reliable means of mapping remote areas of the world, particularly those for which standard aerial photography is unacceptable, for example due to heavy cloud cover.

Although originally motivated by Canadian ice applications, the routine surveillance capabilities that will be provided through *RADARSAT* are expected to provide innovative mechanisms for resource managers to observe the world. A Canadian company, *RADARSAT* International, has been formed to commercialize products from this data stream.

13.5.6 Concluding Remarks

It seems that there are two necessary conditions for ice remote sensing development to flourish:

1. Provision of an adequate level of funding which is likely to be maintained

for at least several years (even though firm commitments may not be in place).
2. Problems of substantial intellectual challenge and scope that will attract top-flight researchers.

The result of this challenge being met by Canada is that diverse, highly capable individuals have joined forces to address the offshore surveillance problem. The results, in spite of the dramatic and unexpected downturn of the original motivating goal, have been successful in their original purposes and have shown relatively rapid and successful adaptation to new goals. This has led to a wealth of expertise and systems capability—the legacy of Canada's sea-ice remote sensing development activities of the past 20 years.

ACKNOWLEDGMENTS

The authors are grateful to Dr. L. Gray (CCRS, Ottawa) for his comments on an early version of this chapter. One of the authors (S. Haykin) is grateful to his research colleague, Dr. T. Bhattacharya, for his valuable inputs relating to the radar vision part of this chapter; he is also indebted to the Natural Sciences and Engineering Research Council for supporting his work on radar vision.

REFERENCES

[1] M. Born and E. Wolf (1959) *Principles of Optics*, Macmillan, New York.

[2] B. Boshash (1989) ''Time-frequency signal analysis,'' in S. Haykin (Ed.) *Advances in Spectral Estimation and Array Processing*, Vol. 1, Prentice-Hall, Englewood Cliffs, NJ.

[3] M. R. Drinkwater, R. Kwok, D. P. Winebrenner, and E. Rignot (1991) ''Multifrequency polarimetric synthetic aperture radar observations of sea ice,'' *J. Geophys. Res.*, **96** (C11), 20, 679–698.

[4] M. Eggers and T. Kuhon (1990) ''Neural network data fusion concepts and applications,'' in *Intl. Joint Conf. on Neural Networks*.

[5] A. K. Gabriel, R. M. Goldstein, and H. A. Zebker (1989) ''Mapping small elevation changes over large areas: differential radar interferometry,'' *J. Geophys. Res.*, **94** (B7), 9183–9191.

[6] R. M. Goldstein and H. A. Zebker (1987) ''Interferometric radar measurements of ocean surface currents,'' *Nature*, **328**, 707–709.

[7] L. C. Graham (1974) ''Synthetic interferometers for topographic mapping,'' in *Proc. IEEE*, **62** (6), 763–768.

[8] A. L. Gray and P. Farris-Manning (1993) ''Repeat-pass interferometry with airborne synthetic aperture radar,'' *IEEE Trans. Geosci. and Remote Sens.*, **31** (1).

[9] S. Haykin (1994) *Neural Networks: A Comprehensive Foundation*, Macmillan, New York.

[10] S. Haykin and T. K. Bhattacharya (1992) ''Adaptive radar detection using supervised learning in time-frequency domain,'' *Seventh Yale Workshop on Adaptive and Learning Systems*.

[11] S. Haykin and T. K. Bhattacharya (1992) ''Adaptive radar detection using supervised learning network,'' *Computational Neuroscience Symp.*, Purdue University, IN.

[12] H. Israelsson and J. Askne (1991) ''Analysis of polarimetric SAR observations of sea ice,'' in *Proc. Intl. Geosci. and Remote Sens. Symp.* 89–92.

[13] V. Kezys and S. Haykin (1993) ''IPIX-A multiparameter *X*-band research radar,'' *26th Intl. Conf. Radar Meteorol.*, 354–356.

[14] Y. B. LeCun, B. Boser, J. S. Denker, D. Henderson, R. E. Howard, W. Hubbard, and I. D. Jackel, 1990. ''Handwritten digit recognition with a back-propagation network,'' in *Advances in Neural Information Processing Systems 2* (D. S. Touretsky, ed.), pp. 396–404, Sam Mateo, CA: Morgan Kaufmann.

[15] F. K. Li and R. M. Goldstein (1990) ''Studies of multibaseline spaceborne interferometric synthetic aperture radars,'' *IEEE Trans. Geosci. and Remote Sens.*, **28** (1) 88–97.

[16] D. R. Lyzenga and J. R. Bennet (1991) ''Estimation of ocean wave spectra using two-antenna SAR systems,'' *IEEE Trans. Geosci. and Remote Sens.*, **29** (3), 463–465.

[17] C. Prati, M. Giani, and N. Leuratti (1990) ''SAR interferometry, a 2D phase unwrapping technique based on phase and absolute value information,'' in *Proc. Intl. Geosci. and Remote Sens. Symp.*, 2043–2046.

[18] C. Prati, F. Rocca, and A. M. Guarnieri (1989) ''Effects of speckle and additive noise on the altimetric resolution of interferometric SAR (ISAR) surveys,'' in *Proc. Intl. Geosci. and Remote Sens. Symp.*, 2469–2472.

[19] R. K. Raney (1971) ''Synthetic aperture radar and moving targets,'' *IEEE Trans. Aerospace and Electron. Syst.*, **AES-7** (3), 499–505.

[20] H. Urkowitz (1964) ''The effect of antenna pattern on the performance of dual-antenna radar airborne moving target indicators,'' *IEEE Trans. Aerospace and Naval Electron.*, **ANE-11**, 218–223.

[21] A. J. Viterbi (1988) ''Wireless digital communication: a view based on three lessons learned,'' *IEEE Communications Mag.*, **24**.

[22] P. M. Woodward (1953) *Probability and Information Theory, with Application to Radar*, Pergamon Press, MacMillan, New York.

[23] H. A. Zebker and R. M. Goldstein (1986) ''Topographic mapping from interferometric synthetic aperture radar observations,'' *J. Geophys. Res.*, **91** (B5), 4993–4999.

INDEX

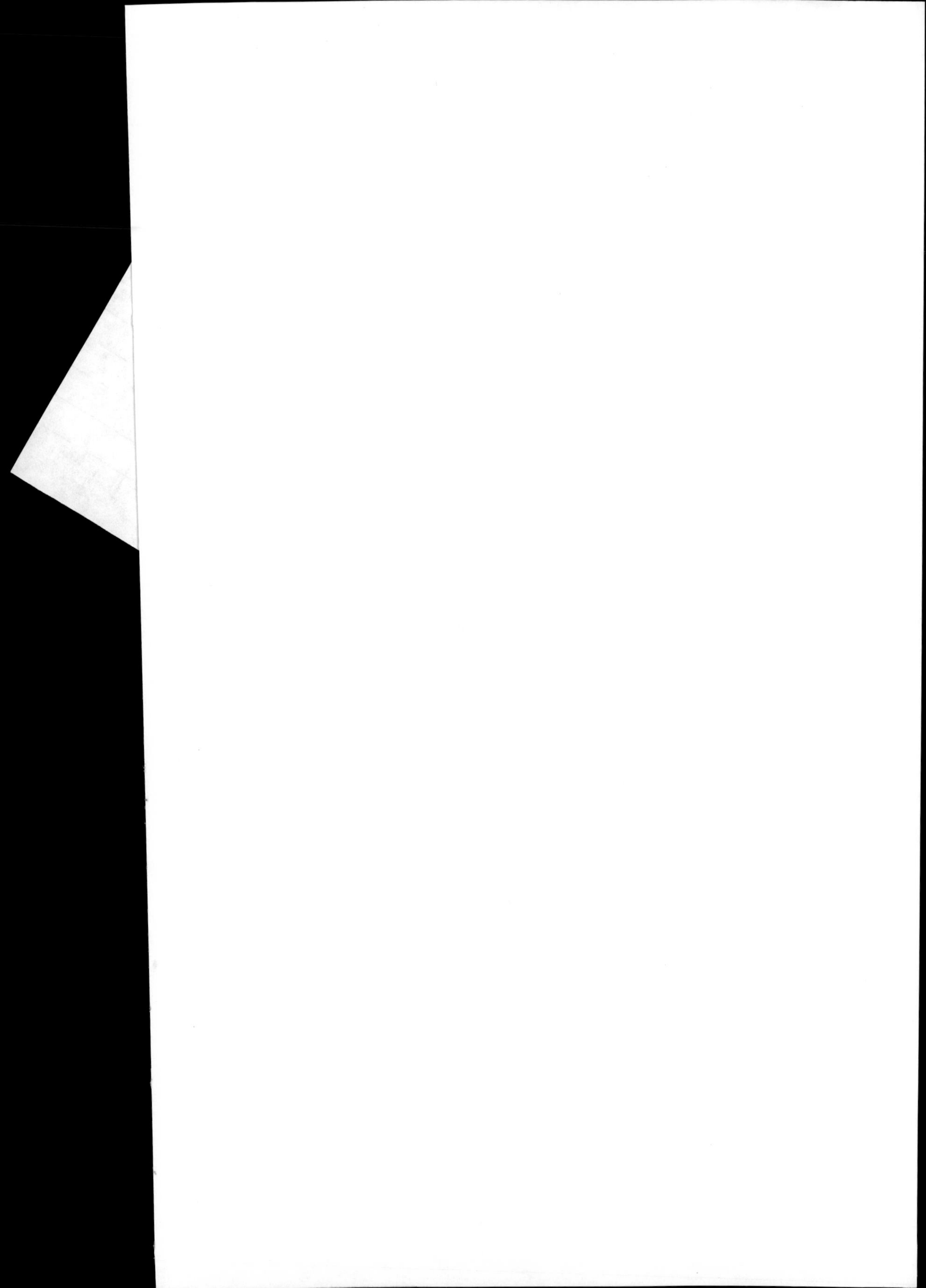